Fred B. Samson   Fritz L. Kr

Ecosystem Management

Selected Readings

With 104 Figures

Springer

Fred B. Samson
U.S. Forest Service, Missoula
Region One
Missoula, MT 59807

Fritz L. Knopf
Natural Resources Science Agency
Fort Collins, CO 80325

The cover photograph is of the boundary between Yellowstone National Park (right) and the Targhee National Forest (left) by Tim Crawford. Used with permission.

Library of Congress Cataloging-in-Publication Data
Ecosystem management: selected readings / Fred B. Samson and Fritz L.
  Knopf, editors.
        p. cm.
    Includes bibliographical references and index.
    ISBN 0-387-94668-3 (hardcover : alk. paper). — ISBN 0-387-94667-5
(pbk. : alk. paper)
    1. Ecosystem management. 2. Biological diversity conservation.
I. Samson, Fred B. II. Knopf, Fritz L.
QH75.E3 1996
333.95—dc20   95-51400

Printed on acid-free paper.

Production managed by Natalie Johnson; manufacturing supervised by Jacqui Ashri.
Typeset by TechType, Inc., Ramsey, NJ.
Printed and bound by Edwards Brothers, Inc., Ann Arbor, MI.
Printed in the United States of America.

9 8 7 6 5 4 3 2 1

ISBN 0-387-94667-5 Springer-Verlag New York Berlin Heidelberg SPIN 10523555 (softcover)
ISBN 0-387-94668-3 Springer-Verlag New York Berlin Heidelberg SPIN 10523628 (hardcover)

Preface

The theory and practice of ecosystem management is pivotal to the debate over how to sustain the health and productivity of our environment. In particular, the role of ecosystems in preserving biological diversity, their contribution to economic growth, and their influence on human well-being is highly controversial (Lubchenco et al. 1991). Traditional resource management does not protect natural values (Sax 1993) or provide for the sustainable production of goods and services (Barnes 1993). Yet a number of researchers and managers further question the ability of science to provide sufficiently powerful tools for the understanding and implementation of ecosystem management (Clark 1996). For example, available data do not permit us to determine ecological limits in most if not all ecosystems (Beier and Rasmussen 1994), and we are unable to define an ecosystem's "range of natural variation," an emerging concept in planning for resource management at the ecosystem level (Samson 1992).

Our central goal in assembling this book is to assist in the dissemination of important litera-ture that can contribute to better informing practicing scientists, resource rofessionals, and students about the ss confronting the development of ecc n management. We have structured this ection of papers around four starting points that appear frequently in the primary research literature. They may be summarized as four challenges: (1) to better understand patterns in biological diversity; (2) restore and promote natural ecological processes; (3) ensure the integrity of ecosystems; and (4) advocate the sustainable use of natural resources. In this book, each of these four topics is addressed by a set of eight key journal articles. The first article in each section provides an overview, followed by case histories and a concluding paper which is a commentary on the difficulty of the issue or assesses its future direction. An article by Risser provides a closing synthesis to this collection.

The authors of the articles in Part 1 — Understand Diversity — speak to the current problems and directions in the conservation of biological diversity. Tilman and Downing argue that preservation of native biodiversity is essential to the maintenance of stable productivity in ecosystems, and the substitution with non-native species should be viewed with caution. Pimm and Gittleman note biological diversity is described by two parameters, α point and β, the turnover of species across space, and that we know too little about where diversity is, particularly β, which is important in establishing ecosystem-level conservation strategies. Four case histories — South America (Mares), two in North America (Sheldon; Knopf and Samson), and for Indo-Pacific countries (Dinerstein and Wikramanayake) — illustrate the need to consider patterns in diversity to establish reasonable and effective conservation strategies. However, efforts to detect patterns and trends in species numbers is not without difficulty, as illustrated by Böhning-Gaese et al. in their comparison of North American breeding bird communities over different spatial and temporal scales. This section concludes (Sisk et al.) with a description of a new analytical tool that provides a preliminary list of countries facing intense conservation problems — a tool that may also find application at smaller spatial scales.

Biological diversity describes the variety of life forms as well as the accompanying ecological processes native to a landscape (Wilcove and Samson 1987). The authors of articles in Part 2 — Restore Ecological Processes — offer insights on ecological processes, focusing on processes that maintain biological diversity. Smith et al. suggest that there is an urgent need to conserve species while simultaneously studying the ecological and evolutionary mechanisms generating patterns in species diversity and isolating mechanisms essential to speciation. Clark and Collins in their respective papers propose two process-dependent models with important implications for plant population dynamics at the metapopulation scale, particularly the ability to predict species distribution or abundance in response to disturbance. Two case histories (Gibson et al. and Motzkin et al.) provide ecological and management perspectives for understanding the roles of ecological processes in conservation of communities and ecosystems. The need to consider ecological processes is extended to animals by Jones et al., who describe organisms as ecosystem engineers. The linkage of ecological to economic systems, another significant feature to management, is outlined by Costanza et al. The concluding paper (Hobbs and Huenneke) considers natural disturbances and their effect on colonization and survival of biological immigrants.

A number of scientists suggest that biological immigrants may pose the most significant threat to the preservation of biological diversity (Culotta 1991). A widely quoted definition of biotic integrity is the ability to maintain and support "a balanced, integrated, adaptive community of organisms having a species composition, diversity, and functional organization comparable to that of the natural habitat of the region" (Karr 1991). The articles in Part 3 — Emphasize Biotic Integrity — begin with a call for more documentation of how biological invasions alter properties of entire ecosystems (Vitousek). Additional case histories show that "no park is an island" (Janzen). Species introductions do have profound consequences on

aquatic communities (Flecker and Townsend), including the exposure of native species to novel disease organisms (Leberg and Vrijenhoek). The introduction of predators may be more important than habitat fragmentation in terrestrial species loss (Short and Turner). Kinsolving and Bain examine a gradient of recovery in the abundance and species richness of fish downstream from a dam. Pyšek and Prach discuss the role of river corridors in encouraging plan invasions — a negative effect on biological diversity at the between-ecosystem scale. The concluding article by Angermeier and Karr argues that biological integrity and not biological diversity is the most effective and meaningful goal for resource management policy.

Part 4 — Promote Ecological Sustainability — begins with Eugene Odum's proposal of the twenty great concepts in ecology for the 1990s. Odum states, "If we are serious about sustainability, we must raise our focus in management and planning to large landscapes and beyond." Four case histories — rain forests and adjoining lands in Guatemala and Indonesia (Salafsky et al.), the Colombian Amazon (Andrade and Rubio-Torgler), Costa Rica (Roth et al.), and the Pacific Northwest of the United States (Hansen et al.) — illustrate potentially sustainable use of wet forests from ecological-economic perspectives. Holling's paper provides direction for the development of programs to evaluate, monitor, and predict community and ecosystem changes across scales, including changes caused by humans. Norton and Ulanowicz argue that monitoring ecological processes alone is adequate to conserve diversity and provide a hierarchical model that emphasizes the importance of scale in sustaining ecosystem health. Part 4 concludes with Daily and Ehrlich's "a framework for estimating population sizes and lifestyles that could be sustained without undermining future generations."

There is widespread agreement about the value of taking an ecosystem-level approach for real world applications in resource management (Grumbine 1994) and "the fact that no one knows what ecosystem management means has not diminished enthusiasm for the concept" (Wilcove and Blair 1995). Where do we begin? A set of useful suggestions is offered in the concluding article by Risser. He provides an interim, integrated strategy, "for organizing information and for guiding those who need to formulate policy and set priorities for research on natural resources" while the interactions of biological diversity, ecosystem processes, integrity and sustainability can be further refined.

The task before resource managers and scientists is to more fully understand the four concepts of diversity, processes, integrity, and sustainability so they function to promote ecosystem management and to strengthen the ties among research, habitat management, and conservation activities by agencies and nations (Samson and Knopf 1993). Many universities, state and national agencies, and conservation organizations are putting together courses and programs to address and implement ecosystem management. We hope this volume will provide students as well as accomplished professionals a starting point to develop, understand, and implement ecosystem management—the emerging paradigm for conservation of natural resources worldwide.

References

Barnes, R. L. 1993. The U. C. C.'s insidious preference for agronomy over ecology in farm lending decisions. *Univ. Colorado Law Rev.* 64:457–512.

Beier, C. and L. Rasmussen. 1994. Effects of whole-ecosystem manipulations on ecosystem internal processes. *Trends Ecol. Evol.* 9:218–223.

Clark, J. 1996. The Great Plains Partnership. *In* F.
B. Samson and F. L. Knopf, eds., Prairie conservation: preserving North America's most endangered ecosystem. Island Press, Covelo, CA (in press).

Cullota, E. 1991. Biological immigrants under fire. *Science* 254:1,444–1,449.

Grumbine, R. E. 1994. What is ecosystem management? *Conser. Biol.* 8:27–39.

Karr, J. 1991. Biological integrity: a long-neglected aspect of water resource management. *Ecol. Appl.* 1:66–84.

Lubchenco, J. A., A. M. Olsen, L. B. Brubacker, S. R. Carpenter, M. M. Holland, S. P. Hubbell, S. A. Levin, J. A. MacMahon, P. A. Matson, J. M. Mellino, H. A. Mooney, H. R. Pullium, L. A. Real, P. J. Regal and P. G. Risser. 1991. The sustainable biosphere initiative: an ecological research agenda. *Ecology* 72:371–412.

Samson, F. B. 1992. Conserving biological diversity in sustainable ecological systems. *Trans. North American Wildlife and Natural Resources Conf.* 57:308–320.

Samson, F. B. and F. L. Knopf. 1993. Managing biological diversity. *Wildl. Soc. Bull.* 21:509–514.

Sax, J. L. 1993. Nature and habitat conservation and protection in the United States. *Ecol. Law Quart.* 20:687–705.

Wilcove, D. S. and R. B. Blair. 1995. The ecosystem bandwagon. *Trends Ecol. Evol.* 10:345.

Wilcove, D. S. and F. B. Samson. 1987. Innovative wildlife management: listening to Leopold. *Trans. North American Wildlife and Natural Resources Conf.* 52:321–329.

Fred B. Samson and Fritz L. Knopf

Contents

Contents

Contributors

G. I. Andrade Fundación Natura Colombia, A.A. 55402, Santafé de Bogotá, D.C., Colombia.

P. L. Angermeier National Biological Survey, Cooperative Fish and Wildlife Research Unit, Department of Fisheries and Wildlife Sciences, Virginia Polytechnic Institute and State University, Blacksburg, VA 24061–0321, USA.

M. B. Bain New York Cooperative Fish and Wildlife Research Unit, Fernow Hall, Cornell University, Ithaca, NY 14853–3001, USA.

K. Böhning-Gaese Abt. für Verhaltensphysiologie, Beim Kupferhammer 8, D-72070 Tübingen, Germany.

J. M. Briggs Division of Biology, Kansas State University, Manhattan, KS 66506, USA.

J. H. Brown Department of Biology, University of New Mexico, Albuquerque, NM 87131, USA.

M. W. Bruford Institute of Zoology, Zoological Society of London, London, U.K.

J. S. Clark Department of Botany, Duke University, Durham, NC 27708, USA.

S. L. Collins Department of Botany and Microbiology, University of Oklahoma, Norman, OK 73019, USA.

R. Costanza Maryland International Institute for Ecological Economics, Center for Environ-mental and Estuarine Studies, The University of Maryland System, Solomons, MD 20688-0038, USA.

G. C. Daily Department of Biological Sciences, Stanford University, Stanford, CA 94305, USA.

E. Dinerstein Conservation Science Program, World Wildlife Fund, 1250 24th St. NW, Washington, D.C. 20037, USA.

J. A. Downing Département de Sciences Biologiques, Université de Montréal, CP, 6128, Succursale, Montréal, Québec H3C 3J7, Canada.

N. E. R. Drake Department of Forestry and Wildlife Management, University of Massachusetts, Amherst, MA, USA.

B. L. Dugelby School of the Environment and Center for Tropical Conservation, Duke University, 3705-C Erwin Rd., Durham, NC 27705, USA.

P. R. Ehrlich Center for Conservation Biology, Department of Biological Sciences, Stanford University, Stanford, CA 94305, USA.

A. S. Flecker Section of Ecology and Systematics, Cornell University, Ithaca, NY 14853, USA.

C. Folke Beijer International Institute of Ecological Economics, S-104 05 Stockholm, Sweden.

S. L. Garman Forest Science Department, Oregon State University, Corvallis, OR 97331, USA.

D. J. Gibson Department of Plant Biology, Southern Illinois University at Carbondale, Carbondale, IL 62901, USA.

J. L. Gittleman Department of Zoology and Graduate Program in Ecology at the University of Tennessee, Knoxville, TN 37996–0810, USA.

A. J. Hansen 17625 Big Elk Meadow Rd., Gallatin Gateway, MT 59730, USA.

R. J. Hobbs CSIRO Division of Wildlife and Ecology, LMB 4, PO Midland, Western Australia 6056, Australia.

C. S. Holling Arthur R. Marshall Laboratory of Ecological Science, Department of Zoology, University of Florida, Gainesville, FL 32611, USA.

L. F. Huenneke Department of Biology, New Mexico State University, Box 30001/Dept 3AF, Las Cruces, NM 88003, USA.

D. H. Janzen Department of Biology, University of Pennsylvania, Philadelphia, PA 19104, USA.

C. G. Jones Institute of Ecosystem Studies (IES), Box AB, Millbrook, NY 12545, USA.

J. R. Karr Institute for Environmental Studies, University of Washington, Seattle, WA 98195, USA.

A. D. Kinsolving Department of Marine and Wildlife Resources, P.O. Box 3730, Pago Pago 96799, American Samoa.

F. L. Knopf Natural Resources Science Agency, Fort Collins, CO 80325, USA.

A. E. Launer Center for Conservation Biology, Department of Biological Sciences, Stanford University, Stanford, CA 94305, USA.

J. H. Lawton NERC Centre for Population Biology, Imperial College, Silwood Park, Ascot, Berks, SL5 7PY, UK.

P. L. Leberg Department of Biology, University of Southwestern Louisiana, Lafayette, LA 70504, USA.

K.-G. Mäler Beijer International Institute of Ecological Economics, S-104 05 Stockholm, Sweden.

M. A. Mares Oklahoma Museum of Natural History and Department of Zoology, University of Oklahoma, Norman, OK 73019, USA.

B. Marks Forest Science Department, Oregon State University, Corvallis, OR 97331, USA.

G. Motzkin Harvard Forest, Harvard University, Petersham, MA 01366, USA.

B. G. Norton School of Public Policy, Georgia Institute of Technology, Atlanta, GA 30332–0345, USA.

E. P. Odum Institute of Ecology, University of Georgia, Athens, GA 30602, USA.

W. A. Patterson III Department of Forestry and Wildlife Management, University of Massachusetts, Amherst, MA, USA.

I. Perfecto School of Natural Resources and Environment, University of Michigan, Ann Arbor, MI 48109, USA.

S. L. Pimm Department of Zoology and Graduate Program in Ecology at the University of Tennessee, Knoxville, TN 37996–0810, USA.

K. Prach University of South Bohemia and Institute of Botany, Academy of Sciences of the Czech Republic, 379 82 Třeboň, Czech Republic.

P. Pyšek Institute of Botany, Academy of Sciences of the Czech Republic, CZ-252 43 Pruhônice, Czech Republic.

B. Rathcke Department of Biology, University of Michigan, Ann Arbor, MI 48109, USA.

P. G. Risser Office of the President, Oregon State University, Corvallis, OR 97331, USA.

D. S. Roth 8 High St. #6, North Grafton, MA 01536, USA.

H. Rubio-Torgler Fundación Natura Colombia, A.A. 55402, Santafé de Bogotá, D.C., Colombia.

N. Salafsky Biodiversity Support Program, c/o WWF, 1250 24th St. NW, Washington, DC 20037, USA.

F. B. Samson USDA Forest Service, Northern Region, P.O. Box 7669, Missoula, MT 59801, USA.

T. R. Seastedt EPO Biology INSTARR, Campus Box 450, University of Colorado, Boulder, CO 80309, USA.

M. Shachak Mitrani Center for Desert Ecology, Blaustein Institute for Desert Research, Ben-Gurion University of the Negev, Sede Boqer 84900, Israel.

A. L. Sheldon Division of Biological Sciences, University of Montana, Missoula, MT 59812-1002, USA.

J. Short CSIRO Western Australian Laboratory, Midland, Western Australia 6056, Australia.

T. D. Sisk National Biological Survey, Department of the Interior, 1849 C St. NW, Washington, DC 20240, USA.

T. B. Smith Department of Biology, San Francisco State University, CA 94132, USA.

K. R. Switky Center for Conservation Biology, Department of Biological Sciences, Stanford University, Stanford, CA 94305, USA.

M. L. Taper Department of Biology, Montana State University, Bozeman, MT 59717, USA.

J. W. Terborgh School of the Environment and Center for Tropical Conservation, Duke University, 3705-C Erwin Rd., Durham, NC 27705, USA.

D. Tilman Department of Ecology, Evolution and Behavior, University of Minnesota, 1987 Upper Buford Circle, St. Paul, MN 55108, USA.

C. R. Townsend Department of Zoology, University of Otago, Dunedin, New Zealand.

B. Turner CSIRO Division of Wildlife and Ecology, LMB 4, PO, Midland, Western Australia 6056, Australia.

R. E. Ulanowicz Chesapeake Biological Laboratory, Box 38, Solomons, MD 20688, USA.

D. L. Urban School of the Environment, Duke University, Durham, NC 27708, USA.

P. M. Vitousek Department of Biological Sciences, Stanford University, Stanford, CA 94305, USA.

R. C. Vrijenhoek Center for Theoretical and Applied Genetics, and Institute of Marine and Coastal Sciences, Rutgers University, New Brunswick, NJ 08903, USA.

L. Wainger Maryland International Institute for Ecological Economics, The University of Maryland, Solomons, MD 20688-0038, USA.

R. K. Wayne Institute of Zoology, Zoological Society of London, Regent's Park London NW1 4RY, U.K.

E. D. Wikramanayake 25, Araliya Mawatha, Sirimal Uyana, Ratmalana, Sri Lanka.

PART 1
Understand Diversity

1
Biodiversity and Stability in Grasslands

David Tilman and John A. Downing

One of the ecological tenets justifying conservation of biodiversity is that diversity begets stability. Impacts of biodiversity on population dynamics and ecosystem functioning have long been debated[1-7], however, with many theoretical explorations[2-6,8-11] but few field studies[12-15]. Here we describe a long-term study of grasslands[16,17] which shows that primary productivity in more diverse plant communities is more resistant to, and recovers more fully from, a major drought. The curvilinear relationship we observe suggests that each additional species lost from our grasslands had a progressively greater impact on drought resistance. Our results support the diversity–stability hypothesis[5,6,18,19], but not the alternative hypothesis that most species are functionally redundant[19-21]. This study implies that the preservation of biodiversity is essential for the maintenance of stable productivity in ecosystems.

The resistance of an ecosystem to perturbation and the speed of recovery, which is called resilience, are two important components of ecosystem stability[6]. Interest in the effects of biodiversity on stability has been heightened by the rapidly accelerating rate of species extinctions[18,22,23]. One view, the diversity–stability hypothesis, holds that species differ in their traits and that more diverse ecosystems are more likely to contain some species that can thrive during a given environmental perturbation and thus compensate for competitors that

are reduced by that disturbance[5,6,7,12,18,19]. This view thus predicts that biodiversity should promote resistance to disturbance. In contrast, the species-redundancy hypothesis asserts that many species are so similar that ecosystem functioning is independent of diversity if major functional groups are present[19-21]. An 11-year study of the factors controlling species composition, dynamics and diversity in successional and native grasslands in Minnesota[16,17] provides a test of the effects of biodiversity on ecosystem response to and recovery from a major perturbation. The study period included the most severe drought of the past 50 years[24] (1987–88), which led to a >45% reduction in above-ground living plant mass and a >35% loss of plant species richness in control plots[24].

Nitrogen is the major nutrient limiting productivity in most terrestrial habitats[25], including these Minnesota grasslands[16]. The species composition, diversity and functioning of these and many other ecosystems depend on the rate of nitrogen supply[17,25,26] and are thus being altered by increased atmospheric nitrogen deposition from agriculture and combustion of fossil fuels[27-29]. In 1982 we established, in four grassland fields, a total of 207 control and experimental plots in which plant species richness was altered through seven different rates of nitrogen addition[16].

We measured resistance to drought by calculating, for each plot, the relative rate of plant community biomass change (dB/Bdt, yr^{-1};

Reprinted with permission from Nature vol. 367, pp. 363-365. Copyright 1994 Macmillan Magazines Ltd.

Fig. 1) from 1986, the year before the drought, to 1988, the peak of the drought. Values closer to zero imply greater drought resistance. For our 207 plots, drought resistance was a significantly ($P < 0.0001$) increasing, but saturating, function of pre-drought plant species richness (Fig. 1). The greatest dependence of drought resistance on species richness occurred in plots with nine or fewer species. The most species-rich plots produced about half of their pre-drought biomass during the drought, whereas the most species-poor plots produced only about one-eighth (Fig. 1).

Other characteristics of plots, such as the rate of nitrogen addition, total above-ground plant biomass, the proportion of total plant biomass from species with the C4 photosynthetic pathway, and differences in these variables among fields, were also correlated with species richness. Species composition and abundances also varied with species richness in these plots[16,17]. More than 90% of the plots that contained four or fewer plant species were dominated (>50% of plot biomass) by *Poa pratensis, Agropyron repens* or *Schizachyrium scoparium*. *Poa* and *Agropyron* are drought-sensitive C3 grasses, and *Schizachyrium Agropyron* are drought-resistant C4 grass. A partial correlation analysis that controlled for all these potentially confounding variables (including the 1986 biomasses of these three species) showed a significant dependence of drought resistance on the natural logarithm of pre-drought species richness ($r_{partial} = 0.21$, $n = 207$, $P < 0.01$). Moreover, when all redundant, nonsignificant variables were removed using backwards elimination, species richness was retained, and its partial correlation with drought resistance was highly significant (Table 1).

The dynamics of individual species in our plots suggest that species richness led to greater drought resistance because species-rich plots were more likely to contain some drought-resistant species. During this two-year drought, the increased growth of these drought-resistant species partially compensated for the decreased growth of other species.

A second component of stability is resilience, or the rate of return to pre-existing conditions after perturbation[6]. We calculated the derivation from pre-drought biomass as the natural

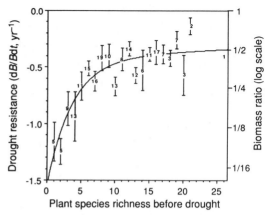

FIGURE 1. Relationship between drought resistance of grassland plots and plant species richness (SR_{86}) preceding a severe drought. Mean, standard error and number of plots with a given species richness are shown. Drought richness was measured as dB/Bdt (yr^{-1}), that is, as 0.5 (ln [biomass$_{1988}$/biomass$_{1986}$]), where biomass$_{1988}$ is at the height of drought and biomass$_{1986}$ is for the year preceding drought. Biomass ratio (biomass$_{1988}$/biomass$_{1986}$; right-hand scale) shows the proportionate decrease in plant biomass associated with the dB/Bdt values. Because the correlation between species richness and drought resistance in our data was no longer significantly ($P \leq 0.05$) positive when all plots with ≤ 8 or ≤ 11 species were ignored, we cannot reject the hypothesis that the relationship may reach a plateau. The solid curve ($dB/Bdt = -1.13e^{-x/3.6} - 0.44e^{-x/110}$, where x is SR_{86}, $r^2 = 0.22$, $P < 0.0001$), which was fitted to all 207 data points, is simply one of many that gave a significantly better fit than a straight line. A simpler equation ($dB/Bdt = 0.304$ ln [SR_{86}] $- 1.21$; $r^2 = 0.21$, $P < 0.0001$) provided an equally good fit.

METHODS. These are described in detail in ref. 16. The 207 plots were located in existing vegetation in four grassland fields in Cedar Creek Natural History Area, Minnesota. Field A had been abandoned for 20 yr, field B for 31 yr and field C for 54 yr in 1988. Each contained 54 plots, each 4 × 4 m. Field D, a prairie opening in native savannah, contained 45 plots, each 2 × 4 m. Nine treatments: no nutrient addition, addition of macro- and micro-nutrients other than nitrogen, and seven treatments that received these nutrients but with seven different rates of nitrogen addition. Field D had five replicates per treatment, the others had six. Vegetation in each plot was sampled by clipping a different 0.3-m² subsection each year, sorting to species, drying and weighing. Species richness is the number of vascular species in a 0.3-m² sample. Biomass is total above-ground living plant mass (g m⁻²).

TABLE 1. Factors influencing drought resistance

Variable	Partial correlation coefficient	P
Field A intercept	−0.56	<0.0001
Biomass of *P. pratensis* (1986)	−0.39	<0.0001
Biomass$_{1986}$	−0.36	<0.0001
ln (SR$_{1986}$)	0.29	<0.0001

Partial correlation of each listed variable with dB/dt, controlling for the other listed variables. Analyses used backwards elimination in multiple regression analyses to retain only variables that had significant ($P < 0.05$) partial correlations with dB/dt. The final multiple regression was highly significant ($F = 48.8$, $n = 207$, $R^2 = 0.48$, $P < 0.0001$). Extensive residual analyses[30] were performed at each step. Candidate variables included ln (SR$_{1986}$), logarithm of experimental plus atmospheric nitrogen addition, biomass$_{1986}$, number of species of C3 and C4 plants in each plot, fraction of biomass comprised of C3 and C4 plants, dummy variables[30] for each field, and the biomass in 1986 of the three most common species in low species-richness (one to four species) plots (*A. repens*, *P. pratensis* and *S. scoparium*). The significant partial correlation for ln (SR$_{1986}$) means that the correlation of species richness with drought resistance does not arise through biomass, species composition or field effects[31]. Other multiple regressions examined change in species richness and biomass preceding drought, detrended annual variation in species richness and biomass preceding drought, and their interactions with SR$_{1982}$ and biomass$_{1982}$. No other multiple regression was significantly better than this one and all showed that partial correlations between species richness and drought resistance were statistically significant. Analyses using other measures of species diversity yielded similar results.

logarithm of the ratio of plot biomass in 1989, 1990, 1991 and 1992 to average pre-drought biomass. For each of the four post-drought years, there were significantly negative intercepts and significantly positive slopes for regressions of these deviations on the natural logarithm of 1989 species richness. These indicate that species-poor plots were still further from their pre-drought biomass than were species-rich plots in each of the four post-drought years.

By 1992, species-rich plots had returned to pre-drought biomass, but the most depauperate plots still had significantly less biomass than their pre-drought average (Fig. 2). When potentially confounding variables were controlled for, there was a significant partial correlation between drought recovery and the natural logarithm of 1989 species richness ($r_{partial} = 0.184$, $P < 0.01$, $n = 207$). Moreover, when all redundant, nonsignificant variables were removed, species richness was retained, and its partial correlation was highly significant (Table 2). Thus, species-poor plots were both more greatly harmed by drought (Fig. 1 and Table 1) and took longer to return to pre-drought conditions (Fig. 2 and Table 2). The stand of native prairie was significantly more resilient than the three successional grasslands (Table 2).

Our results and earlier studies[5,12,14,15] support the diversity–stability hypothesis[5], and show that ecosystem functioning is sensitive to biodiversity. Our results do not support the species-redundancy hypothesis because we always found a significant effect of biodiversity on drought resistance and recovery even when we controlled statistically for the abundances of C3 (often drought sensitive) and C4 (often drought resistant) plant functional groups (Table 1).

Our results show that ecosystem resistance to drought is an increasing but nonlinear function of species richness. This is expected from the mechanism underlying the diversity–stability hypothesis. Functional diversity should be a saturating function of species richness because,

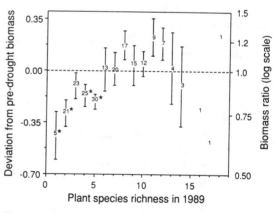

FIGURE 2. Deviation of 1992 biomass from average (1982–1986) pre-drought biomass was measured as in [(biomass$_{1992}$)/(average pre-drought biomass)]. Mean, standard error and number of plots are indicated for each level of species richness. Negative values mean that 1992 biomass was lower than pre-drought average. Biomass ratio is (biomass$_{1992}$)/(average pre-drought biomass). Student's t-tests showed that plots containing 1, 2, 4 or 5 species differed significantly (*$P < 0.05$) from their pre-drought average, but that plots with greater richness did not differ significantly from pre-drought averages.

TABLE 2. Factors influencing drought resistance

Variable	Partial correlation coefficient	P
Field A intercept	-0.39	<0.0001
Field B intercept	-0.30	<0.0001
Field C intercept	-0.20	0.003
ln (SR_{1989})	0.18	0.009
SR_{C3}	-0.18	0.012
Biomass of *Schizachyrium* (1989)	-0.16	0.027
SR_{C4}	-0.14	0.042

Partial correlations of each listed variable with deviation from pre-drought biomass, holding other listed variables constant. These seven variables were retained in multiple regression analysis of 1992 deviation from pre-drought biomass against the same candidate variables used in Table 1, but using 1989 values. Backwards elimination, with residual analysis, was used to retain only significant ($P < 0.05$) variables. The overall regression had $F = 14.0$, $n = 206$, $R^2 = 0.33$, $P < 0.0001$. SR_{C3} is the number of C3 species and SR_{C4} is the number of C4 species in plots in 1989. The significantly positive slope for ln (SR_{1989}) and the significantly negative intercepts for fields A, B and C indicate that species-poor plots in these fields have not yet attained pre-drought biomass, whereas more species-rich plots have. Field D, a native grassland, recovered most rapidly, followed by field C, then B, then A, in order of successional age.

in species-rich ecosystems, additional species are more likely to be similar to existing species[21]. Thus, the progressive loss of species should have progressively greater impacts on ecosystem stability.

In addition to drought, grassland ecosystems experience periodic invasions of insect or mammalian herbivores, unusually late or early frosts, unusually wet or cool years, hail, fire, and other perturbations. Because different species are likely to perform best for particular combinations of these disturbances, the long-term stability of primary production in these and other grasslands should depend on their biodiversity. Although we do not know how the stability of other ecosystems depends on biodiversity, these results lend further urgency to pleas for the conservation of biodiversity.

References

1. Elton, C. S. *The Ecology of Invasions by Animals and Plants* (Chapman & Hall, London, 1958).
2. MacArthur, R. H. *Ecology* **36**, 533–536 (1955).
3. May, R. M. *Stability and Complexity in Model Ecosystems* (Princeton University Press, 1973).
4. Goodman, D. *Q. Rev. Biol.* **50**, 237–266 (1975).
5. McNaughton, S. *J. Am. Nat.* **11**, 515–525 (1977).
6. Pimm, S. L. *Nature* **307**, 321–326 (1984).
7. Schulze, E. D. & Mooney, H. A. *Biodiversity and Ecosystem Function* (Springer, Berlin, 1993).
8. Gardner, M. R. & Ashby, W. R. *Nature* **228**, 784 (1970).
9. Murdoch, W. W. *J. Appl. Ecol.* **12**, 795–807 (1975).
10. Yodzis, P. *Nature* **284**, 544–545 (1980).
11. King, A. W. & Pimm, S. L. *Am. Nat.* **122**, 229–239 (1983).
12. McNaughton, S. *J. Ecol. Monogr.* **86**, 259–294 (1985).
13. Wolda, H. *Am. Nat.* **112**, 1017–1045 (1978).
14. Ewel, J. J., Mazzarino, M. J. & Berlsh, C. W. *Ecol. Appl.* **1**, 289–302 (1991).
15. Frank, D. A. & McNaughton, S. J. *Oikos* **62**, 360–362 (1991).
16. Tilman, D. *Ecol. Monogr.* **57**, 189–214 (1987).
17. Tilman, D. *Plant Strategies and the Dynamics and Structure of Plant Communities* (Princeton Univ. Press, 1988).
18. Ehrlich, P. R. & Ehrlich, A. H. *Extinction. The Causes and Consequences of the Disappearance of Species* (Random House, New York, 1981).
19. Lawton, J. H. & Brown, V. K. in *Biodiversity and Ecosystem Function* (eds Schulze, E. D. & Mooney, H. A.) 225–270 (Springer, Berlin, 1993).
20. Walker, B. H. *Conserv. Biol.* **6**, 18–23 (1991).
21. Vitousek, P. M. & Hopper, D. U. in *Biodiversity and Ecosystem Function* (eds Schulze, E. D. & Mooney, H. A.) 3–14 (Springer, Berlin, 1993).
22. Wilson, E. O. *Biodiversity* (National Academy Press, Washington DC, 1988).
23. Ehrlich, P. R. & Wilson, E. O. *Science* **253**, 758–762 (1991).
24. Tilman, D. & El Haddi, A. *Oecologia* **89**, 257–264 (1992).
25. Vitousek, P. *Am. Nat.* **119**, 553–572 (1982).
26. Pastor, J., Aber, J. D., McClaugherty, C. A. & Melillo, J. M. *Ecology* **65**, 256–268 (1984).
27. Woodin, S. & Farmer, A. *Biol. Conserv.* **63**, 23–30 (1993).
28. Heil, G. W., Werger, J. A., DeMol, W., Vandam, D. & Heijne, B. *Science* **239**, 764–765 (1988).
29. Berendse, F., Aerts, R. & Bobbink, R. in *Landscape Ecology of a Stressed Environment* (eds.

Vos, C. C. & Opdam, P.) 104–121 (Chapman & Hall, London, 1993).

30. Draper, N. R. & Smith, H. *Applied Regression Analysis* (Wiley, New York, 1981).
31. Snedecor, G. W. & Cochran, W. G. *Statistical-Methods* (Iowa State University Press, Ames, 1980).

Acknowledgments

We thank R. Inouye, J. Lawton, C. Lehman, E. McCauley, M. McGinley, S. McNaughton, R. Peters, S. Pimm, P. Reich, M. Ritchie and D. Wedin for comments and the NSF and the Andrew Mellon Foundation for Support.

2
Biological Diversity: Where Is It?

Stuart L. Pimm and John L. Gittleman

To preserve biological diversity, wealthy nations usually save the large and cuddly species that excite the public's imagination[1]. This is not necessarily a bad strategy. The relatively large size of these species means that their population densities are low. Consequently, the minimal numbers required for persistence inhabit large areas and many other less charismatic species are protected under their umbrella. Furthermore, large species are often top predators and thus play crucial, so-called keystone roles in the community's dynamics. Yet such a species-biased approach is not necessarily appropriate for the great majority of animal species that are not furred and feathered, or for plants, or, indeed, for all organisms in species-rich tropical nations deciding on how to allocate land for national parks. The obvious strategy protects areas of greatest diversity. Two recent papers on mammals by Mares in this issue and Pagel et al.[2] illustrate the complex biological issues involved in predicting what diversity might remain after future planners have taken a cookie cutter to their wilderness.

At the outset, estimating the total number of species on the planet is not trivial. Counting one, two, three . . . would be difficult enough, even if all the species were described—and probably the majority are not. Small species escape our notice. The number must be estimated from what we know about general features of ecological communities and their food webs[3]. Understanding how species are distributed across the planet is even more daunting. One tactic documents the patterns in diversity among the furred and feathered, understands their causes, and applies the principles to other groups.

Diversity is described by two parameters: point or α-diversity, (practically, the number of species in a specified area) and β-diversity, the turnover of species across space. Empirical patterns of diversity have a long history[4]. Continent-wide maps of α-diversity based on individual species' ranges, cross-continental comparisons, and their interpretation are not new either[5]. Theory predicts α-diversity to increase with the total number of individuals encompassed[6], and so to increase with both the area sampled (the well-documented species-area relation) and the productivity per unit area. Empirically, we also know that α-diversity is less on remote islands and increases as we move toward the equator. (Total tropical diversity also owes much to the large area of the tropics, a fact easily overlooked on maps with Mercator projection). There are surprises, however. The α-diversity typically peaks along gradients of productivity, declining in the most productive systems[7]; some taxa are more diverse further north[8], and some deserts host unusually diverse plant communities[9]. The longitudinal, westward increase in vertebrate diversity in North America is also large[2,5].

We know less about β-diversity. As a first step, we must understand how large are species' ranges, for there are two extreme scenarios. If ranges are large, then α-diversity is almost independent of the area sampled and the national park cookie cutter can be placed anywhere. Alternatively, total diversity may be high, while α-diversity is low with species' ranges being small and adjoining rather than overlapping. Many parks would be needed to protect diversity. The 523 species of North American mammals are geographically very restricted. The median range is only 1.2% of the area of the continent but range size increases greatly with latitude. Our north temperate experience of many local, but very few global species extinctions following extensive and long-term changes to our environment sends the wrong message to managers of tropical diversity. The gray wolf and grizzly bear have been extirpated from much of Europe and eastern North America but survive locally across ranges that once covered most of three continents. Moreover, even historically, eastern North America supported relatively few species. Tropical species range much less widely and are that much more vulnerable, and there are many more species to lose.

Species ranges tend to follow the major habitat divisions (rain forests, deserts, and so on) for these divisions are themselves defined by their constituent species. In North America, areas with more of the 23 habitats defined by Pagel *et al.* have more species of mammal. Species most commonly occupy only one or two of these habitats. Mares identifies six major habitats in South America. The two of greatest areal extent are drylands and the Amazon lowlands. Theory[6] predicts the diversity of an area to scale as the (area)$^{0.25}$. The drylands are about twice the size of lowlands, ought to contain 18% more species and have 19% more mammals. But what about endemic species, those found only in one habitat? The drylands house 53% more endemic mammalian species and 440% more endemic genera than the lowlands. The concentration of diversity is in the western montane forest. It has 11% of the area, yet 76% as many mammalian species as the lowlands, 63% as many endemics.

The drylands' reputation for being areas of low diversity is false, at least for mammals. Nor do they merely contain a subset of species found in the rain forests. Mares shows that 68% of the mammalian species (and 95% of the genera) in the Amazon lowlands are found in other habitats. Much as we might be appalled by the rates of tropical deforestation, we cannot ignore the drier areas.

Useful as cautionary tales, these results also point to serious gaps in our knowledge. We do not always understand the causes of α-diversity: there are many theories but little consensus. Unless we understand the principles, we have no chance of predicting patterns among other species groups and in other areas where the rate of environmental change may preclude even our describing the species, let alone mapping their ranges. We know far too little about β-diversity to predict its current patterns, or its future, when remnant natural areas are surrounded by highly modified habitats. Part of the problem is familiar: larger remnants increase α-diversity, more remnants increase β-diversity. The optimal allocation of remnants involves knowing both diversities. Even less clear is whether species will survive outside these protected remnants other than human commensals that, like the starling, have followed us worldwide. And to what extent will these commensals penetrate the natural areas? We clearly know too little about where the diversity is, why it is there, and what it will become.

References and Notes

1. R. Tobin, *The Expendable Future* (Duke Univ. Press, Durham, NC, 1990).
2. M. Mares, *Science* **255**, 976 (1992); M. D. Pagel, R. M. May, A. R. Collie, *Am. Nat.* **137**, 791 (1991).
3. R. M. May, *Science* **241**, 1441 (1988).
4. A. R. Wallace, *The Geographical Distribution of Animals* (Macmillan, London, 1876).
5. J. J. Schall and E. R. Pianka, *Science* **201**, 679 (1978).
6. F. W. Preston, *Ecology* **43**, 185 (1962).
7. M. L. Rosenzweig and Z. Abramsky, in *Historical and Geographical Determinants of Community Diversity* (Univ. of Chicago Press, in press).
8. O. Järvinen and R. A. Väisänen, *Oikos* **30**, 496 (1978).
9. R. Rice and M. Westaby, *Vegetatio* **52**, 129 (1983).

3
Neotropical Mammals and the Myth of Amazonian Biodiversity

Michael A. Mares

Data were compiled on the distribution of mammal taxa (883 species, 242 genera, 45 families, and 10 orders) among South America's six major macrohabitats: lowland Amazon forest, western montane forests, Atlantic rain forest, upland semideciduous forest, southern mesophytic forest, and drylands. The drylands are the richest area in numbers of species supported and are more diverse than the other habitats, including the lowland Amazon rain forest, when endemics are considered. An analysis of number of endemic and nonendemic taxa versus size of area found a simple positive linear relationship: the drylands, almost twice as extensive as the Amazon lowlands, support more endemic taxa. Conservation plans that emphasize the wet tropics and fail to consider the drylands as special repositories of mammal diversity will be unable to preserve a significant number of novel taxa.

A great deal of controversy has been focused on the suggestion that the biosphere is approaching a massive, human-induced extinction episode rivaling or surpassing the megaextinctions chronicled in the fossil record[1]. There is evidence that Neotropical ecosystems, especially the lowland Amazon rain forest, contain an unusually rich array of species, whether vertebrates, invertebrates, or plants are considered[2,3]. Because tropical countries are primarily developing nations, efforts to preserve species have assumed importance in the international political arena[4]; issues of environmental integrity have influenced government policies concerning economic development throughout the world.

Biologists have used the biodiversity issue to rally support for increased funding for research in the wet tropics[5], yet there are problems associated with arousing public concern with visions of doomed species. For one thing, data on which such negative scenarios are based may be incorrect[6,7]. Additionally, if species are disappearing at a rate below what has been suggested, the public could perceive biologists as alarmists, unnecessarily predicting a mass extinction—one that neither seems demonstrable with hard data nor ever seems to arrive.

In the literature dealing with the potential loss of diversity, at least one important question has yet to be posed: Is the tropical rain forest uncommonly diverse? Few investigators have examined organismal diversity in South America's deserts, grasslands, or scrublands[8]. I report that when numbers of species and higher taxa of mammals are compared between the Amazon lowlands and the continent's drylands, the Amazon supports fewer taxa at all levels and fewer endemic taxa. Drylands are often considered areas of low diversity, but for mammals they are the most species-rich area on the continent.

One important hypothesis to explain elevated species richness in the Amazon forest postulates the evolution of the biota in Pleistocene refugia[9]. Purportedly, sets of closely related species and subspecies evolved after their previously continuous geographic ranges were fractured as a result of widespread habitat disruptions related to Pleistocene glaciations. The resultant isolates underwent speciation through a classic geographic isolation mechanism. Other causes of elevated species richness have been proposed[10], but whatever the mechanism of species multiplication, these areas contain many closely related species.

Species are made up of individuals that are themselves carriers of genetic information being transferred through time[11]. The potential loss of genetic information from tropical habitats has been invoked as a tragic consequence of extinction[12]. However, the community that contains the most species may not contain the greatest amount of unique genetic information[13]. The argument that some sets of species are more valuable than others is most easily understood with a simple example. Assume that one could save only four of ten species scheduled for extinction. What would be the most effective conservation strategy to employ in choosing which to preserve? Suppose that all ten species were mammals—four rodents, two rabbits, two monkeys, and two bats. Other things being equal, most biologists would choose one species within each higher taxonomic category. Why should this be so?

Although viable hybrids that develop between some plant species may offer unique genetic combinations, when considering animals one attempts to maximize the genetic distance between the species selected for preservation[13]. For example, mitochondrial DNA divergence between two subspecies of the endangered black rhinoceros (Diceros bicornis) was found to be only 0.29%; conservation biologists recommended that they be combined into a single breeding population[14]. The variety of genetic information stored in species representing four different orders of mammals (rodent, rabbit, monkey, and bat) is greater than that contained in four species from a single order (that is, four rodents). Closely related species may share 95% or more of their

nuclear DNA sequences, implying a great similarity in overall genetic information[15]. Choosing two such species would increase by a minuscule amount the quantity of genetic information that is protected. In contrast, species in different higher taxa imply a much greater genetic distance between them[15,16]. By deciding to preserve distantly related taxa, we ensure that genetic diversity, not merely species number, is maximized. This concept is referred to by Pielou[17] as hierarchical diversity; she noted[17] (p. 303), "Suppose we were comparing two communities and that both had the same number of species in the same relative proportions. . . . [I]f in one community all the species belonged to a different genus, it would be reasonable to regard the latter community as the more diverse of the two." When considering Neotropical diversity, the hierarchical diversity of macrohabitats should be considered before continentwide conservation plans are formulated.

I divided South America into six major macrohabitats (Fig. 1): (i) Amazonian lowlands, including the Colombia Chocó, the Pacific lowland rain forest, and all other lowland wet and dry forest below 1500 m (5.34 million km², 30% of the continental land area); (ii) western montane tropical and subtropical forests above 1500 m and below the páramo, extending in a narrow band from central Ven-

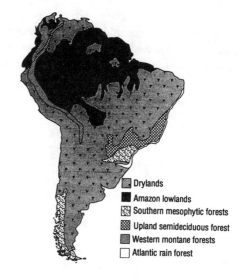

Legend:
- Drylands
- Amazon lowlands
- Southern mesophytic forests
- Upland semideciduous forest
- Western montane forests
- Atlantic rain forest

FIGURE 1. Map of the six major macrohabitats of South America (21).

ezuela southward along the eastern Andean and pre-Andean ranges to northwestern Argentina (0.58 million km^2, 3.2%); (iii) Atlantic rain forest of southeastern Brazil, a narrow strip (0.19 million km^2, 1.0%) confined to the east-facing slopes of the coastal ranges; (iv) upland semideciduous forest of Brazil, Argentina, Uruguay, and Paraguay (0.72 million km^2, 4.0%); (v) southern mesophytic forests of southern Brazil, Uruguay, and Argentina (0.78 million km^2, 4.4%); and (vi) drylands, the largest macrohabitat (10.2 million km^2, 57.3%), including the llanos of Venezuela and Colombia, the cerrado grasslands and caatinga scrublands of Brazil, the chacoan thorn forest, the Argentine pampas, the páramo and puna of the high Andes, and the lowland deserts of Argentina, Chile, and Peru.

Mammal species were assigned to one or more macrohabitats. I examined distribution patterns of species and higher taxa among the macrohabitats, noting endemicity to a macrohabitat. There are several possible sources of error in these analyses. Species distributions and habitat preferences of mammals, although better understood than most other South American organisms, are imperfectly known in many cases. By using broadly defined macrohabitats, I hoped to compensate for this type of error. Thus, for example, the dry and wet tropical forests were combined into the lowland Amazon forest category. Taxonomic problems can also influence these analyses. In some cases, although familial determinations are stable, generic rankings are unclear. However, recent taxonomic research helped clarify some partic-

ularly vexing problems. Additionally, the patterns are so prevalent that it would take a great deal of taxonomic reclassification or a significant change in the habitat preferences of many taxa to alter them. The data analyses yield surprising results (Table 1).

The Amazon lowlands support fewer than 60% (434/732) of the number of species found in all other macrohabitats. Moreover, 68% of the species found in the Amazon lowlands are also distributed in other habitats. If one compares endemic species, there are almost 2.5 (336/138) times as many endemics found in other macrohabitats as in the lowland Amazon rain forest. This pattern is even more pronounced at higher taxonomic categories.

Almost all lowland Amazon rain forest genera (95.4%) also occur in other South American (or Central American) macrohabitats. By contrast, fewer than 60% of the genera occurring in non-Amazonian habitats are also found in the lowland rain forest, and 6.5 (65/10) times as many endemic genera are restricted to non-Amazonian habitats. Essentially all families found in the Amazon are also distributed among the other macrohabitats. Although only one family (the monotypic Callimiconidae) is restricted to the lowland Amazon forest, two marsupial (Caenolestidae and Microbiotheriidae) and five rodent (Myocastoridae, Chinchillidae, Ctenomyidae, Octodontidae, and Abrocomidae) families are endemic to other macrohabitats.

An analysis by macrohabitat is also revealing (Table 2). In comparing the Amazon lowlands and the drylands, the Amazon supports fewer

TABLE 1. Taxonomic diversity and endemism between Amazonian lowland rain forests and all other South American macrohabitats. Endemics are indicated in parentheses.

Taxonomic category	Taxa per macrohabitat			
	Lowland Amazon rain forest (no.)	All other macrohabitats combined (no.)	Amazon taxa shared with other macrohabitats (%)	Taxa shared between non-Amazon and Amazon macrohabitats (%)
Orders	9 (0)	10 (0)	100.0	90.0
Families	36 (1)	42 (3)	94.4	78.6
Genera	151 (10)	227 (65)	95.4*	59.9
Species	434 (138)	732 (336)	68.2*	38.7†

*Eight genera and 13 species are shared with regions outside South America. †Two hundred eighty-three lowland Amazon species are shared with other habitats.

TABLE 2. Taxonomic diversity and endemism among major macrohabitats. Endemics are indicated in parentheses.

Taxonomic category	Taxa per macrohabitat					
	Lowland Amazon rain forest (no.)	Western montane forests (no.)	Atlantic rain forest (no.)	Upland semideciduous forest (no.)	Drylands (no.)	Southern mesophytic forest (no.)
Orders	9 (0)	10 (0)	9 (0)	9 (0)	10 (0)	6 (0)
Families	36 (1)	31 (0)	31 (0)	32 (0)	39 (2)	22 (1)
Genera	151 (10)	128 (7)	102 (2)	106 (0)	183 (44)	58 (4)
Species	434 (138)	332 (87)	170 (19)	192 (5)	509 (211)	94 (14)

species, fewer endemic species, fewer genera, fewer endemic genera, fewer families, and fewer endemic families.

The lowland Amazon rain forest is not the only tropical rain forest in South America. There are extensive montane wet forests in the west and the Atlantic rain forest in the east. Although it might seem logical to combine all rain forests to obtain an estimate of a "rain-forest fauna," an analysis revealed minimal species overlap between the lowland Amazon forest and the Atlantic rain forest. Indeed, not a single species is restricted to the combined area. In effect, there is no mammalian rain forest fauna when restriction to rain forest habitats is the determining factor.

Two forests could be combined: the lowland Amazon forest and the western montane forests. When these two areas are merged, the composite fauna contains 557 species, 170 genera, and 38 families. Moreover, in addition to the endemic taxa shown in Table 2, the combined area yields one additional endemic family (endemic to the two macrohabitats considered as a unit), 17 additional endemic genera, and 75 additional endemic species. This area supports a massive tropical and subtropical forest fauna, with species occurring from lowland wet forests to high, cool, Andean cloud forests and extending from eastern South America to Colombia, then southward to Argentina. Only when this aggregate area is considered does a forest fauna compare favorably with that of the drylands (compare Tables 2 and 3). The forests of southeastern Brazil do not fare as well in combination. Even when joined with the upland semideciduous forest, the Atlantic rain forest is basically a depauperate habitat compared to the drylands (Table 3).

The six macrohabitats differ greatly in size, varying by more than two orders of magnitude. Because area has been considered to play an important role in the conservation of biodiversity[18], I performed a regression analysis on the numbers of species and genera occurring within each macrohabitat against the area of the macrohabitats. Data were examined for possible linear, semilog, and log-log relationships; results of all analyses were comparable, but the best fit is given by a linear model.

I found (Fig. 2A) a statistically significant relation for all species versus area and for all genera versus area. When endemic and nonendemic taxa were considered separately, the results (Fig. 2B) showed a greater level of significance for endemic species versus area and for endemic genera versus area.

The results clearly show that the size of the area and the number of taxa it supports are related and that the relation is mainly the result of the endemic taxa. Drylands support more endemic genera and species than other habitats because they are larger. As area increases, the

TABLE 3. Mammalian faunas for combined woodlands macrohabitats. Endemic taxa are in parentheses.

Taxonomic category	Taxa in combined macrohabitats	
	Lowland Amazon rain forest plus western montane forests (no.)	Atlantic rain forest plus upland semideciduous forest (no.)
Families	38 (2)	33 (0)
Genera	170 (34)	116 (10)
Species	557 (341)	225 (38)

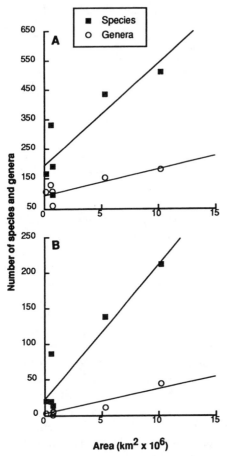

FIGURE 2. Regression plots of taxonomic diversity (total number of species and genera) within the major macrohabitats of continental South America versus area. (A) Total number of species and genera per macrohabitat versus area: species versus area (slope = 34.598, $R^2 = 0.73$, $P \leq 0.05$). (B) Number of endemic species and genera in each macrohabitat versus area: endemic species versus area (slope = 19.109, $R^2 = 0.86$, $P \leq 0.01$); endemic genera versus area (slope = 3.830, $R^2 = 0.88$, $P \leq 0.01$).

served. Endemics, however, are more common in the more extensive macrohabitats, the Amazon lowlands and, especially, the drylands. These data make it clear that, as far as mammal species richness is concerned, the tropical rain forest enjoys no special advantage. Its diversity comes from the same processes that prevail in other places[19].

On the basis of these findings, if one could choose only a single macrohabitat in which to preserve the greatest amount of mammalian biodiversity in South America, one would work in the largely continuous deserts, scrublands, and grasslands. This is exactly the converse of the funding, research, and conservation strategies that have been employed to date. The emphasis on developing additional lowland rain forest parks and reserves may be misguided as far as mammals are concerned; a greater amount of mammalian diversity would be preserved by increasing the number of protected areas in the drylands. Unfortunately, scientists and the ubiquitous popular media have paid scant attention to the need to preserve deserts, grasslands, or scrublands. These dry areas are very likely far more highly threatened than the largely inaccessable rain forests of the lowland tropics[7,8,20].

number of endemic species increases more rapidly than does the number of endemic genera (that is, the slope for species is steeper than that for genera). This is not surprising; genera are more widespread than species, having a geographical distribution that is the sum of the geographic ranges of the component species.

The nonendemics include a broad array of eurytopic taxa, such as bats and carnivores, that range widely throughout all macrohabitats. These are core taxa that will be preserved regardless of which macrohabitats are con-

References and Notes

1. P. R. Ehrlich and E. O. Wilson, *Science* **253**, 758 (1991).
2. E. O. Wilson, Ed., *Biodiversity* (National Academy Press, Washington, DC, 1988).
3. N. Myers, *Conversion of Tropical Moist Forests* (National Academy of Sciences, Washington, DC, 1980).
4. M. A. Mares, *Science* **233**, 734 (1986); J. A. McNeely, *Economics and Biological Diversity: Developing and Using Economic Incentives to Conserve Biological Diversity* (IUCN, Gland, Switzerland, 1988); M. K. Tolba, *Environ. Conserv.* **17**, 105 (1990).
5. M. E. Soulé and K. A. Kohm, Eds., *Research Priorities for Conservation Biology* (Island Press, Washington, DC, 1989).
6. P. M. Fearnside, *Environ. Conserv.* **17**, 213 (1990).
7. W. J. Boecklen and N. J. Gotelli, *Biol. Conserv.* **29**, 63 (1984); W. J. Boecklen, in *Latin American Mammalogy: History, Biodiversity, and Conservation*, M. A. Mares and D. J. Schmidly, Eds.

(Univ. of Oklahoma Press, Norman, OK, 1991), pp. 150–166.

8. K. H. Redford, A. Taber, J. A. Simonetti, *Conserv. Biol.* **4,** 328 (1990).

9. G. T. Prance, Ed., *Biological Diversification in the Tropics* (Columbia Univ. Press, New York, 1982).

10. K. E. Campbell, Jr., and D. Frailey, *Quat. Res.* **21,** 369 (1984); P. A. Colinvaux *et al., Nature* **313,** 42 (1985).

11. M. Eigen *et al., Science* **244,** 673 (1989).

12. M. L. Oldfield, *The Value of Conserving Genetic Resources* (Sinauer, Sunderland, MA, 1989).

13. R. I. Vane-Wright, C. J. Humphries, P. H. Williams, *Biol. Conserv.* **55,** 235 (1991).

14. M. V. Ashley, D. J. Melnick, D. Western, *Conserv. Biol.* **4,** 71 (1990).

15. C. G. Sibley and J. E. Ahlquist, *J. Mol. Evol.* **20,** 2 (1984); R. J. Britten, *Science* **231,** 1393 (1986).

16. J. C. Avise *et al., Annu. Rev. Ecol. Syst.* **18,** 489 (1987); D. Goodman, P. R. Giri, S. J. O'Brien, *Evolution* **43,** 282 (1989); D. P. Mindell and R. L. Honeycutt, *Amer. Rev. Ecol. Syst.* **21,** 541 (1990).

17. E. C. Pielou, *Ecological Diversity* (Wiley, New York, 1975).

18. D. S. Simberloff, *Annu. Rev. Ecol. Syst.* **19,** 473 (1988).

19. The rain forest may prove to be unusually rich for other taxa, especially plants, insects, and fish, but quantitative data comparing the diversity of other groups across macrohabitats are lacking, as are data comparing higher taxonomic level diversity and endemism. The lowland forest may also play a novel role in maintaining gaseous balance in the global atmosphere [G. M. Woodwell *et al., Science* **222,** 1081 (1983); R. P. Detwiler and C. A. S. Hall, *ibid.* **239,** 42 (1988)], but this also remains to be clarified.

20. E. Medina, *Interciencia* **10,** 224 (1985); P. M. Fearnside, *Environ. Conserv.* **17,** 213 (1990); V. Roig, in *Latin American Mammalogy: History, Biodiversity, and Conservation,* M. A. Mares and D. J. Schmidly, Eds. (Univ. of Oklahoma Press, Norman, OK, 1991), pp. 239–279.

21. Map modified from P. Hershkovitz [in *Evolution, Mammals, and Southern Continents,* A. Keast, F. C. Erk, B. Glass, Eds. (State University of New York Albany, 1972), pp. 311–431] and H. Walters [*Die Vegetation der Erde* (Fischer Verlag, Stuttgart, 1968), vol. 2].

22. I thank J. K. Braun for technical and editorial assistance; T. E. Lacher, Jr., M. R. Willig, L. Vitt, J. Caldwell, and L. B. Mares for their comments on the manuscript; M. R. Willig for statistical assistance; and C. Kacmarcik for graphic and computational assistance. Supported by National Science Foundation grant BSR-8906665.

4
Conservation of Stream Fishes: Patterns of Diversity, Rarity, and Risk

Andrew L. Sheldon

Introduction

The rivers and lakes of North America are inhabited by a rich assemblage of fishes. Approximately 700 species regularly spend all or significant portions of their life histories in the fresh waters of the United States and Canada (Lee et al. 1980); the diversity exceeds that of the birds of the same region (Robbins, Bruun, & Zim 1966). The distinctive Mexican fauna (Miller 1966; Miller & Smith 1986), with numerous endemics and an infusion of neotropical taxa, adds many more. Although the diversity of fishes in the United States does not approach that of the great tropical rivers of South America, Africa, and Asia, it far exceeds that of European fresh waters (Lowe-McConnell 1969, 1975).

Freshwater fishes of the United States include survivors of ancient lineages such as the bowfin (Amiidae), gars (Lepisosteidae), sturgeons (Acipenseridae), paddlefish (Polyodontidae), and mooneyes (Hiodontidae). Many families (e.g., minnows, Cyprinidae) are shared with Eurasia, and in some families, including the pikes (Esocidae) and sculpins (Cottidae), genera or species occur throughout the Holarctic. Sunfishes (Centrarchidae) and a catfish family (Ictaluridae) are restricted to North America and the suckers (Catostomidae) nearly

so. Smaller families also confined to North America are pirate perch (Aphredoderidae), trout-perches (Percopsidae), and cavefishes (Amblyopsidae). Darters (Etheostomatini, Percidae) have radiated spectacularly (Page 1983); they contribute about 20% of the species in the total fish fauna north of Mexico and are among the most attractive of all fishes. The cyprinid genus *Notropis* (which may be polyphyletic) is similarly diverse. Gilbert (1976) and Hocutt & Wiley (1986) provide thorough treatments of the origin and distribution of North American fishes. Moyle and Cech (1982) illustrate representatives of most families and summarize their biology.

Minnows dominate the fauna and the Percidae (mostly darters) run a close second in diversity. Other prominent families are the suckers, catfishes, and sunfishes. Killfishes (Cyprinodontidae) and livebearers (Poeciliidae) are widely distributed and are especially characteristic of remnant waters of the southwestern United States. Salmonidae (trout, salmon, whitefish, grayling) and sculpins become increasingly common in the northern United States and Canada. Approximately twenty additional families contribute species, many of which are widespread and numerous.

These fishes are not evenly distributed (Fig. 1). Rivers west of the continental divide contain

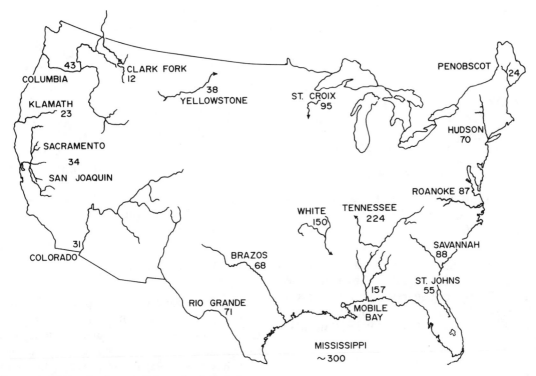

FIGURE 1. Native freshwater fishes of selected river systems.

Conservation Status

relatively few species although their faunas are distinctive and endemism is high (e.g., Stanford & Ward 1986). East of the divide, the Mississippi system is far richer than the smaller Gulf and Atlantic coastal rivers although these in turn have more species than the Pacific drainages. Fish diversity is greatest in the southeastern United States (see McAllister et al. [1986] for a quantitative analysis of patterns of species richness). Diversity decreases northward, especially along the Atlantic coast. Rivers such as the Savannah that drain physiographically diverse areas have more species than rivers such as the St. Johns in Florida that are confined to lowlands.

The Great Lakes and tributaries, which are not considered further in this essay, have 162 native species (Underhill 1986). Most of these occur in streams. The lacustrine component of the fish fauna has been catastrophically altered by introductions, exploitation, and pollution (e.g., Smith 1968; Crowder 1980), resulting in the extinction of three species and one subspecies (Williams & Nowak 1986).

More than half of the freshwater fishes of the United States and Canada receive some legal protection in at least part of their range (Johnson 1987). Far fewer (a total of 75 species and subspecies) are federally designated as endangered or threatened (U.S. Fish & Wildlife Service 1987). Endangered taxa are strongly concentrated in the arid west (Fig. 2). Threatened species (Fig. 3) likewise are predominantly western but include a greater proportion from the diverse southeastern fauna. Many of these fishes, especially the western ones, are extremely localized in springs and fragmented desert drainages. Ono et al. (1983) provide species accounts for these and additional rare species.

The emphasis in the conservation of fishes has been on habitat preservation and single-species management. Given the restricted habitats of many fishes and the economic value of water in arid landscapes, this emphasis is understandable and sensible. Focused plans for

FIGURE 2. Distribution of federally listed endangered fishes. Taxa restricted to a single state (●) are distinguished from those in two or more states (○).

FIGURE 3. Distribution of federally listed threatened fishes. Symbols as in Figure 2.

the management of species and habitats will continue to be an integral part of fish conservation. However, I suggest that a comprehensive view of river systems and their fishes is required if more than a small fraction of the fish fauna is to persist.

Patterns of Distribution and Abundance

Few species of stream fish are widely distributed (McAllister et al. 1986). Abundances in local assemblages approximate lognormal frequency distributions (Loubens 1970; Sheldon 1987). Thus, most species are geographically restricted and numerically rare when they occur. Furthermore, patchy distributions are common. Some threatened species (Fig. 3) apparently exist in only a few widely separate localities. Although this apparent patchiness may be a consequence of general rarity and low detectability (e.g., Starnes et al. 1977), present evidence implies both local rarity and restricted range.

Historical records are adequate to show substantial declines of many stream fishes. For example, the rosy-face shiner, *Notropis rubellus,* remains widely distributed (Lee et al. 1980) and locally common, yet it had disappeared from much of its range in Ohio by midcentury (Trautman 1957). A more extreme case is the harelip sucker *Lagochila,* which vanished from the rivers of nine southeastern and midwestern states before 1900 (Ono et al. 1983).

Rivers as Islands

Species-area relationships and problems of connectedness occupy a prominent place in conservation theory, and stream systems are amenable to this approach. The geometry of drainage area has been related to species richness of fishes (Eadie et al. 1986; Larsen et al. 1986; Swift et al. 1986) and molluscs (Sepkoski & Rex 1974). Discharge, a more direct measure of available stream habitat, was used by Livingstone, Rowland, & Bailey (1982) and Sheldon (1987) in place of area. The observed increase of faunal diversity with increasing drainage area is a result of several processes. Total area of stream surface increases with drainage area, yielding concomitant increases in numbers of individuals and species (Preston 1962). Furthermore, the fusion of small, low-order (Strahler 1957) streams with increasing drainage area forms larger streams, which provide additional habitats. Predictable changes in habitat, composition, and diversity of the fish fauna with increasing stream order are well documented (Thompson & Hunt 1930; Kuehne 1962; Sheldon 1968; Horwitz 1978). Speciation rates may be greater in larger drainages if isolation by distance and intervening high order streams reduce gene flow among populations of small-stream upland forms. Wiley & Mayden (1985) map vicariant fishes separated by large rivers. In summary, a mixture of ecological and evolutionary processes governed by the template of watershed geometry (Leopold, Wolman, & Miller 1964; Richards 1982) is capable of producing a strong positive dependence of fish diversity on drainage area. If diversity depends on area, drastic alteration of drainages can be expected to reduce diversity in streams remote from those directly affected by pollution or impoundment (Sheldon 1987).

Figure 4 contrasts species-area relationship in tributaries of the Mississippi with presently isolated streams draining to the Atlantic and Gulf of Mexico. All the streams lie in regions of fairly similar rainfall, runoff, and physiography. The independent coastal drainages consistently support fewer native fishes than interconnected streams of comparable area in the Mississippi system.

An especially vivid demonstration of insular effects (Fig. 4) is provided by two Mississippi tributaries isolated by waterfalls (Stauffer et al. 1982; Hocutt et al. 1986). Although proximate ecological factors, such as water chemistry and lack of lowland habitats, may be involved, the low diversities of fishes in the New River above Kanawha Falls and the upper Cumberland river above Cumberland Falls are consistent with an hypothesis of extreme insularity. (Both streams contain endemic fishes.) Even small streams

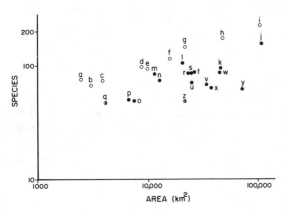

FIGURE 4. Species–drainage area curves for Mississippi tributaries (○), independent Gulf and Atlantic drainages (●), and Mississippi tributaries isolated by major waterfalls (◑). a. Guyandotte b. Big South Fork Cumberland c. Little Kanawha d. Licking e. Big Sandy f. Kentucky g. Green h. Cumberland i. Tennessee j. Mobile k. Santee l. Pearl m. Escambia n. Neuse o. York p. Rappahannock q. Cumberland above falls r. Pascagoula s. Roanoke t. Savannah u. James v. Pee Dee w. Apalachicola x. Potomac y. Susquehanna z. New.

having free interchange with larger systems support many fish species. For example, Cicerello & Butler (1985) reported 73 species from Buck Creek, Kentucky (drainage area 767 km^2), which is a tributary of the Cumberland below the falls. Excluding 15 species associated with the low-gradient downstream reaches, which are influenced by an impoundment on the Cumberland River, and three introduced or recently invading species, leaves a richer fauna than that of the nearby much larger upper Cumberland.

The Riverine Archipelago

A higher order of faunal complexity is imposed on river systems by major watershed divides, the largest lowland rivers, and saltwater barriers. Some of these barriers have been breached by low sea levels during the Pleistocene and stream capture so rivers are not only insular but archipelagic.

Interdrainage differentiation of fish faunas is shown in Figure 5. Faunal dissimilarity was calculated from data in Hocutt & Wiley (1986) and Stauffer et al. (1982) using Preston's (1962) resemblance equation. Preston's index z ranges from 0 (complete identity) to 1.0 (no species in common). Preston (1962) proposed that z = .27 implies "isolates in equilibrium" under his hypothesis of the canonical lognormal distribution of species abundances.

Neighboring rivers on the Atlantic coast and within the Mississippi system are differentiated to a degree that, especially in the coastal rivers, approaches that of "isolates in equilibrium." Gulf coast drainages are more strongly differentiated. (This region is also one in which intraspecific genetic similarities between drainages are low [Bermingham & Avise 1986].) The greatest dissimilarities are between drainages separated by the coastal-Mississippi divide (e.g., the Potomac and the Monongahela, the Mobile and the Tennessee). Kanawha Falls isolates the New River from the lower Kanawha to a comparable degree. The upper and lower Cumberland are less differentiated. (In this calculation I assumed that several species presently found in tributaries below the falls are upper Cumberland species transferred by head-

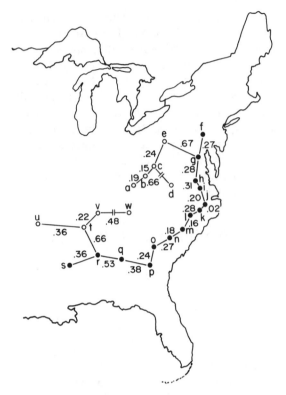

FIGURE 5. Faunal dissimilarity (Preston's z) of nearby drainages in the Mississippi (○) and Gulf Coast-Atlantic (●) regions. a. Big Sandy b. Guyandotte c. Kanawha (excluding New) d. New e. Monongahela f. Susquehanna g. Potomac h. James i. Roanoke j. Tar k. Neuse l. Cape Fear m. Pee Dee n. Santee o. Savannah p. Altamaha q. Apalachicola r. Mobile s. Pearl t. Tennessee u. White v. lower Cumberland w. Cumberland above falls.

ward erosion of the falls [Starnes & Etnier 1986]. Alternatively, these species may be ecologically restricted to the Cumberland Plateau physiographic region.) A single trans-Mississippi comparison (the Tennessee and the White) shows the effect of isolation of the Ozark and Appalachian uplands by lowland and large river habitats.

Faunal distinctions among rivers are being eroded by introductions from other continents (Courtenay & Stauffer 1984) and transfers within North America (Moyle et al. 1986). Stocking of game and forage fishes, aquarium releases, and bait-bucket introductions have all contributed to the breakdown of regional uniqueness.

Proportionally, the effects of introductions are especially noticeable in the depauperate faunas of western drainages. For example, the fauna of the Clark Fork in western Montana (Fig. 1) has been trebled by introductions (Brown 1971), to the clear detriment of some native fishes such as the cutthroat trout *Salmo clarki*.

Introductions are less obvious in the diverse faunas of the southeast. The New River, in which 43% of the fishes are nonnative (Hocutt et al. 1986), is a regional extreme, and Jenkins's (1987) careful analysis indicates from 12 to 32% of the species in 11 other Virginia drainages were introduced. Most of these introductions occurred in the last 30 years and adverse effects on the native fishes are not apparent. However, diverse faunas are not immune to impacts from introduced species, and Lemley (1985) has documented reductions of native fishes in North Carolina streams by introduced green sunfish, *Lepomis cyanellus*.

Wholesale invasions become possible when drainages are linked by navigational or hydro-electrical projects. Catastrophic changes in the Great Lakes (Smith 1968) were triggered by invading species (lamprey, alewife) entering through navigation canals. Balon et al. (1986) predict reciprocal invasions when the Rhine-Danube ship canal is complete. In the United States, two (Mobile, Tennessee) of the richest (Fig. 1) and most distinctive (Fig. 5) fish faunas are now connected by the Tennessee-Tombigbee Waterway (Boschung 1987). Although mainstream impoundments on the Tennessee and lentic environments associated with locks on the Tombigbee may limit dispersal of many stream fishes, the stage is set for invasions, hybridization, species replacement, and ultimately a reduction in total species richness of the combined fauna (Preston 1962).

Conclusions

The common perception of all small non-game fishes as "minnows" and their existence in environments where few persons ever see them makes it difficult to muster public support for their conservation. In conflicts over development of increasingly valuable water resources, the fishes have few advocates. Beyond the serious issue of public awareness, however, lie other problems intrinsic to river systems and their biota.

In any species-rich fauna, most species are rare, and stream fishes are no exception (Sheldon 1987). In such cases reductions in the number of species by environmental stress or habitat fragmentation may be more predictable than the identities of censored species, so faunal management, rather than a focus on single species, is required.

Species-area relationships (Fig. 4) imply that extinctions will follow fragmentation of drainage networks. The impoundment of virtually the entire 1000 km length of the Tennessee River caused the disappearance of the minnow *Hybognathus nuchalis* from the entire system including tributaries not directly affected by reservoirs (Etnier et al. 1979). Other species may have been lost in some tributaries in the 50 years since construction began. Records and collections, as is inevitable when many rare species are involved, are probably inadequate to fully document such extinctions if they occurred. Both theory and experience with other taxa and habitats suggest that extinctions are inevitable.

The between-drainage component of fish diversity is large (Fig. 5), but introductions and extinctions will eventually homogenize the fish fauna to the point where much of the variety and evolutionary history has been lost. Prohibition of interbasin transfers of water and more effective controls on introductions could greatly slow this process.

The geometry of rivers further complicates preservation. Rivers are open, directional systems so protection of any segment requires control over the entire upstream network and the surrounding landscape. There is little likelihood that such protection can be given to very many streams of order 4 or higher, yet it is these streams that support the greatest diversity of fishes. Optimistically, some degree of riparian control and prevention of industrial and agricultural pollution, channelization, and impoundment may be sufficient to maintain diverse assemblages. Still larger streams, such as the Tennessee, have been irreparably modified. Conservation efforts should be focused on the largest reasonably natural drainages in as many major river systems as possible. McDowall (1984) and Maitland (1985) suggest useful criteria for site selection and protection of freshwater fishes. However, the size of North American rivers and the diversity of their fishes make the problems far more difficult. (Buck Creek [Cicerello & Butler 1985] supports more native fishes than the British Isles.)

A comprehensive plan for the conservation of North American fishes lies in the future. A biogeographic overview, as opposed to piecemeal species management, is a necessary component of such a plan although single-species management is compatible with a more inclusive approach since streams containing rare species often support diverse assemblages. The conservation of fishes is also compatible with the protection of other organisms, such as molluscs and crayfishes, and of riparian vegetation, water quality, amenity, and the entire spectrum of values of running-water ecosystems. The evolutionary wealth of fishes that so impressed David Starr Jordan (1922) and earlier scientists is an important component of that value spectrum.

Acknowledgments

Peter Moyle and Jim Williams made helpful comments on the manuscript. DeWayne Williams prepared the figures. Dave Etnier, Herb Boschung, and the Corps of Engineers, U.S. Army, provided information and slides for the oral presentation of this material. I thank them all.

References

Balon, E K., S. S. Crawford, and A. Lelek. 1986. Fish communities of the upper Danube River (Germany, Austria) prior to the new Rhein-Main-Donau connection. Environmental Biology of Fishes **15**:243–271.

Bermingham, E., and J. C. Avise. 1986. Molecular zoogeography of freshwater fishes in the southeastern United States. Genetics **113**:939–965.

Boschung, H. T. 1987. Physical factors and the distribution and abundance of fishes in the upper Tombigbee River system of Alabama and Mississippi, with emphasis on the Tennessee-Tombigbee Waterway. Pages 184–192 *in* W. J. Matthews and D. C. Heins, editors. Community and Evolutionary Ecology of North American Stream Fishes. University of Oklahoma Press, Norman, Oklahoma.

Brown, C. J. D. 1971. Fishes of Montana. Big Sky Books, Bozeman, Montana.

Cicerello, R. R., and R. S. Butler. 1985. Fishes of Buck Creek, Cumberland River drainage, Kentucky. Brimleyana 11:133-159.

Courtenay, W. R., Jr., and J. R. Stauffer, Jr., editors. 1984. Distribution, Biology, and Management of Exotic Fishes. Johns Hopkins University Press, Baltimore.

Crowder, L. B. 1980. Alewife, rainbow smelt and native fishes in Lake Michigan: Competition or predation? Environmental Biology of Fishes 5:225-233.

Eadie, J. M., T. A. Hurley, R. D. Montgomerie, and K. I. Teather. 1986. Lakes and rivers as islands: Species-area relationships in the fish faunas of Ontario. Environmental Biology of Fishes 15:81-89.

Etnier, D. A., W. C. Starnes, and B. H. Bauer. 1979. Whatever happened to the silvery minnow (*Hybognathus nuchalis*) in the Tennessee River? Proceedings of the Southeastern Fishes Council 2(3):1-3.

Gilbert, C. R. 1976. Composition and derivation of the North American freshwater fish fauna. Florida Scientist 39:104-111.

Hocutt, C. H., and E. O. Wiley. 1986. The Zoogeography of North American Freshwater Fishes. John Wiley, New York.

Hocutt, C. H., R. E. Jenkins, and J. R. Stauffer, Jr. 1986. Zoogeography of the fishes of the central Appalachians and central Atlantic Coastal Plain. Pages 161-211 in C. H. Hocutt and E. O. Wiley, editors. The Zoogeography of North American Freshwater Fishes. John Wiley, New York.

Horwitz, R. J. 1978. Temporal variability patterns and the distributional patterns of stream fishes. Ecological Monographs 48:307-321.

Jenkins, R. E. 1987. Introduced fishes in Virginia drainages, with special reference to cryptic introductions and biogeographic study. Abstracts of the 67th Annual Meeting, American Society of Ichthyologists and Herpetologists, 57-58.

Johnson, J. E. 1987. Protected Fishes of the United States and Canada. American Fisheries Society, Bethesda, Maryland.

Jordan, D. S. 1922. Days of a Man. World Book Company, Yonkers-on-Hudson, New York.

Kuehne, R. A. 1962. A classification of streams, illustrated by fish distributions in an eastern Kentucky creek. Ecology 43:608-614.

Larsen, D. P., J. M. Omernik, R. M. Hughes, et al. 1986. Correspondence between spatial patterns in fish assemblages in Ohio streams and aquatic ecoregions. Environmental Management 10:815-828.

Lee, D. S., C. R. Gilberg, D. H. Hocutt, R. E. Jenkins, D. E. McAllister, and J. R. Stauffer, Jr. 1980 et seq. Atlas of North American Freshwater Fishes. North Carolina State Museum of Natural History, Raleigh.

Lemley, A. D. 1985. Suppression of native fish populations by green sunfish in first-order streams of Piedmont, North Carolina. Transactions of the American Fisheries Society 114:705-712.

Leopold, L. B., M. G. Wolman, and J. P. Miller. 1964. Fluvial Processes in Geomorphology. W. H. Freeman, San Francisco.

Livingstone, D. A., M. Rowland, and P. E. Bailey. 1982. On the size of African riverine fish faunas. American Zoologist 22:361-369.

Loubens, G. 1970. Etude de certains peuplements ichtyologiques par des peches au poison (2e note). Cah. ORSTOM (Hydrobiol.) 4:45-61.

Lowe-McConnell, R. H. 1969. Speciation in tropical freshwater fishes. Biological Journal of the Linnean Society 1:51-75.

Lowe-McConnell, R. H. 1975. Fish Communities in Tropical Freshwaters. New York: Longman.

Maitland, P. S. 1985. Criteria for the selection of important sites for freshwater fish in the British Isles. Biological Conservation 31:335-353.

McAllister, D. E., S. P. Platania, F. W. Schueler, M. E. Baldwin, and D. S. Lee. 1986. Ichthyofaunal patterns on a geographic grid. Pages 17-51 in C. H. Hocutt and E. O. Wiley, editors. The Zoogeography of North American Freshwater Fishes. John Wiley, New York.

McDowall, R. M. 1984. Designing reserves for freshwater fish in New Zealand. Journal of the Royal Society of New Zealand 14:17-27.

Miller, R. R. 1966. Geographical distribution of Central American fishes. Copeia 1966:773-802.

Miller, R. R., and M. I. Smith. 1986. Origin and geography of the fishes of central Mexico. Pages 487-517 in C. H. Hocutt and E. O. Wiley, editors. The Zoogeography of North American Freshwater Fishes. John Wiley, New York.

Moyle, P. B., and J. J. Cech, Jr. 1982. Fishes: An Introduction to Ichthyology. Prentice-Hall, Englewood Cliffs, New Jersey.

Moyle, P. B., H. W. Li, and B. A. Barton. 1986. The Frankenstein effect: Impact of introduced fishes on native fishes in North America. Pages 415-426 in R. H. Stroud, editor. Fish Culture in Fisheries Management. American Fisheries Society, Bethesda, Maryland.

Ono, R. D., J. D. Williams, and A. Wagner. 1983. Vanishing Fishes of North America. Stone Wall Press, Washington, D.C.

Page, L. M. 1983. Handbook of Darters. TFH Publications, Neptune City, New Jersey.

Preston, F. W. 1962. The canonical distribution of commonness and rarity. Ecology **43**:185–215, 410–412.

Richards, K. 1982. Rivers—Form and Process in Alluvial Channels. Methuen, New York.

Robbins, C. S., Bruun, B., and H. S. Zim. 1966. Birds of North America. Golden Press, New York.

Sepkoski, J. J., Jr., and M. A. Rex. 1974. Distribution of freshwater mussels: Coastal rivers as biogeographic islands. Systematic Zoology **23**:165–188.

Sheldon, A. I. 1968. Species diversity and longitudinal succession in stream fishes. Ecology **49**:193–198.

Sheldon, A. I. 1987. Rarity: Patterns and consequences for stream fishes. Pages 203–209 *in* W. J. Matthews and D. C. Heins, editors. Community and Evolutionary Ecology of North American Stream Fishes. University of Oklahoma Press, Norman, Oklahoma.

Smith, S. H. 1968. Species succession and fishery exploitation of the Great Lakes. Journal of the Fisheries Research Board of Canada **25**:667–693.

Stanford J. A., and J. V. Ward. 1986. Fishes of the Colorado system. Pages 385–402 *in* B. R. Davies and K. F. Walker, editors. The Ecology of River Systems, Dr. W. Junk Publishers, Dordrecht, the Netherlands.

Starnes, W. C., and D. A. Etnier. 1986. Drainage evolution and fish biogeography of the Tennessee and Cumberland rivers drainage realm. Pages 325–361 *in* C. H. Hocutt and E. O. Wiley, editors. The Zoogeography of North American Freshwater Fishes. John Wiley, New York.

Starnes, W. C., D. A. Etnier, I. B. Starnes, and N. H. Douglas 1977. Zoogeographic implications of the rediscovery of the percid genus *Ammocrypta* in the Tennessee River drainage. Copeia **1977**:783–786.

Stauffer, J. R., Jr., B. M. Burr, C. H. Hocutt, and R. E. Jenkins. 1982. Checklist of the fishes of the central and northern Appalachian Mountains. Proceedings of the Biological Society of Washington **95** 27–47.

Strahler, A. N. 1957. Quantitative analysis of watershed geomorphology. American Geophysical Union Transactions **38**:913–920.

Swift, C. C., C. R. Gilbert, S. A. Borton, G. H. Burgess, and R. W. Yerger. 1986. Zoogeography of the freshwater fishes of the southeastern United States: Savannah River to Lake Pontchartrain. Pages 213–265 *in* C. H. Hocutt and E. O. Wiley, editors. The Zoogeography of North American Freshwater Fishes. John Wiley, New York.

Thompson, D. H., and F. D. Hunt. 1930. The fishes of Champaign County. Bulletin Illinois Natural History Survey **19**:5–101.

Trautman, M. B. 1957. The Fishes of Ohio. University of Ohio Press, Columbia.

Underhill, J. C. 1986. The fish fauna of the Laurentian Great Lakes, the St. Lawrence lowlands, Newfoundland and Labrador. Pages 105–136 *in* C. H. Hocutt and E. O. Wiley, editors. The Zoogeography of North American Freshwater Fishes. John Wiley, New York.

U.S. Fish and Wildlife Service. 1987. Endangered and Threatened Wildlife and Plants. Washington, D.C.

Wiley, E. O., and R. I. Mayden. 1985. Systematics and speciation in phylogenetic systematics, with examples from the North American fish fauna. Annals of the Missouri Botanical Garden **72**:596–635.

Williams, J. D., and R. M. Nowak. 1986. Vanishing species in our own backyard: Extinct fish and wildlife of the United States and Canada. Pages 107–139 *in* I. Kaufman and K. Mallory, editors. The Last Extinction. MIT Press, Cambridge, Massachusetts.

5
Scale Perspectives on Avian Diversity in Western Riparian Ecosystems

Fritz L. Knopf and Fred B. Samson

Riparian vegetation has been referred to as the aorta of an ecosystem because of its significance to the perpetuation of water, fish, wildlife, rangeland, and forest resources (Wilson 1979). Within the United States, this vegetation provides the most productive and valuable wildlife habitat in the Southwest (Johnson & Jones 1977), West (Graul & Bissell 1978), Midwest (Samson 1979), Northwest (Thomas et al. 1979; Raedeke 1988), and Great Plains (Tubbs 1980). The value of riparian areas, especially as they provide habitats for wildlife, has stimulated an entirely new subdiscipline of ecological studies (Johnson & Lowe 1985), and conservation of streamside vegetation has been mandated by a number of executive orders, legislative acts, and internal agency policies (Knopf et al. 1988a).

Interest in conservation of riparian vegetation has focused primarily upon the value of the vegetation in providing habitat for a locally rich avifauna. Although less than 1% of the western landscape of the United States supports riparian vegetation, this vegetation provides habitat for more species of breeding birds than any other vegetation association in western states. Of all species breeding in northern Colorado, 82% occur in riparian vegetation (Knopf 1985). Over half of southwestern species are dependent upon riparian vegetation, and 78 (47%) of the 160 avian species that breed in that region are restricted to it specifically (Johnson et al. 1977). In New Mexico, 46% of the species

breeding in the San Juan Valley (Schmitt 1976) and 49% of the species breeding in the Gila Valley (Hubbard 1971) could disappear from the region if all riparian vegetation were lost. Hubbard (1977) concluded that 16–17% of the breeding avifauna of temperate North America occurs along these two river systems.

Riparian vegetation also provides habitat for many birds during migration (Stevens et al. 1977; Henke & Stone 1979). However, riparian associations may be most critical as wintering habitat for a number of species. Fretwell (1972) emphasized that habitats for over-wintering— not habitats used for breeding—may regulate avian populations. Samson (1979) reminded us of this oversight relative to lowland forest management, and the concept of an "ecological crunch" (Wiens 1979) predicts that avian populations in temperate latitudes are regulated by resource availability during the winter. The recent attention being paid to Neotropical migrant birds (Terborgh 1989) has focused the conservation of a large guild of species upon wintering areas (Hagan & Johnston 1992; Sherry & Holmes 1993). Within riparian systems specifically, birds use the vegetation differently during winter and summer, with winter appearing to be the time of greatest stress due to resource availability and competition (Anderson & Ohmart 1977; Rice et al. 1980).

This paper provides scale perspectives on patterns within avian assemblages of riparian vegetation. Pattern detection is fundamental to explaining processes within ecological frameworks (Wiens 1983, 1989), and we hope to encourage broader perspectives and approaches to conservation of riparian systems as they provide habitat for native birds. Whereas riparian bird studies have been reported primarily within symposia and proceedings (Knopf 1987), our message should be viewed as a selective rather than exhaustive review of published information on riparian bird diversity.

Dimensional Images of Riparian Vegetation as Avian Habitats

Avian assemblages within riparian vegetation can be viewed from three separate spatial perspectives. These perspectives include the site-specific avifauna, relationships among species assemblages within the surrounding landscape/drainage, and the distinctiveness of the riparian assemblage of species within a regional or continental avifauna. These perspectives follow Whittaker's (1975) alpha, beta, and gamma levels of diversity, and they provide very different images of the value of a local tract of riparian vegetation as it contributes to the avian diversity of North America.

Alpha View – The Site Avifauna

The site perspective has dominated riparian conservation thinking. Extremely high avian species richness and evenness (both components of alpha, or site-specific, diversity) have been the leading argument for the need to protect riparian vegetation.

Relative to conservation of riparian vegetation, site diversity can provide misleading views about the ecological value of a specific tract. Some precautions concerning the application of alpha diversity measures to riparian bird assemblages include the following

1. The enhanced richness of bird species in riparian vegetation is often attributed to structural complexity of the vegetation (Anderson & Ohmart 1977; Bull & Skovlin 1982) in locales surrounded by shrub-steppe, grassland, agricultural lands, or desert. The relationship

seems weaker, however, where surrounding uplands are forested (see Salt 1957) and the contribution of "edge effect" (Willson 1974) is difficult to separate. We also note that foliage structure is correlated with the species composition of the vegetative association, and many studies (Franzreb 1978; Holmes & Robinson 1981; Rice et al. 1984; Robinson & Holmes 1984; Rotenberry 1985) have emphasized that birds respond to the presence of a tree species. The correlation of vegetation structure with the alpha diversity of birds appears simplistic and, perhaps, secondary to floristic composition at a site.

2. Disturbances within forests often have differential impacts upon the structural layers of that vegetative association. However, both natural disturbances, such as flooding, and human-induced disturbances such as cattle grazing, within riparian vegetation virtually always result in damage to the woody understory; birds frequenting dense shrub tend to be especially susceptible to disturbance (Klebenow & Oakleaf 1984; Sedgwick & Knopf 1987; Knopf et al. 1988b). Critical changes in populations of these species are masked by population events occurring in the midstory and canopy vegetation (see Knopf & Sedgwick 1987) if the avifauna is monitored using alpha diversity measures.

3. Alpha diversity is often calculated to include information on the relative abundance of species. However, species with the widest geographic distributions – "ecological generalists" – tend to be the most common locally (Bock 1987), thus diluting information on unique species within a local vegetative complex.

4. Most significantly, alpha diversity measures are insensitive to bird species composition (Balda 1975). Brown-headed Cowbirds (*Molothrus ater*) substitute equally for Black-headed Grosbeaks (*Pheucticus melanocephalus*), European Starlings (*Sturnus vulgaris*) for Lazuli Buntings (*Passerina amoena*), and Mourning Doves (*Zenaida mactrichas*). Within each of these arbitrarily defined species pairs, the former occurs across many habitats whereas the latter is restricted to moist-soil sites. Administrative planning to conserve biotic diversity has tra-

ditionally emphasized the "content" of species presence at the expense of "context" information (Noss & Harris 1986; Murphy 1989).

A remaining problem associated with using alpha diversity to evaluate riparian tracts is especially critical relative to conservation philosophy. Managing for alpha diversity puts those species unique to riparian vegetation at a disadvantage. Just as management for endangered species, waterfowl, or cavity-nesting birds is oriented towards specific soil/vegetative associations, so must riparian vegetation be managed for the unique habitat that it provides for native birds (Graul et al. 1976). Evaluations of the significance of riparian vegetation to birds must segregate species relative to their dependency upon riparian ecosystems within broader-than-local contexts.

Beta View — The Landscape Interface

Beta diversity comparisons within a vegetative association are often referred to as gradient analyses (Whittaker 1975). Riparian ecosystems tend to be viewed as links between disjunct islands of forest across watershed gradients. In the accentuated topographical relief of western landscapes, however, they are highly linearized forests transecting watersheds between upland associations of high elevations and very different associations at lower elevations. Within a management unit, conservation of a riparian avifauna requires perspectives and interpretations of that fauna relative to avian assemblages within the entire watershed. Beta diversity considerations pertinent to defining a focus for riparian bird conservation at a locale include the following:

1. Avian assemblages of riparian tracts and adjoining uplands are not independent. Upland bird communities can influence species composition within a riparian site and vice versa. Many species of birds (such as doves) within more xeric landscapes move to riparian sites to obtain water on a regular basis. Other species need riparian vegetation for selected ecological prerequisites (such as nest substrates) but actually forage in surrounding uplands (Pleasants 1979). Desert

uplands surrounding an island of riparian vegetation in Arizona, for example, contribute only 1–1.5% of the species breeding in the riparian avifauna, whereas the riparian avifauna contributed 7–15% of the birds breeding in the adjacent uplands (Szaro & Jakle 1985). Szaro and Jakle concluded that the riparian community effects extended 0.6–1.0 km into the uplands.

2. Interactions between riparian and upland assemblages can vary seasonally. Some permanent resident species of uplands, such as Gila Woodpeckers (*Melanerpes uropygialis*) tend to use riparian areas in the winter more than during the breeding season (R. C. Szaro, personal communication).

3. The number of bird species shared between riparian and adjacent upland sites varies across a watershed. For example, the number of species using both the riparian and upland vegetation at a locale appears greatest at intermediate-elevation sites within the South Platte River drainage of northern Colorado (Knopf 1985). The riparian avifauna appears most unique from contiguous uplands at the headwaters of that drainage and at lower elevations where the stream is of the sixth order (see Kuehne 1962 for a review of stream ordering). At the intermediate elevations, 20–50% of the bird species breeding on a site may be shared between the two vegetative communities. Mixing of avifaunal communities between sites and years is most variable at these intermediate elevations. Avifaunal mixing is lower (8–19%) at other elevations and tends to be relatively consistent between years. This watershed pattern in Colorado can be attributed to the greater annual turnover of species within both riparian and upland vegetation at intermediate elevations.

4. Birds move among locales at a greater rate in riparian than in upland vegetation. Avifaunal similarity among six sites across an altitudinal cline in northern Colorado is twice as great in riparian sites than in adjacent upland sites (Knopf 1985). The consequence of this interchange is that the riparian avifauna is most unique at the upper and lower reaches of a watershed, supporting conclusions from riparian/upland compari-

sons at various elevations that conservation efforts should focus on the two ends of the stream gradient. This generalization probably does not apply within the coastal ranges of the far west, where headwater riparian areas contain many woody species typical of surrounding uplands, and its universality needs to be tested elsewhere.

5. Riparian vegetation may provide critical habitat for birds during seasonal migrations. Wauer (1977) and others have speculated that birds that spend the winter in the Neotropics actually migrate along riparian corridors seasonally. The potential for birds to migrate along riparian corridors in the xeric western landscape is obvious, and current studies are under way along the San Pedro River of Arizona to confirm or reject such use.

6. The destruction of riparian vegetation not only can cause local extinctions (Dobkin & Wilcox 1986) but also surely reduces the ability of populations to recolonize sites. Avian populations in small, remnant islands of forest often are reduced to below minimum levels necessary for continued survival and thereby are subjected to local extinction (see Blake & Karr 1984; Harris 1984). Species turnover patterns within a riparian forest have been likened to patterns typical of island-biogeographic theories (Rice et al. 1983; Gutzwiller & Anderson 1987). Yellow-billed Cuckoos, already extirpated in Washington, have declined in California recently as extensive tracts of riparian vegetation have been either eliminated or fragmented into a series of remote islands (Laymon & Halterman 1989). The smaller the drainage of riparian forest fragment, the more isolated the local avifauna and the greater the likelihood of local extinctions of riparian-obligate species.

Gamma View — Regional Comparisons

Besides viewing riparian systems as significant entities providing habitat for species, conservation biologists (Forman & Baudry 1984; Noss 1987; Simberloff & Cox 1987) frequently cite these systems as examples of corridors when addressing integrated landscape conservation. Just as one can dismiss the assumption that floristic composition within riparian corridors is stable and predictable along a drainage (Szaro 1988), so too can one dismiss the idea of western riparian vegetation connecting two forest islands of similar floristic composition and edaphic characteristics within a region. From an avian perspective especially, riparian systems must be viewed as internally dynamic forest islands of maximized edge.

Considerations of the gamma diversity perspective on the ecology of a local riparian avifauna include the following:

1. Simply (and emphatically) stated, riparian vegetation has not provided habitats for the radiation of many new bird species in North America. At many locales, the riparian-dependent avifauna of western North America can be viewed as extensions of the avifaunas of the North American eastern and western deciduous forests and the semitropical vegetation of Mexico. Ecological generalists such as the American Robin (*Turdus migratorius*) and the Brown-headed Cowbird occur throughout western states and reach their highest local densities in riparian systems and suburban plantings. Species of the eastern deciduous forest such as Orchard Orioles (*Icterus spurius*) have narrower ecological tolerance limits and generally do not occur in the western locations except where riparian vegetation (Tubbs 1980) or other deciduous vegetation (Anderson 1971; Martin 1981) provide habitat on the Great Plains.

2. Riparian bird movements can extend across a region, enabling species to colonize, or recolonize, geographic locales or regions. Biologists must consider the potential for creating dispersal opportunities across historic, zoogeographic barriers for riparian-facultative species. Riparian vegetation can develop rapidly with changing hydrologies both locally and regionally (Wedel 1986). Such developments and degradations occur with changes in historical flow patterns within drainages (Nadler & Schumm 1981) and can result in the movement of many species into a region from distant locations (Knopf & Scott 1990). In northeastern Colorado, almost 90% of the contemporary avifauna has arrived since the development

of a gallery forest along the South Platte River within the last 90 years (Knopf 1986).

3. Riparian-facultative species moving into riparian forests at new locales may represent potential competitors for congeneric species native to a region. The almost universal example of European Starlings displacing native species from nesting cavities illustrates such competitive displacement.

4. Geographic mixing of avifaunas may also have genetic consequences within populations. The faunal mixing seen in eastern Colorado has not only enhanced the local avian assemblage but also has provided a contact zone for some congeneric species that had been geographically separated since the last glacial advance. Some of these species hybridize within the new contact zone (Williams & Wheat 1971; Rising 1983).

Conservation Directions

Riparian vegetation provides habitat for a maximum number of species locally, and such assemblages are easily offered as "display areas" of conservation success. Based upon the foregoing discussion of spatial and temporal scale, however, an emphasis on maximizing the number of species within a management unit will favor the addition of ecological generalists to each assemblage at the expense of regionally unique components—management toward a cosmopolitan avifauna. The biological uniqueness of a riparian tract increases with the number of regionally evolved species (Wilcove 1988). Thus, it follows that management activities to conserve riparian areas should

1. de-emphasize practices promoting alpha diversity locally;
2. emphasize beta diversity applications on management units to promote conservation of biotic integrity;
3. restore ecological processes that favor long-term sustainability of the ecosystem (Samson & Knopf 1982, 1993).

Scale considerations are fundamental to the understanding of avian species biology (Knopf et al. 1990) and community patterns (Wiens et al. 1987) and large-scale perspectives (both spatial and temporal) are especially difficult to evaluate and incorporate into the conservation of landscapes. Incorporating large-scale perspectives into management unit decisions will necessitate specific information about the patterns of the regional avifauna, with the approach characterized by a narrowing of focus from regional to between-habitat diversity perspectives. Questions to be addressed about an avifauna prior to developing a riparian conservation program include the following:

1. What was the historical extensiveness and composition of the riparian tract being considered? Is the vegetative association native or created? What ecological processes ultimately favor long-term sustainability (Samson 1992) of the riparian vegetation?
2. What is the relative importance of riparian vegetation in different upland vegetative communities along the watershed? Does the riparian avifauna make positive contributions to upland avifaunas within a drainage?
3. What "buffer zone" widths are necessary to provide a filter for modulating the effects of human-induced disturbances within upland landscapes across the watershed?
4. Which riparian-dependent species are regionally endemic—species that evolved within that biogeographic domain? A major problem facing conservation of biological diversity has been the lack of conformity between biological and political provinces in North America, leading to accelerated rates of avifaunal mixing of the continent (Knopf 1992).
5. What real or potential role does riparian vegetation play in faunal mixing at the regional level? Will degradation or enhancement activities be severing or creating potential links for the dispersal of individuals within targeted populations? What are the projected biological consequences of these activities?

Inferences arising from this hierarchical series of questions will provide direction while simultaneously generating testable hypotheses. Subsequent testing of those hypotheses will foster a continual refining of conservation activities for a riparian avifauna within a region.

Acknowledgments

This manuscript evolved from an invited presentation at the Streamside Management Symposium on Riparian Wildlife and Forestry Interactions held at the University of Washington, February 11–13, 1987 (Raedeke 1988). It was originally accepted to be published in *Bird Conservation 4* in 1988; it was withdrawn from that volume and updated in September 1993. We thank Robert G. Anthony, David S. Dobkin, William H. Howe, James A. Sedgwick, and Robert C. Szaro for comments on earlier drafts of the manuscript.

References

Anderson, B. W. 1971. Man's influence in hybridization in two avian species in South Dakota. Condor 73:342–347.

Anderson, B. W., and R. D. Ohmart. 1977. Vegetation structure and bird use in the lower Colorado River Valley. Pages 23–34 in R. R. Johnson and D. A. Jones, editors. Importance, preservation and management of riparian habitat: A symposium. General Technical Report RM-166. U.S. Forest Service, Rocky Mountain Forest and Range Experiment Station, Fort Collins, Colorado.

Balda, R. P. 1975. Vegetation structure and breeding bird diversity. Pages 59–80 in D. R. Smith, editor. Symposium on management of forest and range habitats for nongame birds. General Technical Report WO-1. U.S. Forest Service, Washington, D.C.

Blake, J. G., and J. R. Karr. 1984. Species composition of bird communities and the conservation benefit of large versus small forests. Biological Conservation 30:173–187.

Bock, C. E. 1987. Distribution-abundance relationships of some Arizona landbirds: A matter of scale? Ecology 68:124–129.

Bull, E. L., and J. M. Skovlin. 1982. Relationships between avifauna and streamside vegetation. Transactions of the North American Wildlife and Natural Resources Conference 47:496–506.

Dobkin, D. S., and B. A. Wilcox. 1986. Analysis of natural forest fragments: Riparian birds in the Toiyabe Mountains, Nevada. Pages 293–299 in J. Verner, M. L. Morrison, and C. J. Ralph, editors. Wildlife 2000. Modeling habitat relationships of terrestrial vertebrates. University of Wisconsin Press, Madison, Wisconsin.

Forman, R. T. T., and J. Baudry. 1984. Hedgerows and hedgerow networks in landscape ecology. Environmental Management 8:495–510.

Franzreb, K. E. 1978. Tree species used by birds in logged and unlogged mixed-coniferous forests. Wilson Bulletin 90:221–238.

Fretwell, S. D. 1972. Populations in a seasonal environment. Princeton University Press, Princeton, New Jersey.

Graul, W. D., and S. J. Bissell. 1978. Lowland river and stream habitat in Colorado: A symposium. Colorado Chapter of The Wildlife Society and Colorado Audubon Council, Denver, Colorado.

Graul, W. D., J. Torres, and R. Denny. 1976. A species-ecosystem approach for nongame programs. Wildlife Society Bulletin 4:79–80.

Gutzwiller, K. J., and S. H. Anderson. 1987. Short-term dynamics of cavity-nesting bird communities in disjunct floodplain habitats. Condor 89:710–720.

Hagan, J. M., and D. W. Johnston, editors. 1992. Ecology and conservation of Neotropical migrant birds. Smithsonian Press, Washington, D.C.

Harris, L. D. 1984. The fragmented forest. University of Chicago Press, Chicago, Illinois.

Henke, M., and C. P. Stone. 1979. Value of riparian vegetation to avian populations along the Sacramento River System. Pages 228–235 in R. R. Johnson and J. F. McCormick, editors. Strategies for protection and management of floodplain wetlands and other riparian ecosystems. General Technical Report WO-12. U.S. Forest Service, Washington, D.C.

Holmes, R. T., and S. K. Robinson. 1981. Tree species preferences of foraging insectivorous birds in a northern hardwoods forest. Oecologia 48:31–35.

Hubbard, J. P. 1971. The summer birds of the Gila Valley, New Mexico. Nemouria 2:1–35.

Hubbard, J. P. 1977. Importance of riparian ecosystems: Biotic considerations. Pages 14–18 in R. R. Johnson and D. A. Jones, technical coordinators. Importance, preservation and management of riparian habitat: A symposium. General Technical Report RM-166. U.S. Forest Service, Rocky Mountain Forest and Range Experiment Station, Fort Collins, Colorado.

Johnson, R. R., and D. A. Jones, Jr., technical coordinators. 1977. Importance, preservation and management of riparian habitat: A symposium. General Technical Report RM-166. U.S. Forest Service, Rocky Mountain Forest and Range Experiment Station, Fort Collins, Colorado.

Johnson, R. R., and C. W. Lowe. 1985. On the development of riparian ecology. Pages 112–116 in R. R. Johnson, C. D. Ziebell, D. R. Patten, P. F. Ffolliot, and R. H. Hamre, technical coordinators. Riparian ecosystems and their management:

Reconciling conflicting uses. General Technical Report RM-120. U.S. Forest Service, Fort Collins, Colorado.

Johnson, R. R., I. T. Haight, and J. M. Simpson. 1977. Endangered species vs. endangered habitats: A concept. Pages 68–79 in R. R. Johnson and D. A. Jones, technical coordinators. Importance, preservation and management of riparian habitat: A symposium. General Technical Report RM-166. U.S. Forest Service, Rocky Mountain Forest and Range Experiment Station, Fort Collins, Colorado.

Klebenow, D. A., and R. J. Oakleaf. 1984. Historical avifaunal changes in the riparian zone of the Truckee River, Nevada. Pages 203–209 in R. E. Warner and K. M. Hendrix, editors. California riparian systems. University of California press, Berkeley, California.

Knopf, F. L. 1985. Significance of riparian vegetation to breeding birds across an altitudinal cline. Pages 105–111 in R. R. Johnson, C. D. Ziebell, D. R. Patten, P. F. Ffolliot, and R. H. Hamre, technical coordinators. Riparian ecosystems and their management: Reconciling conflicting uses. General Technical Report RM-120. U.S. Forest Service, Fort Collins, Colorado.

Knopf, F. L. 1986. Changing landscapes and the cosmopolitism of the eastern Colorado avifauna. Wildlife Society Bulletin 14:132–142.

Knopf, F. L. 1987. "On publishing symposia" revisited. When are proceedings no longer timely? Wildlife Society Bulletin 15:603–605.

Knopf, F. L. 1992. Faunal mixing, faunal integrity, and the biopolitical template for diversity conservation. Transactions of the North American Wildlife and Natural Resources Conference 57:330–342.

Knopf, F. L., and M. L. Scott. 1990. Altered flows and created landscapes in the Platte River headwaters, 1840–1990. Pages 47–70 in J. M. Sweeney, editor. Static management of dynamic ecosystems. The Wildlife Society, West Lafayette, Indiana.

Knopf, F. L., and J. A. Sedgwick. 1987. Latent population responses of summer birds to a catastrophic, climatological event. Condor 89:869–873.

Knopf, F. L., R. R. Johnson, T. Rich, F. B. Samson, and R. C. Szaro. 1988a. Conservation of riparian ecosystems in the United States. Wilson Bulletin 100:272–294.

Knopf, F. L., J. A. Sedgwick, and R. W. Cannon. 1988b. Guild structure of a riparian avifauna relative to seasonal cattle grazing. Journal of Wildlife Management 52:280–290.

Knopf, F. L., J. A. Sedgwick, and D. B. Inkley. 1990. Regional correspondence among shrub-steppe bird habitats. Condor 92:45–53.

Kuehne, R. A. 1962. A classification of streams, illustrated by fish distribution in an eastern Kentucky creek. Ecology 43:608–614.

Laymon, S. A., and M. D. Halterman. 1989. A proposed habitat management plan for Yellow-billed Cuckoos in California. Pages 272–277 in D. L. Abell, technical coordinator. Proceedings of the California riparian systems conference: Protection, management, and restoration for the 1990's. General Technical Report PSW-110. U.S. Forest Service, Berkeley, California.

Martin, T. E. 1981. Limitation in small habitat islands: Chance or competition? Auk 98:715–734.

Murphy, D. D. 1989. Conservation and confusion: Wrong species, wrong scale, wrong conclusions. Conservation Biology 3:82–84.

Nadler, C. T., and S. A. Schumm. 1981. Metamorphosis of South Platte and Arkansas rivers, eastern Colorado. Physical Geography 2:95–115.

Noss, R. F. 1987. Corridors in real landscapes: A reply to Simberloff and Cox. Conservation Biology 1:159–164.

Noss, R. F., and L. D. Harris. 1986. Nodes, networks, and MUMs: Preserving diversity at all scales. Environmental Management 10:299–309.

Pleasants, B. Y. 1979. Adaptive significance of the variable dispersion pattern of breeding northern orioles. Condor 81:28–34.

Raedeke, K. J., editor. 1988. Streamside management—riparian wildlife and forestry interactions. Contribution No. 59. University of Washington Institute of Forest Resources, Seattle, Washington.

Rice, J., B. W. Anderson, and R. D. Ohmart. 1980. Seasonal habitat selection by birds in the lower Colorado River valley. Ecology 61:1402–1411.

Rice, J., B. W. Anderson, and R. D. Ohmart. 1984. Comparison of the importance of different habitat attributes to avian community organization. Journal of Wildlife Management 48:895–911.

Rice, J., R. D. Ohmart, and B. W. Anderson. 1983. Turnovers in species composition of avian communities in contiguous riparian habitats. Ecology 64:1444–1455.

Rising, J. D. 1983. The Great Plains hybrid zones. Current Ornithology 1:131–157.

Robinson, S. K., and R. T. Holmes. 1984. Effects of plant species and foliage structure on the foraging behavior of forest birds. Auk 101:672–684.

Rotenberry, J. T. 1985. The role of vegetation in avian habitat selection: Physiognomy or floristics? Oecologia 67:213–217.

Salt, G. W. 1957. An analysis of the avifaunas in the Teton Mountains and Jackson Hole, Wyoming. Condor 59:373–393.

Samson, F. B. 1979. Lowland hardwood bird communities. Pages 49–66 in R. M. DeGraaf and K. E.

Evans, editors. Management of North Central and Northeastern forest for nongame birds. General-Technical Report NC-51. U.S. Forest Service, St. Paul, Minnesota.

Samson, F. B. 1992. Conserving biological diversity in sustainable ecological systems. Transactions of the North American Wildlife and Natural Resources Conference 52:308–320.

Samson, F. B., and F. L. Knopf. 1982. In search of a diversity ethic for wildlife management. Transactions of the North American Wildlife and Natural Resources Conference 47:421–431.

Samson, F. B., and F. L. Knopf. 1993. Managing biological diversity. Wildlife Society Bulletin 21:509–514.

Schmitt, C. G. 1976. Summer birds of the San Juan Valley, New Mexico. Publication No. 4. New Mexico Ornithological Society, Albuquerque, New Mexico.

Sedgwick, J. A., and F. L. Knopf. 1987. Breeding bird response to cattle grazing in a cottonwood bottomland. Journal of Wildlife Management 51:230–237.

Sherry, T. W., and R. T. Holmes. 1993. Are populations of Neotropical migrant birds limited in summer or winter? Implications for management. Pages 47–57 in D. M. Finch and P. W. Stangel, editors. Status and management of Neotropical migrating birds. General Technical Report RM-229. U.S. Forest Service, Rocky Mountain Forest and Range Experiment Station, Fort Collins, Colorado.

Simberloff, D., and J. Cox. 1987. Consequences and costs of conservation corridors. Conservation Biology 1:63–71.

Stevens, L., B. T. Brown, J. M. Simpson, and R. R. Johnson. 1977. The importance of riparian habitat to migrating birds. Pages 156–164 in R. R. Johnson and D. A. Jones, editors. Importance, preservation and management of riparian habitat: A symposium. General Technical Report RM-166. U.S. Forest Service, Rocky Mountain Forest and Range Experiment Station, Fort Collins, Colorado.

Szaro, R. C. 1988. Riparian forest and scrubland community types of Arizona and New Mexico. Desert Plants 9:69–138.

Szaro, R. C., and M. D. Jakle. 1985. Avian use of a desert riparian island and its adjacent scrub habitat. Condor 87:511–519.

Terborgh, J. 1989. Where have all the birds gone? Princeton University Press, Princeton, New Jersey.

Thomas, J. W., C. Maser, and J. E. Rodiek. 1979. Wildlife habitats in managed rangelands—the Great Basin of southeastern Oregon. General Technical Report PNW-80. U.S. Forest Service, Portland, Oregon.

Tubbs, A. A. 1980. Riparian bird communities of the Great Plains. Pages 413–433 in R. M. DeGraff and N. B. Tilghman, technical coordinators. Management of western forests and grasslands for nongame birds. General Technical Report INT-86. U.S. Forest Service, Odgen, Utah.

Wauer, R. H. 1977. Significance of Rio Grande riparian systems upon the avifauna. Pages 165–174 in R. R. Johnson and D. A. Jones, editors. Importance, preservation and management of riparian habitat: A symposium. General Technical Report RM-166. U.S. Forest Service, Rocky Mountain Forest and Range Experiment Station, Fort Collins, Colorado.

Wedel, W. R. 1986. Central plains prehistory: Holocene environments and culture change in the Republican River basin. University of Nebraska Press, Lincoln, Nebraska.

Whittaker, R. H. 1975. Communities and ecosystems. MacMillan, New York.

Wiens, J. A. 1979. On competition and variable environments. American Scientist 65:590–597.

Wiens, J. A. 1983. Avian community ecology: An iconoclastic view. Pages 355–403 in A. H. Brush and G. A. Clark, Jr., editors. Perspectives in ornithology. Cambridge University Press, Cambridge, England.

Wiens, J. A. 1989. The ecology of bird communities. Cambridge University Press, Cambridge, England.

Wiens, J. A., J. T. Rotenberry, and B. Van Horne. 1987. Habitat occupancy patterns of North American shrubsteppe birds: The effects of spatial scale. Oikos 48:132–147.

Wilcove, D. S. 1988. National forests. Policies for the future, vol. 2. Protecting biological diversity. The Wilderness Society, Washington, D.C.

Williams, O., and G. P. Wheat. 1971. Hybrid jays in Colorado. Wilson Bulletin 83:343–346.

Wilson, M. F. 1974. Avian community organization and habitat structure. Ecology 55:1017–1029.

Wilson, I. O. 1979. Public forum. Pages 77–87 in O. B. Cope, editor. Grazing and riparian/stream ecosystems. Trout Unlimited, Denver, Colorado.

6
Beyond "Hotspots": How to Prioritize Investments to Conserve Biodiversity in the Indo-Pacific Region

Eric Dinerstein and Eric D. Wikramanayake

Introduction

The forested habitats of the Indo-Pacific region are among the most species-rich on Earth but face severe threats from deforestation (Myers 1988). Erosion of biological diversity in the Indo-Pacific is directly linked to deforestation; with few exceptions, nearly all of this region was once forested. Remaining forests and other natural habitats vary in richness, requiring strategies to establish priorities to conserve the most important areas. Two such approaches, "ecological hotspots" (Myers 1988) and "mega-diversity countries" (Mittermeir & Werner 1990), use lists of plant species or other taxa to identify biologically rich biogeographic units or countries. However, these largely descriptive efforts lack a paradigm to establish conservation priorities at regional, national, and subnational levels. They fail to incorporate the size or location of existing parks, to identify gaps, or to quantify conservation potential and threats for biologically rich countries or protected areas.

We present a new approach to conservation planning to address these concerns: a conservation potential/threat index (CPTI). The index forecasts how deforestation during the coming decade will affect conservation or establishment of forest reserves. The index compares biological richness with reserve size, size of

protected areas, size of remaining forest cover, and deforestation rate. We provide examples for regional, national, and subnational geographical scales to indicate the broad usefulness of the CPTI for setting priorities for conserving biological diversity. Second, we analyze the size of protected areas in the region by country and by biogeographic unit because conservation of biological diversity, from populations to entire ecosystems, will be most successful in large reserves (Schoenwald-Cox et al. 1983; Redford & Robinson 1991). From our analyses we identify where reserves are most needed and how funding can best be invested to finance effective conservation action.

Methods

Geographical Scope

Relevant reviews of protected areas have treated the Indo-Malayan realm and the South Pacific islands separately (IUCN 1986; MacKinnon & MacKinnon 1986), but we considered both regions together in our analysis because funding agencies often administer the two regions as a single unit (for example, the World Bank, the World Wildlife Fund, IUCN). We used national rather than biogeographic units for our analysis because (1) conservation planning and financing by donors, conservation

Reprinted with permission from Conservation Biology vol. 7, pp. 53-65. Copyright 1993 The Society for Conservation Biology and Blackwell Science, Inc.

agencies, and foundations are generally country-specific; (2) projects are usually implemented at the national level; and (3) a number of biogeographic units constitute national or administrative units as well, especially among the island nations in Asia and the South Pacific (see IUCN 1986; MacKinnon & MacKinnon 1986). We include for consideration 23 countries (Appendix A) in the Indo-Malayan realm and the South Pacific (hereafter referred to as the Indo-Pacific countries). Hong Kong, Singapore, and the Maldive Islands were excluded from this analysis because of their negligible forest cover.

Sources for Data

The legal status of protected forested habitats varies considerably among countries. Our base data are from the International Union for the Conservation of Nature and Natural Resources [IUCN] (1990, 1991). The list of protected areas is updated for Malaysia by the Malaysian Forest Department Working Paper (personal communication), for Indonesia by MacKinnon (1990), and for the Philippines by the Integrated Protected Areas System plan (C. Roque, personal communication). For Papua New Guinea, we include the terrestrial wildlife management areas, although they are not recognized by IUCN (1990). We consider small reserves to be those less than 300 km^2, intermediate-sized reserves to be between 300 and 1000 km^2, and large reserves to be 1000 km^2 or more (Appendix A). We base this designation on results of other studies that address reserve size and minimum viable population size for mammals (Redford & Robinson 1991) and on our own analysis of the distribution of reserve size in the Indo-Pacific (see Fig. 1).

Estimates of remaining forested habitat cover are drawn largely from MacKinnon & MacKinnon (1986) for Indo-Malayan countries unless superseded by more recent data (for Indonesia, MacKinnon [1990]; for Malaysia, Forestry Department of Malaysia [personal communication]; for the Philippines, SSC [1988]). Data for the South Pacific countries are from IUCN (1986). In several instances, FAO (1981) data for forest cover in each country were higher than data from MacKinnon & MacKinnon (1986), partly because of

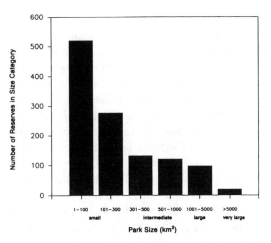

FIGURE 1. Size distribution of protected areas in 23 countries of the Indo-Pacific region.

differences in classification of forest types. We chose to use the MacKinnon & MacKinnon (1986) data in these situations because their estimates were more conservative, were field-based, and used consistent criteria across the Indo-Malayan region.

Deforestation rates for all Indo-Malayan countries, Papua New Guinea, Fiji, and the Solomon Islands are from the World Environment Report (WRI 1990) and McNeely et al. (1990). We lack data on deforestation rates for New Caledonia and Tonga, so we assumed these nations to have the same rate as Papua New Guinea. These rates, particularly for New Caledonia, may be underestimates (IUCN 1986).

Construction of Conservation Potential/Threat Index (CPTI)

We assumed that park coverage (expressed as percentage of country under formal protection) and deforestation rates will remain constant during the next 10 years. In this analysis, forest cover includes all natural habitats with substantial tree cover, such as scrub jungle, open woodland, swamps, and mangroves, and is not restricted to closed-canopy forests. To standardize the analysis, we use percentage of remaining forest cover rather than absolute forest cover to rank countries regardless of their size.

The amount of remaining forested habitat projected in 10 years time was calculated by subtracting the amount of forest lost during the next 10 years from the existing forested area, given a constant rate of deforestation. This estimate is probably conservative because annual deforestation rates are increasing in India, Indonesia, Myanmar, Philippines, Thailand, and Vietnam (WRI 1990; Table 1).

The regional CPTI places the Indo-Pacific countries into one of four categories based upon current efforts at protection and future threats of deforestation. Two categories are created by a horizontal line along the y axis, which delineates the global country average for protected areas (4%) (WRI 1990). Two other categories are created by the intersection of a vertical line along the x axis, which represents an estimate of the minimum amount of multiple-use forest areas that would be required in each country to maintain minimal ecosystem functioning. We have estimated this figure at 20% forest cover outside of protected areas. Services provided might include adequate protection of watersheds, agricultural productivity, stability of local climate, fuelwood supplies, a sustainably-harvested local timber industry, and conservation of some fraction of biological diversity falling outside the formally protected reserves.

Limitations of the Data

We use rates of deforestation rather than rates of degradation or fragmentation to estimate threats to biodiversity, because data for the latter two variables are unavailable. They are likely, however, to be strongly correlated with rates of deforestation. Estimates of remaining habitat in several countries do not differentiate between primary and secondary forest cover because data are unavailable. The index should be used to analyze primary forest cover separately when data are available. We assumed that existing reserves and remaining forest outside reserves would not be significantly degraded; this assumption is no doubt unrealistic. We also recognize that a number of protected areas exist only on paper, others are operational but lack proper management, and that our threat index, at present, does not convey the extent to which parks are protected from logging, poaching, and encroachment. Thus, the predictions of our index must be viewed as the most optimistic available for some of the 23 countries.

Assessing Biological Richness

We focus our analysis at the species level and assume that all taxonomic groups have equal importance. To assess species richness, we tabulated the number of species per country for the following taxonomic groups: mammals, birds, reptiles, amphibians, freshwater fishes, swallowtail butterflies, and vascular plants (see Appendix A; for list of references please contact first author). Our species richness index is the sum of all above taxa. The animal species richness is this sum minus vascular plants. Countries are ranked by species richness and by endemic species.

Invertebrates are represented only by swallowtail butterflies; therefore our database underestimates species richness and endemism for countries covered by moist tropical forests that are rich in invertebrates (such as Indonesia, Papua New Guinea, Malaysia, Philippines). Endemism was indexed by the number of endemic mammals, birds, and vascular plants per country. Information on other taxa was unavailable for all countries. Analyses of generic level diversity are available for part of the region but are not included in this paper.

Evaluation of Funding Levels

We use data from Abramovitz (1991) on amount of funds invested by U.S. institutions during fiscal year 1989 for biodiversity conservation in each country. We express investment as dollars invested per km^2 of remaining forest or forest-associated habitat within each country. This data set does not include contributions by international donors and host country governments. However, it does include contributions from the Global Environment Facility (UNDP 1991).

Application of the CPTI to Specific Countries and Reserves

On a national scale, we assess conservation threats, conservation potential, and biological richness for Indonesia, Malaysia, and the Philippines. We modified the regional CPTI by

TABLE 1. Forest cover, deforestation rates, and protected areas for countries of the Indo-Pacific region.

Country	Country Area (km²)	Remaining Habitat minus Reserves (km²)	Rate of Deforestation	Protected Area (km²)	Number of Terrestrial Reserves	Number of Reserves >300 km²	Number of Reserves >1000 km²
Bangladesh	143,998[a]	8,410[a]	0.9[j]	968[b]	8[b]	1[b]	0[b]
Bhutan	46,620[a]	22,980[a]	0.1[j]	9,243[h]	7[b]	2[b]	1[b]
Brunei	5,765[a]	3,491[a]	1.5[j]	1,224[h]	4[b]	2[b]	0[b]
Myanmar	678,031[a]	255,819[a]	0.3[j]	1,733[h]	2[b]	1[b]	1[b]
Cambodia	181,940[a]	42,899[a]	0.3[j]	0[h]	0[h]	0[h]	0[b]
China	9,597,000[a]	644,069[n]	0.6[l]	219,471[h]	289[h]	75[h]	35[h]
India	3,166,828[a]	547,252[a]	0.3[j]	134,811[h]	359[h]	118[b]	18[h]
Indonesia	1,919,433[a]	801,066[a]	0.5[j]	238,960[h]	162[h]	61[h]	29[h]
Laos	236,725[a]	68,971[a]	1.2[j]	0[h]	0[h]	0[h]	0[h]
Malaysia	332,669[a]	193,161[a]	1.2[j]	11,622[h]	45[h]	7[h]	3[h]
Nepal	141,414[a]	53,813[a]	4.3[j]	9,585[h]	11[b]	6[b]	3[b]
Pakistan	803,941[a]	17,983[a]	0.3[j]	36,550[h]	53[h]	24[h]	11[b]
Philippines	300,000[a]	64,206[a]	1.0[j]	5,508[h]	28[h]	5[h]	1[h]
Sri Lanka	65,610[a]	11,045[a]	3.5[j]	7,837[h]	43[h]	7[h]	1[h]
Taiwan	35,988[a]	10,719[a]	0.6[l]	2,886[h]	5[h]	4[h]	1[h]
Thailand	514,000[a]	132,590[a]	2.7[j]	51,057[h]	83[h]	51[h]	15[b]
Vietnam	329,566[a]	67,468[a]	0.7[j]	8,920[h]	58[b]	7[b]	1[b]
Fiji	18,235[b]	8,110[d]	0.1[m]	53[h]	2[b]	0[b]	0[h]
Papua New Guinea	462,840[b]	381,750[d]	0.1[j]	8,734[i]	14[i]	3[i]	2[i]
Solomon Islands	29,790[e]	24,230[c]	0.1[m]	0[h]	0[h]	0[h]	0[h]
New Caledonia	18,760[e]	16,000[f]	0.1[m]	484[h]	12[b]	0[h]	0[h]
Vanuatu	12,189[o]	9,000[o]	0.1[m]	0[b]	0[h]	0[h]	0[h]
Tonga	747[o]	70[o]	0.1[m]	0[h]	0[h]	0[h]	0[h]

References: (abbreviated references are listed in References):

[a] MacKinnon & MacKinnon (1986) (without desert, semidesert, saltwater marshes).

[b] Data from World Conservation and Monitoring Centre (derived from IUCN 1986) and personal communication.

[c] WCMC data; personal communication. Original extent of closed canopy forests, including tropical coniferous forests in the Asia-Pacific region, compared with remaining extent as judged from most recently available maps and FAO statistics for 1980. Sources: N. M. Collins, J. A. Sayer, T. Whitmore, editors. 1991. The conservation atlas of tropical forests, Asia and the Pacific. MacMillan Press, New York, New York.

[d] WCMC data; personal communication. Source: FAO Rome statistics for remaining cover of tropical closed and open broadleaved, tropical coniferous, and bamboo forests in 1980. FAO. 1988. An interim report on the state of forest resources in the developing countries. Rome, Italy.

[e] IUCN (1986).

[f] S. Davis, S. Droop, P. Gregerson, L. Henson, C. Leon, J. Villa-Lobos, and J. Zantovska. 1986. Plants in danger — what do we know? IUCN, Gland Switzerland. (Only 10% of territory is considered relatively undisturbed rain forest.)

[g] S. H. Pearsall, 1988. An evaluation of biological diversity in the South Pacific: issues, data, institutions, programs. USAID/Nature Conservancy Dept, unpublished.

[h] IUCN (1990).

[i] Protected areas of Papua New Guinea include 2 National Parks and 12 Wildlife Management Areas. Source: Conservation of nature through protected areas system in Papua New Guinea, unpublished manuscript.

[j] Rates from McNeely et al. (1990).

[k] The World Factbook. 1990. Central Intelligence Agency, U.S. Government Printing Office, Washington, D.C.

[l] Deforestation rate estimated for Asian region; from McNeely et al. (1990).

[m] Deforestation rates estimated as being the same as Papua New Guinea (Oceania region); from McNeely et al. (1990).

[n] China Ministry of Forestry Data (1990) (S. Shen personal communication).

[o] IUCN (1991).

adding a horizontal line designating 10% of the land area protected per country; this follows current IUCN recommendations and indicates a major commitment to conservation. A second vertical line indicating 40% remaining forest cover outside reserves reflects the target levels (including plantations and secondary forests) cited in national forest plans in some Indo-Pacific countries. It represents adequate levels of forest cover.

To apply our index at the level of the conservation unit, we assess the conservation potential of the high-priority protected areas of six of the seven biogeographic units of Indonesia— Sumatra, Java and Bali, Kalimantan, Sulawesi, Maluku, and Irian Jaya. Data on endemism are not available for the Lesser Sundas, so they are excluded from the analysis. For each conservation unit, we compare size of reserve (as a proxy for conserving viable populations), amount of remaining forest cover in the biogeographic unit containing the reserve, and number of birds and mammals endemic to the biogeographic unit and present in the reserve. We also consider human population density in each unit. For Maluku, we use data on endemic birds and mammals for the islands rather than for a reserve, because the islands are small and most of the island endemics should be contained in the respective reserves. Data are adapted from MacKinnon (1990). Data on plant diversity and endemism, insect richness, and ecosystem diversity were not available at this geographical site.

Results

Protected Areas in the Indo-Pacific Region

Sixty-eight percent of the protected areas in the Indo-Pacific region are less than 300 km^2, and 90% are less than 1000 km^2 (Fig. 1). The mean size of reserves in the Indo-Pacific region ($\bar{x} = 570.7$ km^2, $n = 1171$, SD $= 1956$) is smaller than for protected areas in Latin America ($\bar{x} = 1448$ km^2, $n = 654$, SD $= 3854$), or Africa ($\bar{x} = 2207$ km^2, $n = 514$, SD $= 6645$) (based on IUCN 1990). Distributions of size of protected areas differ significantly among regions (Kruskal Wallis: $X^2 = 32.88$, df $= 2$, $p < 0.01$).

China, India, Indonesia, and Thailand contain 82% of the large reserves ($n = 122$) and 86% of the terrestrial areas under formal protection in the Indo-Pacific (Table 1). Nine countries contain 1–3 reserves larger than 1000 km^2, and 11 countries lack large protected areas. The number of large protected areas (Pearson $r = 0.83$, df $= 21$, $p < 0.01$) and the

amount protected by country ($r = 0.88$, df $= 21$, $p < 0.01$) are positively correlated with country size. Only Indonesia and India have both extensive reserves and large amounts of forests remaining outside protected areas. Several countries, such as Papua New Guinea, Myanmar, and Laos, have few or no reserves, but have large amounts of unprotected, contiguous forest (Table 1).

The inadequacy of reserve coverage is apparent in comparing the distribution of large protected areas and the distribution of biogeographic units. Only 9 of the 49 major units of the Indo-Pacific recognized by MacKinnon & MacKinnon (1986) and IUCN (1986) contain more than three large reserves (excluding northern China). These data are related to size of the biounit, but it is still disturbing to report that 34 of the biounits contain only one or no large reserve.

Conservation Potentials and Threats

The index shows that countries vary considerably in terms of conservation threat and potential (Fig. 2 and Table 1 for data for construction of CPTI). Relatively few countries fall within Category I. Indo-Chinese countries tend to cluster in or near Category III, as do several of the South Pacific countries. Malaysia falls on the edge of Category I (4.2% of Malaysia is protected). However, we placed Malaysia in Category III because only 2.1% of the biologically rich state of Sarawak, which contains 46% of the remaining forested habitat in Malaysia, is formally protected. Forested reserves in China, Pakistan, Sri Lanka, Philippines, Vietnam, and Bangladesh will be under greatest threat because little unprotected forest will remain in 10 years.

Less than 4% of the protected areas of each country are in large reserves more than 1000 km^2, with the exception of Indonesia, Nepal, Thailand, and Bhutan. Consideration of only large tropical reserves as a fraction of the total protected area within a country results in all countries dropping to 0–2% along the y axis, except Indonesia and Thailand. In addition, most of the large protected forests in tropical countries are lower-montane or montane forests rather than lowland forests (MacKinnon &

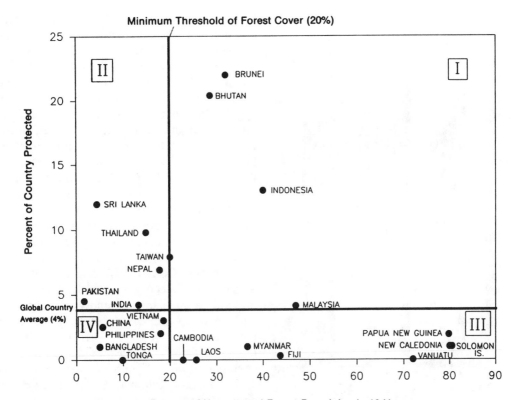

FIGURE 2. Conservation potential/threat index. Unprotected forest that will remain in 10 years was calculated using remaining forest areas and current deforestation rates. Category I: Countries with a relatively large percentage (>4%) of forests under formal protection and that will have a high proportion (>20%) of unprotected forested areas in 10 years. Category II: Countries with a relatively large percent of forests (>4%) under formal protection, but that will have little (>20%) unprotected forests left in 10 years. Category III: Countries with a relatively low percentage (<4%) of forests presently protected. However, under current deforestation rates these countries will still have a relatively large proportion (>20%) of their unprotected forests remaining in 10 years. Category IV: Countries with a relatively low proportion (<4%) of forests presently protected. Under current deforestation rates, in 10 years very little unprotected forest will remain.

MacKinnon 1986). Lowland moist tropical forests are greatly underrepresented in terms of coverage by large reserves in the region and emerge as the most obvious gap in the regional system.

Biological Richness and Remaining Forest Habitat

Indices of species richness and endemism, arranged by country and by category, reveal several general patterns. First, species richness by country indicates that larger countries contain the most species (Fig. 3 and Appendix A). Second, of the groups we tabulated, plants contribute most to overall species richness by country. Third, the number of animal species is strongly correlated with vascular plant species richness (Pearson $r = 0.89$, df $= 21, p < 0.01$). Fourth, endemism is related to both country size and geographical isolation (Appendix A). Endemism is very high in large countries such as Indonesia, China, and India, and in island nations such as Papua New Guinea, Philippines, New Caledonia, Taiwan, Sri Lanka, and Fiji. With the exception of Papua New Guinea, species richness in the South Pacific countries is

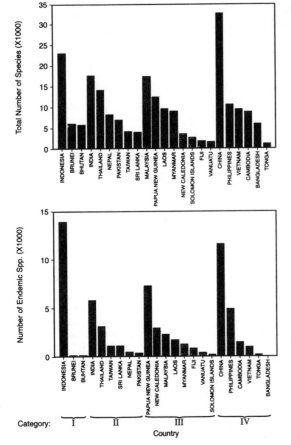

FIGURE 3. Indices of species richness and endemism by category and by country.

moderate to low relative to the mainland countries. The relationship between amount of remaining habitat (including protected and unprotected forests) and species richness (Pearson $r = 0.94, n = 23, p < 0.001$) indicates that most countries with high species richness still have large amounts of forest habitats remaining, except the Philippines and Vietnam. The number of endemic species is positively correlated with remaining habitat (Pearson $r = 0.79, n = 23, p < 0.001$), but countries such as the Philippines and New Caledonia, although lacking extensive remaining primary forest habitat, are high in numbers of endemic species and should be considered in immediate conservation efforts.

At the national level, the biologically rich countries of Indonesia, Malaysia, and the Philippines show high internal variability in status of protection and size of remaining unprotected

forest areas (Fig. 4). At present, most Indonesian units have high potential for conservation, Malaysian units have intermediate potential, and Philippine units are under great threat. The use of 10% protected area and 40% forest cover as additional criteria may seem unrealistic for the Philippine units, Java and Bali, but several subunits of Indonesia and Indonesia itself already achieve or come close to these targets. Use of the 10% line of protection is justified in that most of the 14 biogeographic units considered are richer in species and endemics than some entire countries in the Indo-Pacific region. Indonesia shows perhaps the greatest variation, spanning Java and Bali (3% forest cover, Category II) and Irian Jaya (77% forest cover, Category I). The placement of Sumatra in Category I is misleading in that lowland forest habitat is under great pressure and underrepresented in terms of coverage; most large reserves are lower montane and montane forests (K. MacKinnon, personal communication). Significant areas of Irian Jaya, Maluku, Kalimantan, and Sulawesi are designated for protection, and adequate forest remains (Category I).

Nearly twice as much area is protected in Peninsular Malaysia and Sabah than in Sarawak. Even though Sarawak contains considerably more forested area than the other two states, it is under great threat and many important areas have been logged. The crisis in the Philippines is illustrated by the fact that all subunits, with the exception of Palawan, cluster in or near Category IV. Palawan (Category III) has the highest percentage of remaining forest cover, whereas Mindanao and Luzon have larger areas of forest cover (in km^2), but only a small amount is under formal protection. Data on the ratio of primary versus secondary and selectively logged forest, when available, will refine the priorities identified by these national CPTIs.

Indonesian biogeographic units also vary considerably in biological richness and in the number of large reserves they contain (Appendix B; contact first author for references). Malaysian units are similar in terms of bird and mammal richness, but endemism is greater in Sabah and Sarawak than in Peninsular Malaysia (Appendix B). Inclusion of data on plant

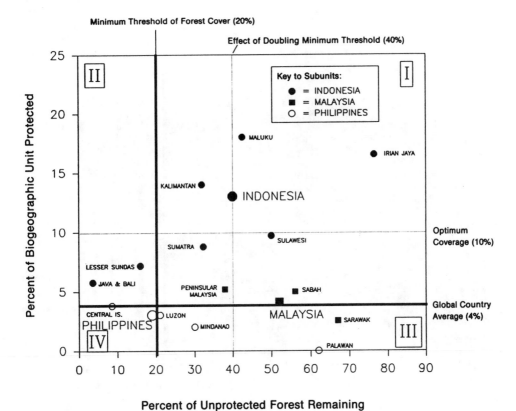

FIGURE 4. Application of conservation potential/threat index to biogeographic and administrative units for Indonesia, Malaysia, and the Philippines. The effect of doubling the minimum threshold of forest cover outside protected areas and increasing protected habitat to optimum levels is illustrated.

endemism and richness would clearly illustrate the conservation importance of Sabah and Sarawak (P. Ashton, personal communication). In the Philippines, Palawan is low in endemism and species richness relative to the four other units (Appendix B); biogeographically, Palawan is more similar to Borneo than to other areas in the Philippines.

At the subnational level, there is considerable variation among the high-priority reserves within Indonesia in terms of reserve size, biological richness, and degree of threat (Fig. 5). The great range in reserve size requires presentation of data on a logarithmic scale. Irian Jaya, Kalimantan, Sulawesi, and Sumatra have large reserves; all but 10 of the 34 reserves exceed the mean reserve size (571 km^2) for the Indo-Pacific region. Among the high-priority conservation units (MacKinnon 1990), the Irian Jaya and Sulawesi reserves are substantially

richer in bird and mammal endemics (Fig. 5). A major threat not illustrated is that about a quarter of Indonesian terrestrial parks ($n = 169$), including some in Figure 5, are not operational, and most are understaffed and underfunded.

Indonesia is the world's fifth most populous nation, but population pressure varies considerably within the units containing the high-priority reserves (Table 2). Use of population density in place of remaining forest cover produces a near mirror image of Figure 5 and is not illustrated here. Data on species richness for plants and insects are vital but currently are not available; they would likely yield different patterns than those observed in Figure 5. Degree of ecosystem or habitat rarity should also be incorporated to highlight those associations underrepresented in the existing network (such as lowland forest and mangroves).

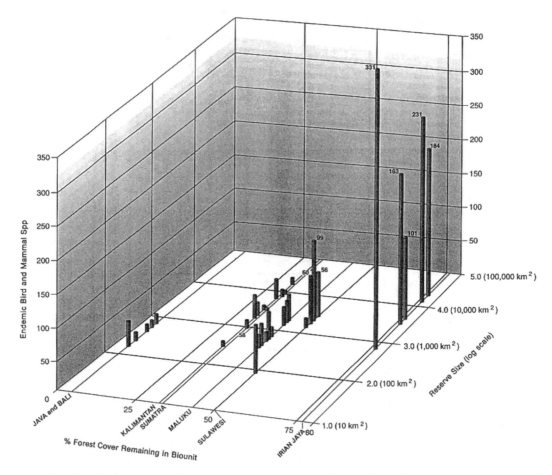

FIGURE 5. Park size, remaining forest cover, and number of endemic birds and mammals among 34 high-priority reserves in Indonesia, as identified by MacKinnon (1990). Park size is graphed on a log scale. Number of endemic birds and mammals represents the number of bird and mammal species endemic to the subunit found within the individual reserve. Total number of endemics is illustrated for the nine most prominent reserves. Reserves are graphed in the sequence in which they appear in Table 2.

Discussion

Reserve Size and Conservation Planning in the Indo-Pacific

A disturbing finding of this analysis is that there are few large conservation units in the Indo-Pacific region. Even more critical is the virtual absence of large reserves in lowland moist tropical forests, the most species-rich habitats. Investment must address this problem immediately because biological diversity will be most completely conserved in large protected areas of more than 1000 km². Small, isolated protected areas of less than 300 km², common

throughout the Indo-Pacific, are inadequate to preserve viable populations of large, keystone species such as top carnivores and larger herbivores and frugivores (Wikramanayake & Dinerstein, unpublished data) or to conserve ecosystem diversity and ecological processes.

Large forest reserves typically protect a variety of habitat types and are better able to withstand incursions and degradation along perimeters than are small parks (Saunders et al. 1991). Small reserves of less than 300 km² or even 100 km², however, should not be dismissed as inadequate and unworthy of financial assistance. For example, some of the reserves in the wet zone of Sri Lanka (such as the Sinharaja Forest Reserve) effectively protect an im-

TABLE 2. Size of high-priority reserves in Indonesia; biological richness and population density in area containing reserve (listed in the order in which they appear in Figure 5).

Park Name	Size (km^2)	Number of Endemic Bird species for Unit	Number of Endemic Mammal species for Unit	% Forest Cover	Population Density (n/km^2)
Java and Bali				3.6	
Gunung Gede	150	21	11		744
Baluran	250	6	1		693
Meru Betiri	500	7	2		693
Bali Barat	570	7	1		506
Ujong Kulon	761	7	4		744
Kalimantan				31.9	
Gunung Palung	300	3	3		22
Gunung Niut	1800	20	7		22
Kutai	2000	1	15		9
Tanjung Puting	3050	2	2		9
Karimun & Bentuang	6000	15	9		22
Sungai Kayan Mentarang	16000	5	4		9
Sumatra				32.3	
Taitai Batti	965	1	10		79
Way Kambas	1235	0	0		47
Barisan Selatan	3650	1	0		228
Gunung Leuser	8080	4	4		149
Kerinci-Selbat	9144	2	1		79
Sulawesi				50	
Tangkoko	89	46	12		132
Rawa Aopa	1500	4	6		48
Morowali	2000	53	7		26
Lore Lindu	2290	65	34		26
Dumoga Bone	3000	46	10		132
Maluku				42.6	25
Kai Besar	370	12	3		
Gunung Sibela	400	26	1		
Wae Bula	600	25	8		
Yamdena	600	11	0		
Taliabu	700	11	0		
Gunung Kapalat Muda	1450	20	3		
Manusela	1800	20	7		
Lolobata	1890	31	1		
Irian Jaya				76.5	4
Arfak	700	278	53		
Tamrau	3313	146	37		
Wasur	4262	74	27		
Mamberamo	14425	191	40		
Lorentz	21500	130	54		

portant forest high in plant endemism, yet these reserves are quite small. New Caledonia is rich in endemic plants but lacks large vertebrates, suggesting the adequacy of small reserves.

Design of protected area networks on a national scale has largely neglected the inclusion of large protected areas, for a range of reasons (see Diamond 1986; Heinen et al. 1987). For example, many reserve boundaries were drawn before conservation biology principles for re-serve design became available to planners. Second, the Indo-Pacific region contains only 13% of the Earth's land area but 50% of its people, creating higher human population densities than in rural areas of Africa or South America. Third, large reserves conflict with other forms of land use, particularly in countries dependent on timber extraction to generate foreign currency (such as Malaysia and Laos). Fourth, extensive deforestation and forest degradation in several countries over the last few centuries have seriously reduced remaining forest cover. Fifth, the customary ownership of forested land, as in South Pacific countries, requires such extensive negotiations with many individuals or clans to establish large tracts of protected forest that only small areas have been proposed for protection. Sixth, the small size of some island nations or island units precludes establishment of large terrestrial reserves.

Because most Indo-Pacific reserves are not large enough to preserve entire ecosystems and maintain minimum viable populations of many larger species, intensive management will be required to deal with demographic, genetic, and environmental threats of extinction associated with isolated populations in small reserves. Large buffer zones must be created adjacent to those protected areas under threat. Unless immediate steps are taken, critical opportunities for expanding or linking existing parks and reserves will be lost. On larger temporal and spatial scales, the long-term prospects for conserving terrestrial habitats will depend on the establishment of a representative regional network of large reserves throughout the Indo-Pacific. This can be accomplished by expanding existing reserves through direct transfer of lands amalgamating or linking smaller reserves where possible, as is occurring in Vietnam (J. MacKinnon, personal communication) and the Philippines. The problems associated with establishing large reserves entirely within a country may be partly overcome by establishing transfrontier reserves (Dinerstein & Wikramanayake, submitted for publication).

Recommendations from the Index

The index yields the following priorities for investing in biodiversity conservation:

Category I is the conservation ideal. Few countries qualify, however, and those that do often have problems protecting established reserves. An important target for conservation financing is to ensure that those high-priority reserves that are essentially "paper parks" become operational as soon as possible.

Category II countries devote a relatively large proportion of the national territory to conservation but suffer from heavy deforestation

outside of protected areas. Countries high in species richness (such as India and Thailand) and high in endemicity (such as India, Thailand, Taiwan, and Sri Lanka) qualify as high-priority countries.

Countries in Category III will have relatively large proportions of forest habitat remaining by the year 2000, and thus have the greatest potential for establishing an integrated protected areas system before forested areas become highly fragmented. Category III countries, such as Malaysia, Papua New Guinea, Laos, Myanmar, New Caledonia, and Solomon Islands, that are high in species richness and endemicity represent important opportunities for establishing national networks of large protected areas. External financing for conservation efforts in Indonesia and the South Pacific has been minimal or nonexistent and should be given highest priority for conservation financing (Table 3; Dinerstein, unpublished data).

Category IV countries are those most threatened, with few protected areas and little of their forests expected to survive the next 10 years. China falls within Category IV because, despite its more than 30 large reserves, the percentage of protected forested area is relatively low. Most reserves are scattered and insularized, resulting in isolated populations; if clumped, the reserves may not serve to preserve all habitat diversity. Biodiversity considerations indicate that key Category IV countries include China, the Philippines, and Vietnam.

In summary, action is required to ensure that countries from Categories II and III move into Category I, and that immediate efforts be taken to halt erosion of biological diversity and to establish parks in Category IV countries. High conservation potential exists in several key areas (such as Indochina, South Pacific, and Sri Lanka) that have received meager support (proposed GEF projects notwithstanding). Bhutan (Category I) is the only Indo-Pacific country in which conservation activities are well-financed.

Sustained financing for ongoing management, expansion of existing areas, and creation of new areas should be addressed, wherever possible, by trust funds. Trust funds are especially suited to countries with low absorptive capacity for conservation programs and funds (such as Bhutan, Laos, and Papua New Guinea). Ini-

TABLE 3. Investment in biodiversity conservation for Indo-Pacific countries by U.S. Institutions, 1989, and planned investment by the Global Environment Facility (GEF) Operations Program beginning in 1992.

Category	Country	Remaining Habitat (km^2)[a]	1989 Funding ($U.S.)[b]	1992 First Tranch GEF[c]	1992 Project Pipeline GED[d]	Dollars/km^2 of Remaining Habitat: U.S. Institutions Only	Dollars/km^2 of Remaining Habitat: GEF Only[e]
I	Bhutan	32,223	249,888	10,000,000		7.8	310.3
	Indonesia	819,255	1,394,244		12,000,000	1.7	14.7
	Brunei	4,715	n.a.			0.0	0.0
II	Nepal	63,398	375,959	3,800,000		5.9	59.9
	Thailand	183,647	699,449			3.8	0.0
	Sri Lanka	18,882	37,800			2.0	0.0
	Pakistan	54,533	65,077			1.2	0.0
	India	682,053	649,752			1.0	0.0
	Taiwan	13,605	n.a.			0.0	0.0
III	Fiji	8,163	116,640	##		14.3	n.a.
	Malaysia	204,783	448,846			2.2	0.0
	Papua New Guinea	390,484	68,952	##	5,000,000	0.2	13.3
	Myanmar	257,552	10,151			0.0	0.0
	Cambodia	42,899	n.a.			0.0	0.0
	Laos	68,971	n.a.	5,500,000		0.0	79.7
	New Caledonia	16,000	n.a.			0.0	n.a.
	Solomon Islands	24,230	0	##		0.0	n.a.
	Vanuatu	9,000	n.a.	##		0.0	n.a.
IV	Philippines	69,714	363,068	20,000,000		5.2	286.8
	China	382,767	666,278			1.7	0.0
	Bangladesh	9,378	15,000			1.6	0.0
	Vietnam	76,388	5,500	3,000,000		0.1	39.3
	Tonga	70	n.a.	##		0.0	n.a.

[a]Remaining habitat equals remaining forest cover outside reserves and protected areas combined.

[b]From Aramovitz (1991).

[c]Data from UNDP/World Bank/UNEP (1991). Report by the Chairman to the participants' meeting to discuss the Global Environmental Facility, vol. 2. World Bank, Washington, D.C. None of the GEF biodiversity projects has been implemented to date.

[d]The first tranche contribution for the South Pacific region will be $8.2 million. Biodiversity projects will be supported in 22 countries, including the 6 covered in this analysis.

[e]UNDP/World Bank/UNEP (1991). GEF projects for Indonesia and Papua New Guinea are not confirmed, and budgets are only illustrative.

[f]GEF projects will have a lifespan of 3 or more years. These data are not annualized because the lengths of some projects are unknown.

tially the trust funds can support training, professional development, and institutional strengthening, with support shifting eventually to protected areas.

Conclusion

Maintenance of the biological integrity of terrestrial ecosystems and communities over the long term will require conservation of large and preferably contiguous tracts of forests. Major new investments in biodiversity conservation should be earmarked for expanding the protected areas system of each country and, where possible, expanding the size of biologically important reserves. Donors must recognize the limited amounts of conservation financing previously available to some high-priority countries identified by our regional analysis and address these deficiencies.

Finally, all Indo-Pacific countries have biological diversity worth conserving. Focusing attention on a few biogeographic areas or countries, as has been the practice to date, is self-defeating in the long term. We encourage planners from each country to experiment with the index approach to setting priorities and assessing threats for conserving populations, species, ecosystems, and ecological processes. A short window of time remains to establish new large protected areas and to extend existing reserves in the region before competing interests preclude such efforts. Second-generation models, tailored to the features of each country, are urgently needed to move conservation planning beyond hotspots.

Acknowledgments

We thank the members of the Conservation Science Program of the World Wildlife Fund and the WWF family for support and encouragement in the development of this paper. T. Agardy, P. Ashton, J. Diamond, G. Hartshorn, D. Hulse, K. MacKinnon, G. Orians, K. Redford, J. Robinson, C. Santiapillai, E. Stevens, J. Sugarjito, and E. O. Wilson improved the manuscript with their comments.

References

Abramowitz, J. N. 1991. Investing in biological diversity. World Resources Institute, Washington, D.C.

Diamond, J. 1986. The design of a nature reserve system for Indonesian New Guinea. Pages 485–503 in M. E. Soulé, editor. Conservation biology: the science of scarcity and diversity. Sinauer Associates, Sunderland, Massachusetts.

FAO. 1981. Tropical forest resources assessment project (GEMS): tropical Africa, tropical Asia, tropical America. FAO/UNEP, Rome, Italy.

Heinen, J. T., B. Kattel, and J. N. Mehta. 1987. National park administration and wildlife conservation in Nepal. Department of National Parks and Wildlife Conservation, Babar Mahal, Kathmandu, Nepal.

IUCN. 1986. Review of the protected areas system in Oceania. International Union for the Conservation of Nature and Natural Resources, Gland, Switzerland.

IUCN. 1990. United Nations list of national parks and protected areas. International Union for the Conservation of Nature and Natural Resources, Gland, Switzerland.

IUCN. 1991. IUCN directory of protected areas in Oceania. International Union for the Conservation of Nature and Natural Resources, Gland, Switzerland.

MacKinnon, J., and K. MacKinnon. 1986. Review of the protected areas system in the Indo-Malayan realm. International Union for the Conservation of Nature and Natural Resources, Gland, Switzerland.

MacKinnon, K. 1990. Biodiversity action plan for Indonesia. Ministry of Population and Environment, Bappenas, and World Bank, Bogor Agricultural University, Bogor, Indonesia.

McNeely, J. A., K. R. Miller, W. V. Reid, R. A. Mittermeier, and T. B. Werner. 1990. Conserving the world's biological diversity. International Union for the Conservation of Nature and Natural Resources, World Resource Institute, CI, World Wildlife Fund—U.S., World Bank, Washington, D.C.

Mittermeier, R. A., and T. B. Werner. 1990. Wealth of plants and animals unites "megadiversity" countries. Tropicus 4:1, 4–5.

Myers, N. 1988. Threatened biotas: "hotspots" in tropical forests. The Environmentalist 8:1–20.

Redford, K. H., and J. G. Robinson. 1991. Park size and the conservation of forest mammals in Latin America. Pages 227–234 in M. Mares and D. Schmidley, editors. Latin American mammals: their conservation, ecology, and evolution.

Saunders, D. A., R. J. Hobbs, and C. R. Margules. 1991. Biological consequences of ecosystem fragmentation: a review. Conservation Biology 5:18–32.

Schoenwald-Cox, C. M., S. M. Chambers, F. MacBryde, and I. Thomas, editors. 1983. Genetics and conservation: a reference for managing wild animal and plant populations. Benjamin/Cummings, Menlo Park, California.

Swedish Space Corporation. 1989. Philippines environment and natural resource management study. The World Bank, Washington, D.C.

UNDP/World Bank/UNEP (United Nations Development Programme, World Bank, and United Nations Environmental Program). 1991. Report by the chairman to the participants' meeting to discuss the Global Environmental Facility, vol. 2. World Bank, Washington, D.C.

WRI. 1990. World resources—1990–91. A guide to the global environment. World Resources Institute, Oxford University Press, New York, New York.

APPENDIX A. Database of biological diversity (for references contact first author).

Country	No. of Mammal spp.	No. of Endemic Mammals	No. of Bird spp.	No. of Endemic Birds	No. of Amphibian spp.	No. of Reptile spp.	No. of Fish spp.	No. of Swallowtail Butterfly spp.	No. of Plant spp.	No. of Endemic Plant spp.
Bangladesh	71	0	340	1	29	129	94	10	5000	–
Bhutan	109	0	429	0	24	19	75	22	5000	47
Brunei	155	0	369	0	76	44	23	35	5000	70
China	394	51	1195	45	265	278	507	99	30000	10000
India	350	38	1200	39	182	453	382	91	15000	5000
Indonesia	515	174	1519	331	270	>600	400	121	20000	11715
Kampuchea	117	1	545	0	28	82	215	22	7571	1175
Laos	157	1	609	3	37	66	244	39	8286	1457
Malaysia	293	16	1200	4	171	294	449	54	15000	1923
Myanmar	300	6	967	3	75	241	196	68	7000	1071
Nepal	125	3	835	1	25	57	102	37	6980	408
Pakistan	150	2	612	1	8	174	156	14	5750	300
Philippines	165	104	541	147	77	212	429	49	8900	3949
Sri Lanka	85	10	340	20	37	158	57	15	3214	887
Taiwan	56	6	276	12	28	48	95	32	3577	900
Thailand	280	8	919	2	101	282	549	56	12000	2742
Vietnam	201	5	586	4	72	212	201	37	8000	800
Fiji	5	0	63	18	3	24	4	1	1500	700
New Caledonia	7	3	68	20	0	29	0	3	3250	2474
Papua New Guinea	187	40	592	195	183	282	150	37	11000	6050
Solomon Islands	47	13	126	40	20	74	0	15	2150	6
Tonga	1	0	21	0	0	6	0	0	770	11
Vanuatu	10	2	53	6	0	22	0	4	1000	150

APPENDIX B. Data on protected areas, forest cover, and biological richness by biogeographic or administrative subunit for Indonesia, Malaysia, and the Philippines.

Subunit Name	Protected Area (km^2)	No. of Parks & Reserves	Remaining Forest Habitat	Remaining Forest w/o Protected Areas	No. of Mammal spp.	No. of Endemic Mammal spp.	No. of Bird spp.	No. of Endemic Bird spp.	Sum of No. of Mammal and Bird spp.	Sum of No. of Endemic Mammal and Bird spp.
Indonesia										
Sumatra	48,311	36	193,086	144,775	194	19	465	9	659	28
Java and Bali	7,375	35	11,444	4,069	133	16	362	25	495	41
Kalimantan	48,792	19	246,524	197,732	201	36	420	25	621	61
Sulawesi	21,451	27	110,850	89,309	114	68	289	92	403	160
Irian Jaya	69,121	17	389,670	320,549	125	73	602	313	727	386
Lesser Sundas	5,935	18	19,341	13,406	41	53	242	73	283	126
Maluku	13,055	10	44,791	31,736	69	12	210	69	279	81
Malaysia										
Peninsular Malaysia	7,400	21	55,100	47,700	208	7	634	3	842	10
Sarawak	2,600	8	84,500	81,900	202	30	511	28	713	58
Sabah	3,900	11	44,400	40,500	199	33	522	28	721	61
Philippines										
Mindanao	728	•	28,380	27,652	25	9	340	94	365	103
Luzon	116	•	23,660	23,544	29	20	408	92	437	112
Palawan	1,000	•	7,410	6,410	25	11	268	26	293	37
Central Island and Visayas	2,400	•	7,440	5,040	8	1	363	95	371	96

The protected areas system of the Philippines is currently undergoing a complete reevaluation under the Integrated Protected Areas System Project (C. Roque, personal communication). Many of the reserves listed in the IUCN (1990) parks directory will be either consolidated, expanded, or reclassified.

7
Avian Community Dynamics Are Discordant in Space and Time

Katrin Böhning-Gaese, Mark L. Taper, and James H. Brown

The threat of global change challenges community ecologists to predict long-term and continental-scale changes in the structure of ecological communities, but the vast majority of studies have been done at very small temporal and spatial scales (Kareiva and Andersen 1988, Tilman 1989, but see Enemar et al. 1984, Holmes et al. 1986, Brown and Heske 1990, Pimm 1991). For example, most studies of competition among species are made in plots of less than 100 m^2 and/or with only one or two years of census data (e.g., Connell 1961, Brown and Kodric-Brown 1979, Rosenzweig et al. 1985, Schmitt 1987, Bock et al. 1992). Although attempting to address global greenhouse warming, studies of effects of elevated atmospheric CO_2 on plant communities are done on similarly small scales (Morison 1990).

But can conclusions that result from studies on small spatial and temporal scales be extrapolated to larger scales? Do ecological communities respond in a similar fashion to variation in abiotic and biotic environmental factors over different scales?

Avian communities provide a good model system to address these questions. Because birds are highly mobile they can potentially respond very rapidly to environmental change in both space and time. Nevertheless, to be able to compare the population dynamics of different species over a wide range of spatial and temporal scales we had to work with a data set collected for a long period of time on a broad spatial scale.

One of the few data sets available on a continental scale is the North American Breeding Bird Survey (BBS), conducted by the U.S. Fish and Wildlife Service and the Canadian Wildlife Service since 1965 (Robbins et al. 1986). The BBS data set is a massive compilation: each year during the breeding season the populations of all breeding birds are censused at approximately 2000 routes throughout North America. Many sites have been censused for more than 20 years. Despite limitations of its sampling design and data (Robbins et al. 1986, Droege 1990, Böhning-Gaese et al. 1993), the BBS represents one of the few data sets available for comparing short-term with long-term population dynamics within and between regions on a continental scale for any organism in North America.

The spatial units of this investigation were regional instead of local avian communities. Local population dynamics are often confounded by sampling error, random demographic fluctuations and immigration and emigration processes. Additionally, by averaging abundances over entire regions, we circumvent many of the sampling problems in the BBS data set (Böhning-Gaese et al. 1993). The regions chosen for the investigation were the physiographic areas or strata (Robbins et al. 1986) of North America. The strata are based largely on

Reprinted with permission from Oikos vol. 70, pp. 121-126. Copyright 1994 Oikos.

Aldrich's (1963) map of the biome types of North America and are ecologically more meaningful than political provinces (e.g., states). In fact, the geographic range boundaries of many species correspond closely to many of the stratum boundaries (Robbins et al. 1986).

In our study we addressed the questions: How similar are the short-term population dynamics among the species of one region? Do species within one region that have similar year-to-year fluctuations in abundance also have similar long-term population trends? Do species with similar population dynamics in one region also have similar population dynamics in other regions?

Methods

Calculation of the Regional Population Dynamics

Each year, the BBS counts birds along approximately 2000 roadside routes distributed across the North American continent north of the U.S.-Mexican border (Robbins et al. 1986). Each route is censused by an experienced volunteer one morning per year during the breeding season in June. A route consists of 50 sampling locations 0.8 km apart. At each location, the observer counts by species all birds heard or seen during a 3-min period.

We restricted our analyses to an ecologically similar group of passerine species: wood warblers (subfamily Parulinae of the Emberizidae), vireos (Vireonidae), gnat-catchers and kinglets (subfamily Sylviinae of the Muscicapidae), titmice and chickadees (Paridae), nuthatches (Sittidae), the brown creeper (*Certhia americana,* Certhiidae), wrens (Troglodytidae) and bluebirds (*Sialia,* Muscicapidae). These species comprise taxonomically defined groups whose members are small, insectivorous, mostly foliage-gleaning songbirds (Ehrlich et al. 1988).

Our goal was to calculate for each stratum and each species the mean number of individuals counted per year, averaging over the routes within the stratum. However, obtaining unbiased averages over multiple routes is not straightforward (Robbins et al. 1980, Geissler

and Noon 1981). Most routes have not been run with high quality continuously from 1965 to 1987. To overcome this problem, we considered only the 20 years from 1968 to 1987, because few routes were sampled before 1968. Additionally, we limited our analysis to routes of high quality: of the routes that were run from 1968 to 1987, we used only those that had no more than two missing counts, no two missing counts in a row, and no missing counts in 1968 and 1987. For analysis the data were transformed to log abundances (ln(count + 1)). More details can be found in Böhning-Gaese et al. (1993).

For each stratum the mean number of individuals per species was calculated for each year, averaging over all the routes within the stratum at which the particular species was observed at least once during the 20 years. Stratum averages were calculated only for species that had been recorded on at least eight routes per stratum. The result were time series for 59 species in up to 22 strata. We restricted the analyses to the six strata of North America with data for more than 25 species (Fig. 1).

Long-Term Population Trends

Long-term population trends were calculated in each stratum for each species by regressing log

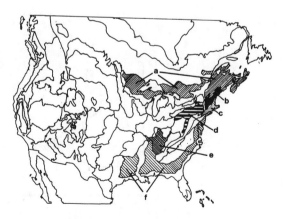

FIGURE 1. The six physiographic regions or strata of North America (Robbins et al. 1986) that were used in the analyses. A: Spruce-Hardwood Forest, b: Central New England, c: Southern New England, d: Allegheny Plateau, e: Highland Rim, f: Upper Coastal Plain.

abundance as a function of time (proc GLM, SAS/STAT 1988). The long-term population trend was the slope of the regression. The dissimilarity of the long-term trends among the species was defined as the difference in the regression coefficient among each pair of species. Thus, two species had similar long-term population trends, whose regressions had similar slopes (e.g., ruby-crowned kinglet (*Regulus calendula*) and boreal chickadee (*Parus hudsonicus*) in the Spruce-Hardwood Forest stratum; Fig. 2b, c). Two species with the same long-term trend had a long-term dissimilarity of 0. If one species had a slope of -1, the other of $+1$, the long-term dissimilarity of the two species was 2.

Short-Term Population Fluctuations

For the short-term population fluctuations we calculated the residuals from the long-term trends. Thus, the short-term fluctuations are the yearly deviations of the abundance of the species from their long-term trends. The similarity of the yearly fluctuations among the species was calculated by the Pearson moment correlation coefficient of the residuals among each pair of species. Thus, two species had similar short-term population fluctuations that had parallel deviations from their long-term trends (e.g., Tennessee warbler (*Vermivora peregrina*) and ruby-crowned kinglet in the Spruce-Hardwood Forest stratum, Fig. 2a, b).

The Pearson moment correlation coefficient ranges from -1 to 1. To be able to compare the long-term population trends and the short-term fluctuations between two species we multiplied the value of the correlation coefficient by -1 and added $+1$. Thus, species with parallel fluctuations got the short-term dissimilarity of 0 [$1 \times (-1) + 1 = 0$] and species with completely dissimilar fluctuations of 2 [$(-1) \times (-1) + 1 = 2$]. Species with the same long-

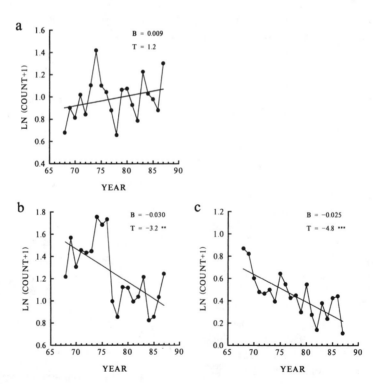

FIGURE 2 . Population dynamics of (a) Tennessee warbler (*Vermivora peregrina*), (b) ruby-crowned kinglet (*Regulus calendula*) and (c) boreal chickadee (*Parus hudsonicus*) in the Spruce-Hardwood Forest stratum from 1968 to 1987. (a) and (b) show very similar short-term population fluctuations (correlation coefficient of the residuals R = 0.763, P = 0.0001, N = 20) but opposite long-term trends. (b) and (c) have similar long-term population trends but no correlation of the short-term fluctuations (R = -0.049, P = 0.838, N = 20). B = slope of the regression, ** = P < 0.01, *** = P < 0.001.

term trends and the same short-term fluctuations had a long-term dissimilarity and a short-term dissimilarity of 0. Species with very different long-term trends and dissimilar short-term fluctuations had a high value, e.g., 2, for the long-term dissimilarity and the maximum value 2 for the short-term dissimilarity.

Similarity of Short-Term Fluctuations and Long-Term Trends

The avian community in each stratum was characterized by two matrices. One matrix described the dissimilarity in short-term fluctuations, the other the dissimilarity in long-term population trends among the species. The correlation between these two matrices was tested statistically using Mantel tests (Smouse et al. 1986). Thus, we were able to compare for all species simultaneously the dissimilarity of their long-term trends with the dissimilarity of their short-term fluctuations.

Similarity of the Population Dynamics Among Different Regions

We compared the dissimilarity matrix for the species in one stratum with the dissimilarity matrices in the other strata. Thereby, the dissimilarity in long-term population trends was compared among the regions as well as the dissimilarity in short-term fluctuations. The correlation between two matrices was again tested using Mantel tests (Smouse et al. 1986).

Results

Similarity of the Short-Term Population Fluctuations Within One Region

There was small but significant similarity in the short-term population fluctuation among the species within a region (Table 1). Depending on the region, 3 to 12% of the correlations were significantly positive, and 2 to 6% significantly negative at the $P = 5\%$ level (compared to 2.5% expected on the basis of chance alone).

In case these low levels of similarity among the species were caused by nonlinear long-term

TABLE 1. Percentage of significantly positive and significantly negative correlations ($P \leq 0.05$) in the short-term population fluctuations among the species in six different strata of North America (compared to 2.5% expected on the basis of chance alone).

Stratum	No. of species	Correlations (%)	
		sig. pos.	sig. neg.
Spruce-Hardwood Forest	38	10.1	2.4
Central New England	30	11.5	5.7
Southern New England	27	3.1	2.8
Allegheny Plateau	33	5.1	2.3
Highland Rim	29	9.6	2.2
Upper Coastal Plain	29	7.6	5.2

population trends we also calculated the residuals after detrending with quadratic regressions, rather than linear ones. This caused a slight decrease in the apparent short-term similarity among species. Nonlinear regressions appear to reflect population dynamics at a scale intermediate between year-to-year fluctuations and decades-long trends.

Did the Species with Similar Short-Term Fluctuations also Have Similar Long-Term Trends?

In none of the regions did species with similar short-term fluctuations have comparable long-term population trends (Table 2). For example, in the Spruce-Hardwood Forest stratum the Tennessee warbler and the ruby-crowned king-

TABLE 2. Mantel matrix correlation coefficients comparing the dissimilarity matrix of the short-term population fluctuations and the dissimilarity matrix of the long-term trends among bird species in six strata of North America. None of the P values are significant after adjusting with the sequential Bonferroni technique (Rice 1989).

Stratum	No. of species	corr. coeff.	P value
Spruce-Hardwood Forest	38	0.08	0.02
Central New England	30	-0.02	0.67
Southern New England	27	-0.08	0.16
Allegheny Plateau	33	-0.11	0.02
Highland Rim	29	-0.07	0.20
Upper Coastal Plain	29	-0.00	0.95

let had very similar short-term population fluctuations (correlation coefficient of the residuals R = 0.763, P = 0.0001, N = 20), but divergent long-term trends (Fig. 2a, b). In the same stratum the long-term population trends of the ruby-crowned kinglet and the boreal chickadee were almost identical, but the two species had no correlation of their short-term fluctuations (R = −0.049, P = 0.8, N = 20; Fig. 2b, c).

Did Species with Similar Population Dynamics in One Region Have Similar Population Dynamics in Other Regions?

Six of the 15 possible comparisons of long-term population trends were significantly positive at the P = 5% level (five after tablewide Bonferroni adjustment (Rice 1989), Table 3). That means that for six pairs of strata the species with similar long-term population trends in one stratum also had similar trends in other strata. Five of these six pairs of regions were geographically close and three had a common boundary (Fig. 1). Four (only two after tablewide Bonferroni adjustment) of the 15 comparisons for the short-term population fluctuations were significant (Table 3). All four pairs were geographically close and three of them had a common boundary (Fig. 1).

Discussion

Our results indicate that avian communities do not respond to environmental variation in a similar fashion over different temporal scales and over different strata. Species within one region responded to year-to-year environmental fluctuations only slightly more similarly than would be expected at random, and these species did not respond similarly to decades-long trends in environmental factors. In addition, species that exhibited similar short- or long-term dynamics within one stratum sometimes showed similar trends in adjacent strata, but only rarely when the strata were widely separated geographically.

We conclude the population dynamics of

TABLE 3. Mantel matrix correlation coefficients comparing the dissimilarity in population dynamics among bird species between six (a–f) strata of North America. Abbr. see Fig. 1. Upper diagonal matrix: correlation of the dissimilarity matrices of the short-term population fluctuations. Lower diagonal matrix: correlation of the dissimilarity matrices of the long-term trends. Exclamation marks indicate significance after adjusting with the sequential Bonferroni technique (Rice 1989).

	a	b	c	d	e	f
a		0.22^x <0.001!^{y} 27^z	0.08 0.22 23	0.03 0.69 25	0.14 0.25 13	0.09 0.47 13
b	0.17 0.004! 27		0.12 0.03 26	0.17 0.002! 26	0.11 0.25 16	0.20 0.05 16
c	−0.03 0.68 23	0.48 <0.001 ! 26		0.16 0.01 26	−0.06 0.58 16	0.01 0.95 15
d	0.42 <0.001! 25	0.22 0.002 ! 26	0.02 0.68 26		−0.07 0.41 18	0.01 0.88 17
e	−0.13 0.34 13	−0.09 0.34 16	−0.09 0.36 16	0.23 0.01 18		0.05 0.44 27
f	0.07 0.54 13	0.07 0.47 16	−0.08 0.44 15	−0.00 0.96 17	0.18 0.002! 27	

x: Mantel matrix correlation coefficient. y: P value. z: no. of species that occurred in both regions.

each species are controlled by a different combination of factors. The species respond to environmental variation in a very individualistic way (Gleason 1917, 1926). Even if the local, year-to-year dynamics of two species are comparable and therefore seem to be regulated by the same factors, they are not driven by the same factors on time scales of several decades or spatial scales encompassing different strata.

For example, the short-term population fluctuations of some European songbird species seem to be driven by the precipitation on their wintering grounds in the African Sahel zone (Hjort and Lindholm 1978, Svensson 1985, Baillie and Peach 1992). But the long-term population trends of these species were probably influenced by habitat changes on their breeding grounds in Europe (Böhning-Gaese 1992).

Are these results specific for avian communities? Are birds especially sensitive to environmental variation? Because birds are highly mobile and the species of small insectivorous passerines that we worked with have short generation times, they can potentially respond very rapidly to environmental change in both space and time. Thus, an advantage of working with birds is that we can see these individualistic responses of species within time periods of 20 years. There is no reason to suppose, however, that other groups of organisms are any less sensitive to environmental change, even if they have low mobility or live long enough to buffer short-term environmental fluctuations. Other groups of organisms with longer lifespans and/ or lower mobility show similar individualistic responses to spatial variability (e.g., Whittaker 1960, Whittaker and Niering 1965, Brown and Kurzius 1987). While such organisms might exhibit greater time lags in response to temporal changes, they should be equally individualistic if observed over sufficiently long periods of time. Indeed, Davis (1986, see also Cole 1982) and Graham (1986, see also Graham and Grimm 1990) observed that species composition of tree and small mammal communities, respectively, changed dramatically from late Pleistocene to contemporary times, with each species shifting its distribution individualistically and with different time lags in response to climatic change.

These individualistic responses of species to environmental change on different temporal and spatial scales make it difficult to deduce the constellation of environmental factors that determine the composition of communities. They will also make it difficult to extrapolate across scales to predict the responses of species and communities to human-caused changes in climate and land use (Peters and Lovejoy 1992).

However, the fact that species respond individualistically to environmental variation does not imply that species do not respond deterministically and predictably to abiotic conditions and to other species, or that communities are not structured. On the contrary, nonrandom patterns of spatial and temporal variation in the abundance of species and the composition of communities can be demonstrated, and the underlying processes can be identified. For example, working with the same bird species in the same strata, Böhning-Gaese et al. (1993) showed that population trends during the last two decades were predictable. A few species with traits indicating high vulnerability to nest predation (low, open nests and high cowbird parasitism) showed significant declines, whereas the majority of species without such traits increased or showed no significant change. A challenge to contemporary community ecology is to reconcile the individualistic nature of species and communities with the deterministic processes that regulate population dynamics and species composition.

Acknowledgments

We thank S. Droege and the U.S. Fish and Wildlife Service for making the BBS data set available, J. Long for providing the Mantel test programs, D. Wiatr for drawing the map and J. S. Brown, J. Burkhardt, B. Gaese, J. R. Gosz, and R. J. O'Connor for comments on earlier versions of the manuscript. The research was supported by a NSF Grant to J.H.B., by a fellowship of the Studienstiftung des deutschen Volkes to K.B.G., and by US-EPA Cooperative Agreement CR-820086 to Dr. Goodman.

References

Aldrich, J. W. 1963. Life areas of North America. — J. Wildl. Manage. 27: 530–531.

Baillie, S. R. and Peach, W. J. 1992. Population limitation in Palaearctic-African migrant passerines. — Ibis 134 suppl.: 120–132.

Bock, C. E., Cruz, A. Jr., Grant, M. C., Aid, C. S. and Strong, T. R. 1992. Field experimental evidence for diffuse competition among southwestern riparian birds. — Am. Nat. 140: 815–828.

Böhning-Gaese, K. 1992. Causes for the decline of European songbirds: an analysis of the migratory bird trapping data of the Mettnau-Reit-Illmitz Program. — J. Ornithol. 133: 413–425.

———, Taper, M. L. and Brown, J. H. 1993. Are declines in North American insectivorous songbirds due to causes on the breeding range? — Conserv. Biol. 7: 76–86.

Brown, J. H. and Kodric-Brown, A. 1979. Convergence, competition, and mimicry in a temperate community of hummingbird-pollinated flowers. — Ecology 60: 1022–1035.

_____ and Kurzius, M. A. 1987. Composition of desert rodent faunas: combinations of coexisting species. – Ann. Zool. Fenn. 24: 227–237.

_____ and Heske, E. J. 1990. Temporal changes in a Chihuahuan desert rodent community. – Oikos 59: 290–302.

Cole, K. L. 1982. Late Quaternary zonation of vegetation in the eastern Grand Canyon. – Science 217: 1142–1145.

Connell, J. H. 1961. The influence of interspecific competition and other factors on the distribution of the barnacle *Chthamalus stellatus*. – Ecology 42: 710–723.

Davis, M. B. 1986. Climatic instability, time lags, and community disequilibrium. – In: Diamond, J. and Case, T. J. (eds), Community ecology. Harper and Row, New York, pp. 269–284.

Droege, S. 1990. The North American Breeding Bird Survey. – In: Sauer, J. R. and Droege, S. (eds), Survey designs and statistical methods for the estimation of avian population trends. U.S. Fish and Wildlife Service, Biological Report 90, Washington, D.C., pp. 1–4.

Ehrlich, P. R., Dobkin, D. S. and Wheye, D. 1988. The birder's handbook: a field guide to the natural history of North American Birds. – Simon and Schuster/Fireside, New York.

Enemar, A., Nilsson, L. and Sjöstrand, B. 1984. The composition and dynamics of the passerine bird community in a subalpine birch forest, Swedish Lapland. – Ann. Zool. Fenn. 21: 321–338.

Geissler, P. H. and Noon, B. R. 1981. Estimates of avian population trends from the North American Breeding Bird Survey. – Stud. Avian Biol. 6: 42–51.

Gleason, H. A. 1917. The structure and development of the plant association. – Bull. Torrey Bot. Club 44: 463–481.

_____ 1926. The individualistic concept of plant association. – Bull. Torrey Bot. Club 53: 7–26.

Graham, R. W. 1986. Response of mammalian communities to environmental changes during the late Quaternary. – In: Diamond, J. and Case, T. J. (eds), Community ecology. Harper and Row, New York, pp. 300–313.

– and Grimm, E. C. 1990. Effects of global climate change on the patterns of terrestrial biological communities. – Trends Ecol. Evol. 5: 289–292.

Hjort, C. and Lindholm, C.-G. 1978. Annual bird ringing totals and population fluctuations. – Oikos 30: 387–392.

Holmes, R. T., Sherry, T. W. and Sturges, F. W. 1986. Bird community dynamics in a temperate deciduous forest: long-term trends at Hubbard Brook. – Ecol. Monogr. 56: 201–220.

Kareiva, P. and Andersen, M. 1988. Spatial aspects of species interactions: the wedding of models and experiments. – In: Hastings, A. (ed.), Community ecology. Springer, New York, pp. 38–54.

Morison, J. I. L. 1990. Plant and ecosystem responses to increasing atmospheric CO_2. – Trends Ecol. Evol. 5: 69–70.

Peters, R. L. and Lovejoy, T. E. (eds) 1992. Global warming and biological diversity. – Yale Univ. Press, New Haven, CT.

Pimm, S. L. 1991. The balance of nature? – Univ. of Chicago Press, Chicago.

Rice, W. R. 1989. Analyzing tables of statistical tests. – Evolution 43: 223–225.

Robbins, C. S., Bystrak, D. and Geissler, P. H. 1986. The Breeding Bird Survey: its first fifteen years, 1965–1979. – U.S. Fish and Wildlife Service Resource Publication 157. Washington, D.C.

Rosenzweig, M. L., Abramsky, Z., Kotler, B. and Mitchell, W. 1985. Can interaction coefficients be determined from census data? – Oecologia 66: 194–198.

SAS/STAT. 1988. SAS/STAT User's Guide, Release 6.03 ed. – SAS Inst., Cary, NC.

Schmitt, R. J. 1987. Indirect interactions between prey: apparent competition, predator aggregation, and habitat segregation. – Ecology 68: 1887–1897.

Smouse, P. E., Long, J. C. and Sokal, R. R. 1986. Multiple regression and correlation extensions of the mantel test of matrix correspondence. – Syst. Zool. 35: 627–632.

Svensson, S. E. 1985. Effects of changes in tropical environments on the North European avifauna. – Ornis Fenn. 62: 56–63.

Tilman, D. 1989. Ecological experimentation: strengths and conceptual problems. – In: Likens, G. E. (ed.), Long-term studies in ecology. Approaches and alternatives. Springer, New York, pp. 136–157.

Whittaker, R. H. 1960. Vegetation of the Siskiyou Mountains. Oregon and California. – Ecol. Monogr. 30: 279–338.

– and Niering, W. A. 1965. Vegetation of the Santa Catalina Mountains, Arizona: a gradient analysis of the south slope. – Ecology 46: 429–452.

8

Identifying Extinction Threats: Global Analyses of the Distribution of Biodiversity and the Expansion of the Human Enterprise

Thomas D. Sisk, Alan E. Launer, Kathy R. Switky, and Paul R. Ehrlich

Relatively pristine habitats around the world are being lost at unprecedented rates (Melillo et al. 1985, Skole and Tucker 1993) as an expanding human population converts them to agriculture, forestry, and urban centers (Hall 1978, Vitousek et al. 1986). As these habitats are altered, untold numbers of species are disappearing before they have been recognized, much less studied (Wilson 1989), and the functioning of entire ecosystems is threatened. This loss of biodiversity, at the very time when the value of biotic resources is becoming widely recognized (Malone 1992), has made it strikingly clear that current strategies for conservation are failing dismally (Ehrlich 1992, Ehrlich and Wilson 1991).

The Center for Conservation Biology at Stanford University has developed a new tool for analyzing anthropogenic threats to biodiversity. It is based on geographic patterns in species distributions, species habitat use, forest loss, and human demographic trends. In this article, we use this tool to identify areas of critical concern, which we define as countries that support high levels of species diversity or endemism and that have a history of rapid forest destruction or face strong pressures from increasing human populations. This article demonstrates the use of low resolution but consistent datasets in continental and global analyses of threats to biodiversity, and it offers a preliminary list of countries deemed facing exceptionally intense conservation problems.

One should be cautious, however, in drawing immediate policy conclusions from our results, because they depend heavily on several underlying assumptions. Foremost among these assumptions is that at comparatively broad geographic levels, biodiversity is correlated with the species diversity of mammals and butterflies, and that the forest loss and human demographic indices used are valid approximations of actual threats of extinction. In addition, our results indicate that there is no single correct way to evaluate threats to biodiversity and to prioritize conservation activities. In many ways, therefore, the process by which we conducted our analyses may be of greater value to those developing conservation policy than the specific results, because it permits similar assessments under different sets of assumptions and scales of analysis.

Problems with Historic Approaches to Conservation

Conservation efforts typically have focused on the species level, yet biodiversity exists in many forms, from genes to ecosystems. Biodiversity also includes dynamic processes such as succession and nutrient cycling. The historic species-

Reprinted with permission from Bioscience vol. 44, pp. 592-602. Copyright 1994 American Institute of Biological Sciences.

level orientation is most likely due to the taxonomic focus of natural historians, but regardless of its origin, it has many shortcomings when applied to conservation problems. For example, activities targeted to a specific species frequently do not include preservation measures for the ecosystems upon which that species depends—witness the ever-increasing reliance of endangered species recovery programs on captive propagation. While the complexities of ecosystems render design and implementation of conservation measures specifically targeting them undeniably difficult, the current overwhelming focus on species misses many of the higher levels of biodiversity. Likewise, preserving populations may be as important as preserving species, because much diversity exists as variation within species, and above all, the services provided to humanity by healthy ecosystems (such as clean air, fresh water, and productive soils) are delivered by local populations (Ehrlich and Daily 1993).

Another difficulty with traditional methods employed for the preservation of biodiversity stems from the paucity of reliable data and the legal means by which organisms receive protection. Government and international agencies typically require extensive knowledge of a species before granting it official protection. The burden of proof is on those who propose protection: a species is assumed to be stable and unthreatened until it is unequivocally shown to be declining and vulnerable (Diamond 1987). The problems associated with this approach are obvious—little information of the quality necessary for such determinations is available for organisms in many regions and for most taxonomic groups. For example, many people, scientists included, address the question of threat through the estimation of extinction probabilities, either via in-depth analyses of the viability of individual populations of target species (e.g., Soulé 1987) or through the ad hoc compilation of species thought to be facing extinction (e.g., Collar and Andrew 1988, Collins and Morris 1985, WCMC 1992).

While these studies and lists can be useful, especially when analyzed collectively with a broad conservation perspective (e.g., Smith et al. 1993), the time required to design and complete scientifically credible field investigations, evaluate the status of target species, and compile complete lists of threatened organisms is prohibitive. As these activities are being conducted, thousands, perhaps even millions, of lesser-known and unstudied organisms are moving toward extinction unnoticed (Ehrlich 1994).

Over the last decade, upwards of 20,000 species have been listed as being at risk of extinction by one or more prominent conservation organizations worldwide (McNeely et al. 1990, Smith et al. 1993, WRI et al. 1990). Meanwhile, various estimates of global extinction rates project annual losses of between 1000 and 30,000 species by the end of this century (Ehrlich and Wilson 1991, Reid and Miller 1989, Wilson 1992). Moreover, estimates of extinction rates are climbing much faster than the rates at which new species are being listed, suggesting that a species-by-species approach is virtually guaranteed to be unsuccessful at preserving biodiversity.

An additional problem with many conservation activities is related to the often unstated assumption that rare species (abundant nowhere) are the most threatened. Given current trends in land use, the rate at which habitats are being lost may, in some cases, be more important for determining risks of extinction than the present commonness or rarity of the species living in them. For example, an uncommon, locally endemic species whose habitat is protected or remote, and therefore buffered from changes in land use, may face a lesser risk of extinction than a more widespread and common species whose entire range is being impacted by human activities. In today's world, where huge tracts of tropical forest are burned each day, a species that is abundant over several thousand square kilometers could be extinguished within a single year.

Equally serious but seldom appreciated failings of current conservation activities are the geographic and taxonomic biases inherent in the information available on biodiversity. The biota of little-studied areas (many developing countries, for example) are poorly known, and the ranges of most species, even in economically developed nations, are not adequately documented (Pimm and Gittleman 1992). Likewise, distributional information is generally available for charismatic taxa, such as primates, large carnivores, and certain birds, but distributional information for most taxa is anecdotal at best. At a broad level, changes in the distribution and integrity of most habitat types are difficult to determine in developed countries, even with the availability of satellite imagery and analysis (Ehrlich 1994). The use of these techniques to quantify habitat loss is much more difficult in developing countries, due to a host of complicating factors, including inadequate ground-truthing, less comprehensive satellite coverage, and the difficulties in identifying and classifying regional habitat types (Fearnside and Salati 1985, Skole and Tucker (1993).

The inadequacy of traditional approaches to identifying conservation priorities is clear, yet conservationists cannot afford to ignore the need for comprehensive, long-term strategies for preserving biodiversity. Intelligent planning requires a logical, consistent approach for making the most of available, albeit limited, data. Conservationists cannot afford to spend their time debating the status of each species separately. Instead, they should concentrate their efforts in geographic areas where especially diverse or unique biota lie in the path of increasing habitat destruction. This article suggests one way of identifying those areas.

The Global Conservation Analysis Project

Over the past five years, we have developed a set of databases that include distribution and habitat information on all species within several large taxonomic groups, along with published estimates of rates of forest loss and changes in the size of human populations. Working within FoxBASE,[1] a commercially available database management program, we developed a relational data structure and a set of programs to query the databases for analyzing threats to biodiversity. The development of these files and programs, together called the Global Conservation Analysis Package, was based on the following assumptions:

---
[1]Fox Software, Perrysburg, Ohio.

- A valid estimate of overall species richness of a country can be made by counting the number of species present from each of a limited number of well-known taxonomic groups.
- A valid estimate of the level of endemism in the biota can be made by counting, for each of the sample taxonomic groups, the number of species that occur only within that country.
- Consideration, in a consistent manner, of all species in the target groups can avoid many biases common to species-by-species preservation measures.
- Measures of recent rates of habitat loss, human population growth, and human population density can serve as indicators of anthropogenic threats to biodiversity.

In this approach, geographic information on species distributions, forest loss, and population pressures are compared to generate measures of historic and current patterns of anthropogenic threats to biodiversity. Areas where rich and/or unique biota coincide with dense, growing human populations and/or with the rapid loss of native forests are those we identify as areas of critical concern and consider to be priority areas for conservation. In these analyses, the vulnerability of biodiversity is determined not by observed declines in the abundances of particular species but by the projected threats to their habitats.

Biodiversity, Forest Loss, and Human Demography

This article draws on four of the databases created for the Global Conservation Analysis Package: all species of mammals, all species of large butterflies, rates of loss of native forest, and human demography. The geographic resolution of the data is at the country level, for which relatively good species distribution and habitat data are available for the entire world. Because many conservation plans are designed and implemented by sovereign nations, and suitable data are not widely available for many subnational units, we selected countries as an appropriate level of resolution for global analyses of conservation priorities. Although biodiversity does not conform to political boundaries, the factors that control its fate often do (Mittermeier 1988).

TABLE 1. Africa. Summary of biodiversity and threat data. Shaded and italicized countries were identified as global and continental areas of critical concern, respectively.

	Number of mammal species (number endemic)	Number of butterfly species (number endemic)	Species richness: by world rank (by region)	Endemism: by world rank (by region)	Population pressure index (annual increase in density)	Population pressure: by world rank (by region)	Forest loss index (annual loss)	Forest loss: by world rank (by region)
Algeria	94 (1)	124 (2)	89 (43)	56 (24)	0.26	95 (35)		
Angola	*269 (2)*	*346 (5)*	*25 (9)*	*43 (16)*	*0.20*	*101 (38)*	*0.0190*	*14 (4)*
Benin	209 (0)	314 (0)	43 (19)	86 (37)	1.38	44 (15)	0.0068	45 (24)
Botswana	187 (0)	138 (0)	66 (32)	86 (37)	0.07	121 (44)	0.0064	48 (26)
Burkina Faso	167 (0)	123 (0)	71 (35)	86 (37)	1.16	51 (18)	0.0083	30 (15)
Burundi	197 (0)	312 (0)	47 (20)	86 (37)	6.67	6 (2)	0.0087	28 (13)
Cameroon	311 (10)	577 (27)	12 (2)	23 (5)	0.85	62 (23)	0.0049	55 (31)
Central African Rep.	266 (3)	359 (2)	24 (8)	48 (16)	0.13	109 (40)	0.0025	73 (37)
Chad	200 (1)	165 (0)	61 (28)	75 (32)	0.10	114 (41)	0.0083	30 (16)
Congo	225 (2)	477 (7)	23 (7)	40 (15)	0.20	101 (38)	0.0021	74 (38)
Egypt	108 (8)	43 (0)	106 (44)	38 (14)	1.33	46 (16)		
Equatorial Guinea	190 (0)	422 (3)	29 (11)	65 (27)	0.37	87 (33)	0.0076	40 (22)
Ethiopia	269 (23)	236 (11)	38 (16)	20 (4)	1.24	48 (17)	0.0039	64 (33)
Gabon	200 (2)	450 (1)	25 (10)	56 (24)	0.10	114 (41)	0.0014	76 (39)
Gambia	116 (0)	109 (0)	82 (40)	86 (37)	2.19	26 (7)	0.0118	24 (11)
Ghana	*236 (0)*	*350 (3)*	*31 (12)*	*65 (27)*	*2.15*	*27 (8)*	*0.0138*	*21 (9)*
Guinea	200 (1)	279 (0)	50 (22)	75 (32)	0.79	66 (26)	0.0150	19 (7)
Guinea-Bissau	125 (0)	109 (0)	79 (39)	86 (37)	0.55	75 (29)	0.0190	14 (4)
Ivory Coast	*224 (1)*	*345 (5)*	*34 (13)*	*48 (16)*	*1.45*	*41 (13)*	*0.0540*	*1 (1)*
Kenya	*328 (12)*	*406 (11)*	*19 (5)*	*28 (7)*	*1.66*	*31 (10)*	*0.0147*	*20 (8)*
Lesotho	134 (0)	178 (3)	71 (35)	86 (37)	1.82	29 (9)	0.0072	43 (23)
Liberia	169 (1)	298 (0)	53 (25)	75 (32)	0.80	65 (25)	0.0200	13 (3)
Libya	98 (5)	34 (0)	109 (46)	43 (16)	0.08	119 (43)		
Malagasy Republic	*95 (73)*	*218 (155)*	*76 (38)*	*5 (1)*	*0.65*	*71 (27)*	*0.0137*	*22 (10)*
Malawi	208 (0)	263 (6)	50 (22)	54 (23)	2.57	22 (5)	0.0059	51 (28)
Mali	198 (0)	128 (0)	63 (29)	86 (37)	0.21	99 (36)	0.0082	32 (17)
Mauritania	116 (1)	101 (0)	85 (41)	75 (32)	0.05	126 (46)		
Morocco	100 (13)	129 (3)	85 (41)	31 (8)	1.41	42 (14)		
Mozambique	230 (2)	230 (1)	50 (22)	56 (24)	0.57	74 (28)	0.0189	16 (6)
Namibia	188 (3)	132 (3)	66 (32)	43 (16)	0.06	123 (45)		
Niger	186 (0)	115 (0)	70 (34)	86 (37)	0.21	99 (36)	0.0081	34 (19)
Nigeria	*268 (5)*	*448 (5)*	*21 (6)*	*35 (12)*	*2.92*	*18 (3)*	*0.0077*	*38 (20)*
Rwanda	213 (0)	329 (0)	38 (16)	86 (37)	9.94	2 (1)	0.0215	8 (2)
Senegal	171 (1)	121 (1)	71 (35)	65 (27)	1.12	53 (19)	0.0086	29 (14)
Sierra Leone	154 (0)	275 (1)	58 (27)	75 (32)	1.58	34 (11)	0.0082	32 (17)
Somalia	203 (10)	132 (1)	63 (29)	34 (10)	0.38	86 (32)	0.0035	68 (34)
South Africa	249 (28)	220 (23)	48 (21)	17 (3)	0.92	61 (22)	0.0047	57 (32)
Sudan	290 (14)	238 (1)	35 (14)	31 (8)	0.33	89 (34)	0.0077	38 (20)
Swaziland	164 (0)	186 (0)	63 (29)	86 (37)	1.47	40 (12)	0.0053	53 (30)
Tanzania	303 (11)	449 (24)	17 (4)	24 (6)	1.02	58 (21)	0.0062	50 (27)
Togo	209 (1)	319 (1)	43 (18)	65 (27)	2.51	23 (6)	0.0065	46 (25)
Tunisia	85 (2)	64 (0)	106 (44)	65 (27)	1.07	55 (20)		
Uganda	*311 (3)*	*491 (9)*	*13 (3)*	*35 (12)*	*2.74*	*21 (4)*	*0.0117*	*25 (12)*
Zaire	422 (25)	705 (47)	6 (1)	15 (2)	0.50	78 (30)	0.0030	70 (35)
Zambia	247 (2)	285 (4)	37 (15)	48 (16)	0.42	82 (31)	0.0030	70 (35)
Zimbabwe	200 (0)	221 (7)	56 (26)	48 (16)	0.82	64 (24)	0.0057	52 (29)

In all datasets, information is coded in a hierarchical numerical format that allows the incorporation and eventual retrieval of data at various geographic levels (global, continental, national, and regional). Importantly, this hierarchical numerical scheme allows the user to adjust the level of resolution to the purpose of the analysis and to the resolution and quality of the available data. It also can accommodate future refinements in that information.

Species Databases

Despite more than a century of active exploration and inventory by biologists, knowledge of species distributions and habitat requirements is incomplete, even for relatively well-studied taxa (May 1988). Information on species distributions and habitat use in developed countries is often available in detail, while in other, even nearby countries, simple species lists have not yet been compiled for most taxa.

Given the impossibility of including data on most kinds of organisms, we chose to focus on two well-known animal taxa, one vertebrate and one invertebrate. Data on all mammal species (probably the best-studied vertebrate group) and on species of the butterfly families Papilionidae, Pieridae, and Nymphalidae (perhaps the best known invertebrate groups) form the core of the species information. These taxa were chosen as representative groups because they have diverse habitat requirements and life-history strategies and, therefore, are likely to provide a better index of overall biodiversity than would any single taxon.

The resulting databases are coarse, but they permit broad, relatively unbiased global analyses. Even though mammals and butterflies are comparatively well-known taxonomic groups, the compilation of these global databases required consultation of more than 140 published references (e.g., Collins and Morris 1985, D'Abrera 1980, 1981, 1984, 1990, 1991, D'Abrera and Classey 1982, Honacki et al. 1982, Nowak and Paradiso 1983).

Forest-Loss Database

In the late 1970s, two landmark efforts were made to assess global rates of habitat loss; both focused on tropical forests. Studies by the National Academy of Sciences (Myers 1980) and the United Nations Food and Agriculture Organization (FAO 1981) provided assessments of the status of tropical forests, including the most detailed deforestation estimates to date. When differences in definitions and underlying assumptions are taken into account, estimated deforestation rates were similar, varying by only 10–15% (Melillo et al. 1985). Updates of

these data suggest that deforestation in the tropics has risen sharply, nearly doubling during the 1980s (Myers 1989). While the magnitude of this increase has been challenged (Sayer and Whitmore 1991), there is widespread agreement that rates are going up.

In 1990, the World Resources Institute published a worldwide compilation of estimated rates of forest loss for all countries during the 1980s (WRI et al. 1990). This information, gathered from many different sources, constitutes a well-referenced, standardized baseline for forest loss and provides some insight into the widely varying estimates of deforestation that are encountered in the literature. The Global Conservation Analysis Package includes forest-loss estimates for 1950–1980 from the FAO report and worldwide forest-loss estimates from the WRI data covering 1980–1989.

Human Demography Database

While data on forest loss provide one measure of recent patterns of habitat disturbance, they are not a comprehensive measure of habitat loss and do not necessarily indicate future trends. Estimates of habitat loss based on rates of deforestation may be particularly misleading in nations that are nearing exhaustion of their timber resources. In addition, reliable data are available only for loss of some forest types in some nations, and many types of habitat alterations (e.g., selective logging and extensive coppicing) are not addressed. And finally, other types of impacts on natural habitats, including grassland plowing, overgrazing of rangelands, wetland drainage, damming of streams, toxification of soils and fresh water, hunting, and introductions of exotic organisms, cannot be inferred from forest-loss data. We therefore employ human demographic trends as imperfect but useful indicators of future habitat conversion and destruction and thus of threats to biodiversity (see also Ehrlich 1994).

Country-specific demography data were gathered from sources including the Population Reference Bureau (1992). Data on population density, current rates of population increase, 50-year population increase, per capita gross national product (GNP), GNP per area, and per capita energy consumption are included in the demography database.

TABLE 2. Asia and Australia. Summary of biodiversity and threat data. Shaded and italicized countries were identified as global and continental areas of critical concern, respectively.

	Number of mammal species (number endemic)	Number of butterfly species (number endemic)	Species richness: by world rank (by region)	Endemism: by world rank (by region)	Population pressure index (annual increase in density)	Population pressure: by world rank (by region)	Forest loss index (annual loss)	Forest loss: by world rank (by region)
Afghanistan	132 (0)	106 (0)	79 (20)	86 (23)	0.68	69 (28)		
Australia	267 (206)	125 (40)	56 (11)	2 (2)	0.02	130 (36)	0.0079	35 (8)
Bangladesh	117 (0)	248 (248)	66 (15)	86 (23)	18.57	1 (1)	0.0065	46 (11)
Bhutan	151 (0)	237 (2)	61 (14)	75 (20)	0.30	91 (30)	0.0045	58 (12)
Brunei	90 (0)	250 (0)	71 (16)	86 (23)	1.30	47 (22)	0.0202	12 (3)
Burma	293 (7)	501 (13)	14 (4)	30 (11)	1.18	50 (24)	0.0040	61 (15)
China	446 (56)	457 (22)	10 (2)	13 (5)	1.59	33 (16)	0.0042	59 (13)
India	*360 (40)*	*514 (47)*	*11 (3)*	*13 (5)*	*5.37*	*8 (4)*	*0.0040*	*61 (15)*
Indonesia	547 (182)	988 (417)	2 (1)	1 (1)	1.54	36 (17)	0.0034	69 (18)
Iran	162 (4)	80 (1)	77 (18)	48 (14)	1.20	49 (23)		
Iraq	102 (0)	41 (0)	106 (26)	86 (23)	1.51	38 (19)		
Israel	88 (2)	49 (0)	109 (27)	65 (17)	3.84	14 (10)		
Japan	107 (27)	116 (2)	85 (21)	21 (8)	0.99	60 (26)		
Jordan	61 (0)	41 (0)	126 (30)	86 (23)	1.35	45 (21)		
Kampuchea	175 (0)	232 (0)	58 (12)	86 (23)	1.11	54 (25)	0.0079	35 (6)
Kuwait	52 (0)	21 (0)	132 (34)	86 (23)	2.48	24 (14)		
Laos	191 (1)	263 (0)	53 (10)	75 (20)	0.54	76 (29)	0.0073	42 (9)
Lebanon	44 (0)	38 (0)	132 (34)	86 (23)	6.87	4 (2)		
Malaysia	305 (19)	450 (6)	17 (5)	25 (9)	1.41	42 (20)	0.0041	60 (14)
Mongolia	136 (2)	p 50 (0)	89 (22)	65 (17)	0.04	128 (35)		
Nepal	185 (3)	198 (0)	60 (13)	56 (16)	3.51	16 (11)	0.0357	4 (1)
New Zealand	7 (0)	11 (0)	135 (36)	86 (23)	0.13	109 (34)	0.0071	44 (10)
North Korea	55 (0)	p 50 (0)	126 (30)	86 (23)	3.50	17 (12)		
Oman	57 (2)	27 (0)	131 (33)	65 (17)	0.18	106 (33)		
Pakistan	180 (4)	114 (0)	71 (16)	48 (14)	4.69	11 (7)	0.0037	65 (17)
Papua New Guinea	214 (47)	303 (56)	46 (8)	12 (4)	0.19	103 (32)	0.0006	78 (19)
Philippines	*195 (110)*	*339 (153)*	*43 (7)*	*4 (3)*	*5.10*	*9 (5)*	*0.0101*	*27 (5)*
Saudi Arabia	94 (0)	76 (0)	102 (25)	86 (23)	0.26	95 (31)		
South Korea	51 (0)	p 50 (0)	126 (30)	86 (86)	4.93	10 (6)		
Sri Lanka	99 (13)	110 (9)	89 (22)	28 (10)	4.02	13 (9)	0.0134	23 (4)
Syria	61 (0)	55 (0)	122 (29)	86 (23)	2.81	19 (13)		
Taiwan	*62 (6)*	*169 (39)*	*89 (22)*	*19 (7)*	*6.32*	*7 (3)*		
Thailand	*319 (319)*	*394 (1)*	*20 (6)*	*35 (12)*	*1.53*	*37 (18)*	*0.0206*	*11 (2)*
Turkey	123 (1)	124 (0)	78 (19)	75 (20)	1.67	30 (15)		
Vietnam	*211 (5)*	*274 (0)*	*48 (9)*	*43 (13)*	*4.58*	*12 (8)*	*0.0079*	*35 (6)*
Yemen	78 (0)	52 (0)	112 (28)	86 (23)	0.69	68 (27)		

Integrating Data Through the Formulation of Indices

To identify areas of critical concern in this analysis, we have generated two indices of biodiversity: species richness (number of species present) and endemism of mammals and butterflies. We have also generated two indices of threat: the rate of forest loss and human population pressure.

Species Richness Index

A general species richness index (IS) was created indicating the mean proportion of the world's mammal and butterfly species in a given country, C:

$$IS_C = 100 \left[\frac{1}{2} \left(\frac{M_C}{M_T} + \frac{B_C}{B_T} \right) \right]$$

where M_C is the number of mammalian species in country C, M_T is the total number of mam-

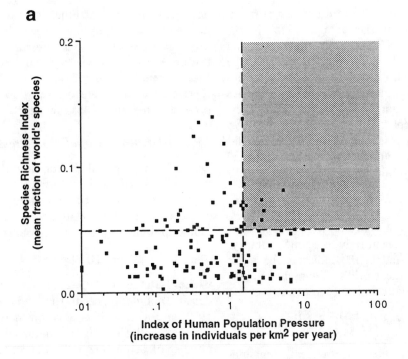

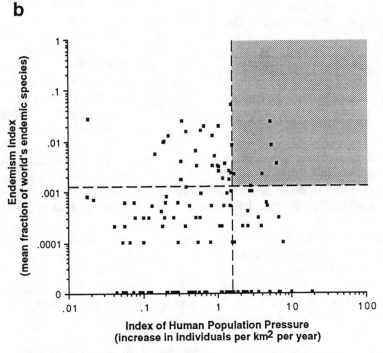

FIGURE 1. The relationship between biodiversity and the threat of habitat loss, as represented by indices developed for 135 continental and large island nations. Countries falling within the shaded regions, referred to in the text as areas of critical concern, have values in the upper quartile for human population pressure (cutoff value = 1.58) and either (a) species richness (cutoff value = 0.048) or (b) endemism (cutoff value = 0.0012).

in country C, M_T is the total number of mammalian species in the database, B_C is the number of butterfly species in country C, and B_T is the total number of butterfly species in the database. A high value for this index identifies a country that supports a high level of species diversity for mammals and/or butterflies. Based on current knowledge, inclusion of two ecologically diverse and distinct taxa should provide a better indicator of overall biological diversity than would any single taxon.

Endemism Index

Similarly, distributional data were used to identify species that occur only in a single country. The endemism index (IE) indicates the mean proportion of the world's mammal and butterfly species that occur only in country C:

$$IE_C = 100 \left[\frac{1}{2} \left(\frac{EM_C}{M_T} + \frac{EB_C}{B_T} \right) \right]$$

where EM_C is the number of endemic mammalian species in country C, and EB_C is the number of endemic butterfly species in country C. A high value for this index identifies a country supporting a distinctive mammal and/or butterfly fauna. To the degree that these groups are typical, the index reflects the distinctiveness of the entire biota.

Forest-Loss Index

We combined FAO data on annual percent loss of tropical forests for 1950–1980 with WRI data on annual percent forest habitat loss during 1980–1989 (converted from decadal loss rate). For countries cited by only one source (most temperate-zone countries), the single available annual rate was used as the index. For tropical countries having two independent estimates, we assumed that the two values were equally valid measures of the relative rate of deforestation among countries and gave them equal weighting when calculating the index of forest loss (IF):

$$IF_C = \tfrac{1}{2} \left(APL_{1950-1980} + APL_{1980-1989} \right)$$

where APL is annual percentage lost.

While this index is based solely on forest loss, it represents a useful (and often the only) estimate of overall habitat loss in each country, because conversion of other habitat types has not been adequately quantified. This index provides a measure of past pressures on biotic resources and may serve as an indicator of future trends. A high value for this index represents a high rate of forest loss (as expressed in percentage lost per year).

Index of Human Population Pressures

Data on human population density (PRB 1992) were combined with information on each country's natural rate of population increase to create an index of the relative rate of increase in human population density for country C:

$$IP_C = HPD \times NRPI$$

where HPD is the 1992 human population density (population per 100 ha) and NRPI is the 1992 natural rate of population increase. The index is equivalent to the annual increase in number of people per 100 ha (1 km²). A high value for this index is assumed to represent greater pressure on natural systems due to high or rapidly increasing human population density. We looked at several other methods for approximating relative human population pressure, including indices based on population density alone, the natural rate of increase alone, and on either population density or the natural rate of increase, whichever was highest. While none of the formulations fully capture the summed impacts of humans on their environment, there was general agreement on the number of countries selected as areas of critical concern and their identities (range: 16–18 countries; similarity of country lists: 0.52–0.79, Jaccard index). Results based solely on the rate of population increase differed most from the other formulations.

None of the human population indices examined incorporate the critical factors of resource consumption and technology (Ehrlich and Ehrlich 1991, Ehrlich and Holdren 1971), the international impacts of the demands of developed nations for natural resources from developing countries, or the greater ability of developed nations to ameliorate the pressures they put on natural systems worldwide. Thus, our index of population pressures does not attempt to represent the impact of a given country on global biodiversity but rather the level of the threat of habitat destruction in that country, whatever its source. We believe that the index adopted here is a coarse, but useful, indicator of the scale of the human enterprise and a valid first approximation of the relative intensity of future environmental impacts.

Methods for Identifying Priority Areas

We use these indices of species diversity, species endemism, forest loss, and human population pressures to identify countries that support relatively high species diversity or endemism and that have undergone or are likely to undergo relatively large-scale perturbation of natural habitats. All countries recognized in June 1991 were included in the databases, but for this analysis we concentrate on continental nations and omit small islands. The numerous political changes in the former Soviet Union preclude the inclusion of the newly independent republics because country-level data are not yet available. The analysis includes the 135 countries for which adequate data on species richness, endemism, and human population pressures exist; forest-loss data are available for 80 of these nations. A discussion of the potential effects of using the incomplete forest-loss data is presented with the results.

The indices for each country were compared to those of other countries on the same continent (continental scale analyses) and to those of all other countries (global scale analysis). Countries within the top 25% of any index (i.e., those with the highest biodiversity, endemism, rates of forest loss, or population pressure) were considered to be critical areas for that particular index. In this analysis, a country was classified as an area of critical concern if it fell within the top quartile for either biodiversity or endemism and either forest loss or population pressure.

This system of ranking and use of quartiles was adopted after it became apparent that attempting to infer precise meanings from the absolute values of the indices was both inappropriate—the purpose of this analysis being to compare countries to one another—and mis-

leading, because biodiversity is essential to the welfare of people in all countries and therefore worthy of protection regardless of how it ranks globally.

For the purpose of identifying areas where high diversity faces intense threat of destruction, retaining for further consideration those countries that fall within the upper quartile for each index constitutes a conservative analytical approach. Rankings were used only to distinguish countries within the top quartile for each parameter, and all subsequent analysis is based on the number of parameters for which each country ranked in the top quartile, not on specific ranks or index values. Selection of a 25% cutoff is arbitrary, but one that we feel is appropriate, because no country experiencing extreme threats and supporting an exceptionally high level of biodiversity is likely to be omitted.

By considering virtually all species in each sample group and all continental and large-island nations, this approach eliminates the need for a priori selection of threatened species and candidate hot-spots. It builds on the analytical approach first advanced by Myers in his analysis of regional conservation hot-spots (Myers 1988, 1990). Through the identification of regions of exceptionally high plant species diversity and endemism, and the consideration of estimated rates of habitat alteration, Myers identified 18 hot-spots around the planet. A somewhat different approach was taken by Mittermeier (1988), who focused on mammal diversity and endemism but did not consider explicitly the current threats of habitat loss. A third synthesis of global conservation priorities was carried out by the International Council for Bird Preservation (Bibby et al. 1992). Their analysis focused on small areas of high bird endemism, and they presented a ranking of the countries that contain the highest numbers of endemic birds listed as threatened by the World Conservation Monitoring Center. These three studies, often cited in conservation literature, examined different taxa and employed different means of characterizing biotic value and threats. Our approach combines advantages from each: we draw on country-specific data on diversity and endemism, and we attempt to quantify threats to native habitats.

Areas of Critical Concern

Continental Analyses

The number of countries, and therefore the number of potential areas of critical concern, varies considerably among continents (Tables 1–5). The percentage of countries identified as areas of critical concern was highest in Australia-Asia (14%; 5 of 36 countries) and Africa (13%; 6 of 46), indicating that rich biota and increasing threats often coincide on those continents. In our analysis, few of the biologically richest countries in South America (8%; 1 of 13), North/Central America (8%; 1 of 13), or Europe (4%; 1 of 27) faced the highest threats from anthropogenic pressures.

In only a few cases did the lack of comparable forest-loss data potentially influence the results of the continental analysis. For Africa, only one country lacking forest-loss data, Morocco, was identified as being in the top quartile in either species richness or endemism. For Australia and Asia, only two countries identified as being relatively high in either of the biodiversity indices, Japan and Taiwan, lacked forest-loss data, and Taiwan was identified as an area of critical concern on the basis of the population index. In both North/Central America and South America, no countries lacking forest-loss data ranked in the top quartile for either of the biodiversity indices. The complete lack of comparable forest-loss data for Europe is a potential problem for that continental analysis—all nine countries identified as having relatively high biodiversity values lacked forest-loss data. Of these nine countries

TABLE 3. Europe. Summary of biodiversity and threat data. The italicized country was identified as a continental area of critical concern. Forest loss data were not available.

	Number of mammal species (number endemic)	Number of butterfly species (number endemic)	Species richness: by world rank (by region)	Endemism: by world rank (by region)	Population pressure index (annual increase in density)	Population pressure: by world rank (by region)
Yugoslavia	110 (3)	132 (1)	79 (1)	54 (1)	0.18	106 (15)
France	93 (0)	147 (0)	82 (2)	86 (6)	0.42	82 (10)
Italy	87 (1)	144 (3)	85 (3)	56 (2)	0.19	103 (14)
Bulgaria	93 (0)	117 (0)	89 (4)	86 (6)	0.00	133 (25)
Romania	95 (0)	125 (0)	89 (4)	86 (6)	0.10	114 (19)
Austria	90 (0)	119 (0)	95 (6)	86 (6)	0.09	117 (20)
Greece	88 (1)	112 (3)	95 (6)	56 (2)	0.08	119 (21)
Spain	87 (1)	122 (3)	95 (6)	56 (2)	0.12	111 (16)
Germany	89 (0)	100 (0)	99 (9)	86 (6)	−0.23	133 (25)
Hungary	87 (0)	102 (0)	99 (9)	86 (6)	−0.22	133 (25)
Switzerland	*78 (1)*	*123 (3)*	*99 (9)*	*56 (2)*	*0.50*	*78 (7)*
Czechoslovakia	87 (0)	96 (0)	102 (12)	86 (6)	0.25	97 (13)
Poland	94 (0)	86 (0)	102 (12)	86 (6)	0.49	80 (8)
Andorra	66 (0)	86 (0)	109 (15)	86 (6)	0.40	84 (11)
Finland	66 (0)	72 (0)	112 (16)	86 (6)	0.04	128 (24)
Liechtenstein	64 (0)	73 (0)	112 (16)	86 (6)	1.14	52 (4)
Portugal	63 (0)	78 (0)	112 (16)	86 (6)	0.11	113 (18)
Sweden	68 (0)	67 (0)	112 (16)	86 (6)	0.11	113 (18)
Belgium	65 (0)	64 (0)	117 (20)	86 (6)	0.66	70 (6)
Luxembourg	60 (0)	61 (0)	117 (20)	86 (6)	0.46	81 (9)
Netherlands	66 (0)	53 (0)	117 (20)	86 (6)	1.85	28 (2)
Norway	65 (0)	66 (0)	117 (20)	86 (6)	0.05	126 (23)
Denmark	61 (0)	53 (0)	122 (24)	86 (6)	0.12	111 (16)
San Marino	54 (0)	69 (0)	122 (24)	86 (6)	1.64	32 (2)
United Kingdom	66 (0)	41 (0)	122 (24)	86 (6)	0.71	67 (5)
Irish Republic	26 (0)	24 (0)	134 (27)	86 (6)	0.30	91 (12)
Albania	78 (0)	105 (0)	102 (12)	86 (6)	2.20	25 (1)

TABLE 4. North America and Central America. Summary of biodiversity and threat data. Shaded and italicized countries were identified as global and continental areas of critical concern, respectively.

	Number of mammal species (number endemic)	Number of butterfly species (number endemic)	Species richness: by world rank (by region)	Endemism: by world rank (by region)	Population pressure index (annual increase in density)	Population pressure: by world rank (by region)	Forest loss index (annual loss)	Forest loss: by world rank (by region)
Belize	173 (0)	392 (0)	38 (8)	86 (12)	0.27	94 (11)	0.0063	49 (9)
Canada	172 (5)	150 (2)	69 (10)	42 (7)	0.02	130 (13)		
Costa Rica	*219 (8)*	*617 (19)*	*14 (2)*	*27 (4)*	*1.51*	*38 (6)*	*0.0372*	*3 (1)*
Cuba	43 (14)	88 (12)	117 (11)	26 (3)	1.04	57 (8)	0.0009	77 (12)
Dominican Republic	24 (0)	96 (0)	126 (12)	86 (12)	3.56	15 (3)	0.0048	56 (10)
El Salvador	171 (1)	397 (0)	38 (8)	75 (10)	7.59	3 (1)	0.0254	6 (3)
Guatemala	197 (4)	450 (11)	27 (5)	33 (5)	2.76	20 (4)	0.0178	17 (6)
Haiti	18 (0)	102 (4)	126 (12)	56 (8)	6.70	5 (2)	0.0208	10 (5)
Honduras	187 (1)	420 (0)	31 (7)	75 (10)	1.57	35 (5)	0.0250	7 (4)
Mexico	446 (138)	471 (46)	9 (1)	6 (1)	1.02	58 (9)	0.0103	26 (7)
Nicaragua	187 (2)	452 (0)	28 (6)	65 (9)	1.07	55 (7)	0.0271	5 (2)
Panama	229 (5)	595 (4)	14 (2)	38 (6)	0.59	73 (10)	0.0076	40 (8)
United States	374 (96)	292 (22)	21 (4)	9 (2)	0.22	98 (12)	0.0026	72 (11)

only Switzerland was identified as an area of critical concern on the basis of the population index, and in light of the drastic political changes occurring in eastern Europe, it is probable that other countries would have been so identified, had habitat-loss data been available.

Global Analyses

In the worldwide analysis, 135 countries were ranked globally for each index (80 for forest loss) and, as in the continental analyses, those falling within the top 25% for either biodiversity or endemism and either forest loss or population pressure were considered global areas of critical concern (see Tables 1–5). Figure 1 plots the two biodiversity indices against the level of threat, as measured by the index of human population pressure. Areas of critical concern appear in the upper right quadrants, where both biodiversity and threat are high. Many countries with higher levels of biodiversity (upper left quadrant) and others with higher threat (lower right quadrants) do not qualify as areas of critical concern because they do not fall into the upper quartiles along both axes. Of the 18 global areas of critical concern, 6 are from Africa and 6 from Australia-Asia; 4 are from North/Central America, and 2 are from South America. Lack of data on

forest loss for 55 countries potentially affected the outcome of four countries: Chile, Japan, Morocco, and Taiwan — each ranked in the top quartile for one or both biodiversity indices. Of these countries, only Taiwan was included in the list of global areas of critical concern, again on the basis of human population pressures.

While use of the Global Conservation Analysis Package is effective in extracting broad trends from existing data, its reliance on rankings makes it sensitive to both the size and the composition of the pool of countries included in each analysis. Although there is a high degree of overlap in the continental and global analyses — 11 out of 14 countries identified as continental areas of critical concern were also identified at the global scale — the differences between the results illustrate this sensitivity (Table 6, first two columns).

Somewhat paradoxically, seven of the countries on the global list were not on the continental list. In addition, three countries identified at the continental level were not identified as such at the global level. Consideration of the relative values of the indices in each analysis, and recognition of region-wide trends in biodiversity and human pressures, explain these apparent contradictions. For example, the countries of Central America are similar in their comparatively high levels of species richness

TABLE 5. South America. Summary of biodiversity and threat data. Shaded and italicized countries were identified as global and continental areas of critical concern, respectively.

	Number of mammal species (number endemic)	Number of butterfly species (number endemic)	Species richness: by world rank (by region)	Endemism: by world rank (by region)	Population pressure index (annual increase in density)	Population pressure: by world rank (by region)	Forest loss index (annual loss)	Forest loss: by world rank (by region)
French Guiana	190 (3)	360 (6)	38 (10)	40 (9)	0.02	130 (13)		
Argentina	285 (41)	267 (9)	31 (8)	16 (6)	0.14	108 (9)	0.0051	54 (4)
Bolivia	303 (19)	842 (101)	7 (5)	11 (5)	0.19	103 (8)	0.0015	75 (8)
Brazil	455 (85)	954 (201)	4 (3)	3 (1)	0.34	88 (5)	0.0027	65 (6)
Chile	112 (12)	84 (34)	95 (13)	18 (7)	0.32	90 (6)		
Colombia	*379 (22)*	*1276 (185)*	*1 (1)*	*7 (2)*	*0.60*	*72 (2)*	*0.0174*	*18 (3)*
Ecuador	291 (21)	1120 (138)	5 (4)	10 (4)	0.85	62 (1)	0.0210	9 (2)
Guyana	227 (0)	368 (2)	29 (7)	75 (12)	0.07	121 (11)	0.0002	79 (9)
Paraguay	147 (4)	318 (2)	55 (11)	43 (10)	0.30	91 (7)	0.0393	2 (1)
Peru	381 (42)	1212 (139)	3 (2)	8 (3)	0.39	85 (4)	0.0036	67 (7)
Suriname	207 (1)	355 (1)	35 (9)	65 (11)	0.06	123 (12)	0.0001	80 (10)
Uruguay	75 (0)	174 (0)	82 (12)	86 (13)	0.09	117 (10)		
Venezuela	311 (8)	757 (32)	8 (6)	21 (8)	0.52	77 (3)	0.0040	61 (5)

and environmental threats (forest loss and human population growth). Because we established the upper quartile as a cutoff for each index, and the number of countries included in the North/Central America continental pool was relatively low (13), it is not surprising that only one country, Costa Rica, fit our criteria and was identified as a continental area of critical concern. Guatemala, while having index values close to those of Costa Rica, does not rank in the top quartile for any of the four indices, and thus was not identified as a continental area of critical concern. However, when all 135 countries included in the global analysis are ranked together, Guatemala ranks in the top quartile, globally, for all four indices, and qualified as one of the 18 global areas of critical concern.

Similarly, of the 46 African countries in the continental analysis, the Malagasy Republic ranks first for endemism and is in the top quartile for forest loss (it is not in the top quartile for either species richness or population pressure). But worldwide, the Malagasy forest-loss rate ranks 22nd of 80 nations considered, just below the upper quartile. Thus, despite a very high endemism rank (fifth in our analysis), the Malagasy Republic did not qualify as a global area of critical concern.

At both geographic scales, drawing from several different datasets dilutes the importance of any single parameter in identifying conservation priorities. Therefore, careful consideration of ranks and relative cutoff levels is key to the formulation and interpretation of any list. Each index provides a simplified view of a complex dataset, while their integration highlights important differences in continental patterns in both biodiversity and threat.

Caveats and Conclusions

The areas of critical concern identified in this article constitute a list of global conservation priority areas that is as valid as any other currently available and better justified than most. However, one of our fears in writing this article was that conservationists might overemphasize the results while ignoring our emphasis on methods and the assumptions underlying them. Therefore, before presenting our conclusions, we offer three cautionary notes.

First, the value of the results rests on the assumption that the taxonomic groups used as indicators track total species diversity. We have work in progress that will examine the correlation between species richness of mammals, butterflies, and passerine birds, drawing on both the Global Conservation Analysis Package[2] and field research in four tropical countries (e.g., Méndez et al. 1994). Currently, however, this issue remains unresolved and hotly debated (e.g., Curnutt et al. 1994, Prendergast et al. 1993, Smith et al. 1993). If regions of high biological diversity for one taxon show little overlap with those for others, any approach that employs a narrow subset of the biota as an indicator for diversity in general may be severely flawed. However, because some simplified measure of biodiversity is needed now, we suggest that indices derived from measures of species richness and endemism for two or more distinct and ecologically diverse taxa (such as mammals and butterflies) provide a reasonable first approximation of overall diversity.

Second, our results depend upon how well deforestation and our index of human population pressure actually measure future threats to biodiversity. At present, we know of no data, available on a global scale, that better reflect the level of threat at the country level. However, the indices employed here are, at best, approximations of the actual threats facing a wide range of habitats in each country.

Third, allocation of the scarce resources available for conservation should not be based on a single criterion or result. We have pointed out the problems with previous efforts to identify priority areas for conservation, as well as the limitations of our own analyses. Because no single approach is likely to capture all critical areas, policy makers should view results collectively, considering the data and analytical approaches upon which the results rest, and pursue additional analyses to resolve conflicting conclusions and recommendations.

Accepting these caveats, it is instructive to compare the lists of global conservation priorities identified in this article with those from three earlier studies (Table 6). A cursory examination suggests that the countries China, India, Colombia, and Ecuador, which were identified in all four studies, and the Malagasy Republic and the Philippines, which were identified in our study and two of the others, are key to the conservation of global biodiversity.

These countries pose very different conservation challenges. Colombia, Ecuador, and the Malagasy Republic are developing nations with small but expanding economies, where properly focused conservation efforts might lead to disproportionate gains in the preservation of biodiversity. For example, development of viable agricultural and economic opportunities in settled regions of these countries, coupled with a leveling of the growth rate of the human population, might eliminate the desperate pressure for expansion of subsistence farming into undeveloped regions (Fearnside 1987, Myers 1992). Many of these areas are unsuitable for agriculture, despite their ability to support some of the most biologically diverse habitats in the world.

China and India, on the other hand, are densely populated, industrializing nations. There, conservation efforts should probably emphasize halting human population growth, deployment of technologies to minimize the environmental impacts of industrialization, and traditional conservation approaches, such as direct acquisition of land for reserves and improved management (Dinerstein and Wikramanayake 1993).

The most important novel contribution of our analysis may be the identification of nations with moderately high biodiversity that are facing extreme threats. More than half of the global areas of critical concern (10 of 18 countries) were not identified in any of the three studies previously discussed. These countries (including Angola, Nigeria, Taiwan, and Guatemala) may constitute the current front lines of the extinction crisis, yet their importance for conservation has been often overlooked.

Regional analyses are a logical next step in the formulation of conservation strategies. A recent study combining measures of diversity and threat at the subnational level for Indonesia, Malaysia, and the Philippines, demonstrated how regional analyses might guide conservation planning (Dinerstein and Wikramanayake 1993). Similar efforts around the world are likely to be made more tractable by the availability of the databases and analytical software of the Global Conservation Analysis Package.

---

[2]In preparation.

TABLE 6. Comparison of the areas of critical concern (global and continental analyses) with conservation priorities identified by three widely cited studies employing different methods. Areas are listed alphabetically by continent, with countries selected by one or more of the other studies appearing in bold.

	Areas of critical concern		Regional hot-spots*	Megadiversity countries†	Range-restricted birds‡
	Global	Continental			
Africa					
Angola	+	+	−	−	−
Ghana	+	+	−	−	−
Ivory Coast	+	−	+	−	−
Kenya	+	+	−	−	−
Malagasy Republic	−	+	+	+	−
Nigeria	+	+	−	−	−
South Africa	−	−	+	−	−
Tanzania	−	−	+	−	−
Uganda	+	+	−	−	−
Asia–Australia					
Australia	−	−	+	+	−
China	+	−	+§	+	+
India	+	+	+^{s}	+	+
Indonesia	−	−	+	+	−
Malaysia	−	−	+	+	−
Philippines	+	+	+	−	+
New Caledonia	−	−	+	−	−
Sri Lanka	+	−	+	−	−
Taiwan	+	+	−	−	−
Thailand	+	+	−	+	−
Vietnam	−	+	−	−	−
Europe					
Switzerland	−	+	−	−	−
North America/Central America					
Costa Rica	+	+	−	−	−
Guatemala	+	−	−	−	−
Honduras	+	−	−	−	−
Mexico	−	−	−	+	+
Nicaragua	+	−	−	−	−
United States	−	−	+	−	+
South America					
Brazil	−	−	+	+	+
Chile	−	−	+	−	−
Colombia	+	+	+	+	+
Ecuador	+	−	+	+	+
Peru	−	−	+‖	+	+

*Myers 1988, 1990. In some cases, only portions of the indicated countries are identified as hot-spots.
†Mittermeier 1988.
‡Bibby et al. 1992.
§Myers (1988, 1990) groups parts of India, China, and Nepal into a hot-spot called E. Himalaya.
‖Myers (1988, 1990) groups parts of Colombia, Ecuador, and Bolivia into a hot-spot called W. Amazon Uplands.

Ultimately, informed conservation decisions can be made only after assessing the distribution of biological diversity and the threats it faces. This assessment requires the integration of large, seemingly incongruous datasets, which can then be employed in multiple analyses at different geographic scales. The approach presented here, based on comprehensive datasets and explicit assumptions, constitutes an important contribution to understanding the distribution of Earth's biodiversity and its relationship to the increasing anthropogenic influences that will determine its future.

Acknowledgments

The authors thank Erica Fleishman for her many contributions to the development of the Global Conservation Analysis Package databases, and they gratefully acknowledge the enormous number of hours of library research and data entry by CCB research assistants past and present: Russell Bell, Sonia Chalfin, Jonathan Hoekstra, Suresh Jesuthasan, Daniel Mandell, Arthur Molina, José Romero, David Steffan, Michael Thornton, and John Wakely. John MacKenzie, Eric Macklin, and Charles Metzler provided computer programming expertise, and work by Marybeth Buechner, Eric Fajer, and Bruce Wilcox was essential to the early stages of this project. Suggestions from Gretchen Daily, Katy Human, and Fraser Smith were very helpful in the later stages of manuscript preparation, as were comments by Norman Myers, Stuart Pimm, and an anonymous reviewer. Financial support for the development of the databases and software that underlie this work was provided by the World Wildlife Fund—United States and the William and Flora Hewlett Foundation.

References

Bibby, C. J., N. J. Collar, M. J. Crosby, M. F. Heath, C. Imboden, T. H. Johnson, A. J. Lange, A. J. Sutterfield, and S. J. Thirgood. 1992. *Putting Biodiversity on the Map: Priority Areas for Global Conservation*. International Council for Bird Preservation, Cambridge, UK.

Collar, N. J., and P. Andrew. 1988. *Birds to Watch: The ICBP World Checklist of Endangered Birds*. Technical Publication 8. International Council for Bird Preservation, Cambridge, UK.

Collins, N. M., and M. G. Morris. 1985. *Threatened Swallowtail Butterflies of the World: The IUCN Red Data Book*. International Union for Nature and the Conservation of Natural Resources, Cambridge, UK.

Curnutt, J., J. Lockwood, H. Luh, P. Nott, and G. Russell. 1994. Hotspots and species diversity. *Nature* 367: 326–327.

D'Abrera, B. 1980. *Butterflies of the Afrotropical Region*. Lansdowne Editions, East Melbourne, Australia.

———. 1981. *Butterflies of the Neotropical Region*. Lansdowne Editions in association with E. W. Classey, Melbourne, Australia.

———. 1984. *Butterflies of South America*. Hill House, Victoria, Australia.

———. 1990. *Butterflies of the Australian Region.*

3rd ed. Hill House, Victoria, Australia.

———. 1991. *Butterflies of the Holarctic Region*. Hill House, Victoria, Australia.

D'Abrera, B., and E. W. Classey. 1982. *Butterflies of the Oriental Region*. Hill House, Victoria, Australia.

Diamond, J. M. 1987. Extant unless proven extinct? Or, extinct unless proven extant? *Conserv. Biol.* 1: 77–79.

Dinerstein, D., and E. D. Wikramanayake. 1993. Beyond "hotspots": how to prioritize investments to conserve biodiversity in the Indo-Pacific region. *Conserv. Biol.* 7: 53–66.

Ehrlich, P. R. 1992. Population biology of checkerspot butterflies and the preservation of global diversity. *Oikos* 63: 6–12.

———. 1994. Energy use and biodiversity loss. *Philos. Trans. R. Soc. Lond. B. Biol. Sci.* 344: 99–104.

Ehrlich, P. R., and G. C. Daily. 1993. Population extinction and saving biodiversity. *Ambio* 22: 64–68.

Ehrlich, P. R., and A. H. Ehrlich. 1991. *Healing the Planet*. Addison-Wesley Publishing Co., Reading, MA.

Ehrlich, P. R., and J. Holdren. 1971. The impact of population growth. *Science* 171: 1212–1217.

Ehrlich, P. R., and E. O. Wilson. 1991. Biodiversity studies: science and policy. *Science* 253: 758–762.

Fearnside, P. M. 1987. Deforestation and international economic development projects in Brazilian Amazonia. *Conserv. Biol.* 1: 214–221.

Fearnside, P. M., and E. Salati. 1985. Explosive deforestation in Rondonia, Brazil. *Environ. Conserv.* 12: 355–356.

Food and Agriculture Organization of the United Nations (FAO). 1981. *Tropical Forest Resources Assessment Project*. Technical Report 2 (four volumes). FAO, Rome, Italy.

Hall, A. V. 1978. Endangered species in a rising tide of human population growth. *Trans. R. Soc. S. Afr.* 43: 37–49.

Honacki, J. H., K. E. Kinman, and J. W. Koeppl, eds. 1982. *Mammal Species of the World*. Allen Press and The Association of Systematics Collections, Lawrence, KS.

Malone, T. F. 1992. The world after Rio. *Am. Sci.* 80: 530–532.

May, R. M. 1988. How many species are there on Earth? *Science* 241: 1441–1449.

McNeely, J. A., K. R. Miller, W. V. Reid, R. A. Mittermeier, and T. B. Werner. 1990. *Conserving the World's Biological Diversity*. International Union for Nature and the Conservation of Natural Resources, World Resources Institute, Conservation International, World Wildlife Fund—United States, World Bank, Washington, DC.

Melillo, J. M., C. A. Palm, R. A. Houghton, G. M. Woodwell, and N. Myers. 1985. A comparison of two recent estimates of disturbance in tropical forests. *Environ. Conserv.* 12: 37–40.

Méndez, C., T. D. Sisk, and N. M. Haddad. 1994. Beyond birds: multitaxonomic monitoring provides broad measure of tropical biodiversity. *Proceedings of the First International Congress of Wildlife Management,* The Wildlife Society, Washington, DC.

Mittermeier, R. A. 1988. Primate diversity and the tropical forest: case studies from Brazil and Madagascar and the importance of the megadiversity countries. Pages 145–153 in E. O. Wilson, ed. *Biodiversity,* National Academy Press, Washington, DC.

Myers, N. 1980. *Conversion of Tropical Moist Forests.* National Academy of Science, Washington, DC.

———. 1988. Threatened biota: hot-spots in tropical forests. *Environmentalist* 8: 1–20.

———. 1989. *Deforestation Rates in Tropical Forests and their Climatic Implications.* Friends of the Earth, London, UK.

———. 1990. The biodiversity challenge: expanded hot-spot analysis. *Environmentalist* 10: 243–256.

———. 1992. Tropical forests: the policy challenge. *Environmentalist* 12: 15–27.

Nowak, R. M., and J. L. Paradiso. 1983. *Walker's Mammals of the World.* 4th ed. Johns Hopkins University Press, Baltimore, MD.

Pimm, S. L. and J. L. Gittleman. 1992. Biodiversity: where is it? *Science* 255: 940.

Population Reference Bureau (PRB). 1992. *World Population Data Sheet.* Population Reference Bureau, Washington, DC.

Prendergast, J. R., R. M. Quinn, J. H. Lawton, B. C. Eversham, and D. W. Gibbons. 1993. Rare species, the coincidence of diversity hotspots and conservation strategies. *Nature* 365: 335–337.

Reid, W. V., and K. R. Miller. 1989. *Keeping Options Alive: The Scientific Basis for Preserving Biodiversity.* World Resources Institute, Washington, DC.

Sayer, J. A. and T. C. Whitmore. 1991. Tropical moist forests: destruction and species extinction. *Biol. Conserv.* 55: 199–213.

Skole, D., and C. Tucker. 1993. Tropical deforestation and habitat fragmentation in the Amazon: satellite data from 1978 to 1988. *Science* 260: 1905–1910.

Smith, F., R. May, R. Pellew, T. Johnson, and K. Walter. 1993. Estimating extinction rates. *Nature* 364: 494–496.

Soulé, M. E. 1987. *Viable Populations for Conservation.* Cambridge University Press, Cambridge, UK.

Vitousek, P. M., P. R. Ehrlich, A. H. Ehrlich, and P. A. Matson. 1986. Human appropriation of the products of photosynthesis. *BioScience* 36: 368–373.

Wilson, E. O. 1989. Threats to biodiversity. *Sci. Am.* 261: 108–112.

———. 1992. *The Diversity of Life.* Harvard University Press, Cambridge, MA.

World Conservation Monitoring Center (WCMC). 1992. *Global Biodiversity, Status of the Earth's Living Resources.* Chapman Hall, London, UK.

World Resources Institute (WRI), United Nations Environment Programme, and United Nations Development Programme. 1990. *World Resources, 1990-91.* Oxford University Press, Oxford UK.

PART 2
Restore Ecological Processes

9

The Preservation of Process:
The Missing Element of
Conservation Programs

T. B. Smith, M. W. Bruford, and R. K. Wayne

Introduction

With the loss of biodiversity has come increasing debate as to how best to preserve it. Conservation strategies have largely focused on patterns of diversity, or how to maximize the number of species that can be saved within a particular geographic region (Bibby *et al.*, 1992; McNeely *et al.*, 1990). While many of these plans are to be applauded most nevertheless fall short of addressing many of the fundamental ecological and evolutionary processes that produced the patterns in the first place. Emphasis on pattern rather than process is in many respects understandable, rapid population declines and diminishing habitats requires swift action, often before all relevant data are collected. However, there is an urgent need to combine strategies to protect patterns of species richness with studies directed toward understanding fundamental ecological and evolutionary mechanisms (Pickett, Parker & Fiedler, 1992; Redford, 1992; Vane-Wright, Humphries & Williams, 1990). If we are to conserve biodiversity the ecological and evolutionary mechanisms generating genetic diversity and the isolating mechanisms critical for speciation must also be preserved. Protecting regions solely on the basis of species richness without also conserving important biotic processes is, in some respects, an extension of the single species approach to conservation, an approach which has been vehemently rejected by many biologists.

Here we discuss some of the potential pitfalls of emphasizing only patterns of high endemism and species richness, and suggest a two-tiered approach to conserving biodiversity. This approach first identifies regions of high biodiversity, but then combines this information with knowledge of ecological, evolutionary and genetic mechanisms necessary to promote and preserve natural dynamics.

While tropical rainforests are generally regarded as containing great species richness, the evolutionary mechanisms which have generated this tremendous diversity are anything but clear. Early work stressed the importance of Pleistocene refugia in generating new species (Haffer, 1969). Although the Pleistocene refugia hypothesis was not without controversy (i.e. Endler, 1982; Mayr & O'Hara, 1986), many conservation plans were designed as if the hypothesis were scientifically unchallenged. In fact, recent work in Central and South America provides little evidence that speciation in any group of organisms was actually tied to isolation in Pleistocene refugia (Patton, Myers & Smith, 1990; Flenley, 1993). With respect to mammals, even the assertion that the greatest diversity is found in lowland forests is being

Reprinted with permission from Biodiversity Letters vol. 1, pp. 164-167. Copyright 1993 Blackwell Science Ltd.

challenged (Mares, 1992). Similar questions have been raised for African rainforests (Endler, 1982). Additionally, recent work on British plants and animals has shown that species-rich areas often do not coincide for different taxa, and that many rare species do not occur in areas of the highest species diversity (Prendergast *et al.*, 1993). We are currently investigating the role that ecotones (between forest and savanna) play in generating genetic diversity in sub-Saharan Africa. While there is little debate that some centrally located regions show greater species richness, small isolated ecotone populations by being more subject to drift and directional selection may exhibit greater interpopulation variation. In this way peripheral ecotonal populations may provide an important source of new genetic variation which ultimately enriches more centrally located populations. Centrally located rainforest areas are presently the main focus of many conservation efforts, while ecotonal regions are rapidly being lost to overgrazing, burning and wood harvesting with little or no effort toward their conservation. It is a serious mistake for conservation organizations to focus solely on protecting areas of high species richness and not to support research on regenerative ecological and evolutionary processes.

In a recent review Redford (1992) discusses the critical importance of ecological extinctions on a microgeographic scale, where natural forest dynamics and community structure are potentially destroyed by declines of important species. His arguments raise some important considerations. Do we focus on protecting remnant habitats that may contain small populations of endemic species, but where major seed dispersers are either extinct or at such low densities as to be ecologically dysfunctional? Or do we concentrate our efforts on preserving habitats which are not on the verge of being destroyed, and where populations of species important in maintaining community dynamics have not declined? There are no easy answers, and ultimately there are many important considerations, but it is clear that greater emphasis should be placed on conserving fundamental ecological processes.

Conservation strategies must also examine ecological processes at macrogeographic levels.

Preserving single habitats without understanding their relationships to others is fraught with pitfalls. An illustration is provided by some species of Costa Rican sphingid moths which are seasonally migratory requiring both lowland dry forest and montane cloud forest (Janzen, 1987). Obviously attempting to conserve these species by preserving only one of these habitats would be a mistake.

Molecular genetic techniques also provide important data for making informed conservation management decisions. However, some conservation biologists view such techniques as being of either marginal importance or too esoteric to be counted as a priority in conservation strategies. In fact, examining the phylogenetic status of taxonomic units in conservation programs and measuring the genetic structure of populations of rare and endangered species have generated much important data. Species number and the rate at which the number of species changes with area are common indicators of biodiversity. In poorly-documented regions, a team of taxonomic experts may be dispatched to rapidly assess the number of species in a wide range of taxonomic groups. Visits to several localities by such groups provide a snapshot of biodiversity over a large region. However, this approach ignores processes that generate biodiversity and does not identify regions with long separate histories of evolution. Nor does it rank different areas according to the phylogenetic uniqueness of their flora and fauna. Molecular genetic techniques not only provide an estimate of the number of distinct forms in an area but provides a measure of how different they are. Phylogenetic analyses of molecular data also allow inference of the historical processes that have generated biodiversity. All else being equal, preservation of an area that contains taxa representing long, distinct lineages is more likely to preserve a greater degree of evolutionary heritage than one with a similar number of species but containing a succession of closely allied forms (May, 1990; Vane-Wright *et al.*, 1990). This is especially true of island faunas that, by their nature, contain species in proportion to their area, but may include many species which are dramatically divergent and represent phylogenetically long distinct lineages. Molecular phylogenetic analyses may also provide a framework to understand the evolution of complex traits, allowing one to discern whether a given adaptation reflects a recent common heritage or an unique evolutionary innovation meriting special preservation (Brooks & McLennan, 1991; Harvey & Pagel, 1991).

Recently, (Kremen, 1992) has presented an alternative to the counting-species by taxonomic-teams approach. She suggests selecting a single group, a target taxon, that is likely to be an indicator of biodiversity in other unrelated taxa. The best taxonomic groups to choose are those that have shown an endemic radiation and, consequently, are likely to have specialized forms for a wide variety of unique habitats. Detailed study of the taxonomy and distribution of species within a target taxon may provide a general index of biodiversity at a given locality, while changes in biodiversity among localities indicate habitat diversity. Morphologic and molecular analysis of a target taxon reveals the complex historical forces that generated biodiversity in the region and provides a taxonomic weight to each species (Kremen, in press; May, 1990; Vane-Wright et al., 1990). A phylogenetic tree of a target taxon when compared to the geographical distribution of species provides a hierarchy for conservation that allows preservation of the maximum amount of evolutionary diversity. In contrast, a total species approach confuses species number and taxonomic distinction. Such an approach may result in phylogenetically distinct species being lost and miss important centres of evolutionary diversity.

An extreme example (Daugherty et al., 1990) concerns the tuatara (genus Sphenodon) which exists on islands off the coast of New Zealand. This group represents the only extant genus of an entire reptilian order and was until recently regarded as one species, S. punctatus. Molecular analyses revealed that the species has a complex evolutionary history, and actually should be divided into three distinct groups. One taxon, which was described in the last century as S. guntheri, and exists on one island only, was found to have three fixed allele differences from S. punctatus, supporting its recognition as a distinct species. A genetically divergent population of S. punctatus was also recognized as a separate taxon. A conservation scheme needs to be designed to preserve these distinct units. Had the tuatara been used as a target taxon, the conservation of the three separate units would have probably been preserved and possibly other distinct forms which showed a coincident pattern of speciation.

The conservation of species found in centres of biodiversity may preserve only a small fraction of the phenotypic and genetic variability found throughout their geographic range. In fact, if a centre of diversity is dominated by species with many distinct peripheral populations, conservation plans that conserve the centre of diversity alone may preserve only a narrow slice of the variability in each species. Consequently, a plan to maximize the preservation of species numbers alone may conflict with one designed to preserve the maximum amount of genetic information found in all populations constituting the flora and fauna of a region.

Molecular genetic analysis can identify biogeographic divisions that exist within each species that would not be evident using other means. Several species may show coincident phylogenetic divisions across their range indicating the existence of biogeographic units having similar species composition but separate evolutionary histories. For example, a sharp phylogenetic division exists between conspecific populations in the gulf and Atlantic coasts of Florida (Avise, 1992). Divisions were found in all eighteen taxa examined and include salt and fresh water fish, birds, reptiles and invertebrates such as oysters and horseshoe crabs. The strong within-species genetic discontinuity between Gulf and Atlantic populations suggest they are distinct biotic realms and should be conserved as separate units reflecting their unique and independent evolutionary heritage (Avise, 1992). Mistakes have already been made. For example, the male founders of the endangered dusky seaside sparrow captive breeding program were drawn from gulf populations that, although close geographically to conspecific Atlantic populations, were from a phylogenetically divergent stock (Avise & Nelson, 1989).

The measurement of the loss of genetic variation in small captive and wild populations is an often discussed application of molecular genetic techniques. However, it is becoming increasingly clear that the loss of genetic variation is rarely the primary threat to endangered species (e.g. Lande, 1988). Perhaps of greater interest is the effect of habitat fragmentation, translocation and reintroduction on the genetic landscape of species. Captive breeding may involve a small sampling of the genetic diversity within species and reintroduction of captive raised individuals or the translocation of individuals may change the genetic composition within a species. Stocking of streams with hatchery raised individuals is one area for concern (Waples & Teel, 1990).

A striking recent example concerns genetic divisions among populations of Grant's gazelle in Kenya and Tanzania. Analysis of their mitochondrial DNA by Peter Arctander and colleagues (pers. comm.) has shown dramatic phylogenetic divisions between closely situated populations in Tsavo East and Amboseli National Parks. Considering the high mobility of gazelles, the lack of significant topographic barriers and the similarity of grassland habitats in these parks, such profound microgeographic subdivisions were unexpected. That they exist suggests a need to consider these areas as separate conservation units with regard to Grant's gazelle. If this species were endangered, reintroduction or augmentation plans would need to consider the existing genetic landscape so as to preserve its historical structure. Additional genetic surveys need to be done on other species in the same areas to determine if similar discontinuities exist within them. In sum, molecular analyses of endangered species permit a detailed mapping of their genetic landscape. This information can then be used to design plans that maximize the preservation of genetic diversity.

An integrative approach to the loss of biodiversity requires the use of ecological, demographic, morphological and molecular genetic data to ensure that the maximum amount of evolutionary information contained in populations existing within a region is preserved. In particular, molecular techniques need to be utilized. The techniques are often simple and straightforward and provide information with clear, tractable implications for conservation. However, molecular biologists with conserva-

tion interests are often frustrated by the lack of financial support. Some governmental scientific funding councils regard conservation genetics as applied rather than pure research and often do not fund conservation genetic projects. Similarly, many non-governmental organizations seem obsessed with funding faunal inventories and may even explicitly exclude genetic analyses from consideration. Obtaining funding to investigate ecological processes which have direct application to conservation is also difficult. Until studies of pattern and process are recognized as integral parts of a whole, we will likely lack coherent, effective conservation programs.

References

Avise, J.C. (1992) Molecular population structure and the biogeographic history of a regional fauna: a case history with lessons for conservation biology. *Oikos*, **63**, 62–76.

Avise, J.C. & Nelson, W.S. (1989) Molecular genetic relationships of extinct Dusky Seaside Sparrow. *Science*, **243**, 646–648.

Bibby, C.J., Collar, N.J., Crosby, M.J., Heath, M.F., Imboden, C., Johnson, T.H., Long, A.J., Stattersfield, A.J. & Thirgood, S.J. (1992) *Putting biodiversity on the map: priority areas for global conservation*. International Council for Bird Preservation, Cambridge.

Brooks, D.R. & McLennan, D.A. (1991) *Phylogeny, ecology, and behavior. A research program in comparative biology*. University of Chicago Press, Chicago.

Daugherty, C.H., Cree, A., Hay, J.M. & Thompson, M.B. (1990) Neglected taxonomy and continuing extinctions of tuatara (*Spenodon*). *Nature*, **347**, 177–179.

Endler, J.A. (1982) Problems in distinguishing historical from ecological factors in biogeography. *Am. Zool.* **22**, 441–452.

Flenley, J. (1993) The origins of diversity in tropical rain forests. *Trends Ecol. Evol.* **8**, 119–120.

Haffer, J. (1969) Speciation in Amazonian forest birds. *Science*, **165**, 131–137.

Harvey, P.H. & Pagel, M.D. (1991) *The comparative method in evolutionary biology*. Oxford University Press, Oxford.

Janzen, D. (1987) When, and when not to leave. *Oikos*, **49**, 241–243.

Kremen, C. (1992) Accessing the indicator properties of species assemblages for natural areas monitoring. *Ecol. Applic.* **2**, 203–217.

Kremen, C. (in press) Biological inventory using target taxa: A case study of the butterflies of Madagascar. *Ecol. Applic.*

Lande, R. (1988) Genetics and demography in biological conservation. *Science,* **241**, 1455–1460.

Mares, M.A. (1992) Neotropical mammals and the myth of Amazonian biodiversity. *Science,* **255**, 976–979.

May, R. (1990) Taxonomy as destiny. *Nature,* **347**, 129–130.

Mayr, E. & O'Hara, R.J. (1986). The biogeographical evidence supporting the Pleistocene forest refuge hypothesis. *Evolution,* **40**, 55–67.

McNeely, J.A., Miller, K.R., Reid, W.V., Mittermeier, R.A. & Werner, T.B. (1990) *Conserving the World's biological diversity.* International Union for the Conservation of Nature, Gland.

Patton, J.L., Myers, P. & Smith, M.F. (1990) Vicariant versus grandiant models of diversification: The small mammal fauna of eastern Andean slopes of Peru. *Vertebrates in the tropics* (ed. by G. Peter and R. Huntterer). Museum Alexander Koenig, Bonn.

Pickett, S.T.A., Parker, V.T. & Fiedler, P.L. (1992) The new paradigm in ecology: implications for conservation biology above the species level. *Conservation biology: the theory and practice of nature conservation, preservation, and management* (ed. by P. L. Fiedler and S. K. Jain), Chapman and Hall, New York.

Prendergast, J.R., Quinn, R.M., Lawton, J.H., Eversham, B.C. & Gibbons, D.W. (1993) Rare species, the coincidence of diversity hotspots and conservation strategies. *Nature,* **365**, 335–337.

Redford, K.H. (1992) The empty forest. *BioScience,* **42**, 412–422.

Vane-Wright, R.I., Humphries, C.J. & Williams, P.H. (1990) What to protect? Systematics and the agony of choice. *Biol. Conserv.* **55**, 235–254.

Waples, R.S. & Teel, D.J. (1990) Conservation genetics of Pacific salmon I. Temporal changes in allele frequency. *Science,* **241**, 1455–1460.

10
Disturbance and Population Structure on the Shifting Mosaic Landscape

James S. Clark

Introduction

The development of an analytical theory of plant demography has been frustrated by problems of complexity related to spatial heterogeneity in population structure and limiting resource (Schaffer and Leigh 1976, Pacala and Silander 1985, Pacala 1988), asymmetric competition (large individuals control a disproportionate share of the resource base) (Harper 1977, Weiner 1985, 1986), and time-dependent parameters such as birth and death rates (Clark 1991a). The problem of applying to plants the population models developed for mobile organisms, such as animals or plankton in chemostats, has long been recognized (Schaffer and Leigh 1976, Pacala 1989). Plant recruitment, growth, and mortality respond to extremely localized resource levels rather than to landscape averages. There is also an uncertainty associated with the timing of recruitment events, which occur episodically, even in "undisturbed" assemblages (Harper 1977, Spurr and Barnes 1980). Different species require different types of disturbances or "regeneration niches" (sensu Grubb 1977), and it is likely that the pattern of these events has important implications for population dynamics at the metapopulation scale (Comins and Noble 1985, Warner and Chesson 1985, Hastings and Wolin

1989, Pacala 1989). Local dynamics are further complicated by larger scale "disturbances" that cause adult mortality, permit regeneration, and synchronize population dynamics across landscapes (Shugart 1984, Turner et al. 1989). Thus analytical treatment of plant demography is restricted largely to annuals (reviewed in Watkinson 1986 and Pacala 1989) or to dynamics within single generations of perennials (Hara 1984, 1985, Clark 1990).

Lack of a tractable model that applies to the demography of perennial plants makes it unclear how population structure depends on local dynamics and disturbance regimes. It appears likely, for example, that the way in which disturbance probability changes with plant age should influence population dynamics (Pickett and White 1985, Clark 1989, Cohen and Levin 1991). Recent field studies have focused on determining how this probability changes over time (Johnson 1979, Romme 1982, Johnson and Van Wagner 1985, Foster 1988, Baker 1989, Clark 1989), and there has been some speculation concerning effects on population structure (Johnson 1979, Clark 1989). Such questions might be addressed with forest "gap" models (Botkin et al. 1972, Shugart 1984) and the appropriate experimental design. But there exists no analyzable model to investigate the consequences of disturbance regimes for structure at the scale of the metapopulation.

Specifically, how should densities and age classes be distributed in the "shifting mosaic" landscape that develops in forests that may rarely experience large exogenous disturbances (Watt 1947, Bormann and Likens 1979)? How does this population structure evolve following larger disturbances? How do dynamics within local cohorts influence metapopulation structure? What type of structure is to be expected in landscapes ranging from those that experience infrequent disturbances to those characterized by infrequent disturbances to those characterized by frequent disturbance? What constitutes an "intermediate disturbance regime" (Connell 1978, Huston 1979) for a shifting mosaic landscape, and what are its implications? How does the structure of a population change at local and regional scales en route to regional dominance or extinction of a species? The answers to these questions represent an important step toward understanding persistence and species coexistence, for it is the age and density structure of a population that determines whether a species can persist. Moreover, interpreting effects of fluctuating environments and climate change requires a theory that establishes the population structure to be expected in the absence of these sources of variability.

Here I present a stochastic model of plant population dynamics to describe and analyze this mosaic structure. The model contains somewhat more complexity than is typically included in analytical population models to accommodate the inherently stochastic nature of plant populations and the mix of spatial and temporal scales. I nonetheless strive to retain the simplicity necessary for analysis and understanding. Population dynamics of perennial plants are necessarily stochastic because populations often consist of cohorts that date from the occurrence of a regeneration niche (Grubb 1977, Harper 1977, Spurr and Barnes 1980). I focus on the distribution of regeneration niches, which occur stochastically in time and space and can be quantified in natural assemblages, and on subsequent thinning on those patches as individuals increase in size. The model structure is similar to that contained in gap models (Botkin et al. 1972, Shugart 1984) in assuming a mosaic of different-aged patches. The products of this theory are probability

distributions of patch age classes, of cohort densities, and of plant age classes. Patch age classes are an important component of population structure, because cohorts may pass through a predictable sequence of stages that determine many ecosystem attributes, such as leaf area (Waring and Schlesinger 1985), primary production (Sprugel 1984, 1985), nutrient cycling (Vitousek and Reiners 1975, Pastor and Post 1986), and gap area (Oliver 1981, Clark 1991b). Plant density approaches a point equilibrium nowhere on this shifting mosaic landscape, so distributions of densities are a useful means for summarizing population size. Age structure depends on distributions of patches and their densities; without knowledge of the mosaic structure, it is impossible to represent age structure in a fashion that permits analysis of the factors that produce it. Results have application to questions of landscape patterns of disturbance, succession, life history, and spatial and temporal heterogeneity.

Theoretical Development

The landscape consists of a mosaic of patches, each supporting a cohort of plants that became established a yr ago. Disturbances serve as regeneration niches (sensu Grubb 1977); they occur episodically, resulting in adult mortality and the initiation of a new cohort of small individuals at high density, with patch age a being reset to zero. The model does not depend on assumptions regarding why disturbances occur (e.g., whether they are considered "allogenic" vs. "autogenic" [Runkle 1985]). Between disturbances on any given patch, plants increase in size, and density $x(a)$ decreases at rate $dx(a)/da$. The structure of this metapopulation is summarized by distributions of (a) patch age classes $\omega(a)$, (b) cohort densities $f(x)$, and (c) plant age classes $f_p(a)$.

Disturbances can occur at two spatial scales (Fig. 1). Type r disturbances are sufficiently small that the population that becomes established on the patch can be treated as a single cohort. Type r disturbances are nested within a landscape that may experience a second disturbance process s that affects larger areas of age t. A type s patch contains many type r patches,

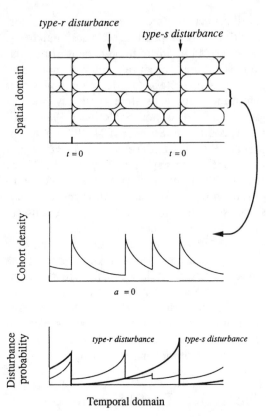

FIGURE 1. Model of a shifting mosaic population that experiences two superimposed and interdependent disturbance processes. A small-scale type r process (e.g., canopy gap formation) is nested within a larger type s process (e.g., fire occurrence) (above). Type s disturbances synchronize type r patches. On any given patch, recruitment is episodic, followed by thinning (center). The occurrence of disturbances that allow recruitment occurs stochastically with probability that may increase with time since the last disturbance (below).

each of age $a \le t$ (Fig. 1). An example would be a landscape that experiences a gap-phase process r and fire regime s. In this case, process r describes a shifting mosaic gap process that develops with time t since the last fire s.

I begin with the simplest case by solving for the structure of this metapopulation that is composed of a mosaic of patches created by a type r (e.g., "gap") disturbance process. The analysis explores how a single-species metapopulation is influenced by local and regional processes. I then generalize this result to include a disturbance process s that synchronizes type r patches across broader portions of a landscape for an

early- and late-successional species. Several assumptions are used to simplify the analysis:

1. The location of a given patch relative to others on the landscape does not influence its dynamics.
2. Dispersal is high and seed production is uninfluenced by local cohort dynamics.
3. Thinning within a patch is deterministic and driven by plant growth.
4. Disturbances destroy existing adults, and they create regeneration opportunities. Because patches are out of phase, the landscape is a mosaic of different-aged patches.
5. Disturbance can be described as a stationary renewal process (Clark 1989).

Assumption (1) is partially relaxed in a later section, *The shifting mosaic with large-scale disturbances,* where type r patches are nested within type s patches. Assumption (2), an assumption my model shares with gap simulation models, simplifies the analysis. I discuss how the decrease in propagule availability at low densities is likely to affect model predictions. Assumption (3) is a reasonable way to simplify local dynamics in view of the ways in which thinning rates compensate for initial density (Harper 1977). Thinning rate depends on growth of the individual plants within the cohort (Harper 1977, Aikman and Watkinson 1980, Norberg 1988). Although densities of seeds and seedlings are highly variable, densities tend to converge to this thinning equation from a range of initial densities as a result of the response of thinning rate to crowding (Harper 1977, Westoby 1984, Clark 1990, 1991b). Assumption (5) is a stationarity assumption, which says that the distribution of disturbance events does not depend on the time at which we observe the process. This assumption is partially relaxed in the section, *The shifting mosaic with large-scale disturbances,* where type r disturbances depend on the elapsed time since the last type s disturbance.

The Shifting Mosaic Structure (the Type r Process)

Metapopulation structure across this landscape depends on the rate of thinning within each patch and on the age distribution of patches. Per capita mortality rates in crowded even-aged plant cohorts tend to be nearly constant following canopy closure and before plants approach maximum size (Harper 1977, Peet and Christensen 1980), because of the effect of individual plant growth on population thinning (Clark 1990). As growth rates of plants decline, so too do density-dependent mortality rates (Fig. 2). At this time when density-dependent mortality is decreasing, density-independent mortality is on the rise, and canopy gaps begin to appear. The extent to which density-dependent vs. density-independent factors dominate in older cohorts, however, will vary with species. I therefore make the simplifying assumption that a local cohort thins at per capita rate λ_p,

$$x(a) = xe^{-\lambda_p a} \qquad (1)$$

where the coefficient x has units of density and incorporates the plant-height : crown-breadth ratio of the species (Norberg 1988, Clark 1990; Fig. 3a).

Age distribution of patches. The age distribution of patches is equivalent to the age distribution of cohorts, because we have thus far assumed that plants become established imme-

diately following patch formation, i.e., a disturbance. This distribution is the proportion of the landscape occupied by age class a, termed the "recurrence time." For a process that began in the distant past, the distribution of recurrence times tends to

$$\omega_r(a) = \frac{S_r(a)}{E[A_r]} \qquad (2)$$

(Cox 1962), where $S_r(a)$ describes the survivor function for the disturbance regime, and $E[A_r]$ is the time A_r expected to elapse between disturbances (Fig. 3c). $S_r(a)$ is the probability that a patch will survive at least a yr without another disturbance and can be estimated from disturbance history data (Johnson and Van Wagner 1985, Clark 1989).

Distribution of cohort densities. The distribution of cohort densities can be derived using the age distribution of patches together with the fact that the thinning Eq. 1 establishes a relationship between patch age and stand density. Let the age at which a patch supports density $x(a)$ be $a(x)$ (Fig. 3d). Then the recurrence time can be solved in terms of density through the inversion $\omega_r[a(x)] = \omega_r(x)$ (Fig. 3e). The second ingredient needed for a distribution of densities is the proportion of time $a(x)$ during which density actually is x, given by the reciprocal of the thinning rate

$$\Delta(x) = \left| \frac{da}{dx(a)} \right| \qquad (3)$$

(Appendix I). The distribution of densities across the landscape then is the production of $\omega_r(x)$ and $\Delta(x)$:

$$f_r(x) = \omega_r(x)\,\Delta(x) \qquad (4)$$

(see Fig. 3f and Appendix I).

For example, consider a species that regenerates immediately following patch formation and thins according to Eq. 1. Further assume that patch formation (disturbance) has constant probability λ_r, with survivor function $S_r(a) = e^{-\lambda_r a}$ and expected interval between the formation of new patches $E[A_r] = 1/\lambda_r$. From Eq. 2, the distribution of recurrence times

$$\omega_r(a) = \lambda_r e^{-\lambda_r a}, \qquad (5)$$

is exponential. Solving for a in Eq. 1 yields

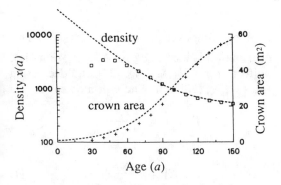

FIGURE 2. The relationship between exclusive crown area projection A and density x in a *Picea* stand. Data points are indicated by squares. Stand densities move horizontally toward the thinning line until the canopy closes, at which point densities follow the exponential model, until plants approach maximum size. Dashed lines are fitted values to the model from Clark (1990). The crown-area projection A is predicted from the same parameter values using the relation $1/x(dx/da) = -1/A(dA/da)$ (Clark 1991b).

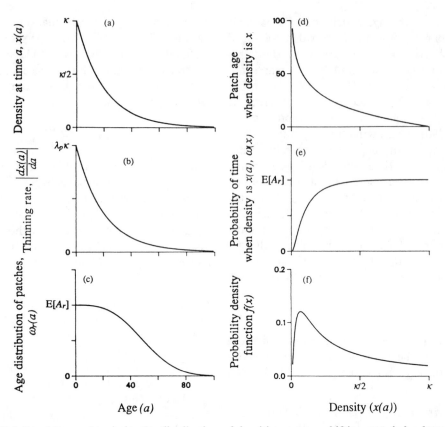

FIGURE 3. Relationships used to derive the distribution of densities across a shifting mosaic landscape having constant disturbance probability that increases with age: (a) survivorship of the local cohort; (b) the rate of change in density; (c) the probability that a randomly selected patch is of age a; (d) the inverse of (a), i.e., the age of a patch when density is x; (e) distribution ω_r (a) from (c) solved in terms of density x; (f) the probability that a randomly selected patch supports density x.

$$a(x) = \frac{\ln\left[\dfrac{x}{x(a)}\right]}{\lambda_p},$$

which is substituted in Eq. 5 to give

$$\omega_r(x) = \omega_r \exp[-\lambda_r a(x)] = \lambda_r \left(\frac{x}{\kappa}\right)^{\lambda_r/\lambda_p}$$

The time spent at a given density is determined from the derivative of Eq. 1, which is used in Eq. 3 to give $\Delta(x) = [\lambda_p x(a)]^{-1}$. The appropriate substitutions in Eq. 4 yield the probability density function

$$f(x) = \begin{cases} \dfrac{\lambda_r}{\lambda_p x}\left(\dfrac{x}{\kappa}\right)^{\lambda_r/\lambda_p} & 0 < x \le \kappa \\ 0 & \text{otherwise} \end{cases} \qquad (6)$$

$f(x)$ is the frequency of density x (Fig. 3f). That Eq. 6 is a probability density function is shown

by integrating over $(0, \kappa)$, $\int_0^\kappa f(x)dx = 1$. This distribution is increasing for $\lambda_r > \lambda_p$ (Fig. 4a), uniform for $\lambda_r = \lambda_p$ (Fig. 4b), and decreasing for $\lambda_r < \lambda_p$ (Fig. 4c). The mth moment of x is

$$E[X^m] = \frac{\lambda_r \kappa^m}{m\lambda_p + \lambda_r},$$

having expectation

$$E[X] = \frac{\lambda_r \kappa}{\lambda_p + \lambda_r}, \qquad (7)$$

and coefficient of variation

$$\text{cv}[X]$$
$$= (\lambda_r + \lambda_p)\sqrt{\frac{1}{\lambda_r}\left[\frac{1}{2\lambda_p + \lambda_r} - \frac{\lambda_r}{(\lambda_r + \lambda_p)^2}\right]}, \quad (8)$$

where X is an observed value of x. The variability of patch densities is maximized at intermediate disturbance frequencies, as can be demonstrated from the variance

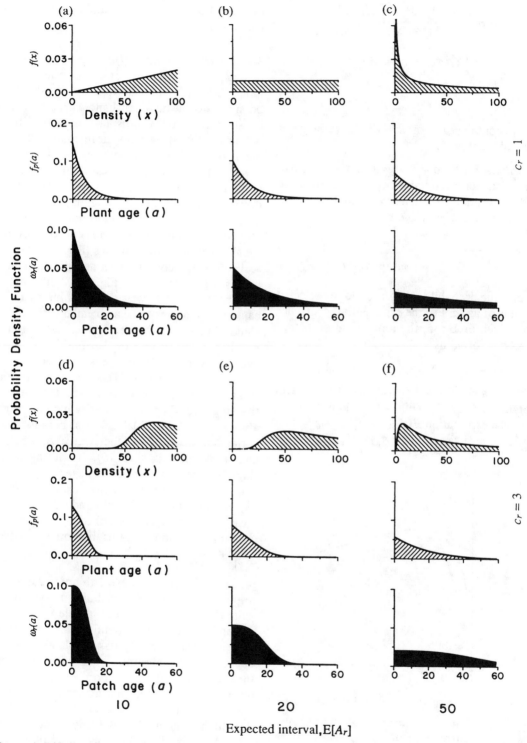

FIGURE 4. Distributions of density $f(x)$ and cohort age classes $f_p(a)$ on a shifting mosaic landscape having six different disturbance regimes, characterized by patch age classes $\omega_r(a)$. The distributions of density are given by Eq. 8. Figure shows examples using two different shape parameter values, $c_r(c_r = 1$, above, and $c_r = 3$, below) and three values of the expected interval, $E[A_r]$: $E[A_r] = 10$, left, $E[A_r] = 20$, center, $E[A_r] = 50$, right. The thinning rate used here is $\lambda_p = 0.05$.

$$\text{Var}[X] = \frac{\lambda_r x^2}{2\lambda_p + \lambda_r} - \frac{\lambda_r^2 x^2}{(\lambda_p \lambda_r)^2}.$$

Setting $\delta \text{Var}[X]/\delta \lambda_r = 0$ yields

$$\lambda_r^* = \frac{(\lambda_p + \lambda_r^*)^3}{(2\lambda_p + \lambda_r^*)^2}, \qquad (9)$$

which contains a root λ_r^* where variance is maximized (Fig. 5).

Distribution of Cohort Age Classes. The age class distribution contains two components, the age distribution of patches $\omega_r(a)$ and the plant density on patches of age a,

$$f_P(a) = \frac{\omega_r(a)x(a)}{\displaystyle\int_0^\infty \omega_r(a)x(a)\, da}. \qquad (10)$$

$f_P(a)$ is the probability that a randomly selected plant is of age a. For the exponential recurrence time given by Eq. 5, the representation of age class a is solved as

$$f_P(a) = (\lambda_r + \lambda_p)\exp[-a(\lambda_r + \lambda_p)], \qquad (11)$$

which is a simple exponential distribution with parameter $(\lambda_r + \lambda_p)$ (Fig. 4a–c). This distribution has expectation $E[A_p] = (\lambda_r + \lambda_p)^{-1}$ and coefficient of variation $cv(A_p) = 1$, where A_p is

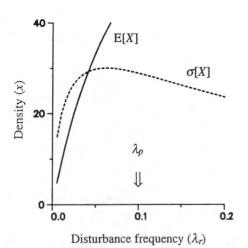

FIGURE 5. The maximum density variance at "intermediate" disturbance frequencies λ_s. Expected density $E[X]$ increases with λ_s, as young, dense stands account for a greater proportion of the metapopulation. Coefficient of variation simultaneously decreases, however, as the population becomes more homogeneous, resulting in an intermediate maximum for standard deviation on density $\sigma[X]$.

the age of a plant selected randomly from this metapopulation.

Local and Regional Effects on Landscape Structure. The structure on this mosaic population, which is characterized by the simplest possible stochastic disturbance process (i.e., constant probability), displays dynamics that will prove robust over the more complex treatments that follow. It is therefore worth considering some of these dynamics at this stage.

Increasing thinning rate λ_p has the effect of decreasing the average age (Fig. 6b) and density (Fig. 6a) of the metapopulation and the less obvious consequence of increasing the coefficient of variation on densities (Fig. 6c). At the scale of a single cohort, lower densities are characteristic of older stands (Fig. 3a). At the metapopulation scale, however, high thinning rates cause low densities to be associated with young stands (Fig. 6a, b). Despite the fact that thinning rate has no effect on the distribution of patch age classes, the average tree is younger.

Although the model assumes deterministic local dynamics, it is worth noting that higher thinning rates do not necessarily increase variability *within* individual cohorts, because that variability depends on the degree of density compensation (Clark 1991b). If within-cohort thinning rates were density and age independent with rate parameter λ_p, for example, then higher mortality rate would indeed cause increased variability as described by the coefficient of variation for this binomial process, $\exp\left[\frac{\lambda_p a}{2}\right] \sqrt{1/x} \, (1 - e^{-\lambda_p a})$. If density compensation is high, however, thinning rates are responsive to random variability in local mortality (Fig. 2), and variance in cohort densities is constrained. The binomial case does not hold, because mortality risk of plants depends on whether neighboring plants have survived. Indeed, given that λ_p is positively related to plant growth rate and that higher growth rates increase density compensation (Clark 1991b), then high thinning rate is likely to attend decreased variability within cohorts, until density compensation declines late in stand development. Thus, increased thinning rates do not necessarily increase within-patch variability, although they do increase among-patch variability.

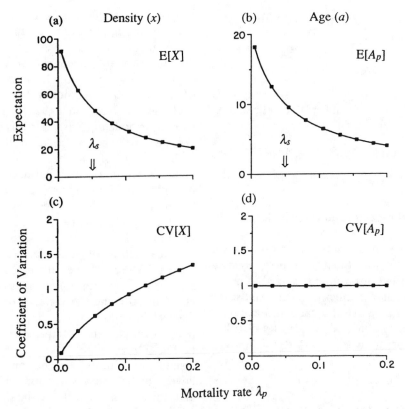

FIGURE 6. Effect of local mortality rate on metapopulation structure. Both density (a) and expected age (b) decrease with increasing mortality rate. The coefficient of variation on density increases (c), however, while that for age is unaffected by mortality rate (d).

At the metapopulation scale, the parameter of the exponential age distribution is not equal to the mortality rate, but is greater than the mortality rate in an amount equal to the disturbance frequency (Eq. 11). Although thinning rate λ_p drives expected density and age in the same direction (i.e., $\partial E[X]/\partial\lambda_p$ and $\partial E[A_p]/\partial\lambda_p$ are both negative for all x and a, respectively) (Fig. 6), the frequency of regeneration opportunities λ_r drives expected density and age in opposite directions (Fig. 4a–c). More frequent regeneration opportunities increase the expected density while decreasing the expected age. Density variance is maximized at intermediate disturbance frequencies (Fig. 5). In general, higher thinning rates λ_p increase variability across the landscape, while more frequent regeneration opportunities (higher λ_r) render the metapopulation more uniform.

Changing Disturbance Probability. These analytical results for constant disturbance probability aid understanding of the more complex structure that arises from more flexible (and realistic) disturbance distributions. These flexible distributions are analyzed here to establish what implications, if any, age-dependent disturbance probability holds for population structure. Changing disturbance probability is accomplished by a power transformation of the time scale using a Weibull distribution $S_r(a) = \exp[-(\lambda_r a)^{c_r}]$. The corresponding distribution of patch age classes is

$$\omega_r(a) = \frac{1}{E[A_r]}\exp[-(\lambda_r a)^{c_r}], \qquad (12)$$

where c_r is a dimensionless shape parameter, $E[A_r] = \Gamma(1/c_r + 1)/\lambda_r$ is the expectation of the Weibull distribution, and $\Gamma(\cdot)$ is the gamma function. The moments of $\omega_r(a)$ are derived in Appendix II. Increasing c_r describes a probability of disturbance that increases with time since the last disturbance. Solving for $a(x)$ from Eq. 12 and making the appropriate substitutions in Eq. 4 results in the distribution of cohort densities

$$f(x) = \frac{1}{\lambda_p E[A_r]x} \exp\left\{-\left[\frac{\lambda_r}{\lambda_p}\ln\left(\frac{\varkappa}{x}\right)\right]^{c_r}\right\},$$

$$0 < x \leq \varkappa. \qquad (13)$$

Eq. 6 is a special case of Eq. 13, where $c_r = 1$. The age distribution from Eq. 10 can be written as

$$f_p(a) = \frac{\exp[-(\lambda_r a)^{c_r} - \lambda_p a]}{\int_0^\infty \exp[-(\lambda_r a)^{c_r} - \lambda_p a]\ da}. \qquad (14)$$

For $c_r = 1$, this expression is equivalent to Eq. 11.

Although we cannot solve for $f_p(a)$, nor for the moments of $f(x)$, the structure represented by these distributions can be analyzed to determine how local dynamics and the landscape disturbance regime affect local and metapopulation structure. The mode of density distribution Eq. 13 occurs where $df(x)/dx = 0$, given by

$$x^* = \begin{cases} \varkappa \exp\left[-\left(\frac{\lambda_p}{\lambda_r}\right)^{c_r/(c_r-1)} c_r^{1/(1-c_r)}\right] & c_r \neq 1 \\ 0 & c_r = 1 \lambda_p > \lambda_r \\ \varkappa & c = 1 \lambda_p < \lambda_r \end{cases}$$

$$(15)$$

For $\lambda_p \approx \lambda_r$, this mode occurs near $\varkappa/2$. The rate at which local populations thin (λ_p) influences the regional density structure according to

$$\frac{\partial f(x)}{\partial \lambda_p} = \frac{c_r \Lambda^{c_r} - 1}{\lambda_p^2 E[A_r]x} \exp(-\Lambda^{c_r}),$$

where $\Lambda = (\lambda_r/\lambda_p)\ln(\varkappa/x)$. Increasing thinning rate decreases the expected density, having a positive effect on the representation of lower density cohorts and a negative effect on higher density cohorts. This sensitivity to thinning rate is greatest at intermediate densities x_p, where

$$x_p = \varkappa \exp\left(-\frac{\lambda_p}{\lambda_r} c_r^{-1/c_r}\right). \qquad (16)$$

The densities most sensitive to thinning rate λ_p, when λ_p is high (low), are rare and greater (less) than the modal density for the landscape. When thinning rates are in the range of the disturbance frequency, local population dynamics

have their strongest influence on the densities that are among those most commonly observed at the metapopulation level and little effect on densities that are infrequent (very low or very high). All else being equal, the landscape density of the metapopulation will be lower and the coefficient of variation higher for greater λ_p. As with the exponential case, kurtosis is lowest and skewness ≈ 0 when $\lambda_p \approx \lambda_r$.

The effects of the regional disturbance regime are summarized by parameters λ_r and c_r. As in the previous exponential case, increased disturbance frequency decreases the expected patch age and variance, but the coefficient of variation on patch ages is independent of the rate at which these patches are formed λ_r (Appendix II). For a given disturbance rate constant (i.e., constant λ_r), increased c_r has the effect of reducing the variance on patch ages, despite the fact that patches are still out of phase. Compare for example $\omega_r(a)$ distributions in Fig. 4b vs. e. This effect on patch ages results, in turn, in distributions of densities having higher modal densities, higher mean, and lower variance. Increasing c_r decreases expected plant age when disturbance is frequent (Fig. 4a vs. d) and vice versa (Fig. 4c vs. f). Increased c_r decreases variability in age structure, and it reduces kurtosis. Thus, reducing the variance on patch ages has effects that are different from either of the effects observed by changing thinning rates or disturbance frequency alone.

The Shifting Mosaic with Large-Scale Disturbances

In order to generalize population dynamics to include large-scale type s disturbances, it is necessary to simplify the shifting mosaic structure and to incorporate the dependency of type r disturbances on time t since the last type s disturbance (Fig. 1). In the last section, I claimed that the patch age distribution was given by the survivor function divided by the expected interval in Eq. 2. In fact, Eq. 2 is a special case of a time-dependent process

$$\omega_r(a, t) = m_r(t - a)S_r(t),$$

where $m_r(a)$ is the renewal density (Cox 1962), the expected number of disturbances to occur on a landscape a yr following the last type s disturbance. As the time since the last type s disturbance becomes remote, $m_r(t - a)$ tends to $1/E[A_r]$, yielding Eq. 2. The renewal density is related to $S_r(a)$ as

$$m(t) = \sum_{i=1}^{\infty} - \frac{d}{dt} S_i(t),$$

where $-dS_i/dt \, \Delta t$ is the probability that the ith disturbance occurs on $(t, t + \Delta t)$ and $S_0(t) = S_r(t)$ (e.g., Cox 1962). This relationship can be used together with the assumption that type s disturbances are much larger than type r disturbances (so much so that type r dynamics can be treated deterministically) to derive and analyze population structure across this larger landscape. The recruitment rate $\beta_k(t)$ of species k on a type s patch is proportional to the renewal density,

$$\beta_k(t) = \varkappa m_r(t).$$

The density of species k on a single type r patch at time t following occurrence of the most recent type s disturbance is

$$x_\varkappa(t) = \varkappa \int_0^t m_r(t - a) S_r(a) S_k(a) \, da. \quad (17)$$

By direct analogy to the type r case (Eq. 4), the distribution of densities across a landscape of type s patches is

$$f(x_k) = \omega_s(x_k) \, \Delta(x_k),$$

where $\omega_s(x_k)$ is the recurrence density for the type s process. The age class distribution on a single type s patch changes with time t,

$$f_k(a, t) = \frac{m_r(t - a) S_r(a) S_k(a)}{\int_0^t m_r(t - a) S_r(a) S_k(a) \, da}. \quad (18)$$

and that for a landscape of type s patches at elapsed time T since the type s process began is

$$g_{kT}(a) = \frac{\int_0^T \omega_s(t) f_k(a, t) \, dt}{\int_0^T \int_0^t \omega_s(t) f_k(a, t) \, da \, dt}. \quad (19)$$

I subsequently use this approach in analytical and numerical examples to demonstrate effects

of local (thinning and small disturbances) and regional (large disturbances) processes on meta-population structure.

An analytical example. Consider the simplest case of constant probability of type r and type s disturbances. The age distribution of type s patches is given by

$$\omega_s(t) = \lambda_s e^{-\lambda_s t},$$

and type r disturbances are occurring at constant rate $m_r(t) = \lambda_r$. The density of species k at time t is solved from Eq. 17 as

$$x_k(t) = \frac{\varkappa \lambda_r}{\lambda_p + \lambda_r} [1 - e^{-t(\lambda_p + \lambda_r)}],$$

with corresponding rate equation

$$\frac{dx_k}{dt} = \varkappa \lambda_r - x_k(\lambda_p + \lambda_r).$$

Solving for time t in terms of density and substituting in $\omega_s(t)$ gives a recurrence density

$$\omega_s(x_k) = \lambda_s \left[1 - \frac{x_k(\lambda_p + \lambda_r)}{\varkappa \lambda_r} \right]^{\frac{\lambda_s}{\lambda_p + \lambda_r}}.$$

The product of this quantity and the reciprocal of dx_k/dt (see Eq. 3) yields the probability density function

$$f(x_k) = \frac{\lambda_s}{\varkappa \lambda_r} \left[1 - \frac{x_k(\lambda_p + \lambda_r)}{\varkappa \lambda_r} \right]^{\frac{\lambda_s}{\lambda_p + \lambda_r} - 1}.$$

$$0 \leq x_k \leq \frac{\varkappa \lambda_r}{\lambda_p + \lambda_r}$$

This distribution differs from Eq. 6 in assuming deterministic type r dynamics. It is the distribution of densities across type s patches. The age distribution on a single type s patch is solved from Eq. 18 as

$$f_k(a, t) = \frac{(\lambda_p + \lambda_r) e^{-a(\lambda_p + \lambda_r)}}{1 - e^{-t(\lambda_p + \lambda_r)}} \qquad 0 \leq a \leq t$$

This result tends to the type r case (Eq. 11) as type s disturbances become rare,

$$\lim_{t \to \infty} f_k(a, t) = (\lambda_p + \lambda_r) e^{-a(\lambda_p + \lambda_r)}.$$

This limiting result also applies to cases of more complex recruitment patterns, provided that recruitment rate tends to some nonzero value with time since the last type s disturbance (Fig.

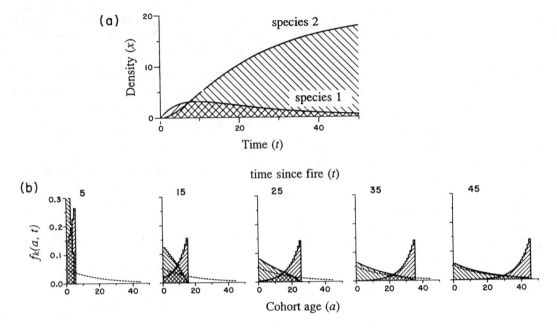

FIGURE 7. Contrasting density and age structure of early- (1) and late- (2) successional species as a function of time since the last large (type s) disturbance. (a) Density of species 1 increases, then decreases following disturbance, while that of species 2 increases. (b) The variance on age increases for both species over time, but distributions are skewed in opposite directions. As time since the last large disturbance becomes remote, the age distribution of species 2 approaches the limiting exponential distribution, given by the broken line.

7). I made some analytical headway with the landscape age distribution (Eq. 19),

$$g_{kT}(a) = \frac{\lambda_s(\lambda_p + \lambda_r)e^{-a(\lambda_p + \lambda_r)}}{1 - e^{-T\lambda_s}} \int_0^T \frac{e^{-\lambda_s t} \, dt}{1 - e^{-t(\lambda_p + \lambda_r)}} \, dt,$$

which tends to

$$g_{k\infty}(a) = \lambda_s(\lambda_p + \lambda_r)e^{-a(\lambda_p + \lambda_r)} \int_0^\infty \frac{e^{-\lambda_s t} \, dt}{1 - e^{-t(\lambda_p + \lambda_r)}} \, dt,$$

as T becomes large.

A Numerical Example. The assumptions of constant disturbance probability can be relaxed to accommodate recruitment rates and small-scale disturbances that depend on large (type s) disturbances. I use the example of fire for this large disturbance, but the model could be applied to other disturbances, such as large blowdowns. To illustrate the effect of these interacting disturbance processes, I consider two species that differ in the time t at which they begin to colonize a site following the most recent type s disturbance. These species are an early-successional species 1 and a late-successional species 2. Species 1 finds regener-

ation niches following some large catastrophic disturbance that results in mortality of many adults, whereas species 2 finds increasingly more regeneration niches with time since this catastrophic disturbance. This second species might be a gap colonizer, because canopy gaps tend to be rare until early-successional species attain maximum size. If the recruitment rates for species k is given by $\beta_k(t)$, where k is either 1 or 2, then these trends in recruitment rates are described by $d\beta_1(t)/dt \leq 0$ and $d\beta_2(t)/dt \geq 0$. To limit the number of subscripts, I represent the thinning and recruitment rates of both species by single parameters λ_p and $\varkappa$, respectively.

In this example, I assume that the distribution of type s patch ages can be described by a Weibull distribution with recurrence time

$$\omega_s(t) = \frac{1}{E[T_s]} \exp[-(\lambda_s t)^{c_s}], \qquad (20)$$

where $E[T_s]$ is time expected to elapse between type s disturbances, and T_s is an observed interval between disturbances. The Weibull distribution has been shown to fit disturbance history data (Johnson 1979, Baker 1989, Clark

1989). The density of regeneration niches for the early-successional species 1 decreases with time. For simplicity I assume this decrease is exponential, being proportional to

$$\beta_1(t) = e^{-pt}.$$

Let the increase in regeneration niches for species 2 be its complement:

$$\beta_2(t) = 1 - e^{-pt}.$$

For species 1, the effects of regional disturbance regime, represented by λ_s and c_s, are qualitatively similar at this scale to those of λ_r and c_r at the finer scale (compare density distributions in Fig. 4 with those for species 1 in Fig. 8). For species 2, the effects are opposite. Increasing intervals between large disturbances have the effect of shifting the mode of the early-successional species 1 to lower densities and the late-successional species 2 toward high density (Fig. 8). As the density of the early-successional species decreases, the distribution becomes more platykurtic and subsequently more leptokurtic, as it becomes extinct in most stands. The density distribution of the late-successional species also becomes initially more platykurtic, as the mode shifts toward higher densities, and it then becomes more leptokurtic and negatively skewed. Both species exhibit the maximum variation in densities at intermediate times since the last large disturbance.

Age distributions are J-shaped for both species (Fig. 8). These landscape age class distributions are a composite of the age class distributions (Eq. 18) of the individual sites (Fig. 7b), with the relative contribution of the distributions at different times since fire determined by the recurrence time of the fire regime. The age distributions of both species have low expectation when disturbances are frequent, because all patches are of young age, and thus cohorts are necessarily young. Species 1 always has a component of older individuals, because it begins to reproduce sooner after the large disturbance. For the early-successional species, results demonstrate the widely observed J-shaped distributions that result from landscape surveys actually composed of local distributions having modes >0 (Hough 1932, West et al. 1981; Fig. 7b). Age class distributions $g_1(a)$ for this species approximate the age distributions $\omega_s(t)$ of the

patches themselves to an extent determined by the degree to which recruitment can continue with time since the last disturbance. The age distribution of species 1 becomes increasingly uniform where disturbance is infrequent, as total density declines. This is because young patches account for a smaller proportion of the total landscape. These results agree with the regional syntheses of West et al. (1981), where composite distributions of older stands were increasingly platykurtic and positively skewed.

Species 2 maintains the J-shaped character because recruitment of young age classes continues. When disturbances are infrequent ($\lambda_s \ll \lambda_p$) the landscape distribution of age class of the late-successional species approaches an exponential distribution with parameter λ_p (Fig. 8).

Discussion

One of the most popular concepts of forest dynamics to emerge in recent years is that of the "shifting mosaic" (Watt 1947, Bormann and Likens 1979, Peet 1981, Shugart 1984). The so-called old-growth forest consists of a patchwork of small cohorts, each having a unique history that depends on the episodic occurrence of disturbances. Processes exogenous to the stand are not required to introduce heterogeneity, because variability is produced by the plants themselves. On any given piece of ground the dynamics consist of recruitment phases followed by periods of thinning. These local dynamics are produced by simulation models of small patches that clearly show this two-phase process (e.g., Shugart 1984:Fig. 5.4). Only when the output from a large number of patches is averaged do recruitment and mortality appear somewhat constant.

In this paper, I derived the structure of a "shifting mosaic" metapopulation, and I examined the effects of local population dynamics and of regional disturbance regimes on that structure. The local dynamics are represented by changing age and density, which depend on growth rates of individual plants within small cohorts. The regional disturbance regime is described by a distribution of events that result in adult mortality and the initiation of new cohorts. The structure of this metapopulation is

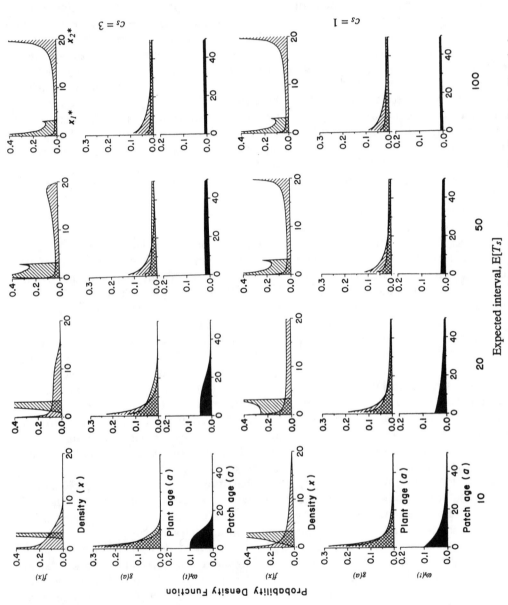

FIGURE 8. Dynamics of two species across a landscape where large disturbances are distributed as $\omega_s(t)$ and where dynamics on any given patch are as shown in Fig. 7. Distributions of densities $f(x)$, cohort age classes $g(a)$, and patch age classes $\omega_s(t)$ are shown for two values of c_s and four different expected intervals between disturbances $E[T_s]$. X_1^* and X_2^* (upper right-hand graph) represent the upper integration limit for density distributions, analogous to x in Eq. 6.

summarized by distributions of cohort (patch) age classes, plant age classes, and cohort densities. This highly idealized class of models serves to isolate several important relationships.

Effects of Local Dynamics

The link between growth of plants and cohort mortality rates (Clark 1991*a*) can be used together with results presented here to demonstrate implications of local dynamics for metapopulation structure. The faster plants grow, the more rapidly they thin. This effect of growth rate on mortality is obvious in all forestry yield tables (reviewed in Clark 1991*b*), and it is commonly observed in populations of herbaceous plants (reviewed in Harper 1977). This negative effect of growth rate on density has effects at the metapopulation level, influencing not only expected values, but also population heterogeneity. The directions and magnitudes of these growth-rate effects depend on the scale at which the population is observed.

Growth rates may have opposing effects on within- vs. among-cohort variability. High growth rate and attendant high thinning rate results in low expected age and density of this shifting mosaic metapopulation (Fig. 6a, b). At the same time, the coefficient of variation on plant age is unchanged (Fig. 6d), but that of density is increased (Fig. 6c). The faster plants grow, however, the greater may be the adjustment of density-dependent mortality rates to stochastic variability in mortality rates (Clark 1991*b*). It is therefore possible that higher growth rates result in more predictable thinning curves. The implication is that increased growth rate can increase local homogeneity while at the same time rendering the metapopulation more heterogeneous. Where large-scale disturbances episodically synchronize patch dynamics across large areas, higher thinning rates imply lower density and lower age for early-successional and late-successional species (Fig. 8).

The Effect of Disturbance Regime

The general patterns in age structure across a gradient from high to low regeneration-niche frequency range from high-density young stands to a mix of densities and age classes to low-density stands dominated by older individuations. Constant disturbance probability results in an exponential age structure with parameter $(\lambda_r + \lambda_p)$ (Fig. 4a–c). Disturbance frequency "substitutes" for mortality rate in the age distribution (Eq. 11), and more frequent disturbance has an effect similar to higher mortality rate. This represents an important result, because exponential age distributions appear to commonly arise in real forests (e.g., Hough 1932, Leak 1965, 1975, Van Wagner 1978), and the parameter of the distribution is shown here to be the sum of two potentially very different sources of mortality, the "thinning rate," which is largely a consequence of competition, and the "disturbance frequency," which can contain a component that is exogenous to the local stand dynamics and which may affect much larger areas. Moreover, while one component (λ_r) of this parameter tends to be associated with recruitment, the other (λ_p) is not.

For density distributions, however, more frequent disturbance has effects opposite of those produced by increased thinning rate; as disturbance becomes more frequent, density increases, and the metapopulation becomes more homogeneous (Fig. 4). Infrequent regeneration niches (e.g., Fig. 4c) produce distributions of densities with a mode near zero, positive skewness, and low variance. Many plants become old and senescent before the next regeneration niche becomes available. Where mortality and regeneration-niche time scales are roughly equivalent (Fig. 4b), the mode occurs at intermediate densities, skewness may approach zero, and variance is high. These stands consist of both young and old cohorts at high and low densities, respectively. Where regeneration niches occur frequently (Fig. 4a), densities are high, skewness is negative, and variance is again low. Where the agent responsible for the regeneration niche also destroys mature individuations, as is often the case, these frequent disturbances result in a transient phase of young individuals at high density that will not reach reproductive maturity before the next disturbance occurs (see below, *Intermediate disturbance and the link to life history theory*). The simple distributions presented here likely overrepresent juvenile classes for species that are rare. A partial antidote to the high variability of juvenile classes would be to ignore them entirely, as is done in gap simulation models, and simply focus on structure of the remaining population.

Increasing Disturbance Probability with Age

Results presented here support earlier speculation (e.g., Johnson 1979) that the shape of the disturbance distribution (represented by parameter c) has important effects on population structure. The simplest case of constant disturbance probability has been advocated on the basis of some empirical studies (Van Wagner 1978) and is assumed in most stochastic population models.

It appears in forests, however, that the probability both of gaps (Kohyama 1987, Suzuki and Tsukahara 1987, Foster 1988) and of fire (Heinselman 1973, Johnson 1979, Romme 1982, Baker 1989, Clark 1989) often increases with time since the last such event. This increasing probability is described here by parameter c_r or $c_s > 1$. As a result of the more precipitous rise in disturbance probability that occurs with large c_r, disturbances tend to occur with higher probabilities at a certain age, and the variance on disturbance intervals declines. For example, stands of some early-successional tree species tend to fall apart at senescence, with gap area increasing rather suddenly as growth rates of trees decline. The population effect of increased c_r depends on the relationship between thinning rate (λ_p) and expected disturbance interval ($E[A_r]$) and on the scale at which the population is observed. In fact, the influence of time-dependent disturbance probability on age structure changes qualitatively from species having low thinning rate ($\lambda_p \ll 1/E[A_r]$) to those having high thinning rate ($\lambda_p \gg 1/E[A_r]$). A clustering of high disturbance probabilities at particular ages (i.e., $c_r \gg 1$) within a shifting mosaic metapopulation has the effects of decreasing the expected age and increasing the expected density for populations characterized by low thinning rates relative to the expected disturbance interval (compare Fig.

4a and d). This response occurs despite the fact that the expected disturbance interval is held constant. However, the coefficient of variation decreases for both age and density.

At high thinning rates ($\lambda_p \gg 1/E[A_r]$, e.g., 50-yr intervals in Fig. 4), increasing parameter c_r increases both the expected age and density (compare Fig. 4c and f), but the coefficients of disturbance intervals within patches (increasing c_r) results in decreased variance on patch age classes, which in turn decreases the coefficient of variation on plant densities and on plant age classes of long- and short-lived species. Despite the fact that patches are out of phase with one another, the increasing predictability of patch age produces younger and denser metapopulations of populations having low thinning rates, older and denser metapopulations of populations having high thinning rates, and more homogeneous metapopulations regardless of thinning rate.

Superimposed and Interdependent Disturbance Processes

On many landscapes, several types of disturbances that provide regeneration niches for different types of species may operate at vastly different scales. An obvious example is fire, which may occur over a spatial scale that is large, and treefalls, which occur on the same piece of ground, albeit at different times (e.g., Spies and Franklin 1989). Moreover, the occurrence of treefalls themselves depends on occurrence of fire, because only when trees that regenerate following fire become large will treefall produce the canopy gaps required for regeneration by a gap species (Peet and Christensen 1980, Oliver 1981, Peet 1981). Early-successional species, such as *Betula papyrifera, Pinus banksiana, P. resinosa, P. contorta,* and *Populus* spp., may colonize soon after such a disturbance at high density and subsequently thin. Other species, such as *Acer saccharum, Tilia americana,* and *Fagus grandifolia,* germinate beneath a closed canopy and maintain low metabolic rates until a gap in the canopy results in increased resources that allow "recruitment" to the canopy. Different sizes and types of gaps may favor different species

(Connell 1979, Runkle 1982, Brokaw 1985, Canham 1988). Although this description is oversimplified, it captures a general pattern that has been observed so consistently that it serves as a basis for prevailing concepts of "succession," "regeneration niche," and "gap-phase replacement."

On a large disturbed area where occurrence of regeneration niches depends on time since this large disturbance, the age distributions of early- and late-successional species are skewed in opposite directions (Fig. 7b). Across a landscape of such large disturbances, however, both distributions are positively skewed, with the early-successional species having a longer "tail" (Fig. 8). The structure of an early-successional population responds to an increase in disturbance frequency in a fashion qualitatively similar to a gap species on a simple shifting mosaic (Fig. 4): density increases and age decreases. Increasing time between large disturbances results in a metapopulation containing few old and low-density stands of early-successional species. The age class distribution approaches that of the disturbance regime itself, with the regional population exhibiting high spatial variability (compare $g(a)$ and $\omega_s(t)$ distributions in Fig. 8). On any given disturbed area, a late-successional species eventually assumes a limiting exponential distribution with the parameter equal to the thinning rate plus the frequency with which canopy gaps occur (Eq. 11; Fig. 7b). Thus, the empirical observation of exponential age-class distributions in "late-successional" species (Leak 1965, 1975) represents the limiting distribution of a more general relationship that accommodates recruitment and mortality from both sources. Both species display maximum variance on stand densities at intermediate disturbance frequencies (Fig. 8, see next section).

The effects of variance in the intervals on large-scale disturbances (compare $c_s = 1$ vs. $c_s = 3$ in Fig. 8) also depend on thinning rate, in addition to whether a species is early vs. late successional. If such disturbances are rare, this variance has little effect on population structure (e.g., $E[T_s] = 100$, $c_s = 1$ vs. $c_s = 3$ in Fig. 8). For long-lived species, or when such disturbances are more frequent (e.g., $E[T_s] = 10$ in Fig. 8), however, decreasing the variance on

expected disturbance interval decreases the expected age and density of late-successional species, and it increases the density and decreases the age of early-successional species. The coefficient of variation on densities is decreased, while the variance is maximized at intermediate frequencies for both species.

It is therefore important to consider the time-dependence of disturbance probability for disturbances that occur frequently, and it is of diminishing importance for disturbances that are rare relative to the time scale of the thinning process. Examples of disturbances that occur with sufficient frequency that this time-dependence is important include gap processes and fire in some temperate forests. Thus, the relative time scales for cohort thinning and disturbance frequency, and the temporal correlation in the mosaic landscape both have important consequences for population structure. The effects of both of these considerations are complex and, in some cases, unexpected. An understanding of the representation of density and age classes and of metapopulation heterogeneity requires knowledge of the time development of cohort densities and changing disturbance probability at several temporal and spatial scales.

Intermediate Disturbance and the Link to Life History Theory

In a companion paper (Clark 1991a) I showed that the probability that a plant will be reproductively mature at the time of the next recruitment opportunity ζ is maximized at "intermediate" disturbance frequencies. This intermediate frequency is one having expectation that is greater than the maturation time a_1 and less than the longevity a_2, which is correlated with maturation time (Loehle 1988), i.e., $a_2 = \alpha a_1$. Because many tree species that occur in temperate forests share similar life histories, this intermediate disturbance frequency is also expected to maximize species diversity. Given the focus on population structure taken here (as opposed to life history optimization in Clark 1991a), it is reasonable to ask whether such an "intermediate" disturbance frequency also exists that maximizes reproductive potential from the standpoint of population structure. Instead of maximizing the reproductive probability ζ discussed in Clark (1991*a*), one candidate for maximization here is the density of reproductive individuals x_r, which includes all individuals of the age interval $(a_1, \alpha a_1)$. This density can be derived as the product of expected density $E[X]$ and the proportion of the population that is of reproductive age,

$$x_r = E[x] \int_{a_1}^{\alpha a_2} f_p(a) \, da.$$

For the exponential case, expected density is given by Eq. 7. The fraction of the population that is reproductively mature is derived from Eq. 11 as

$$\int_{a_1}^{\alpha a_2} f_p(a) \, da = \exp[-(\lambda_r + \lambda_p)a_1]$$
$$- \exp[-(\lambda_r + \lambda_p)\alpha a_1]. \quad (24)$$

The density of reproductive individuals is thus

$$\frac{\lambda_r x}{\lambda_p + \lambda_r} \{\exp[-(\lambda_r + \lambda_p)a_1] - \exp[-(\lambda_r + \lambda_p)\alpha a_1]\}.$$

This result is proportional to ζ for the same disturbance regime, given by Eq. 10 in Clark (1991*a*), i.e.,

$$x_r = \varkappa\zeta,$$

and so is maximized at the same disturbance frequency (Fig. 9).

The fact that the disturbance regime that maximizes reproductive probability is also that which produces the age and density structure containing the maximum density of reproductive individuals suggests (1) an internal consistency of the theory of population dynamics on a mosaic landscape (i.e., the individual perspective does not yield results different from a population perspective), (2) a close link between optimal life history from an individual perspective and the mosaic structure of a metapopulation, and (3) a specific definition for the "intermediate disturbance hypothesis" as one that maximizes both the reproductive potential of individual plants and the reproductive portion of the metapopulation. As disturbance becomes less frequent, the reproductive fraction of the population increases, but more local populations go extinct (Fig. 9); the population consists of increasingly larger and older individuals within lower density stands. At the other

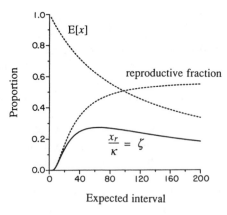

FIGURE 9. Comparison of predictions for the optimal life history from Clark (1991*a*) with that for maximum density of reproductive individuals from theory presented here. The reproductive fraction x_r (solid curve) is the product of expected density $E[x]$ and the fraction of the population that is of reproductive age (broken curves). This product is equal to the optimal maturation time ζ from Clark (1991*a*), when density is measured in units of $\varkappa$, i.e., $\zeta = x_1/\varkappa$. The maturation time is $a_1 = 30$ and thinning rate is $\lambda_p = 0.01$.

end of the spectrum are populations nearing extinction because mortality agents occur too frequently. These stands consist of individuals that are too young to reproduce, and they occur at high density only because they are young and thus small. Thus, the high-density stage that occurs when disturbance is frequent is a transient phenomenon, because the seed source rapidly diminishes with this loss of reproductive individuals from the population.

The maximum among-cohort variability in density classes also occurs when disturbance frequency is in the range of thinning rate (Figs. 4, 5, 8). This conclusion applies to the simple shifting mosaic (Fig. 4), and to both early- and late-successional species on the more complex landscape that experiences large-scale disturbances, despite the fact that disturbance frequency is driving the density distribution of early- and late-successional species in opposite directions (Fig. 8). Thus, at a range of spatial scales, the approximate equivalence of these time scales maximizes variability, and it coincides with the relationship where persistence is predicted to be most likely. The turnover of species along disturbance gradients (e.g.,

Huston 1979, Tilman 1988) derive at least in part from this relationship between time scales. These relationships are important in view of the frequent regional changes in forest tree composition that have occurred in temperate zones just over the last 10,000 yr (Wright 1974, Davis 1981, Payette and Filion 1985, Jacobson et al. 1987).

Population Dynamics as a Mosaic

There is often a concern in population theory that deterministic results might be modified in some qualitative way upon the introduction of random or time-dependent variability in demographic parameters such as birth or death rates. A number of studies show that this type of variability may produce unexpected results in some models (Abrams 1984, Chesson 1984), while in others the main effect of limited stochasticity is to simply replace a fixed-point equilibrium with a probability distribution having expectation in the neighborhood of that point equilibrium (May 1973, Turrelli 1981). For many applications, it is reasonable to ignore this random component of birth and death rates, because the effect of variation appears obvious or because variation in the system under consideration is deemed insignificant (it may be small in magnitude, or dependencies may be weak).

The type of stochasticity treated here cannot so easily be ignored as it represents the basic demography that characterizes many populations of perennial plants. This variation is that which results from the uncertainty associated with the timing of recruitment events. Even if all environmental variables could be held constant, a survey of stem densities and age classes across any landscape would generate a frequency distribution, because of the discrete nature of individual plants, because plants are constantly increasing in size, and because increasing individual size for a population of sessile organisms necessitates decreasing density with time. As plants approach maximum size, growth slows and density is reduced to a point where canopy gaps begin to appear. It is at this point that recruitment rate increases. Variability is produced by the population itself. The result is a metapopulation that is a collection of cohorts (Watt 1947, Bormann and Likens 1979), each possessing a unique history and thus structure.

In order to address this process, I have treated the metapopulation as a mosaic of local populations. Plant densities, competitors, and resources do not approach a constant density anywhere on a landscape, recruitment is confined to short periods, and a long history of observation implicates the importance of the distribution of regeneration niches on landscapes (e.g., Grubb 1977). Taller plants preempt more light and larger root systems may permit control of a disproportionate amount of belowground nutrients. The establishment of a seedling is most strongly dependent on the occurrence of "safe sites" (Harper 1977) or "regeneration niches" (e.g., Watt 1947, Grubb 1977). "The vast majority of tree species are dependent either on fire, flooding, or windthrow to provide suitable seedbeds for their establishment" (Spurr and Barnes 1980: 71). Different species utilize different regeneration niches, and it is the availability of regeneration niches that poses an important limiting factor on a population's success.

Competition influences the probability of surviving until that next recruitment opportunity occurs. A strong competitor can insure local extinction of a weaker competitor for the same regeneration niche by increasing that competitor's mortality rate to an extent where it does not survive to reproductive maturity or until the next recruitment opportunity. Thus, theory presented here places competition within the context of survivorship of established cohorts within patches of finite area.

Finally, the occurrence of these opportunities for recruitment (gaps) often depends on the dynamics of the overstory, so it might also be viewed as one component of competition. A more complete model might generate the distribution of regeneration niches from the assemblage itself, with the occurrence of each type of regeneration niche influenced by the size and density of each species within the assemblage. I view this approach as less useful at the level of analytical models because of its complexity, the lack of information on direct and higher order effects of species combinations on the disturbance regime, and the problem of understanding how the assemblage interacts with exogenous factors to affect the distributions of disturbances such as blowdowns, fires, and landslides. Moreover, the gap process in temperate forests may be relatively insensitive to species composition (see Runkle 1985, Clark 1991a). This approach is also more applicable to real situations, because the distributions of disturbances and plant survivorship can be quantified in forests without understanding all of the dependencies of that disturbance occurrence on the physical environment and biotic factors. These simple analytical results provide a perspective for complementary numerical models that assume a similar structure but contain more complexity.

Acknowledgments

For their helpful discussions and/or reviews of the paper I thank P. Abrams, H. Caswell, P. Chesson, P. Munholland, S. Pickett, D. Royall, D. Tilman, H. E. Wright, and three anonymous reviewers. This research was supported by a fellowship from the Graduate School of the University of Minnesota, and NSF grants BSR-8715251 and BSR-8818355. Limmological Research Center contribution number 377 and New York State Education contribution number 608.

References

Abrams, P. A. 1984. Variability in resource consumption rates and the coexistence of competing species. Theoretical Population Biology 25:106–124.

Aikman, D. P., and A. R. Watkinson. 1980. A model for growth and self-thinning in even-aged monocultures of plants. Annals of Botany 45:419–427.

Baker, W. L. 1989. Effect of scale and spatial heterogeneity on fire-interval distributions. Canadian Journal of Forest Research 19:700–706.

Bormann, F. H., and G. E. Likens. 1979. Pattern and process in a forested ecosystem. Springer-Verlag, New York, New York, USA.

Botkin, D. F., J. F. Janak, and J. R. Wallis. 1972. Some ecological consequences of a computer model of forest growth. Journal of Ecology 60:849–872.

Brokaw, N. V. 1985. Treefalls, regrowth, and community structure in tropical forests. Pages 53–69 in S. T. A. Pickett and P. S. White, editors. The ecology of natural disturbance and patch dynamics. Academic Press, New York, New York, USA.

Canham, C. D. 1988. Growth and canopy architecture of shade-tolerant trees: response to canopy gaps. Ecology 69:789–795.

Chesson, P. L. 1984. The storage effect in stochastic population models. Lecture Notes in Biomathematics 54:76–89.

Clark, J. S. 1989. Ecological disturbance as a renewal process: theory and application to fire history. Oikos 56:17–30.

_____ . 1990. Integration of ecological levels: individual plant growth, population mortality, and ecosystem process. Journal of Ecology 78:275–299.

_____ . 1991a. Disturbance and tree life history on the shifting mosaic landscape. Ecology 72:1102–1118.

_____ . 1991b. Relationships between individual plant growth and the dynamics of populations and ecosystems. In D. DeAngelis and L. Gross, editors. Populations, communities, and ecosystems: an individual perspective. Routledge, Chapman, and Hall, New York, New York, USA, in press.

Cohen, D., and S. A. Levin. 1991. Dispersal in patchy environments: the effects of temporal and spatial structure. Theoretical Population Biology, in press.

Comins, H. N., and I. R. Noble. 1985. Dispersal, variability, and transient niches: species coexistence in a uniformly variable environment. American Naturalist 126:706–723.

Connell, J. H. 1978. Diversity in tropical rain forests and coral reefs. Science 199:1302–1310.

_____ . 1979. Tropical rain forests and coral reefs as open non-equilibrium systems. Pages 141–163 in R. M. Anderson, B. O. Turner, and L. R. Turner, editors. Population dynamics. Blackwell, Oxford, England.

Cox, D. R. 1962. Renewal theory. Methuen, London, England.

Davis, M. B. 1981. Quaternary history and the stability of forest communities. Pages 132–153 in D. C. West, H. H. Shugart, and D. B. Botkin, editors. Forest succession: concepts and application. Springer-Verlag, New York, New York, USA.

Foster, D. R. 1988. Species and stand response to catastrophic wind in central New England, U.S.A. Journal of Ecology 76:135–151.

Grubb, P. J. 1977. The maintenance of species richness in plant communities. The importance of the regeneration niche. Biological Review 52:107–145.

Hara, T. 1984. Dynamics of stand structure in plant monocultures. Journal of Theoretical Biology 110:223–239.

_____ . 1985. A model for mortality in a self-thinning plant population. Annals of Botany 55:667–674.

Harper, J. L. 1977. Population biology of plants. Academic Press, London, England.

Hastings, A., and C. L. Wolin. 1989. Within-patch dynamics in a metapopulation. Ecology 70:1261–1266.

Heinselman, M. L. 1973. Fire in the virgin forests of the Boundary Waters Canoe Area, Minnesota. Quaternary Research 3:329–382.

Hough, A. F. 1932. Some diameter distributions in forest stands of northwestern Pennsylvania. Journal of Forestry 30:933–943.

Huston, M. A. 1979. A general hypothesis of species diversity. American Naturalist 113:81–101.

Jacobson, G. L., T. Webb, and E. C. Grimm. 1987. Patterns and rates of vegetation change during the deglaciation of eastern North America. Pages 277–288 in W. F. Ruddiman and H. E. Wright, editors. North America and adjacent oceans during the last deglaciation. The geology of North America, K-3. Geological Society of America, Boulder, Colorado, USA.

Johnson, E. A. 1979. Fire recurrence in the subarctic and its implications for vegetation composition. Canadian Journal of Botany 57:1374–1379.

Johnson, E. A., and C. E. Van Wagner. 1985. The theory and use of two fire history models. Canadian Journal of Forest Research 15:214–220.

Kohyama, T. 1987. Stand dynamics in a primary warm-temperate rain forest analyzed by the diffusion equation. Botanical Magazine (Tokyo) 100:305–317.

Leak, W. B. 1965. The J-shaped probability distribution. Forest Science 11:405–409.

_____. 1975. Age distribution in virgin red spruce and northern hardwoods. Ecology 56:1451–1454.

Loehle, C. 1988. Tree life history strategies: the role of defenses. Canadian Journal of Forest Research 18:209–222.

May, R. M. 1973. Stability and complexity in model ecosystems. Princeton University Press, Princeton, New Jersey, USA.

Norberg, Å. 1988. Theory of growth geometry of plants and self-thinning of plant populations: geometric similarity, elastic similarity, and different growth modes of plant parts. American Naturalist 131:220–256.

Oliver, C. D. 1981. Forest development in North America following major disturbance. Forest Ecology and Management 3:153–168.

Pacala, S. W. 1988. Competitive equivalence: the coevolutionary consequences of sedentary habit. American Naturalist 132:576–593.

_____. 1989. Plant population dynamic theory. Pages 54–67 in J. Roughgarden, R. M. May, and S. A. Levin, editors. Perspectives in ecological theory. Princeton University Press, Princeton, New Jersey, USA.

Pacala, S. W., and J. A. Silander. 1985. Neighborhood models of plant population dynamics. I. Single-species models of annuals. American Naturalist 125:385–411.

Pastor, J., and W. M. Post. 1986. Influence of climate, soil moisture, and succession on forest carbon and nitrogen cycles. Biogeochemistry 2:3–27.

Payette, S., and L. Filion. 1985. White spruce expansion at the tree line and recent climatic change. Canadian Journal of Forest Research 15:241–251.

Peet, R. K. 1981. Changes in biomass and production during secondary succession. Pages 324–338 in D. C. West, H. H. Shugart, and D. B. Botkin, editors. Forest succession: concepts and applications. Springer-Verlag, New York, New York, USA.

Peet, R. K., and N. L. Christensen. 1980. Succession: a population process. Vegetatio 43:131–140.

Pickett, S. T. A., and P. S. White. 1985. Patch dynamics: a synthesis. Pages 371–384 in S. T. A. Pickett and P. S. White, editors. The ecology of natural disturbance and patch dynamics. Academic Press, Orlando, Florida, USA.

Romme, W. H. 1982. Fire and landscape diversity in subalpine forests of Yellowstone National Park. Ecological Monographs 52:199–221.

Runkle, J. R. 1982. Patterns of disturbance in some old-growth mesic forests of eastern North America. Ecology 63:153–1546.

_____. 1985. Disturbance regimes in temperate forests. Pages 17–34 in S. T. A. Pickett and P. S. White, editors. The ecology of natural disturbance and patch dynamics. Academic Press, Orlando, Florida, USA.

Schaffer, W. M., and E. G. Leigh. 1976. The prospective role of mathematical theory in plant ecology. Systematic Botany 1:209–232.

Shugart, H. H. A. 1984. Theory of forest dynamics: the ecological implications of forest succession models. Springer-Verlag, New York, New York, USA.

Spies, T. A., and J. F. Franklin. 1989. Gap characteristics and vegetation response in coniferous forests of the Pacific Northwest. Ecology 70:543–545.

Sprugel, D. G. 1984. Density, biomass, productivity, and nutrient-cycling changes during stand development in wave-regenerated balsam fir forests. Ecological Monographs 54:165–186.

_____. 1985. Natural disturbance and ecosystem energetics. Pages 335–352 in S. T. A. Pickett and P. S. White, editors. The ecology of natural disturbance and patch dynamics. Academic Press, New York, New York, USA.

Spurr, S. H., and B. V. Barnes. 1980. Forest ecology. Wiley, New York, New York, USA.

Suzuki, E., and Tsukahara, J. 1987. Age structure and regeneration of old growth *Cryptomeria japonica* forests on Yakushima Island. Botanical Magazine (Toyko) **100**:223–241.

Tilman, D. 1988. Plant strategies and the dynamics and structure of plant communities. Princeton University Press, Princeton, New Jersey, USA.

Turelli, M. 1981. Niche overlap and invasion of competitors in random environments. I. Models without demographic stochasticity. Theoretical Population Biology **20**:1–56.

Turner, M. G., V. H. Dale, and R. H. Gardner. 1989. Predicting across scales: theory development and testing. Landscape Ecology **3**:245–252.

Van Wagner, C. E. 1978. Age-class distribution and the forest fire cycle. Canadian Journal of Forest Research **8**:220–227.

Vitousek, P. M., and W. A. Reiners. 1975. Ecosystem succession and nutrient retention: a hypothesis. BioScience **25**:376–381.

Waring, R. H., and W. H. Schlesinger. 1985. Forest

ecosystems: concepts and management. Academic Press, Orlando, Florida, USA.

Warner, R. R., and P. L. Chesson. 1985. Coexistence mediated by recruitment fluctuations: a field to the storage effect. American Naturalist **125**:769–787.

Watkinson, A. R. 1986. Plant population dynamics. Pages 137–184 *in* M. J. Crawley, editor. Plant ecology. Blackwell Scientific, Oxford, England.

Watt, A. S. 1947. Pattern and process in the plant community. Journal of Ecology **35**:1–22.

Weiner, J. 1985. Size hierarchies in experimental populations of annual plants. Ecology **66**:743–752.

_____ . 1986. How competition for light and nutrients affects size variability in *Ipomoea tricolor* populations. Ecology **67**:1425–1427.

West, D. C., H. H. Shugart, and J. W. Ranney. 1981. Population structure of forests over a large area. Forest Science **27**:701–710.

Westoby, M. 1984. The self-thinning rule. Advances in Ecological Research **14**:167–225.

Wright, H. E. 1974. Landscape development, forest fires, and wilderness management. Science **186**:487–495.

Appendix I

The Effect of Thinning Rate on the Distribution of Densities

Consider a cohort of plants that are rapidly growing (and thus, rapidly thinning) soon after a regeneration niche becomes available ($a = 1$). The high thinning rate dictates that the population actually spends a small fraction of a year at a given density x. In contrast, at some later time $a \geqslant 1$ plants are thinning less rapidly (Fig. 1). The "fraction" of time $a(x)$ actually spent at a given density is much greater in an amount determined by the thinning rate. This result is demonstrated as follows:

Define discrete probability density functions U and V such that

$$U(x) = \Pr[u - 1 \leq x(a) < u]$$

and

$$V[a(x)] = \Pr[v - 1 \leq a(x) < v],$$

where u and v are values of density and age, respectively (Fig. A.1). U is the probability that

a randomly selected patch supports density x. V is the probability that a randomly selected patch is a yr old. We wish to solve for $U(x)$ and to establish its relationship to the continuous distribution $f(x)$.

$U(x)$ is the product of two events, the probability that a random patch is of an age $a(x)$, i.e., that age at which the patch supports density x, and the portion of age $a(x)$ at which density actually is x,

$$U(x) = V[a(x)] \left| \frac{a(u - 1) - a(u)}{v - (v - 1)} \right|. \quad (I.1)$$

The quotient in Eq. I.1 is the relative proportion of the age interval ($v = 1$, v) at which density is between $u - 1$ and u (Fig. A.1). During the remainder of the interval ($v - 1$, v) density is not between $u - 1$ and u. Expanding $a(x)$ in a Taylor series about ($v - 1$) and ignoring all terms of order ≥ 2, we can write

$$a(u) = (v - 1) + \frac{da}{dx} [u - x(v - 1)] + o(x^2),$$

and

$$a(u - 1) = (v - 1) + \frac{da}{dx} [(u - 1) - x(v - 1)] + o(x^2),$$

where $o(x^2)$ represents all terms of order ≥ 2. Substituting in Eq. I.1 yields

$$U(x) = V[a(x)] \left| \frac{da}{dx} \frac{u - 1 - u}{v - (v - 1)} \right|$$

$$= V(a) \left| \frac{da}{dx} \right| .$$

Given that $\omega(a)$ is the limiting distribution for $V(a)$, then $\Delta(x) = |da/dx|$ and $f(x)$ is the continuous analogue of $U(x)$.

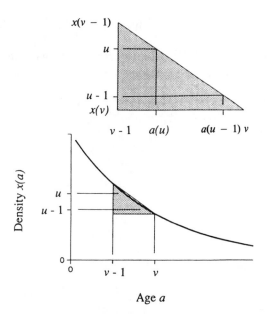

FIGURE A.1. Geometric interpretation of the contribution of thinning rate to the distribution of cohort densities. The cohort thins at a rate that is approximated by the slope of the shaded triangle. Labeled values of density and age are references in Appendix I.

Appendix II

Derivation of Moments for the Distribution of Patch Ages

Given the distribution of patch ages (recurrence times)

$$\omega_r(a) = \frac{\lambda_r}{\Gamma(1/c + 1)} \exp[-(\lambda_r a)^c] ,$$

the mth moment of $\omega_r(a)$ is written as

$$E[a^m] = \int_0^\infty a^m \omega_r(a) \, da$$

$$= \frac{\lambda_r}{\Gamma(1/c + 1)} \int_0^\infty a^m \exp[-(\lambda_r a)^c] \, da.$$

The substitution $u = (\lambda_r a)^c$ yields

$$\frac{1}{\lambda_r^m c \Gamma(1/c + 1)} \int_0^\infty u^{\left(\frac{m+1}{c}\right) - 1} e^{-u} \, du.$$

The integral expression is now a gamma function with parameter $\frac{m + 1}{c}$. Using this fact, together with the property $\Gamma(\alpha + 1) = \alpha \Gamma(\alpha)$, the mth moment is written as

$$\frac{\Gamma\left(\frac{m + 1}{c}\right)}{\lambda_r^m \Gamma(1/c)},$$

yielding expectation

$$E[A_r] = \frac{\Gamma(2/c)}{\lambda_r \Gamma(1/c)},$$

and coefficient of variation

$$cv[A'] = \frac{\Gamma^{1/2}(1/c)}{\Gamma(2/c)}\left[\Gamma(3/c) - \frac{\Gamma^2(2/c)}{\Gamma(1/c)}\right]^{1/2}.$$

By use of well-known properties of the gamma function, it can be shown that these results collapse to the exponential for $c = 1$, i.e., $E[A'] = 1/\lambda_r$ and $cv[A_r] = 1$, and that $c = 2$ decreases this expectation to $E[A_r] = \frac{1}{\lambda_r\sqrt{\pi}}$.

11
Fire Frequency and Community Heterogeneity in Tallgrass Prairie Vegetation

Scott L. Collins

Introduction

Pattern in plant communities is a function of many biotic and abiotic factors and these factors operate on different species at different spatial scales (Pielou 1960, Greig-Smith 1979, Legendre and Fortin 1989). One product of the nonrandom distribution of species at different spatial scales is community heterogeneity. In this paper, spatial heterogeneity is the mean degree of dissimilarity in species composition from one point to another in a community (Inouye et al. 1987, Collins 1990). Heterogeneity differs from pattern analysis in that the latter is a measure of the degree of spatial autocorrelation, or distance decay in the case of gradients, in a community (Palmer 1988, Legendre and Fortin 1989). Like the measurement of pattern, the measurement of spatial heterogeneity is scale dependent; larger samples are likely to include a larger subset of species in a community, and thus, increase the degree of similarity among plots, than are smaller samples.

One common mechanism producing spatial heterogeneity in communities is disturbance (e.g., Collins 1989, Chaneton and Facelli 1991, Glenn et al. 1992). Studies of disturbance effects on heterogeneity, however, are often confounded by scale problems. Two theoretical models, the disturbance heterogeneity model (DHM, Kolasa and Rollo 1991), and the inter-mediate disturbance hypothesis (IDH, Connell 1978) are relevant. These models make contrasting predictions concerning the relationship between heterogeneity and disturbance.

The DHM (Kolasa and Rollo 1991) is a between-patch heterogeneity model based on analyses that combine measurements from both disturbed and undisturbed patches. In this case, disturbance increases community heterogeneity in areas where the size of the disturbance is small relative to the size of the community (e.g., the effects of gaps in forests). As size of the disturbed area increases relative to the total area of interest, the DHM predicts a quadratic relationship in which heterogeneity is highest in communities where 50% of the area is disturbed (Kolasa and Rollo 1991). One important assumption of the DHM providing the basis of the quadratic relationship is that spatial heterogeneity within disturbed patches is similar to heterogeneity within undisturbed patches.

The IDH is a within-patch model that predicts a quadratic relationship between disturbance frequency and species diversity. Given that diversity is positively correlated with heterogeneity (Whittaker and Levin 1977), this model predicts that within-patch heterogeneity will be highest at intermediate frequencies of disturbance.

Whereas spatial heterogeneity is measured over a number of points in space at a single point in time, temporal heterogeneity is measured over several points in time at a single point in space (Kolasa and Rollo 1991). A linkage between space and time is an important component of many models of landscape dynamics (Delcourt et al. 1983, Urban et al. 1987, Turner et al. 1989), yet few studies have documented a statistical space–time relationship between community variables in terrestrial systems. Kolasa and Rollo (1991) state, however, that there is a positive relationship between spatial and temporal heterogeneity in an area over time.

In this study, I used community composition data collected over a 9-yr period in quadrats permanently located in experimental management units to test hypotheses derived from the IDH and DHM models. First, I determined if within-patch heterogeneity showed a quadratic relationship with disturbance frequency as predicted by the IDH. Secondly, I determined if burned and unburned grasslands exhibited similar degrees of spatial heterogeneity as assumed by the DHM. The relationship between disturbance and heterogeneity may differ at larger spatial scales (Glenn et al. 1992). As noted above, the DHM predicts a quadratic relationship between among-site heterogeneity and number of sites disturbed in a region. Thirdly, I determined if there was a positive relationship between spatial and temporal heterogeneity as suggested by Kolasa and Rollo (1991).

Methods

Study Site

The study was conducted at Konza Prairie Research Natural Area (KPRNA), a 3487-ha tallgrass prairie located in Riley and Geary counties, northeastern Kansas, USA. The climate is continental with hot summers, cold winters, strong winds, and low humidities. Mean annual temperature is $\approx 13°C$, and mean annual precipitation, which varies drastically from year to year, is 835 mm (1951–1980), most of which occurs in May–June and September. From 1981 to 1990, growing season precipitation (April–September) ranged from a low of 434 mm in 1988 to a high of 849 mm in 1986. As

part of the Long-term Ecological Research Program (LTER), KPRNA is divided into a series of replicated management units subjected to different frequencies of spring (mid-April) burning (Hulbert 1985).

Data Collection

Vegetation data have been collected on eight study sites since 1981. In 1983, an additional six sites were added, and five more sites were added in 1984 for a total of 19 sites in 12 management units. In each of the 19 study sites, species composition was measured in permanently located 10-m^2 circular quadrats. Five quadrats were evenly spaced (12.5 m) along each of four 50-m long transects within a study site (total = 20 quadrats per site). Because each site is subjected to a controlled disturbance regime, a site as used here represents a disturbed (burned) or undisturbed (unburned) patch. In each year, cover of species in each quadrat was estimated in May, July, and September using a modified Daubenmire scale (1 = 0–1% cover, 2 = 2–5% cover, 3 = 6–25% cover, 4 = 26–50% cover, 5 = 51–75% cover, 6 = 76–95% cover, 7 = 96–100% cover). Although this cover scale yields relatively coarse estimates of cover in a single quadrat, it is a highly repeatable method for rapidly sampling a large number of quadrats over a short period of time (total = 380 quadrats). For each species, Daubenmire scale values were converted to the midpoint value per quadrat for analysis.

Data Analysis

As noted above, spatial heterogeneity is measured as the mean dissimilarity in species composition, among samples at a site within a year (Inouye et al. 1987, Collins 1989, Scheiner 1990). In this study, percent dissimilarity (PD) was defined as:

$$PD = 1 - PS$$
$$PS = 1 - 0.5 \sum_{t=1}^{s} |p_a - p_b|,$$

where PS is percent similarity, p_a is the proportional cover of species p in quadrat a, p_b is the proportional cover of species p in quadrat b, and s is the total number of species (Whittaker 1975). As the degree of difference in composition among quadrats increases, spatial heterogeneity, as measured by PD, increases. Percent dissimilarity was calculated for all possible two-way combinations of quadrats at a site, resulting in a matrix of 190 values. Within-site heterogeneity is the average of the 190 values.

Linear and polynomial regressions were used to test the hypothesis from the IDH that there was a quadratic relationship between disturbance frequency and heterogeneity. For this analysis within-site heterogeneity was measured as the heterogeneity of a site ($n = 19$) in 1990, the most recent year of sampling. Disturbance frequency was measured as the number of times the site had been burned since the experimental protocol was established in 1972 (max = 19 times).

To test the assumption of the DHM that heterogeneity within undisturbed sites is similar to heterogeneity within disturbed sites, within-site heterogeneity of management units burned annually, every other year, every 4 yr, and unburned were compared using Kruskal-Wallis one-way nonparametric analysis of variance and Dunn's multiple-comparison test (Neave and Worthington 1988). Statistical comparisons were made for all years from 1983 to 1990 (too few sites for comparison in 1981).

Spearman rank correlation was used to test the hypothesis that a relationship existed between the within-site heterogeneity at time t, and the temporal change in species composition at a site from time t to time $t + 1$. That is, does the degree of within-site heterogeneity in one year affect the amount of change in composition from one year to the next at that site? Again, within-site heterogeneity is measured as average dissimilarity among the 20 quadrats at a site. Temporal variation in composition is dissimilarity in species composition at a site from one year to the next. In this case, temporal variation in a single PD value based on average cover values ($n = 20$) for all species at a site at time t and time $t + 1$. Data for these analyses were segregated by burning frequency.

Because measurement of heterogeneity may be scale dependent, I analyzed the larger scale effects of fire frequency on heterogeneity among different sites, as well. To do so, I calculated the mean cover value for each species among the 20 quadrats at each site. Percent dissimilarity then was calculated for all pairwise comparisons among all sites at Konza, resulting in matrices of 28 values for 8 sites in 1981, 91 values for 14 sites in 1983, and 171 values for 19 sites in 1984 to 1990. Regional heterogeneity is the mean percent dissimilarity for all pairwise comparisons in each year. Spearman rank correlation and polynomial regression were then used to determine if there was a linear or quadratic relationship, respectively, between regional heterogeneity and the proportion of sites burned each year.

Results

Heterogeneity in 1990 was significantly negatively related to burning frequency (Fig. 1). The quadratic relationship ($r^2 = 0.42$, $F = 5.9$, df $= 16$, $P = .02$) had a higher r^2 but lower P value than the linear relationship ($r^2 = 0.39$, $F = 10.9$, df $= 16$, $P = .004$). More importantly, the quadratic relationship was opposite that predicted by the IDH. That is, heterogeneity was lowest at an intermediate frequency of disturbance. This result is important given the significant positive relationship between heterogeneity and species diversity (measured as the Shannon-Wiener index, H'; $r_s = 0.90$, df $= 17$, $P < .0001$) and total species richness ($r_s = 0.78$, df $= 19$, $P < .01$).

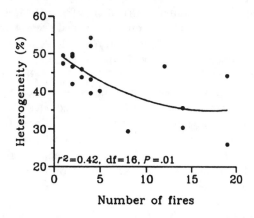

FIGURE 1. Relationship between number of times a site has been burned between 1972 and 1990 and site heterogeneity ($= 1 -$ percent similarity) in species composition, as measured in 1990.

TABLE 1. Average heterogeneity in species composition ($=1$ − percent similarity) of vegetation on management units with different burning frequencies. Values in the same rows with similar superscripts were not significantly different based on Kruskal-Wallis (K-W) one-way analysis of variance and Dunn's multiple-comparisons tests.

| | Burn frequency | | | | | | | | | |
| | Annual | | 2-yr | | 4-yr | | Unburned | | | |
Year	%	n	%	n	%	n	%	n	K-W	P
1981	30.6	2	...		44.2	4	42.1	2	2.0	
1983	25.3[a]	4	...		33.1[ab]	4	40.2[b]	6	11.0	<.01
1984	24.4[a]	4	26.6[a]	2	34.8[b]	7	40.8[b]	6	12.9	<.01
1985	22.6[a]	4	19.8[a]	2	33.6[b]	7	37.9[b]	6	14.2	<.01
1986	26.9[a]	4	27.5[a]	2	37.8[b]	7	45.4[b]	6	13.8	<.01
1987	29.5[a]	4	30.4[a]	2	39.2[ab]	7	48.0[b]	6	10.8	<.02
1988	33.1[a]	4	25.9[a]	2	44.8[ab]	7	51.6[b]	6	11.5	<.01
1989	33.7[a]	4	38.5[ab]	2	42.4[ab]	7	46.6[b]	6	10.0	<.02
1990	38.9[a]	4	38.0[ab]	2	44.6[ab]	7	48.4[b]	6	4.4	.05

Mean heterogeneity on annual burned grasslands was always less than on unburned grasslands (Table 1). Sites burned every other year were often less heterogeneous than unburned sites and both were sometimes less heterogeneous than 4-yr burn sites (Table 1). The decrease in heterogeneity with burning results from increased cover of C_4 grasses following fire. Cover of the dominant C_4 grass *Andropogon gerardii* was significantly negatively related to heterogeneity ($r_s = -0.48$, df = 19, $P = .05$).

The relationship between heterogeneity in a site at time t (spatial heterogeneity) and change in species composition from time t to time $t + 1$ (temporal heterogeneity) was significant and positive on annually burned sites ($r_s = 0.44$, df = 30, $P < .01$), and on sites burned every 4 yr ($r_s = 0.65$, df = 40, $P < .01$) (Fig. 2). A similar trend occurred on sites burned every other year ($r_s = 0.33$, df = 10, $.05 < P < .10$). There was no relationship between spatial heterogeneity and temporal variation on unburned sites.

At the regional level, there was no linear ($r_s = 0.03$, df = 7, $P > .70$) or quadratic ($r^2 = 0.09$, $F = 0.3$, df = 7, $P = .75$) relationship between heterogeneity and the proportion of sites burned in a year.

Discussion

Results from this study indicate that for tallgrass prairie (1) heterogeneity was lowest on sites with intermediate burning frequencies, (2) heterogeneity on burned grassland was significantly lower than on unburned grassland at a small but not at a larger scale, (3) there was a positive relationship between spatial and temporal heterogeneity in two of the four burning treatments, and (4) heterogeneity was positively correlated with species richness and diversity. Patterns of heterogeneity at Konza are consistent with other analyses of disturbance effects on prairie heterogeneity. For example, small-scale heterogeneity (0.25 m^2) in a grassland burned one time was significantly lower than in unburned grassland in Oklahoma (Collins 1989). However, at a larger scale, there were no significant differences in among-site heterogeneity between burned and unburned sites in the region of Konza Prairie on sites with an unknown disturbance history (Glenn et al. 1992). Thus, at larger spatial scales, factors including site history, regional climate, edaphic factors, and topography may have a greater impact on heterogeneity than fire frequency.

The intermediate disturbance hypothesis predicts that diversity, and in turn, heterogeneity will be highest at intermediate disturbance frequencies. Just the opposite was found in this study. In general, frequent burning changes species composition (Gibson and Hulbert 1987) by eliminating fire-intolerant species and increasing productivity of fire-adapted matrix-forming species, such as the C_4-grass *Andropogon gerardii* (Knapp 1985, Collins 1987,

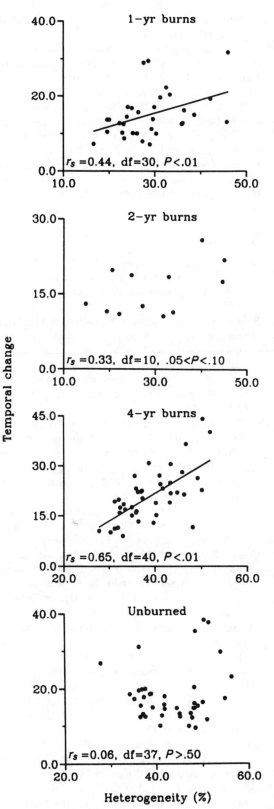

Tester 1989). As productivity increases at Konza Prairie, species richness decreases (S. L. Collins and J. M. Briggs, *personal observation*), perhaps as a result of competitive exclusion (Wilson and Shay 1990). As production of a few species increases, heterogeneity decreases. Eventually, however, frequent burning results in chronic soil nitrogen deficiency (Seastedt and Ramundo 1990). Only species tolerant of burning and low nutrient conditions can exist on frequently burned sites. Assuming that N availability is patchily distributed (e.g., Tilman 1982) then compositional heterogeneity will increase as soil resource heterogeneity increases under long-term burning regimes.

The significant relationship between spatial and temporal heterogeneity on annually burned and 4-yr burning sites is intriguing. The pattern on 2-yr burn sites was nearly significant, as well (.05 < P < .10). All relationships were positive, suggesting that high within-site heterogeneity yields high temporal heterogeneity at least under certain disturbance regimes. Indeed, previous analyses of community dynamics have demonstrated a high degree of spatial and temporal variability in species distribution and abundance at Konza Prairie (Collins and Glenn 1990, 1991). Temporal variability is a measure of change in species composition over time (immigration plus extinction of species). Although immigration rates are constant on different sites at Konza (Collins and Glenn 1991), as species richness increases over time on a site, there is a significant increase in local rates of extinction (Glenn and Collins 1992). These patterns of immigration and extinction produce compositional change, which yields high temporal heterogeneity. The lack of a significant spatial-temporal relationship on unburned sites is probably a function of secession as these sites become dominated by long-lived woody perennials such as *Symphoricarpos orbiculatus, Rhus glabra, Rosa arkansana,* and *Ceanothus herbaceus.*

FIGURE 2. Spearman rank correlation of within-site heterogeneity (=1 − percent similarity) in species composition in a given year and proportional change in species composition from one year to the next on sites that are annually burned, burned every 2 yr, burned every 4 yr, or unburned.

Other disturbances (e.g., grazing) are likely to have an effect on heterogeneity opposite that of fire (Gessaman and McMahon 1984). Grazing by ungulates, for example, may increase heterogeneity in grasslands by decreasing dominance and increasing diversity (Belsky 1983, Bakker et al. 1984, Collins 1990). This result would be system and scale dependent (Milchunas et al. 1988, Milchunas and Lauenroth 1989). For instance, Glenn et al. (1992) found that grazed sites had lower within-community but higher among-community heterogeneity compared to burned sites. Highest regional heterogeneity occurred among sites with a combination of burning and grazing.

Currently, generalities regarding heterogeneity are limited because of the small number of studies that have quantified patterns of heterogeneity or measured the mechanisms that create and maintain heterogeneity in communities (Legendre and Fortin 1989, Kolasa and Rollo 1991). Perhaps the relationship between disturbance and heterogeneity reflects a complex response of the system to climatic variation and evolutionary history of disturbance as suggested in the models of Denslow (1980) and Milchunas et al. (1988). Simple predictive models of heterogeneity in response to increasing size of disturbance (e.g., Connell 1978, Kolasa and Rollo 1991) provide a necessary first step for a predictive approach to scale-related questions on patterns of community heterogeneity. Such models are often based on the tenuous assumption that heterogeneity within a disturbance is equal to heterogeneity in adjacent undisturbed areas. Predicting community heterogeneity in response to disturbance will depend on community composition and productivity, evolutionary history, and the type and frequency of disturbance. As Kolasa and Rollo (1991) noted, there is indeed a heterogeneity of heterogeneity. A focus on variation, however, in addition to averages, will help to increase our understanding of factors that affect patterns of spatial heterogeneity and temporal variability in plant communities.

Acknowledgments

I thank Lynn Francis, Susan Glenn, Deborah Goldberg, Bruce Hoagland, Jurik Kolasa, Dan Milchunas, Ernie Steinauer, and two anonymous reviewers for many helpful comments on previous drafts of the manuscript. Earlier data sets were collected by Lloyd Hulbert, Ross Sherwood, Marc Abrams, David Gibson, and Gary Merrill. Data from Konza Prairie Research Natural Area were collected as part of the Konza Prairie LTER program (NSF grants DEB-8012166 and BSR-8514327), Division of Biology, Kansas State University. Data and supporting documentation are stored in data sets PVC01 and PVC02 in the Konza Prairie Research Natural Area data bank. Additional support for data analysis and manuscript preparation was provided by NSF grant BSR-9007450 to Scott Collins and Susan Glenn.

References

Bakker, J. P., J. de Leeuw, and S. E. van Wieren. 1984. Micro-patterns in grassland vegetation created and sustained by sheep-grazing. Vegetatio 55:153–161.

Belsky, A. J. 1983. Small-scale pattern in grassland communities in the Serengeti National Park, Tanzania. Vegetatio 55:141–151.

Chaneton, E. J., and J. M. Facelli. 1991. Disturbance effects on plant community diversity: spatial scales and dominance hierarchies. Vegetatio 93:143–155.

Collins, S. L. 1987. Interaction of disturbances in tallgrass prairie: a field experiment. Ecology 68:1243–1250.

_____ . 1989. Experimental analysis of patch dynamics and community heterogeneity in tallgrass prairie. Vegetatio 85:57–66.

_____ . 1990. Patterns of community structure during succession in tallgrass prairie. Bulletin of the Torrey Botanical Club 117:397–408.

Collins, S. L., and S. M. Glenn. 1990. A hierarchical analysis of species abundance patterns in grassland vegetation. American Naturalist 135:633–648.

Collins, S. L., and S. M. Glenn. 1991. Importance of spatial and temporal variation in species regional abundance and distribution. Ecology 72:654–664.

Connell, J. H. 1978. Diversity in tropical rain forests and coral reefs. Science 199:1302–1310.

Delcourt, H. R., P. A. Delcourt, and T. Webb, III. 1983. Dynamic plant ecology: the spectrum of vegetational change in space and time. Quaternary Science Review 1:153–175.

Denslow, J. S. 1980. Patterns of plant species diversity during succession under different disturbance regimes. Oecologia (Berlin) 46:18–21.

Gessaman, J. A., and J. A. MacMahon. 1984. Mammals in ecosystems: their effects on the composition and production of vegetation. Acta Zoologica Fennici 172:11–18.

Gibson, D. J., and L. C. Hulbert. 1987. Effects of fire, topography and year-to-year climatic variation on species composition in tallgrass prairie. Vegetatio 72:175–185.

Glenn, S. M., and S. L. Collins. 1992. Effects of scale and disturbance on rates of immigration and extinction of species in prairies. Oikos 63:273–280.

Glenn, S. M., S. L. Collins, and D. J. Gibson. 1992. Disturbances in tallgrass prairie: local versus regional effects on community heterogeneity. Landscape Ecology, *in press*.

Greig-Smith, P. 1979. Pattern in vegetation. Journal of Ecology 67:755–780.

Hulbert, L. C. 1985. History and use of Konza Prairie Research Natural Area. Prairie Scout 5:63–93.

Inouye, R. S., N. J. Huntly, D. Tilman, J. R. Tester, M. Stillwell, and C. Zinnel. 1987. Old-field succession on a Minnesota sand plain. Ecology 68:12–26.

Knapp, A. K. 1985. Effect of fire and drought on the ecophysiology of *Andropogon gerardii* and *Panicum virgatum* in a tallgrass prairie. Ecology 66:1309–1320.

Kolasa, J., and C. D. Rollo. 1991. Introduction: the heterogeneity of heterogeneity: a glossary. Pages 1–23 *in* J. Kolasa and S. T. A. Pickett, editors. Ecological heterogeneity. Springer-Verlag, New York, New York, USA.

Legendre, P., and M.-J. Fortin. 1989. Spatial pattern and ecological analysis. Vegetatio 80: 107–138.

Milchunas, D. G., and W. K. Lauenroth. 1989. Three-dimensional distribution of plant biomass in relation to grazing and topography in the shortgrass steppe. Oikos 55:82–86.

Milchunas, D. G., O. E. Sala, and W. K. Lauenroth. 1988. A generalized model of the effects of grazing by large herbivores on grassland community structure. American Naturalist 132:87–106.

Neave, H. R., and P. L. Worthington. 1988. Distribution-free tests. Unwin Hyman, London, England.

Palmer, M. W. 1988. Fractal geometry: a tool for describing spatial patterns of plant communities. Vegetatio 75:91–102.

Pielou, E. C. 1960. A single mechanism to account for regular, random, and aggregated populations. Journal of Ecology 48:575–584.

Scheiner, S. M. 1990. Affinity analysis: effects of sampling. Vegetatio 86:175–181.

Seastedt, T. R., and R. A. Ramundo. 1990. The influence of fire on belowground processes of tallgrass prairie. Pages 99–117 *in* S. L. Collins and L. L. Wallace, editors. Fire in North American tallgrass prairies. University of Oklahoma Press, Norman, Oklahoma, USA.

Tester, J. R. 1989. Effects of fire frequency on oak savanna in east-central Minnesota. Bulletin of the Torrey Botanical Club 116:134–144.

Tilman, D. 1982. Plant strategies and the dynamics and structure of plant communities. Princeton University Press, Princeton, New Jersey, USA.

Turner, M. G., V. H. Dale, and R. H. Gardner. 1989. Predicting across scales: theory development and testing. Landscape Ecology 3:245–252.

Urban, D. L., R. V. O'Neill, and H. H. Shugart. 1987. Landscape ecology. BioScience 37:119–127.

Whittaker, R. H. 1975. Communities and ecosystems. Second edition. MacMillan, New York, New York, USA.

Whittaker, R. H., and S. A. Levin. 1977. The role of mosaic phenomena in natural communities. Theoretical Population Biology 12:117–139.

Wilson, S. D., and J. M. Shay. 1990. Competition, fire, and nutrients in a mixed-grass prairie. Ecology 71:1959–1967.

12
Management Practices in Tallgrass Prairie: Large- and Small-Scale Experimental Effects on Species Composition

David J. Gibson, T. R. Seastedt, and John M. Briggs

Introduction

Many studies from grasslands have reported how differing management techniques affect production levels and species composition (e.g., Ehrenreich & Aikman 1963; Wells 1980; Parr & Way 1988). In most studies the main emphasis has been on a single treatment (e.g., mowing, grazing or burning) under either highly controlled small-scale, experimental conditions (Hover & Bragg 1981; Collins 1987; Cox 1988) or less rigorous large-scale descriptive field studies (e.g., Abrams & Hulbert 1987; Gibson & Hulbert 1987). There are inherent strengths and weaknesses to both these approaches. Experimental studies, usually carried out at only one site or in small plots, may reflect local conditions; conversely, large-scale field observations usually lack statistical rigour (Hurlbert 1984) and treatment effects may be obscured by large-scale landscape heterogeneity (e.g. Gibson 1988a). Recently, there has been a trend towards multiscale analyses incorporating both types of approach (e.g. Briggs, Seastedt & Gibson 1989; Collins & Glenn 1990; Glenn, Collins & Gibson 1992). This approach recognizes that many ecological phenomena, such as disturbance events (Pickett *et al.* 1989), are scale-dependent (Allen & Starr 1982).

Multiscale observations or experiments are lacking for many applied ecological phenomena. Here, such a multiscale approach is taken by studying the effects of soil type/fertilizer, burning and mowing on species composition within both large- (*c.* 10 000 m²) and small-scale (100 m²) experimental plots. The working null hypothesis was that any response of the plant community to prairie management techniques would be observed at both levels of analysis, i.e. the response to the treatment should be independent of the spatial scale at which it is measured.

Materials and Methods

Study Site

The study was conducted on Konza Prairie Research Natural Area (NPRNA), Manhattan, Kansas, USA. Konza prairie is a 3487 ha native tallgrass prairie site in the Flint Hills region of north-eastern Kansas and consists of several soil types located along a topoedaphic gradient from freely drained uplands to moist lowlands. Dominant species include the warm-season (C_4) grasses *Andropogon gerardii*, *A. scoparius* and *Sorghastrum nutans,* and various forbs including *Aster ericoides, Ambrosia psilostachya* and *Artemisia ludoviciana*. Nomenclature follows the Great Plains Flora Association (1986).

The Flint Hills region is one of the few unploughed regions of surviving tallgrass prairie (Reichman 1987). Steep slopes and rocky terrain precluded ploughing and the area has been traditionally grazed by cattle or cut for hay (Aldous 1935). Management practices are aimed at maximizing the production and abundance of palatable and nutritious native species for cattle grazing (Aldous 1935; Owensby & Anderson 1967; Owensby & Smith 1979). Prescribed burning is necessary to maintain high production levels and keep out woody species (Aldous 1934; Hulbert 1973, Bragg & Hulbert 1976; Adams, Anderson & Collins 1982). A combination of burning and fertilizer can offer the best pastures in terms of production and range quality (Owensby & Smith 1979).

Large-scale plots were located along the southern edge of the Geary County portion of Konza Prairie. The small-scale plots were located 3 km to the north in a lowland Irwin soil area immediately to the west of the KPRNA headquarters. Both areas were ungrazed throughout the duration of the study.

Large-Scale Plots

Burning and mowing were initiated in 27 soil-treatment combinations in 1972 and 1974 respectively, soon after Konza Prairie was first acquired in 1971 (Gibson 1989). Prior to that time the area had been burned approximately every other year and grazed by cattle for about 100 years since settlement. Forty-five large-scale plots, c. 200×500 m ($10\,000$ m^2) each, were established on one of five soil types subject to April burning at 1-, 2- and 4-year intervals, annual burning at three seasons, no burning or mowing at three seasons: 27 soil-treatment combinations in all (Table 1). Burning was conducted in March, April or November. Mowing occurred in March, July or November, with the hay either left or removed (baled) depending on the treatment. The most common soils are the Florence (clayey-skeletal, montmorillonitic, mesic, Udic Argiustolls) and Tully (fine, mixed, mesic Pachic Argiustolls) Series and the majority of burning and mowing treatments were located on these. Tully soils have a significantly higher pH (6.3), nitrate (0.5 μg g^{-1}), and total nitrogen content (329.1 μg g^{-1}) than Florence soils (pH = 6.2, nitrate = 0.3 μg g^{-1}, total nitrogen = 370.7 μg g^{-1}) (S. L. Collins & J. M. Briggs, unpublished). Other

TABLE 1. Summary of 27 soil-treatment combinations in large-scale studies to determine the effects of fire, mowing and soil type in tallgrass prairie vegetation. Values indicate the number of replication sample sites per soil-treatment combination

Treatment combinations	Soil Series				
	Tully	Irwin	Sogn	Florence	Benfield
Annual burn					
March	2			1	
Late April	6*	1		3	
November	2			1	
2-year cycle	3			3	
4-year cycle	2			2	
Unburned	2	2	1	3	1
Mown & baled:					
March				1	
Mown & left:					
July	1	1	1		
Mown & baled:					
July	2	1		1	
Mown & left:					
November	1				
Mown & baled:					
November		1		1	

*Includes one replicate plot each of burned with and against the prevailing wind.

soil series sampled were Sogn (loamy, mixed, mesic Lithic Haplustolls), Benfield (fine, mixed, mesic Udic Argiustolls) and Irwin (fine, mixed, mesic Pachic Argiustolls); all are chernozems classified within the Mollisol order (Jantz et al. 1975). Nomenclature for soils follows the Agency for International Development, United States Department of Agriculture, Soil Conservation Service (Soil Survey Staff 1990). These soil-treatment combinations are representative of grazing management regimes in the tallgrass prairie (Aldous 1935). The number of replicate soil-treatments varied from one to five (Table 1) as a function of the availability of suitable topoedaphic landscape units and is thus necessarily an unbalanced design.

Percentage cover of all vascular plants was recorded in 1983 using a modified Daubenmire (1959) scale (Abrams & Hulbert 1987) in 20 circular plots of 10 m^2 evenly distributed within each of the 45 plots. Each plot was sampled in late spring, mid-summer and late summer, with the maximum cover value attained in any season by each species within a plot being retained for subsequent analysis.

Small-Scale Plots

Sixty-four plots of 100 m^2 were established in 1986 to study more closely the effects of burning, mowing and soil fertility. The plots were arranged according to a split-plot design with groups of 16 plots arranged into four blocks. Half of the plots within each block were burned annually in early May; the rest were left unburned. Burning intensity and duration was the same as in the large-scale plots. Within each block, eight of the plots were mowed with the hay being removed; the rest were left unmowed. The plots were mowed twice a year (early and mid-summer) in 1986 and 1987, and once a year (late June) thereafter. Within a block, one of four levels of fertilizer treatment was applied to individual plots; N, 10 g m^{-2} nitrogen in the form of ammonium nitrate; P, 1 g m^{-2} phosphorus as superphosphate (P_2O_5); N + P; and an unfertilized control. By 1989, these treatments had led to a significant ($P < 0.05$) increase in soil nitrate and ammonium levels in fertilized plots compared with controls (nitrate = 2.5 μg g^{-1} in fertilized plots and 0.2 μg g^{-1} in control plots; ammonium = 4.7 μg g^{-1} and 3.3 μg g^{-1}, respectively. Ammonium levels were also significantly lower in burned (3.6 μg g^{-1}) compared with unburned (4.4 μg g^{-1}) plots. In July 1989, species cover of all vascular plants was estimated as for the large-scale plots in a 10-m^2 circular plot centred in each treatment plot.

Data Analysis

Canopy coverage data for the large-scale plots were analyzed using TWINSPAN (Hill 1979), a polythetic divisive classification technique, to obtain an objective classification of the 45 large-scale plots (representing 27 soil-treatment combinations). This procedure is frequently used to classify plant communities in order to provide an ecological understanding of management practices (see for example Fox & Murphy 1990).

Data from the small-scale experimental plots were analyzed using ANOVA according to a split-plot design. The least significant difference (LSD) pairwise comparison test was used to compare treatment means following any significant ($P < 0.05$) ANOVA. Canopy coverage data were arcsin transformed, and species richness data were square root transformed prior to analysis to reduce heteroscedasity (Sokal & Rohlf 1981).

Results

Large-Scale Plots

TWINSPAN classified the large-scale plots based on species abundance (Fig. 1). After three levels of division, the large-scale plots were assigned to groups with 2- to 14-member units. The first and major division was on the basis of the soil subgroups; plots on lowland Tully soil series were separated from plots on upland Florence soils. The four Irwin soil plots and two Sogn soil plots were grouped with the Tully soils (Irwin and Tully soils being both Pachic Argiustolls; Sogn a Lithic Haplustoll) and the Benfield soil plot grouped with the Florence soils (the latter two are both Udic Argiustolls). The four indicator species associated with this first division were all linked with the upland soils. The second division separated plots mainly on the basis of burning regime (Fig. 1) with annually burned and plots burned every 2 or 4 years being separated from unburned plots. This separation was most evident on the Florence and Benfield soils. In both cases *Poa pratensis* was an indicator species reflecting higher cover in unburned situations. A separation of plots according to mowing treatment was only apparent at the third level of division. On Tully, Irwin and Sogn soils, plots mowed in July were separated from unburned, 4-year-burned and November-mowed plots on the basis of higher cover values for *Bromus inermis* and *Oxalis stricta*.

Cover and richness of plant life-form groups differed between the TWINSPAN vegetation types (which were based only on cover of individual species) (Table 2). Although the primary divi-

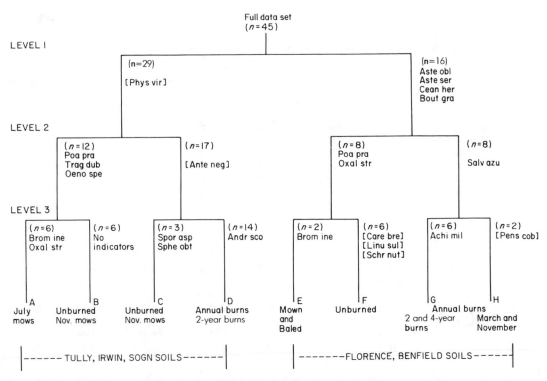

FIGURE 1. Results of TWINSPAN analysis of large-scale plot × species data, expressed as a dendrogram, showing hierarchical subdivisions of the data set to give eight end-groups (A . . . H) at level 3 of the division process. Sample groups distinguished at each level are shown (n = number of sample plots per group), together with indicator pseudospecies for each division. Where no indicators were distinguished by the program, preferential pseudospecies for the group are given in square brackets (where frequency in the groups exceeded 50%). Species codes: Achi mil, *Achillea millefolium*; Andr sco, *Andropogon scoparius*; Ante neg, *Antennaria neglecta*: Aste obl, *Aster oblongifolius*; Aste ser, *Aster sericeus*; Bout gra, *Bouteloua gracilis*: Brom ine, *Bromus inermis*; Care bre, *Carex brevior*; Cean her, *Ceanothus herbaceous*; Linu sul, *Linum sulcatum*; Oeno spe, *Oenothera speciosa*; Oxal str, *Oxalis stricta*; Pens cob, *Penstemon cobaea*; Phys vir, *Physalis virginiana*; Poa pra, *Poa pratensis*; Salv azu, *salvia azurea*; Schr nut, *Schrankia nuttallii*; Spor asp, *Sporobolus asper*; Sphe obt, *Sphenopholis obtusata*; Trag db, *Tragopogon dubius*.

sion of the plots into two initial groups, shown in the upper portion of the TWINSPAN dendrogram, indicates a principal gradient in species composition based on soil type, there were no differences between the cover and richness of plants in life-form groups with respect to soil type. The cover of C_4 grasses did not differ between vegetation groups. However, the cover of C_3 grasses was lower in groups composed of annually burned plots. In addition, the cover of woody species, the number of woody and exotic species, and diversity were also reduced in annually burned plots. The number of native species was significantly lower in annually burned plots on lowland Tully soils compared

with unburned or July-mown plots. This pattern was not observed on the Florence/Benfield soils. Mown plots had a higher cover of C_3 grasses, annual grasses, and annual/biennial forbs (herbaceous non-grasses), and a greater number of annual grasses and exotic species.

Overall, analysis of the large-scale plots identified soil type as the principal factor allowing discrimination of vegetation types. Burning followed by mowing regimes were of secondary and tertiary importance, respectively. Soil type, however, did not determine the abundance and richness of species within life-form groups.

TABLE 2. Mean cover and richness of plant life-forms from large-scale plots units classified according to TWINSPAN vegetation groups. Mean values followed by the same letter in a row are not significantly different ($P < 0.05$) (least significant difference test)

	Twinspan vegetation group						
	A	B	C	D	E	F	G/H
Cover							
C_4 perennial grasses	180.7[a]	164.4[a]	176.6[a]	199.6[a]	193.0[a]	164.6[a]	191.0[a]
C_3 perennial grasses	88.2[a]	61.0[a]	16.6[b]	1.2[c]	66.5[a]	33.2[b]	1.5[c]
Cyperaceae & Juncaceae	1.4[c]	11.9[abc]	30.2[a]	19.3[ab]	8.7[abc]	2.9[bc]	2.2[c]
Annual grasses	21.5[a]	<0.1[b]	<0.1[b]	0.0[b]	13.5[a]	1.2[b]	<0.1[b]
Perennial forbs	16.3[a]	25.5[a]	32.0[a]	28.4[a]	28.5[a]	37.3[a]	30.9[a]
Annual/biennial forbs	14.7[a]	0.3[b]	0.2[b]	3.0[b]	12.8[a]	1.1[ab]	0.4[b]
Woody plants	0.18[cd]	5.1[bcd]	15.4[a]	6.4[bc]	1.5[bc]	4.3[bcd]	7.2[abcd]
Total cover	265.4[a]	246.7[a]	281.0[a]	257.9[a]	350.5[a]	244.7[a]	233.1[a]
Richness							
C_4 perennial grasses	7.0[c]	6.8[c]	7.3[bc]	6.6[c]	9.0[abc]	8.7[bc]	10.6[a]
C_3 perennial grasses	5.2[a]	4.8[a]	4.3[a]	2.7[b]	5.5[a]	4.5[a]	3.9[a]
Cyperaceae & Juncaceae	3.0[a]	3.0[a]	3.0[a]	3.7[a]	2.0[a]	2.7[a]	3.6[a]
Annual grasses	1.0[a]	0.2[bc]	0.3[bc]	0.0[c]	1.0[c]	0.5[b]	0.1[c]
Perennial forbs	27.7[ab]	26.5[b]	28.7[ab]	22.6[c]	33.0[a]	30.2[ab]	31.4[a]
Annual/biennial forbs	7.3[abc]	4.5[bc]	4.7[bc]	3.1[c]	8.0[abc]	9.2[a]	3.3[c]
Woody plants	1.3[c]	2.5[bc]	2.0[bc]	1.5[c]	5.4[a]	2.5[bc]	1.8[bc]
Native species	46.3[b]	46.0[b]	48.7[ab]	39.9[c]	54.5[ab]	56.0[a]	53.4[a]
Exotic species	6.2[a]	2.3[bc]	1.7[bc]	1.0[c]	6.0[a]	2.5[b]	1.3[c]
Total species	52.8[ab]	48.3[b]	50.3[ab]	40.8[c]	60.0[a]	58.5[ab]	54.6[ab]

Small-Scale Plots

The cover of species life-form groups and individual species responded to several of the treatments. Either alone as a main effect or in an interaction C_4 grasses attained significantly higher cover in response to burning or mowing, but not both (Table 3), and decreased cover in response to nitrogen fertilization when unburned (Table 4). The cover and richness of the woody species (i.e., *Amorpha canescens*, *Rosa arkansana* and *Symphoricarpus orbiculatus*) (Table 4) were reduced by mowing and burning, and by nitrogen fertilizer when unmowed. The cover of forbs was significantly lower on plots that were burned, except when mown (Table 3). Forb species richness, in contrast, was lowest in unmown and unburned plots.

Sixty-two vascular species were recorded in the small-scale plots, with 28 frequent enough (present in >10% of the 64 plots) to warrant statistical analysis. Of these, 12 showed a significant response (ANOVA, $P < 0.05$) to one or more of the experimental treatments. None of these significant treatment effects reflected a

TABLE 3. Small-scale experimental plots: mean cover and species richness of species and species groups showing a significant interaction of mowing (M = mown, U = unmown) and burning (B = burned, U = unburned) treatments. Mean values are significantly different ($P < 0.05$) when followed by a different upper case superscript within a mowing treatment or different lower case superscript within a burning treatment (least significant difference test)

		Mowing	
	Burn	U	M
Cover			
Andropogon scoparius	U	16.94[Bb]	34.41[Aa]
	B	29.09[Aa]	18.97[Bb]
Antennaria neglecta	U	0.06[Ab]	3.00[Aa]
	B	1.03[Aa]	1.19[Ba]
Asclepinas viridis	U	0.97[Ba]	1.03[Aa]
	B	2.84[Aa]	0.19[Ab]
C_4 grasses	U	67.70[Bb]	93.50[Aa]
	B	111.40[Aa]	88.40[Ab]
Forbs	U	102.10[Aa]	15.50[Ab]
	B	68.70[Ba]	23.70[Ab]
Richness			
Forbs	U	6.50[Bb]	7.50[Aa]
	B	7.90[Aa]	7.60[Aa]
C_4 grasses	U	4.13[Bb]	5.70[Aa]
	B	4.80[Ab]	5.50[Aa]

TABLE 4. Small-scale experimental plots: mean cover and species richness of species and species groups showing a significant interaction of mowing (M = mown, U = unmown) or burning (B = burned, U = unburned) treatments with fertilizer (C = control, P = phosphorus, N = nitrogen, N + P = nitrogen & phosphorus) treatment. Mean values are significantly different (at $P < 0.06$ for *B. curtipendula* otherwise at $P < 0.05$) when followed by different upper case superscripts in rows within a mowing treatment or a different lower case superscript in columns within a fertilizer treatment (least significant difference test)

		Fertilizer			
		C	P	N	N + P
Covering	Mowing				
Andropogon gerardii	U	35.31[Aa]	30.69[Aa]	14.19[Ba]	15.44[Bb]
	M	22.94[ABb]	21.88[ABc]	15.19[Ba]	24.75[Ac]
Panicum virgatum	U	16.00[Ba]	14.90[Ba]	33.50[Aa]	32.80[A]
	M	2.00[Ab]	2.20[Ab]	2.30[Ab]	3.94[Ab]
Achillea millefolium	U	0.06[Ab]	0.06[Ab]	0.00[Aa]	0.00[Aa]
	M	0.37[Aa]	0.31[Aa]	0.00[Ba]	0.00[Ba]
Solidago canadensis	U	8.44[Ca]	11.25[BCa]	15.37[Ba]	33.19[Aa]
	M	0.00[Ab]	0.00[Ab]	0.00[Ab]	0.00[Ab]
Bouteloua curtipendula	U	3.94[Aa]	5.69[Aa]	13.43[Aa]	0.00[Bb]
	M	9.56[Aa]	9.44[Aa]	8.56[Aa]	9.44[Aa]
C_4 grasses	U	101.90[Aa]	93.10[Aa]	88.80[Aa]	74.30[Bb]
	M	84.60[Aa]	84.30[Aa]	89.80[Aa]	105.10[Aa]
Woody species	U	5.61[Aa]	3.71[Aa]	0.10[Ba]	0.10[Ba]
	M	0.00[Ab]	0.10[Ab]	0.00[Aa]	0.00[Aa]
	Burning				
Solidago canadensis	U	6.60[Ba]	9.41[Ba]	9.71[Ba]	29.40[Aa]
	B	5.30[Aa]	5.30[Aa]	7.70[Aa]	6.90[Ab]
C_4 grasses	U	89.70[Aa]	84.20[Aa]	78.70[ABa]	69.70[Bb]
	B	96.80[Aa]	93.30[Aa]	100.00[Aa]	109.60[Aa]
Woody species	U	5.60[Aa]	3.80[Aa]	0.00[Ab]	0.10[Ab]
	B	0.00[Ba]	0.00[Ba]	0.10[Aa]	0.00[Ba]
Richness					
Woody species	U	0.40[Aa]	0.40[Aa]	0.00[Ab]	0.00[Aa]
	B	0.00[Ba]	0.00[Ba]	0.10[Aa]	0.00[Ba]

main effect response to burning. Overall, more species responded to mowing or fertilizer, either alone as a main effect (mowing = 6 species, fertilizer = 6 species) or in an interaction (3 species), than with burning (burning alone = 0 species, burning × mowing = 1 species, burning × fertilizer = 0 species, 3 way interaction = 1 species).

Several species responded to the fertilizer treatment with reduced cover values following the application of nitrogen either alone or in combination with phosphorus (Table 5). Other species showed this response only in unmown plots (e.g., *Andropogon gerardii,* Table 4). *Panicum virgatum,* which was reduced in cover on mown plots, was the only species to show an increase in cover on unmown plots fertilized by nitrogen (Table 4). Species richness, of both all species and forbs, was reduced significantly in plots which received nitrogen (Table 5).

Mowing acted to reduce total cover and the cover of individual species (*Ambrosia psilostachya* and *Aster ericoides*) (Table 6). Several species were reduced (e.g., *Andropogon gerardii, Panicum virgatum*), while *Solidago canadensis* was eliminated entirely, when mown and fertilized (Table 4). *Achillea millefolium* had its highest cover values in plots that were mown but not fertilized with nitrogen. Burning also acted as a moderator for fertilizer responses (Table 4); *Solidago canadensis* had higher cover values in unburned plots, especially plots treated with both nitrogen and phosphorus. The C_4 grasses as a group had generally lower cover in unburned plots in the presence of nitrogen.

TABLE 5. Small-scale experimental plots: mean cover and species richness of species and species groups showing a significant fertilizer effect. Mean values within a row are significantly different ($P < 0.05$) when followed by a different superscript (least significant difference test)

	Fertilizer treatment			
	Control	Phosphorus	Nitrogen	N & P
Cover				
Antennaria				
neglecta	2.19[A]	3.00[A]	0.06[B]	0.03[B]
Carex brevior	0.06[B]	0.13[AB]	0.00[B]	0.22[A]
Dalea candida	3.16[A]	1.19[AB]	0.09[B]	0.19[B]
Woody spcies	2.81[A]	1.91[AB]	0.03[B]	0.03[B]
Richness				
Forbs	9.38[A]	7.94[A]	6.19[B]	6.06[B]
All species	15.69[A]	14.19[A]	12.06[B]	11.75[B]

Discussion

Applicability to Management Practices

Both the large- and small-scale plots indicated that a distinct assemblage of species and species life-forms results from each combination of management practices. As in other studies (e.g. Gibson & Hulbert 1987), soil type played a major role in determining the nature of the plant communities (Fig. 1). This was principally a topographic effect, as the separation of upland (Florence) from primarily deep lowland soils (Tully) indicated. Within the upland/lowland soil contrast, the species groups were identified according to burning and mowing regime. Mowing and burning were similar treatments in that grasses, especially warm-season (C_4) grasses, were favoured at the expense of woody species and forbs.

The cover of eight of 28 common prairie species, the cover of C_4 grasses and woody species, and the richness of forbs and total species richness responded to fertilizer in the small-scale plots. Competition for limiting soil nutrients is ascribed a major role in recent models of plant community development (Tilman 1982, 1988). However, in the small-scale plots the fertilizer only elicited an increase in cover for two species, *Panicum virgatum* and *Solidago canadensis,* and then only when un-

TABLE 6. Small-scale experimental plots: mean cover of species and total cover showing only a significant mowing effect. Mean values within a row are significantly different ($P < 0.05$ except *A. psilostachya* where $P < 0.07$)

	Mowing treatment	
Cover	Unmown	Mown
Ambrosia psilostachya	2.53	0.25
Andropogon bladhii	0.94	11.72
Aster ericoides	27.31	0.30
Total cover	178.67	111.52

mown (Table 4). In response to the fertilizer, other species (*Andropogon gerardii, Achillea millefolium* and *Bouteloua curtipendula*), C_4 grasses and woody species decreased in cover, but only when in combination with either mowing or burning. The decrease in species richness in response to nitrogen fertilizer may well be due to a disproportionate increase in mortality of smaller or shade-intolerant species in more productive low-light sites (Kirchner 1977; Gibson 1988b; Goldberg & Miller 1990; Carson & Pickett 1990). A decrease in species richness in tallgrass prairies has been linked to litter build-up following a disruption of the natural disturbance regime (Collins 1978; Gibson & Hulbert 1987; Collins & Gibson 1990). Nitrogen is a major limiting factor in tallgrass prairie, especially under frequent fire conditions (Seastedt, Briggs & Gibson 1991).

Scale of Investigation

This study allowed a comparison between responses to treatments at two spatial scales; i.e. 100 m^2 and 10 000 m^2 small- and large-scale plots. The burning and mowing treatments imposed on the small-scale plots and the large-scale plots had been in place for different periods of time: 3 years on the small-scale plots and 11 (burning) and 9 (mowing) years, respectively, on the large-scale plots. However, system response to fire at least has been shown to be rapid (Gibson 1988a) and this difference is not considered to be a problem in interpreting the results of this study. A comparison of results from different spatial scales is relevant especially in the light of recent interpretations of grassland community structure. Collins & Glenn (1990) propose that grasslands show self-similarity in response to similar disturbance (i.e. show fractal properties). If so, then similar responses to the treatments should have been evident at both of the spatial scales tested in this study.

In the large-scale plots the primary classification of treatment plots distinguished the vegetation of upland and lowland soils. The second level of resolution was on the basis of burning frequency, particularly between unburned areas and annually burned areas. The resolution of mowing effects was not discernible at this level of analysis until after soil type and burning had been considered (Fig. 1). This indicates that the effects of burning and mowing on grassland plant communities are not resolvable at the same hierarchical scale of analysis when considered at this larger spacial scale. In contrast, the small-scale plots failed to indicate a significant effect of burning on the cover of common species. This is somewhat surprising given the ubiquity of species responses to fire in grassland (e.g., Adams, Anderson & Collins 1982; Abrams & Hulbert 1987; Biondini, Steuter & Grygiel 1989; but see also Wilson & Shay 1990) and the production responses from this experiment (Seastedt, Briggs & Gibson 1991). Indeed, mowing and fertilizer assumed a more important role in determining species cover responses than did fire. Wilson & Shay (1990) similarly found fire to be less important than nutrients in controlling species composition in mixed-grass prairie. Fertilizer can be considered a somewhat similar treatment to the various soil types investigated in the large-scale plots; although the latter also incorporate important soil moisture differences. The relative importance of mowing and burning is, therefore, reversed with respect to the large and small plots.

Thus, within a soil type (small-scale plots), mowing effects are greater than burning, but between soil types (large-scale plots) burning effects are greater than mowing. This inconsistency in the resolution of common management treatments at different scales is contrary to Collins & Glenn's (1990) fractal model for grasslands, indicating a need to consider scale-dependent causal mechanisms. Small plots subject to different but adjacent treatments offer the potential for dispersal from one plot to another (e.g., Peterson, Kaufman & Kaufman 1985; Evans 1991). However, between-treatment dispersal distances would be greater in large plots, thus increasing the independence of individual treatments. Indeed, consistency at various spatial scales under experimental conditions should only occur if the underlying mechanisms (e.g. soil depth, seed dispersal) responsible for the variables of interest remain constant over those scales; a condition that is unlikely to be met under field conditions (Krummel et al. 1987).

Acknowledgments

The late Lloyd C. Hulbert initiated the mowing and burning treatments, and collected the data from Old Konza. R.A. Ramundo supervised the treatments on the small-scale plots, and C. Gray of the KSU Soil Testing Laboratory conducted soil analyses. W.P. Carson, S.L. Collins, and S.M. Glenn provided suggestions that improved the original manuscript. C.M. Bundrick provided valuable statistical advice, and A. Goldin, M. Ransom and C. Rice provided advice on soil taxonomy. This research was funded by NSF grant BSR 85–14327 for Long-Term Ecological Research to Kansas State University. Data and documentation are stored in the Konza Prairie Research Natural Area Data Bank. Konza Prairie is owned by the Nature Conservancy and administered by Kansas State University.

References

Abrams, M.D. & Hulbert, L.C. (1987) Effect of topographic position and fire on species in tallgrass prairie in northeast Kansas. *American Midland Naturalist* 117, 442–445.

Adams, D.E., Anderson, R.C. & Collins, S.L. (1982) Differential response of woody and herbaceous species to summer and winter burning in an Oklahoma grassland. *Southwestern Naturalist, 27,* 55–61.

Aldous, A.E. (1934) *Effect of burning on Kansas bluestem pastures.* Kansas State College of Agriculture and Applied Science, Agricultural Experiment Station, Technical Bulletin 38.

Aldous, A.E. (1935) *Management of Kansas permanent pastures.* Kansas State College of Agriculture and Applied Science, Agricultural Experiment Station, Bulletin 272.

Allen, T.F.H. & Starr, T.B. (1982) *Hierarchy: Per-*

spectives for Ecological Complexity. University of Chicago Press, Chicago.

Biondini, M.E., Steuter, A.A. & Grygiel, C.E. (1989) Seasonal fire effects on the diversity patterns, spatial distribution and community structure of forbs in the Northern Mixed Prairie, USA. *Vegetatio* 85, 21–31.

Bragg, T.B. & Hulbert, L.C. (1976) Woody plant invasion of unburned Kansas bluestem prairie. *Journal of Range Management,* 29, 19–24.

Briggs, J.M., Seastedt, T.R. & Gibson, D.J. (1989) Comparative analysis of temporal and spatial variability in above-ground production in a deciduous forest and prairie. *Holartic Ecology,* 12, 130–136.

Carson, W.P. & Pickett, S.T.A. (1990) Role of resources and disturbance in the organization of an old-field plant community. *Ecology,* 71, 226–238.

Collins, S.L. (1987) Interaction of disturbances in tallgrass prairie: a field experiment. *Ecology,* 67, 87–94.

Collins, S.L. & Gibson, D.J. (1990) Effects of fire on community structure in tallgrass and mixed-grass prairie. *Effects of Fire on Tallgrass Prairie Ecosystems* (eds S.L. Collins & L.L. Wallace), pp. 81–98. University of Oklahoma Press, Norman, USA.

Collins, S.L. & Glenn, S.M. (1990) A hierarchical analysis of species' abundance patterns in grassland vegetation. *American Naturalist,* 135, 633–648.

Cox, J.R. (1988) Seasonal burning and mowing impacts on *Sporobolus wrightii* grasslands. *Journal of Range Management,* 41, 12–15.

Daubenmire, R.F. (1959) A canopy-coverage method of vegetational analysis. *Northwest Science,* 33, 43–66.

Ehrenreich, J.H. & Aikman, J.M. (1963) An ecological study of the effect of certain management practices on native prairie in Iowa. *Ecological Monographs,* 33, 113–130.

Evans, E.W. (1991) Experimental manipulation of herbivores in native tallgrass prairie: responses of aboveground anthropods. *American Midland Naturalist,* 125, 37–46.

Fox, A.M. & Murphy, K.J. (1990). The efficacy and ecological impacts of herbicide and cutting regimes on the submerged plant communities of four British rivers. II. A multivariate analysis of the effects of management regimes on macrophytic communities. *Journal of Applied Ecology,* 27, 541–548.

Gibson, D.J. (1988a) Regeneration and fluctuation of tallgrass prairie vegetation in response to burning frequency. *Bulletin of the Torrey Botanical Club,* 115, 1–12.

Gibson, D.J. (1988b) The maintenance of plant and soil heterogeneity in dune grassland. *Journal of Ecology,* 76, 497–508.

Gibson, D.J. (1989) Hulbert's study of factors effecting botanical composition of tallgrass prairie. *Prairie Pioneers: Ecology, History and Culture* (eds. T.B. Bragg & J. Stubbendieck), pp. 115–133. *Proceedings of the Eleventh North American Prairie Conference,* University of Nebraska Printing, Lincoln, USA.

Gibson, D.J. & Hulbert, L.C. (1987) Effects of fire, topography and year-to-year climatic variation on species composition in tallgrass prairie. *Vegetatio,* 72, 175–185.

Glenn, S., Collins, S.L. & Gibson, D.J. (1992) Disturbances in tallgrass prairie: local and regional effects on community heterogeneity. *Landscape Ecology* (in press).

Goldberg, D.E. & Miller, T.E. (1990) Effects of different resource additions on species diversity in an annual plant community. *Ecology* 71, 213–225.

Great Plains Flora Association (1986) *Flora of the Great Plains.* University Press of Kansas, Lawrence. 1392 p.

Hill, M.O. (1979) *TWINSPAN: A FORTRAN Program for arranging Multivariate Data in an Ordered Two-way Table by Classification of the Individuals and Attributes.* Department of Ecology and Systematics, Cornell University, Ithaca.

Hover, E.I. & Bragg, T.B. (1981) Effect of season of burning and mowing on an Eastern Nebraska Stipa-Andropogon prairie. *American Midland Naturalist,* 105, 13–18.

Hulbert, L.C. (1973) Management of Konza prairie to approximate pre-white-man fire influences. *Third Midwest Prairie Conference Proceedings* (ed. L.C. Hulbert). Kansas State University, Manhattan, Kansas.

Hurlbert, S.H. (1984) Pseudoreplication and the design of field experiments. *Ecological Monographs,* 54, 187–211.

Jantz, D.R., Harner, R.F., Rowland, H.T. & Gier, D.A. (1975) *Soil survey of Riley County and part of Geary County, Kansas.* United States Department of Agriculture Soil Conservation Service.

Kirchner, T.B. (1977) The effects of resource enrichment on the diversity of plants and arthropods in a shortgrass prairie. *Ecology,* 58, 1334–1344.

Krummel, J.R., Gardner, R.H., Sugihara, G., O'Neill, R.V. & Coleman, P.R. (1987) Landscape patterns in a disturbed environment. *Oikos,* 48, 321–324.

Owensby, C.E. & Anderson, K.L. (1967) Yield response to time of burning in the Kansas Flint Hills. *Journal of Range Management,* 20, 12–16.

Owensby, C.E. & Smith, E.F. (1979) Fertilizing and

burning Flint Hills bluestem. *Journal of Range Management,* **32,** 254–258.

Parr, T.W. & May, J.M. (1988) Management of roadside vegetation: the long-term effects of cutting. *Journal of Applied Ecology,* **25,** 1073–1087.

Peterson, S.K., Kaufman, G.A. & Kauffman, D.W. (1985) Habitat selection by small mammals of the tall-grass prairie: experimental patch choice. *Prairie Naturalist,* **17,** 65–70.

Pickett, S.T.A., Kolasa, J., Armesto, J.J. & Collins, S.L. (1989) The ecological concept of disturbance and its expression at various hierarchical levels. *Oikos,* **54,** 131–139.

Reichman, O.J. (1987) *Konza Prairie: a Tallgrass Natural History.* University of Kansas Press, Lawrence. 248 pages.

Seastedt, T.R., Briggs, J.M. & Gibson, D.J. (1991) Controls of nitrogen limitation in tallgrass prairie. *Oecologia,* **87,** 72–80.

Soil Survey Staff. (1990) *Keys to Soil Taxonomy,* 4th edn. SMSS Technical monograph No. 6. Blacksburg. 422 pages.

Sokol, R.R. & Rohlf, F.J. (1981) *Biometry,* 2nd edn. W.H. Freeman & Company, New York.

Tilman, D. (1982) *Resource Competition and Community Structure.* Princeton University Press, Princeton, NJ.

Tilman, D. (1988) *Dynamics and Structure of Plant Communities.* Princeton University Press, Princeton, NJ.

Wells, T.C.E. (1980) Management options for lowland grassland. *Amenity Grassland: An Ecological Perspective* (eds I.H. Rorison & R. Hunt), pp. 163–173. Wiley, Chichester.

Wilson, S.D. & Shay, J.M. (1990) Competition, fire, and nutrients in a mixed-grass prairie. *Ecology,* **71,** 1959–1967.

13

Fire History and Vegetation Dynamics of a *Chamaecyparis Thyoides* Wetland on Cape Cod, Massachusetts

Glenn Motzkin, William A. Patterson III, and Natalie E. R. Drake

Introduction

Recent attempts to conserve 'natural community diversity' have resulted in the creation of many nature reserves selected for their exemplary and often unusual vegetation associations. An understanding of the ecological and historical processes that have allowed these unusual associations to develop is required in order to identify appropriate conservation objectives for these natural areas, and to develop management strategies for achieving these objectives. Frequently, such an understanding requires investigations at a variety of temporal scales in order to relate recent trends to longer-term dynamics. This is particularly true in regions with a history of intensive human settlement, and for communities dominated by long-lived tree species.

In this study we investigate long- and short-term vegetation dynamics in an Atlantic white cedar (*Chamaecyparis thyoides*) wetland on Cape Cod, Massachusetts. Although many cedar wetlands are now preserved as unique natural communities (Whigham 1987), developing conservation and management objectives for these wetlands has proved to be difficult, because previous investigations (Little 1950; Hickman & Neuhauser 1977) that have focused solely on modern vegetation dynamics have

produced conflicting assertions about successional trends and life-history characteristics of the two tree species (Atlantic white cedar and red maple *Acer rubrum*) that dominate most cedar wetlands. Differences in land-use practices between presettlement Indian populations and postsettlement Europeans in eastern North America (Cronon 1983; Patterson & Sassaman 1988) suggest that it may not be possible to infer long-term processes from current vegetation patterns. Therefore we employed fine-resolution pollen analysis (Green 1983; Green & Dolman 1988) of peat sediments and age-structure analyses of existing trees to relate recent vegetation trends to processes occurring over longer time periods.

Atlantic White Cedar Wetlands in the North-Eastern United States

Throughout their distribution range in eastern North America, *Chamaecyparis thyoides* wetlands are uncommon, having decreased historically in areal extent and in biological diversity (Laderman 1989). Current estimates indicate that there are less than 5300 ha of *C. thyoides* wetlands remaining in the glaciated north-eastern United States (Motzkin 1991). Much of the historic decrease is attributable to human activity through conversion to agricultural, industrial, or commercial uses, and through selective logging of cedar. Flooding by beavers and sea level rise have been responsible for additional losses.

Natural autogenic successional processes, in which, in the absence of disturbance, *C. thyoides* is replaced by more shade-tolerant species, have been implicated in the loss of cedar wetlands by some investigators (Buell & Cain 1943; Little 1950). Little (1950) states that *C. thyoides* typically forms even-aged stands that establish after disturbances such as fire (especially under flooded conditions when seed reserves in the upper organic layers remain protected), windthrow, temporary flooding and salt-spray damage.

According to Little (1950) *C. thyoides* is less shade-tolerant than *Acer rubrum* which reproduces beneath established cedar canopies to form all-aged populations that may replace cedar when overstories break up with increasing stand age. Little's model of autogenic succession in *C. thyoides* wetlands raises concerns about the loss of cedar dominance in sites protected from disturbance and has led to silvicultural practices that emphasize cutting and the use of prescribed fire (Korstian & Brush 1931; Noyes 1939; Little 1950; Johnson 1980). In the North-east, small clearcuts within existing stands are recommended for regenerating even-aged patches of cedar (Little 1950; Zampella 1987; Roman, Good & Little, in press).

Hickman & Neuhauser (1977) present a different interpretation of successional trends in cedar wetlands of New Jersey, however. They report no maple or cedar regeneration beneath undisturbed cedar canopies and state that during establishment cedar may be more shade-tolerant than maple. Contrary to Little (1950), Hickman & Neuhauser (1977, p. 35) assert that cedar should not be considered "in any general way subclimax to maple." Resolution of this apparent conflict has important implications for the management of *C. thyoides* wetlands.

In this study we investigated successional trends in an old-growth stand at the Marconi Atlantic White Cedar Swamp (MAWCS), a 5-ha wetland at Cape Cod National Seashore. The MAWCS is currently protected from fire and timber cutting and is therefore potentially threatened by the successional trends identified

by Little (1950). This study addresses the following questions:

1. How have autogenic and allogenic processes influenced successional trends at the Marconi Atlantic White Cedar Swamp during the past ~1000 years.
2. How have disturbance frequency and type changed since European settlement?
3. Is there evidence that either cedar or maple continuously establish beneath existing overstorey canopies?
4. Is Atlantic white cedar likely to persist at the site in the foreseeable future?
5. What conservation and management objectives emerge from an understanding of the long-term dynamics of this system?

Study Site

The Marconi Atlantic White Cedar Swamp is located in South Wellfleet, Massachusetts (Fig. 1). It occupies a kettle depression in outwash sand and gravel deposits which support oak–

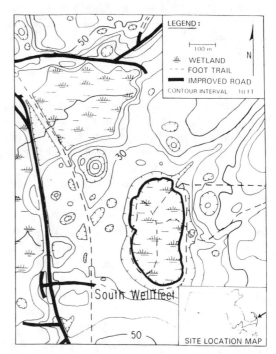

FIGURE 1. Location map for the Marconi Atlantic White Cedar Swamp (MAWCS), Cape Cod National Seashore, South Wellfleet, Massachusetts.

pitch pine (*Quercus* spp.–*Pinus rigida*) forests on the adjacent upland. Maximum depth of organic sediments in the Swamp, which has no surface inflow or outflow, is 7 m. The climate of the region is humid continental, with a maritime influence. Precipitation averages *c.* 103 cm year^{-1} (Patterson *et al.* 1985).

Historical records suggest that Europeans settled Wellfleet *c.* AD 1640 (Altpeter 1937). By the late seventeenth century, forests on outer Cape Cod had been so heavily exploited that several towns adopted ordinances prohibiting the cutting of wood (Altpeter 1937; McCaffrey 1973). Agricultural activity peaked in 1860, with *c.* 30% of the land area being used for crop and pasture (Altpeter 1937). Pitch pine that now occurs on the upland surrounding the Marconi Swamp established following agricultural abandonment between the late nineteenth and mid-twentieth centuries.

Data from Belling (1977) indicate that *C. thyoides* occurred in Massachusetts as early as 6800 BP, but its distribution and relative abundance since that time are uncertain. Cedar first arrived at the Marconi site *c.* 3000 years ago. Within 500–1000 years of its arrival, cedar dominated the site and continued to do so until *c.* 1000 years ago when it declined in Belling's pollen profiles, and pitch pine and red maple increased. A charcoal layer coincides with these changes in pollen suggesting that a fire swept through the area, resulting in an increase in pitch pine in the uplands and a decrease in the relative importance of cedar in the Swamp. Cedar pollen percentages increased again in the profile at approximately the time of European settlement (Belling 1977).

Atlantic white cedar in Massachusetts was much valued for wood products (Emerson 1850), and it is likely that there was at least some removal of cedar from the Marconi site in the eighteenth and nineteenth centuries. Cedar has been the dominant tree species at the Swamp since at least the mid-19th century, however, and much of the current stand is comprised of mature trees as old as 200 years. For the past 30 years, the Swamp has been protected except for limiting cutting to build a boardwalk for public access.

Only three overstorey species currently occur in the Marconi Swamp; cedar dominates, red maple is common and pitch pine is present as scattered individuals. Understorey species include *Vaccinium corymbosum, Rhododendron viscosum, Clethra alifolia, Ilex verticillata, I. glabra, Leucothoe racemosa* and *Gaylussacia frondoas. Sphagnum* spp. dominate the ground cover with *Maianthemum canadense, Woodwardia virginica* and *Osmunda cinnamomea* also present.

Methods

Vegetation Characterization

A 20-m × 20-m grid was established throughout the Swamp in 1988 and sampled for overstorey cover using variable-radius plots (Mueller-Dombois & Ellenberg 1974). All live and dead trees were tallied by species at each of 162 points on the grid. On 57 of the 400-m^2 grid cells (30% sample), the relevé method (Mueller-Dombois & Ellenberg 1974) was used to estimate percentage cover for all vascular plant species, with stems grouped in height classes defined according to Küchler's Physiognomic Classification system (Küchler 1967). Species were assigned cover scores according to the Braun–Blanquet scale (Mueller-Dombois & Ellenberg 1974). Cover values were used to calculate importance values, as follows. For each species on each relevé, Braun–Blanquet cover values were summed for all the strata within which the species occurred. The number of strata within which the species occurred was then subtracted from the summation, with the value 1 added to the total (see Clark & Patterson 1985). This procedure results in a minimum value of 1 assigned to a species that occurs in only one stratum with a cover value of 1 (single occurrence, minimal cover). Because of the nonlinear nature of the Braun–Blanquet cover class values, this procedure emphasizes the occurrences of species within strata (more than their cover) and increases the importance of species that occur in multiple strata relative to those that occur in a single stratum.

Patterns of structural variability within the Marconi Swamp were also evaluated through interpretation of 1984 colour-infrared aerial photographs (scale: 1:25 000). Nomenclature follows Seymour (1989).

Age-Structure Analysis

Ten sample plots were located in the three distinct stands identified by our vegetation characterization (see Results), and diameter at breast height (d.b.h.) and crown classification (dominant, codominant, intermediate or suppressed) were recorded for all stems >5 cm d.b.h. Increment cores for age determination were taken within 30 cm of the ground from all trees >5 cm d.b.h. Sample plots were 20 m × 20 m (five plots in stand I, two in stand II, and one in stand III), except in two plots in stand III which comprised dense stands of small stems where 10-m × 10-m plots were employed (Fig. 2).

Within each plot, height and basal diameter of all woody stems >3 cm tall but <5 cm d.b.h. were sampled in 10–15 randomly located 4-m² subplots. The small size of tree stems in

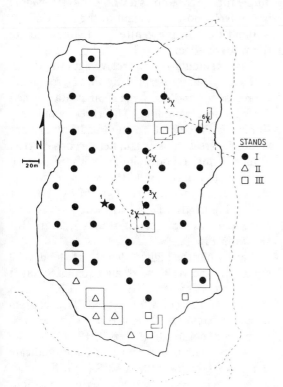

FIGURE 2. Sample locations in the MAWCS. Small circles, triangles and squares represent relevés categorized by stand (see Results and Fig. 3). Large, open polygons are intensive sample plots, and dashed lines represent trails. The sites from which the peat core (*) and surface samples (X) for pollen and charcoal analyses were extracted are also indicated.

the understorey subplots made coring for age determinations impractical. To determine the age distribution of stems >5 cm d.b.h., we cut, aged, and recorded the basal diameter of 55 maple and 24 cedar stems. Regression analyses revealed significant correlations between age and diameter for both species. Equations generated were used to estimate the ages of tree stems from the subplots.

Paleoecological Investigations

To reconstruct the history of vegetation previously occupying the site, we examined written accounts and maps on file at Cape Cod National Seashore and the Wellfleet Historical Society. We also obtained a peat core 10 m in diameter and 91 cm long from a point near the centre of the Swamp (Fig. 2). The core was sampled at 1-cm intervals, and subsamples selected at 2–3-cm intervals were examined microscopically for pollen and charcoal content after preparation using a modification of the acetolysis technique (Faegri & Iversen 1975). Although we could not separate the pollen of *Chamaecyparis* from that of red cedar *Juniperus virginiana,* there is currently little *Juniperus* growing in the forests surrounding the Swamp. This fact, plus the occurrence of *Chamaecyparis* wood throughout the core, suggests that most of the Cupressaceae pollen occurring in the peat samples was derived from cedar growing in the Swamp itself.

Surface samples of organic material were obtained from the Swamp to evaluate the relationship between current vegetation and modern pollen rain. Samples were obtained from five locations for which we had overstorey basal area data (see Fig. 2). We selected locations that ranged from nearly pure stands of dense, mature cedar to openings that contained up to one-third of their basal area as red maple.

The area of charcoal fragments on microscope slides was estimated using the point-count method of Clark (1982), with values expressed as the ratio of charcoal area to fossil pollen (Ch:P).

Peat sections 5 cm long were removed at 33–38, 60–65, and 85–90 cm and dated by ^{14}C analysis at Beta Analytic, Inc., Miami, Florida. Pollen indicators [i.e., increases in *Ambrosia, Rumex, Plantago* and Gramineae for European settlement (*c.* AD 1650) and a decline in *Castanea* (*c.* AD 1910) (Clark & Patterson 1984)] were used to date peat from the upper portion of the core. Detrended correspondence analysis (Hill 1979; Gauch 1982) of pollen data was used to examine the relationship between fire occurrence and vegetation change.

Results

Vegetation Characterization

Analysis of aerial photographs indicated the presence of three structurally distinct stands (Fig. 3). The stand that occupies most of the wetland (stand I) consists of large, open-grown cedars as well as forest-grown cedars and maples of approximately the same height. A smaller stand (II) with cedar of greater density and smaller diameter occurs in the southwestern portion of the Swamp. In the southeastern portions of the Swamp are areas of dense but shorter cedar (stand III). Additional areas of dense (and apparently young) cedar were identified during field sampling. These areas are very small and are not recognizable as cedar on aerial photographs. They appear instead as canopy gaps or areas with low, dense vegetation (Fig. 3). Two relevés (one sampled intensively) were taken in such an area and are categorised as stand III in Fig. 2. Individuals in stands II and III lack the large, open-grown crowns typical of stems in stand I.

TABLE 1. Average basal area for stands I, II and III at the Marconi Atlantic White Cedar Swamp

Species	Status	Basal area (m²/ha)		
		stand I (*n* = 110)	stand II (*n* = 7)	stand III (*n* = 16)
Chamaecyparis thyoides	live	40.8	36.6	42.2
	dead	7.9	8.9	6.1
Acer rubrum	live	9.9	11.0	5.2
	dead	0.9	0.3	0.9
Pinus rigida	live	1.3	0.0	1.2
	dead	0.1	0.0	0.1
Other	live	0.2	0.0	0.6
Average Total Basal Area		61.6	56.8	56.3

The three stands are similar with respect to total basal area and its distribution among species (Table 1). The ratio of cedar to red maple for the Swamp as a whole is 4.3:1. Maple is relatively more important at the edges of the wetland than in the centre, particularly along the western edge.

Table 2 summarizes the relevé data for stands defined by photointerpretation and ground reconnaissance and for vegetation strata defined during relevé sampling. The three stands are similar with respect to summed importance values for trees, mosses and herbaceous species. Stands I and III are similar with respect to shrub importance, with the summed importance values for stand II two-thirds those of stands I and III. The distribution of relevés categorized as representing stands I, II or III is shown in Fig. 2.

Stand I has the highest density of stems >20 cm d.b.h. and very few smaller stems. Stand II is dominated by stems 10–20 cm d.b.h., with few stems >25 cm d.b.h. Stand III is characterized by stems 5–10 cm d.b.h., with very few stems >20 cm d.b.h. (Motzkin 1990).

Variable-radius-plot and relevé data indicate that the vegetation of MAWCS is fairly homogeneous with respect to species composition. Although the three stands are structurally distinct (Fig. 3; Motzkin 1990), they are quite uniform floristically.

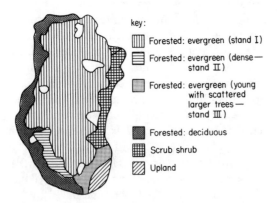

key:

▨ Forested: evergreen (stand I)

▤ Forested: evergreen (dense— stand II)

▨ Forested: evergreen (young with scattered larger trees— stand III)

■ Forested: deciduous

▦ Scrub shrub

▨ Upland

FIGURE 3. Vegetation of the MAWCS based upon interpretation of 1984 colour–infrared aerial photographs. Unshaded areas represent canopy gaps with low, dense vegetation.

Age-Frequency Distributions

Increment cores were obtained for 586 live cedar trees and 135 live maples. Regression

TABLE 2. Average species importance values by stand

	Stand		
	I	II	III
Trees			
Chamaecyparis thyoides	9.0	10.2	11.2
Acer rubrum	8.8	8.4	7.8
Pinus rigida	0.1	0.6	0.0
Shrubs			
Clethra alnifolia	3.4	3.8	2.6
Smilax spp.	2.3	1.2	2.4
Vaccinium corymbosum	6.4	6.6	7.8
Rhododendron viscosum	4.5	4.6	4.8
Leucothoe racemosa	3.5	1.4	2.2
Ilex glabra	2.0	2.2	2.6
Gaylussacia frondosa	1.0	0.0	0.6
Rhus radicans	1.4	0.2	0.8
Amelanchier spp.	0.2	0.0	0.0
Gaylussacia baccata	2.1	0.0	1.0
Ilex verticillata	2.4	1.0	2.4
Unknown	0.8	0.6	0.6
Aronia spp.	1.1	0.0	1.2
Kalmia angustifolia	0.5	0.0	1.2
Lyonia ligustrina	0.0	0.0	0.2
Nemopanthus mucronata	0.1	0.0	0.0
Decodon verticillatus	0.0	0.4	0.0
Herbs			
Trientalis borealis	0.8	0.6	0.0
Woodwardia virginica	2.0	0.8	1.2
Osmunda cinnamomea	0.4	0.4	0.0
Aralia nudicaulis	0.1	0.0	0.0
Carex spp.	0.4	0.4	0.4
Grass spp.	0.1	0.0	0.0
Bartonia virginca	0.1	0.0	0.0
Mitchella repens	0.1	0.0	0.0
Mosses			
Sphagnum spp.	6.4	6.4	5.2
Other mosses	3.5	3.6	4.0
Sum of average importance values			
Trees	17.9	19.2	19.0
Shrubs	31.7	22.0	30.4
Herbs	4.0	2.2	1.6
Mosses	9.9	10.0	9.2
Total	63.5	53.4	60.2

analyses of age vs. d.b.h. for stems with sound cores showed a significant positive correlation for both cedar ($R^2 = 0.68$, $P = 0.0001$) and maple ($R^2 = 0.43$, $P = 0.0001$). Regression equations (for cedar, age = 17.8 + 3.79 × d.b.h.; and for maple, age = 29.5 + 3.28 × d.b.h.) were used to estimate the ages of live stems for which sound cores could not be obtained. For small stems (<5 cm d.b.h.), age and basal diameter (*BD*) are positively correlated (for cedar, age = 1.2 + 12.0*BD*, $R^2 =$ 0.91, $P = 0.0001$; for maple, age = 2.34 + 6.84*BD*, $R^2 = 0.80$, $P = 0.0001$).

Age–frequency distributions, in 10-year age classes, were developed for cedar and maple in each stand (Fig. 4). These graphs include over-storey stems for which ages from sound cores were obtained, estimates of ages of live stems for which no sound cores were obtained, and estimates of ages of understorey stems. Inclusion of ages estimated from regression equa-

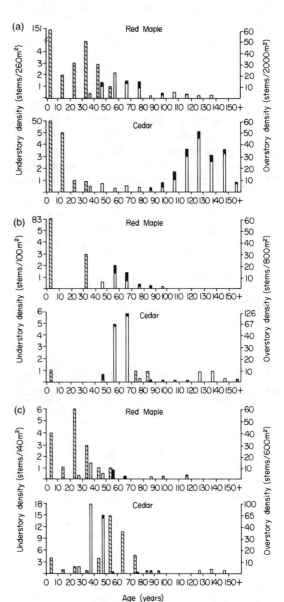

FIGURE 4. Age–frequency distributions for cedar and red maple in (a) stand I, (b) stand II and (c) stand III: (□) overstory ages from cored stems; (▧) estimated understory ages; (■) estimated overstory ages.

tions does not appear to change the general pattern of age–frequency distributions, but instead emphasizes trends evident from aged trees.

Stand I

This stand occupies most of the Swamp and is dominated by cedars 100–150 years old (Fig. 4a). Density of cedars older than 100 years is higher here than in the other stands, although for red maple older than 100 years density is much lower than for cedar (Fig. 4).

Age–frequency distributions for the five intensive sample plots in this mature stand reveal two distinct subtypes. Stand Ia has abundant cedar in the 100–150-year-old age classes but no cedar that has regenerated in the past 80–90 years (Table 3). The few maple stems that occur here are scattered throughout several age classes. Stand Ib includes areas with lower densities of cedar in the oldest age classes and some cedar in several of the younger age classes. The regeneration, 40–80 years ago, of maple stems in stand Ib appears to correspond with the regeneration of cedar (Table 3).

Stand II

The stand in the south-western portion of the Swamp is dominated by cedars of intermediate age (50–70 years old) with a few maples of the same age. The density of cedar is approximately 5.5 times that of maple for these age classes (Table 3). There is no evidence of continuous recruitment of either species, but there are a few open-grown cedar stems that correspond in age to the 100–150-year-old stems observed in stand I. Remains of old, cut stumps are found throughout this stand.

Stand III

Trees in this stand are largely even-aged (Fig. 4c), with more cedar than maple in the 30–50-year-age classes. A few 100–150-year-old cedar are scattered throughout. Most understorey cedar stems were estimated to be about the same age as overstorey stems (Fig. 4c).

Paleoecology of the Swamp

Sediment dating by pollen indicators (i.e., the rise in agricultural indicators at ~28 cm = c. AD 1650) and by ^{14}C dating (c. AD 1500 at 35.5 cm; c. AD 1220 at 62.5 cm; and c. AD 860 at 87.5 cm) yields estimated peat accumulation rates of 0.079–0.09 cm year^{-1}. Thus each 1-cm sample represents c. 12 years, with gaps of between about 12 and 24 years between samples. This resolution seems adequate for characterizing fluctuations in the abundance of cedar, which,

TABLE 3. Cedar and red maple stem density (stems/ha) by age class for stands at the Marconi Atlantic White Cedar Swamp

Age class	Cedar				Maple			
	Ia	Ib	II	III	Ia	Ib	II	III
0–10 years	250	2667	125	222	5750	5833	8834	306
10–20 years	–	278	00	83	125	56	–	56
20–30 years	–	56	–	236	–	167	–	469
30–40 years	–	97	13	2050	125	256	292	369
40–50 years	–	67	88	250	25	267	75	225
50–60 years	–	33	838	1233	25	231	263	175
60–70 years	–	50	1575	917	13	125	175	42
70–80 years	–	42	134	450	25	108	63	–
80–90 years	25	17	96	100	13	17	38	8
90–100 years	75	25	25	83	13	33	13	17
100–110 years	113	75	25	–	–	50	–	–
110–120 years	188	183	25	–	25	17	–	25
120–130 years	350	192	113	67	–	25	–	–
130–140 years	288	67	125	100	38	–	–	–
140–150 years	288	117	25	17	–	–	–	–
150+ years	63	42	13	–	–	–	–	–

in the absence of disturbance, can live to ages of 200–300 years or more.

Pollen stratigraphy shows that cedar has been present at the Marconi Swamp for the past 1100 years. Cupressaceae pollen percentages fluctuate from <5% to >80% prior to the increase in agricultural weed pollen at a depth of ~28 cm in the core (Fig. 5). The abundances of pine, oak, *Sphagnum* and several shrub and fern species also fluctuate widely in the lower two-thirds of the core. Charcoal:pollen ratios vary from 0 to >26 000.

Shortly before the time of settlement (c. AD 1650), Cupressaceae percentages rise to levels that consistently exceed 80% during postsettlement time. Values for all other types except agricultural weeds have declined since settlement. Pollen percentages generally reflect the patterns of land-use outlined earlier. In contrast to the lower portion of the core, there is little charcoal in the upper 20 cm.

Cupressaceae pollen percentages for surface samples (Fig. 5) are uniformly high despite the fact that cedar basal area at the locations from which the samples were taken vary (Table 4) as do distances from the edge of the Swamp (Fig. 2). Sample 6 (Fig. 2), for example, was taken from an opening (stand III) near the upland but has a cedar pollen percentages >70. This suggests that cedar pollen is overrepresented in small openings in the otherwise cedar-dominated swamp and that low cedar percentages during prehistoric time represented swampwide declines in cedar and not simply small openings that might result from the destruction of one or a few overstorey trees.

Vegetation changes associated with presettlement fire occurrence and postsettlement fire suppression are evident in the results of detrended correspondence analysis of the pollen data (Fig. 6). Cedar pollen percentages decrease with increasing axis 1 values, whereas values for *Sphagnum* and fern allies increase. Most presettlement samples have values higher than 40 on axis 1. A subset of these samples (group B) has axis 1 values <100, with samples having generally higher cedar percentages and low-to-moderate *Sphagnum* and/or fern percentages compared to those farther to the right on axis one. Group C is represented by 10 samples that often comprise two or three adjacent samples (38/41, 47/50/53, 61/65/68) that are

shifted far to the right on axis 1 compared to levels above and below them. These samples often have particularly large *Sphagnum* and Ch:P values and low cedar pollen percentages, perhaps indicating open conditions associated with stand-replacing fires. Level 77, which is not associated with high Ch:P values, occurs in the extreme lower right-hand corner of Fig. 6. Cedar and pine percentages for this sample are at their lowest presettlement levels in the profile, whereas *Sphagnum* percentages reach a maximum value of 82.5 indicating a sudden change in environmental conditions (Tallis 1964). This sample may represent a disturbance other than fire (e.g., a major windstorm).

High charcoal values at 30 and 32 cm occur after an increase in cedar pollen and just prior to the time that agricultural indicators increase, but they are not associated with a shift in the sample locations to the right on Figure 6. These values may represent a fire that occurred on the upland with little effect on the wetland itself. This interpretation is consistent with the fact that cedar pollen percentages continue to rise through the period represented by the fire and there are no other distinct changes in wetland species. Pine percentages decline sharply at this time, however, and *Myrica* (an upland shrub) percentages increase.

Postsettlement pollen spectra (group A; 0–30 cm plus the surface samples) are concentrated on the extreme left of axis 1. Only level 62 among the presettlement spectra is grouped with levels representing postsettlement time. All of these samples have abundant cedar pollen, little charcoal and few *Sphagnum*/fern spores.

Discussion

Evidence for Episodic Regeneration

Each of the stands identified by our analysis have age distributions that span only a few decades indicating that establishment of both cedar and maple is episodic rather than continuous. Stand I is dominated by even-aged, mature cedars. Even where cohorts of younger stems occur (stand Ib), the age structure of the cohorts indicates distinct episodes of establishment rather than continuous recruitment. These episodes probably correspond with dis-

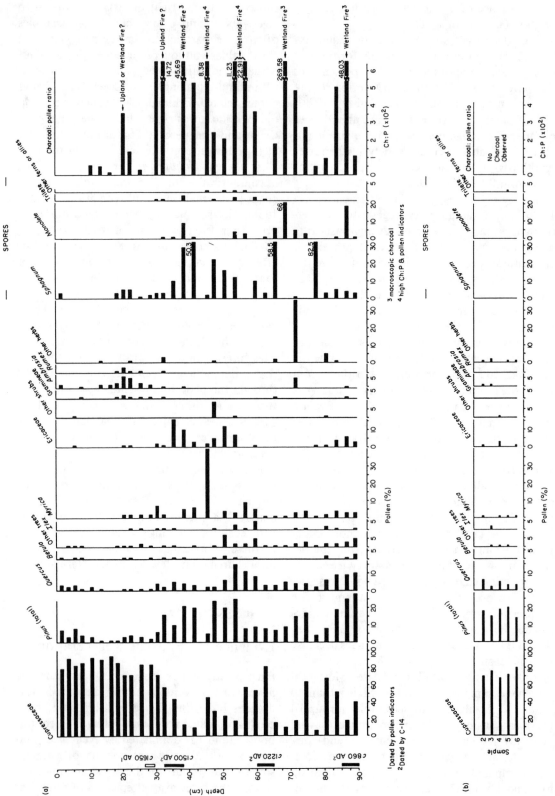

FIGURE 5. Pollen percentages and charcoal pollen ratios for samples from the top 90 cm of peat in (a) the MAWCS and (b) for surface samples

TABLE 4. Cedar and maple pollen percentages for modern surface samples from the Marconi Atlantic White Cedar Swamp with associated basal areas from variable radius plot sampling

Surface sample	Pollen (%)		Basal area (m²/ha)	
	cedar	red maple	cedar	red maple
1 (top of core)	79.2	–	56.4	10.4
2	70.8	–	55.2	4.6
3	76.4	–	47.2	11.5
4	79.8	–	29.9	10.4
5	67.4	0.8	36.8	9.2
6	71.4	0.2	54.1	5.8

turbances—especially selective cutting. Periodic small blowdowns cannot be ruled out, but we saw little evidence in the form of downed stems to suggest that large, old trees were dying individually or in small groups to form these gaps. Stand Ib has lower densities of 100–150-year-old stems than stand Ib (Table 3) suggesting the removal by selective cutting of overstorey stems. The regeneration, 40-to-80 years ago, of maple in stand Ib corresponds with an increase in cedar stems, further suggesting that selective cutting during the first half of the

twentieth century allowed both cedar and especially maple to become established. Selective cutting of only parts of the Swamp would be consistent with nineteenth- and twentieth-century ownership patterns which show the Swamp divided among as many as 12 different owners at various times.

The ability of red maple to sprout and grow quickly after a disturbance like cutting and the corresponding inability of cedar to reproduce vegetatively could result in a distribution of younger stems similar to that which occurs in stand Ib. Korstian & Brush (1931) describe the development after selective logging of two-aged stands similar to those encountered in portions of stand I. In contrast, abundant cut stumps and an absence of maple throughout stands II and III suggest that episodes of cedar rather than maple establishment followed brief periods of more intensive cutting in the area now occupied by these stands.

The relationship between disturbance events and episodes of cedar and maple regeneration supports Hickman & Neuhauser's (1977) observations that neither cedar nor maple successfully establish beneath closed canopies of mature cedar (e.g. as in stand Ia). Maple does not occur in all-aged populations in any of the stands we sampled. Disturbance in the form of selective cutting is a likely explanation for the occurrence of younger maple in some portions of stand Ib. The even-aged cohorts of maple that we have described at Marconi are not consistent with the all-aged populations that would be expected from Little's (1950) explanation for cedar–maple interactions in New Jersey.

Fire and the Paleoecology of Cedar

The intent of our analysis was to characterize the vegetation and disturbance regime of the Swamp in the period immediately preceding the advent of European settlers. This period is often viewed by the National Park Service as a benchmark from which subsequent vegetation change can be measured. The results of our fine-resolution pollen analysis (Fig. 5) indicate a significant change in plant communities in response to changing disturbance regimes following European settlement. Fires were

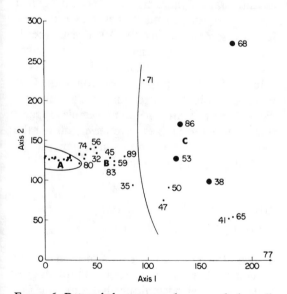

FIGURE 6. Detrended correspondence analysis ordination of pollen percentage data for surface and stratigraphic samples from the MAWCS. Enclosed circles represent samples that have abundant charcoal and little Cupressaccae pollen. See text for descriptions of groups A, B and C.

common prior to settlement. Most probably began in the highly flammable oak–pitch-pine forests on the upland and burned as intense surface or crown fires until they entered the Swamp. Cedar foliage is highly flammable, and wind-driven fires could have burned through the crowns of cedar despite the presence of standing water in the wetland. In addition, some fires may have burned at times when conditions were dry enough to allow the surface of the wetland to burn. A layer of charcoal 5 cm thick at 65 cm below the surface of the wetland indicates that one such fire burned c. AD 1170. Thinner layers of macroscopically visible charcoal at 38 and 87 cm (c. AD 1470 and AD 870, respectively) indicate other fires that almost certainly burned through the swamp. High Ch:P ratios and associated changes in pollen assemblages at 45 cm and 53–57 cm suggest that additional prehistoric fires burned in the Swamp c. AD 1390 and 1265.

Evidence that fires burned into the Swamp is strengthened by the fact that declines in cedar pollen percentages sometimes coincide with major charcoal peaks (especially at 45 and 57 cm). At other times, fires may have burned either through young cedar or other shrubby vegetation (e.g. at 38 cm). Despite the occurrence of at least five major fires in the Swamp during the last 1000 years, there is no evidence that cedar was completely extirpated from the wetland for even brief periods. Fire episodes were, in fact, often followed by the regeneration of a new cedar stand (e.g. at 38 and 86 cm). High presettlement pollen percentages for shrub species corresponding with low percentages for cedar suggest that the Swamp at times supported an open cedar stand quite different from the one at the site today. High percentages for monolete fern spores (perhaps representing *Woodwardia virginica,* which is found in the Swamp today) coincide with high presettlement charcoal values. This provides additional evidence for the opening of the wetland surface to high light levels for at least a short time following prehistoric fires. No fern spores were found in any of the modern surface samples, whereas values in excess of 60% coincide with the severe fire that burned into the Swamp c. AD 1170. At no time during the several hundred years prior to European settlement did cedar pollen percentages attain the consistently high levels (70–80%) found in surface samples.

High postsettlement cedar pollen percentages are accompanied by very low Ch:P values, which have never exceeded 500 during the past c. 200 years. Values do exceed 400 at 20 cm when pollen percentages for indicators of agricultural activity (*Ambrosia,* Gramineae, *Plantago* and *Rumex*) peak. This probably represents the period of maximum agricultural activity on Cape Cod (c. AD 1860). Upland pollen percentages show little response to this fire, but cedar declines about 10% and then recovers. A coincident rise in *Sphagnum* spores, which also fluctuate in response to fire in presettlement sediments, suggests that this postsettlement fire may have impacted at least a portion of the Swamp. The lack of macroscopically visible charcoal and of any response in fern spore percentages suggests a fire of different character than the ones that burned into the Swamp prior to settlement, however. The charcoal may represent the burning of slash from selective logging of cedar or a fire set near the Swamp to clear pastures of burn agricultural refuse.

Of particular interest in the interpretation of modern stand dynamics is the fact that pollen percentages for red maple never exceed 0.5% for any sample during the 1000-year record. Maples are insect pollinated, and percentages are normally quite low, but we have observed values > 5% for samples taken from Cape Cod swamps dominated by red maple. The very low values for the Marconi Swamp samples suggest that mature red maples have never, during the last 1000 years, been more important in the Swamp than they are today.

Allogenic Factors and the Development of Cedar

The results of our investigation of both current vegetation dynamics and vegetation trends over a 1000-year period indicate that allogenic factors have strongly influenced the vegetation of the Marconi Atlantic White Cedar Swamp. Age structure analyses of the modern vegetation suggest that cedar and maple establishment occurs only during distinct episodes following disturbance events. During the past 150 years, timber cutting has been the primary disturbance influencing the regeneration of both cedar and maple. Light thinnings apparently favoured red maple establishment over cedar in some areas, whereas more-intensive cutting resulted in the regeneration of even-aged stands dominated by cedar. Areas lacking evidence of cutting lack significant regeneration of either species. Prior to European settlement, fire rather than cutting influenced vegetation patterns at the Swamp. Fires burning through the wetland destroyed existing cedar stands but allowed for subsequent cedar regeneration from seed stored in the upper organic soil horizon or from surviving mature trees. Presettlement fire frequency was apparently great enough to prevent dense stands of mature cedar (like the one that currently occupies much of the Swamp) from persisting for more than brief periods.

Previous studies of succession in Atlantic white cedar swamps have focused on the influence of autogenic processes (Little 1950; Hickman & Neuhauser 1977). The results of our investigation suggest that although disturbance regimes were different before and after European settlement, allogenic factors have been the dominant force influencing succession in the Marconi Atlantic White Cedar Swamp.

The Future Course of Vegetation Development

Current National Park Service policies of wildfire suppression and preservation of natural vegetation rule out fire and timber cutting as factors likely to influence future vegetation development in the Marconi Atlantic White Cedar Swamp. Unlike wilderness areas in western North America, the Marconi Swamp is in an area where lightning fires are rare (Patterson *et al.* 1985). With no active prescribed burning program in place at the Seashore, only uncontrolled fires of human origin are likely to occur. With increasing fragmentation of the forest due to recreational and residential development in the area surrounding the Swamp, it is unlikely that fires will reach a size that could threaten the Swamp.

Never, during the last 1000 years, has the interval between fires exceeded the expected lifespan of cedar (200–300 years), and it is thus difficult to predict the course of future vegetation development. Factors other than fire that might influence vegetation development include windthrow and salt-spray damage associated with storm winds (Little 1950). Although these disturbances may occur, there are no studies of their influences on vegetation succession in cedar wetlands. A disturbance-free period of up to several hundred years is possible, however, and our discussion of future vegetation trends focuses on the effects of autogenic succession over such a time period.

In the absence of disturbance it is unlikely that significant amounts of cedar or maple will regenerate beneath the existing overstorey. As the mature, 100–150-year-old stand ages, individual trees may die creating small canopy gaps (Little 1950). Portions of the mature stand lacking young maple or cedar but with abundant shrubs will experience increases in shrub cover. Subsequent regeneration of either cedar or maple at these shrub-dominated sites will depend upon gap size and local site conditions. Elsewhere in the mature stand, past cutting has allowed the establishment of younger cohorts of cedar and maple. In these areas established younger cedar and/or maple rather than shrubs will fill canopy gaps. Most areas have been thinned only lightly, favouring the establishment of maple. Thus, maples will probably increase in importance relative to cedar throughout much of the present mature stand. Cedar appears to be longer-lived than red maple, however. Cedars >150 years old are scattered throughout the Swamp, with the oldest cedar more than 200 years old. The oldest red maple that we aged was only 140 years. Thus, although red maple may increase in importance relative to cedar as existing mature stands break up, cedar may eventually again increase in relative importance as the young stems of both cedar and maple age and maples die at a younger age.

Stands II and III are currently 30–50 and 50–70 years old, respectively. These stands could survive for another 100–200 years before attaining the age where stand break-up might be expected. Unless disturbed by fire, windthrow, cutting, or pathogens, these stands will probably remain intact during the time that the existing mature stand is breaking up. Thus, even if cedar dominance is greatly reduced through the break-up of the existing old stand (stand I), it is unlikely that cedar will become completely extirpated from the Marconi site within the next several centuries. It therefore seems reasonable to conclude that although modern disturbance regimes are quite unlike those of prehistoric times, cedar is not immediately threatened by a management strategy that favours protection over active manipulation of the vegetation. Cedar is likely to survive for a very long time at Marconi, but in a community that is different in structure and species abundances than the one existing at the site prior to settlement of New England by Europeans. A similar conclusion may apply to many natural areas preserved for the unique species or communities that they contain.

From ecological as well as management perspectives, an important observation that emerges from this investigation is that the vegetation of the Marconi Swamp has been extremely dynamic during the entire 1000-year period that we examined. This suggests that attempting to preserve the stand structure as it now occurs may not be an appropriate or practical conservation objective. Rather, priority should perhaps be placed on protecting a sufficient range of community types so that the disturbance of any particular site may be considered acceptable as a functional process in natural area preservation.

Acknowledgments

This study was supported in part by a grant from the National Park Service to WAP and by a grant from the Massachusetts Natural Heritage and Endangered Species Program to GM and WAP. We thank Bill Wilson for discussions of the ideas presented, and Charles Laing for preparation of Fig. 5. David Foster and an anonymous reviewer provided valuable comments on earlier drafts of this paper.

References

Altpeter, L.S. (1937) *A history of the forests of Cape Cod*. MSC thesis, Harvard University, Cambridge, MA.

Belling, A.J. (1977) *Postglacial migration of* Chamaccyparis thyoides (L). B.S.P. (southern white cedar) in the northeastern United States. PhD thesis, New York University, New York.

Buell, M.F. & Cain, R.L. (1943) The successional role of southern white cedar, *Chamaecyparis thyoides,* in southeastern North Carolina. *Ecology,* 24, 85–93.

Clark, J.S. & Patterson, W.A. III (1984) Pollen, Pb-210, and opaque spherules: An integrated approach to dating and sedimentation in the intertidal environment. *Journal of Sedimentary Petrology,* 54, 1251–1265.

Clark, J.S. & Patterson, W.A. III (1985) The development of a tidal marsh: Upland and oceanic influences. *Ecological Monographs,* 55, 189–217.

Clark, R.L. (1982) Point count estimation of charcoal in pollen preparations and thin sections of sediments. *Pollen et Spores,* 24, 523–535.

Cronon, W. (1983) *Changes in the Land.* Hill and Wang, New York.

Emerson, G. (1850) *Trees and Shrubs of Massachusetts.* Little & Brown, Boston.

Faegri, K. & Iversen, J. (1975) *Textbook of Pollen Analysis.* Hafner Press, New York.

Gauch, H.G. (1982) *Multivariate Analysis in Community Ecology.* Cambridge University Press, Cambridge.

Green, D.G. (1983) The ecological interpretation of fine resolution pollen records. *New Phytolologist,* 94, 459–477.

Green, D.G. & Dolman, G.S. (1988) Fine resolution pollen analysis. *Journal of Biogeography,* 15, 531–541.

Hickman, J.C. & Neuhauser, J.A. (1977) Growth patterns and relative distribution of *Chamaecyparis thyoides* and *Acer rubrum* in Lebanon State Forest, New Jersey. *Bartonia,* 45, 30–36.

Hill, M.O. (1979) DECORANA—*A FORTRAN Program for Detrended Correspondence Analysis and Reciprocal Averaging.* Cornell University, Ithaca, NY.

Johnson, J.W. (1980) Atlantic white cedar. *Forest Cover Types of the United States and Canada* (ed. F.H. Eyre), p. 75. Society of American Foresters, Washington.

Korstian, C.F. & Brush, W.D. (1931) *Southern white cedar.* USDA Technical Bulletin no. 251. Washington, DC.

Küchler, A.W. (1967) Vegetation Mapping. The Ronald Press Co. New York.

Laderman, A.D. (1989) *The ecology of Atlantic white cedar wetlands: a community profile.* United States Fish and Wildlife Service Biological Report 85 (7.21), Washington, DC.

Little, S. (1950) Ecology and silviculture of white

cedar and associated hardwoods in southern New Jersey. *Yale University School of Forestry Bulletin,* **56**, 1–103.

McCaffrey, C.A. (1973) *An ecological history of the Province Lands, Cape Cod National Seashore.* Report, Amherst, MA.

Motzkin, G. (1990) *Age structure and successional status of the Marconi Atlantic White Cedar Swamp, Cape Cod National Seashore, So. Wellfleet, Massachusetts.* MSc thesis, University of Massachusetts, Amherst.

Motzkin, G. (1991) *Atlantic white cedar wetlands of Massachusetts.* Massachusetts Agricultural Experiment Station Research Bulletin Number 731, Amherst.

Mueller-Dombois, D. & Ellenberg, H. (1974) *Aims and Methods of Vegetation Ecology.* John Wiley & Sons, New York.

Noyes, J.H. (1939) *Silvicultural management of southern white cedar in Connecticut.* MF thesis, Yale University, New Haven, NH.

Patterson, W.A., III & Sassaman, K.E. (1988) Indian fires in the prehistory of the Northeast. *Holocene Human Ecology in Northeastern North America.* (ed. G.P. Nicholas), pp. 107–135. Plenum Publishing Corporation, New York.

Patterson, W.A., III, Saunders, K.E., Horton, L.J.

& Foley, M.K. (1985) Fire management options for coastal New England forests: Acadia National Park and Cape Cod National Seashore. *Proceedings—Symposium and Workshop on Wilderness Fire* (Technical Coordination by J.E. Lotan, B.M. Kilgore, W.C. Fischer & R.W. Mutch), pp. 360–365. USDA Forest Service General Technical Report INT-182.

Roman, C.T., Good, R.E. & Little, S. (in press) Ecology of Atlantic white cedar swamps in the New Jersey Pinelands. *Management for Wetlands: Series on Tasks for Vegetation Science* (eds J. Kvet, D. Whigham & R. Good). Junk, The Hague.

Seymour, F.C. (1989) *The Flora of New England,* 2nd edn. Phytologia Memoirs V. Plainfield, New Jersey.

Tallis, J.H. (1964) Studies on Southern Pennine peats. III. The behaviour of *Sphagnum. Journal of Ecology,* **52**, 345–353.

Whigham, D.F. (1987) Ecosystem processes and biogeographical considerations. *Atlantic White Cedar Wetlands* (ed. A.D. Laderman), pp. 295–311. Westview Press, Boulder, CO.

Zampella, R.A. (1987) Atlantic white cedar management in the New Jersey Pinelands. *Atlantic White Cedar Wetlands* (ed. A.D. Laderman), pp. 295–311. Westview Press, Boulder, CO.

14
Organisms as Ecosystem Engineers

Clive G. Jones, John H. Lawton, and Moshe Shachak

Interactions between organisms are a major determinant of the distribution and abundance of species. Ecology textbooks (e.g., Ricklefs 1984, Krebs 1985, Begon et al. 1990) summarise these important interactions as intra- and interspecific competition for abiotic and biotic resources, predation, parasitism and mutualism. Conspicuously lacking from the list of key processes in most text books is the role that many organisms play in the creation, modification and maintenance of habitats. These activities do not involve direct trophic interactions between species, but they are nevertheless important and common. The ecological literature is rich in examples of habitat modification by organisms, some of which have been extensively studied (e.g. Thayer 1979, Naiman et al. 1988). However, in general, population and community ecology have neither defined nor systematically identified and studied the role of organisms in the creation and maintenance of habitats. There is not even a word, or words, in common use to describe the process. We will call the process *Ecosystem Engineering* and the organisms responsible *Ecosystem Engineers*.

Beaver (*Castor canadensis*) are familiar examples of organisms acting as ecosystem engineers. By cutting trees and using them to construct dams they alter hydrology, creating wetlands that may persist for centuries. "These activities retain sediments and organic matter in the channel, . . . modify nutrient cycling and decomposition dynamics, modify the structure and dynamics of the riparian zone, influence the character of water and materials transported downstream, and ultimately influence plant and animal community composition and diversity" (Naiman et al. 1988).

However, beaver are by no means the only ecosystem engineers. As we will show, a vast array of species have effects that are fundamentally similar, albeit often on more modest spatial and temporal scales. Yet there is no common language to describe what ecosystem engineers do, no formal structure to model their effects, and no general theory round which to organise understanding of the process. Examples, which are developed more formally below (for others, see Table 1), include not only beaver and their dams but also gophers, ants and termites that move soil, woodpeckers that drill holes, alligators that make wallows, rock-eating snails, trees, corals, seagrass beds and *Sphagnum* blanket bogs.

The purposes of this article are fourfold: (i) to define and to give examples of ecosystem engineering by organisms; (ii) to develop a conceptual framework that explains and classifies its effects; (iii) to show how organismal engineering differs from related concepts (e.g. 'keystone species' Paine 1969, Krebs 1985); (iv) and to identify questions for further work on organisms as ecosystem engineers. First we define what we mean by an ecosystem engineer,

TABLE 1. Examples of organisms acting as ecosystem engineers. Classification according to Fig. 1. Additional examples are discussed in the text.

Organism	Habitat	Activity	Impact	Refs.
Case 2 (allogenic)				
American alligators, *Alligator mississippiensis*	Everglades National Park	create wallows	retain water in droughts; provide refuges for fish, fisheating birds, etc.	Finlayson & Moser (1991)
Rabbits, *Oryctolagus cuniculus*, badgers, *Meles meles*	Europe	dig extensive burrows (rabbit warrens, badger setts)	burrows occupied by other species, e.g. fox, *Vulpes vulpes*, and by many invertebrates	Southern (1964); Neal & Roper (1991)
Case 3 (autogenic)				
Marine phytoplankton	Gulf of Maine	blooms of phytoplankton particles scatter and absorb light in upper layers of water column	Enhance warming of surface waters that may initiate development of thermocline	Townsend et al. (1992)
Microalgae in sea ice	Antarctica	scatter and absorb light within ice and underlying seawater; reduce strength of ice	enhance melting and break up of ice	Buynitskiy (1986); Arrigo et al. (1991)
Freshwater phytoplankton	Lake St. George, Ontario	intercept light in upper water column; small algal spp. more effective than large spp.	light interception leads to shallower mixing depth, lower metalimnetic temperatures and lower heat content of water column	Mazumder et al. (1990)
Cyanobacteria and other nonvascular plants	desert and semi-desert soils	exude mucilaginous organic compounds	glue the organisms, organic matter and soil particles together to form a microphytic crust; change infiltration, percolation, retention and evaporation of water; reduce soil erosion; affect seedling emergence	West (1990)
Bog moss, *Sphagnum* spp.	Northern and western Britain	build 'blanket' and 'raised' bogs via accumulated peat	major changes in hydrology, pH, and topography	Tansley (1949)
Submerged macrophytes	freshwater lakes, ponds and rivers	grow to create weed beds	attenuate light; steepen vertical temperature gradient; retard flow; enhance sedimentation; oxygenate rhizosphere	Carpenter & Lodge (1986)
Forest trees (broad-leaved and coniferous)	Hubbard Brook Experimental Forest, New Hampshire	shed branches and trunks into streams	create debris dams; alter morphology and stability of stream channels, storage and transport of dissolved organic matter and sediments; different tree species may create dams which differ in persistence	Likens & Bilby (1982); Hedin et al. (1988)
Higher plants	ubiquitous	dead leaves etc. accumulate a litter	alter microenvironment of soil; change surface structure, affecting drainage, and transfer of heat and gasses; act as physical barrier for seeds and seedlings; numerous impacts on structure and composition of plant communities	Facelli & Pickett (1991)
Terrestrial plants in 29 families, with >1,500 species	ubiquitous	grow structures (modified leaves, leaf axils etc.) that impound water	create small aquatic habitats, supporting a highly specialised insect fauna	Fish (1983)

(continued)

TABLE 1. (*continued*)

Organism	Habitat	Activity	Impact	Refs.
Case 4				
Marine meiofauna (protozoa and representatives of many invertebrate phyla)	ubiquitous	biodeposition, bioturbation, porewater circulation, and faecal pellet production	change physical, chemical and biological properties of sediments; change direction and magnitude of nutrient fluxes; increase oxygenation of sediments	Reichelt (1991)
Marine burrowing macrofauna	ubiquitous	burrow into and redistribute sediments; bioturbation; burrow ventilation	create dynamic sediment mosaics; actively transport solutes into burrows; increase oxygenation of sediments; stimulate microflora; increase decomposition rates	Anderson & Kristensen (1991); de Wilde (1991); Meadows & Meadows (1991b)
Marine zooplankton	ubiquitous	fiter living, dead organic and inorganic (e.g. clay) particles, and concentrate into faecal pellets	sinking faecal pellets important in vertical transport and exchange of elements and organic compounds in oceans	Dunbar & Berger (1981); Wallace et al. (1981); Fowler & Knauer (1986)
Fiddler crab, *Uca pugnax*	New England salt marsh	dig burrows	increase soil drainage and oxidation-reduction potential; increase decomposition rates; increase primary production at intermediate tidal heights	Bertness (1985)
European periwinkle, *Littorina littorea*	New England rocky beach	bulldoze sediments from hard substrates	prevent sediment accumulation and hence growth and establishment of algal canopy; algae are case 3 engineers and further increase sedimentation rates; faunal composition markedly different with and without snails	Bertness (1984a)
Snails, *Euchondrus* spp.	Negev desert	eat endolithic lichens and the rock they grow in	increase rate of nitrogen cycling, soil formation and rock erosion	Shachak et al. (1987); Jones & Shachak (1990)
Bagworm caterpillars, *?Penestoglossa* sp.	Golden Gate Highlands, South Africa	eat endolithic lichens and construct larval shelters ('bags') from quartz crystals	small increase in erosion rate, nutrient cycling and soil formation	Wessels & Wessels (1991)
Mound-building termites, Isoptera	widespread in tropics and subtropics	mound and subterranean gallery construction; redistribution of soil particles	change mineral and organic composition of soils; alter hydrology and drainage	Elmes (1991)
Earthworms, Lumbricidae, Megascolecidae	ubiquitous	burrowing, mixing and casting	change mineral and organic composition of soils; affect nutrient cycling; alter hydrology and drainage; affect plant population dynamics and community composition	Lal (1991); Thompson et al. (1993)
Blind mole rats, *Spalax ehrenbergi*	Israel	digging and tunnelling	move large quantities of soil; increase aeration; create distinctive ecosystem	Heth (1991)
Mole rats, Bathyergidae (several genera)	South African lowland fynboss	digging and tunnelling	create impressive, cratered landscapes, with effects on soil formation, plant productivity and species composition	Richardson et al. (in press)

(*continued*)

TABLE 1. (*continued*)

Organism	Habitat	Activity	Impact	Refs.
Prairie dogs, *Cynomys* spp.	North American short and mixed grass prairie	continual intense disruption by burrowing, creating soil mounds	change physical and chemical properties of soil persisting for 100–1000s of years	Whicker & Detling (1988)
Pocket gophers, *Geomys bursarius*	North American grasslands and arid shrublands	construct tunnels and move soil to surface mounds	alter patterns and rates of soil development, nutrient availability and microtopography; change plant demography, diversity and primary productivity; affect behaviour and abundance of other herbivores	Huntly & Inouye (1988); Moloney et al. (1992)
Indian crested porcupine, *Hystrix indica*	Negev desert	digging for food	dig up to 2–3 holes m^{-2}; diggings accumulate organic matter, runoff water; create favourable sites for seed germination	Yair & Rutin (1981); Gutterman (1982)
Elephants, *Loxodonta africana*	East African woodland and savannah	physical disturbance and destruction of trees and shrubs	widespread vegetation changes; alteration of fire regime; effects on food supply and population dynamics of other animals; ultimately changes in soil formation, riparian zones, and biogeochemical cycling	Naiman (1988)
Case 5 (autogenic) and case 6 (allogenic) (examples combining elements of both)				
Crustose coralline algae, *Porolithon, Lithophyllum*	coral reefs	overgrow and cement together detritus on outer algal ridge of barrier reef	break force of water and protect corals against major wave action; effect via own bodies (case 5) and secretion of 'cement' (case 6)	Anderson (1992)
Ribbed mussels, *Geukensia demissa*	Rhode Island *Spartina* salt marsh	secrete byssal threads, and form dense mussel beds	on marsh edge, dense beds of mussels (case 5) and byssal threads (case 6) bind and protect sediments and prevent physical erosion and disturbance, e.g. by storms	Bertness (1984b)

before providing examples and a conceptual framework for what is, and is not, ecosystem engineering.

Definitions

Ecosystem engineers are organisms that directly or indirectly modulate the availability of resources (other than themselves) to other species, by causing physical state changes in biotic or abiotic materials. In so doing they modify, maintain and/or create habitats.

The direct provision of resources by an organism to other species, in the form of living or dead tissues is not engineering. Rather, it is the stuff of most contemporary ecological research, for example plant-herbivore or predator-prey interactions, food web studies and decomposition processes.

Autogenic engineers change the environment via their own physical structures, i.e. their living and dead tissues. *Allogenic engineers* change the environment by transforming living or non-living materials from one physical state to another, via mechanical or other means.

Armed with these definitions we can now proceed to consider some detailed examples. we are not, at this juncture, concerned with the magnitude or scale of the impacts of engineers on communities and ecosystems. We are interested solely in discovering properties that all engineers have in common. We address the scale and magnitude of their effects later.

Classification of Organisms as Engineers

Table 1 summarises examples of organisms as ecosystem engineers. The table is illustrative, not exhaustive. Additional examples are discussed at greater length in the text.

All the examples of which we are aware can be assigned to one of five possible cases (Fig. 1), or to a combination of two or more of these cases. As in many other areas of ecology, the diversity of biological processes means that precise pigeon-holing is sometimes difficult. The boundaries between types of engineering are occasionally fuzzy and, in the real world, separating engineering from other ecological processes may also be difficult, simply because these non-trophic interactions always co-occur with trophic interactions. We discuss some difficult cases as we proceed. The majority of examples are, however, easy to classify. The legend to Fig. 1 explains the conventional notation used to describe them. For clarity, it is easiest to introduce the arguments using beaver and their dams.

Beaver conform to case 4 in Fig. 1. That is they are allogenic engineers, taking materials in the environment (in this case trees, but in the more general case it can be any living or non-living materials), and turning them (engineering them) from physical state 1 (living trees) into physical state 2 (dead trees in a beaver dam). This act of engineering then creates a pond, and it is the pond which has profound effects on a whole series of resource-flows used by other organisms. The critical step in this process is the transformation of trees from state 1 (living) to state 2 (a dam). This transformation modulates the supply of other resources, particularly water, but also sedi-

ments, nutrients etc. A critical characteristic of ecosystem engineering is that it must change the availability (quality, quantity, distribution) of resources utilised by other taxa, excluding the biomass provided directly by the population of allogenic engineers. Engineering is not the direct provision of resources in the form of meat, fruits, leaves, or corpses. Beaver are not the direct providers of water, in the way that prey are a direct resource for predators, or leaves are food for caterpillars.

Now consider the autogenic equivalents of beaver (Fig. 1, case 3). Simple examples are the growth of a forest or a coral reef. Trees and corals are direct sources of food and living space for numerous organisms, but the production of branches, leaves or living coral tissue does not constitute engineering. Rather, it conforms to case 1 in Fig. 1 (the direct provision of resources). However, the development of the forest or the reef results in physical structures which do change the environment and modulate the distribution and abundance of other resources. This modulation constitutes autogenic engineering. Trees alter hydrology, nutrient cycles and soil stability, as well as humidity, temperature, windspeed and light levels (see Holling 1992); corals modulate current speeds, siltation rates and so on. It is obvious, but surprisingly rarely explicitly stated, that numerous inhabitants of the habitats so created are dependent upon the physical conditions modulated by the autogenic engineers, and upon resource flows which they influence but do not directly provide; without the engineers, most of these other organisms would disappear.

One further example may help to clarify the distinction between case 1 and case 3. The growth of seagrass beds modulates ocean currents, which in turn may alter sedimentation rates and hence food supplies for other organisms with substantial effects upon their performance (e.g. growth and survival in the clam *Mercenaria mercenaria* (Irlandi and Peterson 1991)). The direct provision of food or living space by seagrasses (case 1) is not critical for *Mercenaria* but the clam's survival is nonetheless dependent upon these plants.

It could be argued that the growth of tree-trunks, branches, reefs or similar substantial biological structures (case 1, Fig. 1) itself con-

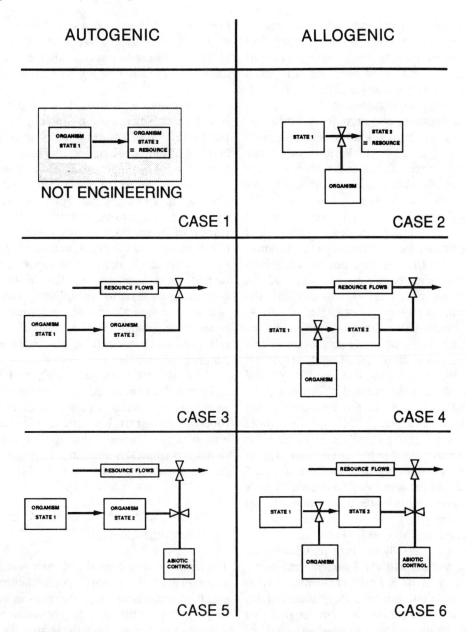

AUTOGENIC | ALLOGENIC

FIGURE 1. Conceptual models of autogenic and allogenic engineering by organisms. For definitions and examples see text and Table 1. The symbol ⋈ defines points of modulation. For example, allogenic engineers transform living or non-living materials from state 1 (raw materials) to state 2 (engineered objects and materials), via mechanical or other means. The equivalent (state 2) products of autogenic engineering are the living and dead tissues of the engineer. These products of both autogenic and allogenic engineering then modulate the flow of one or more resources to other species (cases 2–4) or modulate a major abiotic controller (e.g. fire), which in turn modulates resource flows (cases 5–6). Case 1, the direct provision of resources by one species to another is not engineering, and involves no modulation of resource flows.

stitutes ecosystem engineering. Inclusion or exclusion is a matter of choice. We have chosen to exclude it (whether it is the provision of food or of 'architecture' (Southwood et al. 1979 and Lawton 1983)) when the structures are considered solely and directly as resources, because this differs in kind from the remaining cases in Fig. 1. To qualify for our definition of an ecosystem engineer, an organism must modulate the supply of other resources for other species, rather than be the direct provider of resources. The growth of biological structures is thus a necessary but not a sufficient requirement for autogenic engineering.

We can illustrate our arguments further by considering the simplest kind of allogenic engineering (case 2, Fig. 1). Various organisms make holes in tree trunks and branches, some quickly (woodpeckers), others more slowly (rot fungi). They transform wood without holes into wood with holes, and indirectly provide resources for other creates, nesting and roosting cavities for birds and bats for instance. The holes are the resource, not the organisms that make them. Notice that if some of the holes fill up with water (Kitching 1983), the little ponds so created are examples of case 4, and are conceptually identical to beaver dams.

Inevitably there are some grey areas in this classification and at the risk of being pedantic it is worthwhile considering just one. The natural hollows and cavities formed at branch junctions and root bases as the tree grows, and which may subsequently fill up with water ('pans' sensu Kitching (1983)) are not the same as rot- or woodpecker-holes and they do not belong in case 4; rather they conform to case 3 (trees are now autogenic engineers because their biomass gives rise to water-filled hollows). Numerous examples of these 'phytotelmata' are summarised by Fish (1983) (Table 1).

It is not universally the case that plants are allogenic and animals are autogenic engineers. Enhanced rates of physical and chemical weathering of rocks into soil by algae or higher plants (Bloom 1978) constitute allogenic engineering (case 4). Plants may act either as autogenic or as allogenic engineers, and provide some of the most complex cases of ecosystem engineering, to which we now turn.

Autogenic (case 5) and allogenic (case 6) engineering are at the extremes of continua that merge with cases 3 and 4 respectively. Cases 5 and 6 have the common property that autogenic or allogenic engineers interact with, and modulate powerful abiotic forces, for instances fires or hurricanes. Examples in cases 5 and 6 are distinguished from cases 3 and 4 by the extreme magnitude of the processes modulated by the engineers, and by the fact that these major abiotic forces are themselves fundamental modulators of the distribution and abundance of resources.

Fire provides a particularly interesting case. It is logical, albeit unconventional, to regard the production of combustible living and dead biomass as autogenic engineering (case 5). Different species of plants produce different qualities and quantities of living and dead fuel, modulating the magnitude, intensity and duration of fire and, in turn, profoundly altering the supply of resources for many other species (Christensen 1985). High grassland productivity in Serengeti-Mara in the 1960s markedly increased the incidence of fire, resulting in conversion of savannah woodland to an alternative state – grassland – which is now maintained by elephants (Dublin et al. 1990). The effects of elephants as allogenic engineers are summarised in Table 1.

Plants also act as allogenic engineers (case 6). In Puerto Rico *Dacryodes excelsa* trees are able to withstand hurricanes because their extensive roots and root grafts bind and stabilise bedrock and superficial rocks; this species therefore dominates tropical mountain forests where hurricanes are common (Basnet et al. 1992).

Difficult Cases: Pollinators, Gall Formers and Cows

The richness of biological processes means that a completely satisfactory, comprehensive yet exclusive definition of ecosystem engineering may be impossible to achieve, although numerous examples are easy to classify (Table 1). Given the diversity of species interactions in nature, efforts to classify many other ecological phenomena suffer similar problems 'at the margin'.

Pollinators and gall-formers present an interesting challenge for our definition of engineering. Both have profound effects on the growth of plant tissue; in so doing, pollinators modulate the supply of resources for seed predators, and gall-formers create structures that are used not only by themselves for shelter and food, but also by inquilines (e.g. Askew 1975). Both types of interaction conform broadly to case 2 (Fig. 1). However, we do not find it helpful to regard pollinators as engineers, not least since self-pollination is case 1 and not engineering. But it is possible to regard gall-formers as engineers for inquilines; they physically modify plant-tissues, and create new habitats and resources for other organisms. The distinction between pollinators and gall-formers is, however, a fine one.

Our definition of engineering also embraces other, unexpected ecological phenomena, for example dung masses produced by large herbivores. Cows turn grass into cow pats, which are then colonised by a rich community of invertebrates, dependent upon the pats for food and shelter (Mohr 1943). The physical structure and environment provided by the droppings is at least as important to its inhabitants as the concentration of food resources (Elton 1966). It does not stretch the definition of engineering too far to regard cows as allogenic engineers, turning grass into cow pats (case 2). We would be the first to admit, however, that it is an unconventional perspective. Similar remarks apply to faecal pellet production by oceanic zooplankton (Table 1). For these and other borderline cases, the common sense way to view the issue is to ask whether understanding the ecological interactions is enhanced by recognising the engineering dimension.

Human Analogues

The parallels between ecological and human engineers are, not surprisingly, very close. Humans are tool-using organisms that specialise in engineering. While human engineering often has intent or purpose, it is probably true to say that the major reason why humans have such adverse effects on the environment is because of the unintended consequences of our activities as engineers. Indeed people are now the primary agents of environmental change in most areas of the world (Naiman 1988, Likens 1992). Many human activities, from dam-building and skyscraper construction to forest clearance and the dredging and canalization of water courses, conform exactly to cases in Fig.

1, in which humans are allogenic engineers, altering the physical environment and modulating the flow of resources to other species.

Construction of nesting boxes for birds and hives for bees are examples of case 2. Ploughing by farmers and the construction of dams and reservoirs by water engineers provide examples of case 4. Building harbours and sea walls to reduce storm damage from waves are examples of case 6. Humans mimic autogenic effects, using tools to construct glasshouses and build air conditioning plants (mimicking case 3), and by bulldozing fire breaks to counteract fire (mimicking case 5). We classify human engineering activities as heavy or light, construction, civil, heating, plumbing, and air-conditioning to name but a few. Organisms do all these jobs, and from a functional perspective we see no fundamental difference between human and non-human engineering.

Related Concepts

The idea that some organisms alter the physical structure of their environment, with impacts on their own and other populations is not new. But earlier work either focuses on particular species and habitats and lack generality, or takes a more general view but fails clearly to define ecological engineering, or to distinguish it from other processes. For example, in an important set of reviews dealing with animal influences on ecosystem dynamics, co-ordinated by Naiman (1988), engineering and direct trophic effects are interwoven. Within this series of papers, Huntly and Inouye (1988) explicitly describe pocket gophers *Geomys bursarius* as "soil engineers" because of their role as earth-movers. Gophers are, indeed, excellent examples of allogenic (case 4) engineers (Table 1).

Ecologists in general have paid surprisingly little attention to how environments are created and maintained; most appear content to follow Andrewartha and Birch (1954) in recognising "a place to live" and "weather" as two of the four essential features of species' environments, without formally considering the role of engineering in habitat modification, creation and maintenance.

Bioturbators

In marine benthic environments the activities of large burrowing animals are known to play a dominant role in determining the physical structure of sediments, altering habitat suitability for other species (Rhoads and Young 1970, Thayer 1979, Lopez and Levinton 1987, Meadows and Meadows 1991a, see also Table 1). Rhoads and Young (1970) called the process 'trophic amensalism' when large deposit feeders create unstable sediments, restricting the presence of suspension feeders and attachment by sessile epifauna. The term amensalism is reasonable, because the effect is asymmetrical and is a form of competitive exclusion (Lawton and Hassell 1981), but the mechanism is not trophic and clearly differs from normal exploitation competition for food. 'Trophic amensalism' is actually another good example of case 4 allogenic engineering brought about by bioturbation of sediments.

Patch Dynamics

Numerous studies recognise the importance of patches of bare or different substrates in otherwise closed communities (e.g. Dayton 1971, Wiens 1976, Paine 1979, Pickett and White 1985). Patches may be created by physical disturbance (waves, fire, landslips) or by the activities of organisms (grazing, predation or engineering), or by interactions between engineering, trophic and physical processes. For some applications of patch dynamic theory the way in which patches are created may be less important than their existence. In general, however, we believe that it is desirable to recognise engineering as one of several distinct ways in which patches are created and maintained, particularly since the factors that control patch formation by engineers are often different from those controlling patch formation by abiotic forces.

Animal and Plant Artifacts

In their *Theory of Environment* Andrewartha and Birch (1984) define a framework for examining all the processes that impinge upon a single species population. This target species occupies the centrum of a web of directly and indirectly acting components, and Andrewartha and Birch point out that a link in the web may be another living organism or its artifact or residue. The Theory of Environment therefore clearly allows for engineering, without explicitly identifying it as a defined modifier in the web, or as a process worthy of study in its own right.

Meadows and Meadows (1991a) and Meadows (1991) review the environmental impacts of animal burrows and burrowing animals, and Hansell (1993) the ecological consequences of animal burrows and nests. Many of these artifacts (e.g., meiofaunal burrows, megapode nests, termite mounds and mole rat colonies) have landscape level effects, and serve to concentrate and redistribute resources for other species — that is they are classic examples of ecological engineering (Table 1).

Meadows (1991) points out that there are "underlying similarities between the impact of [burrowing] animals from different terrestrial and aquatic habitats on environmental change and modification." He provides a formal system to quantify the impacts of burrowing, distinguishing between animals with large per capita but geographically restricted effects (e.g. badgers) and those species with small per capita effects that nevertheless, because of their abundance and distribution, have impacts on entire landscapes (e.g. earthworms). Hansell (1993) recognises that the "services and substances of the builders create a new range of habitat niches which can be exploited by a wide variety of specialists" and suggests that "the presence of nest builders and burrowers can . . . significantly contribute to species diversity in habitats." The examples provided by both authors all conform to either case 2 (other species use the nests and burrows) or case 4 (species respond to changes in distribution and abundance of resources). Their work therefore differs from ours only in its more restricted focus.

Extended Phenotypes

The importance of animal artifacts is also recognised by Dawkins (1982), as an example of species' extended phenotypes. Dawkins writes (p.200): "A beaver dam is built close to the lodge, but the effect of the dam may be to flood an area thousands of square meters in extent. As to the advantage of the pond from the beaver's point of view, the best guess seems to be that it increases the distance the beaver can travel by water, which is safer than travelling by land, and easier for transporting wood. . . . If this interpretation is right the lake may be regarded as a huge extended phenotype." Dawkins recognises other products of animal engineering as extensions of species' phenotypes and hence subject to natural selection, including caddis-fly cases, termite mounds and birds' nests.

He also points out that not every example of what we are now calling allogenic engineering can be regarded as an extended phenotype, because impacts on the environment are of no consequence to the engineer's fitness and hence are not subject to natural selection. A good example would be water-filled footprints made by an ungulate. The distinction between engineering that is subject to natural selection (because it is an extended phenotype) and engineering that is not ('accidental' engineering) appears to be unimportant in terms of its shorter-term ecological consequences; all types of engineering modify and modulate resource flows for other organisms. But there may be interesting longer-term differences, particularly in the nature of the feed-back loops that operate on "extended phenotype' versus 'accidental' forms of engineered artifacts. We return to this point later.

Keystone Species

Keystone species (Paine 1969, Krebs 1985, Daily et al. 1993) play a critical role in determining community structure. By definition, removal of keystone species causes massive changes in species composition and other ecosystem attributes. The critical links are usually regarded as trophic and therefore within the realm of traditional ecological thinking. For example, removing top predators has a cascading effect throughout the foodweb, altering species composition and hence physical structure and nutrient cycling at lower levels (Estes and Palmisano 1974, Carpenter et al. 1987).

But critical effects frequently involve engi-

neering, for example via disturbance (e.g. bioturbators (Thayer (1979), above; case 4). In the frequently cited example of sea-otters *Enhydra lutris,* removal of otters leads to an increase in sea urchins (*Strongylocentrotus* sp.) and hence to the disappearance of kelp beds, which in turn changes wave action and siltation rates, with profound consequences for other inshore flora and fauna (Estes and Palmisano 1974). Kelp are autogenic engineers (case 3); removal of kelp by urchins is, amongst other things, allogenic engineering (case 4). In other words, in this familiar example, the species traditionally regarded as the keystone (sea otter) has major effects because it changes the impact of one engineer (urchin) on another (kelp), with knock-on effects on other species in the web of interactions. The equally well known impact of kangaroo rats (*Dipodomys* spp.) on desert vegetation occurs because the rodents not only eat seeds but also cause considerable physical disturbance. By burrowing and moving large quantities of soil they create many shallow pits and little mounds, which facilitate decomposition and the establishment of annual plants (case 4 engineering) (Brown and Heske 1990).

Direct effects of keystone species via their role as engineers have recently been reported by Daily et al. (1993). Red-naped sapsuckers *Sphyrapicus nuchalis* (a type of woodpecker) act as keystone species in two ways in Colorado subalpine meadows. Their nesting holes drilled in aspens, *Populus tremuloides,* are essential nesting sites for two species of swallows (case 2); the swallows are missing from the community in the absence of sapsuckers. Feeding holes drilled by the sapsuckers in willows, *Salix* spp., also make sap flows available to several birds, mammals and insects (directly changing the distribution and abundance of this resource for other species – and therefore again conforming to case 2).

It is theoretically possible (though we think it will be uncommon in practice) for a keystone species to exert its effects entirely trophically, without also acting as an engineer, or without changing the engineering role of other species in the web. On the other hand, many engineers are keystone species even though they play relatively minor roles in community food webs.

Krebs concludes his textbook review of keystone species as follows: "Keystone species may be relatively rare in natural communities, *or they may be common but not recognised* (our emphasis). At present, few terrestrial communities are believed to be organised by keystone species, but in aquatic communities keystone species may be common". We believe that such views probably reflect a consensus among ecologists. They persist because we have failed to recognize the role of ecosystem engineers as keystone species. It is trite, but true, that a forest is a forest because it has trees, which not only provide food and living space but which also autogenically engineer the forest climate, and modulate the flows of many other resources to forest inhabitants, both above- and below-ground. Many single species of trees in temperate or boreal forests (with low tree-species richness) are both keystone species and significant ecosystem engineers.

Our views are very close to Holling's (1992) *Extended Keystone Hypothesis,* in which he argues that "all terrestrial ecosystems are controlled and organised by a small set of key plant, animal, and abiotic processes that structure the landscape at different scales." We would add two points. First a critical, but not exclusive controlling mechanism is some form of engineering; and second, we believe that keystone engineers occur in virtually all habitats on earth, not just terrestrial ones.

"Top Down" vs. "Bottom Up," Asymmetrical and Indirect Effects

Traditional population models focus on reciprocally coupled pairs of interactions, interspecific competition (-/-), enemy-victim (-/+) and so on (Williamson 1972). Highly asymmetrical competitive interactions (amensalism; 0/-) are common, possibly the norm, in some situations (Lawton and Hassell 1981). Enemy-victim interactions may also be asymmetrical (donor-controlled; 0/+), that is prey abundance controls predator abundance, but not vice versa (Lawton 1989, Hawkins 1992), with a growing debate in ecology about the relative importance of such 'bottom up' vs 'top down' effects

(Hunter and Price 1992). Ecological engineering adds enormously to the catalogue of important, highly asymmetrical species interactions, because engineers impact upon many taxa (positively or negatively), but there may often be no direct, reciprocal effects of the impacted species upon the population of engineers.

One example will suffice. Beaver beneficially influence the abundances of aquatic biota, but not vice versa; that is they have massive 'bottom up' effects (o/-) that benefit numerous other aquatic organisms. Their activities are also detrimental to terrestrial species living upstream (and perhaps downstream) from the dam (o/-) just as bioturbators exclude sessile epifauna requiring stable substrates (see above). Generally we expect both the positive and negative effects of engineers to be highly asymmetrical.

This is not to say that there cannot be any feedbacks from organisms in the engineered habitat, back to the engineer. Undoubtedly there are, although feedback pathways are probably often rather long, indirect, and frequently slow. They remain virtually unstudied. For some engineers, it is difficult to imagine any reciprocal effects. For instance, the insect habitants of abandoned birds' nests (e.g. some tineid moths in the genus *Monopis* and staphylinid beetles in the genus *Microglotta* (Walsh and Dibb 1954, Emmet 1979)) probably never encounter the builder. Gophers, in contrast, engineer soil and change vegetation composition, biomass and productivity (Table 1); in turn, grasshopper populations become more abundant in the vicinity of gopher mounds (Huntly and Inouye 1988). Two feedbacks may operate on gophers. The first is reasonably well documented, positive and relatively direct, via plants that are food for gophers; soil disturbance favours the plant species that gophers prefer to eat. Second, it is at least conceivable, but untested, (D. Tilman, pers. comm.) that grasshoppers compete with gophers for food, providing a longer, negative feed-back loop on gopher numbers.

We predict that if feed-backs exist at all between engineers and the organisms they affect, they will characteristically be indirect, involving several intermediate processes and species.

Spatial and Temporal Scales

The impact of an ecological engineer depends upon the spatial and temporal scale of its actions. Water filled woodpecker holes and beaver dams may both be examples of case 4 engineering, but there is not much doubt about which is the more significant ecological phenomenon. Six factors scale the impact of engineers. They are:

i. Lifetime per capita activity of individual organisms.
ii. Population density.
iii. The spatial distribution, both locally and regionally, of the population.
iv. The length of time the population has been present at a site.
v. The durability of constructs, artifacts and impacts in the absence of the original engineer.
vi. The number and types of source flows that are modulated by the constructs and artifacts, and the number of other species dependent upon these flows.

Thus, the most obvious ecological engineering is attributable to species with large per capita effects, living at high densities, over large areas for a long time, giving rise to structures that persist for millennia and which affect many resource flows — for instance mima mounds created by fossorial rodents, including gophers (Cox and Gakahu 1985, 1986, Cox et al. 1987, Naiman 1988). Autogenic engineers may also have massive effects; as Holling (1992) succinctly states: "To a degree, . . . the boreal forest 'makes its own weather' and the animals living therein are exposed to more moderate and slower variation in temperature and moisture than they would otherwise be." Boreal forest trees have large per capita effects on hydrology and climatic regimes, occur at high densities over large areas, and live for decades. But their impacts as autogenic engineers may have a relatively short memory if the forest is logged.

Organisms with small individual impacts can also have huge ecological effects, providing that they occur at sufficiently high densities over large areas, for sufficient periods of time. Burrowing meiofauna and bogforming *Sphag-*

num mosses (Table 1) are good examples. Accumulated *Sphagnum* peat may persist for hundreds to thousands of years after the death of the living moss.

Ecological engineers may also enhance and speed up large scale physical processes, including geological erosion and weathering (Yair and Rutin 1981, Krumbein and Dyer 1985, Hoskin et al. 1986). Examples of rock-eating snails and caterpillars are listed in Table 1. Worldwide, but especially in the tropics, heavily undercut costal cliffs of sedimentary rock are apparently being eroded by tides and storms. In fact the process is greatly accelerated by two groups of engineers, both with low per capita effects, but very abundant. Cyanobacteria (*Hyella* spp.) bore the rock and are food for chitons which rasp away the rock to reach them, apparently speeding up coastal erosion by an order of magnitude or more (Krumbein and Dyer 1985). Similarly, organisms whose shells, body parts and dead tissues help form sedimentary rocks, coal and soil (e.g. molluscs, diatoms and many higher plants) create structures whose effects on ecosystems persist for eons.

Engineering impacts are often greatest when the resource flows that are modulated are utilised by many other species, or when the engineer modulates abiotic forces that affect many other species. Not surprisingly, engineering that effects soils, sediments, rocks, hydrology, fire and hurricanes provides some of the most striking examples.

We know of very few field manipulation experiments designed to quantify the impact of ecosystem engineers by removing or adding species. Studies by Bertness (1984a,b, 1985) are excellent examples of manipulative experiments on both allogenic (cases 4 and 6) and autogenic (case 5) engineers (Table 1). A recent study by Hall et al. (1993) shows the potential power of field manipulations for disentangling per capita impacts from population impacts (although in the present context it is not ideal because predation and disturbance [case 4 engineering] effects are confounded). Edible crabs (*Cancer pagurus*) hunting for prey dig pits in shallow subtidal areas of the west coast of Scotland. The pits are conspicuous features of the seabed topography, yet exclusion of crabs from areas of the sea bed for twelve months failed to reveal any landscape-level effects of crab pits, either on substrate structure and composition (particle sizes, organic carbon etc.) or faunal diversity, composition and abundance. Crabs appear not to be abundant enough to significantly alter community structure, either by predation or by engineering, despite large and visually conspicuous individual impacts. This example contrasts markedly with the substantial effects of fiddler crabs (*Uca pugnax*) on productivity, decomposition, oxygenation and drainage in a New England salt marsh, revealed by experimental removal of crabs (Bertness 1985) (Table 1).

Another extremely poorly researched problem is the way in which the persistence of the products or effects of engineering influence population, community and ecosystem processes. If engineers make long-lived artifacts, then their effects will usually, also, be long lived. But ephemeral products can also have long-term impacts. For example, faecal pellets produced by marine zooplankton (Table 1) decompose relatively quickly, but not before they have sunk into the deep ocean, removing nutrients from surface waters for millenia.

A useful thought experiment is to consider taking the engineers away and imagining the consequences. In many cases their impacts are ephemeral, operating on timescales shorter than, or similar to, the lifetime of the organism itself (e.g. the nests of small passerine birds). But in other cases, the engineers leave monuments with impacts that extend many lifetimes beyond their own — mima mounds, termite nests, buffalo wallows, beaver dams, peat, sedimentary rocks and so on. These persistent effects must greatly slow down rates of ecological change, and impose considerable buffering and inertia on many ecological systems and processes. Rates of decay, the 'half-lives' of the products of ecological engineers, and their contributions to population, community and ecosystem stability, resistance and resilience (Pimm 1984) deserve much more attention from theoretical and experimental ecologists.

Evolutionary Effects

Earlier, we distinguished between engineered artifacts subject to natural selection as extended phenotypes, and 'the rest' — by-products of some other activity that are not themselves directly subject to selection; water-filled ungulate hoof prints were given as an example. Extended phenotype engineering, by definition, creates structures or effects that directly influence individual fitness (or colony fitness in social insects), for instance a beaver dam, woodpecker hole or termite mound. But the evolutionary effects of extended phenotype engineering, other 'accidental' engineering, and of organisms in the engineered habitat upon the engineer are far from straightforward, and generally unstudied. For example, the extended phenotype may be subject to selection from the physical environment, implying no biotic feedbacks. Or it may be subject to selection from polyphagous predators (e.g. nest-robbing snakes) that are in no way dependent upon the engineered habitat for their existence. On the other hand, engineering might generate habitats with species populations that, ultimately, feed back positively or negatively upon the engineers, via predation, disease, competition, or the invasion of additional species of engineers.

Some engineering undoubtedly has had evolutionary effects on other organisms. One of the best documented examples in the fossil record is a decline, from the Devonian onwards, in the diversity of immobile suspension feeders living on soft marine substrata, as mobile taxa diversified (Thayer 1979). Thayer attributes these major changes in the structure of marine benthic communities to the evolution of 'biological bulldozers' — bioturbators or engineers — that disturb sediments (see above), fouling, overturning and burying immobile suspension feeders, which are now largely confined to hard substrates. Thayer also speculates that by increasing the turnover rate of nutrients in sediments of continental shelves, bulldozers may have contributed to the Mesozoic diversification of phytoplankton (coccoliths, diatoms and dinoflagellates) and, via trophic linkage, to diversification of zooplankton (radiolaria and foraminifera).

An intriguing, but rarely considered problem is the degree to which engineering by other taxa might similarly have changed major patterns in the radiation and extinction of earth's biota. To the extent that engineers shape and modify most, possibly all, habitats on earth (see below), and given the trite but true observation that all organisms are adapted to their environment, engineering in some form or other must have driven, or contributed to, the evolution of myriads of species. But the extent to which major patterns of evolution might have been different if some types of ecological engineering had not evolved, or had taken a different form, is almost entirely unknown.

Questions

We finish with a haphazard list of open questions.

Are there any ecosystems on earth that have not been physically engineered by one or more organisms to a significant degree? A cautious, preliminary answer is no, there are not. We initially thought that it would be difficult to identify evidence of ecosystem engineering by biota in the open waters of oceans or large lakes, or in snow fields and ice packs, for instance. But the examples in Table 1 show that this prediction was wrong. We currently cannot identify any habitat on earth that is not engineered in some way by one or more species.

How many species (or what proportion of species) in various communities have a clearly defined and measurable impact as engineers? Is it 10%, 1% or 0.1%? What are the relative frequencies of the five classes of engineering identified in Fig. 1, say in terms of the numbers of species acting as engineers? Cases 5 and 6 are presumably rather rare; but how much rarer are they than the other types of engineering? Is the predominance of examples involving burrowing animals in case 4 real (we could easily have included still more examples) or is it because this form of engineering is particularly easy to see?

Are the most physically structured ecosystems (or subsystems, e.g. soil or sediments) the ones in which engineers are most important? How much of the structure have they created and modified?

How may other species are impacted by engineers in any ecosystem? What happens to species richness if we remove or add engineers? How much of the effects of keystone species are due to engineering versus trophic effects? Earlier, we speculated that few keystone effects are

purely trophic; is this hypothesis correct? How do engineering and trophic relations interact?

Should conservationists and nature reserve managers pay more attention to the role of ecological engineers in maintaining ecosystem integrity, or do managers largely know which the important species are, without having put a name to the idea, or without having recognising the common themes identified in this paper?

How should we model engineering? The biological details in each case will be complicated, and there are at least five kinds of engineering; but there are also several sorts of interspecific competition, various ways of being a herbivore and a rich catalogue of enemy-victim interactions, none of which has stopped theoreticians from developing appropriate families of relatively simple models to understand and to predict the dynamics of such interactions. There is, in principle, no reason why we cannot write down the equation:

$$d\text{mayfly}/d\text{beaver} = F(x,y,z),$$

where mayfly populations respond to changes in beaver numbers on long time-scales, and where the response is influenced by various key variables, including feedbacks to beaver from other components in the engineered habitat. Interesting theoretical questions centre on the generation time of the engineer, the half-life of whatever it is that is engineered, the rate of restoration of non-engineered habitat, the generation times of impacted species, and their various interactions. There are intriguing problems of nested time-scales, delayed responses, donor-control, long chains of indirect interactions and so on, that might usefully be explored using relatively simple models.

Extending these arguments, there is no reason in principle why processes driven by engineering should not be coupled to the rich diversity of trophic linkages to create not simply descriptions and models of foodwebs (e.g. Pimm et al. 1991), but of *interaction webs,* that more accurately reflect interactions in communities and ecosystems. The conceptual framework brought together by Carpenter (1988) under the title of 'complex interactions' should clearly embrace engineering as one component.

Intriguingly, once the need to define and study interaction webs is recognised, it also becomes apparent that ecological engineering as we have defined it is only part of the picture. We have focused on physical state changes wrought on biotic and abiotic materials by organisms. But physical engineering is not the only form of ecosystem engineering that organisms carry out. Chemical and transport engineering are two obvious other forms that we consider will conceptually fit into the same general classification scheme. For brevity, we have not examined them here.

Last, but by no means least, what new insights will the concept of engineering bring to ecology? This review attempts to define and classify the phenomenon. But definition and classification are merely a small beginning, not an end. Are there major patterns in the distribution and abundance of organisms that might be explained, at least in part, by ecosystem engineering? What are they? What predictions might we make that would not have been made without the conceptual framework provided here, or something akin to it? We do not, currently, know the answers to these questions. But if the notion of organisms as ecosystem engineers results in an accumulation of "just-so" stories, it will not have been particularly useful.

Acknowledgments

This work was supported by the Mary Flagler Cary Charitable Trust, the US-Israel Binational Science Foundation, and the NERC Centre for Population Biology, UK. Numerous colleagues have allowed us to sharpen and refine our ideas about engineering by discussing problems with them, and have generously provided us with examples and references; to all of them we express our sincere thanks. Contribution to the programme of the Institute of Ecosystem Studies. Publication 171 from the Mitrani Center.

References

Anderson, F. O. and Kristensen, E. 1991. Effects of burrowing macrofauna on organic matter decomposition in coastal marine sediments. – Symp. zool. Soc. Lond. 63: 69–88.

Anderson, R. A. 1992. Diversity of eukaryotic algae. – Biodiv. Conserv. 1: 267–292.

Andrewartha, H. G. and Birch, L. C. 1954. The distribution and abundance of animals. – University of Chicago Press, Chicago.

_____ and Birch, L. C. 1984. The ecological web. More on the distribution and abundance of Animals. – University of Chicago Press, Chicago.

Arrigo, K. R., Sullivan, C. W. and Kremer, J. N. 1991. A bio-optical model of Antarctic sea ice. – J. Geophys. Res. 96 C6: 10581–10592.

Askew, R. R. 1975. The organisation of chalcid-dominated parasitoid communities centred upon endophytic hosts. – In: Price, P. W. (ed.), Evolutionary strategies of parasitic insects and mites. Plenum, New York, pp. 130–153.

Basnet, K., Likens, G. E., Scatena, F. N. and Lugo, A. E. 1992. Hurricane Hugo: damage to a tropical rain forest in Puerto Rico. – J. Trop. Ecol. 8: 47–55.

Begon, M., Harper, J. L. and Townsend, C. R. 1990. Ecology. Individuals, populations and communities, Second ed. – Blackwell Scientific, Boston.

Bertness, M. D. 1984a. Habitat and community modification by an introduced herbivorous snail. – Ecology 65: 370–381.

_____ 1984b. Ribbed mussels and *Spartina alterniflora* production in a New England salt marsh. – Ecology 65: 1794–1807.

_____ 1985. Fiddler crab regulation of *Spartina alterniflora* production on a New England salt marsh. – Ecology 66: 1042–1055.

Bloom, A. L. 1978. Geomorphology. – Prentice Hall, New York.

Brown, J. H. and Heske, E. J. 1990. Control of a desert-grassland transition by a keystone rodent guild. – Science 250: 1705–1707.

Buynitskiy, V. K. 1968. The influence of microalgae on the structure and strength of Antarctic sea ice. – Oceanology 8: 771–776.

Carpenter, S. R. (ed.). 1988. Complex interactions in late communities. – Springer, New York.

_____ and Lodge, D. M. 1986. Effects of submerged macrophytes on ecosystem processes. – Aquat. Bot. 26: 341–370.

_____ , Kitchell, J. F., Hodgson, J. R., Cochran, P. A., Elser, J. J., Elser, M. M., Lodge, D. M., Kretchmer, D., He, X. and von Ende, C. N. 1987. Regulation of lake primary productivity by food web structure. – Ecology 68: 1863–1876.

Christensen, N. L. 1985. Shrubland fire regimes and their evolutionary consequences. – In: Pickett, S. T. A. and White, P. S. (eds.), The ecology of natural disturbance and patch dynamics. Academic Press, Orlando, FL, pp. 85–100.

Cox, G. W. and Gakahu, C. G. 1985. Mima mound microtopography and vegetation pattern in Kenyan savannas. – J. Trop. Ecol. 1: 23–36.

_____ and Gakahu, C. G. 1986. A latitudinal test of the fossorial rodent hypothesis of Mima mound origin. – Z. Geomorphologie 30: 485–501.

———, Lovegrove, B. G. and Siegfied, W. R. 1987. The small stone content of mima-like mounds in the South African Cape region. — Catena 14: 165–176.

Daily, G. C., Ehrlich, P. R. and Haddad, N. M. 1993. Double keystone bird in a keystone species complex. — PNAS 90: 592–594.

Dawkins, R. 1982. The extended phenotype. — Oxford University Press, Oxford.

Dayton, P. K. 1971. Competition, disturbance, and community organisation: the provision and subsequent utilization of space in a rocky intertidal community. — Ecol. Monogr. 41: 351–389.

Dublin, H. T., Sinclair, A. R. E. and McGlade, J. 1990. Elephants and fire as causes of multiple stable states in the Serengeti-Mara woodlands. — J. Anim. Ecol. 1990: 1147–1164.

Dunbar, R. B. and Berger, W. H. 1981. Fecal pellet flux to modern bottom sediment of Santa Barbara Basin (California) based on sediment trapping. — Geol. Soc. Amer. Bull. pt. 1 92: 212–218.

Elmes, G. W. 1991. Ant colonies and environmental disturbance. — Symp. zool. Soc. Lond. 63: 1–13.

Elton, C. S. 1966. The pattern of animal communities. — Methuen, London.

Emmet, A. M. (ed.). 1979. A field guide to the smaller British Lepidoptera. — British Entomological and Natural History Society, London.

Estes, J. A. and Palmisano, J. F. 1974. Sea-otters: their role in structuring nearshore communities. — Science 185: 1058–1060.

Facelli, J. M. and Pickett, S. T. A. 1991. Plant litter: its dynamics and effects on plant community structure. — Bot. Rev. 57: 1–32.

Finlayson, M. and Moser, M. (eds.). 1991. Wetlands. — International Waterfowl and Wetlands Research Bureau, Oxford.

Fish, D. 1983. Phytotelmata: flora and fauna. — In: Frank, J. H. and Lounibos, L. P. (eds), Phytotelmata: terrestrial plants as hosts for aquatic insect communities. Plexus, Medford, NJ, pp. 1–27.

Fowler, S. W. and Knauer, G. A. 1986. Role of large particles in the transport of elements and organic compounds through the oceanic water column. — Prog. Oceanog. 16: 147–194.

Gutterman, Y. 1982. Observations on the feeding habits of the Indian crested porcupine (*Hystrix indica*) and the distribution of some hemicryptophytes and geophytes in the Negev desert highlands. — J. Arid. Env. 5: 261–268.

Hall, S. J., Robertson, M. R., Basford, D. J. and Fryer, R. 1993. Pitdigging by the crab *Cancer pagurus*: a test for long-term, large-scale effects on infaunal community structure. — J. Anim. Ecol. 62: 59–66.

Hansell, M. H. 1993. The ecological impact of animal nests and burrows. — Funct. Ecol. 7: 5–12.

Hawkins, B. A. 1992. Parasitoid-host food webs and donor control. — Oikos 65: 159–162.

Hedin, L. O., Mayer, M. S. and Likens, G. E. 1988. The effect of deforestation on organic debris dams. — Verh. Internat. Verein. Limnol. 23: 1135–1141.

Heth, G. 1991. The environmental impact of subterranean mole rats (*Spalax ehrenbergi*) and their burrows. — Symp. zool. Soc. Lond. 63: 265–280.

Holling, C. S. 1992. Cross-scale morphology, geometry, and dynamics of ecosystems. — Ecol. Monogr. 62: 447–502.

Hoskin, C. M., Reed, J. K. and Mook, D. H. 1986. Production and off-bank transport of carbonate sediment, Black Rock, southwest Little Bahama Bank. — Mar. Geol. 73: 125–144.

Hunter, M. D. and Price, P. W. 1992. Playing chutes and ladders: heterogeneity and the relative roles of bottom-up and top-down forces in natural communities. — Ecology 73: 724–732.

Huntly, N. and Inouye, R. 1988. Pocket gophers in ecosystems: patterns and mechanisms. — BioScience 38: 786–793.

Irlandi, E. A. and Peterson, C. H. 1991. Modification of animal habitat by large plants: mechanisms by which seagrasses influence clam growth. — Oecologia 87: 307–318.

Jones, C. G. and Shachak, M. 1990. Fertilization of the desert soil by rock-eating snails. — Nature 346: 839–841.

Kitching, R. L. 1983. Community structure in water-filled treeholes in Europe and Australia—comparisons and speculations. — In: Frank, J. K. and Lounibos, L. P. (eds), Phytotelmata: terrestrial plants as hosts for aquatic insect communities. Plexus, Medford, NJ, pp. 205–222.

Krebs, C. J. 1985. Ecology. The experimental analysis of distribution and abundance. Third ed. — Harper and Row, New York.

Krumbein, W. E. and Dyer, D. B. 1985. This planet is alive—weathering and biology, a multi-faceted problem. — In: Drever, J. I. (ed.), The chemistry of weathering. D. Reidel Publishing Co., pp. 143–160.

Lal, R. 1991. Soil conservation and biodiversity. — In: Hawksworth, D. L. (ed.), The biodiversity of microorganisms and invertebrates: its role in sustainable agriculture. CAB International, Wallingford, pp. 89–103.

Lawton, J. H. 1983. Plant architecture and the diversity of phytophagous insects. — Ann. Rev. Ent. 28: 23–39.

——— 1989. Food webs. — In: Cherrett, J. M. (ed.), Ecological concepts. Blackwell Scientific Publications, Oxford, pp. 43–78.

_____ and Hassell, M. P. 1981. Asymmetrical competition in insects. – Nature 289: 793–795.

Likens, G. E. 1992. The ecosystem approach: its use and abuse. – Ecology Institute, Oldendorf/Luhe.

_____ and Bilby, R. E. 1982. Development, maintenance, and role of organic-debris dams and routing in forest drainage basins. USDA Forest Service General Technical Report PNW141. USDA Forest Service, Pacific Northwest Forest and Range Experimental Station, pp. 122–128.

Lopez, G. R. and Levinton, J. S. 1987. Ecology of deposit-feeding animals in marine sediments. – Quart. Rev. Biol. 62: 235–260.

Mazumder, A., Taylor, W. D., McQueen, D. J. and Lean, D. R. S. 1990. Effects of fish and plankton on lake temperature and mixing depth. – Science 247: 312–315.

Meadows, P. S. 1991. The environmental impact of burrows and burrowing animals – conclusions and a model. – Symp. zool. Soc. Lond. 63: 327–338.

_____ and Meadows, A. (eds.). 1991a. The environmental impact of burrowing animals and animal burrows. – Clarendon Press, Oxford.

_____ and Meadows, A. 1991b. The geotechnical and geochemical implications of bioturbation in marine sedimentary ecosystems. – Symp. zool. Soc. Lond. 63: 157–181.

Mohr, C. O. 1943. Cattle droppings as ecological units. – Ecol. Monogr. 13: 276–298.

Moloney, K. A., Levin, S. A., Chiariello, N. R. and Buttel, L. 1992. Pattern and scale in a serpentine grassland. – Theoret. Pop Biol. 41: 257–276.

Naiman, R. J. 1988. Animal influences on ecosystem dynamics. – BioScience 38: 750–752.

_____, Johnston, C. A. and Kelley, J. C. 1988. Alteration of North American streams by beaver. – BioScience 38: 753–762.

Neal, E. G. and Roper, T. J. 1991. The environmental impact of badgers (*Meles meles*) and their sets. – Symp. zool. Soc. Lond. 63: 89–106.

Paine, R. T. 1969. A note on trophic complexity and community stability. – Am. Nat. 103: 91–93.

_____ 1979. Disaster, catastrophe, and local persistence of the sea palm *Postelia palmaeformis*. – Science 205: 685–687.

Pickett, S. T. A. and White, P. S. (eds), 1985. The ecology of natural disturbance and patch dynamics. – Academic Press, Orlando.

Pimm, S. L. 1984. The complexity and stability of ecosystems. – Nature 307: 321–326.

_____, Lawton, J. H. and Cohen, J. E. 1991. Food web patterns and their consequences. – Nature 350: 669–674.

Reichelt, A. C. 1991. Environmental effects of meiofaunal burrowing. – Symp. zool. Soc. Lond. 63: 33–52.

Rhoads, D. C. and Young, D. G. 1970. The influence of deposit-feeding organisms on sediment stability and community trophic structure. – J. Mar. Res. 28: 150–178.

Richardson, D. M., Cowling, R. M., Bond, W. J. Stock, W. D. and Davis, G. W. in press. Links between biodiversity and ecosystem function: evidence from the Cape Floristic Region. – In: Davis, G. W. and Richardson, D. M. (eds), Biodiversity and ecosystem function in Mediterranean-type ecosystems. Springer-Verlag, Heidelberg.

Ricklefs, R. E. 1984. Ecology, Second ed. – Chiron Press, New York.

Shachak, M., Jones, C. G. and Granot, Y. 1987. Herbivory in rocks and the weathering of a desert. – Science 236: 1098–1099.

Southern, H. N. (ed.). 1964. The handbook of British mammals, First ed. – Blackwell Scientific, Oxford.

Southwood, T. R. E., Brown, V. K., and Reader, P. M. 1979. The relationship of plant and insect diversities in succession. – Biol. J. Linn. Soc. 12: 327–348.

Tansley, A. G. 1949. Britain's green mantle. – George Allen and Unwin, London.

Thayer, C. W. 1979. Biological bulldozers and the evolution of marine benthic communities. – Science 203: 458–461.

Thompson, L., Thomas, C. D., Radley, J. M. A., Williamson, S. and Lawton, J. H. 1993. The effect of earthworms and snails in a simple plant community. – Oecologia 95: 171–178.

Townsend, D. W., Keller, M. D., Sieracki, M. E. and Ackleson, S. G. 1992. Spring phytoplankton blooms in the absence of vertical water column stratification. – Nature 360: 59–62.

Wallace, G. T., Jr., Mahoney, O. M., Dulmage, R., Storti, F. and Dudek, N. 1981. First-order removal of particulate aluminum in oceanic surface water. – Nature 293: 729–731.

Walsh, G. B. and Dibb, J. R. (eds.). 1954. A coleopterist's handbook. – The Amateur Entomologists' Society, London.

Wessels, D. C. J. and Wessels, L.-A. 1991. Erosion of biogenically weathered Clarens sandstone by lichenophagous bagworm larvae (Lepidoptera; Pyschidae). – Lichenol. 23: 283–291.

West, N. E. 1990. Structure and function of microphytic soil crusts in wildland ecosystems of arid to semi-arid regions. – Adv. Ecol. Res. 20: 180–223.

Whicker, A. D. and Detling, J. K. 1988. Ecological consequences of prairie dog disturbances. – BioScience 38: 778–785.

Wiens, J. A. 1976. Population responses to patchy environments. – Annu. Rev. Ecol. Systm. 7: 81–120.

de Wilde, A. W. J. 1991. Interactions in burrowing communities and their effects on the structure of marine ecosystems. — Symp. zool. Soc. Lond. 63: 107–117.

Williamson, M. 1972. The analysis of biological populations. — Edward Arnold, London.

Wood, T. G. and Sands, W. A. 1978. The role of termites in ecosystems. — In: Brian, M. V. (ed.), Production ecology of ants and termites. Cambridge University Press, Cambridge, pp. 245–292.

Yair, A. and Rutin, J. 1981. Some aspects of the regional variation in the amount of available sediment produced by isopods and porcupines, northern Negev, Israel. — Earth Surf. Proc. Land, 6: 221–234.

15
Modeling Complex Ecological Economic Systems: Toward an Evolutionary, Dynamic Understanding of People and Nature

Robert Costanza, Lisa Wainger, Carl Folke, and Karl-Göran Mäler

Recent understanding about system dynamics and predictability that has emerged from the study of complex systems is creating new tools for modeling interactions between anthropogenic and natural systems. A range of techniques has become available through advances in computer speed and accessibility and by implementing a broad, interdisciplinary systems view.

Systems are groups of interacting, interdependent parts linked together by exchanges of energy, matter, and information. Complex systems are characterized by strong (usually nonlinear) interactions between the parts, complex feedback loops that make it difficult to distinguish cause from effect, and significant time and space lags, discontinuities, thresholds, and limits. These characteristics all result in scientists' inability to simply add up or aggregate small-scale behavior to arrive at large-scale results (Rastetter et al. 1992, von Bertalanffy 1968). Ecological and economic systems both independently exhibit these characteristics of complex systems. Taken together, linked ecological economic systems are devilishly complex.

Although almost any subdivision of the universe can be thought of as a system, modelers of systems usually look for boundaries that minimize the interaction between the system under study and the rest of the universe in order to make their job easier. The interactions between ecological and economic systems are many and strong. So, splitting the world into separate economic and ecological systems is a poor choice of boundary.

Classical (or reductionist) scientific disciplines tend to dissect their subject into smaller and smaller isolated parts in an effort to reduce the problem to its essential elements. To allow the dissection of system components, it must be assumed that interactions and feedbacks between system elements are negligible or that the links are essentially linear so they can be added up to give the behavior of the whole (von Bertalanffy 1968). Complex systems violate the assumptions of reductionist techniques and therefore are not well understood using the perspective of classical science. In contrast, systems analysis is the scientific method applied across many disciplines, scales, resolutions, and system types in an integrative manner.

In economics, for example, a typical distinction is made between partial equilibrium analysis and general equilibrium analysis. In partial equilibrium analysis, a subsystem (a single market) is studied with the underlying assumption that there are no important feedback loops from other markets. In general equilibrium analysis, on the other hand, the totality of markets are studied to bring out the general interdependence in the economy. The large-scale, whole-economy, general equilibrium effects are usually quite different from the sum of the constituent small-scale partial equilibrium effects. Add to this observation the further complication that in reality equilibrium is never achieved, and one can begin to see the limitations of classical, reductionist science in understanding complex systems.

Economic and ecological analysis needs to shift away from implicit assumptions that eliminate links within and between economic and natural systems because, due to the strength of the real-world interactions between these components, failing to link them can cause severe misperceptions and indeed policy failures (Costanza 1987). Because reductionist thinking fails in the quest to understand complex systems, new concepts and methods must be devised.

Achieving a comprehensive understanding that is useful for modeling and prediction of linked ecological economic systems requires the synthesis and integration of several different conceptual frames. As Levins (1966) has described this search for robustness, "we attempt to treat the same problem with several alternative models each with different simplifications. . . Then, if these models, despite their different assumptions, lead to similar results we have what we call a robust theorem which is relatively free of the details of the model. Hence our truth is the intersection of independent lies" (p. 423).

Existing modeling approaches can be classified according to a number of criteria, including scale, resolution, generality, realism, and precision. The most useful approach within this spectrum of characteristics depends on the specific goals of the modeling exercise. We describe here a few examples of how one might match model characteristics with several of the possible modeling goals relevant for ecological economic systems, and we claim that a better appreciation of the range of possible model characteristics and goals can help to match characteristics and goals.

Complex-systems analysis offers great potential for generating insights into the behavior of linked ecological economic systems. These insights will be needed to change the behavior of the human population toward a sustainable pattern, one that works in synergy with the life-supporting ecosystems on which it depends. The next step in the evolution of ecological economic models is to fully integrate the two fields and not just transfer methods between them. Clark's (1976, 1981, 1985) bioeconomics work was the start of this recognition of the importance of linking the mutually interacting subparts. But much work remains to be done to bring the two fields and the technology that supports them to the point where their models can adequately interact. Transdisciplinary collaboration and cooperative synthesis among natural and social scientists will be essential (Norgaard 1989).

Computers and Modeling

Until computers became available, the equations that described the dynamics of systems had to be solved analytically, severely limiting the level of complexity (as well as the resolution) of the systems that could be studied and the complexity of the dynamics that could be examined for any particular system. Table 1 shows the limits of analytical methods in solving various classes of mathematical problems in general.

Only relatively simple linear systems of algebraic or differential equations can, in general, be solved analytically. The problem is that most complex, living systems (like economies and ecosystems) are decidedly nonlinear, and efforts to approximate their dynamics with linear equations have been of only limited usefulness. In addition, complex systems often exhibit discontinuous and chaotic behavior (Rosser 1991) that can only be adequately represented with numerical methods and simulations using computers.

We differentiate here between the use of linear systems of equations to model complex-system dynamics (which does not work well) versus the use of linear systems to understand systems structure (which may work reasonably well). Integrating these views of structure and dynamics is a key item for research on complex ecological economic systems.

In recent years, computers have become not only faster but also much more accessible. This

TABLE 1. The limits of analytical methods in solving mathematical problems (after von Bertalanffy 1968). The thick solid line divides the range of problems that are solvable with analytical methods from those that are difficult or impossible using analytical methods and require numerical methods and computers to solve. Systems problems are typically nonlinear and fall in the range that requires numerical methods. It should be noted that whereas some special problems that fall in the areas labeled impossible in the table are actually possible to solve using analytical methods (frequently requiring special tricks), in general one cannot depend on a solution being available. Computers have guaranteed that a solution can be found in all the cases listed in the table.

Equations	Linear			Nonlinear		
	One equation	Several equations	Many equations	One equation	Several equations	Many equations
Algebraic	Trivial	Easy	Difficult	Very difficult	Very difficult	Impossible
Ordinary differential	Easy	Difficult	Essentially impossible	Very difficult	Impossible	Impossible
Partial differential	Difficult	Essentially impossible	Impossible	Impossible	Impossible	Impossible

ease of access has allowed researchers to develop methods to allow adaptive, evolutionary, dynamic solutions. For example, Holland and Miller (1991) describe how recent computer and machine learning (a form of artificial intelligence) advances have spawned "artificial adaptive agents," computer programs that can simulate evolution and acquire sophisticated behavioral patterns. In these programs, individual agents (e.g., processes, elements, and pieces of computer code) in networks of interacting agents reproduce themselves in the next time period based on some measure of their performance in the current time period. The system exhibits changing group behavior over time and mimics evolution. To exhibit this adaptive behavior, the actions of the agents must be assigned values, and the agents must act to increase these values over time. Algorithms like these can provide a realistic representation of ecological and economic processes.

Another useful technique is metamodeling, in which more general models are developed from detailed ones. Richard Cabe, Jason Shogren, and their colleagues (1991) have developed this technique to link models of agricultural production and economic behavior that could not normally be used together because, for one, they run at different time and space scales. Their models, which cover the entire midwestern farm belt of the United States,

provide a method for a quick and cost-efficient evaluation of ecological economic policies.

Computer hardware advances such as CRAY supercomputers and Connection Machines (massively parallel supercomputers) facilitate the modeling of complex systems using advanced numerical computation algorithms (e.g., finite difference and finite element routines, cellular automata algorithms, and emerging methods that employ at least a modicum of artificial intelligence). For example, parallel computers make high spatial resolution and regional and global ecological economic models computationally feasible (Costanza et al. 1990, Costanza and Maxwell 1991) and allow the types and resolution of evolutionary and metamodeling approaches to expand dramatically. These new capabilities, linked with a more realistic and pluralistic view of the various roles and limitations of models in understanding and decision making, can dramatically increase the effectiveness of modeling.

Purposes of Models

Models are analogous to maps. Like maps, they have many possible purposes and uses, and no one map or model is right for the entire range of uses (Levins 1966, Robinson 1991). It is inappropriate to think of models or maps as anything but crude, although in many cases absolutely essential, abstract representations of complex territory. Their usefulness can best be judged by their ability to help solve the navigational problems faced. Models are essential for policy evaluation, but they are often also misused because there is "the tendency to use such models as a means of legitimizing rather than informing policy decisions. By cloaking a policy decision in the ostensibly neutral aura of scientific forecasting, policy-makers can deflect attention from the normative nature of that decision . . . " (Robinson in press).

In the case of modeling ecological economic systems, purposes can range from developing simple conceptual models to provide a general understanding of system behavior, to detailed realistic applications aimed at evaluating specific policy proposals. It is inappropriate to judge this whole range of models by the same criteria. At minimum, the three criteria of realism (simulating system behavior in a qualitatively realistic way), precision (simulating behavior in a quantitatively precise way), and generality (representing a broad range of systems' behaviors with the same model) are necessary. Holling (1964) first described the fundamental trade-offs in modeling among these three criteria. Later, Holling (1966) and Levins (1966) expanded and further applied this classification. No single model can maximize all three of these goals, and the choice of which objectives to pursue depends on the fundamental purposes of the model. Several examples in the literature of ecological and economic models demonstrate the various ways in which trade-offs are made among realism, precision, and generality.

High-generality Conceptual Models

In striving for generality, models must give up some realism and/or precision. They can simplify relationships and/or reduce resolution. Simple linear and nonlinear economic and ecological models tend to have high generality but low realism and low precision (Brown and Swierzbinski 1985, Clark and Monroe 1975, Kaitala and Pohjola 1988, Lines 1989, 1990b).

Examples include Holling's four-box model (Holling 1987), the ecological economy model of Brown and Roughgarden (1992), most conceptual macroeconomic models (Keynes 1936, Lucas 1975), economic growth models (Solow 1956), and the evolutionary games approach. For example, the ecological economy model (Brown and Roughgarden 1992) contains only three state variables (labor, capital, and natural resources), and the relationships among these variables are highly idealized. But the purpose of the model was not high realism or precision but rather to address some basic questions about the limits of economic systems in the context of their dependence on an ecological life-support base.

High-precision Analytical Models

Often one wants high precision (quantitative correspondence between data and model) and is willing to sacrifice realism and generality. One strategy here is to keep resolution high but to simplify relationships and deal with short time frames. Some models strive to strike a balance between mechanistic small-scale models that trace small fluctuations in a system and more general whole-system approaches that remove some of the noise from the signal but do not allow the modeler to trace the source of system changes. The alternative some ecologists have devised is to identify one or a few properties that characterize the system as a whole (Wulff and Ulanowicz 1989). For example, Hannon and Joiris (1987) used an economic input-output model to examine relationships between biotic and abiotic stocks in a marine ecosystem; they found that this method allowed them to show the direct and indirect connection of any species to any other and to the external environment in this system at high precision (but low generality and realism). Also using input-output techniques, Duchin's (1988, 1992) aim was to direct development of industrial production systems to efficiently reduce and recycle waste in the manner of ecological systems. Large econometric models (Klein 1971) used for predicting short-run behavior of the economy belong to this class of models, because they are constructed to fit existing data as closely as possible, at the sacrifice of generality and realism.

High-realism Impact-analysis Models

When the goal is to develop realistic assessments of the behavior of specific complex systems, generality and precision must be relaxed. High-realism models are concerned with accurately representing the underlying processes in a specific system, rather than with precisely matching quantitative behavior or being generally applicable. Dynamic, nonlinear, evolutionary systems models at moderate to high resolution generally fall into this category. Coastal physical-biological-chemical models (Wroblewski and Hofmann 1989), which are used to investigate nutrient fluxes and contain large amounts of site-specific data, fall into this category, as do micromodels of behavior of particular business activities. Another example is a model of coastal landscape dynamics (Costanza et al. 1990), which includes high spatial and temporal resolution and complex nonlinear process dynamics. This model divides a coastal landscape into 1-square-kilometer cells, each of which contains a process-based dynamic ecological simulation model. Flows of water, sediments, nutrients, and biomass from cell to cell across the landscape are linked with internal ecosystem dynamics to simulate long-term successional processes and responses to various human impacts in a realistic way. But the model is site specific and of only moderate numerical precision.

Moderate-generality and Moderate-precision Indicator Models

In many types of systems modeling, the desired outcome is to accurately determine the overall magnitude and direction of change, trading off realism for some moderate amount of generality and precision. For example, aggregate measures of system performance such as standard gross national product, environmentally adjusted net national product (or green NNP), which includes environmental costs (Mäler 1991), and indicators of ecosystem health (Costanza et al. 1992) fit into this category. The

microcosm systems employed by Taub (1989) allow some standardization for testing ecosystem responses and developing ecosystem performance indices. Taub (1987) notes, however, that many existing indicators of change in ecosystems are based on implicit ecological assumptions that have not been critically tested, either for their generality, realism, or precision.

Scale and Hierarchy

In modeling complex systems, the issues of scale and hierarchy are central (O'Neill et al. 1989). Some claim that the natural world, the human species included, contains a convenient hierarchy of scales based on interaction-minimizing boundaries: scales ranging from atoms to molecules to cells to organs to organisms to populations to communities to ecosystems (including economic and/or human-dominated ecosystems) to bioregions to the global system and beyond (Allen and Starr 1982, O'Neill et al. 1986). By studying the similarities and differences among different kinds of systems at different scales and resolutions, one might develop hypotheses and test them against other systems to explore their degree of generality and predictability.

The term *scale* in this context refers to both the resolution (spatial grain size, time step, or degree of complication of the model) and extent (in time, space, and number of components modeled) of the analysis. The process of *scaling* refers to the application of information or models developed at one scale to problems at other scales. In both ecology and economics, primary information and measurements are generally collected at relatively small scales (i.e., small plots in ecology or individuals or single firms in economics), and that information is then often used to build models at radically different scales (i.e., regional, national, or global). The process of scaling is directly tied to the problem of aggregation (the process of adding or otherwise combining components), which in complex, nonlinear, discontinuous systems (like ecological and economic systems) is far from a trivial problem (O'Neill and Rust 1979, Rastetter et al. 1992). For example, in applied economics, basic data sets are usually derived from national accounts that contain data that are linearly aggregated over individuals, companies, or organizations. Sonnenschein (1974) and Debreu (1974) have shown that, unless one makes strong and unrealistic assumptions about the individual units, the aggregate (large-scale) relations between variables have no resemblance to the corresponding relations on the smaller scale.

Rastetter et al. (1992) describe and compare three basic methods for scaling that are applicable to complex systems. All of their methods are attempts to use information about the nonlinear small-scale variability in the large-scale models. They list partial transformations of the fine-scale mathematical relationships to coarse scale using a statistical expectations operator that incorporates the fine-scale variability; partitioning or subdividing the system into smaller, more homogeneous parts (i.e., spatially explicit modeling); and calibration of the fine-scale relationships to coarse-scale data when this data is available. They go on to suggest a combination of these methods as the most effective overall method of scaling in complex systems.

A primary reason for aggregation error in scaling complex systems is the nonlinear variability in the fine-scale phenomenon. For example, Rastetter et al. (1992) give a detailed example of scaling a relationship for individual leaf photosynthesis as a function of radiation and leaf efficiency to estimate the productivity of the entire forest canopy. Because of nonlinear variability in the way individual leaves process light energy, one introduces significant aggregation error by simply using the fine-scale relationships among photosynthesis, radiation, and efficiency along with the average values for the entire forest to get total forest productivity.

One must somehow understand and incorporate this nonlinear fine-scale variability into the coarse-scale equations using some combination of the three methods mentioned above. The statistical expectations method implies deriving new coarse-scale equations that incorporate the fine-scale variability. The problem is that incorporation of this variability often leads to equations that are extremely complex and cumbersome (Rastetter et al. 1992). The partitioning method implies subdividing the forest into many relatively more homogeneous levels or zones and applying the basic fine-scale equa-

tions for each partition. This approach requires a method for adjusting the parameters for each partition, a choice of the number of partitions (the resolution), and an understanding of the effects of the choice of resolution and parameters on the results. The recalibration method implies simply recalibrating the fine-scale equations to coarse-scale data. It presupposes that coarse-scale data are available (as more than simply the aggregation of fine-scale data). In many important cases, however, this coarse-scale data is either extremely limited or is not available. Thus, although a judicious application of all three aggregation methods is necessary, from the perspective of complex systems modeling, the partitioning approach seems to hold particular promise, because it can take fullest advantage of emerging computer technologies and databases.

From the scaling perspective, hierarchy theory is a potentially useful tool for partitioning systems in ways that minimize aggregation error. According to hierarchy theory, nature can be partitioned into naturally occurring levels, which share similar time and space scales and which interact with higher and lower levels in systematic ways. Each level in the hierarchy experiences the higher levels as constraints and the lower levels as noise. For example, individual organisms experience the ecosystem they inhabit as a slowly changing set of constraints, and the operation of their component cells and organs is what matters most to them. However, Norton and Ulanowicz (1992) suggest that what appears to be noise at a lower level could be turned into significant perturbations on the higher level. This change can happen when a critical mass of components participate in a trend, a behavioral pattern, that affects the slower processes at the higher level. The rapid and extensive human uses of fossil fuels could be seen as such a trend, causing perturbations at the global atmospheric level, which might feed back and radically alter the framework of action at the lower level.

Shugart et al. (1991) explains the relationship between scales: "Clearly, natural patterns in environmental constraints contribute substantially to the spatial pattern and temporal dynamics of particular ecosystems . . . these patterns, especially temporal ones, may resonate with natural frequencies of plant growth forms (i.e., phenology and longevity) to amplify environmental patterns" (p. 232). The simplifying assumptions of hierarchy theory may ease the problem of scaling by providing a common (but somewhat generalized) set of rules that could be applied at any scale in the hierarchy.

Fractals and Chaos

The concept of fractals (Mandelbrot 1977) can be seen as another related approach to the problem of scaling, based on the fundamental principle of self-similarity between scales. This concept implies a regular and predictable relationship between the scale of measurement (here meaning the resolution of measurement) and the measured phenomenon. For example, the measured length of a coastline is an increasing function of the resolution at which it is measured. At higher resolutions, one can recognize and measure more of the small-scale bays and indentations in the coast and the total length measured increases.

The relationship between length and resolution usually follows a regular pattern that can be summarized in the following equation:

$$L = k \, s^{(1-D)}$$

where L equals the length of the coastline or other fractal boundary, s equals the size of the fundamental unit of measure or the resolution of the measurement, k equals a scaling constant, and D equals the fractal dimension.

Phenomena that fit this equation are said to be self-similar because, as resolution is increased, one perceives patterns at the smaller scale similar to those at the larger scale. This convenient scaling rule has proven useful in describing many kinds of complex boundaries and behaviors (Mandelbrot 1983, Milne 1991, Olsen and Schaffer 1990, Sugihara and May 1990, Turner et al. 1989). One test of the principle of self-similarity is that it can be applied to produce computer-generated shapes that have a decidedly natural and organic look to them (Mandelbrot 1977).

Certain nonlinear dynamical systems models exhibit behaviors whose phase plots ($x[t]$ versus $x[t-dt]$) are fractals. These chaotic attractors,

as they are called, are one of four possible pure types of attractors that can be used to classify system dynamics. The other three are point attractors (indicating stable, non-time varying behavior), periodic attractors (indicating periodic time behavior), and noisy attractors (indicating stochastic time behavior). Real-system behavior can be thought of as representing some combination of these four basic types.

The primary questions about the range of applicability of fractals and chaotic-systems dynamics to the practical problems of modeling ecological economic systems are the influence of scale, resolution, and hierarchy on the mix of behaviors one observes in systems. This problem is key for extrapolating from small-scale experiments or simple theoretical models to practical applied models of ecological economic systems.

Resolution and Predictability

The significant effects of nonlinearities raise some interesting questions about the influence of resolution (including spatial, temporal, and component) on the performance of models, in particular on their predictability. For example, the relationship between the degree of complication (the number of components included) and the predictability of models is an important input to model design. Hofmann (1991) discusses this concern in the context of scaling coastal models to the global scale. The difficulty of using aggregate models that integrate over many details of finer resolution models is that the aggregated models may not be able to represent biological processes on the space and time scales necessary. Hofmann suggests that coupled detailed models (in which the output of one model becomes the input for another) may be a more practical method for scaling models to larger systems.

Costanza and Maxwell (in press) analyzed the relationship between spatial resolution and predictability and found that, although increasing resolution provides more descriptive information about the patterns in data, it also increases the difficulty of accurately modeling those patterns. There may be limits to the predictability of natural phenomenon at particular resolutions, and scaling rules that determine how both "data" and "model" predictability change with resolution.

Predictability (Colwell 1974) measures the reduction in uncertainty about one variable given knowledge of others using categorical data. One can define spatial autopredictability (P_a) as the reduction in uncertainty about the state of a pixel in a scene, given knowledge of the state of adjacent pixels in that scene, and spatial cross-predictability (P_c) as the reduction in uncertainty about the state of a pixel in a scene, given knowledge of the state of corresponding pixels in other scenes. P_a is a measure of the internal pattern in the data, whereas P_c is a measure of the ability of a model to represent that pattern.

Some limited testing of the relationship between resolution and predictability (by resampling land-use map data at different spatial resolutions) showed a strong linear relationship between the log of P_a and the log of resolution (measured as the number of pixels per square kilometer). This fractal-like characteristic of self-similarity with decreasing resolution implies that predictability, like the length of a coastline, may be best described using a unitless dimension that summarizes how it changes with resolution. One can define a fractal predictability dimension (D_P) in a manner analogous to the normal fractal dimension that summarizes this relationship. D_P allows convenient scaling of predictability measurements taken at one resolution to other resolutions.

Cross-predictability (P_c) can be used for pattern matching and testing the fit between map scenes. In this sense, it relates to the predictability of models versus the internal predictability in the data revealed by P_a. Although P_a generally increases with increasing resolution (because more information is being included), P_c generally falls or remains stable (because it is easier to model aggregate results than fine-grain ones). Thus, we can define an optimal resolution for a particular modeling problem that balances the benefit in terms of increasing data predictability (P_a) as one increases resolution, with the cost of decreasing model predictability (P_c). Figure 1 shows this relationship in generalized form.

These results may be generalizable to all forms of resolution (spatial, temporal, and

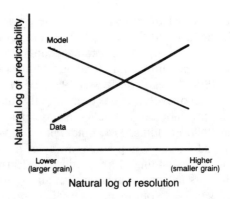

FIGURE 1. Hypothetical relationship between resolution and predictability of data and models, plotted on log-log axes (from Costanza and Maxwell in press). Data predictability is a measure of the internal pattern in the data (e.g., the degree to which the uncertainty about the state of landscape pixels is reduced by knowledge of the state of adjacent pixels in the same map). Model predictability is a measure of the correspondence between data and models (e.g., the degree to which the uncertainty about the state of pixels is reduced by knowledge of the corresponding state of pixels in a model of the system). In general, data predictability rises with increasing resolution (because more internal patterns are perceived), whereas model predictability falls (because it becomes more difficult to match the high-resolution patterns). Particular types of models and data sets would fall on different lines, and certain types of models would require certain types of data. An optimal resolution occurs where the data and model predictability lines intersect.

number of components) and may shed some light on chaotic behavior in systems. When looking across resolutions, chaos may be the low level of model predictability that occurs as a natural consequence of high resolution. Lowering model resolution can increase model predictability by averaging out some of the chaotic behavior, at the expense of losing detail about the phenomenon. For example, Sugihara and May (1990) found chaotic dynamics for measles epidemics at the level of individual cities, but more predictable periodic dynamics for whole nations.

Evolutionary Approaches

In modeling the dynamics of complex systems, it is impossible to ignore the discontinuities and surprises that often characterize these systems and the fact that they operate far from equilibrium in a state of constant adaptation to changing conditions (Holland and Miller 1991, Kay 1991, Lines 1990, Rosser 1991, 1992). The paradigm of evolution has been broadly applied to both ecological and economic systems (Arthur 1988, Boulding 1981, Lindgren 1991, Maxwell and Costanza in press) as a way of formalizing understanding of adaptation and learning behaviors in nonequilibrium dynamic systems. The general evolutionary paradigm posits a mechanism for adaptation and learning in complex systems at any scale using three basic interacting processes: information storage and transmission, generation of new alternatives, and selection of superior alternatives according to some performance criteria.

The evolutionary paradigm is different from the conventional optimization paradigm popular in economics in at least four important respects (Arthur 1988): evolution is path dependent, meaning that the detailed history and dynamics of the system are important; evolution can achieve multiple equilibria; there is no guarantee that optimal efficiency or any other optimal performance will be achieved, due in part to path dependence and sensitivity to perturbations; and lock-in (survival of the first rather than survival of the fittest) is possible under conditions of increasing returns. Arthur (1988) notes, "conventional economic theory is built largely on the assumption of diminishing returns on the margin (local negative feedbacks)," but life itself can be characterized as a positive feedback, self-reinforcing, autocatalytic process (Günther and Folke in press, Kay 1991), and we should expect increasing returns, lock-in, path dependence, multiple equilibria, and suboptimal efficiency to be the rule rather than the exception in economic and ecological systems.

Cultural versus Genetic Evolution

In biological evolution, the information storage medium is the genes, the generation of new alternatives is by sexual recombination or ge-

netic mutation, and selection is performed by nature according to criteria of fitness based on reproductive success. The same process of change occurs in ecological, economic, and cultural systems, but the elements on which the process works are different. For example, in cultural evolution the storage medium is the culture (the oral tradition, books, film, or other storage medium for passing on behavioral norms), the generation of new alternatives is through innovation by individual members or groups in the culture, and selection is again based on the reproductive success of the alternatives generated. Reproduction is carried out by the spread and copying of the behavior through the culture rather than by biological reproduction.

One may also talk of economic evolution, a subset of cultural evolution dealing with the generation, storage, and selection of alternative ways of producing things and allocating that which is produced. The field of evolutionary economics has grown up in approximately the last decade based on these ideas (cf. Day 1989, Day and Groves 1975). Evolutionary theories in economics have already been successfully applied to problems of technical change, to the development of new institutions, and to the evolution of means of payment.

For large, slow-growing animals like humans, genetic evolution has a built-in bias toward the long run. Changing the genetic structure of a species requires that characteristics (phenotypes) be selected and accumulated by differential reproductive success. Behaviors learned or acquired during the lifetime of an individual cannot be passed on genetically. Genetic evolution is therefore usually a relatively slow process requiring many generations to significantly alter a species' physical and biological characteristics.

Cultural evolution is potentially much faster. Technical change is perhaps the most important and fastest-evolving cultural process. Learned behaviors that are successful, at least in the short term, can be almost immediately spread to other members of the culture and passed on in the oral, written, or video record. The increased speed of adaptation that this process allows has been largely responsible for *Homo sapiens'* amazing success at appropriating the resources of the planet. Vitousek et al. (1986) estimate that humans directly control from 25% to 40% of the total primary production of the planet's biosphere, and this control is beginning to have significant effects on the biosphere, including changes in global climate and in the planet's protective ozone shield.

The costs of this rapid cultural evolution, therefore, are potentially significant. Like a car that has increased speed, humans are in more danger of running off the road or over a cliff. Cultural evolution lacks the built-in long-run bias of genetic evolution and is susceptible to being led by its hyperefficient short-run adaptability over a cliff into the abyss.

Another major difference between cultural and genetic evolution may serve as a countervailing bias, however. As Arrow (1962) has pointed out, cultural and economic evolution, unlike genetic evolution, can to some extent employ foresight. If society can see the cliff, perhaps it can be avoided.

Although market forces drive adaptive mechanisms (Kaitala and Pohjola 1988), the systems that evolve are not necessarily optimal, so the question remains: What external influences are needed, and when should they be applied to achieve an optimal economic system via evolutionary adaptation? The challenge faced by ecological economic systems modelers is to first apply the models to gain foresight and then to respond to and manage the system feedbacks in a way that helps avoid any foreseen cliffs (Berkes and Folke in press). Devising policy instruments and identifying incentives that can translate this foresight into effective modifications of the short-run evolutionary dynamics is the challenge (Costanza 1987).

Evolutionary Criteria

A critical problem in applying the evolutionary paradigm in dynamic models is defining the selection criteria a priori. In its basic form, the theory of evolution is circular and descriptive (Holling 1987). Those species or cultural institutions or economic activities survive that are the most successful at reproducing themselves. But we only know which ones were more successful after the fact. To use the evolutionary

paradigm in modeling, we require a quantitative measure of fitness (or more generally performance) to drive the selection process.

Several candidates have been proposed for this function in various systems, ranging from expected economic utility to thermodynamic potential. Thermodynamic potential is interesting as a performance criterion in complex systems because even simple chemical systems can be seen to evolve complex nonequilibrium structures using this criterion (Nicolis and Prigogine 1977, 1989, Prigogine 1972), and all systems are (at minimum) thermodynamic systems (in addition to their other characteristics). Therefore, thermodynamic constraints and principles are applicable across both ecological and economic systems (Eriksson 1991).

This application of the evolutionary paradigm to thermodynamic systems has led to the development of far-from-equilibrium thermodynamics and the concept of dissipative structures (Prigogine 1972). An important research question is to determine the range of applicability of these principles and their appropriate use in modeling ecological economic systems.

Many dissipative structures follow complicated transient motions. Schneider and Kay (in press) propose a way to analyze these chaotic behaviors and note that, "Away from equilibrium, highly ordered stable complex systems can emerge, develop and grow at the expense of more disorder at higher levels in the system's hierarchy." It has been suggested that the integrity of far-from-equilibrium systems has to do with the ability of the system to attain and maintain its (set of) optimal operating point(s) (Kay 1991). The optimal operating point(s) reflect a state where self-organizing thermodynamic forces and disorganizing forces of environmental change are balanced. This idea has been elaborated and described as "evolution at the edge of chaos" by Bak and Chen (1991) and Kauffman and Johnson (1991).

The concept that a system may evolve through a sequence of stable and unstable stages leading to the formation of new structures seems well suited to ecological economic systems. For example, Gallopin (1989) stresses that to understand the processes of economic impoverishment "The focus must necessarily shift from the static concept of poverty to the

dynamic processes of impoverishment and sustainable development within a context of permanent change. The dimensions of poverty
cannot any longer be reduced to only the economic or material conditions of living; the
capacity to respond to changes, and the vulnerability of the social groups and ecological systems to change become central" (p. 394).

In a similar fashion, Robinson (1991) argues
that sustainability calls for maintenance of the
dynamic capacity to respond adaptively, which
implies that we should focus more on basic
natural and social processes than on the particular forms these processes take at any time.
Berkes and Folke (in press) have discussed the
capacity to respond to changes in ecological
economic systems, in terms of institution building, collective actions, cooperation, and social
learning. These activities might enhance the
capacity for resilience (increase the capacity to
recover from disturbance) in interconnected
ecological economic systems.

The Holling Model

One broad conceptual application of these
ideas to ecological and economic systems, with
the goal of maximal generality, is the model of
Holling (1987, 1992). Holling proposes four
basic functions common to all complex systems
and a spiraling evolutionary path through
them (Figure 2). The functions (boxes) are:
exploitation (e.g., r-strategists, pioneers, op-
portunists, and entrepreneurs), conservation
(e.g., K-strategists, climax ecosystems, consol-
idation, and rigid bureaucracies), release (e.g.,
fire, storms, pests, and political upheavals),
and reorganization (e.g., accessible nutrients
and abundant natural resources). Within this
model, systems evolve from the rapid colonization and exploitation phase, during which they
capture easily accessible resources, to the conservation stage of building and storing increasingly complex structures. Examples of the ex-

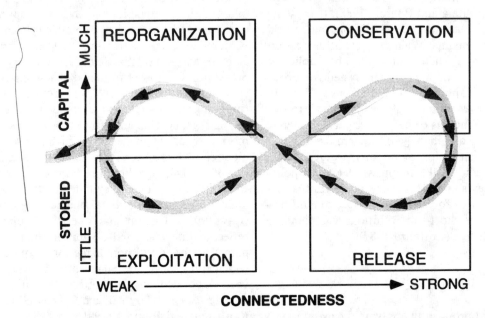

FIGURE 2. The four general system functions and the flow of events between them (from Holling 1987, 1992).
The arrows show the speed of that flow in the ecosystem cycle; arrows close to each other indicate a rapidly
changing situation and arrows far from each other indicate a slowly changing situation. The cycle reflects
changes in two attributes: on the Y axis, the amount of accumulated capital (nutrients and carbon) stored in
variables that are dominant keystone variables at the moment, and, on the X axis, the degree of
connectedness among variables. The exit from the cycle indicated at the left of the figure indicates the stage
where a flip is most likely into a less- or more-productive and organized system, that is, devolution or
evolution as revolution.

ploitation phase are early successional ecosystems colonizing disturbed sites or pioneer societies colonizing new territories. Examples of the conservation phase are climax ecosystems or large, mature bureaucracies.

The release or "creative destruction" (Schumpeter 1950) phase represents the breakdown of mature structures via aperiodic events such as fire, storms, pests, or political upheavals. The released structure is then available for reorganization and uptake in the exploitation phase. The amount of ongoing creative destruction that takes place in the system is critical to its behavior. The conservation phase can often build elaborate and tightly bound structures by severely limiting creative destruction (the former Soviet Union is a good example), but these structures become brittle and susceptible to massive and widespread destruction. If some moderate level of release is allowed to occur on a more routine basis, the destruction is on a smaller scale and leads to a more resilient system. It could be argued that patterns of behavior with moderate levels of ongoing creative destruction evolved in those local communities and human cultures that managed to survive for thousands of years or more.

Creative destruction, in terms of shocks or surprises, seems to be crucial for system resilience and integrity. Similarly, it has been argued that episodic events, such as the Chernobyl accident, the Rhine chemical spill, and the death of seals in the North Sea, are shocks to the social-cultural value system and may stimulate positive change toward more resilient ecological economic systems (Berkes and Folke in press).

Fire climax systems, such as the pine forests of Yellowstone National Park, are a good example of the range of possibilities for creative destruction. In its unmanaged state, Yellowstone burned over extensive areas relatively often, but the high fire frequency kept the amount of fuel insufficient to create extremely destructive fires. The more-frequent, small- to moderate-size fires released nutrients stored in the litter and supported a spurt of new growth without destroying all the old growth. On the other hand, when fires were suppressed and controlled, fuel built up to high levels and (because control and suppression

are never perfect—remember the former Soviet Union), when the fire did come it wiped out much of the forest.

The Holling four-box model may serve as a minimal ecological economic model aimed at generality (at the expense of precision and realism). It raises some interesting questions about the relationships among diversity, stability, resilience, control, creativity, surprise, and evolution in ecological and economic systems that are ripe for further analyses.

Evolutionary Game Theory

Evolutionary game theory is the combination of traditional game theory and evolutionary models. The evolution of evolutionary game theory is itself quite interesting, because it relied on several interacting disciplines. Game theory began with von Neuman (1928) as a mathematical exercise for analyzing parlor games. It continued with von Neuman and Morgenstern (1944), who developed the theory for applications in economics. The central idea is that a game consists of a number of players who all act rationally (i.e., use the information available to select strategies that will maximize their expected payoff). The situation when all players have picked their optimal strategies and no one can expect to increase their payoff is called a Nash equilibrium. Conventional game theory with rational players has had an enormous impact on economics, and an increasing number of economic problems are being studied using its tools.

Game theory was imported into evolutionary biology (Maynard-Smith and Price 1973, Maynard-Smith 1979, 1982) to improve understanding of biological processes. There was, however, an important change in the interpretation. Whereas economists used static, rational strategic choices on the part of the players, Maynard-Smith introduced evolution by identifying strategies with genes and the payoff with reproductive success. Reproductive strategies with high payoffs would be expected to have a proportionally higher representation in the population. A Nash equilibrium corresponds to an evolutionarily stable strategy (i.e., a strategy that would be immune to invasion by other strategies).

After this further development within evolutionary biology, evolutionary game theory was then reimported into economics when it became clear that it could be used for an improved understanding of the evolution of various economic institutions (e.g., means of payments and property rights) and of technical processes in production. Economists interested in evolutionary game theory have even started reapplying it to biological evolution (Selten 1980), and political scientists and others have developed it for the interdisciplinary analysis of the evolution of cooperation in both economic and ecological systems (Axelrod 1984).

Through research within four disciplines — mathematics, biology, economics, and political science — evolutionary game theory has developed as a rather important tool for understanding these social and biological processes in isolation. To address ecological economic systems, conventional game theory must be integrated with evolutionary game theory. This integration would include the analysis of games in which different subsets of the players have different time horizons, payoff structures, and objectives. For example, some organisms within the system may choose strategies not only with regard to reproductive success but also with regard to other goals. Some progress has been made along these lines (e.g., Banerjee and Weibull 1991), and it seems to be a fruitful area for future research.

Summary of Questions and Opportunities

Based on this synthesis, the major researchable questions and opportunities in modeling complex ecological economic systems can be divided into three broad, interdependent categories. These categories are listed below.

Application of the Evolutionary Paradigm to Modeling Ecological Economic Systems

The evolutionary paradigm provides a general framework for complex ecological economic systems dynamics. It incorporates the elements of uncertainty, surprise, learning, path dependence, multiple equilibria, suboptimal performance, lock-in, and thermodynamic constraints. In applying the evolutionary paradigm, a key feature is the choice of the measure (or multiple measures) of performance on which the system's selection process will work. Several such measures have been proposed and partially tested, but additional research and testing in this area may have a high payoff. An important research question is the range of applicability of nonequilibrium thermodynamic principles and their appropriate use in modeling ecological economic systems. Key methods include adaptive computer simulation models and integrated conventional/evolutionary game theory.

Scale and Hierarchy Considerations in Modeling Ecological Economic Systems

The key questions involve exactly how hierarchical levels interact with each other and how to further develop the three basic methods of scaling (statistical expectations, partitioning, and recalibration) for application to complex ecological economic systems. Additional questions concern the range of applicability of fractals and chaotic-systems dynamics to the practical problems of modeling ecological economic systems. In particular, what is the influence of scale, resolution, and hierarchy on the mix of behaviors one observes in systems? This question is key for extrapolating from small-scale experiments or simple theoretical models to practical applied models of ecological economic systems at regional and global scales.

The Nature and Limits of Predictability in Modeling Ecological Economic Systems

The significant effects of nonlinearities raise some interesting questions about the influence of resolution (including spatial, temporal, and component) on the performance of models, in particular on their predictability. There may be limits to the predictability of natural phenomena at particular resolutions, and fractal-like rules that determine how both data and model predictability change with resolution. To test these limits, we need better measures of model correspondence with reality and long-term, aggregate-system performance that incorporate the three conflicting criteria of generality, realism, and precision.

Acknowledgments

The Beijer International Institute for Ecological Economics, a research institute of the Royal Swedish Academy of Sciences, provided support for the preparation of this article, which was presented at the workshop on Complex Ecological Economic Systems Modeling held at the Beijer Institute in Stockholm, Sweden, 27–29 July 1992. We thank C. S. Holling, J. Kay, R. E. Ulanowicz, J. Bartholomew, and four anonymous reviewers for detailed comments and suggestions on earlier drafts. We are also indebted to the other participants at the Beijer workshop (G. Brown, C. Clark, P. Dasgupta, F. Duchin, K.-E. Eriksson, S. O. Funtowicz, G. Gallopin, C. Lindgrend, B. Hannon, E. Hofmann, H. Isomäki, R. Kaufmann, V. Kaitala, M. Lines, T. Maxwell, J. McGlade, J. Robinson, C. Perrings, J. B. Rosser, J. Roughgarden, J. Shogren, and F. Taub) for their valuable input and feedback on the article. The article also benefited from interactions with faculty from the University of Maryland's Multiscale Experimental Ecosystem Research Center, which is devoted to research on many of the topics we have identified.

References

Allen, T. F. H., and T. B. Starr. 1982. *Hierarchy*. University of Chicago Press, Chicago.

Arrow, K. 1962. The economic implications of learning by doing. *Review of Economic Studies*. 29:155–173.

Arthur, W. B. 1988. Self-reinforcing mechanisms in economics. Pages 9–31 in P. W. Anderson, K. J. Arrow, and D. Pines, eds. *The Economy as an Evolving Complex System*. Addison-Wesley, Redwood City, CA.

Axelrod, R. 1984. *Evolution of Cooperation*. Basic Books, New York.

Bak, P., and K. Chen. 1991. Self-organized criticality. *Sci. Am.* 264:46.

Banerjee, A., and J. W. Weibull. 1991. Evolutionary selection and rational behavior. Research Papers in Economics 4, Department of Economics, University of Stockholm, Sweden.

Berkes, F., and C. Folke. In press. Investing in cultural capital for a sustainable use of natural capital. In A. M. Jansson, C. Folke, R. Costanza, and M. Hammer, eds. *Investing in Natural Capital: The Ecological Economic Approach to Sustainability*.

Boulding, K. E. 1981. *Evolutionary Economics*. Sage, Beverly Hills, CA.

Brown, G. M., and J. Roughgarden. 1992. An ecological economy: notes on harvest and growth. Beijer Discussion Paper Series 12, Beijer International Institute of Ecological Economics, Stockholm, Sweden.

Brown, G. M., and J. Swierzbinski. 1985. Endangered species, genetic capital and cost-reducing R&D. Pages 111–127 in D. O. Hall, N. Myers, and N. S. Margaris, eds. *Economics of Ecosystems Management*. Dr. W. Junk Publ., Dordrecht, The Netherlands.

Cabe, R., J. Shogren, A. Bouzaher, and A. Carriquiry. 1991. Metamodels, response functions, and research efficiency in ecological economics. Working paper 91-WP 79, Center for Agricultural and Rural Development, Iowa State University, Ames.

Clark, C. W. 1976. *Mathematical Bioeconomics*. John Wiley & Sons, New York.

———. 1981. Bioeconomics of the ocean. *BioScience* 31:231–237.

———. 1985. *Bioeconomic Modelling and Fisheries Management*. John Wiley & Sons, New York.

Clark, C. W., and G. R. Munro. 1975. The economics of fishing and modern capital theory: a simplified approach. *J. Environ. Econ. Manage.* 2:92–106.

Colwell, R. K. 1974. Predictability, constancy, and contingency of periodic phenomena. *Ecology* 55:1148–1153.

Costanza, R. 1987. Social traps and environmental policy. *BioScience* 37:407–412.

Costanza, R., and T. Maxwell. 1991. Spatial ecosystem modeling using parallel processors. *Ecol. Model.* 58:159–183.

———. In press. Resolution and predictability: an approach to the scaling problem. *Landscape Ecol.*

Costanza, R., B. Norton, and B. J. Haskell, eds. 1992. *Ecosystem Health: New Goals for Environmental Management*. Island Press, Washington, DC.

Costanza, R., F. H. Sklar, and M. L. White. 1990. Modeling coastal landscape dynamics. *BioScience* 40:91–107.

Day, R. H. 1989. Dynamical systems, adaptation and economic evolution. MRG Working Paper no. M8908, University of Southern California, Los Angeles.

Day, R. H., and T. Groves, eds. 1975. *Adaptive Economic Models*. Academic, New York.

Debreu, G. 1974. Excess demand functions. *Journal of Mathematical Economics* 1:15–23.

Duchin, F. 1988. Analyzing structural change in the economy. Pages 113–128 in M. Ciaschini, ed. *Input-Output Analysis: Current Developments*. Chapman and Hall, New York.

_____ . 1992. Industrial input-output analysis: implications for industrial ecology. *Proc. Natl. Acad. Sci.* 89:851–855.

Eriksson, K.-E. 1991. Physical foundations of ecological economics. Pages 186–196 in L. O. Hansson and B. Jungen, eds. *Human Responsibility and Global Change*. University of Göteborg Press, Göteborg, Sweden.

Gallopin, G. C. 1989. Global impoverishment, sustainable development and the environment: a conceptual approach. *International Social Science Journal* 121:375–397.

Günther, F., and C. Folke. In press. Characteristics of nested living systems. *Journal of Biological Systems*.

Hannon, B., and C. Joiris. 1987. A seasonal analysis of the southern North Sea ecosystem. *Ecology* 70:1916–1934.

Hofmann, E. E. 1991. How do we generalize coastal models to global scale? Pages 401–417 in R. F. C. Mantoura, J. M. Martin, and R. Wollast, eds. *Ocean Margin Processes in Global Change*. John Wiley & Sons, New York.

Holland, J. H., and J. H. Miller. 1991. Artificial adaptive agents in economic theory. *American Economic Review* 81:365–370.

Holling, C. S. 1964. The analysis of complex population processes. *Can. Entomol.* 96:335–347.

_____ . 1966. The functional response of invertebrate predators to prey density. *Mem. Entomol. Soc. Can.* 48.

_____ . 1987. Simplifying the complex: the paradigms of ecological function and structure. *European Journal of Operational Research* 30:139–146.

_____ . 1992. Cross-scale morphology, geometry and dynamics of ecosystems. *Ecol. Monogr.* 62:447–502.

Kaitala, V., and M. Pohjola. 1988. Optimal recovery of a shared resource stock: a differential game model with efficient memory equilibria. *Nat. Res. Model.* 3:91–119.

Kauffman, S. A., and S. Johnson. 1991. Coevolution to the edge of chaos: coupled fitness landscapes, poised states, and coevolutionary avalanches. *J. Theor. Biol.* 149:467–505.

Kay, J. J. 1991. A nonequilibrium thermodynamic framework for discussing ecosystem integrity. *Environ. Manage.* 15:483–495.

Keynes, J. M. 1936. *General Theory of Employment, Interest and Money*. Harcourt Brace, London, UK.

Klein, L. R. 1971. Forecasting and policy evaluation using large-scale econometric models: the state of the art. Pages 133–177 in M. D. Intriligator, ed. *Frontiers of Quantitative Economics*. North-Holland Publ., Amsterdam, The Netherlands.

Levins, R. 1966. The strategy of model building in population biology. *Am. Sci.* 54:421–431.

Lindgren, K. 1991. Evolutionary phenomena in simple dynamics. Pages 295–312 in C. G. Langton, C. Taylor, J. D. Farmer, and S. Rasmussen. *Artificial Life*. Addison-Wesley, Redwood City, CA.

Lines, M. 1989. Environmental noise and nonlinear models: a simple macroeconomic example. *Economic Notes* 19:376–394.

_____ . 1990. Stochastic stability considerations: a nonlinear example. *International Review of Economics and Business* 37:219–233.

Lucas, R. E. 1975. An equilibrium model of the business cycle. *Journal of Political Economy* 83:1113–45.

Mäler, K.-G. 1991. National accounts and environmental resources. *Environmental and Resource Economics* 1:1–15.

Mandelbrot, B. B. 1977. *Fractals: Form, Chance and Dimension*. W.H. Freeman, San Francisco, CA.

_____ . 1983. *The Fractal Geometry of Nature*. W.H. Freeman, San Francisco, CA.

Maxwell, T., and R. Costanza. In press. An approach to modelling the dynamics of evolutionary self-organization. *Ecol. Model.*

Maynard-Smith, J. 1979. Game theory and the evolution of behavior. *Proc. R. Soc. Lond. B* 205:475–488.

_____ . 1982. *Evolution and the Theory of Games*. Cambridge University Press, New York.

Maynard-Smith, J., and G. R. Price. 1973. The logic of animal conflict. *Nature* 246:15–18.

Milne, B. T. 1991. Lessons from applying fractal models to landscape patterns. Pages 199–235 in M. G. Turner and R. Gardner, eds. *Quantitative Methods in Landscape Ecology*. Springer-Verlag, New York.

Nicolis, G., and I. Prigogine. 1977. *Self-organization in Non-equilibrium Systems*. John Wiley & Sons, New York.

_____ . 1989. *Exploring Complexity*. W.H. Freeman, New York.

Norgaard, R. B. 1989. The case for methodological-pluralism. *Ecological Economics* 1:37–57.

Norton, B. G., and R. E. Ulanowicz. 1992. Scale and biodiversity policy: a hierarchical approach. *Ambio* 21:244–249.

Olsen, L. F., and W. M. Schaffer. 1990. Chaos versus noisy periodicity: alternative hypotheses for childhood epidemics. *Science* 249:499–504.

O'Neill, R. V., D. L. DeAngelis, J. B. Waide, and T. F. H. Allen. 1986. *A Hierarchical Concept of Ecosystems.* Princeton University Press, Princeton, NJ.

O'Neill, R. V., A. R. Johnson and A. W. King. 1989. A hierarchical framework for the analysis of scale. *Landscape Ecol.* 3:193–205.

O'Neill, R. V., and B. Rust. 1979. Aggregation error in ecological models. *Ecol. Model.* 7:91–105.

Prigogine, I. 1972. Thermodynamics of evolution. *Physics Today* 23:23–28.

Rastetter, E. B., A. W. King, B. J. Cosby, G. M. Hornberger, R. V. O'Neill, and J. E. Hobbie. 1992. Aggregating fine-scale ecological knowledge to model coarser-scale attributes of ecosystems. *Ecological Applications* 2:55–70.

Robinson, J. B. 1991. Modelling the interactions between human and natural systems. *International Social Science Journal* 130:629–647.

_____. Of maps and territories: the use and abuse of socio-economic modelling in support of decision-making. *Technological Forecast and Social Change.*

Rosser, J. B. 1991. *From Catastrophe to Chaos: A General Theory of Economic Discontinuities.* Kluwer, Amsterdam, The Netherlands.

_____. 1992. The dialogue between the economic and ecologic theories of evolution. *Journal of Economic Behavior and Organization* 17:195–215.

Schneider, E. D., and J. J. Kay. In press. Life as a manifestation of the second law of thermodynamics. *International Journal of Mathematical and Computer Modelling.*

Schumpeter, J. A. 1950. *Capitalism, Socialism and Democracy.* Harper & Row, New York.

Selten, R. 1980. A note on evolutionary stable strategies in asymmetrical animal conflicts. *J. Theoret. Biol.* 84:93–101.

Shugart, H. H. 1989. The role of ecological models in long-term ecological studies. Pages 90–109 in G. E. Likens, ed. *Long-Term Studies in Ecology: Approaches and Alternatives.* Springer-Verlag, New York.

Shugart, H. H., G. B. Bonan, D. L. Urban, W. K. Lavenroth, W. J. Parton, and G. M. Hornberger.

1991. Computer models and long-term ecological research. Pages 221–239 in P. G. Risser, ed. *Long-term Ecological Research: An International Perspective.* John Wiley & Sons, New York.

Solow, R. M. 1956. A contribution to the theory of economic growth. *Quarterly Journal of Economics* 70:65–94.

Sonnenschein, H. 1974. Market excess demand functions. *Econometrica* 40:549–563.

Sugihara, G., and R. M. May. 1990. Nonlinear forecasting as a way of distinguishing chaos from measurement error in time series. *Nature* 344:734–741.

Taub, F. B. 1987. Indicators of change in natural and human impacted ecosystems: status. Pages 115–144 in S. Draggan, J. J. Cohrssen, and R. E. Morrison, eds. *Preserving Ecological Systems: The Agenda for Long-term Research and Development.* Praeger, New York.

_____. 1989. Standardized aquatic microcosm: development and testing. Pages 47–92 in A. Boudou and F. Ribeyre, eds. *Aquatic Ecotoxicology: Fundamental Concepts and Methodologies.* vol II. CRC Press, Boca Raton, FL.

Turner, M. G., R. Costanza, and F. H. Sklar. 1989. Methods to compare spatial patterns for landscape modeling and analysis. *Ecol. Model.* 48:1–18.

Vitousek, P., P. R. Ehrlich, A. H. Ehrlich, and P. A. Matson. 1986. Human appropriation of the products of photosynthesis. *BioScience* 36:368–373.

von Bertalanffy, L. 1968. *General System Theory: Foundations, Development, Applications.* George Braziller, New York.

von Neuman, J. 1928. Sur Theorie der Gesellschaftsspiel. *Mathematische Analen* 100:295–320.

von Neuman, J., and O. Morgenstern. 1944. *Theory of Games and Economic Behavior.* Princeton University Press, Princeton, NJ.

Wulff, F., and R. E. Ulanowicz. 1989. A comparative anatomy of the Baltic Sea and Chesapeake Bay ecosystems. Pages 232–256 in F. Wulff, J. G. Field, and K. H. Mann, eds. *Network Analysis of Marine Ecology: Methods and Applications.* Springer-Verlag, NY.

Wroblewski, J. S., and E. E. Hofmann. 1989. U.S. interdisciplinary modeling studies of coastal-offshore exchange processes: past and future. *Prog. Oceanogr.* 23:65–99.

16
Disturbance, Diversity, and Invasion: Implications for Conservation

Richard J. Hobbs and Laura F. Huenneke

Introduction

Preservation of natural communities has historically consisted of measures protecting them from physical disturbance. Timber harvests and livestock grazing are usually excluded from preserves, and fire suppression has been practiced — within the U.S. system of national parks, for example. Ecologists and conservationists have come to recognize, however, that many forms of disturbance are important components of natural systems. Many plant communities and species are dependent on disturbance, especially for regeneration (Pickett & White 1985). Preserves should be large enough to allow the natural disturbance regime to operate and to support a mosaic of patches in different stages of disturbance, successional recovery, and community maturation (Pickett & Thompson 1978). In addition, both theory (the intermediate disturbance hypothesis, Connell 1978) and growing empirical evidence suggest that moderate frequencies or intensities of disturbance foster maximum species richness. To preserve biotic diversity and functioning natural ecosystems, then, conservation efforts must include explicit consideration of disturbance processes.

Disturbance acts in plant communities in another way, however, by promoting invasions by non-native and weedy plant species (Ewel 1986; Hobbs 1989, 1991; Rejmánek 1989). Invasive species have recently gained notoriety as major conservation and management concerns in natural ecosystems (see MacDonald et al. 1989; Soulé 1990; Westman 1990). The control of non-native plants has become one of the most expensive and urgent tasks of managers in several U.S. national parks, in island preserves such as the Galapagos, and elsewhere. Invasive plants can reduce or displace native species, both plant and animal, and may even alter ecosystem function (Vitousek 1986; Schofield 1989). They have become recognized, therefore, as significant conservation concerns.

Disturbance thus presents a conundrum to conservation management: the continued existence of particular species or communities often requires disturbance of some type — and hence disturbance regimes must be integrated with management plans — but disturbance may simultaneously lead to the degradation of natural communities by promoting invasions. Here we examine this problem by discussing the types of disturbance important in maintaining plant species diversity and those that encourage invasions. We identify particular cases where conflicts are most likely to arise. Our examples are drawn primarily from grassland vegetation, although we discuss other ecosystem types such as shrublands and woodlands. We close by

Reprinted with permission from Conservation Biology vol. 6, pp. 324–337. Copyright 1992 The Society for Conservation Biology and Blackwell Science, Inc.

suggesting guidelines for evaluating the proper role of disturbance in the management of a natural area or preserve.

Theoretical Background

There has been considerable debate on the definition of disturbance, and on what does and does not constitute a disturbance to any given community or ecosystem (see Rykiel 1985; van Andel & van den Berg 1987). Definitions of disturbance vary, from Grime's (1979) view of disturbance as a process removing or damaging biomass, to White and Pickett's (1985) definition of "any relatively discrete event in time that disrupts ecosystem, community or population structure and changes resources, substrate availability, or the physical environment." Petraitis et al. (1989) expand the definition further to include any "process that alters the birth and death rates of individuals present in the patch" by directly killing individuals or by affecting resource levels, natural enemies, or competitors in ways that alter survival and fecundity. Temporal and spatial scale are clearly important in our recognition of the "discreteness" of a disturbance event, as nearly any ecological or biogeochemical process might fall under the last, most inclusive definition. Pickett et al. (1989) define a disturbance as a change in structure caused by factors external to the hierarchical level of the system of interest; this is necessary to distinguish disturbance from other changes in the system. In our discussion below, we will include both direct disturbances (those affecting the survivorship of individuals directly) and indirect disturbance (those affecting resource levels or other conditions that then influence individuals in the patch). Disturbances to plant communities thus include such events as fires, storms, and floods; but other changes such as altered grazing regimes or nutrient inputs would also be classed as disturbance if they affected resource levels and demographic processes.

Within a given patch, the response of any community to a disturbance (or to the disturbance regime, characterised by the natural distribution of disturbance sizes, frequencies, intensities, and timing) is determined by the attributes of component species. Disturbance frequency is also important; the time interval between successive disturbances can have significant effects on community response. This is because species composition changes with time since disturbance, and many species require some time after disturbance to reach reproductive maturity. If a second disturbance occurs before they reach that stage, there will not be any propagules available to recolonize the patch. The response of a community to disturbance is then predicted on the basis of the life history responses of those species available for recruitment or invasion (Noble & Slatyer 1980; Moore & Noble 1990).

There has been increasing recognition among ecological researchers of the importance of natural disturbance in the function of terrestrial ecosystems. Pickett and White (1985) provided a comprehensive review of the role of disturbance in the dynamics of many ecosystem types. They distinguished several components of natural disturbance regimes, including frequency, intensity, and size of disturbance (White & Pickett 1985), each acting in a distinctive way on communities and populations. Petraitis et al. (1989) presented a more detailed analysis of these components, and recognized that hypotheses about the relationship of disturbance and community response can be sorted into two groups: those postulating selective mortality or action for a specific target group, and those dealing with random or catastrophic mortality. Petraitis et al. (1989) suggested that selective mortality could maintain species diversity or richness at some equilibrium level, while random mortality would prevent the establishment of community equilibrium (for example, preventing the dominance of one superior competitor and the exclusion of other species). They pointed out that both equilibrium and nonequilibrium models of communities predict greatest species richness at intermediate levels of disturbance. Various versions of this "intermediate disturbance hypothesis" thus predict a similar result—highest species numbers when disturbances occur at intermediate frequencies or with intermediate intensities (Fig. 1)—despite different underlying theories of community function (see Fox 1979; Huston 1979; Sousa 1984). These arguments are often

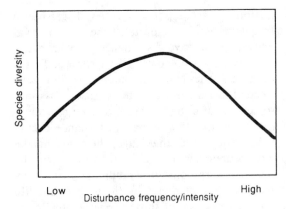

FIGURE 1. The intermediate disturbance hypothesis, which indicates that species diversity within a given patch should be highest at intermediate frequencies or intensities of disturbance (after Connell 1978).

based on the fact that only a few species (ruderals) can persist in the face of frequent, severe disturbance; only a few species (the longest-lived, best competitors, and those able to regenerate without disturbance) can persist over the long term in the absence of disturbance; but many species (including some representatives of each of these, plus intermediates) can find some place to survive in a region comprising patches in various stages of recovery, arising at some intermediate frequency.

How is "intermediate" defined? It is perhaps easiest to relate the frequency of discrete events to the longevity of major species in the system. Approximately half the lifespan of the dominant species has been used as one estimation of intermediate disturbance frequency (Hobbs et al. 1984). Definitions of intermediate intensity may have fewer external referents, however; intensity can be evaluated in terms of percentage of individuals killed, or the degree of structural or resource alteration caused.

The above discussion concentrates on within-patch diversity (alpha diversity), but disturbance is also important for creating or maintaining diversity between patches or at the landscape level (beta diversity). By creating patches of different ages and successional stages, disturbance affects structural and habitat diversity as well as overall species diversity. While we concentrate on within-patch diversity in this review, the role of disturbance in creating landscape mosaics should also be noted (see Turner 1987).

While disturbance is important for main-taining diversity both within communities and at a landscape level, it has become increasingly recognized that disturbance may also have undesirable effects. Particularly important is the recognition that disturbance may act to increase the likelihood of invasion of a community. For invasion to occur there must be available propagules of an invasive species capable of dispersing into a given plant community, and there then has to be a suitable microsite for germination and establishment to occur. That is, there has to be a suitable invasion "window" (Johnstone 1986). Disturbance usually acts primarily by affecting the availability of suitable microsites, although some forms of disturbance may affect the availability of invasive propagules. For instance, non-native herbivores may bring seed into an area either on their coats or in feces. Here we will discuss primarily the effect of disturbance on microsite availability.

The spatial and temporal distribution of disturbance in a region or an ecosystem gives rise to the disturbance mosaic of an area. Pickett and Thompson (1978) pointed out that the recurrence of disturbances necessitates the preservation of a "minimum dynamic area," or an area large enough to contain within it multiple patches in various stages of disturbance or recovery such that internal recolonization can contribute to the maintenance of the overall ecosystem. The dynamics of patch disturbance and of biotic exchanges among patches, which determine the pattern of recovery, are of major concern in defining the minimum critical size of ecosystems (see Lewin 1984), the size required to maintain characteristic species composition and system function. With increasing fragmentation of natural areas, it is likely that these minimum areas are now to be found only within the largest conservation units, and disturbance regimes and biotic exchanges between patches are liable to be significantly altered in smaller remnant areas (Hobbs 1987; Saunders et al. 1991). In particular, invasions are likely to become more important. How shall managers respond to or compensate for the changed nature of disturbance? We approach this question by surveying the major types of disturbance and reviewing their effects on plant species diversity and invasions. We mostly consider grasslands, but we also include illustrative examples from other vegetation types.

Empirical Evidence

1. Fire

The central role of fire in maintaining the open nature of the vegetation has been acknowledged for many grasslands, particularly in mesic regions. Further, research has documented that fire can stimulate or maintain high primary productivity. In tall-grass prairies of North America, fire enhances productivity by removing the thick litter layer and altering the microclimate and nutrient content of surface soil (see Knapp & Seastedt 1986). Fire also influences species diversity and the characteristic structure of these prairie communities. Classical work on fire ecology of prairies (Kucera & Koelling 1964; Abrams et al. 1986) found that annual burning favored tall warm-season grasses and resulted in low abundance of typical prairie forbs after 5–10 years. Biennial burning resulted in the highest community diversity with mixed grasses and forbs. Areas with long fire-free periods resembled unburned areas in their heavy litter accumulation and decline in grasses.

Fires may favor the dominant "matrix" prairie grasses and thus can actually decrease diversity (Collins 1987). Apparently most prairie fires stimulate individual grasses and do not kill them; few openings are created for the establishment of new individuals or species. As we have noted previously, however, species diversity comprises two main components: species density or alpha diversity within a patch, and patch diversity or the number of types of different patches or microhabitats. Glenn-Lewin and ver Hoef (1988) reported that grasslands vary in the degree to which these two contribute to overall diversity. In three grasslands, patch diversity rather than species density was the major contributor to overall community diversity. Fire (and other disturbances) may create a heterogeneous patch structure, even if within patches it serves to decrease species density.

Life history, of course, determines the vulnerability and response of plants to fire. In annual grasslands in California, fire had only temporary effects on botanical composition (forbs increased and grass dominance decreased for a brief time). Here the restructuring of the community each autumn with germination quickly swamps any temporary effect on the seed bank or on germination conditions (Heady 1972).

Suppression of fires in ecosystems dominated by fire-adapted species can cause severe disruption of community and ecosystem processes, which may have implications for the conservation of native, fire-tolerant species. For example, Cowling et al. (1986) found that fire suppression has been responsible for the conversion of a South African open, grassy veld to a vegetation now dominated by undesirable non-native shrubs. They suggested frequent prescribed fires as the best mechanism for restoring the original open nature of the vegetation and for maintaining populations of the region's endemic geophytes. Strang (1973) similarly suggested that fire was an expensive but necessary part of reversing the conversion of moist grassland in south-central Africa to brush. Fire can also be used more precisely to favor the performance of one species over another. For example, in an attempt to restore prairie on the site of an abandoned agricultural site, fire was used successfully to create openings in a turf of non-native *Poa* species and to enhance the colonization and expansion of true prairie species (Curtis & Partch 1948).

As early work in tall-grass prairie confirmed, the overall fire regime rather than any single fire is the critical factor in determining community response. Fires of differing intensity or occurring in different seasons are likely to affect species diversity in a variety of ways by altering the potential of individual species to regenerate. Hobbs et al. (1984) provide an example of how fire intervals can alter the diversity of species that are able to regenerate in heathland, and hence affect overall community diversity. An intermediate fire frequency resulted in the highest species diversity.

Fire has been discussed as a factor that can increase the likelihood of invasions (Christensen & Burrows 1986). Fire acts to remove much of the plant canopy and usually has a short-term fertilizing effect on the soil; hence both light and nutrient availability can be increased temporarily. Zedler and Scheid (1988) discuss the invasion of coastal chaparral by *Carpobrotus edulis* following fire. There is clear evidence, however, that not all fires result in increased invasion and that variations in fire regime can affect the extent of invasion. Hobbs and Atkins (1990) have illustrated how invasion of *Banksia* woodlands differs between fires burned in spring versus autumn. In some cases, fire per se does not affect the degree of invasion, or will do so only when combined with some other type of disturbance, such as mechanical disturbance of the soil or nutrient input. For instance, Hester and Hobbs (1992) studied burned and unburned shrubland patches within an area of remnant vegetation in the Western Australian wheatbelt and found that invasion by non-native annuals was restricted to the remant edges, even following burning. In adjacent woodland, the abundance of non-native species actually declined following the fire. Following another fire in the same area, this time in heathland vegetation, invasion increased only where the fire impinged on a roadverge that had been subject to prior disturbance during road grading. This interaction is important when management of roadside vegetation corridors is considered (see Loney & Hobbs, 1991; Panetta & Hopkins 1991).

Because species vary in their response to fires, fire may favor one set of species over another; these relationships can explain the balance between native and non-native species in some fire-impacted systems. Where native species are sensitive to fire (because fuel loads were such that fires in the native ecosystem were of low frequency and intensity), fire can enhance the invasion of non-native fire-tolerant species. When these fire-tolerant species contribute to increased fuel loads and inflammability, the disturbance regime can be shifted toward more frequent and intense fires; these fires further enhance the dominance of non-native over native species. Just such a cycle has enhanced invasion of woody species in South African Mediterranean systems, and of annual grasses into other Mediterranean-climate regions (MacDonald et al. 1989). Similarly, invasion of fire-tolerant grasses in dry Hawaiian lowlands has had severe effects on native species (Hughes et al. 1991).

2. Grazing

Grazing animals are conspicuous and important features of many grasslands; it has long

been known that some plants are tolerant of grazing while others are not, and that grazing alters the appearance, productivity, and composition of grasslands. Milchunas et al. (1988) have reviewed the effects of grazing by large herbivores on differing types of grassland and relate these to the intermediate disturbance hypothesis. They suggest that grazing constitutes a disturbance only where the evolutionary history of grazing is short. This has also been discussed by Naveh and Whittaker (1980) and Peet et al. (1983). We suggest, however, that in any situation a significant change in grazing regime will constitute a disturbance. Thus, imposition of grazing animals (or different herbivores) on a system not previously subject to that type or level of grazing will constitute a disturbance. So, too, will the removal of grazing from a system with a long grazing history. Species diversity will be affected by the direction of change in grazing regime relative to the historical regime (Ranwell 1960; White 1961; van der Maarel 1971; Milchunas et al. 1990; Dolman & Sutherland 1991). Numerous authors have reported maximum species diversity under intermediate levels of grazing (Zeevalking & Fresco 1977; Milchunas et al. 1988; Puerto et al. 1990).

The most detailed understanding of how grazing affects community structure comes from the chalk grasslands of Britain and northern Europe; these infertile sites support a diverse mixture of grasses and forbs, with species adapted to openings of different kinds and scales (Grubb 1976). These communities, although admittedly artifacts of human activity (clearing, fires, or grazing), have long been prominent features of the landscape; today they are of major conservation value both for their diversity and for particular rare species. Repeatedly it has been demonstrated that grazing is an important factor in the maintenance of chalk grassland diversity; the cessation of grazing leads to dominance of a few grasses, and even to incursions by shrubs or other woody species (Wells 1969). Entire components of the flora may be lost; for example, During and Willems (1986) blamed the loss of most lichens and the impoverishment of the bryophte flora in Dutch chalk grasslands on the absence of grazing.

Grazing maintains high species diversity in

other grasslands, as well. Grazing management is an important and successful technique for preserving diversity and conservation value of old grasslands and pastures in England (Hopkins & Wainwright 1989). Sykora et al. (1990) found that grassland on embankments in the Netherlands was converted to woody scrub in the absence of grazing; under light grazing, a species-poor grassland resulted from competition from a few competitive grasses. Under more intensive grazing, those grasses did not dominate, and a more diverse grassland was maintained. Mediterranean-climate grasslands may respond similarly to grazing management; in a California grassland on serpentine substrate, cessation of livestock grazing enhanced the dominance of non-native annual grasses and led to a rapid decline in abundance of the diverse native annual forb flora (Huenneke et al., unpublished data).

One straightforward effect of grazing is the elimination of trees and shrubs invading mesic grasslands. Without grazing, many North American prairie sites have been converted to woodland. Similarly, there are also documented cases of grazing preventing or reversing the succession of African savanna to woodland. For example, Smart et al. (1985) found that in Uganda the exclusion of elephants was even more important than fire suppression in encouraging acacia invasion, leading to the loss of many species including the original grassland dominants. In these regions, a long evolutionary history of grazing has led to the dominance of grassland plants adapted to and tolerant of grazing pressure. Caldwell's work (for example, Caldwell et al. 1981) has documented the many physiological traits that affect a plant's tolerance of grazing losses.

In contrast, regions with no recent history of grazing are often dominated by plants that lack these tolerance mechanisms. Extreme examples are presented by oceanic islands with no native mammalian herbivores, where the introduction of livestock or other grazers has usually been catastrophic in its effect on native vegetation — for example, the effect of feral goats on island floras (Coblentz 1978) and of introduced herbivores in the Galapagos on native vegetation (Hamann 1975, 1979). A less obvious but still major impact has been made on regions with few native grazers (at least since post-

Pleistocene time), such as the intermountain West (Mack & Thompson 1982).

In semi-arid grasslands in the American Southwest, species diversity has declined and, in many cases, the physiognomy of the vegetation has been altered from perennial grassland to shrub-dominated desert scrub. The chief question of range management and ecology is the determination of the proper utilization rate: what level of grazing will maximize productivity and maintain the grassland's general character? Unfortunately, it is not known what utilization level maximizes plant species diversity or productivity, or whether the same level maximizes both. Westoby et al. (1989) outlined the differences in grazing management that would result from considering grazing in an equilibrial, successional context versus a non-equilibrial series of alternate states; working with the second mental model requires much more active management to "seize opportunities and to evade hazards."

Given grazing's impact on community structure, it has been used as a management tool in conservation applications. One example is a grassland restoration project, where an abandoned, species-poor pasture in Holland was being grazed by cattle; seed inputs from cattle feces (together with openings created by grazing) contributed significantly to increasing species diversity (Bulow-Olsen 1980). In another case, sheep were used to restore abandoned fields (Gibson et al. 1987), again by importing seeds and creating openings for recruitment. Several grasses in the Middle East, wild cereal ancestors of conservation interest, are negatively affected by heavy grazing but also vulnerable to competition from tall perennial grasses. Therefore the two sets of species alternate on lightly grazed or protected sites (Noy-Meir 1990). Upland British grassland species of conservation value vary in their response. Some benefit from removal of grazing, while others are negatively influenced by the resulting increase in grass (Rawes & Welch 1972). Wells (1969) commented that grazing (or mowing) during the season when the dominant grass species is growing most rapidly is usually the most effective way to maintain diversity in chalk grassland. He stated that the cessation of grazing is the major conservation problem in those grasslands, eliminating many forbs and causing increases of litter and woody species. He added, however, that grazing should be timed to avoid the sensitive phases in the life cycle of species vulnerable to grazing.

This raises an important point: Effects of grazing are species-specific. That is, two species in the same community may vary in their response to grazing or to a specific grazing regime. For example, in an English high-elevation grassland on limestone, after sheep were excluded from one site, several rare shrubs benefited from protection, but one species declined (Elkington 1981). The optimal design of grazing management may thus be difficult. Vinther (1983) found that a mesic meadow was maintained as open meadow if it was heavily grazed—because tree seedlings were killed by browsing—or if it was not grazed at all—because seedlings couldn't establish in the dense herb layer. Intermediate grazing levels allowed woody regeneration and loss of the meadow's open character. Unfortunately, these same intermediate levels of grazing are those maximizing the richness of herbaceous species in the short term. An alternate means of preventing woody plant encroachment would then be necessary to allow continued management for maximum species diversity.

Grazing's impact presents an interesting contrast to mowing, which is often suggested as an alternative to grazing management. Mowing can reduce the growth of competitively dominant grasses, allowing the persistence of less competitive species, but it does not create openings for recruitment of seedlings as grazing does. Sykora et al. (1990) emphasized the different results of the two, with grazing creating more microsites for establishment and greater heterogeneity, while providing seed dispersal in animal feces, hooves, and coats. As van den Bos and Bakker (1990) pointed out, grazers do not use an entire area evenly but always prefer some spots to others, so they create greater heterogeneity than does mowing. There is also a difference in the form in which nutrients are returned or retained in the system (Rizand et al. 1989). Grazing is thus an amalgam of different effects. Clearly, if mowing is to be used by managers in preference to grazing, more sophisticated methods involving variations in mowing time and pattern and degree of mulch removal should be investigated.

Grazing animals may frequently be implicated in the invasion of natural communities. Grazers may import non-native plant propagules into native vegetation, but they may also act to provide microsites for invasion. In particular, where grazing alters the vegetation structure or is accompanied by soil disturbance (trampling, digging, and so forth), conditions are modified in such a way that invading species can become established. For instance, Cross (1981) showed that grazing by the non-native sika deer facilitated the invasion of oak woodlands by *Rhododendron ponticum* by removing the herbaceous understory and providing more safe sites for establishment. The arrival of large numbers of livestock following European settlement has been implicated in the decline in native perennial grasses and their replacement with non-native annual grasses in several grassland areas in North America and Australia (Moore 1970; Mack 1981, 1989). Braithwaite et al. (1989) suggested that water buffalo activities aid in the establishment of *Mimosa pigra* in northern Australia. Pickard (1984) implicated grazing disturbance as one of the major factors influencing invasion on Lord Howe Island in the South Pacific.

3. Soil Disturbances

In grasslands, as in most plant community types, soil disturbance creates openings for establishment, frequently of weedy or ruderal species. It is unclear whether temporary increases in nutrients and other resources are directly responsible for this enhancement of establishment or whether reduced competition from neighboring plant canopies and roots is more important, and it is usually difficult to separate the two effects. Where such disturbance has long been a component of the ecosystem, there is likely a substantial fraction of the flora that is specialized or adapted to establishment there. Thus in the Mediterranean region, where human agricultural and other activity has long created such soil disturbance, there is a large and successful group of weedy species. These are the colonists and invaders that have become so pervasive in disturbed sites elsewhere in the world, where agricultural activity has a much shorter history and where few native species are adapted to such a habitat (Naveh 1967; Hobbs & Hopkins, 1990).

Plowing is said to diminish species richness, especially that of dicots, in lowland grasslands (Fuller 1987). Even so, particular species may require plowing to persist (Preston & Whitehouse 1986). Smaller-scale disturbances may be equally important in providing opportunities both ecological and evolutionary; for example, in tall-grass prairies, mounds created by badger excavations support a distinctive and diverse flora of "fugitive" prairie plants that live only on those mounds (Platt 1975). This distinctive group of species contributes substantially to the overall diversity of those prairies, particularly to overgrazed ones in which the background or matrix is relatively species-poor. Other disturbances by prairie dogs, buffaloes, and gophers also have significant effects on prairie diversity (Coppock et al. 1983; Collins & Barber 1985; Huntly & Inouye 1988; Whicker & Detling 1988; Martensen et al. 1990). Mounds of bare soil formed by the activity of pocket gophers act in Californian annual grasslands to provide substrates for seedling establishment in an environment of lower density and altered microclimate and soil nutrient status (Hobbs & Mooney 1985; Koide et al. 1987). Coffin and Lauenroth (1988) used a modeling approach and found that the effect of soil disturbances (ant mounds and mammal burrows) on a shortgrass community was chiefly a function of disturbance frequency and secondarily of disturbance size.

While soil disturbances, especially by animals, often have important effects on the dynamics of native plant communities, there are also numerous examples of such soil disturbances facilitating invasion by non-native species. Disturbance by gophers was found by Hobbs and Mooney (1985, 1991) to be an important factor in the invasion of serpentine grassland by *Bromus mollis* and other non-native annual grasses following years of above-average rainfall. *Bromus mollis* became established in greater abundance on gopher mounds than in undisturbed grassland, and was virtually absent from areas where gophers were excluded. *Bromus mollis* was able to disperse seeds onto gopher mounds more effectively than some of the native species because of its taller inflorescence, and it then survived better

on the more open microhabitat than in the undisturbed grassland.

Experiments in which artificial soil disturbances were created have had mixed results, with the effects varying among different plant communities. Hobbs and Atkins (1988) found that some communities were more readily invaded than others, and that soil disturbance did not necessarily increase the ease with which non-native species could become established or survive. Disturbance had the largest effect in the communities that were already more susceptible to invasion.

Why does soil disturbance facilitate invasion? Disturbance may act primarily by providing a rougher surface on which seeds can lodge; in other words, the disturbance increases the availability of safe sites (Hobbs & Atkins 1988). Hobbs and Mooney (1985) found that plants of both native and non-native species grew much larger on gopher mound microhabitats than in undisturbed grassland, but Koide et al. (1987) found that nutrient availability was actually lower in gopher mound soils than in undisturbed soil. Hence removal of competitors may be the major factor in this case.

4. Nutrient Inputs

Another type of disturbance, which is often less obvious, is a change in the input and cycling of nutrients in an ecosystem. Input of additional nutrients, particularly nitrogen and phosphorus, in low-fertility sites can be as devastating as eutrophication in freshwater ecosystems. Fertilization has contributed to a marked decline in species richness in British and Dutch grasslands (Willis 1963; Bakker 1987; Fuller 1987). Grasses are often the species to respond and to dominate under nutrient enrichment, to the detriment of broadleaved plants. During and Willems (1986) suggested that continued input of pollutants and nitrogen were partially responsible for the floristic impoverishment of nonvascular flora in Dutch chalk grassland. Input of atmospheric nitrogen was apparently to blame for the increasing dominance of one grass species and the loss of many forbs and other grasses, regardless of management (mowing, grazing, burning) in chalk grassland (Bobbink & Willems 1987). Certainly the problem of increased deposition of nutrients from the atmosphere is likely to be chronic and widespread.

Gough and Marrs (1990) suggested that high phosphorus levels in the soil of abandoned pastures precluded the reestablishment of species-rich grassland there. Natural or successional losses of phosphorus were too slow from a management perspective; incursion of scrub or woody species apparently increased levels of extractable phosphorus. They suggested that managers use cropping (cutting and removing above-ground biomass each season) or heavy leaching to lower soil-extractable phosphorus levels more quickly. Marrs (1985) reported a similar effort to reduce soil fertility in a site where managers were attempting to reestablish an acid heathland. In another twenty-two-year experiment with cutting, Rizand et al. (1989) found that retaining clippings on the site kept phosphorus availability high, with a possible negative influence on species composition, compared with removal of clippings or with grazing. Green (1972) pointed out early on that chalk grassland, dune grassland, and heath were all seral, low-fertility ecosystems with high conservation value. He suggested more study of nutrient budgets on those systems and pointed out that grazing, burning, and mowing all decreased the likelihood of nutrient accumulations. In North American old fields, nutrient-enriched fields supported lower species richness and retained a weedy annual, largely non-native flora (Carson & Barrett 1988) rather than the perennial grasses typical of fields of equivalent age.

In ecosystems with predominantly nutrient-poor soils, addition of nutrients can constitute a major disturbance, which has been shown in many examples to facilitate invasion by non-native species. Huenneke et al. (1990) have shown that a serpentine grassland dominated by annual forbs can be transformed in two years into one dominated by non-native grasses by the addition of nutrients, particularly nitrogen and phosphorus. Hobbs et al. (1988) produced similar results and showed that survival of non-native grasses was significantly enhanced on fertilized plots, while that of native forbs was reduced. In both these cases, invasion was not related directly to soil disturbance and, in fact, Hobbs et al. (1988) found that subsequent gopher disturbance actually reduced the dominance of non-native grasses and allowed the re-establishment of native forbs.

Nutrient input has also been shown to facilitate invasion of Australian plant communities. Heddle and Specht (1975) reported increased abundances of non-native herbaceous species in areas of heathland that had received fertilizer. Other studies have indicated a strong relationship between the degree of invasion by non-native species and soil nutrient levels, particularly of phosphorus (Cale & Hobbs 1991; Hester & Hobbs 1992). Experiments where nutrients were added to plots within a number of different plant communities in Western Australia showed that increased nutrients resulted in increased growth of non-native species in some plant communities but not others (Hobbs & Atkins 1988). Of particular interest was the finding that a combination of soil disturbance and nutrient addition had the greatest effect in enhancing the establishment and growth of non-native species.

5. Trampling

Like the other disturbances we have discussed, trampling can create openings in vegetation that provide opportunities for new individuals to become established, and it can slow the growth of dominant species sufficiently to allow the persistence of less vigorous species. Again, intermediate levels of trampling seem most effective at maintaining high species richness because of the suppression of competitive dominants (Liddle 1975). The season or timing of trampling has a significant effect on the chance, rate, and species composition of recovery (Harrison 1981). There are species-specific responses to trampling: in one study most but not all species were negatively affected (Crawford & Liddle 1977): invertebrates seem far more sensitive than plants (Duffey 1975). We have encountered little information on the effects of trampling on invasions, although trampling effects are frequently considered together with those of grazing.

6. Fragmentation

The fragmentation and insularization of ecosystems is not a disturbance within an individual system but a landscape-level disturbance resulting in the rearrangement of the landscape matrix. By influencing edge effects and the likelihood of movement of nutrients, propagules, and fauna from adjacent patches, fragmentation affects disturbance regimes in individual patches of remnant vegetation (Hobbs 1987; Saunders et al. 1991). How does fragmentation affect the species composition and richness of grasslands? Simberloff and Gotelli (1984) surveyed patches of prairie and found that "archipelagoes" of small grassland patches supported more species than did single large patches of equivalent total area. Thus small patch size does not constrain total species richness. Quinn and Robinson (1987) and Robinson and Quinn (1988) used an experimental approach to this question, subdividing annual grassland into fenced patches separated by heavily grazed zones; species richness was substantially higher in the more subdivided treatments. Single species frequently came to dominate single plots, so a region with a greater number of patches supported both more dominant species (alternate dominants in different plots) and more edge species (growing along the greater perimeter). Murphy (1989) has pointed out that Robinson and Quinn's (1988) study was carried out at an inappropriate scale and in a grassland that is dominated by non-native annuals. However, the point that fragmentation will lead to an increase in edge species is important. From a conservation management perspective, one would want to know just which species are being favored by edge effects. A higher total species richness could be primarily due to an increased number of ruderal or weedy species of low conservation value (as found, for example, for invertebrates by Webb & Hopkins [1984]), or to a higher number of legitimate community members.

7. Interaction of Disturbances

Of course, most ecosystems experience multiple disturbances and are shaped by multiple factors. In many cases the results are not merely additive, and disturbances can act synergistically. For example, grazing reduced fuel loads, reduced fire frequency, and allowed the invasion of woody species into many regions of semi-arid grassland (such as the historical ex-

pansion of pinyon-juniper vegetation into western U.S. grasslands; Wright et al. 1979). In an experimental study, Collins (1987) found that fire significantly increased species diversity in grazed tallgrass prairie but not on ungrazed grassland; in some respects the effects of grazing and fire were additive. Collins and Gibson (1990) have further illustrated how grazing, fire, and small-scale soil disturbance all affect the matrix structure of these grasslands differently, and hence can interact to increase community diversity. Leigh et al. (1987) found that rabbit populations increased on burned areas of subalpine vegetation, while Noy-Meir (1988) found that elevated populations of voles had the greatest effects on grasslands where other grazing was minimized. Sykora et al. (1990) suggested that fire in Dutch grasslands increased nutrients and thus increased the likelihood of "ruderalization"—increasing dominance by a few grasses leading to a decline in diversity. Hodgkin (1984) found that woody encroachment increased soil fertility and changed the nature of British dune grassland. It was suggested that the myxomatosis-caused decline in rabbit populations had resulted in the increased establishment of woody vegetation, and that the resulting scrub had increased soil nutrients to the point that weedy plant species were favored.

Invasion by the nitrogen-fixing *Myrica faya* onto young laval flows in Hawaii has been shown to alter the nature of ecosystem development following volcanic eruptions (Vitousek et al. 1987; Vitousek & Walker 1989). In this and other cases, such as that of *Mimosa pigra* in Australia (Braithwaite et al. 1989), the invading plants themselves constitute a major disturbance to the systems they are invading.

Good Disturbances Turned Bad: Conflicts

Are there cases where disturbance is a necessary component of ecosystem and community dynamics, but also enhances the likelihood of invasion? From the foregoing, it would seem that virtually any type of disturbance can facilitate invasion under certain circumstances. Invasion is, after all, simply a subset of the possible recolonization response to disturbance. As an example, Griffin et al. (1989) have shown how periodic flooding can lead to the invasion of arid zone river systems by *Tamarix aphylla*. It is not the type of disturbance but rather certain aspects of its action in a particular system that shift the result toward enhancement of invasions at the expense of natives (see McIntyre et al. 1988). For example, it is not fire per se but the combination of fire with other disturbance, or the adoption of a fire regime inappropriate to the life histories of native plants, that favors non-native fire-tolerant species at the expense of natives. A primary consideration, then, must be the suite of adaptations and life histories found in the native plants, particularly those of conservation value.

The relationship between soil disturbance and invasion is also complex, and mechanical disturbance in the absence of nutrient addition may not necessarily lead to enhanced invasion (see Hobbs 1989). Frequently, however, physical disturbance and nutrient enrichment coincide, as when rabbits scrape the soil and defecate at the same time, or when disturbance enhances nitrogen mineralization. An important problem for systems with a naturally low nutrient status is the gradual nutrient enrichment that can occur via atmospheric input, windblown fertilizer, or input from livestock feces (Landsberg et al. 1990; Cale & Hobbs 1991). An increased baseline nutrient status will have important implications for the whole ecosystem, but in the short term it may exacerbate the likelihood of invasions by weedy pest species.

Conclusions

No system can remain immune from certain disturbances (such as nutrient input from the atmosphere); in the future, few areas will even be protected from direct human activity. Some disturbance types can be modified by on-site management (fire and grazing regimes) while others cannot (floods, storms). Human-induced disturbances such as road construction can also be minimized. "Natural disturbance regimes" may be desirable but are often imprac

ticable in the altered settings of contemporary reserves. We need to acknowledge the actual disturbance regime operating currently in a reserve, and the current propagule rain, which determines the importance of coping with likely invasions. Further, managers need to take an active role in designing the disturbance regime, tailoring it to the landscape, the biotic community, and their specific conservation goals.

Denslow (1980) hypothesized that any natural community would be richest in species adapted to establishment in the type of patch most commonly created by disturbance. For example, where large scale disturbances are the norm, most species will establish there and species richness will decline through time and succession. In contrast, in an ecosystem where small-scale disturbances are normal, most species will establish in small scale gaps or in undisturbed sites, and diversity will increase with time after a large disturbance. Total diversity of native species at the landscape level will be greatest when disturbance occurs at its historical frequency and in the historical pattern (Fig. 2). Changes in the size of the frequency, as well as the type, of disturbance will mean that most native species will no longer be well adapted for recruitment or establishment.

In addition, even when disturbance regimes have not been significantly altered, the availability of weedy or invasive species may alter system response to disturbance. Management must consider not only alterations of the original disturbance regime, but also alteration in the pool of potential responding species (in other words, the availability of colonists or pests).

The response of invaders to disturbance is an extreme case of an underlying, unavoidable conflict—any disturbance and any management regime will be good for some species and bad for others. The decision may be easy (although the techniques for management may not be) when the choice is between natives and non-native pest species. The dilemma is thornier when non-natives have some appeal of their own (for instance in terms of grazing value), and it is still more difficult when the choice is between one set of native species and another. In the end, the wisest choice may be to use a diversity of management strategies, to en-

FIGURE 2. Any change in the historical disturbance regime of an ecosystem may alter species composition by reducing the importance of native species, by creating opportunities for invasive species, or both.

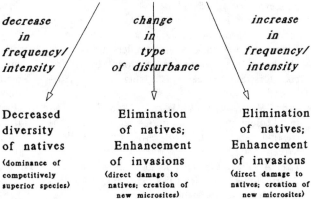

Natural Disturbance Regime
maintains native species diversity
(historical type, frequency, intensity of disturbance)

decrease in frequency/ intensity	*change in type of disturbance*	*increase in frequency/ intensity*
Decreased diversity of natives (dominance of competitively superior species)	**Elimination of natives; Enhancement of invasions** (direct damage to natives; creation of new microsites)	**Elimination of natives; Enhancement of invasions** (direct damage to natives; creation of new microsites)

courage different species in different parts of the reserve or in different reserves within a region. There is no single optimal strategy; managers must make decisions based on the likely costs and benefits in terms of maintenance of diversity versus invasion by non-natives.

We draw the following conclusions from this review of scientific research on disturbance and species composition and diversity in grasslands.

• Disturbance plays an integral role in structuring plant communities, but some types or combinations of disturbance can increase the potential of invasion by non-native species.
• Background levels of disturbance, resource availability, and the pool of potential species in any ecosystem all differ now from primeval condition. This is true even in the largest parks and reserves (see Chase 1987). It is not enough to say that the original disturbance regime is the desired state.
• Species vary in their response to disturbance, requiring managers to make deliberate choices of which taxa to favor.
• Managers may have to choose between specific conservation targets, such as preventing the spread of invasive species, and the more general goal of maintaining overall species diversity.
• Nearly all systems are likely to be nonequilibrial in the future; we must be activists in determining which species to encourage and which to discourage. We cannot just manage

passively, or for maximal diversity, but must be selective and tailor management to specific goals.

Acknowledgments

We thank Juli Armstrong, Richard Groves, and two referees for constructive comments on the draft manuscript.

References

Abrams, M. D., A. K. Knapp, and L. C. Hulbert. 1986. A ten-year record of aboveground biomass in a Kansas tallgrass prairie: effects of fire and topographic position. American Journal of Botany 73:1509–1515.

Bakker, J. P. 1987. Restoration of species-rich grassland after a period of fertiliser application. Pages 185–200 in J. van Andel, J. P. Bakker, and R. W. Snaydon, editors. Disturbance in grasslands: causes, effects and processes. Junk, Dordrecht.

Bobbink, R., and J. H. Willems. 1987. Increasing dominance of *Brachypodium pinnatum* in chalk grasslands: a threat to a species-rich ecosystem. Biological Conservation 40:301–314.

Braithwaite, R. W., W. M. Lonsdale, and J. A. Estbergs. 1989. Alien vegetation and native biota in tropical Australia: the impact of *Mimosa pigra*. Biological Conservation 48:189–210.

Bulow-Olsen, A. 1980. Changes in the species composition in an area dominated by *Deschampsia flexuosa* as a result of cattle grazing. Biological Conservation 18:257–270.

Caldwell, M. M., J. H. Richards, D. A. Johnson, R.

S. Nowak, and R. S. Dzurec. 1981. Coping with herbivory: photosynthetic capacity and resource allocation in two semiarid *Agropyron* bunchgrasses. Oecologia **50**:14–24.

Cale, P., and R. J. Hobbs. 1991. Condition of roadside vegetation in relation to nutrient status. Pages 353–362 in D. A. Saunders and R. J. Hobbs, editors. Nature conservation 2: the role of corridors. Surrey-Beatty, Chipping Norton, Australia.

Carson, W. P., and G. W. Barrett. 1988. Succession in old-field plant communities: effects of contrasting types of nutrient enrichment. Ecology **69**:984–994.

Chase, A. 1987. Playing god in Yellowstone. The destruction of America's first national park. Harcourt Brace Jovanovich, New York.

Christensen, P. E., and N. D. Burrows. 1984. Fire: an old tool with a new use. Pages 57–66 in R. H. Groves and J. J. Burdon, editors. Ecology of biological invasions: an Australian perspective. Australian Academy of Science, Canberra, Australia.

Coblentz, B. E. 1978. The effects of feral goats (*Capra hircus*) on island ecosystems. Biological Conservation **13**:279–286.

Coffin, D. P., and W. K. Lauenroth. 1988. The effect of disturbance size and frequency on a shortgrass plant community. Ecology **69**:1609–1617.

Collins, S. L. 1987. Interaction of disturbances in tallgrass prairie: a field experiment. Ecology **68**:1243–1250.

Collins, S. L., and S. C. Barber. 1985. Effects of disturbance on diversity in mixed-grass prairie. Vegetation **64**:87–94.

Collins, S. L., and D. J. Gibson. 1990. Effects of fire on community structure in tallgrass and mixed-grass prairie. Pages 81–98 in S. L. Collins and L. L. Wallace, editors. Fire in North American tallgrass prairies. University of Oklahoma Press, Norman, Oklahoma.

Connell, J. H. 1978. Diversity in tropical rain forests and coral reefs. Science **199**:1302–1310.

Coppock, D. L., J. K. Detling, J. E. Ellis, and M. I. Dyer. 1983. Plant-herbivore interactions in a North American mixed-grass prairie. 1. Effects of black-tailed prairie dogs on intraseasonal aboveground plant biomass and nutrient dynamics and species diversity. Oecologia (Berlin) **56**:1–9.

Cowling, R. M., S. M. Pierce, and E. J. Moll. 1986. Conservation and utilization of South Coast Renosterveld, an endangered South African vegetation type. Biological Conservation **37**:363–377.

Crawford, A. K., and M. J. Liddle. 1977. The effect of trampling on natural grassland. Biological Conservation **12**:135–142.

Cross, J. R. 1981. The establishment of *Rhododen-dron ponticum* in the Killarny oakwoods, S. W. Ireland. Journal of Ecology **69**:807–824.

Curtis, J. T., and M. L. Partch. 1948. Effect of fire on the competition between blue grass and certain prairie plants. American Midland Naturalist **39**:437–443.

Denslow, J. S. 1980. Patterns of plant species diversity during succession under different disturbance regimes. Oecologia **46**:18–21.

Dolman, P., and W. Sutherland. 1991. Historical clues to conservation. New Scientist **1749**:22–25.

Duffey, E. 1975. The effects of human trampling on the fauna of grassland litter. Biological Conservation **7**:255–274.

During, H. J., and J. H. Willems. 1986. The impoverishment of the bryophyte and lichen flora of the Dutch chalk grasslands in the thirty years 1953–1983. Biological Conservation **36**:143–158.

Elkington, T. T. 1981. Effects of excluding grazing animals from grassland on sugar limestone in Teesdale, England. Biological Conservation **20**:25–35.

Ewel, J. 1986. Invasibility: lessons from South Florida. Pages 214–230 in H. A. Mooney and J. A. Drake, editors. Ecology of biological invasions of North America and Hawaii. Springer-Verlag, New York.

Fox, J. F. 1979. Intermediate-disturbance hypothesis. Science **204**:1344–1345.

Fuller, R. M. 1987. The changing extent and conservation interest of lowland grasslands in England and Wales: a review of grassland surveys 1930–84. Biological Conservation **40**:281–300.

Gibson, C. W. D., T. A. Watt, and V. K. Brown. 1987. The use of sheep grazing to recreate species-rich grassland from abandoned arable land. Biological Conservation **42**:165–183.

Glenn-Lewin, D. C., and J. M. ver Hoef. 1988. Scale, pattern analysis, and species diversity in grasslands. Pages 115–129 in H. J. During, M. J. A. Werger, and J. H. Willems, editors. Diversity and pattern in plant communities. SPB Academic Publishing, The Hague, The Netherlands.

Gough, M. W., and R. H. Marrs. 1990. A comparison of soil fertility between semi-natural and agricultural plant communities: implications for the creation of species-rich grassland or abandoned agricultural land. Biological Conservation **51**:83–96.

Green, B. H. 1972. The relevance of seral eutrophication and plant competition to the management of successional communities. Biological Conservation **4**:378–384.

Griffin, G. F., D. M. Stafford Smith, S. R. Morton, G. E. Allan, and K. A. Masters. 1989. Status and implications of the invasion of tamarisk (*Tamarix aphylla*) on the Finke River, Northern Territory, Australia. Journal of Environmental Management **29**:297–315.

Grime, J. P. 1979. Plant strategies and vegetation processes. Wiley, New York.

Grubb, P. J. 1976. A theoretical background to the conservation of ecologically distinct groups of annuals and biennials in the chalk grassland ecosystem. Biological Conservation **10**:53–76.

Hamann, O. 1975. Vegetational changes in the Galapagos Islands during the period 1966–73. Biological Conservation **7**:37–59.

Hamann, O. 1979. Regeneration of vegetation on Santa Fe and Pinta Islands, Galapagos, after the eradication of goats. Biological Conservation **15**:215.

Harrison, C. 1981. Recovery of lowland grassland and heathland in southern England from disturbance by seasonal trampling. Biological Conservation **19**:119–130.

Heady, H. F. 1972. Burning and the grasslands in California. Proceedings of the Twelfth Annual Tall Timbers Fire Ecology Conference.

Heddle, E. M., and R. L. Specht. 1975. Dark Island Heath (Ninety-Mile Plain, South Australia). VIII. The effects of fertilizers on composition and growth. Australian Journal of Botany **23**:151–164.

Hester, A. J., and R. J. Hobbs. 1992. Influence of fire and soil nutrients on native and non-native annuals at remnant vegetation edges in the Western Australian wheatbelt. Journal of Vegetation Science **3**:101–108.

Hobbs, R. J. 1987. Disturbance regimes in remnants of natural vegetation. Pages 233–240 in D. A. Saunders, G. W. Arnold, A. A. Burbidge, and A. J. M. Hopkins, editors. Nature conservation: the role of remnants of native vegetation. Surrey Beatty, Chipping Norton, Australia.

Hobbs, R. J. 1989. The nature and effects of disturbance relative to invasions. Pages 389–405 in J. A. Drake, H. A. Mooney, F. di Castri, R. H. Groves, F. J. Kruger, M. Rejmánek, and M. Williamson, editors. Biological invasions. A global perspective. Wiley, Chichester, England.

Hobbs, R. J. 1991. Disturbance as a precursor to weed invasion in native vegetation. Plant Protection Quarterly **6**:99–104.

Hobbs, R. J., and L. Atkins. 1988. The effect of disturbance and nutrient addition on native and introduced annuals in the Western Australian wheatbelt. Australian Journal of Ecology **13**:171–9.

Hobbs, R. J., and L. Atkins. 1990. Fire-related dynamics of a *Banksia* woodland in south-west Western Australia. Australian Journal of Botany **38**:97–110.

Hobbs, R. J., and A. J. M. Hopkins. 1990. From frontier to fragments: European impact on Australia's vegetation. Proceedings of the Ecological Society of Australia **16**:93–114.

Hobbs, R. J., and H. A. Mooney. 1985. Community and population dynamics of serpentine grassland annuals in relation to gopher disturbances. Oecologia (Berlin) **67**:342–351.

Hobbs, R. J., and H. A. Mooney. 1991. Effects of rainfall variability and gopher disturbance on serpentine annual grassland dynamics in N. California. Ecology **72**:59–68.

Hobbs, R. J., S. L. Gulmon, V. J. Hobbs, and H. A. Mooney. 1988. Effects of fertilizer addition and subsequent gopher disturbance on a serpentine annual grassland community. Oecologia (Berlin) **75**:291–295.

Hobbs, R. J., A. U. Mallik, and C. H. Gimingham. 1984. Studies on fire in Scottish heathland communities. III. Vital attributes of the species. Journal of Ecology **72**:963–976.

Hodgkin, S. E. 1984. Scrub encroachment and its effects on soil fertility on Newborough Warren, Anglesey, Wales. Biological Conservation **29**:99–119.

Hopkins, A., and J. Wainwright. 1989. Changes in botanical composition and agricultural management of enclosed grassland in upland areas of England and Wales, 1970–86, and some conservation implications. Biological Conservation **47**:219–235.

Huenneke, L. F., S. P. Hamburg, R. Koide, H. A. Mooney, and P. M. Vitousek. 1990. Effects of soil resources on plant invasion and community structure in Californian serpentine grassland. Ecology **71**:478–491.

Hughes, R. F., P. M. Vitousek, and J. T. Tunison. 1991. Effects of invasion by fire-enhancing C4 grasses on native shrubs in Hawaii Volcanoes National Park. Ecology **72**:743–746.

Huntly, N., and R. Inouye. 1988. Pocket gophers in ecosystems: patterns and mechanisms. BioScience **38**:786–793.

Huston, M. 1979. A general hypothesis of species diversity. American Naturalist **113**:81–101.

Johnstone, I. M. 1986. Plant invasion windows: a time-based classification of invasion potential. Biological Reviews **61**:369–394.

Knapp, A. K., and T. R. Seastedt. 1986. Detritus accumulation limits productivity of tallgrass prairie. BioScience **36**:662–668.

Koide, R., L. F. Huenneke, and H. A. Mooney. 1987. Gopher mound soil reduces growth and affects ion uptake of two annual grassland species. Oecologia (Berlin) **72**:284–290.

Kucera, C. L., and M. Koelling. 1964. The influence

of fire on composition of central Missouri prairie. American Midland Naturalist **72**:142–147.

Landsberg, J., J. Morse, and P. Khanna. 1990. Tree dieback and insect dynamics in remnants of native woodlands on farms. Proceedings of the Ecological Society of Australia **16**:149–165.

Leigh, J. H., D. J. Wimbush, D. H. Wood, M. D. Holgate, A. V. Slee, M. G. Stanger, and R. I. Forrester. 1987. Effects of rabbit grazing and fire on a subalpine environment. I. Herbaceous and shrubby vegetation. Australian Journal of Botany **35**:433–464.

Lewin, R. 1984. Parks: how big is big enough? Science **225**:611–612.

Liddle, M. J. 1975. A selective review of the ecological effects of human trampling on natural ecosystems. Biological Conservation **7**:17–36.

Loney, B., and R. J. Hobbs. 1991. Establishment, maintenance and rehabilitation of vegetation corridors. Pages 299–311 in D. A. Saunders and R. J. Hobbs, editors. Nature conservation 2: the role of corridors. Surrey-Beatty, Chipping Norton, Australia.

MacDonald, I. A. W., L. L. Loope, M. B. Usher, and O. Hamann. 1989. Wildlife conservation and the invasion of nature reserves by introduced species: a global perspective. Pages 215–255 in J. A. Drake, H. A. Mooney, F. di Castri, R. H. Groves, F. J. Kruger, M. Rejmánek, and M. Williamson, editors. Biological invasions: a global perspective. Wiley, Chichester, England.

Mack, R. N. 1981. Invasion of *Bromus tectorum* L. into western North America: an ecological chronicle. Agro-Ecosystems **7**:145–165.

Mack, R. N. 1989. Temperate grasslands vulnerable to plant invasions: characteristics and consequences. Pages 155–179 in J. A. Drake, H. A. Mooney, F. di Castri, R. H. Groves, F. J. Kruger, M. Rejmánek, and M. Williamson, editors. Biological invasions: a global perspective. Wiley, Chichester, England.

Mack, R. N., and J. N. Thompson. 1982. Evolution in steppe with few large, hooved mammals. American Naturalist **119**:757–773.

Marrs, R. H. 1985. Techniques for reducing soil fertility for nature conservation purposes: a review in relation to research at Roper's Heath, Suffolk, England. Biological Conservation **34**:307–332.

Martensen, G. D., J. H. Cushman, and T. G. Whitham. 1990. Impact of pocket gopher disturbance on plant species diversity in a shortgrass prairie community. Oecologia (Berlin) **83**:132–138.

McIntyre, S., P. Y. Ladiges, and G. Adams. 1988. Plant species-richness and invasion by exotics in relation to disturbance of wetland communities on the Riverine Plain, NSW. Australian Journal of Ecology **13**:361–373.

Milchunas, D. G., W. K. Lauenroth, P. L. Chapman, and M. K. Kazempour. 1990. Community attributes along a perturbation gradient in a shortgrass steppe. Journal of Vegetation Science **1**:375–384.

Milchunas, D. G., O. E. Sala, and W. K. Lauenroth. 1988. A generalized model of the effects of grazing by large herbivores on grassland community structure. American Naturalist **132**:87–106.

Moore, A. D., and I. R. Noble. 1990. An individualistic model of vegetation stand dynamics. Journal of Environmental Management **31**:61–81.

Moore, R. M. 1970. Australian grasslands. Australian National University Press, Canberra, Australia.

Murphy, D. D. 1989. Conservation and confusion: wrong species, wrong scale, wrong conclusions. Conservation Biology **3**:82–84.

Naveh, Z. 1967. Mediterranean ecosystems and vegetation types in California and Israel. Ecology **48**:445–459.

Naveh, Z., and R. H. Whittaker. 1980. Structural and floristic diversity of shrublands and woodlands in northern Israel and other Mediterranean areas. Vegetatio **41**:171–190.

Noble, I. R., and R. O. Slatyer. 1980. The use of vital attributes to predict successional changes in plant communities subject to recurrent disturbances. Vegetatio **43**:5–21.

Noy-Meir, I. 1988. Dominant grasses replaced by ruderal forbs in a vole year in undergrazed Mediterranean grasslands. Journal of Biogeography **15**:579–587.

Noy-Meir, I. 1990. The effect of grazing on the abundance of wild wheat, barley and oat in Israel. Biological Conservation **51**:299–310.

Panetta, F. D., and A. J. M. Hopkins. 1991. Weeds in corridors: invasion and management. Pages 341–351 in D. A. Saunders and R. J. Hobbs, editors. Nature conservation 2: the role of corridors. Surrey-Beatty, Chipping Norton, Australia.

Peet, R. K., D. C. Glenn-Lewin, and J. Walker Wolf. 1983. Prediction of man's impact on plant species diversity. Pages 41–54 in W. Holzner, M. J. A. Werger, and I. Ikusima, editors. Man's impact on vegetation. Junk, The Hague, The Netherlands.

Petraitis, P. S., R. E. Latham, and R. A. Niesenbaum. 1989. The maintenance of species diversity by disturbance. Quarterly Review of Biology **64**:393–418.

Pickard, J. 1984. Exotic plants on Lord Howe Island: distribution in space and time. Journal of Biogeography **11**:181–208.

Pickett, S. T. A., and J. N. Thompson. 1978. Patch dynamics and the design of nature reserves. Biological Conservation 13:27–37.

Pickett, S. T. A., and P. S. White, editors. 1985. The ecology of natural disturbance and patch dynamics. Academic Press, Orlando, Florida.

Pickett, S. T. A., J. Kolasa, J. J. Armesto, and S. L. Collins. 1989. The ecological concept of disturbance and its expression at various hierarchical levels. Oikos 54:129–136.

Platt, W. J. 1975. The colonisation and formation of equilibrium plant species associations on badger disturbances in a tall-grass prairie. Ecological Monographs 45:285–305.

Preston, C. D., and H. L. K. Whitehouse. 1986. The habitat of *Lythrum hyssopifolium* in Cambridgeshire, its only surviving English locality. Biological Conservation 35:41–62.

Puerto, A., M. Rico, M. D. Matias, and J. A. Garcia. 1990. Variation in structure and diversity in Mediterranean grasslands related to trophic status and grazing intensity. Journal of Vegetation Science 1:445–452.

Quinn, J. F., and G. R. Robinson. 1987. The effects of experimental subdivision on flowering plant diversity in a California annual grassland. Journal of Ecology 75:837–855.

Ranwell, D. S. 1960. Newborough Warren, Anglesey. III. Changes in vegetation on parts of the dune system after the loss of rabbits by myxomatosis. Journal of Ecology 48:385–397.

Rawes, M., and D. Welch. 1972. Trials to recreate floristically rich vegetation by plant introduction in the Northern Pennines, England. Biological Conservation 4:135–140.

Rejmánek, M. 1989. Invasibility of plant communities. Pages 369–388 in J. A. Drake, H. A. Mooney, F. di Castri, R. H. Groves, F. J. Kruger, M. Rejmánek, and M. Williamson, editors. Biological invasions: a global perspective. Wiley, Chichester, England.

Rizand, A., R. H. Marrs, M. W. Gough, and T. C. E. Wells. 1989. Long-term effects of various conservation management treatments on selected soil properties of chalk grassland. Biological Conservation 49:105–112.

Robinson, G. R., and J. F. Quinn. 1988. Extinction, turnover and species diversity in an experimentally fragmented California annual grassland. Oecologia (Berlin) 76:71–82.

Rykiel, E. J. 1985. Towards a definition of ecological disturbance. Australian Journal of Ecology 10:361–365.

Saunders, D. A., R. J. Hobbs, and C. R. Margules. 1991. Biological consequences of ecosystem fragmentation. Conservation Biology 5:18–32.

Schofield, E. K. 1989. Effects of introduced plants and animals on island vegetation: examples from the Galapagos Archipelago. Conservation Biology 3:227–238.

Simberloff, D., and N. Gotelli. 1984. Effects of insularization on species richness in the prairie-forest ecotone. Biological Conservation 29:27–46.

Smart, N. O. E., J. C. Hatton, and D. H. N. Spence. 1985. The effect of long-term exclusion of large herbivores on vegetation in Murchison Falls National Park, Uganda. Biological Conservation 33:229–245.

Soulé, M. E. 1990. The onslaught of alien species, and other challenges in the coming decades. Conservation Biology 4:233–239.

Sousa, W. P. 1984. The role of disturbance in natural communities. Annual Review of Ecology and Systematics 15:353–391.

Strang, R. M. 1973. Bush encroachment and veld management in south-central Africa: the need for a reappraisal. Biological Conservation 5:96–104.

Sykora, K. V., G. van der Krogt, and J. Rademakers. 1990. Vegetation change on embankments in the southwestern part of the Netherlands under the influence of different management practices (in particular sheep grazing). Biological Conservation 52:49–81.

Turner, M. G. editor. 1987. Landscape heterogeneity and disturbance. Springer, New York.

van Andel, J., and J. P. van den Bergh. 1987. Disturbance of grasslands. Outline of theme. Pages 3–13 in J. van Andel, J. P. Bakker, and R. W. Snaydon, editors. Disturbance in grasslands: causes, effects and processes. Junk, Dordrecht.

van den Bos, J., and J. P. Bakker. 1990. The development of vegetation patterns by cattle grazing at low stocking density in the Netherlands. Biological Conservation 51:263–272.

van der Maarel, E. 1971. Plant species diversity in relation to management. Pages 45–63 in E. Duffey and A. S. Watt, editors. The scientific management of animal and plant communities for conservation. Blackwell Scientific Publications, Oxford, England.

Vinther, E. 1983. Invasion of *Alnus glutinosa* in a former grazed meadow in relation to different grazing intensities. Biological Conservation 25:75–89.

Vitousek, P. M. 1986. Biological invasions and ecosystem properties: can species make a difference? Pages 163–176 in H. A. Mooney and J. A. Drake, editors. Ecology of biological invasions of North America and Hawaii. Springer, New York.

Vitousek, P. M., and L. R. Walker. 1989. Biological invasion by *Myrica faya* in Hawaii: plant demography, nitrogen fixation, ecosystem effects. Ecological Monographs **59**:247–265.

Vitousek, P. M., L. R. Walker, L. D. Whiteaker, D. Mueller-Dombois, and P. A. Matson. 1987. Biological invasion by *Myrica faya* alters ecosystem development in Hawaii. Science **238**:802–804.

Webb, N. R., and P. J. Hopkins. 1984. Invertebrate diversity on fragmented *Calluna* heathland. Journal of Applied Ecology **21**:921–933.

Wells, T. C. E. 1969. Botanical aspects of conservation management of chalk grasslands. Biological Conservation **2**:36–44.

Westman, W. E. 1990. Park management of exotic plant species: problems and issues. Conservation Biology **4**:251–259.

Westoby, M., B. Walker, and I. Noy-Meir. 1989. Range management on the basis of a model which does not seek to establish equilibrium. Journal of Arid Environments **17**:235–240.

Whicker, A. D., and J. K. Detling. 1988. Ecological consequences of prairie dog disturbances. BioScience **38**:778–784.

White, D. J. B. 1961. Some observations on the vegetation of Blakeney Point, Norfolk, following the disappearance of rabbits in 1954. Journal of Ecology **49**:113–118.

White, P. S., and S. T. A. Pickett. 1985. Natural disturbance and patch dynamics: an introduction. Pages 3–13 in S. T. A. Pickett and P. S. White, editors. The ecology of natural disturbance and patch dynamics. Academic Press, Orlando, Florida.

Willis, A. J. 1963. Braunton Burrows: the effects on the vegetation of the addition of mineral nutrients to dune soils. Journal of Ecology **51**:353–374.

Wright, H. A., L. F. Neuenschwander, and C. M. Britton. 1979. The role and use of fire in sagebrush-grass and pinyon-juniper plant communities. USDA Forest Service General Technical Report INT-58.

Zedler, P. H., and G. A. Scheid. 1988. Invasion of *Carpobrotus edulis* and *Salix lasiolepis* after fire in a coastal chaparral site in Santa Barbara County, California. Madroño **35**:196–201.

Zeevalking, H. J., and L. F. M. Fresco. 1977. Rabbit grazing and species diversity in a dune area. Vegetatio **35**:193–196.

PART 3
Emphasize Biotic Integrity

17
Biological Invasions and Ecosystem Processes: Towards an Integration of Population Biology and Ecosystem Studies

Peter M. Vitousek

Introduction

Much of the recent progress in ecosystem ecology can be traced to studies which have examined the responses of ecosystems to disturbance (Odum 1969). For example, early studies of forest clear-felling (cf. Hesselman 1917, in Stålfelt (1960) demonstrated that soil nutrient availability is usually enhanced in harvested sites. More recently, studies at the Hubbard Brook Experimental Forest (Likens et al. 1970, Bormann and Likens 1979), in Sweden (Tamm et al. 1974, Wiklander 1981), and elsewhere documented that forest cutting alters watershed-level hydrology and nutrient losses; longer-term measurements have documented the reestablishment of biotic regulation of water and nutrient cycling during secondary succession (Bormann and Likens 1979). Studies of a wide range of harvested sites have provided a geographic perspective for patterns in nutrient losses following cutting (Vitousek and Melillo 1979), and detailed experimental studies have evaluated physical and microbial mechanisms controlling variations in loss (Vitousek and Matson 1984, 1985, Matson et al. 1987). The development and testing of theories concerning the regulation of nutrient cycling in forest ecosystems (Vitousek and Reiners 1975, Bormann and Likens 1979, Vitousek and Walker 1987) have been driven in large part by these studies of clearcutting.

Natural disturbances such as fire (Christensen and Muller 1975, Raison 1979) and periodic insect or pathogen outbreaks (Swank et al. 1981, Matson and Boone 1984, Matson and Waring 1984) can affect ecosystems in ways similar to acute anthropogenic disturbance. Viewed on a longer time scale, however, these can also be viewed as cyclic phenomena driven by processes internal to ecosystems (fuel accumulation, nutrient deficiency). Natural disturbance of this sort actually may be integral to the normal functioning of many ecosystems (Holling 1981).

Experimental studies of ecosystem-level responses to disturbance have yielded substantial benefits to applied as well as basic ecology (if indeed the dichotomy has much meaning). For example, studies of lake eutrophication in the Experimental Lakes Area of Ontario are both a convincing analysis of *why* increased phosphorus loading causes eutrophication and a widely comprehensible and politically influential demonstration of how the process works (Schindler et al. 1973, Schindler 1989).

Biological invasions by exotic species represent a wholly different kind of disturbance from those described above, but they too could yield both basic and practical results. Population biologists and community ecologists have long used responses to invasion as a means of analyzing population processes (cf. Elton 1958, Moulton and Pimm 1983, Mooney and Drake

1986). If an introduced species can in and of itself alter ecosystem-level processes such as primary or secondary productivity, hydrology, nutrient cycling, soil development, or disturbance frequency, then clearly the properties of individual species can control the functioning of whole ecosystems (Vitousek 1986). Such species then would provide a useful framework for integrating population and physiological processes into ecosystem studies.

Practically, an exotic species which altered ecosystem properties would not merely compete with or consume native species—it would alter the fundamental rules of existence for all organisms in the area. It could have significant social or economic effects if it altered any of the "ecosystem services" (cf. Ehrlich and Mooney 1983) that affect humanity (such as the regulation of water quality). Finally, biological invasions by exotic species may provide a model for evaluating the possible effects of the release of genetically altered organisms (Regal 1986).

Studies of biological invasions are not the only way to examine the ecosystem-level importance of individual species. Many studies have attempted to determine the importance of particular species without actually altering species composition, and a smaller number have experimentally added or removed species and determined ecosystem-level consequences. The latter approach has been very successful with animals; they can often be enclosed or exclosed by the experimenter. However, the same approach is not applicable to abundant perennial plants—for them, removal itself represents a disturbance with ecosystem-level consequences similar to those of clearcutting. On the other hand, studies of ecosystem-level consequences of biological invasions by exotic species can be done with any group of organisms, and the unprecedented mobility of humans and our associated species ensures that there is more than enough material for study.

I will describe an instance in which an exotic plant species clearly alters ecosystem-level characteristics, and then review briefly a number of other invasions which appear to alter ecosystem properties. Finally, I will discuss how the ecosystem-level effects of invading species could be used to provide the raw material for integrated studies of population biology and ecosystem ecology.

Myrica faya in Volcanic Regions in Hawai'i

Isolated oceanic islands have long been recognized as being unusually susceptible to biological invasion (Wallace 1880, Elton 1958); their biota is highly endemic, usually not very diverse compared with continental areas, disharmonic in species composition (often lacking in major groups such as mammals or ants), and not well adapted to the increased frequency of disturbance which generally accompanies invasion by *Homo sapiens* (Vitousek et al. 1987, Loope and Mueller-Dombois, in press). The Hawaiian Islands are Earth's most isolated archipelago; their native biota are relatively well characterized and quite low in overall species diversity (Carlquist 1980, Mueller-Dombois et al. 1981, Stone and Scott 1985). Biological invasions of Hawai'i are frequent and often highly successful. For example, the native flora consists of about 1100 species—and an additional 4600 exotic plants have been identified there, of which perhaps 800 are invasive and at least 86 represent serious threats to native species or ecosystems (Smith 1985). One species of mammal (a bat) is native; at least 18 more have become established after introduction by humans (Stone 1985).

My colleagues and I have been studying the ecosystem-level consequences of an ongoing biological invasion by *Myrica faya* Ait., an actinorrhizal nitrogen fixer, in a young volcanic region of Hawai'i Volcanoes National Park (HVNP) (Vitousek et al. 1987, Turner and Vitousek 1987). Kilauea Volcano has erupted frequently in historic times, and ^{14}C dating has extended the chronology of lava flows and ash deposits back several thousand years (Lockwood and Lipman 1980, J. P. Lockwood, pers. comm.). Ecosystems developing on recent volcanic substrates in HVNP are relatively low in nitrogen (Vitousek et al. 1983), as is true of primary succession in general (Walker and Syers 1976, Robertson and Vitousek 1981,

Vitousek and Walker 1987). Symbiotic nitrogen fixing plants should have a distinct competitive advantage early in primary succession (Walker and Syers 1976, Gorham et al. 1979, Tilman 1982), and indeed they dominate early stages of many primary seres (Stevens and Walker 1970). However, no native symbiotic nitrogen fixers occur early in primary rainforest succession in Hawai'i, despite the occurrence of legumes (particularly *Acacia koa*) later in succession.

Biological invasion by *Myrica faya* adds a symbiotic nitrogen fixer to nitrogen-deficient sites, so it has great potential to alter ecosystem-level properties and processes. *Myrica* was introduced to Hawai'i in the late 1800s, before which the actinorrhizal symbiosis had been absent from the native flora. It was first observed in HVNP in 1961; by 1977 it covered 600 ha despite intensive control efforts (Whiteaker and Gardner 1985). Control was then abandoned, and by 1985 *Myrica* was present in varying densities over 12,200 ha in HVNP and 34,365 ha in the Hawaiian Islands (Whiteaker and Gardner 1985). *Myrica* occurs in sites ranging from < 15 yr-old volcanic cinder to closed-canopy native rainforest, but its cover is greatest in open-canopied seasonal montane rainforest and in forests thinned but not destroyed by volcanic ashfall. This pattern of dispersal reflects *Myrica's* physiology and mode of seed dispersal (La Rosa et al. 1985, Vitousek et al. 1987). *Myrica* grows very slowly in the shade of a closed forest canopy; it is bird-dispersed, and seed inputs are very highly concentrated under potential perch trees. Consequently, it colonizes most heavily in sites with both perch trees and open canopies.

We predicted that biological invasion by *Myrica faya* would alter primary successional ecosystems in Hawai'i by increasing the amount and biological availability of fixed nitrogen (Vitousek et al. 1987). In order to establish this prediction, three requirements must be met: 1) nitrogen must be limiting to plant and/or microbial activity; 2) nitrogen fixation by *Myrica* must alter ecosystem-level nitrogen budgets substantially; and 3) nitrogen fixed by *Myrica* ultimately must be available to other organisms.

All three requirements are met in open-canopied sites created by volcanic cinder-fall (Vitousek et al. 1987). The first (nitrogen limitation) was determined by fertilizing 26- and 195-yr-old sites with nitrogen, phosphorus, and all other nutrients (including micronutrients but excluding N and P) in factorial combination. Growth of *Metrosideros polymorpha*, the dominant native tree, was doubled by additions of nitrogen; no other nutrient or combination of nutrients had a significant main or interactive effect. Additions of nitrogen (alone) to an open-canopied site yielded an even larger growth increment, while added nitrogen had no effect on growth in an approximately 2000-yr-old site where native nitrogen availability was much greater (Vitousek et al. 1983, 1987). Clearly nitrogen availability does limit primary production in young volcanic sites.

The second requirement (a significant alteration in the system-level nitrogen budget) was evaluated by measuring nitrogen fixation by *Myrica*, fixation by lichens and native non-symbionts, and inputs through rainfall. These measurements were carried out in sites with substantial populations of *Myrica*, in sites with very little *Myrica*, and in one plot where *Myrica* was excluded experimentally. Measurement of fixation by *Myrica* was a 4-step process based on the acetylene reduction assay for nitrogenase (Bergersen 1980). We measured moles of C_2H_2 reduced per mole of $^{15}N_2$ fixed, moles of C_2H_2 reduced per gram of nodule (diurnally and seasonally on three contrasting sites), grams of nodule per individual *Myrica* of several size classes in three sites (Turner and Vitousek 1987), and finally the population and size distribution of *Myrica* in several sites. These calculations yielded an estimated 18 kg ha^{-1} yr^{-1} of nitrogen fixed by *Myrica* in a heavily colonized open-canopied site (Vitousek et al. 1987). Of course, this multiplicative combination of measurements raises the possibility of propagating errors; we are now pursuing alternative estimates based on ^{15}N natural abundance (Shearer and Kohl 1986).

Fixation by *Myrica* is quantitatively more important (in sites where it is abundant) than are other sources of fixed nitrogen. Rainfall added at most 5 kg ha^{-1} yr^{-1} of nitrogen to these sites; native nitrogen fixers (lichens with blue-green algal symbionts like *Stereocaulon volcani, Nostoc* in bryophyte mats, and decomposers of wood and *Metrosideros* leaf litter) added about 0.5 kg ha^{-1} yr^{-1} more (Vitousek et al. 1987). Invasion by *Myrica* can therefore quadruple inputs of fixed nitrogen.

Finally, the availability of *Myrica*-fixed nitrogen was examined by measuring pool sizes of inorganic nitrogen and net nitrogen mineralization in soil under *Myrica*, under *Metrosideros*, and in the open. Soil and forest floor under *Myrica* had significantly higher pool sizes and mineralization of available nitrogen than soil elsewhere; therefore we concluded that biological invasion by *Myrica* alters the availability as well as the quantity of nitrogen in young volcanic sites (Vitousek et al. 1987).

All three requirements were met; consequently biological invasion by *Myrica faya* has been shown to alter ecosystem-level properties of young volcanic sites in HVNP by adding fixed nitrogen. The population-level processes which permit and/or facilitate its invasion, together with the physiological characteristics which cause it to alter nitrogen budgets, therefore have important consequences to local ecosystems. In turn, we should now be able to observe how the altered ecosystem-level properties feed back to affect population and physiological processes of the native biota.

Invasions and Ecosystem Alterations

How often do biological invasions alter ecosystem-level properties and/or processes? Several recent reviews have addressed this question (Vitousek 1986, Ramakrishnan and Vitousek, in press, MacDonald et al., in press). Much of the available information is anecdotal and unavoidably biased towards successful invasions (cf. Simberloff 1986), in this case those which do alter ecosystems. I believe that the majority of successful invasions do not alter large-scale ecosystem properties and processes in a meaningful way. Nonetheless, some (such as *Myrica faya*) clearly do, and a tentative classification of the ways in which biological invaders can alter ecosystems may be useful. I suggest in-

vaders can change ecosystems where they 1) differ substantially from natives in resource acquisition or utilization; 2) alter the trophic structure of the invaded area; or 3) alter disturbance frequency and/or intensity.

Resource Acquisition and Utilization

Myrica faya fits into the first category—by fixing atmospheric nitrogen, it expands the resource base for the entire ecosystem, with consequences that could go well beyond its own growth. Another clear example is provided by invasions of salt-cedar (*Tamarix* spp.) in riparian areas of the semi-arid southwestern United States. *Tamarix* is a phreatophyte (rooted into ground water) which does not actively regulate its transpiration; as a consequence it can desiccate watercourses and marshes (Neill 1983). Deeply rooted invading plants can also alter the resource base of an ecosystem by bringing nutrients to the surface where they may be available to a range of organisms (Hodgkin 1984).

Not all changes in resource base increase productivity; in California and Australia, the exotic ice-plant *(Mesembryanthemum crystallinum)* accumulates salt from throughout the rooting zone and thereby reduces soil fertility (Vivrette and Muller 1977, Kloot 1983). It may also increase soil erosion (Halvorson, in press). Plants which produce low-quality acid litter also could reduce overall soil nutrient availability (Pastor et al. 1984). This effect has been documented clearly in tree plantations (cf. Nihlgård 1972, Perala and Alban 1982); whether it occurs as a consequence of natural invasions remains to be documented, but the invasion and rapid growth of *Pinus radiata* in areas of Australian *Eucalyptus* woodland (Chilvers and Burdon 1983) could cause such changes.

Invading species which differ from natives in their efficiency of resource utilization also could alter ecosystem-level processes effectively. One way in which one plant species could be more efficient than another is through differences in life-form; for example, perennials maintain internal storage pools of energy and nutrients which they can utilize in subsequent growing seasons, while annuals have only seed storage and current photosynthesis and nutrient uptake. Adding a perennial to a system dominated by annuals could therefore alter ecosystem properties. The invasion of floating aquatic plants into open-water habitats represents an analogous change. For example, the water-fern *Salvinia molesta* has altered productivity and water chemistry substantially in Africa, India, and Papua New Guinea (Mitchell et al. 1980, Thomas 1981). Invaders which differ from natives in photosynthetic pathway (C_3, C_4, CAM) could also change ecosystems (Ramakrishnan and Vitousek, in press), although I am not aware that any such changes have been documented unequivocally as being due to differing photosynthetic pathways.

Trophic Structure

Experimental studies and examinations of biological invasions have demonstrated that manipulating the uppermost level of a trophic pyramid can have ecosystem-level consequences disproportionate to the amounts of energy and/or nutrients involved (Paine 1966, Dayton et al. 1984). This effect may be responsible in part for the observation that it is easier to document ecosystem-level consequences of biological invasions by animals than plants (Vitousek 1986). Animal invasions are particularly disruptive on oceanic islands; these often lacked *any* large generalist herbivore before human settlement. Additions of pigs, goats, and cattle have altered islands dramatically; consequent changes in soil erosion, nutrient cycling, and subsequent invasibility have been identified (Stone 1985, Stone et al., in press, Merlin and Juvik, in press).

The effects of animal invaders on oceanic islands may be so severe because islands often have only two trophic levels, producers and decomposers (excluding specialist herbivores and their carnivores). Adding large generalist herbivores without their predators therefore can depress producer populations and/or standing crop. Subsequent addition of a predator (in Hawai'i, human hunters) often results in greater plant cover (including that of natives) in accessible areas (Jacobi and Scott 1985, Stone et al., in press).

Alterations in trophic structure per se are not the only reason for severe ecosystem-level consequences of animal invasions on oceanic islands. The flora of such islands is often lacking in chemical and mechanical defenses against herbivores (Carlquist 1980); animal invasion can therefore cause more damage than might occur simply by adding another trophic level to a continental area. Invasions by animals can be extremely disruptive in continental regions; European wild boars in the Great Smoky Mountains of the south-eastern United States provide one clear example (Bratton 1975, Singer et al. 1984). In this case, however, the ecosystem-level effect is probably due to physical disturbance (see below), and the invasion may be facilitated by human removal of carnivores (wolves, puma) in the region.

Disturbance Frequency and Intensity

Biological invaders which alter the disturbance regime of an invaded area can have significant ecosystem-level consequences; natural disturbance regulates both population and ecosystem-level properties in many, perhaps most, ecosystems (Matson and Waring 1984, Pickett and White 1985, Vitousek and Denslow 1986). Invading animals may change ecosystems because they *are* agents of disturbance (in the sense of destruction of biomass—Grime 1979); this effect may be particularly marked on oceanic islands where native plants are poorly protected against grazing. Moreover, the feeding habit of certain animals is particularly destructive—the rooting activity of pigs is most likely responsible for their very striking effects on soils and nutrient cycling (Singer et al. 1984, Vitousek 1986).

Biological invasions also alter ecosystems through their influence on fire regimes. Exotic grasses have invaded semi-arid shrublands and woodlands in many areas; they often produce considerably more aboveground litter than native species. This litter can increase the probability, extent, and severity of fires (Parsons 1972, Smith 1985, Christensen and Burrows 1986, Mack 1986). Moreover, many of these grasses are adapted to rapid seeding or sprouting after fire, while native plants often are not; relative dominance by exotic grasses and the probability of subsequent fires consequently increases after each fire.

This classification of ecosystem-level effects of invading species is tentative, and there may well be examples of ecosystem alteration that are not encompassed within it. Further research on the ways in which invaders can alter ecosystems and the frequency with which they do so would be most useful.

Integrating Population and Ecosystem Ecology

A few biological invaders can be clearly shown to have altered ecosystem-level properties and processes through their own activities. The number of examples is relatively small, but serious attempts to evaluate such effects have been few. Moreover, the lack of detailed background information in most sites and the coarseness of most ecosystem-level measurements make it difficult or impossible to detect small or subtle effects of biological invaders; only major changes can be identified clearly. The fact that there *are* examples in which plant and animal invasions do alter ecosystems is unambiguous evidence that some individual species affect the properties of some ecosystems.

It is perhaps better established that the properties of ecosystems affect population-level processes, including the invasibility of particular communities (Orians 1986). In the case of *Myrica faya* in Hawai'i Volcanoes National Park, nitrogen-limited open-canopied forests represent the primary habitat for colonization. The presence of trees provides perches and perhaps an alternative food source for birds, while substantial light penetration to the soil surface permits rapid growth of *Myrica*. These conditions are realized in seasonal montane forest ecosystems and volcanic-ash damaged rainforest (Vitousek et al. 1987).

More generally, biological invasions by plants are often concentrated in human-disturbed sites (cf. Allan 1936, Egler 1942, Elton 1958), and it appears that many invaders are more successful on more fertile soils (Gerrish and Mueller-Dombois 1980, Bridgewater and Backshall 1981). Invasions of all kinds are more often successful on oceanic islands then continents (Elton 1958), and this is true of invasion into parks and preserves as well as disturbed areas (Loope and Mueller-Dombois, in press).

A demonstration that individual species affect ecosystem-level properties and that ecosystem properties in turn affect species populations does not in and of itself integrate population biology and ecosystem ecology — but it does provide raw material for such an integration. Where individual species invade and alter ecosystems, population-level processes *become* ecosystem-level processes — and where ecosystem properties are changed by invasions, the basic rules of existence for all organisms also change. Native species could be equally important in controlling ecosystems, but their effects are often more difficult to demonstrate. The value in studying invading exotic species is that changes in ecosystems can be observed directly and manipulated experimentally.

Acknowledgments

I thank P. A. Matson, D. Mueller-Dombois, C. P. Stone, L. R. Walker, and L. Whiteaker for their collaboration in the study of *Myrica faya* in Hawai'i. L. F. Huenneke, J. Armstrong, and R. Riley commented critically on an earlier version of this manuscript, and C. Nakashima prepared the manuscript for publication. Research and manuscript preparation were supported by NSF grant BSR-8415821 to Stanford University.

References

Allan, H. H. 1936. Indigene versus alien in the New Zealand plant world. — Ecology 17:187–193.

Bergersen, F. J. (ed). 1980. Methods for evaluating biological nitrogen fixation. — Wiley, New York.

Bormann, F. H., and Likens, G. E. 1979. Pattern and process in a forested ecosystem. — Springer, New York.

Bratton, S. P. 1975. The effect of the European wild boar, *Sus scrofa*, on gray beech forest in the Great Smoky Mountains. — Ecology 56:1356–1366.

Bridgewater, P. B. and Backshall, D. J. 1981. Dynamics of some Western Australian ligneous formations with special reference to the invasion of exotic species. — Vegetatio 46:141–148.

Carlquist, S. J. 1980. Hawaii, a Natural History. — Pacific Tropical Botanical Garden, Lawai, Kauai, Hawaii.

Chilvers, G. A. and Burdon, J. J. 1983. Further studies on a native Australian eucalypt forest invaded by exotic pines. — Oecologia (Berl.) 59:239-245.

Christensen, P. E. and Burrows, N. D. 1986. Fire: an old tool with a new use. — In: Groves, R. H. and Burdon, J. J. (eds), Ecology of biological invasions: An Australian perspective. Australian Academy of Sciences, Canberra, pp. 97-105.

Christensen, N. L. and Muller, C. H. 1975. Effects of fire on factors controlling plant growth in *Adenostoma* chaparral. — Ecol. Monogr. 45:29-55.

Dayton, P. K., Currie, V., Gerrodette, T., Keller, B. D., Rosenthal, R. and Ven Tresca, D. 1984. Patch dynamics and stability of some California kelp communities. — Ecol. Monogr. 54:253-289.

Egler, F. E. 1942. Indigene versus alien in the development of arid Hawaiian vegetation. — Ecology 23:14-23.

Ehrlich, P. R. and Mooney, H. A. 1983. Extinction, subsituation, and ecosystem services. — Bioscience 33:248-253.

Elton, C. S. 1958. The ecology of invasions by animals and plants. — Methuen, London.

Gerrish, G. and Mueller-Dombois, D. 1980. Behavior of native and non-native plants in two tropical rainforests on Oahu, Hawaiian Islands. — Phytocoenologia 8:237-295.

Gorham, E., Vitousek, P. M., and Reiners, W. A. 1979. The regulation of element budgets over the course of terrestrial ecosystem succession. — Ann. Rev. Ecol. Syst. 10:53-84.

Grime, J. P. 1979. Plant strategies and vegetation processes. — Wiley, New York.

Halvorson, W. C. Alien plants at Channel Islands National Park. — In: Stone, C. P., Smith, C. W., and Tunison, J. T. (eds), Alien plant invasions in Hawai'i: Management and research in near-native ecosystems. Cooperative National Park Resources Study Unit. University of Hawaii, Honolulu. (In press).

Hesselman, H. 1917. Studier över saltpeterbildningen i naturliga jordmåner och dess betydelse i växtekologiskt avseende. — Medd. Stat. skogsforskningsanst. 12:297.

Hodgkin, S. E. 1984. Scrub encroachment and its effects on soil fertility on Newborough Warren, Anglesey, Wales. — Biol. Conserv. 29:99-119.

Holling, C. S. 1981. Forest insects, forest fires, and resilience. — In: Mooney H. A., Bonnicksen T. M., Christensen N. L., Lotan J. E. and Reiners W. A. (coordinators), Fire regimes and ecosystem properties. U.S. Forest Service General Technical Report WO-26, pp. 445-464.

Jacobi, J. D. and Scott, J. M. 1985. Status, research,

and management needs of the native Hawaiian biota. — In: Stone, C. P. and Scott, J. M. (eds), Hawai'i's terrestrial ecosystems: Preservation and management. University of Hawaii Press for Cooperative National Park Resources Study Unit, Honolulu, pp. 3-22.

Kloot, P. M. 1983. The role of common iceplant *(Mesembryanthemum crystallinum)* in the deterioration of medic pastures. — Aust. J. Ecol. 8:301-306.

LaRosa, A. M., Smith, C. W. and Gardner, D. E. 1985. Role of alien and native birds in the dissemination of fire tree *(Myrica faya* Ait. — Myricaceae) and associated plants in Hawaii. — Pacific Sci. 39:372-378.

Likens, G. E., Bormann, F. H., Johnson, N. M., Fisher, D. W. and Pierce, R. S. 1970. Effects of forest cutting and herbicide treatment on nutrient budgets in the Hubbard Brook ecosystem in New Hampshire. — Ecol. Monogr. 40:23-47.

Lockwood, J. P. and Lipman, P. W. 1980. Recovery of datable charcoal beneath young lavas: lessons from Hawaii. — Bull. Volcanol. 43:609-615.

Loope, L. and Mueller-Dombois, D. Characteristics of invaded islands. — In: Drake, J., diCastri, F., Groves, R., Kruger, F., Mooney, H. A., Rejmanek, M. and Williamson, M. (eds), Biological invasions: a global perspective. Wiley (in press).

MacDonald, I. A. W., Loope, L. L., Usher, M. B. and Hamann, O. Wildlife conservation and the invasion of nature reserves by exotic species: a global perspective. — In: Drake, J., diCastri, F., Groves, R., Kruger, F., Mooney, H. A., Rejmanek, M. and Williamson, M. (eds), Biological invasions: a global perspective. Wiley (in press).

Mack, R. N. 1986. Alien plant invasion into the Intermountain West: a case history. — In: Mooney, H. A. and Drake, J. A. (eds), Ecology of biological invasions of North America and Hawaii. Springer, New York, pp. 191-213.

Matson, P. A. and Boone, R. D. 1984. Natural disturbance and nitrogen mineralization: waveform dieback of mountain hemlock in the Oregon Cascades. — Ecology 65:1511-1516.

_____ and Waring, R. H. 1984. Effects of nutrient and light limitation on mountain hemlock: susceptibility to laminated root rot. — Ecology 65:1517-1524.

_____ , Vitousek, P. M., Ewel, J. J., Mazzarino, M. J. and Robertson, G. P. 1987. Nitrogen transformations following tropical forest felling and burning on a volcanic soil. — Ecology 68:490-1502.

Merlin, M. D. and Juvik, J. O. Relationships between native and alien plants on islands with and without wild ungulates. — In: Stone, C. P., Smith, C. W. and Tunison, J. T. (eds), Alien plant invasions in Hawai'i: Management and research in near-native ecosystems. Cooperative National Park Resources Study Unit. University of Hawaii, Honolulu. (In press).

Mitchell, D. S., Petr, T. and Viner, A. B. 1980. The water-fern *Savinia molesta* in the Sepik River, Papua New Guinea. — Environ. Conserv. 7:115–122.

Mooney, H. A. and Drake, J. (eds) 1986. Biological invasions of North America and Hawaii. — Springer, New York.

Moulton, M. P. and Pimm, S. L. 1983. The introduced Hawaiian avifauna: biogeographic evidence for competition. — Am. Nat. 121:669–690.

Mueller-Dombois, D., Bridges, K. W. and Carson, H. L. (eds) 1981. Island ecosystems: Biological organization in selected Hawaiian communities. — Hutchinson Ross, Stroudsburg, PA.

Neill, W. M. 1983. The tamarisk invasion of desert riparian areas. — Educ. Bull. 83-4, Desert Protective Council, Spring Valley, CA.

Nihlgård, B. 1972. Plant biomass, primary production and distribution of chemical elements in a beech and a planted spruce forest in South Sweden. — Oikos 23:69–81.

Odum, E. P. 1969. The strategy of ecosystem development. — Science 164:262–270.

Orians, G. H. 1986. Site characteristics favoring invasions. — In: Mooney, H. A., and Drake, J. (eds), Ecology of biological invasions of North America and Hawaii. Springer, New York, pp. 133–148.

Paine, R. T. 1966. Food web complexity and species diversity. — Am. Nat. 100:65–75.

Parsons, J. J. 1972. Spread of African pasture grasses to the American tropics. — J. Range Manage. 25:12–17.

Pastor, J., Aber, J. D., McClaugherty, C. A. and Melillo, J. M. 1984. Above-ground production and N and P cycling along a nitrogen mineralization gradient on Blackhawk Island, Wisconsin. — Ecology 65:256–268.

Perala, D. A., and Alban, D. H. 1982. Biomass, nutrient distribution and litterfall in *Populus, Pinus* and *Picea* stands on two different soils in Minnesota. — Plant Soil 64: 177–192.

Pickett, S. T. A. and White, P. S. (eds), 1985. The ecology of natural disturbance and patch dynamics. — Academic Press, New York.

Raison, R. J. 1979. Modification of the soil environment by vegetation fires, with particular reference to nitrogen transformations: a review. — Plant Soil 51:73–108.

Ramakrishnan, P. S. and Vitousek, P. M. Ecosystem-level consequences of biological invasions. — In: Drake, J., diCastri, F., Groves, R., Kruger, F.,

Mooney, H. A., Rejmanek, M., and Williamson, M. (eds), Biological invasions: a global perspective. Wiley, (in press).

Regal, P. J. 1986. Models of genetically engineered organisms and their ecological impact. — In: Mooney, H. A. and Drake, J. A. (eds), Ecology of biological invasions of North America and Hawaii. Springer, New York, pp. 111–129.

Robertson, G. P. and Vitousek, P. M. 1981. Nitrification potentials in primary and secondary succession. — Ecology 62:376–386.

Schindler, D. W. 1989. Experimental perturbations of whole lakes as tests of hypotheses concerning ecosystem structure and function. — Oikos 57:25–41.

_____ , Kling, H., Schmidt, R. V., Prokopowich, J., Frost, V. E., Reid, R. A. and Capel, M. 1973. Eutrophication of Lake 227 by addition of phosphate and nitrate: the second, third, and fourth years of enrichment 1970, 1971, and 1972. — J. Fish. Res. Bd Can. 30:1415–1440.

Shearer, G. and Kohl, D. H. 1986. N_2 fixation in field settings: estimations based on natural ^{15}N abundance. — Aust. J. Plant Physiol. 13:699–756.

Simberloff, D. 1986. Introduced insects: a biogeographic and systematic perspective. — In: Mooney, H. A. and Drake, J. A. (eds), Ecology of biological invasions of North America and Hawaii. Springer, New York, pp. 3–26.

Singer, F. J., Swank, W. T. and Clebsch, E. E. C. 1984. Effects of wild pig rooting in a deciduous forest. — J. Wildl. Manage. 48:464–473.

Smathers, G. A. and Gardner, D. E. 1979. Stand analysis of an invading firetree (*Myrica faya* Aiton) population, Hawaii. — Pacific Sci. 33: 239–255.

Smith, C. W. 1985. Impact of alien plants on Hawaii's native biota. — In: Stone, C. P. and Scott, J. M. (eds), Hawaii's terrestrial ecosystems: Preservation and management. Cooperative National Park Resources Study Unit, Univ. of Hawaii, Honolulu, pp. 180–250.

Stålfelt, M. G. 1960. Växtekologi. — Translated as "Stålfelt's plant ecology", 1972, Halstead Press, New York.

Stevens, P. R. and Walker, T. W. 1970. The chronosequence concept and soil formation. — Quart. Rev. Biol. 45:333–350.

Stone, C. P. 1985. Alien animals in Hawaii's native ecosystems: towards controlling the adverse effects of introduced vertebrates. — In: Stone, C. P. and Scott, J. M. (eds), Hawai'i's terrestrial ecosystems: Preservation and management. Cooperative National Park Resources Study Unit, Univ. of Hawaii, Honolulu, pp. 251–297.

_____ , Higashino, P. K., Tunison, J. T., Cuddihy,

L. W., Anderson, S. J., Jacobi, J. D., Ohashi, T. J. and Loope, L. L. Success of alien plants after feral goat and pig removal. – In: Stone, C. P., Smith, C. W. and Tunison, J. T. (eds), Alien plant invasions in Hawai'i: Management and research in near-native ecosystems. Cooperative National Park Resources Study Unit, Univ. of Hawaii, Honolulu, (in press).

_____ and Scott, J. M. (eds) 1985. Hawai'i's terrestrial ecosystems: Preservation and management. – Cooperative National Park Resources Study Unit, Univ. of Hawaii, Honolulu.

Swank, W. T., Waide, J. B., Crossley, D. A. and Todd, R. L. 1981. Insect defoliation enhances nitrate export from forested ecosystems. – Oecologia (Berl.) 51:297–299.

Tamm, C. O., Holmen, H., Popovic B. and Wiklander, G. 1974. Leaching of plant nutrients from soils as a consequence of forestry operations. – Ambio 3:211–221.

Thomas, K. J. 1981. The role of aquatic weeds in changing the pattern of ecosystems in Kerala. – Environ. Conserv. 8:63–66.

Tilman, D. 1982. Resource competition and community structure. – Princeton Univ. Press, Princeton, NJ.

Turner, D. R. and Vitousek, P. M. 1987. Nodule biomass in the nitrogen-fixing alien *Myrica faya* in Hawaii Volcanoes National Park. – Pacific Sci. 41:186–190.

Vitousek, P. M. 1986. Biological invasions and ecosystem properties: can species make a difference? – In: Mooney, H. A. and Drake, J. (eds), Biological invasions of North America and Hawaii. Springer, pp. 163–176.

_____ and Denslow, J. S. 1986. Nitrogen and phosphorus availability in treefall gaps in a lowland tropical rainforest. – J. Ecol. 74:1167–1178.

_____ , Loope, L. L. and Stone, C. P. 1987. Introduced species in Hawaii: Biological effects and opportunities for ecological research. – Trends Ecol. Evol. 2:224–227.

_____ and Matson, P. A. 1984. Mechanisms of nitrogen retention in forest ecosystems: a field experiment. – Science 225:51–52.

_____ and Matson, P. A. 1985. Disturbance, nitrogen availability, and nitrogen losses in an intensively managed loblolly pine plantation. – Ecology 66:1360–1376.

_____ and Melillo, J. M. 1979. Nitrate losses from disturbed forests: patterns and mechanisms. – For. Sci. 25:605–619.

_____ and Reiners, W. A. 1975. Ecosystem succession and nutrient retention: a hypothesis. – Bioscience 25:376–381.

_____ and Walker, L. R. 1987. Colonization, succession, and resource availability: ecosystem-level interactions. – In: Gray, A., Crawley, M. and Edwards, P. J. (eds), Colonization, succession, and stability. Blackwell, Oxford, pp. 207–223.

_____ , Van Cleve, K., Balakrishnan, N. and Mueller-Dombois, D. 1983. Soil development and nitrogen turnover in montane rain forest soils in Hawai'i. – Biotropica 15:268–274.

_____ , Walker, L. R., Whiteaker, L. D., Mueller-Dombois, D. and Matson, P. A. 1987. Biological invasion by *Myrica faya* alters ecosystem development in Hawaii. – Science 238:802–804.

Vivrette, N. J. and Muller, C. H. 1977. Mechanism of invasion and dominance of coastal grassland by *Mesembryanthemum crystallinum*. – Ecol. Monogr. 47:301–318.

Walker, T. W. and Syers, J. K. 1976. The fate of phosphorus during pedogenesis. – Geoderma 15:1–19.

Wallace, A. R. 1880. Island life. – Macmillan, London.

Whiteaker, L. D., and Gardner, D. E. 1985. The distribution of *Myrica faya* Ait. in the State of Hawai'i. Cooperative National Park Resources Study Unit Tech Report 55. Dept of Botany, Univ. of Hawaii.

Wiklander, G. 1981. Rapporteur's comment on clearcutting. – In: Clark, F. E. and Rosswall, T. H. (eds), Nitrogen cycling in terrestrial ecosystems: Processes, ecosystem strategies, and management implications. Ecol. Bull. (Stockholm) 33:642–647.

18
No Park Is an Island: Increase in Interference from Outside as Park Size Decreases

Daniel H. Janzen

Introduction

There is evident conservation value to numerous small habitat preserves (parks) when large protected areas are impractical (Simberloff 1982). However, such a management policy brings to mind a caution that is often unappreciated for parks and other forms of conserved pristine vegetation, and becomes ever more appropriate as park size decreases. The smaller the patch (island) of habitat that is viewed as pristine, the greater the effect of the surrounding secondary successional vegetation and croplands as a source of 1) animals and seeds that enter the park and interact with the residents, and 2) food subsidy for residents capable of foraging outside of the pristine forest. Every field biologist working in or around apparently pristine vegetation is aware to some degree of this problem, but none of us wish to fully acknowledge its potential impact on biotic interactions and our ecological/evolutionary interpretations of them. It is a particularly inconvenient acknowledgment for studies of biotic interactions at levels of organization greater than pairs of individuals.

The problem arises because a park differs in one very substantial biological manner from a conventional island surrounded by water, a very different habitat. The conventional island is not surrounded by a habitat rich in organisms competent to forage extensively on the island (excluding beach inhabitants) and prone to bombard the island with juveniles quite capable of taking up residence. Likewise, the conventional island's residents are generally not inclined to forage heavily in the surrounding ocean. Oceanic birds nesting on small islands, and hermit crabs, are obvious exceptions. However, these sorts of animals are generally not included in discussions of how species-packing on islands relates to island size or to discussions of ecological interactions such as competition, mutualism, etc. In strong contrast, a park, and especially its naturally-occurring successional habitats in pristine vegetation (tree falls, landslides, large mammal trails, watercourse edges, etc.), is bombarded by an enormous multi-species seed shadow generated by the exploded populations of successional plants and associated animals of roadsides, abandoned croplands, brushy pastures, logged forest, etc. Furthermore, the animals of both natural disturbed sites and closed-canopy forest have many reasons to forage outside the park in these anthropogenic habitats.

Since the phenomenon I describe does not seem to threaten extinction, and may result in increased densities of desired species, it might seem so esoteric as to be not worrisome to park management (though its significance to more esoteric studies seems evident). However, parks differ quantitatively from zoos and botanical gardens in the interactions that are present in the park; parks function to conserve interactions, and the conservation of organisms is a happy by-product. A belt of food-rich secondary succession around a small park may be as destructive to these interactions as would be the elimination of the large carnivores within the park. The evolutionary ecology of large tree replacement in tree falls within the park may be rendered uninterpretable as much by heavy seed rain from outside the park as by deliberate planting in tree falls by park managers.

To move from the vague to the specific, below I briefly comment on the primary and secondary succession in a tropical tree fall, a habitat where the phenomenon over which I worry is particularly evident. The question is simple. What determines the successional outcome in a particular tree fall in an essentially pristine forest? The answer is likewise simple. The outcome is determined by what plants were there when the tree fell, what plants arrive later, and how these plants interact. In a small park, all these processes are influenced by more than the organisms of the closed-canopy forest and its natural disturbance sites.

Replace a Guapinol

In the eastern portion of Santa Rosa National Park in northwestern lowland Costa Rica (25 km south of La Cruz, 0–350 m a.s.l., see Boza and Mendoza 1981 for a general description), there is a 10 ha patch of apparently original forest (though the mahoganies, *Swietenia macrophylla*, were selectively removed in the 1940's). This forest is bounded on two sides by old pastures and in other directions grades into 10–80-yr old secondary successional forest. In the essentially pristine closed-canopy forest, the general canopy is 20–40 m in height; its undulation is accentuated by the very broken topography of ravines and small plateaus. At least 80% of the crowns of large trees are the following species: *Hymenaea courbaril* (Leguminosae), *Quercus oleoides* (Fagaceae), *Luehea speciosa* (Tiliaceae), *Brosimum alicastrum* and *Ficus obtusifolia* (Moraceae), *Zuelania guidonia* (Flacourtiaceae), *Manilkara zapota* and *Mastichodendron capiri* (Sapotaceae) (plant names used in this paper are in accordance with

Janzen and Liesner 1980). There are another 30 species of rare large trees and sub-canopy level trees. *H. courbaril* or guapinol and *Q. oleoides* or roble encino are the most common large trees and one could easily describe the forest as being a guapinol-encino association, with the guapinol more common on the moist sites and the roble encino more common on the drier sites (the nearby pasture area was once a nearly pure roble encino forest, cf. Boucher 1981, and the guapinol forest may once have been little more than a fringe on the once much more extensive roble encino forest). The western edge of the guapinol-encino forest is cut through by the main blacktop road through the Park and the forest is generally referred to there as "bosque húmedo" or "bosque siempre verde". These names derive from the nearly evergreen pristine forest, which stands in strong contrast to the highly deciduous nature of the secondary successional forest that covers much of the remainder of the Park.

In late December 1978, one of the severe wind storms associated with the beginning of the rain-free dry season (December to mid-May) blew down a 31 m tall guapinol (three other large guapinoles were blown down within 1000 m of this one in the same storm). In falling, the guapinol left a gap in the canopy, produced a small patch of bare soil, exposed a large area of understory to direct insolation, and crushed a smaller area of understory vegetation. My question is, can the succession leading to the eventual replacement of this guapinol by another large tree or trees of the same or other species be studied as an example of succession in pristine forest of this type, and can the results be generalized to the evolutionary ecology questions currently fashionable about tropical forest dynamics? Has the succession in this tree fall been as altered by processes outside the pristine forest as if a forester had deliberately seeded the tree fall with a desirable tree species? For me, the best way to answer such questions, in the context of the subject of this essay, is to examine the guapinol tree fall at present, 3.5 growing seasons after the tree fell. My intent is to identify some of those interactions that may have been important so far and that may have been influenced by the area outside of the pristine forest.

Before examining this particular tree fall, more background is needed on the area. Santa Rosa National Park, 10800 ha in extent, was a cattle ranch for at least 200 yr. During that time, at least 30% of the forest was cleared and planted to grass. At least 70% of the remaining forest was lumbered and then allowed to return to secondary woody succession. The successional woody vegetation and pastures were variously grazed and browsed by cattle until 1978. The pastures are highly interdigitated with forest of all ages, and are presently gradually returning to woody vegetation (but the return is slowed by dry season grass fires). There has been essentially no hunting since 1971, when the Park was established.

Anatomy of a Guapinol Tree Fall

The tree had 12.5 m of straight and unbranched 68.8 cm DBH bole, with a broadly conical crown 20.5 m in depth (distance from the first major fork to the uppermost layer of leaves). It was healthy and the crown symmetrical and intact. When the wind uprooted the guapinol at the beginning of the dry season, the soil was still moist. However, no rain fell for nearly 6 months on the newly bared soil, litter and understory vegetation. When uprooted, it produced a 2.3-m-tall hemispherical but thin mound of dirt and guapinol roots on the west side of an 80 cm deep hemispherical pit from which the root crown was torn. The soil was deep latosol with old volcanic ash intrusions. The guapinol fell such that the bole crushed a few understory saplings but the crown did not tear away portions of adjacent crowns of large trees. This was partly an accident of the way it fell and partly because, as is generally the case with this species of tree, the crown was free of large vines connecting it with other crowns. The gap in the canopy was 124 m^2 in extent. This tree fall and the succession in it is representative of numerous other tree falls that I have examined in pristine forest in the Park.

The tree fall led to a strong and heterogeneous change in insolation at ground level. The freshly bared litter-free soil of the pit and mound was in direct sunlight or below open sky from about 0800 to 1600 hours (and most of it still was 3.5 yr later). The (previously) understory plants at the margins of the soil pit also received direct sunlight, as did much of the litter beneath them. The litter at this site received the most intense sunlight of any of that exposed by the tree fall because the understory vegetation tends to be most sparse around the bases of large trees such as this one. Immediately after the tree fell, the bole was partly covered by the crowns of 1–3 m tall understory shrubs and treelets whose crowns had merely been pushed to one side as the bole passed by. These small woody plant crowns shaded the litter somewhat and quickly responded to the increased insolation by producing more leafy crowns. By August 1982, these crowns generated as dense a shade, if not more so, as occurred before the tree fell. Where the crown of the guapinol landed, the understory plants were severely crushed and also covered with slowly decomposing medium-sized to large guapinol branches. Here, the overstory canopy was still intact, and therefore there was little or no change in insolation. By August 1982, the area where the crown fell was nearly bare of plants and littered with rotting vegetation.

By August 1982, 3.5 yr after the guapinol fell, the three new vegetation types are very different from each other in composition but strongly reflect the initial differences described above.

The Soil Pit and Mound

This area of intense disturbance is highly heterogeneous in vegetation regeneration. On the bare soil, a few individuals of fast-growing "colonizing" species have taken root (Tab. 1) but they have not yet produced an approximation of a partly-closed canopy at any height. (I will later discuss *Cecropia peltata* as a detailed example). In addition to the plants that have grown on the bare soil from seedlings, 16 species of perennial vines that were present before the tree fell have grown into the open area over the bared soil or over the immediately adjacent vegetation (Tab. 1). Each of these species of vines is represented by only 1–2 individuals. Of similar biology to the vines, but

TABLE 1. Vegetation of the soil pit and its insolated immediate surroundings at ground level

Plant species	Number of individuals	Height of tallest (cm)	Seed dispersal agents
Rooted from seed on the bared soil:			
Cecropia peltata (Moraceae)	4	800	birds, bats, coatis, monkeys, rodents
Adiantum conccinum (Adiantaceae)	5	25	?
Eupatorium quandrangulare (Compositae)	2	72	wind
Verbesina gigantea (Compositae)	2	145	wind
Cordia linnaei (Boraginaceae)	1	148	birds
Tetracera volubilis (Dilleniaceae)	1	6	birds
Cassia hayesiana (Leguminosae)	1	62	birds
Hamelia patents (Rubiaceae)	4	110	birds
Miconia argentea (Melastomataceae)	7	123	birds
Vismia baccifera (Guttiferae)	1	188	bats?
Solanum americanum (Solanaceae)	1	202	bats?
Piper marginatum (Piperaceae)	6	48	bats
Piper pseudo-fulligineum (Piperaceae)	3	87	bats
Pityrogramma calomelanos (Gymnogrammaceae)	2	36	?
Borreria ocimoides (Rubiaceae)	32	3	?
Philonotis sp. (Bartramaceae)	?	1	?
Prockia crucis (Flacourtiaceae)	1	110	birds
Vines growing across the bare soil and pit edge from old root stocks:			
Pithecoctinium crucigerum (Bignoniaceae)	2	20	wind
Gouania polygama (Rhamnaceae)	1	130	wind
Tetracera volubilis (Dilleniaceae)	1	30	birds
Serjania atrolineata (Sapindaceae)	1	20	wind
Paulinia cururu (Sapindaceae)	2	10	birds, coatis
Desmodium axillare (Leguminosae)	1	8	sticktight on mammal
Lygodium venustum (Schizaeaceae)	1	22	?
Petraea volubilis (Verbenaceae)	1	18	wind
Centrosema pubescens (Leguminosae)	1	40	explosive fruit
Discorea convolvulacea (Dioscoreaceae)	1	23	wind
Macfadyena unguis-cati (Bignoniaceae)	1	6	wind
Passiflora platyloba (Passifloraceae)	1	240	large mammals, rodents
Forsteronia spicata (Apocynaceae)	1	110	wind
Cissus rhombifolia (Vitaceae)	2	30	birds, mammals
Sicydium tamnifolium (Cucurbitaceae)	2	100	birds
Ipomoea squamosa (Convolvulaceae)	1	200	wind
Established prior to tree fall, self-supporting, not reproducing in forest understory:			
Annona purpurea (Annonaceae)	1	260	large mammals, rodents
Hymenaea courbaril (Leguminosae)	3	120	large mammals, rodents
Allophyllus occidentalis (Sapindaceae)	1	180	birds, large mammals
Sapium thelocarpum (Euphorbiaceae)	1	230	birds
Solanum accrescens (Solanaceae)	1	310	bats?
Malvaviscus arboreus (Malvaceae)	1	330	birds, mammals
Swartzia cubensis (Leguminosae)	1	110	birds, mammals
Spondias mombin (Anacardiaceae)	1	350	mammals
Astronium graveolens (Anacardiaceae)	1	130	wind
Lasiacis sorghidea (Gramineae)	2	110	?
Olyra latifolia (Gramineae)	2	60	rodents
Streptochaeta spicata (Gramineae)	2	80	sticktight on mammal

(continued)

TABLE 1. (*continued*)

Plant species	Number of individuals	Height of tallest (cm)	Seed dispersal agents
Panicum trichoides (Gramineae)	1	0	?
Scleria pterota (Cyperaceae)	1	90	?
Baccharis trinervis (Compositae)	1	210	wind
Manilkara zapota (Sapotaceae)	1	120	mammals
Established prior to tree fall, reproductive in forest understory:			
Rourea glabra (Connaraceae)	2	100	birds
Hirtella racemosa (Chrysobalanaceae)	3	240	birds, rodents
Eugenia aff. *Oerstediana* (Myrtaceae)	2	300	birds
Ocotea veraguensis (Lauraceae)	1	140	birds, rodents
Piper pseudo-fulligineum (Piperaceae)	1	160	bats

more stationary, there were 14 species of saplings (juveniles) of self-supporting plants which had been rooted at the site when the tree fell but which would not have reached reproductive maturity in the shaded understory (as is also the case with the above mentioned vines) (Tab. 1). These plants have responded to the insolation by enlarging their crowns into the isolated area. Finally, there were 5 species of shrubs and treelets that normally reproduce in the understory but responded to the increased light by increased foliation and by sexual reproduction (Tab. 1).

The Newly Insolated Understory Vegetation Near the Fallen Bole

In strong contrast to the open and easily penetrable vegetation in the area of the soil pit, after 3.5 yr of growth the vegetation along the fallen guapinol bole is a nearly impenetrable stand of 1–10 cm diameter stems (measured 1 m above the ground) supporting a very dense canopy 2.5–4 m above the ground with emergents reaching 6 m. Large perennial vines course through the foliage at the level of the canopy. The following species constitute at least 90% of the volume of leafy vegetation: self-supporting—*Astronium graveolens* (Anacardiaceae), *Cecropia peltata, Brosimum alicastrum, Castilla elastica* and *Trophis racemosa* (Moraceae), *Trema micrantha* (Ulmaceae), *Cordia linnaei* (Boraginaceae), *Genipa americana* (Rubiaceae), *Casearia sylvestris* (Flacourtiaceae), *An-*

nona reticulata (Annonaceae), *Picramnia quaternaria* (Simaroubaceae), *Acacia collinsii* (Leguminosae); perennial vines—*Machaerium kegelii* and *Desmodium axillare* (Leguminosae), *Gouania polygama* (Rhamnaceae), *Paulinia cururu* (Sapindaceae), *Callichlamys latifolia, Pithecoctinium crucigerum* and *Cydista aequinoctialis* (Bignoniaceae), DHJ 12205 (Malpighiaceae), *Passiflora pulchella* (Passifloraceae). It is striking that only one of the above self-supporting species—the single 5-m tall *Brosium alicastrum* sapling—is a member of the 30–40 m canopy of this forest when mature. In contrast to the self-supporting plants leaning into the area around the soil pit at the base of the fallen guapinol, the trees near the bole have symmetrical crowns and straight vertical central axes.

The Crushed Understory Where the Guapinol Crown Fell

At this date, there appears to have been little or no replacement of the understory plants killed and branches stripped off by the guapinol crown when it fell. In view of the deep shade cast by the overstory canopy at this site, it will probably be scores of years before the vegetation again resembles normal understory.

Origin and Survival of the Plants in the Tree Fall

Where did they come from? Irrespective of whether the seeds arrived before or after the guapinol fell, every species of plant I have mentioned above as associated with the tree fall occurs commonly in the Park on roadsides, pasture edges, old fencerows, regenerating fields and other sites of primary and secondary anthropogenic succession. They also occur in what is in all likelihood pristine forest and its associated natural disturbance sites. Except for the two large trees, *Brosimum alicastrum* (ojoche) and *Hymenaea courbaril* (guapinol), and two large woody vines (*Callichlamys latifolia* and *Machaerium kegelii*) all of these species have large breeding populations in the anthropogenic portions of the Park vegetation. There are many more individuals, breeding individuals, and breeding individuals with extra-large flower and fruit crops in the anthropogenic vegetation than in the guapinol-encino forest and its tree falls. While the proportions of seeds in the seed rain from the two different origins need not reflect the absolute numbers of seeds produced in both habitats, it probably approximates it, at least in direction. It is therefore very likely that the seeds which produced the plants in the guapinol tree fall came from that portion of their population in anthropogenic vegetation. The nearest portion of this vegetation is approximately 60 m from the area of impact of the guapinol crown.

Not only is it likely that the seeds came from vegetation other than natural tree falls, but the number and patterns of arrival of these seeds are likely to be very different from those that would arrive as seed rain from natural tree falls. The species and numerical relations of the plants in the tree fall surely must be related to the numbers and timings of seeds that arrive. It follows from this that the proportions, ages, and fates of the plants in the guapinol tree fall cannot be viewed as "natural" in any sense of the word. That is to say, there is no reason to believe that for any given plant in the guapinol tree fall, the array of competitors and mutualists it now experiences approximates the array with which it on average interacted during its evolution.

Who brought them there? Just considering the 52 species of plants in the immediate vicinity of the soil pit and mound at the base of the guapinol, 62% of the plants are dispersed by birds and/or mammals (in addition, some of the wind-dispersed seeds are secondarily dispersed by seed predator rodents). Individuals of all of these vertebrates have foraging ranges large enough to encompass both the segment of pristine forest containing the guapinol tree fall and large areas of anthropogenic secondary woody succession. Of greater significance for this story, all of these animal species have populations sustained in great part by insects, fruits, flowers, foliage and seeds harvested from the anthropogenic vegetation. It is certain that their densities are much higher for the general region than would be the case were they being sustained only by the products of a large expanse of guapinol-encino forest. The anthropogenic forest can have highly diverse effects on the seed rain into the guapinol tree fall. On the one hand, it may be such a high-quality foraging area that animals that would normally frequent tree falls in pristine forest will find them comparatively poor as foraging sites and therefore the seed rain from these animals will be diminished. On the other hand, the anthropogenic vegetation may generate so many seeds or such intense portions of seed shadows that the tree fall receives far more seeds of this or that species than it ever would if far from such a source area. The actual situation for any given tree fall or class of tree falls cannot be determined any other way than through empirical observation.

How are plant-plant interactions affected by herbivores? Animals do more than bring seeds to the tree fall; by preying on dispersed seeds, trampling, and browsing, they further influence the outcome of this multi-specific vegetative melee. For example, one of the reasons why about 1 m^2 of the soil pit is bare of plants after 3.5 yr is that this portion of the site is crossed by a peccary trail. The peccaries both trample seedlings and browse certain species of foliage (cf. *Cecropia* below). *Liomys salvini* mice forage nightly in the tree fall. These heteromyid rodent seed predators both consume and disperse the seeds of almost every species of plant mentioned in this paper. For example, some individuals can live for months on a pure diet of guapinol seeds. The foliage of the plants in the guapinol tree fall is fed on by caterpillars, beetles, orthopteroids, true bugs and a miscellany of other insects. Just as with the plants, the individual insects are members of large populations distributed over the appropriate resource base. The densities of these insects, and those of the vertebrates mentioned above, are very different in the guapinol tree fall from what they would be were the tree fall part of a large expanse of intact forest. Furthermore, since the large "ocean" of anthropogenic vegetation generates carnivores that feed on the herbivores, as well as the herbivores themselves, it cannot be predicted at this stage of knowledge whether the herbivore density will be increased or decreased by the nearby anthropogenic vegetation.

Cecropia Peltata as an Example

To be more specific, I will examine *Cecropia peltata* (guarumo) in the context of the above generalizations. I chose this tree because it is familiar to many readers and to travellers in the tropics, and because it is a conspicuous member of the tree fall flora. However, a similar story can be told for each species that I have recorded growing in this particular tree fall.

Three of the four *Cecropia* saplings growing in the area of the soil pit are from seeds that germinated in the early wet season (May–June) of 1979 on the crest of the soil-root mound. One is 2.5 growing seasons of age, and unoccupied by ants. The three of equal age are 4, 7 and 8 m in height and the tallest will probably flower in 1983 for the first time. All three tall *Cecropia* are occupied by healthy colonies of *Azteca* ants, free of vines, and herbivorized to the degree normal for trees of this size.

The seeds of these *Cecropia* arrived at the site in the feces of a small to medium-sized bat (e.g., *Carollia, Artibeus*) or bird (e.g., *Chiroxiphia, Ramphastos*), or a non-volant mammal (e.g., *Tayassu, Capucinus, Nasua, Ototylomys*) that ate the fruit from the tree or ground below. The source tree was very likely one of the hundreds of adults growing along the roadside and in the several km^2 of late secondary succession within several km of the tree fall.

All of the animals that could defecate *Cecropia* seeds in the tree fall feed heavily not only on the abundant *Cecropia* fruits, but also on the abundant other fruits (and insects) of this vegetation. For example, the *Carollia* bats that could have defecated the seeds feed heavily on *Piper amalago* fruits (Fleming 1981) which occur at exceptionally high density in anthropogenic late secondary succession in the Park. In short, the *Cecropia* seed rain into the guapinol tree fall is generated by a density of *Carollia* and other frugivores far higher than could be supported by the pristine guapinol-encino forest with its sparse and highly pulsed seed, fruit and insect yield. For example, the guapinol trees produce fruit crops only at 3–5 yr intervals (Janzen 1978) and the understory fruiting shrubs (e.g., *Hirtella racemosa* (Chrysobalanaceae), *Ourata lucens* (Ochnaceae), *Psychotria nervosa* (Rubiaceae), *Mouriri myrtilloides* (Melastomataceae), *Picramnia quaternaria* (Simaroubaceae)) have very small crops during only certain times of year. Except when one of the large moraceous trees or *Zuelania guidonia* (Flacourtiaceae) fruits in the guapinol-encino forest, the density of small juicy fruits is very low compared to late secondary succession.

Not only the common animals of secondary succession, such as *Carollia*, are affected by the large amounts of this vegetation. In the vicinity of the tree fall, birds that are generally thought of as deep forest birds, such as the long-tailed manakin *(Chiroxiphia linearis)*, feed heavily on the small fruits of the forest understory shrubs listed above (Janzen unpubl., Foster 1976). However, these birds also make long foraging forays into secondary succession (and more rarely nest there) to feed on fruits at abundant *Cecropia, Hamelia patens* (Rubiaceae), *Casearia corymbosa* and *Casearia sylvestris* (Flacourtiaceae), *Trema micrantha* (Ulmaceae), *Clidemia octona* (Melastomataceae), *Alophyllus occidentalis* (Sapindaceae), etc. If the density of these birds is raised by increased fruit availability, they will be at a higher density in pristine forest near late secondary succession than in pristine forest alone. Not only does the secondary succession contain more fruiting species (and fruit through more of the year) than does the pristine forest, but when understory shrubs of the pristine forest are exposed to insolation along roadsides and other kinds of edges, they often bear exceptionally heavy fruit crops.

It is probably not an accident that the four *Cecropia* seedlings that survived to sapling status in the guapinol tree fall soil pit area were all on the very steep-sided mound. Peccaries are fond of *Cecropia* seedlings (5–30 cm tall, months before they have acquired an obnoxious ant colony) and I suspect that they ate every seedling that appeared on the bare soil of the pit or its edges. While the peccary troup in this area forages throughout the pristine forest (and seeks shade and water there), there is far more browse and fallen fruit forage for them in the secondary succession than in the pristine forest. Their visitation rate and their density are undoubtedly higher in the guapinol tree fall because of the nearby secondary succession.

To mature in the guapinol tree fall, these *Cecropia* saplings had to do more than just arrive and survive herbivory. They had to be found by one or more *Azteca* ant queens. The founding queens originated in the same large adult *Cecropia* as produced the seeds, and the density of founding queens in the vicinity of old secondary succession is much higher than in large expanses of pristine forest, even if it does contain an occasional mature *Cecropia* in an old tree fall.

A major value of the *Azteca* ant colony is in keeping the tree free from climbing vines (Janzen 1969, 1973). The ocean of late secondary succession generates enormous numbers of seeds of the species of vines that are a threat to *Cecropia* saplings (all those listed in Tab. 1 except *Callichlamys latifolia* and *Machaerium kegelii*). The ants are also functional in removing pyraustine pyralid larvae that do severe damage by rolling and consuming leaves. However, they do not remove the insects that specialize at avoiding the patrolling ants (just as is the case with acacia-ants and acacias, Janzen 1967). For example, there are large populations of three butterflies (*Historis odius, istoris acheronta, Colabura dirce*), one skipper (Hesperiidae) and one chrysomelid beetle (probably *Coelomera atrocaerulea*) that feed solely on the leaves of the population of *Cecropia* in late secondary succession. The ants cannot thoroughly remove these five species of herbivores. It would not be surprising to find that the density of *Cecropia* in natural tree falls in

pristine guapinol-encino forest is too low to sustain one or more of these species.

The *Azteca* ants are usually effective at deterring the *Atta cephalotes* leaf-cutter ants from harvesting *Cecropia* leaves at Santa Rosa. However, for other less well-protected plants in the tree fall, it is an important fact that the anthropogenic secondary succession maintains a very high density of large *Atta* colonies. Whether leaf-cutters forage in any given tree fall depends on the proximity of a leaf cutter nest, and this in turn is related to nest density. The density of leaf-cutter nests is related to both the quality of the habitat and the amount of colonization by new queens. This is in turn related to the number of large mature colonies in the general vicinity. The particular guapinol tree fall under scrutiny here is within foraging range of a large leaf cutter colony. Of the three guapinol saplings growing at the margin of the original soil pit and therefore potential candidates to replace the one that fell, two had over 90% of their 1982 leaf crop removed by leaf cutters in May, and by the end of August 1982 these leaves had not been replaced. Such defoliations have a strong potential for influencing the competitive interactions between *Cecropia*, guapinol, and other woody plants in the tree fall.

The most bizarre "herbivore" threat to *Cecropia* saplings in the Park is the large lineated woodpecker *(Dryocopus lineatus)* that tears open the stems to prey on the ant colony. It weakens the stems so badly that they are permanently deformed and often break off at the damage point in a windstorm. The large number of *Cecropia* in late secondary succession may be instrumental in leading to the bird learning to exploit this food source. Further, the late secondary succession contains very large numbers of dying trunks and stems in which this bird forages heavily for wood-mining insects; its density is likely to be much higher in the presence of such a food source than in pristine forest.

Once the *Cecropia* in the guapinol gap attain flowering size, they will still be influenced by the other *Cecropia* in the secondary succession. *C. peltata* (like other species of *Cecropia*) is dioecious and therefore must outcross. It is also believed to be wind-pollinated. The large number of trees in secondary succession must create an enormous group pollen shadow that must greatly facilitate pollination of the female trees in tree falls. However, the importance of this process will depend on the way wind moves between the pristine forest and the secondary successional forest, and how far *Cecropia* pollen moves on it. It is possible that for a female *Cecropia* to be successfully pollinated in a tree fall it may have to have a male of equal reproductive state so close as to be within the same tree fall, and therefore the *Cecropia* trees in nearby secondary succession are irrelevant to the tree fall *Cecropia* except as sources of animal-dispersed seed and herbivores.

If a *Cecropia* tree attains fruiting status in the guapinol tree fall, the same processes that put its seed into the tree fall also apply in dispersing its seeds to new tree falls. Its seeds are likely to be more broadly dispersed within and between habitats than were there no secondary succession in the vicinity. Whether its fruits will be more thoroughly eaten (removed) will depend on the relative fruit abundance of *Cecropia* and other plants in the secondary succession and to what degree this attracts the animals that would otherwise be getting their fruit in the guapinol tree fall.

If one examines third- or greater order interactions of the *Cecropia* with the other members of the habitat, the connections are clear but their numerical values even more difficult to divine than would be those of the above discussion. For example, as mentioned above, a major challenge to *Cecropia* saplings is being used as a vine trellis. *Passiflora platyloba* is one such vine; at Santa Rosa it is fed on by the caterpillars of three species of heliconiine butterflies all of which have large populations in the secondary succession around the pristine forest but also oviposit on the *P. platyloba* in the forest tree falls. One of these, *Heliconius hecale*, is a prominent visitor (and probable pollinator) to the flowers of *Hamelia patens*. This shrubby treelet occurs in the tree fall and is common in the anthropogenic secondary succession, where partly insolated plants bear huge fruit crops. These fruit crops are heavily fed on by the long-tailed manakins mentioned earlier as an example of one of the birds that lives in the understory of the pristine forest but forages

for fruits in both tree falls and the anthropogenic secondary succession. A manakin could well have been the bird that defecated *Cecropia* seeds in the guapinol tree fall.

The above considerations of *Cecropia* biology bring to mind the possibility that *Cecropia* is not even a natural member of guapinol (or other large tree) tree falls in pristine guapinol-encino forest at Santa Rosa National Park. It may well have been a tree of much larger disturbance sites such as new river terraces created by exceptional rainy seasons, cliff edges, and landslide scars. In these sites, just as in anthropogenic secondary succession in logged areas, abandoned fields and roadsides, the local density could have been high enough for high-quality cross-pollination, coupled with sufficient site duration for a long life as a seed-producing adult (*C. peltata* can live to at least 12–20 yr of age at Santa Rosa, as determined by counting the cycles of long and short internodes associated with the long rainy and dry seasons). If the above supposition was in fact the case, the natural sub-populations of *Cecropia* would have hit forest tree falls in their vicinity with seeds, just as they would hit other larger and more long-lived disturbance sites. The degree to which the resulting *Cecropia* adults in tree falls were part of the breeding population, as opposed to being as dead as is a rainforest tree seedling that comes from a seed carried to the sea in a river and washed up on an island beach, depends on the numerical values of all the various interactions mentioned earlier for the *Cecropia* in the guapinol tree fall.

At Santa Rosa there are no areas of pristine forest far enough from anthropogenic secondary succession to test the above hypothesis on *Cecropia* population structure. However, such forest exists in the rainforest inland from Llorona in northern Corcovado National Park on the Osa Peninsula of southwestern Costa Rica. Here, *Cecropia peltata* is a common member of the succession in anthropogenic disturbance sites as well as along rivers, in landslide scars, etc. However, it is generally absent from natural tree falls several kilometers into the forest from abandoned fields and roadsides. When a large portion of the soil of this forest is bared by clearing or by a major storm, it remains free of *Cecropia* seedlings for

1–3 yr. On the other hand, when old fields are recleared or "primary" forest near old fields is cleared, the bare ground is immediately colonized by seedlings of *Cecropia, Piper, Trema* and other fast-growing trees. Presumably these appear from a soil seed bank rather than immediate seed-rich fecal input. Such an observation is hardly new in human experience, and probably was a major driving force in the cultural evolution of true slash and burn agriculture. Where it seems to need reappreciation is in understanding treefall vegetation composition in pristine forest within foraging and dispersal range of the animals and plants of anthropogenic secondary succession. For example, Holthuijzen and Boerboom (1982) studied the seed bank in the soil of "pristine Surinam lowland rainforest" and found an average of 73 *Cecropia* seeds per square meter of litter and topsoil. However, their description of the study site says "the nearest *Cecropia* stand, bordering a forest road, was about two km from the sample area"; this puts their "pristine" rainforest sample well within daily movement range of the host of animals that would feed on the anthropogenic roadside *Cecropia* fruits.

We do not begin to know enough of the biology of tropical animals and plants to know what that foraging range might be (and it surely varies with the species). However, distances of 1–5 km are certainly within that range. The *Carollia* in the vicinity of the guapinol tree fall regularly forage 1–3 km (Fleming and Heithaus 1981). The peccaries in a preserve of similar vegetation 50 km to the south move over an area 1–10 km in diameter during the day and year (C. Vaughan, pers. comm.). *Heliconius hecale* moves 0.1 to 2 km in vegetation like that at Santa Rosa on a daily and life-span basis in search of flowers, mates, and oviposition sites (L. Gilbert, pers. comm.). I have found the large mammal-dispersed seeds of *Mastichodendron capiri*, which had to originate in the guapino-lencino forest or further away, in the secondary succession 2.5 km from the guapinol tree fall.

On a Larger Scale

The problem I discuss here exists on a scale much larger than one tropical park and its

secondary succession. The patterns and densities of migratory birds in, within, and out of the tropics have undoubtedly been strongly altered by anthropogenic alteration of extra-tropical and tropical regions far from any specific tropical "pristine" segment of vegetation (Keast and Morton 1980). The "pristine" forests we study in the best-preserved portions of Neotropical parks are about as natural as will be the forest parks of Uganda 10000 yr from now after all the big game animals have been shot out (cf. Janzen and Martin 1982). I think we have no choice but to abandon any hope of studying the ecology of large ecosystems as systems in evolutionary equilibrium and to deal with ecosystems as only quite rough approximations of evolutionary steady states. Such a view brings to mind a generally unstudied trait of ecosystems, the degree of which the ecologically interacting organisms have been moved out of evolutionary equilibrium, and how much the "inefficiency" of the phenotypes results from the new levels and directions of resource flow.

To bring up the *Cecropia* example again, to what degree are some animals of the guapinol tree fall subsisting on *Cecropia* seed rain from the secondary succession? Might we expect the next million years of *Cecropia* life in anthropogenic secondary succession to generate a *Cecropia* fruit type that generates a much more directed seed shadow into some type of regularly recurring anthropogenic vegetation type (e.g., roadsides) and away from tree falls in pristine forest where survivorship of first or second generation offspring is essentially nil? Is the "natural" *Cecropia* genotype already severely contaminated with genes that have been highly successful in anthropogenic secondary succession but have low fitness in tree fall succession in pristine forest?

In Closing

Natural secondary and primary disturbance sites in pristine forest are likely to pick up or contain a fauna and flora that will be the same as or strongly influenced by anthropogenic successional habitats within 5 km (or perhaps more) of the sites. While I have not discussed this, the animal visitors to a pristine forest

canopy fruit crop are subject to the same
process. For me this means that small parks
(small islands of relatively pristine vegetation in
a sea of secondary succession) may well maintain nearly all the *species* that they always had,
but the densities and ecological relationships
(hence selective pressures) may be altered
strongly. It is hard to avoid the conclusion that
in some circumstances, it may be much better to
surround a small patch of primary forest with
species-poor vegetation of non-invasive species
of low food value (e.g., grain fields, closely
cropped pastures, cotton fields, sugar cane)
than to surround it with an extensive area of
secondary succession rich in plants and animals
that will invade the pristine forest.

From the conservation standpoint, one is
tempted to say that invasion of secondary successional species doesn't matter as long as the
habitat and its species are conserved. However,
there are two reasons why it matters. The first is
speculative, but probably real. If, for example,
a tree like *Cecropia* becomes a prominent part
of guapinol tree falls at Santa Rosa, it could
well alter the competitive successes of the various canopy-level tree species that normally
develop in the tree fall. This could well change
the species composition of the canopy, and
even something as prominent as the guapinol
might disappear. The second reason is a recognition of the nature of a park or other habitat
preserve. As mentioned in the introduction, all
the *species* at Santa Rosa could be maintained
(at enormous expense) in zoos, botanical gardens and game preserves. But a park is not so
much a living museum of the species as of their
interactions. We have to ask if the function of
a park or other primary forest preserve is to
conserve species we all like or if its function is
the much more complex act of conserving that
set of interactions we view as characteristic of
pristine forest. The interactions are often what
we put up park boundaries for. If you alter a
species' interactions, you have altered it as
much as if you changed its color, diet or teeth.
What we always seem to bemoan is the extinction of species from natural areas, when we
should be just as worried about the addition of
species. That *Cecropia* tree may not belong in
the Santa Rosa guapinol tree fall any more than
an *Eucalyptus* sapling belongs there.

Acknowledgments

This study was supported by NSF GB 80-11558 and
by Servicios de Parques Nacionales de Costa Rica.
The manuscript was constructively criticized by W.
Hallwachs and G. Stevens. L. D. Gomez aided in
plant determinations.

References

Boucher, D. H. 1981. Seed predation by mammals
and forest dominance by *Quercus oleoides*, a
tropical lowland oak.—Oecologia (Berl.) 49:
409–414.
Boza, M. A. and Mendoza, R. 1981. The national
parks of Costa Rica.—INCAFO, S.A., Madrid,
Spain.
Fleming, T. H. 1981. Fecundity, fruiting pattern,
and seed dispersal in *Piper amalago* (Piperaceae),
a bat-dispersed tropical shrub.—Oecologia (Berl.)
51:42–46.
_____ and Heithaus, E. R. 1981. Frugivorous bats,
seed shadows, and the structure of tropical forests.—Biotropica (supplement) 13:45–53.
Foster, M. S. 1976. Nesting biology of the long-
tailed manakin.—Wilson Bull. 88:400–420.
Holthuijzen, A. M. A. and Boerboom, J. H. A.
1982. The *Cecropia* seedbank in the Surinam
lowland rain forest.—Biotropica 14:62–68.
Janzen, D. H. 1967. Interaction of the bull's-horn
acacia (*Acacia cornigera* L.) with an ant inhabitant
(*Pseudomyrmex ferruginea* F. Smith) in eastern
Mexico.—Univ. Kansas Sci. Bull. 47:315–558.
_____ 1969. Allelopathy by myrmecophytes: the ant
Azteca as an allelopathic agent of *Cecropia*.—
Ecology 50:146–153.
_____ 1973. Dissolution of mutalism between *Cecropia* and *Azteca* ants.—Biotropica 5:15–28.
_____ 1978. Seeding patterns of tropical trees.—In:
Tomlinson, P. B. and Zimmerman, M. H. (eds),
Tropical trees as living systems. Cambridge Univ.
Press, New York, pp. 83–128.
_____ and Liesner, R. 1980. Annotated check-list of
plants of lowland Guanacaste Province, Costa
Rica, exclusive of grasses and non-vascular cryp-
tograms.—Brenesia 18:15–90.
_____ and Martin, P. S. 1982. Neotropical anachro-
nisms: the fruits the gomphotheres ate.—Science
215:19–27.
Keast, A. and Morton, E. S. (eds) 1980. Migrant
birds in the Neotropics.—Smithsonian Inst. Press,
Washington, D.C.
Simberloff, D. 1982. Big advantages of small ref-
uges.—Natural History 91(4):6–14.

19
Community-Wide Consequences of Trout Introduction in New Zealand Streams

Alexander S. Flecker and Colin R. Townsend

Introduction

Twenty-five years ago not one of these rivers had the least interest for the angler . . . the rod of the fisherman never cast a shadow on their waters; every one of these mighty rivers, every one of the thousand creeks and streams that flow into them . . . were tenantless and profitless to the sportsman (Spackman 1892, as cited in McDowall 1990a). . . . brown trout have since been stocked in almost every conceivable lake, river, or stream, such that the present naturalized population encompasses every suitable ecological niche within the confines of New Zealand (MacCrimmon and Marshall 1968).

One of the important challenges linking ecology and conservation is understanding how introduced species influence natural communities (e.g., Elton 1958, Diamond and Case 1986, Mooney and Drake 1986, Drake et al. 1989, Coblentz 1990, Soulé 1990, Pimm 1991). Biological invasions have been of interest to ecologists for decades (e.g., Elton 1958); however, the attention paid to introduced species is today of unprecedented proportions. A clear message from the growing literature is that our knowledge of invasion processes is limited, and serious shortcomings exist in our ability to establish reliable invasion "forecasting rules" (e.g., Mooney and Drake 1986, Drake et al. 1989, Simberloff 1990, Ramakrishnan 1991, Townsend 1991). Regrettably, the staging ground for advances in invasion ecology will occur within the continuing homogenization of the world's biotic realms.

Ecological studies of species introductions largely emphasize how invaders directly affect particular resident species (Simberloff 1990). Much of this work is strongly biased towards conspicuous species or species producing dramatic ecological disruptions (Simberloff 1981, 1985, 1990, Bertness 1984, Herbold and Moyle 1986). In contrast, relatively little is known about subtle effects of invaders on community organization, some of which may be indirect and cascade through multiple trophic levels. There is increasing evidence that indirect interactions are common in natural communities (e.g., Kerfoot and Sih 1987, Pimm 1991, Strauss 1991, Flecker 1992a), although effects are often not readily apparent and can be difficult to measure. Thus, the consequences of introduced species could be considerably underestimated by failing to adequately appreciate the importance of subtle impacts of biological invasions (see Pimm 1991, Townsend 1991, Allan and Flecker 1993).

In this study, we experimentally examined multiple trophic-level effects of introduced brown trout (*Salmo trutta* L.) in New Zealand tussock grassland streams. Brown trout were first introduced by European settlers into Australasia during the 1860s and have been subse-

Reprinted with permission from Ecological Applications vol. 4, pp. 798–807. Copyright 1994 the Ecological Society of America.

quently transferred to all continents of the world except Antarctica (MacCrimmon and Marshall 1968). In New Zealand, some 60 × 10^6 fish were liberated by 1920; today, brown trout are among the most widely distributed of the country's freshwater fishes (McDowall 1990a). There is growing evidence that trout have been a major source of population decline of native fishes and have completely replaced resident species in some locations (e.g., Mc-Dowall 1984, 1990a, b, Minns 1990, Townsend and Crowl 1991, Crowl et al. 1992). For example, Minns (1990) found from his extensive analysis of <4200 stream sites throughout New Zealand, that native fishes generally did not co-occur with exotic species. An issue that remains virtually unexplored, however, is the significance of such broadscale species substitutions to community organization.

The question of focus in this paper is: Does the replacement of native fishes by brown trout have potential community-level consequences in New Zealand tussock streams? Although species substitutions within the guild of insectivorous fishes have been documented, it is unclear whether the impacts are restricted to a single food web component, or whether they are propagated through multiple trophic levels. Understanding the community implications of species replacements is important in shaping our view of stream conservation. Clearly, species are not fully independent entities, and biotic and abiotic connections are essential for the maintenance of the integrity of running water systems.

Study Site

The study was conducted in the Shag River, a stony-bottom stream located in Otago Province on the South Island of New Zealand. The Shag River originates in the Kakanui Mountains at an elevation of ≈1200 m and flows southeasterly for 84 km before reaching the Pacific Ocean. The study site is a third-order reach located at 600 m (170°24′ E, 45°11′ S) and varies in width from 2 to 4 m. The stream is open and unshaded, passing through grassland dominated by tussocks (mainly *Festuca* sp. and *Chionochloa rigida*) and shrubs (e.g., *Discaria toumatou, Olearia* sp., *Coprosma* sp., and

Aciphylla spp.). The surrounding area has been grazed by sheep and cattle for <120 yr (Thompson 1949). The climate is typical of the central Otago region in that flows are relatively stable during the summer months when, in general, little rainfall occurs.

Fish Fauna and Food Web Complexity

New Zealand streams have a species-poor fish fauna. Characteristically, only two native fish species inhabit the upper reaches of the Shag River, the common river galaxias (Galaxiidae: *Galaxias* cf. *vulgaris*) and long-finned eel (Anguillidae: *Anguilla dieffenbachii*). Galaxiids are southern hemisphere salmoniform fishes, and, in New Zealand, some species are referred to as native or Maori trout. Both galaxiids and eels are predators of invertebrates, although eels become piscivorous once they attain a large body size (McDowall 1990a). In addition, the Shag River contains introduced brown trout (Salmonidae: *Salmo trutta*), which were first released into the lower reaches in 1869 by the Otago Acclimatisation Society (Otago Acclimatisation Society 1964). It is unclear when trout first invaded higher elevation sites, because several groups of small waterfalls may have acted as barriers to dispersal. The first known introduction into the study reach was recorded in the 1930s and came from a different genetic stock than the original population (D. Scott, *personal communication*). Like galaxiids, the diet of brown trout is composed mainly of invertebrates; however, larger trout also prey on small galaxiids when present (Crowl et al. 1992). Diet overlap between trout and common river galaxias has led some authors to speculate that interspecific competition might occur between these species (e.g., Cadwallader 1975, Crowl et al. 1992). The occurrence of both trout and galaxiids in the Shag River contrasts with many streams in the region in which the two taxa do not coexist (Townsend and Crowl 1991, Crowl et al. 1992).

Trophic structure in the Shag River during the summer months is relatively simple. Less than 10 taxa of fish and invertebrates comprise the vast bulk of individuals and biomass in the community. Many species can be classified into feeding guilds or functional groups, although some undergo ontogenetic shifts in diet. Virtually all benthic invertebrates found in the Shag River can be considered either browsers (i.e., feed on algae and detritus), filter feeders, or predators (Cowie 1980, Ryder 1983). Thus, there are few food web components, especially if trophically analogous taxa are grouped together as "trophic species" (see Martinez 1991).

Methods

Experimental Design

Our major objective was to explore whether the replacement of native *Galaxias* fishes by introduced brown trout has potential community-wide consequences in New Zealand tussock streams. We conducted two field experiments that compared the effects of the different fish species on lower trophic levels. We aimed to examine: (a) whether stream fishes were strong determinants of patterns of invertebrate abundance, (b) how the fish species might differentially affect insect assemblages, and (c) whether impacts of fishes on insects might cascade to the bottom trophic level (i.e., the algal-detritus resource base).

The first experiment was conducted in January 1992, during the early weeks of the austral summer. Three treatments were established using in situ flow-through stream channels: (1) no fish, (2) galaxiid enclosure, and (3) juvenile brown trout enclosure. Experimental channels were constructed from 5 m lengths of PVC plastic sewer pipe (38 cm inner diameter) cut longitudinally. Each stream channel contained a set of removable screens (7 mm mesh) at each end, permitting free immigration and emigration of invertebrates, but preventing the passage of fishes. The bottom of each channel was covered with a layer of gravel overlaid by cobbles. Substrata were collected from the stream and scrubbed in order to remove attached invertebrates and periphyton. The three experimental treatments were replicated in four randomized complete blocks for a total of 12 channels. Blocks were located in a 0.8-km section of stream, and each block was separated by at least 100 m.

Eight *Galaxias* or juvenile trout were added to each of the experimental enclosures. We attempted to evenly size the two fish species used in the experiment. Galaxias were slightly longer than the juvenile trout (galaxias fork length $\bar{X} \pm 1$ se = 90.7 ± 1.8 mm; trout fork length 85.3 ± 1.3 mm; $P < 0.02$). Conversely, trout biomass (7.4 ± 0.4 g) was slightly greater than galaxias biomass (6.2 ± 0.4 g), although these differences were not significant ($0.05 < P < 0.10$). Fish biomass represented values from the higher end of the natural range observed in Otago streams (A. D. Huryn and T. A. Crowl, *unpublished data*). Galaxiids were collected in the study reach, whereas juvenile trout were captured in a reach farther downstream containing high trout densities. Fish were collected using a backpack electrofisher and kept in field holding pens for ≈ 24 h before being added to enclosure channels.

We allowed an initial 10-d period for colonization by algae and invertebrates before stocking fish in any of the channels. We then added fish to the appropriate treatments, placed nylon net covers (3.5 mm mesh) over all channels, and ran the experiment for another 10–11-d period. End screens were brushed gently once each day in order to prevent the mesh from clogging. The experiment was taken down in blocks over a 2-d period for logistical reasons.

At the end of the experiment, a set of fine-mesh nets (200 μm) was placed at both ends of all channels within a block to prevent the escape of invertebrates. Before collecting invertebrates, we measured the size of distinct grazing bands that had developed on the channel walls. Insects were washed from the substrates into the end nets and preserved in 95% ethanol mixed with rose bengal as a colorant. In addition, a 24.6 cm^2 circle of periphyton was removed with a brush from the top surface of five randomly chosen cobbles from each channel. Algal samples were filtered in the field onto glass fiber filters (Whatman GF/C), wrapped in foil, and immediately preserved on ice.

Invertebrate and algal samples were processed in the laboratory. Invertebrate samples were washed through a 1-mm sieve, which retained the vast majority of insects. Smaller invertebrates (< 10 mm body length) were subsampled ($\approx 15\%$ of individuals), whereas larger individuals (≥ 10 mm body length) were sorted from the entire channel collection. Invertebrates were identified (generally to genus or species) using available keys (e.g., Winterbourn and Gregson 1989). Filtered algal samples were analyzed for chlorophyll a and ash-free dry mass (AFDM), as measures of algal standing crop. Chlorophyll was extracted in 90% ethanol and corrected for phaeophytin following methods described by Nusch (1980). After pigment extraction, ash-free dry mass was determined from each algal sample. Samples were dried in an oven at 60°C for 24 h, weighed, ashed in a muffle furnace at 550° for 2 h, and reweighed.

The second experiment was conducted in late summer (March 1992). This experiment differed from the previous one in several ways; however, the main fish treatments were similar. Three fish treatments were established in each of four blocks, corresponding to those described above. Unlike the earlier experiment, channels were divided into two equally sized sections (2.5 m each) using coarse mesh (3.5 mm) partitions placed at the midpoint. In each upstream channel section, an ambient nutrient treatment was maintained, whereas nutrients were added to the downstream sections. Here, we report results from the ambient nutrient sections only, because these provide the comparisons pertinent to the main objectives in this paper. Details about procedures of this experiment are the same as those described above unless noted otherwise.

Statistical Analyses

The experiments were analyzed as randomized complete block designs, with blocks representing different locations in the stream (Anderson and McLean 1974). Data were log transformed for taxa in which the variances were not homogeneous. Differences between treatments in the number of invertebrates and algal standing crop (i.e., AFDM and chlorophyll a) were analyzed using ANOVA (randomized complete block model). In addition, two a priori orthogonal contrasts (Kirk 1982) were performed that tested for: (1) fish predation effect (fish exclusion vs. galaxias plus trout

treatments) and (2) predator species effect (galaxias vs. trout). Because contrasts were both orthogonal and planned, we used a contrast (rather than experiment-wise) alpha value of 0.05 as the critical value for considering treatments significantly different (Kirk 1982).

There were few common taxa. We established as a criterion for analyzing a taxon separately that a mean density of <50 individuals/m^2 be found in at least one treatment. We also analyzed large invertebrate predators (i.e., ≥ 10 mm body length) even when they failed to meet the above criterion. All statistical analyses were performed using Statistix Version 1.1 (1986).

Results

Effects on Invertebrate Abundance and Biomass

Density and biomass of colonizing invertebrates were strongly influenced by the experimental treatments (Tables 1–4). In both experiments, densities of total insects were consistently lowest in channels with trout. In early summer (Experiment 1), a significant fish effect was observed (no fish vs. fish contrast, $P < 0.014$, Fig. 1a). Relative to the exclusion treatment, we found fewer total insects when either fish species was present. Nevertheless, there was a substantially greater reduction in insect numbers in trout ($\bar{X} \pm 1$ SE = 2685.3 $\pm$ 404.0 individuals/m^2) compared to galaxias channels (3952.6 $\pm$ 456.8 individuals/m^2) ($P < 0.018$, Fig. 1a).

By late summer (Experiment 2), we no longer detected a significant effect of fish per se (i.e., fish vs. no fish contrast, $P < 0.155$), but continued to observe strong overall differences among treatments (Tables 3 and 4). This was because there were highly significant reductions in total insect densities in trout ($\bar{X} \pm 1$ SE = 552.1 $\pm$ 99.7 individuals) compared to galaxias (987.4 $\pm$ 141.4 individuals/m^2) treatments ($P < 0.010$, Fig. 2a).

In neither experiment were significant differences in biomass observed when fish and no fish treatments were compared. Insect biomass was remarkably similar between fish exclusion and galaxias channels (Tables 1–4, Fig. 3). In

TABLE 1. Summary of insect densities (number/m^2, mean ± 1 SE) for different fish treatments from Shag River Experiment 1. No Fish = Fish exclusion channels; Galaxias = Galaxias enclosure channels; Trout = Trout enclosure channels.

	No fish	Galaxias	Trout
Deleatidium	2952.2 ± 255.6	2528.7 ± 392.1	1635.7 ± 265.2
Olinga	947.8 ± 88.2	789.0 ± 68.5	621.9 ± 108.8
Helicopsyche	70.8 ± 17.6	117.2 ± 30.9	30.7 ± 10.8
Total grazers	4003.2 ± 242.5	3462.8 ± 403.3	2317.4 ± 360.6
Coloburiscus	115.6 ± 31.0	158.2 ± 47.0	110.2 ± 25.5
Austrosimulium	167.4 ± 42.5	87.4 ± 20.8	74.2 ± 34.8
Stenoperla	8.2 ± 1.6	8.3 ± 2.3	4.2 ± 1.4
Archichaulodes	2.6 ± 0.6	4.2 ± 1.6	1.2 ± 0.4
Hydrobiosid larvae	11.6 ± 1.1	9.6 ± 2.3	6.2 ± 0.7
Total predators	24.2 ± 2.8	24.8 ± 4.8	12.8 ± 2.2
Total invertebrates	4502.0 ± 190.2	3952.6 ± 456.8	2685.3 ± 404.0
Invertebrate biomass (g/m^2)	2.90 ± 0.07	2.90 ± 0.26	1.71 ± 0.28
Species richness	25.5 ± 1.0	24.2 ± 1.0	23.2 ± 0.6
Chlorophyll *a* (μg/cm^2)	0.36 ± 0.06	1.18 ± 0.32	1.89 ± 0.82
AFDM (mg/cm^2)	4.0 ± 0.6	9.9 ± 2.3	16.2 ± 5.9

contrast, marked reductions in insect biomass were apparent in trout channels, with biomass generally ≈ 50–60% of that observed for the other two treatments (Experiment 1: $P <$ 0.007, Tables 1 and 2; Experiment 2: $P <$ 0.026, Tables 3 and 4).

TABLE 2. Summary of probabilities for ANOVAs testing for overall treatment effects and orthogonal contrasts for Shag River Experiment 1. ANOVAs were performed on log-transformed data for taxa indicated with an asterisk.

		Contrast	
	Overall	No fish vs. fish	Galaxias vs. trout
Deleatidium	.002	.003	.004
Olinga	.191	.121	.322
Helicopsyche	.099	.925	.039
Total grazers	.007	.009	.015
Coloburiscus	.364	.551	.207
Austrosimulium	.094	.038	.736
Stenoperla	.111	.279	.066
Hydrobiosid larvae*	.033	.028	.082
*Archichaulodes**	.017	.266	.008
Total large predators	.017	.105	.010
Total invertebrates	.010	.014	.018
Invertebrate biomass*	.011	.059	.007
Species richness	.083	.047	.263
Chlorophyll *a**	.034	.014	.460
AFDM*	.023	.010	.266

Few invertebrate species were common, and only five taxa displayed densities $<$ 50 individuals/m^2 over both experiments (the mayflies *Deleatidium* spp. and *Coloburiscus humeralis*, the caddisflies *Olinga feredayi* and *Helicopsyche* sp., and black fly larvae *Austrosimulium* sp.) (Tables 1 and 3). Almost all taxa were less abundant in late summer, after many larval insects had emerged. Effects of the experimental treatments varied according to taxon and date (Tables 1–4). Perhaps the most striking feature was that densities of all taxa were lowest in trout channels compared to the other treatments, although, in some cases, differences were not statistically significant. In contrast, few generalizations could be made to collectively describe the responses of individual taxa to galaxias. In Experiment 1, for example, the most abundant taxon, *Deleatidium* spp., comprised $<$ 70% of browsing insects, and exhibited patterns similar to total invertebrates (i.e., densities in no fish channels $<$ galaxias $<$ trout, Table 1). On the other hand, the common caddis *Helicopsyche* displayed the greatest densities in the galaxias channels ($\bar{X} \pm$ 1 SE = 117.2 ± 30.9 individuals/m^2) relative to both the fish exclusion (70.8 ± 17.6 individuals/m^2) and trout enclosure (30.7 ± 10.8 individuals/m^2) treatments. Black fly larvae (*Austrosimulium*) displayed yet another pattern, with

TABLE 3. Summary of insect densities (number/m², mean ± 1 SE) for different fish treatments from Shag River Experiment 2. No Fish = Fish exclusion channels; Galaxias = Galaxias enclosure channels; Trout = Trout enclosure channels.

	No fish	Galaxias	Trout
Deleatidium	232.0 ± 35.0	202.7 ± 44.8	120.5 ± 25.8
Olinga	617.3 ± 134.3	714.9 ± 117.5	381.3 ± 83.3
Total grazers	873.7 ± 167.3	987.3 ± 104.9	518.2 ± 98.5
Stenoperla	14.5 ± 3.8	13.8 ± 3.7	10.5 ± 5.4
Total predators	16.3 ± 3.4	14.5 ± 3.9	11.3 ± 5.2
Total invertebrates	933.9 ± 162.2	987.4 ± 141.4	552.1 ± 99.7
Invertebrate biomass (g/m²)	2.04 ± 0.68	2.32 ± 0.74	1.21 ± 0.38
Species richness	16.5 ± 1.3	14.0 ± 2.0	13.5 ± 1.9
Chlorophyll *a* (μg/cm²)	0.90 ± 0.14	0.75 ± 0.19	2.00 ± 0.51
AFDM (mg/cm²)	9.7 ± 1.6	7.8 ± 2.8	18.2 ± 3.3

comparable reductions of black flies in both fish treatments irrespective of the predator species (Tables 1 and 2). Finally, trout markedly reduced the number of invertebrate predators in relation to the galaxias treatment ($P <$ 0.017, Tables 1 and 2); however, there were no significant effects of fish per se, because invertebrate predator densities were similar between galaxias enclosures ($\bar{X} \pm 1$ SE = 24.8 ± 4.8) and fish exclusions (24.2 ± 2.8 individuals/m²).

Only two taxa (*Deleatidium* and *Olinga*) were common in the late summer experiment, and these displayed qualitatively similar patterns

(Tables 3 and 4). Strong overall treatment effects were observed for both taxa, yet there were no significant differences when contrasting the fish and no fish treatments. Instead, both taxa displayed significant density reductions in trout channels relative to the galaxias treatment. In contrast to the earlier experiment, no significant differences were ob-

TABLE 4. Summary of probabilities for ANOVAs testing for overall treatment effects and orthogonal contrasts for Shag River Experiment 2. ANOVAs were performed on log-transformed data for taxa indicated with an asterisk.

		Contrast	
	Overall	No fish vs. fish	Galaxias vs. trout
Deleatidium	.038	.052	.050
Olinga	.032	.433	.013
Total grazers	.020	.294	.008
Stenoperla	.715	.607	.544
Total large predators	.622	.466	.540
Total invertebrates	.019	.155	.010
Invertebrate biomass*	.059	.433	.026
Species richness	.011	.004	.506
Chlorophyll *a**	.051	.420	.022
AFDM*	.066	.636	.026

a.

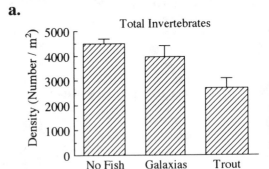

b.

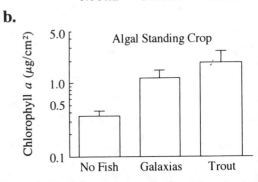

FIGURE 1. Total invertebrate density (a) and algal standing crop (b) from the three fish treatments in Experiment 1 (January 1992) (means and 1 SE). Probability values for ANOVAs are given in Table 2.

a.

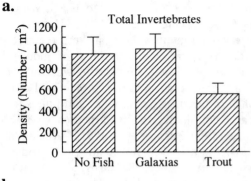

b.

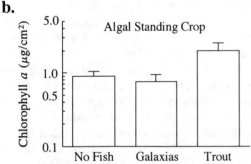

FIGURE 2. Total invertebrate density (a) and algal standing crop (b) from the three fish treatments in Experiment 2 (March 1992) (means and 1 SE). Probability values for ANOVAs are given in Table 4.

Total Invertebrate Biomass

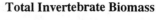

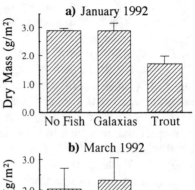

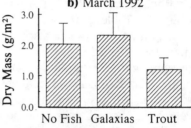

FIGURE 3. Total invertebrate biomass from each of the three treatments in the January experiment (a) and March experiment (b) (means and 1 SE). Probability values for ANOVAs are given in Tables 2 and 4.

served for large invertebrate predators, although densities were again lowest in trout enclosure channels.

Invertebrate Diversity

Approximately 20–25 insect taxa colonized each of the channels during the early summer experiment. Slight differences in species richness were observed among treatments, although the effect was not significant in the overall analysis ($P < 0.083$, Tables 1 and 2). Nevertheless, the small reduction in species richness in channels with fish was significant when contrasted to the fish exclusions (i.e., no fish vs. fish contrast, $P < 0.047$), with a typical decrease of 1–2 insect species.

Similar trends were observed in the late summer experiment, although fewer species of insects were collected (12–17 species per channel) (Tables 3 and 4). In this case, the overall analysis was significant ($P < 0.011$), and a significant fish effect was observed on species richness ($P < 0.004$), but no differences were found in relation to fish species.

Effects on Algal Standing Crop

In both experiments, algal standing crops and total invertebrate density exhibited inverse patterns (Figs. 1 and 2; Tables 1–4). In early summer, algal standing crop from trout channels was <4–$5 \times$ that of fish exclusions, whereas values in galaxias enclosures were in-

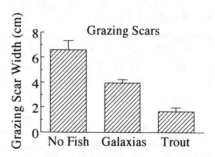

FIGURE 4. Width of algal grazing band from the three experimental treatments in the January experiment (means and 1 SE). Probability values are given in *Effects on algal standing crop.*

termediate between trout and no fish treatments (Table 1, Fig. 1b). In this experiment, a significant fish effect on algal standing crop was observed (chlorophyll a:$P < 0.034$; AFDM: $P < 0.023$); however, fish species did not differ significantly in their effect on algal standing crop (Table 2). In the late summer experiment, strong treatments differences were once again observed; algal standing crop was greater in trout channels than in galaxias channels (chlorophyll a: $P < 0.022$; AFDM: $P < 0.026$) (Fig. 2b, Tables 3 and 4).

Finally, grazing insects left distinct grazing scars (i.e., areas cleared of algae) that differed strongly among treatments (Fig. 4). Insects thoroughly grazed stones and channels when fish were excluded, resulting in wide grazing bands along the channel walls. This contrasted sharply with significantly narrower grazing bands in channels with fish (no fish vs. fish contrast, $P < 0.001$). Moreover, grazing band width was highly dependent on fish species; the width of scars measured in the trout enclosures was about half that in galaxias channels (galaxias vs. trout contrast, $P < 0.008$). These observations suggest that foraging activity of grazing insects may be differentially constrained by the two fish species. This could result if grazers are foraging different distances from refugia in relation to varying predation hazard.

Discussion

Trout provide one of the best illustrations of the willful introduction of fishes in running water communities (MacCrimmon and Marshall 1968, MacCrimmon 1971). Historically, questions about their ecological impacts were not thoroughly explored due to social, political, and scientific factors. Recently, a number of reviews have evaluated stream fish introductions (e.g., Courtenay and Stauffer 1984, Moyle 1986, Moyle et al. 1986, Williams et al. 1989, Crowl et al. 1992). The general focus has been on the repercussions of fish invasions for indigenous fish assemblages via a suite of ecological and genetic mechanisms. Such invasions are believed to be one of the major sources of population decline for native stream fishes

(e.g., Williams et al. 1989, Moyle and Williams 1990, Allan and Flecker 1993). However, in stream communities there has been remarkably little effort to extend this line of inquiry to other trophic levels. Contrasting with this is the rich literature on multiple trophic-level effects of introduced species in lakes and ponds (e.g., Zaret and Paine 1973, Carpenter 1988, Mc-Queen 1990, Spencer et al. 1991, Carpenter and Kitchell 1992, DeMelo et al. 1992), some of which has formed the basis of the current controversy about whether the introduction of piscivorous species is an effective management tool for controlling algal productivity in lakes (see Carpenter et al. 1985, Gulati et al. 1990).

Here we found differences in the interactions between common river galaxias and brown trout involving lower trophic levels. In our experiments, brown trout caused a consistent reduction in insect abundance, relative to both galaxias and the absence of fish. These effects were manifested in the bottom trophic level where the greatest algal standing crops were measured when trout were present. In contrast, the impacts of native galaxias were considerably more complex and variable. These included density reductions (*Deleatidium* and *Austrosimilium*), increases (*Helicopsyche*), or no effects at all (e.g., large invertebrate predators), depending on the insect species and the seasonal timing of the experiment. Likewise, algal responses reflected the abundance of insect grazers (Figs. 1 and 2). Thus, the consequences of species substitutions in the guild of insectivorous fishes can propagate to the bottom trophic level in this tussock stream community.

We suggest that at least two mechanisms are important in explaining the "top-down" effects of fish that we observed. First, insect density was significantly influenced by fish treatment (Tables 1 and 2). Grazing insects, in particular, displayed strong reductions in response to fish. Further, grazers comprised the major component of the insect assemblage in terms of numbers of individuals. In addition to top-down effects on grazer density, fish appeared to modify insect foraging behavior. This was manifested as differences among treatments in the width of insect grazing scars. Large grazing scars were found in the complete absence of fish (Fig. 4). Moreover, grazing scars were significantly wider in galaxiid compared to trout channels. These patterns suggest that fish constrain activity of grazing invertebrates, whereby the distances that insects venture from refugia differ according to fish species. This interpretation would be consistent with findings that fish modify stream invertebrate behaviors and limit the amount of time foragers spend on exposed surfaces outside of refugia (e.g., Kohler and McPeek 1989, Culp et al. 1991, Huang and Sih 1991, McIntosh and Townsend 1994).

Evidence for Community-wide Effects of Invaders

Whether impacts of invasive species are generally propagated throughout entire communities is a subject of current debate. Certainly, some classes of invaders are likely to wreak havoc in natural systems, especially taxa that constitute new habitat, modify existing habitat, or represent ecological novelties in "naive" communities (Vitousek 1986, Simberloff 1990, Pimm 1991). Based on a literature review, Simberloff (1981, 1990) concluded that there is little evidence that introductions generally result in community-level effects (also see Ebenhard 1988, Lodge 1993). Simberloff argued that in the vast majority of cases, natural communities are resistant to invasions, which add rather than subtract species to communities, and fail to precipitate a chain of secondary extinctions. Furthermore, he contended that innocuous introductions are hardly noticed and rarely reported by the casual observer and reasoned that the bias in this largely anecdotal literature may be towards overestimating the prevalence of community-level impacts.

Several authors have questioned Simberloff's bold assertion (e.g., Herbold and Moyle 1986, Pimm 1991). Herbold and Moyle (1986) maintained that complex effects of introduced species are rarely detected, because they are seldom examined. They suggested that, to the contrary, this has resulted in underestimations of the impacts of introduced species. Moreover, Simberloff's review (1981) considered "effects" as extinctions only, thus overlooking changes in

densities which may be the more common community-level consequence of invasions. In his assessment of the invasion literature, Pimm (1991) arrived at the conclusion that species introductions do indeed affect community composition in most cases. He generated a series of predictions about community resistance to invasion based on food web theory, relying on a selected subset of studies to support his arguments. Likewise, Vitousek (1986) examined the evidence that biological invasions can strongly modify ecosystem function. After reviewing a variety of pertinent studies, he concluded unequivocally that ecosystem properties can be altered by invasions in otherwise intact ecosystems. Linked with changes in ecosystem function were effects of invaders on community structure.

This debate serves to underscore that surprisingly few data exist that address the consequences of species introductions at the community level (Simberloff 1990). In particular, relatively little is understood about invasions in which a species substitution occurs within an already existent guild in a community. The direct and indirect effects of species replacements involving ecological analogs may be comparatively subtle and therefore considerably more difficult to document.

How Unique are New Zealand Streams?

Can we expect to find multiple trophic-level effects of trout elsewhere? A number of authors have suggested that invasions on island ecosystems are often particularly successful. At least two explanations have been proposed: (1) food webs are frequently characterized as relatively simple on islands, and (2) island assemblages are often "naive," having little prior history with many ecological novelties when long isolated from mainlands (Pimm 1991). Classic examples of catastrophic introductions on islands include whole-scale declines of avifaunas from the Australian brown tree snake in Guam (Savidge 1987) and from invasions by mammalian predators in Hawaii (Ebenhard 1988).

Certainly New Zealand streams are species

poor. If, indeed, low-diversity food webs are where community-wide consequences of predators are most likely to be manifested (Strong 1992, but see Pimm 1991), then the strong impacts of trout observed in our New Zealand stream may not be surprising. However, it is difficult to generalize about the relative complexity of food webs on islands as opposed to continental land masses. While the number of species on continents may be greater, we know little about the complexity of trophic interactions in different types of streams. Likewise, results from experimental studies in streams on the influence of fish predation are too equivocal to establish whether multi-trophic level effects are generally limited to running waters with few community constituents. Nevertheless, Power (1990a) recently described a spectacular four-tiered trophic cascade from a riverine food web in California, suggesting that strong community-level effects of predators can propagate in systems considerably more complex than the stream studied here.

Furthermore, it is hard to judge the degree to which trout represent ecological novelties in New Zealand stream communities. Both trout and galaxiids are insectivorous fishes, belonging to the order Salmoniformes. However, galaxiids are generally small, less aggressive, and considerably more nocturnal than trout. This contrasts with running water communities in North America where the introduction of trout has resulted in species and subspecies substitutions with other native salmonids, often via hybridization (e.g., Williams et al. 1989, Echelle 1991, Goodman 1991). Thus, in North American trout streams, it may not be easy to find species replacements resulting in striking community-level changes. Yet in some regions, trout have been introduced where drift-feeding fishes do not naturally occur, such as mid-elevation streams of the northern Andes (Flecker 1992b). In streams where a previously nonexistent guild has become established, community-level effects of invasions are likely to be more readily detectable.

Challenges in Detecting Community-level Effects

Even if community-level effects of biological invasions are commonplace, numerous obstacles may impede our ability to discern their impacts. Notorious problems of spatial and temporal heterogeneity in streams have plagued investigators searching for mechanisms explaining patterns of distribution and abundance. Experimental studies in running waters have often required levels of replication that are ruled out for logistical reasons (see Allan 1984).

Challenges in detecting community-wide effects of introduced species are further complicated by the confounding influences of land use changes. In New Zealand, the onslaught of biological invasions associated with European settlement was accompanied by the widespread conversion of wildlands to pasture (Crosby 1986). Indeed, extensive landscape changes may have been important in setting the stage for successful introductions (see Elton 1958, Orians 1986). Moreover, substantial modifications in nutrient dynamics are likely to have ensued with the introduction of ruminant grazers and heavy fertilizer inputs. Such broad-scale changes, whose local effects are not well understood, can only exacerbate problems in documenting impacts of species replacements.

Finally, we may never be able to understand fully the community-level changes that have followed poorly documented extinctions. One can only speculate on the contribution of trout to the extinction of the New Zealand grayling (Prototroctidae: *Prototroctes oxyrhynchus*). This endemic fish species was once so ubiquitous that early European settlers described their taking by the cartload (McDowall 1990a). Population declines were noted soon after the introduction of trout; by the turn of the century the species had largely disappeared, and it became extinct in the 1920s (McDowall 1990a,c). Unlike any living fish species in New Zealand today, the grayling was a highly specialized herbivore which scraped algae from stone surfaces. Evidence from other regions of the world indicates that such algivorous fishes can play a prominent functional role (e.g., Power et al. 1985, Power 1990b, Flecker 1992a). Interactions between trout, galaxiids and other trophic levels would almost certainly have been modified by the presence of an abundant and comparatively large vertebrate grazer. The demise of the grayling adds a further problematic dimension in attempts to reconstruct how trout introduction has affected the dynamics of New Zealand stream communities.

Acknowledgments

We thank Chris Arbuckle, Kirsty Barr, Fenella Harper, Vivian Huryn, Sunny Power, Stephen Scott, Trina Shierlaw, and others for assistance in the field. Helpful suggestions were provided by Todd Crowl, Alex Huryn, Phil Lester, Mike Scarsbrook, and the University of Otago Stream Team. Angus McIntosh gave us particularly important advice on chlorophyll analyses and other aspects of the study. Comments on an earlier draft of the manuscript were made by Dave Allan, Steve Bartell, J. A. Drake, Bobbi Peckarsky, Sunny Power, and the Cornell Aquatic Entomology Group. Special thanks to Val Allen, Murray McKenzie, and the Boys in Blue for their patience and essential logistical support. G. MacDonald and D. Andrew kindly granted permission to work on their property. This research was supported by grants from the University of Otago.

References

Allan, J. D. 1984. Hypothesis testing in ecological studies of aquatic insects. Pages 484–507 *in* V. H. Resh and D. M. Rosenberg, editors. The ecology of aquatic insects. Praeger, New York, New York, USA.

Allan, J. D., and A. S. Flecker. 1993. Biodiversity conservation in running waters. BioScience **43**:32–43.

Anderson, V. L., and R. A. McLean. 1974. Design of experiments. Marcel Dekker, New York, New York, USA.

Bertness, M. D. 1984. Habitat and community modification by an introduced herbivorous snail. Ecology **65**:370–381.

Cadwallader, P. L. 1975. Feeding relationships between galaxiids, bullies, eels, and trout in a New Zealand river. Australian Journal of Marine and Freshwater Research **26**:299–316.

Carpenter, S. R., editor. 1988. Complex interactions in lake communities. Springer-Verlag, New York, New York, USA.

Carpenter, S. R., and J. F. Kitchell. 1992. Trophic cascade and biomanipulation: interface of research and management—a reply to the comment by DeMelo et al. Limnology and Oceanography **37**:208–213.

Carpenter, S. R., J. F. Kitchell, and J. R. Hodgson. 1985. Cascading trophic interactions and lake productivity. BioScience **35**:634–639.

Coblentz, B. E. 1990. Exotic organisms: a dilemma for conservation biology. Conservation Biology **4**:261–265.

Courtenay, W. R., Jr., and J. R. Stauffer, Jr., editors. 1984. Distribution, biology, and management of exotic fishes. Johns Hopkins University Press, Baltimore, Maryland, USA.

Cowie, B. 1980. Community dynamics of the benthic fauna in a west coast stream ecosystem. Dissertation. University of Canterbury, Christchurch, New Zealand.

Crosby, A. W. 1986. Ecological imperialism: the biological expansion of Europe, 900–1900. Cambridge University Press, Cambridge, England.

Crowl, T. A., C. R. Townsend, and A. R. McIntosh. 1992. The impact of introduced brown and rainbow trout on native fish: the case of Australasia. Reviews in Fish Biology and Fisheries **2**:217–241.

Culp, J. M., N. E. Glozier, and G. J. Scrimgeour. 1991. Reduction of predation risk under the cover of darkness: avoidance responses of mayfly larvae to a benthic fish. Oecologia **86**:163–169.

DeMelo, R., R. France, and D. J. McQueen. 1992. Biomanipulation: hit or myth? Limnology and Oceanography **37**:192–207.

Diamond, J., and T. J. Case. 1986. Overview: introductions, extinctions, exterminations, and invasions. Pages 65–79 *in* J. Diamond and T. J. Case, editors. Community ecology. Harper & Row, New York, New York, USA.

Drake, J. A., H. A. Mooney, F. di Castri, R. H. Groves, F. J. Kruger, M. Rejmánek, and M. Williamson, editors. 1989. Biological invasions: a global perspective. John Wiley and Sons, Chichester, England.

Ebenhard, T. 1988. Introduced birds and mammals and their ecological effects. Swedish Wildlife Research (Viltrevy) **13**:1–107.

Echelle, A. A. 1991. Conservation genetics and genic diversity in freshwater fishes of Western North America. Pages 141–154 *in* W. L. Minckley and J. E. Deacon, editors. Battle against extinction: native fish management in the American West. University of Arizona Press, Tucson, Arizona, USA.

Elton, C. S. 1958. The ecology of invasions by animals and plants. Methuen, London, England.

Flecker, A. S. 1992*a*. Fish trophic guilds and the structure of a tropical stream: weak direct vs. strong indirect effects. Ecology **73**:927–940.

———. 1992*b*. Fish predation and the evolution of invertebrate drift periodicity: evidence from neotropical streams. Ecology **73**:438–448.

Herbold, B., and P. B. Moyle. 1986. Introduced species and vacant niches. American Naturalist **128**:751–760.

Huang, C., and A. Sih. 1991. Experimental studies

on behaviorally mediated, indirect interactions through a shared predator. Ecology **71**:1515–1522.

Goodman, B. 1991. Keeping anglers happy has a price: ecological and genetic effects of stocking fish. BioScience **41**:294–299.

Gulati, R. D., E. H. R. R. Lammens, M.-L. Meijer, and E. van Donk, editors. 1990. Biomanipulation–tool for water management. Kluwer Academic, Dordrecht, The Netherlands.

Kerfoot, W. C., and A. Sih, editors. 1987. Predation: direct and indirect impacts on aquatic communities. University Press of New England, Hanover, New Hampshire, USA.

Kirk, R. E. 1982. Experimental design. Second edition. Brooks/Cole, Belmont, California, USA.

Kohler, S. L., and M. A. McPeek. 1989. Predation risk and the foraging behavior of competing stream insects. Ecology **70**:1811–1825.

Lodge, D. M. 1993. Species invasions and deletions: community effects and responses to climate and habitat change. Pages 367–387 in P. M. Kareiva, J. G. Kingsolver, and R. B. Huey, editors. Biotic interactions and global change. Sinauer, Sunderland, Massachusetts, USA.

MacCrimmon, H. R. 1971. World distribution of rainbow trout (*Salmo gairdneri*). Journal of the Fisheries Research Board of Canada **28**:663–704.

MacCrimmon, H. R., and T. L. Marshall. 1968. World distribution of brown trout, *Salmo trutta*. Journal of the Fisheries Research Board of Canada **25**:2527–2548.

Martinez, N. D. 1991. Artifacts or attributes? Effects of resolution on the Little Rock Lake food web. Ecological Monographs **61**:367–392.

McDowall, R. M. 1984. Exotic fishes: the New Zealand experience. Pages 200–214 in W. R. Courtenay, Jr. and J. R. Stauffer, Jr., editors. Distribution, biology, and management of exotic fishes. Johns Hopkins University Press, Baltimore, Maryland, USA.

———. 1990a. New Zealand freshwater fishes: a natural history and guide. Heinemann, Auckland, New Zealand.

———. 1990b. When galaxiid and salmonid fishes meet–a family reunion in New Zealand. Journal of Fish Biology 37 (Supplement A):35–43.

———. 1990c. Freshwater fishes and fisheries of New Zealand: the angler's Eldorado. Critical Reviews in Aquatic Sciences **2**:281–341.

McIntosh, A., and C. R. Townsend. 1994. Interpopulation variation in mayfly anti-predator tactics: differential effects of contrasting predatory fish. Ecology **75**:2078–2090.

McQueen, D. J. 1990. Manipulating lake community structure: where do we go from here? Freshwater Biology **23**:613–620.

Minns, C. K. 1990. Patterns of distribution and association of freshwater fish in New Zealand. New Zealand Journal of Marine and Freshwater Research **24**:31–44.

Mooney, H. A., and J. A. Drake, editors. 1986. Ecology of biological invasions of North America and Hawaii. Ecological Studies 58. Springer-Verlag, New York, New York, USA.

Moyle, P. B. 1986. Fish introductions into North America: patterns and ecological impact. Pages 27–43 in H. A. Mooney and J. A. Drake, editors. Ecology of biological invasions of North America and Hawaii. Ecological Studies 58. Springer-Verlag, New York, New York, USA.

Moyle, P. B., H. W. Li, and B. A. Barton. 1986. The Frankenstein effect: impact of introduced fishes on native fishes in North America. Pages 415–426 in R. H. Stroud, editor. Fish culture in fisheries management. American Fisheries Society, Bethesda, Maryland, USA.

Moyle, P. B., and J. E. Williams. 1990. Biodiversity loss in the temperate zone: decline of the native fish fauna of California. Conservation Biology **4**:275–284.

Nusch, E. A. 1980. Comparison of different methods for chlorophyll and phaeopigment determination. In H. Ria, editor. The measurement of photosynthetic pigments in fresh waters and the standardization of methods. Archiv für Hydrobiologie, Ergebnisse der Limnologie **14**:14–36.

Orians, G. H. 1986. Site characteristics favoring invasions. Pages 133–148 in H. A. Mooney and J. A. Drake, editors. Ecology of biological invasions of North America and Hawaii. Ecological Studies 58. Springer-Verlag, New York, New York, USA.

Otago Acclimatisation Society. 1964. Centennial year and 98th annual report. Otago Acclimatisation Society, Dunedin, New Zealand.

Pimm, S. L. 1991. The balance of nature? Ecological issues in the conservation of species and communities. University of Chicago Press, Chicago, Illinois, USA.

Power, M. E. 1990a. Effects of fish in river food webs. Science **250**:411–415.

———. 1990b. Resource enhancement by indirect effects of grazers: armored catfish, algae and sediment. Ecology **71**:897–904.

Power, M. E., W. J. Matthews, and A. J. Stewart. 1985. Grazing minnows, piscivorous bass, and stream algae: dynamics of a strong interaction. Ecology **66**:1448–1456.

Ramakrishnan, P. S., editor. 1991. Ecology of biological invasion in the tropics. International Scientific, New Delhi, India.

Ryder, G. 1983. An examination of the river continuum concept in New Zealand. B.Sc. Honors Thesis. University of Otago, Dunedin, New Zealand.

Savidge, J. A. 1987. Extinction of an avifauna by an introduced snake. Ecology 68:660–668.

Simberloff, D. 1981. Community effects of introduced species. Pages 53–81 in M. H. Niteki, editor. Biotic crises in ecological and evolutionary time. Academic Press, New York, New York, USA.

———. 1985. Predicting ecological effects of novel entities: evidence from higher organisms. Pages 152–161 in H. O. Halvorson, editor. Engineered organisms in the environment: scientific issues. American Society for Microbiology, Washington, D.C., USA.

———. 1990. Community effects of biological introductions and their implications for restoration. Pages 128–136 in D. R. Towns, C. H. Daugherty, and I. A. E. Atkinson, editors. Ecological restoration of New Zealand islands. Conservation Sciences Publication Number 2. Department of Conservation, Wellington, New Zealand.

Soulé, M. E. 1990. The onslaught of alien species, and other challenges in the coming decades. Conservation Biology 4:233–239.

Spencer, C. N., B. R. McClelland, and J. A. Stanford. 1991. Shrimp stocking, salmon collapse, and eagle displacement. BioScience 41:14–20.

Statistix. 1986. Statistix. Version 1.1. NH Analytical Software, St. Paul, Minnesota, USA.

Strauss, S. Y. 1991. Indirect effects in community ecology: their definition, study and importance. Trends in Ecology and Evolution 6:206–210.

Strong, D. R. 1992. Are trophic cascades all wet? Differentiation and donor control in speciose ecosystems. Ecology 73:747–754.

Thompson, H. M. 1949. East of the Rock and Pillar: a history of the Strath Taieri and Macraes Districts. Capper, Christchurch, New Zealand.

Townsend, C. R. 1991. Exotic species management and the need for a theory of invasion ecology. New Zealand Journal of Ecology 15:1–3.

Townsend, C. R., and T. A. Crowl. 1991. Fragmented population structure in a native New Zealand fish: an effect of introduced brown trout? Oikos 61:347–354.

Vitousek, P. M. 1986. Biological invasions and ecosystems properties: can species make a difference? Pages 163–176 in H. A. Mooney and J. A. Drake, editors. Ecology of biological invasions of North America and Hawaii. Ecological Studies 58. Springer-Verlag, New York, New York, USA.

Williams, J. E., J. E. Johnson, D. A. Hendrickson, S. Contreras-B., J. D. Williams, M. Navarro-M., D. E. McAllister, and J. E. Deacon. 1989. Fishes of North America endangered, threatened, or of special concern: 1989. Fisheries 14:2–20.

Winterbourn, M. J., and K. L. D. Gregson. 1989. Guide to the aquatic insects of New Zealand. Bulletin of the New Zealand Entomological Society 9. Auckland, New Zealand.

Zaret, T. M., and R. T. Paine. 1973. Species introduction in a tropical lake. Science 182:449–455.

20

Variation Among Desert Topminnows in Their Susceptibility to Attack by Exotic Parasites

Paul L. Leberg and Robert C. Vrijenhoek

Introduction

Of the problems facing managers of biodiversity, none may be as difficult to correct as exotic species (Coblentz 1990; Soulé 1990). Exotics can directly affect the viability of native species through competition, predation, and habitat alteration (Simberloff 1981; Coblentz 1990; Soulé 1990), and they pose an additional threat if they harbor parasites and pathogens that also attack native species. For example, oxen probably introduced rinderpest to Africa, where the disease caused high mortality in wild and domestic ungulates (Ford 1971). Indigenous North and South American people were decimated by the introduction of smallpox by European colonists (Crosby 1972). Parasites associated with an introduced sturgeon (*Acipenser stellaturs*) devastated populations of a native species (*A. nudiventris*) in the Aral Sea (Bauer & Hoffman 1976). Introduced mosquitoes (*Culex pipiens*) facilitated transmission of avian malaria to forest-dwelling birds of Hawaii and led to the extinction of many species (Warner 1968).

Increased disease susceptibility may be one of the most serious consequences of the loss of genetic diversity in small remnant populations and captive populations of an endangered species (O'Brien & Evermann 1988; Scott 1988; Lyles 1990). Variability at immune response loci of the host appears to be necessary to sustain a coevolutionary struggle with a rapidly evolving microparasite (those with short generation times relative to the host; such as viruses, bacteria, protozoa, and many helminths) (Hamilton 1980; Anderson & May 1982). But the role that genetic variation plays in host responses to exotic diseases has received little empirical attention.

In this study, we evaluate the susceptibility of several genetic lineages of topminnows in the genus *Poeciliopsis* (Cyprinodontiformes: Poeciliidae) to infection by a novel parasite, *Gyrodactylus turnbulli* (Trematoda: Monogenea). This natural ectoparasite of the common guppy, *Poecilia reticulata* (Cyprinodontoformes: Poeciliidae), is native to northern South America and adjacent Caribbean islands. Guppies, however, have obtained a worldwide distribution in tropical and subtropical areas (Welcomme 1981), which has led to a corresponding increase in the distribution of guppy gyrodactylids (Harris 1986). Presently, we have no records of guppies or gyrodactylids in the state of Sonora, Mexico (home of the *Poeciliopsis* strains used in this study), but guppies have been introduced into several other Mexican states (Hendrickson 1983; Contreras-B. & Escalante-C. 1984) and into Arizona (Minckley 1973). Furthermore, mosquitofish (Poeciliidae: *Gambusia*), which also are known to host gy-

Reprinted with permission from Conservation Biology vol. 8, pp. 419–424. Copyright 1994 The Society for Conservation Biology and Blackwell Science, Inc.

rodactylids (Rogers & Wellborn 1965), have been introduced into Arizona and Sonora (Minckley 1973; Hendrickson 1983).

Methods

The Monogenean Parasites

Gyrodactylids infect a large number of marine and freshwater teleosts (Malmberg 1970). The absence of reports of gyrodactylids infecting *Poeciliopsis* should not be taken as evidence that these fish lack gyrodactylids. Because of their small size (~0.5 mm) the parasites can easily go undetected or unreported. Careful examination of our laboratory stocks did not reveal gyrodactylids.

The *G. turnbulli* used in this study were obtained from naturally infected guppies originally collected in Trinidad by J. Endler. A colony of the parasites was maintained on laboratory stocks of Trinidadian guppies for several years before we obtained them. The parasite has a short generation time (2.4 days at 25° C) as it grows and reproduces directly on the fish host. Using a dissecting microscope, we could easily count parasites on anesthetized fish (for methods see Scott 1982). Fish mortality due to gyrodactylid infection appears to be direct, but parasite infestations might also contribute to lethal bacterial and fungal infections. *Gyrodactylus* infections are known to cause mortality in guppies in the wild (Lyles 1990).

The Fish Hosts

Topminnows of the genus *Poeciliopsis* are small, freshwater fish that range from southern Arizona in the United States, through western Mexico and Central America, to northern parts of South America (Rosen & Bailey 1963). Like many stream-dwelling fishes in the arid regions of Arizona and northwestern Mexico, the habitats of these topminnows are rapidly disappearing due to river impoundment, diversion, and groundwater pumping (Minckley & Deacon 1991). The decline of *P. occidentalis* in Arizona is also attributable to introduced mosquitofish (*Gambusia affinis*, and possibly *G. holbrooki*), which are effective predators on newborn and juvenile topminnows (Meffe et al. 1983; Meffe 1985).

Poeciliopsis from northwestern Mexico exhibit sexual and clonal reproduction modes. Natural hybridization between the sexual species *P. monacha* and *P. lucida* produced the hybridogenetic biotype *P. monacha-lucida* (Schultz 1969). Reproduction in *P. monacha-lucida* (abbreviated as *ML*) is hemiclonal—the females transmit, without recombination, only the haploid *M* genome to their eggs. The *L* genome is discarded during oogenesis, and it must be regained in each generation by matings with *P. lucida* males. We used two sympatric hemiclones, *ML/VII* and *ML/VIII*, which occur in the Arroyo de Jaguari tributary of the Arroyo de Jaguari tributary of the Rio Fuerte (Roman numerals identify allozyme genotypes of the maternal *M* genomes) (Vrijenhoek et al. 1978). The paternal *L* genomes of two *ML* strains were "standardized" by inseminating their mothers with sperm from an isogenic strain of *P. lucida* (S68–4 PC; see below). Thus, genetic differences in the susceptibility of the two hemiclones must be encoded by their respective *M* genomes. We also examined strains of two sexual species derived from the same locality: a randomly bred stock (VQ89–7 PB) of *P. monacha* collected by Vrijenhoek and Quattro in 1989, a randomly bred stock (WQ90–7 PC) of *P. lucida* collected by Weeks and Quattro in 1990, and the inbred strain (S68–4 PC) of *P. lucida* (used as the paternal genome of the hybridogens) collected by Schultz in 1968 and maintained in the laboratory for more than 40 generations of brother-sister matings.

In the first two of three experiments described below, individual fish were maintained in separate 4-liter buckets. Water temperatures were maintained at approximately 27° C. An air-stone was placed in each bucket, and fish were fed commercial fish food *ad libidum* daily. The three experiments described below were performed from October 1 to November 30, 1991. All studies with *G. turnbulli* took place in a quarantine chamber that was separate from our fish-culture facility.

Experiment 1

To determine differences in host susceptibility and mortality, we manually infected fish with a single parasite. Parasitic infections and census methods followed Scott (1982). Seven adult females from each of the following strains were used: *P. lucida* (outcrossed), *P. monacha* (outcrossed), *ML/VII*, and *ML/VIII*. Five uninfected individuals of each strain served as controls for natural mortality. The controls were treated identically to infected fish in all aspects of care, handling, and censusing.

Experiment 2

To give the parasites a better chance at establishment so that we could characterize their population growth, we manually infected fish with two parasites each. We used outcrossed *P. lucida* ($n = 8$), inbred *P. lucida* ($n = 8$), *ML/VII* ($n = 11$), and *ML/VIII* ($n = 11$). *Poeciliopsis monacha* were not included due to an unrelated decline in the laboratory stock. Half of the *P. lucida* in both the inbred and outcrossed treatments were males.

Experiment 3

To determine whether *G. turnbulli* would infect *Poeciliopsis* through casual contact with free-swimming guppies, ten *ML/VII*'s, ten *ML/VIII*'s, ten *P. lucida* (outbred), and two *P. monacha* (outbred) were placed together in an 80-liter aquarium with a single infected guppy. We performed two trials using this design. In the first trial, the infected guppy carried four parasites; in the second, the guppy had five. After five days, the fish were euthanized by immersion for 10 minutes in ice-water ($0°C$), and tissues were removed for electrophoretic identifications. Only four gene loci [*Ldh-1*, *Pgd*, *Idh-2*, and *Ck-A* (formerly *Mp-3*)] were necessary for complete identification of these strains (for methods, see Vrijenhoek et al. 1978).

Statistics

Fisher's Exact Test was used to evaluate differences among strains in mortality rates and parasite prevalence (i.e., percentage of the host population infected at any point in time). The Kruskal-Wallis Test was used to evaluate dif-

ferences among fish strains in parasite loads per infected individual. Probabilities of Type-I errors are indicated by an α.

Results

Experiment 1: Infection of Fish with One Parasite

No *P. monacha, P. lucida*, or hemiclone *ML/VIII* individuals carried parasites at the first census period (7 days after infection). In contrast, five of the seven *ML/VII* individuals were infected. One *ML/VII* died before its parasite load could be assessed on day 7. Inspection of its carcass revealed associated *Gyrodactylus*, but an accurate census was not possible because the carcass had deteriorated. Of the six *ML/VII*s living on day 7, the proportion infected ($p = 5/6 = 0.0833$) was significantly greater than the zero-level prevalence in each of the other three strains ($\alpha = 0.001$). The five infected fish had a mean abundance of 4.3 (± 2.1 SE) parasites each. The two *ML/VII*s with the highest parasite loads on day 7 did not survive to the census on day 14. Only four (of seven) *ML/VII*s survived to day 14, and none were infected. The mortality of *ML/VII*s was significantly higher than that of the other strains ($\alpha = 0.037$).

One exposed *P. lucida* died one day after it was infected, but its parasite load was not determined due to advanced deterioration of the carcass. We doubt that its parasite population could have increased to lethal numbers in less than 24 hours. Its death probably resulted from anesthesia or handling stress. With the exception of a single *P. lucida*, all control fish survived the 14-day trial period. Again, we suspect that handling stress or anesthesia resulted in the death of the single *P. lucida*, because this species is not as robust as *P. monacha* or the hybridogenetic strains in the laboratory.

Experiment 2: Infection of Fish with Two Parasites

We censused the parasites at 5, 7, and 9 days after infection to obtain a better picture of parasite population growth. Significant differences in the prevalence of infection existed among the four strains for all three census periods (Table 1, $\alpha \leq 0.04$). The proportion of *ML/VII* infected was always higher than that of *ML/VIII* or *P. lucida*. *ML/VIII* individuals were refractory to the parasite, but several *P. lucida* were infected. Although the prevalence of infections on outbred *P. lucida* appeared to be high at the census on day 5 ($p = 0.50$), the differences between inbred and outbred strains were not significant ($\alpha \geq 0.608$) for any of the census periods. One infected *P. lucida* and one infected *ML/VII* died between day 5 and day 7.

The abundance of parasites on infected fish was significantly higher for *ML/VII* than *P. lucida* for all three censuses ($\alpha < 0.04$). Parasite numbers on infected *ML/VII*s increased from a mean of 4.8 to 10.3 per fish during the course of the experiment (Table 1). In contrast, parasite populations did not increase on infected inbred or outbred *P. lucida*; instead, they decreased.

Experiment 3: Exposure to Infected Guppies

There was no mortality of *Poeciliopsis* five days after they were placed in aquaria with infected guppies. Both guppies died during the 5-day period. Only two *Poeciliopsis* were infected, one *ML/VII* individual in each tank. Each fish had two parasites. The results of this experiment are in agreement with earlier experiments in which *ML/VII* was the strain most susceptible to infection. Clearly, this strain can be infected through casual contact with an infected guppy.

Discussion

The parasite *Gyrodactylus turnbulli* was not transmitted to either of the sexual species via casual contact with infected guppies. Similarly, one sexual species, *P. monacha*, was not successfully parasitized following manual infections of individuals with a single parasite. Although the other sexual species, *P. lucida*, was not successfully parasitized when individuals were infected with a single parasite, infection with two parasites resulted in low-level infections. However, the parasite populations failed to grow on *P. lucida*. This low-level susceptibility of *P. lucida* appeared to be unaffected by its history of inbreeding or outbreeding.

Significant differences in resistance to *G. turnbulli* existed between hemiclones *ML/VII* and *ML/VIII*. In all three experiments, no parasites were evident on *ML/VIII*. In contrast, *G. turnbulli* was capable of infecting *ML/VII* through casual contact with infected guppies. The manual infections revealed that roughly one-half of the *ML/VII*s were infected with parasites at the first census period. When *ML/VII* individuals were exposed to two parasites per fish, the parasites proliferated in just 9 days to a mean abundance of 10.3 parasites per infected fish. Apparently, these heavy infections contributed to the increased mortality of this strain. The controls used in Experiment 1, coupled with results of other laboratory studies (Schultz & Fielding 1989; Wetherington et al. 1989), revealed no evidence that *ML/VII* is inherently less viable than *ML/VIII* under normal laboratory conditions.

Differences in susceptibility between hemi-

TABLE 1. Infection of fish with two parasites.

Strain	N_0	Day 5			Day 7			Day 9		
		N	P	A	N	P	A	N	P	A
P. lucida										
outbred	8	8	0.50	2.0 (± 0.4)	7	0.14	2	7	0.14	2
inbred	8	8	0.25	1	8	0.13	1	8	0	0
Total	16	16	0.38	1.2 (± 0.3)	15	0.13	1.5 (± 0.5)	15	0.07	2
ML/VII	11	11	0.45	4.8 (± 1.2)	10	0.40	7.0 (± 1.7)	10	0.40	10.3 (± 2.6)
ML/VIII	11	11	0	0	11	0	0			

The initial number of fish (N_0), number of fish surviving (N), prevalence of infected fish (P), and mean abundance (A) of parasites per infected fish (*SE*, in parens) are given for each of the census periods.

clones *ML/VII* and *ML/VIII* must result from specific alleles or gene/environment interactions carried by their respective *M* genomes, because their paternal *L* genomes were identical in these experiments. Most of the genetic differences between the *M* genomes of *ML/VII* and *ML/VIII* are believed to have been "frozen" from extant variation in *P. monacha* when these strains arose via hybridization (Vrijenhoek et al. 1978; Wetherington et al. 1989). Although our laboratory stock of *P. monacha* was refractory to infection by the present strain of *G. turnbulli*, susceptible genotypes probably existed in the natural population of *P. monacha* that gave rise to *ML/VII* and *ML/VIII*. Alternatively, the susceptibility of *ML/VII* may be due to mutations that accumulated in its *M* genome subsequent to its origin.

Similar experiments with inbred guppies revealed interstrain variation for resistance to a related parasite, *Gyrodactylus bullatarudis* (Madhavi & Anderson 1985). Crosses between strains suggested that resistance was dominant over susceptibility. In contrast, partial resistance to *G. turnbulli* associated with the *lucida* genome used in this study appeared to be recessive in the *ML/VII* hybrids. Not all *ML/VII*'s were infected, however, despite their genetic uniformity. Perhaps variable susceptibility in this strain results from complex multigenic interactions between *monacha* and *lucida* genomes and the environment to which these hybrids are exposed.

Loss of genetic variation due to seven generations of sib-matings resulted in an increased susceptibility of guppies to *G. turnbulli* (Lyles 1990). But susceptibility of *Poeciliopsis* to this parasite cannot be attributed to inbreeding or genotypic uniformity among individuals. The isogenic strain of *P. lucida* was no more susceptible than an outbred strain. Also, hemiclone *ML/VIII* was genetically uniform among individuals and yet completely refractory to infection. In contrast, *ML/VII* was equally uniform and yet highly susceptible. Both hemiclonal forms are highly heterozygous due to their hybrid constitution (Vrijenhoek et al. 1978). Furthermore, in nature, the paternal *L* genomes of *ML* hybrids is substituted in each generation, giving the hybrids access to all the

allelic variability in the gene pool of *P. lucida*. Although paternal variation contributes to phenotypic diversity among *ML* individuals, this variation is not transmitted to the progeny. Only the heritable *M* genome can be molded by natural selection to produce a more resistant *ML* population. Although the nonheritable paternal variation cannot be molded by natural selected among *ML* individuals, it can be selected in the coexisting *P. lucida* population. To the extent that *P. lucida* co-evolves with its parasites, *lucida* variation expressed by the *L* genome of *ML* hybrids may also contribute to the increased fitness of hemiclonal strains.

Conservation Implications

Introduction of guppies and associated gyrodactylids into northwestern Mexico could result in elimination of susceptible *Poeciliopsis* lineages (sexual or asexual). By maintaining a reservoir of infections in guppies, the parasite used in this study could decimate a partially susceptible strain like *ML/VII*. On the other hand, invading parasites might be more diverse genetically and, thus, might be able to infect *ML/VIII* and sexual species as well. Susceptibility of *ML/VII* to the present strain of *G. turnbulli* does not mean that this hemiclone would be an easy victim for each novel disease organism to which it is exposed. The different *M* genomes of *ML/VII* and *ML/VIII* could confer differential susceptibility to a variety of diseases. For example, *Poeciliopsis* in this region harbor several native parasites: neascus larvae of a strigeid trematode (Vrijenhoek 1975), larvae of the trematode *Uvulifer* sp. (Lively et al. 1990), and larvae of two heterochelid nematode species, *Contracaecum* spp. (Vrijenhoek 1978).

Frequency-dependent processes appear to affect relationships between *Poeciliopsis* clones and their natural parasites (Lively et al. 1990). *Poeciliopsis monacha* co-exists with two sperm-dependent clones (*MML/I* and *MML/II*) of the triploid gynogenetic biotype *P. 2 monacha-lucida*. Sperm from *P. monacha* males is necessary for activation of triploid embryogenesis, but inheritance is strictly clonal (Schultz 1969).

Parasite loads of *Uvulifer* sp. were highest on clone *MML/I* in habitats where this clone was most frequent; they were correspondingly lower in habitats where the alternate clone (*MML/II*) was more frequent. Frequency-dependent selection by co-evolved native parasites might help to stabilize clonal diversity in these unisexual fish complexes. Nevertheless, introduction of an exotic parasite like *Gyrodactylus* could temporarily tip such a balance, leading to elimination of one or more clones.

Although we cannot predict the consequences of introduced parasites for these fish populations, we suspect that some clones would be lost because of their inability to respond rapidly to new challenges. Triploid gynogenetic forms of *Poeciliopsis* should be more vulnerable than the diploid hybridogens, because the triploids do not benefit from paternal variability. We feel it would be a terrible loss to see these unusual reproductive complexes diminished in this or any other manner. They have taught us so much about the benefits of genetic diversity in biotically and physically heterogeneous environments (Schultz & Fielding 1989; Vrijenhoek 1993). We urge the appropriate Mexican wildlife agencies to regularly monitor these streams for possible introductions of *Poecilia reticulata* and other unwanted exotics such as *Gambusia*. Early detection of introduced exotics would facilitate their eradication before the exotics and their diseases have had a chance to spread. Similarly, introduced populations of guppies and mosquitofish in Arizona should be carefully examined to see if they pose a threat of transmitting exotic diseases to native, federally endangered topminnows (*Poeciliopsis occidentalis*).

Acknowledgments

We thank Anne Houde and Denise Calaprice for providing the parasites used in this study. Clark Craddock and Stephen Karl provided helpful criticisms of an earlier manuscript. This is contribution no. D-67175-1-93 of the New Jersey Agricultural Experiment Station and No. 93-10 of the Institute of Marine and Coastal Sciences, supported by state funds and NSF grant nos. DEB88-05361 and BSR-9123943.

References

Anderson, R. M., and M. May. 1982. Coevolution of hosts and parasites. Parasitology 85:411–426.

Bauer, O. N., and G. L. Hoffman. 1976. Helminth range extension by translocation of fish. Pages 163–172 in L. A. Page, editors. Wildlife diseases. Plenum Press, New York.

Coblentz, B. E. 1990. Exotic organisms: A dilemma for conservation biology. Conservation Biology 4:261–265.

Contreras-B., S., and M. A. Escalante-C. 1984. Distribution and known impacts of exotic fishes in Mexico. Pages 102–130 in W. R. Courtenay, Jr., and J. R. Stauffer, Jr., editors. Distribution, biology, and management of exotic fishes. Johns Hopkins University Press, Baltimore, Maryland.

Crosby, A. W., Jr. 1972. The Columbian exchange. Biological and cultural consequences of 1492. Greenwood Press, Westport, Connecticut.

Ford, J. 1971. The role of the trypanosomiases in African ecology: A study of the tsetse-fly problem. Clarendon Press, Oxford, England.

Hamilton, W. D. 1980. Sex versus non-sex versus parasite. Oikos 35:282–290.

Harris, P. D. 1986. Species of *Gyrodactylus* von Nordmann 1932 (Monogenea Gyrodactylidae) from poeciliid fishes, with a description of *G. turnbulli* sp. nov. from the guppy, *Poecilia reticulata* Peters. Journal of Natural History 20:183–191.

Hendrickson, D. A. 1983. Distribution records of native and exotic fishes in Pacific drainages of northern Mexico. Journal of the Arizona-Nevada Academy of Sciences 18:33–38.

Lively, C. M., C. Craddock, and R. C. Vrijenhoek. 1990. The Red Queen hypothesis supported by parasitism in sexual and clonal fish. Nature 344:864–866.

Lyles, A. M. 1990. Genetic variation and susceptibility to parasites: *Poecilia reticulata* infected with *Gyrodactylus turnbulli*. Ph.D. thesis. Princeton University, Princeton, New Jersey.

Madhavi, R., and R. M. Anderson. 1985. Variability in the susceptibility of the fish host, *Poecilia recticulata*, to infection with *Gyrodactylus bullatarudis* (Monogenea). Parasitology 91:531–544.

Malmberg, G. 1970. The excretory systems and the marginal hooks as a basis for the systematics of *Gyrodactylus* (Trematoda, Monogenea). Arkiv für Zoologi 23:1–235.

Meffe, G. K. 1985. Predation and species replacement in American southwestern fishes: A case study. Southwestern Naturalist 30:173–187.

Meffe, G. K., D. A. Hendrickson, W. L. Minckley, and J. N. Rinne. 1983. Factors resulting in decline of the endangered Sonoran topminnow *Poeciliopsis occidentalis* (Atheriniformes: Poeciliidae) in the United States. Biological Conservation 25:135–159.

Minckley, W. L. 1973. Fishes of Arizona. Arizona Game & Fish Department, Phoenix, Arizona.

Minckley, W. L., and J. E. Deacon. 1991. Battle against extinction: Native fish management in the American West. University of Arizona Press, Tucson, Arizona.

O'Brien, S. J., and J. F. Evermann. 1988. Interactive influence on infectious disease and genetic diversity in natural populations. Trends in Ecology and Evolution 3:254–259.

Rogers, W. A., and T. L. Wellborn, Jr. 1965. Studies of *Gyrodactylus* (Trematoda: Monogenea) with descriptions of five new species form the southeastern U.S. Journal of Parasitology 51:977–982.

Rosen, D. E., and R. M. Bailey. 1963. The poeciliid fishes (Cyprinodontiformes), their structure, zoogeography, and systematics. Bulletin of the American Museum of Natural History 126:1–176.

Schultz, R. J. 1969. Hybridization, unisexuality, and polyploidy in the teleost *Poeciliopsis* (Poeciliidae) and other vertebrates. American Naturalist 103:605–619.

Schultz, R. J., and E. Fielding. 1989. Fixed genotypes in variable environments. Pages 32–38 in R. Dawley and J. Bogart, editors. Evolution and ecology of unisexual vertebrates. Bulletin 466. New York State Museum, Albany, New York.

Scott, M. E. 1982. Reproductive potential of *Gyrodactylus bullatatudis* (Monogenea) on guppies (*Poecilia reticulata*). Parasitology 85:217–236.

Scott, M. E. 1988. The impact of infection and disease on animal populations: Implications for conservation biology. Conservation Biology 2:40–56.

Simberloff, D. 1981. Community effects of introduced species. Pages 53–81 in M. A. Nitecki, editors. Biotic crises in ecological and evolutionary time. Academic Press, New York.

Soulé, M. E. 1990. The onslaught of alien species, and other challenges in the coming decades. Conservation Biology 4:233–239.

Vrijenhoek, R. C. 1975. Effects of parasitism on the esterase patterns of fish eyes. Comparative Biochemistry and Physiology 50B:75–77.

Vrijenhoek, R. C. 1978. Genetic differentiation among larval nematodes infecting fishes. Journal of Parasitology 64:790–798.

Vrijenhoek, R. C. 1993. The origin and evolution of clones versus the maintenance of sex in *Poeciliopsis*. Journal of Heredity 84:388–395.

Vrijenhoek, R. C., R. A. Angus, and R. J. Schultz. 1978. Variation and clonal structure in a unisexual fish. American Naturalist 112:41–55.

Warner, R. E. 1968. The role of introduced diseases in the extinction of the endemic Hawaiian avifauna. Condor 70:101–120.

Welcomme, R. L. 1981. Register of international transfers of inland fish species. Food and Agriculture Organization Fisheries Technical Paper 213:1–120.

Wetherington, J. D., S. C. Weeks, K. E. Kotora, and R. C. Vrijenhoek. 1989. Genotypic and environmental components of variation in growth and reproduction of fish hemiclones (*Poeciliopsis*: Poeciliidae). Evolution 43:635–645.

21

A Test of the Vegetation Mosaic Hypothesis: A Hypothesis to Explain the Decline and Extinction of Australian Mammals

Jeff Short and Bruce Turner

Introduction

Most extinctions and major range contractions within the Australian fauna have taken place among mammals from arid and semiarid areas, within the the size range 0.035–5.5 kg (the critical weight range [CWR] of Burbidge & McKenzie 1989). Included are large mice and rats (Eutheria: Muridae), bandicoots and bilbies (Marsupialia: Perameloidea), bettongs, hare-wallabies, and nail-tail wallabies (Marsupialia: Macropodoidea), quolls (Marsupialia: Dasyuroidea), and possums (Marsupialia: Phalangeroidea). One possible explanation for the decline and extinction of these species relates to the loss of a fire regime imposed by Aborigines pursuing their traditional lifestyle (Bolton & Latz 1978; Kitchener et al, 1980; Johnson & Roff 1982; Allen 1983; Burbidge 1985; Burbidge et al. 1988; Burbidge & McKenzie 1989; Johnson & Southgate 1990).

In the vast spinifex deserts of central and Western Australia (Figure 1a), Aborigines made extensive use of fire to modify the landscape (Gould 1971). Frequent, small-scale fires set by Aborigines are believed to have created a vegetation with a fine-grained mosaic of different seral stages. When Aborigines moved out of these deserts and adopted less traditional lifestyles, this regime was replaced by one of infrequent, large-scale fires caused by lightning strike. The effect is believed to have created a juxtaposition of food in recently burned spinifex and shelter in long-unburned spinifex that was too coarse-grained to be used by medium-sized mammals. This change in the spatial distribution of resources is believed to have been a major factor in the extinction of the medium-sized mammals (Bolton & Latz 1978; Johnson & Roff 1982; Burbidge et al. 1988).

We postulate that if medium-sized mammals require an ecotone between two or more habitat types or seral stages to meet their needs for food and shelter, then they will be disadvantaged as habitat becomes increasingly uniform (coarse-grained). This should be reflected by a decrease in density, condition of individuals, and reproductive success as the intermingling of vegetation types or seral stages is diminished.

This study aimed to test the hypothesis that population parameters of medium-sized mammals respond to scale of the vegetation mosaic in such a way that their persistence at that site may be threatened. We measured density, condition, and reproduction of golden bandicoots (*Isoodon auratus*), northern brush-tailed possums (*Trichosurus vulpecula arnhemensis*), and burrowing bettongs (*Bettongia lesueur*) at 24 sites on Barrow Island that covered a gradient in diversity of vegetation associations. Sites had mixes of vegetation types that varied from

Reprinted with permission from Conservation Biology vol. 8, pp. 439–449. Copyright 1994 The Society for Conservation Biology and Blackwell Science, Inc.

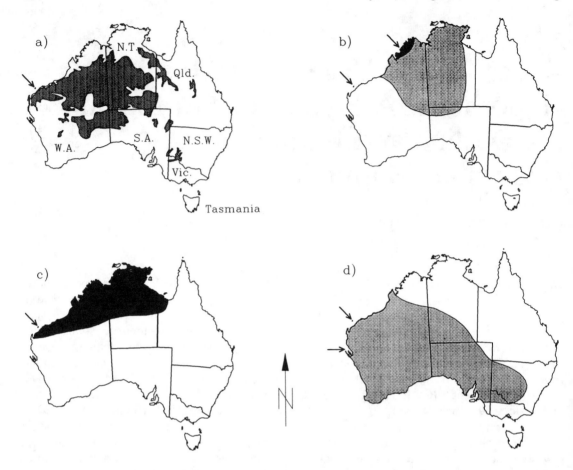

Figure 1. (a) The distribution of spinifex grasslands (*Triodia* and *Plectrachne* spp.) in Australia (Australian Surveying and Land Information Group 1990). The arrow marks Barrow Island. W. A. = Western Australia; N. T. = Northern Territory; Qld = Queensland; N. S. W. = New South Wales; Vic = Victoria; S. A. = South Australia. (b) The former (lightly shaded) and present (heavily shaded or arrowed) range of the golden bandicoot (*Isoodon auratus*). The northern arrow indicates Augusta Island; the southern arrow indicates Barrow Island. (c) The former (lightly shaded) and present (heavily shaded or arrowed) range of the northern brush-tailed possum (*Trichosurus vulpecula arnhemensis*). The arrow indicates Barrow Island. (d) The former (lightly shaded) and present (heavily shaded or arrowed) range of the burrowing bettong (*Bettongia lesueur*) (Strahan 1983; Burbidge et al. 1988). The northern arrow indicates Barrow Island; the southern arrow indicates Bernier and Dorre Islands.

fine-grained to coarse-grained and encompassed the range of variation present on the island.

Barrow Island (Fig 1*a*) lies 60 km off the northwest coast of Australia and has a vegetation dominated by spinifex (*Triodia* spp.). Islands such as Barrow provide opportunities for the study of medium-sized mammals that are now absent or rare on the mainland. Barrow Island is now the only remaining site where these three species of medium-sized mammal occur together. Barrow Island did not support an Aboriginal population after its separation from the mainland by rising sea levels around 8000 years before the present. Despite this, its 14 species of mammals (Butler 1970, 1975) have persisted. Barrow Island appears to burn infrequently (in intervals of more than 50 years) due to lightning strikes, and much of the island burns at once. Hence, its mosaic scale derives from topographic diversity and edaphic variation.

A major oil-field operation on Barrow Island has enhanced the naturally occurring mosaic on the island since 1964 by mechanically removing large amounts of late seral-stage vegetation (dominant on Barrow) and allowing early seral-stage vegetation to take its place. This mimics the changes wrought by fire—the removal of climax vegetation and its temporary replacement by a different suite of species—and provides a further opportunity to test the hypothesis that medium-sized mammals require a fine-grained mosaic. If the scale of mosaic is important, then this mix of early and late successional stages should be reflected in some way in measurable parameters of animal populations. We compared density, condition, and reproductive success in the half of the island dominated by the oil-field operation with those in the other half where disturbance was much less.

Methods

The Fauna

The golden bandicoot has a distribution that formerly took in the desert and tropical woodland of the northwestern third of Western Australia and much of the Northern Territory (Figure 1b). It is the smallest member of its genus, with a body weight up to 670 g reported from the Kimberley (McKenzie 1983). Individuals spend the day in nests concealed within dense vegetation, made of flattened heaps of sticks and debris with no obvious entrance. They are known to dig burrows in sandy soil during hot weather. The golden bandicoot is nocturnal and forages for termites, ants, centipedes, moths, insect larvae, small reptiles, roots, and tubers (McKenzie 1983).

The golden bandicoot has suffered a major range contraction and now has only sparse, scattered populations in the far northwest of Western Australia in addition to the Barrow Island and Augusta Island populations (Friend 1990). It was widely distributed in central Australia until the 1930s; the most recent museum record is at the Granites in the Northern Territory in 1952 (Johnson & Southgate 1990).

The brush-tailed possum in northern Australia (Figure 1c) lacks the distinctive brush tail and is considerably smaller (1.1–1.6 kg) than southern forms (1.5–4.0 kg) (Kerle et al. 1991). It dens in tree hollows in northern Australia but spends much of its time foraging on the ground. On Barrow Island, possums spend the day under cap rock or in termite mounds. The northern subspecies is described as common but limited (Kerle 1983). Winter (1979) viewed it as sparsely distributed with occasional pockets of higher density, but he believed it to be secure within that range. It has declined in the desert portion of its range over the last 50 years due to a combination of long periods with below-average rainfall, deterioration of habitat quality in key refuge areas from grazing by livestock and rabbits, and predation by dingoes and introduced predators (Kerle et al. 1992).

The burrowing bettong is one of nine species of rat-kangaroos of the family Potoroidae, within the superfamily Macropodoidea. It is distinctive among the macropods as being the only species to construct and live in warrens. Rat-kangaroos are small in size (<4 kg), nocturnal, and omnivorous. They dig for part of their food: roots, tubers, underground fungi, and invertebrates, but they also feed on fruit, flowers, and the leaves of forbs and shrubs.

The bettongs, along with other medium-sized mammals, have suffered major contractions in range since European settlement. The burrowing bettong is now confined to three islands off the coast of Western Australia: Barrow, Bernier, and Dorre (Short & Turner 1993; Figure 1d).

Study Area

Barrow Island lies approximately 60 km off the northwest coast of Western Australia, 1300 km north of Perth (Figure 1a). It is a limestone island of approximately 233 km^2, and it experiences a monsoonal climate (Gentilli 1972) moderated by its oceanic position. Mean annual rainfall is approximately 326 mm ($n = 21$), with most falling in February-March and May-June. A significant component of annual rainfall comes from cyclones, which commonly dump between 100 and 320 mm of rain in a few days. Cyclones crossed the coast in the vicinity of Barrow Island at the rate of 1.3 per annum in the 21-year period to 1989. Hence, rainfall on Barrow Island is often dominated by comparatively few high-intensity falls: 71% of annual rainfall falls in two distinct periods each year, one in January-March and the other in May-August.

The island has been a fauna and flora reserve since 1910 and a producing oil-field since 1964. The oil lease covers the entire island, but the major oil-field infrastructure occupies about 100 km² of the island. In 1987 there were approximately 700 wells, 10 separator stations, 1000 km of pipeline, several hundred kilometers of road, and an infrastructure of work-

shops, warehouses, tank farms, administrative buildings, and accommodation (Figure 2a). An extensive grid of seismic lines (long-narrow strips of land cleared for seismic measurements) covers the entire island in a regular grid.

Buckley (1983) described the vegetation of Barrow Island. The major types in decreasing order of areal extent were *Triodia wiseana* hummock grassland on limestone uplands, *Triodia angusta* hummock grassland on watercourses and lowland loams, *Triodia pungens* hummock grassland on red sand, *Spinifex lon-*

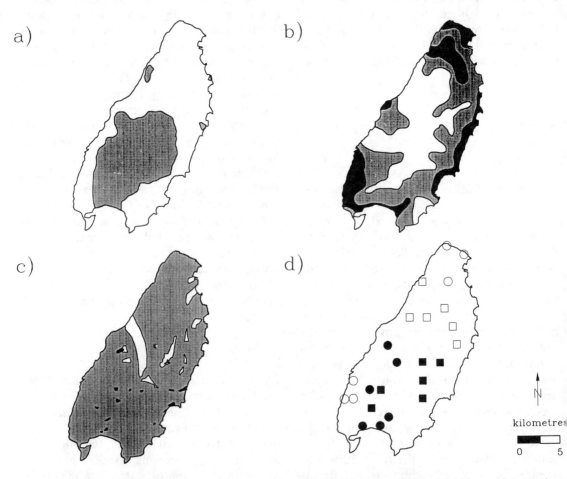

FIGURE 2. (a) Barrow Island, with area of major oil field activity (light stipple). (b) Isobars of vegetation diversity from Shannon-Wiener indices (SWI) of vegetation types. Areas with high diversity (SWI > 1.2), moderate diversity (0.6 > SWI > 1.2), and low diversity (SWI < 0.6) are shown as heavy stipple, light stipple, and unshaded, respectively. (c) Barrow Island, with areas that burned in the 1961 fire (light stipple) and subsequent fires (dark stipple). (d) Barrow Island, with the location of 24 trapping grids. The 24 blocks were divided among four treatments: high diversity—high disturbance (solid circles), high diversity—low disturbance (open circles), low diversity—high disturbance (solid squares), and low diversity—low disturbance (open squares).

gifolius assemblages on white calcareous fore-dunes; short forb communities on floodout flats; and limited areas of mangroves (*Avicennia marina*) and salt flats near the mouths of intermittent sandy watercourses.

The major source of a vegetation mosaic on Barrow Island is the intermingling of different vegetation types, and the mix of areas disturbed by oil-field activities with nearby undisturbed areas. Vegetation and landform types form patches of varying size. Therefore, some areas have a mix of several vegetation types that intermingle in a small area: a fine-grained mosaic. In contrast, other areas are dominated by a single vegetation or landform type: a coarse-grained mosaic. The interior of Barrow Island is a shallowly dissected limestone upland clothed with *Triodia wiseana* and scattered *Ficus platypoda*. There is little diversity of habitat. In contrast, coastal areas are a mix of three to six vegetation types and hence are highly diverse.

There are also large differences across the island in the degree of disturbance caused by the oil-field operations. The oil-field operations remove patches and strips of climax vegetation over substantial areas for wells, pipelines, roads, and "borrow pits." Regeneration on disturbed sites, or around their margins, commences with a vegetation similar to that induced by burning: a species-rich mix of dicots and young palatable *Triodia*.

The fire history of Barrow Island is largely unknown prior to the 1960s. There was a major fire in 1864 (Cox 1977). Captain Jarman, a visiting mariner, reported: "we fired it [the island] from end to end, making a splendid sight in the strong breeze blowing at the time, which caused a perfect sea of flame to traverse the Island at a most astonishing pace." Aerial photography taken in 1943 shows no evidence of burns in the previous decade. Ninety percent of the island burned in 1961, prior to initial oil drilling in 1964 (Figure 2c). Since then there have been over a dozen minor fires (Figure 2c).

Experimental Design

Population densities, physical condition of individuals, and fecundity of bandicoots, pos-

sums, and bettongs were measured at 24 trapping grids. We used a three-factor orthogonal design: two levels of mosaic scale (coarse and fine), two levels of disturbance (areas within and beyond the oil-field), and two seasons (dry and wet). We posed the question of how mosaic scale, disturbance due to the oil-field, and season of year affect population density, physical condition of individuals, and reproductive success.

Scale of vegetation mosaic was assessed in two ways. Shannon-Wiener diversity indices (Krebs 1985:521) of vegetation type and the length of interface between vegetation types (ecotone length) were calculated for areas with a radius of 1 km centered in trapping grids. Vegetation boundaries were derived from aerial photographs of the island (1971 black and white at a scale of 1:40,000) and vegetation categories from Buckley (1983).

The locations of the 24 trapping grids are shown in Fig. 2d. Grid locations were chosen from a random set of 50 blocks of 1 km^2 used in spotlight surveys (Short & Turner 1991). The trapping grids were centered on the six blocks with the highest and the six blocks with the lowest diversity indices outside of the oil field, and on the six blocks with the highest and six blocks with the lowest diversity indices within the oil field. Half the grids were trapped in January and half in June 1989. Three of six grids from each treatment, selected randomly, were trapped at each season. January is the end of the dry season (15% of annual rain falls in the six months from July to December) and June the end of the wet season.

Twelve cage traps were spaced in two rows (80 m apart) of six traps (65 m apart). Traps were set for four nights per grid. Measurements of trapped animals included weight, sex, reproductive status, and head and foot length. The number of animals caught per trapping grid was converted to an estimated density of captures per trap (Caughley 1977:20). Physical condition was indexed as the cube root of body weight divided by length of hind foot.

The bandicoot's home range is generally less than 10 ha (Heinsohn 1966; Gordon 1974; Lobert 1990; Copley et al. 1990); the possum's is 1–11 ha (Kerle 1984); and the bettong's is about 120 ha (Short & Turner 1990). These ranges are small relative to the area around each trapping grid assessed for mosaic scale (314 ha) and to the size of the island (23,300 ha).

Analysis

Differences in abundance between treatment (expressed as estimated density of capture per 48 trap nights) were analyzed for each species by three-factor analysis of variance (two levels of each factor and three replicate grids per treatment).

Condition of bandicoots was analyzed in the same way. Condition of possums and bettongs was analyzed by one-way analysis of variance due to the absence of condition values at some trapping grids (no animals of the particular species caught).

The impact of mosaic scale, disturbance, and season on reproduction (number of females with pouch young) in the three species was tested by Chi-square analysis of 2 × 2 contingency tables.

Results

Measurements of Mosaic Scale

Figure 3 shows the curvilinear relationship between vegetation diversity and ecotone length on Barrow Island. Both measures are highly correlated. Isobars of vegetation diversity are shown in Figure 2b. Higher values of this index denote areas of greater vegetation diversity; lower values denote areas of lesser diversity.

Table 1 summarizes the characteristics of each treatment. The 12 grids in areas characterized as fine-grained mosaic had significantly higher Shannon-Wiener indices (mean: 1.44

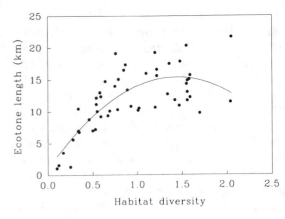

FIGURE 3. The relationship between vegetation diversity and ecotone boundary length for the 50 blocks on Barrow Island. The relationship is curvilinear ($y = 0.95 + 20.07x - 6.98x^2$; $r = 0.76$). Fine-grained mosaics are characterized by high habitat diversity and a longer length of ecotone per unit area; coarse-grained mosaics are characterized by low habitat diversity and shorter ecotone length.

versus 0.37; $f = 58.2$, $p < 0.001$) and ecotone lengths (mean: 14.2 versus 9.4 km; $f = 21.6$. $p < 0.001$) than the 12 grids in areas characterized as coarse-grained mosaic.

Similarly, the 12 grids in areas of high oil field activity had a significantly greater number of oil wells (7.6 versus 0.2 km^{-2}; $f = 123.3$, $p < 0.001$) and length of road (3.1 versus 0.7 km^{-2}, $f = 117.0$, $p < 0.001$) than grids beyond the oil field. This difference was likely to have been reflected in a greater mix of early and late seral stages of vegetation (a finer mosaic).

Abundance

In total, 383 bandicoots, 71 possums, and 38 bettongs were caught on the 24 grids in 1152 trap nights (Table 2). If a grid is assumed to

TABLE 1. Differences in vegetation mosaic between trapping grids.*

Treatment	Shannon-Wiener Index	Habitat Richness	Ecotone Length (km)	Number of Oil Wells	Length of Roads (km)
High Diversity-High Disturbance	1.10	2.8	13.4	23.2	9.5
High Diversity-Low Disturbance	1.78	5.0	14.9	0.5	1.9
Low Diversity-High Disturbance	0.22	2.0	8.7	24.8	10.1
Low Diversity-Low Disturbance	0.51	2.8	10.1	0.7	2.8

*Measurements were made for areas of 1 km radius (3.14 km^2) around each of the 24 trapping grids. Values are the means of the six blocks in each treatment.

TABLE 2. Results from three grids of 12 cage traps for four nights in four major treatments at two seasons on Barrow Island.

	Captures (recaptures)			Trap success
Treatment	Bandicoots	Possums	Bettongs	(%)
January				
High Diversity-High Disturbance	40 (22)	6 (1)	5 (1)	52.1
High Diversity-Low Disturbance	44 (30)	3 (0)	2 (0)	54.9
Low Diversity-High Disturbance	43 (22)	12 (0)	4 (0)	56.3
Low Diversity-Low Disturbance	38 (23)	6 (5)	3 (0)	52.1
All grids	165 (97)	27 (6)	14 (1)	53.8
June				
High Diversity-High Disturbance	63 (19)	16 (3)	6 (1)	75.0
High Diversity-Low Disturbance	61 (22)	6 (2)	9 (2)	70.8
Low Diversity-High Disturbance	53 (33)	9 (6)	7 (5)	78.5
Low Diversity-Low Disturbance	41 (23)	10 (4)	2 (1)	56.3
All grids	218 (97)	44 (15)	24 (9)	70.7

sample an area of 4.7 ha (120 × 390 m), then this gives raw indices of abundance of 3.3 ha^{-1} for bandicoots, 0.6 ha^{-1} for possums, and 0.3 ha^{-1} for bettongs—suggesting minimum population estimates for the three species on Barrow Island of 77,000, 14,000, and 7,000, respectively. All species were widespread on the island. Bandicoots, possums, and bettongs were caught on 100%, 75%, and 63% of grids, respectively.

Analysis of variance of the abundance of animals in each treatment indicate that there were no significant effects of mosaic scale for any species (Table 3). Disturbance due to the oil-field operation similarly had no detectable effect on the abundance of any of the three species. Season of year was important for ban-

TABLE 3. Analysis of variance (F ratios and significance level) of the effect of disturbance, mosaic scale, and season on the abundance of bandicoots, possums, and bettongs expressed as estimated density of captures per 48 trap nights.

	Species		
Treatment	Bandicoot	Possum	Bettong
Disturbance	0.68 (n.s.)	0.48 (n.s.)	0.34 (n.s.)
Mosaic Scale	2.42 (n.s.)	0.08 (n.s.)	0.34 (n.s.)
Season	7.53 ($p < 0.05$)	1.95 (n.s.)	0.94 (n.s.)
D × Ms	0.30 (n.s.)	0.96 (n.s.)	0.34 (n.s.)
Ms × S	1.41 (n.s.)	0.31 (n.s.)	0.34 (n.s.)
D × S	0.21 (n.s.)	0.02 (n.s.)	0.34 (n.s.)
D × Ms × S	0.13 (n.s.)	0.49 (n.s.)	0.94 (n.s.)

$F_{1,16} (0.05) = 4.49$

dicoots. They were nearly 60% more abundant in winter than in the previous summer, which can be attributed to the large breeding pulse that we observed in January (see below). Bandicoots caught in June included a higher percentage of smaller animals (44% less than 300 g, compared to 33% in January) and a lower percentage of recaptures (31% versus 37%), suggesting that the increase in trap success measured an increase in abundance rather than a greater willingness to enter traps.

Condition

The physical condition of bandicoots showed a significant interaction between season and oil field (Table 4). Bandicoots were in significantly better condition in grids beyond the oil field in January, but this was reversed in June.

The physical condition of possums showed no relationship to mosaic scale or to disturbance (Table 4), but possums were in significantly better condition in January.

The condition of bettongs showed no trend with mosaic scale or disturbance (Table 4); it was significantly higher in January.

Reproduction

Reproduction in the bandicoot was strongly seasonal (Table 5). Of females greater than 250 g, 82% carried pouch young in January, compared to only 4% in June. The minimum weight of a bandicoot with pouch young was 205 g.

TABLE 4. Analysis of variance (F ratios and significance level) of the effect of disturbance, mosaic scale, and season on the condition of bandicoots, possums, and bettongs.

Treatment	Species		
	Bandicoot	Possum	Bettong
Disturbance	0.03 (n.s.)	0.61 (n.s.)	0.14 (n.s.)
Mosaic			
Scale	0.48 (n.s.)	3.57 (n.s.)	0.27 (n.s.)
Season	0.31 (n.s.)	5.20 ($p <$ 0.05)	18.4 ($p <$ 0.001)
D × Ms	0.58 (n.s.)	–	–
Ms × S	4.34 (n.s.)	–	–
D × S	5.25 ($p <$ 0.05)	–	–
D × Ms × S	0.03 (n.s.)	–	–

$F_{1,16}$ (0.05) = 4.49

There was no significant association between mosaic scale and the proportion of bandicoot females carrying young, but disturbance due to the oil field did appear to have an effect (Table 5). Of females in disturbed areas, 31% (14 of 45) carried pouch young, compared to 54% (30 of 56) in little-disturbed areas.

The number of pouch young per female was not significantly related to maternal body weight ($f = 1.71$, n.s.), to maternal condition ($f = 1.32$, n.s.), or to the number of bandicoots caught on the grid ($f = 0.23$, n.s.). However, there was a strong negative association between the number of pouch young per female and an index of the size of young (hindfoot length; slope of regression significantly different from zero; $f = 6.88$, $p < 0.05$), indicating an attrition of pouch young during the course of

TABLE 5. Chi-squared analysis of the effect of disturbance, vegetation mosaic, and season on reproduction in bandicoots, possums, and bettongs.

Treatment	Species		
	Bandicoot	Possum	Bettong
Disturbance	5.11 ($p <$ 0.05)	13.5 ($p <$ 0.01)	0.02 (n.s.)
Mosaic Scale	0.65 (n.s.)	0.08 (n.s.)	0.28 (n.s.)
Season	63.2 ($p <$ 0.001)	0.04 (n.s.)	0.02 (n.s.)

pouch life. Females may have as many as four young at birth (commonly two), but it seems that only one or two survive to independence.

There was no significant association between the number of female possums with pouch young and mosaic scale or season of year (Table 5). However, significantly more females in high disturbance areas (15 of 16) carried pouch young than in low disturbance areas (1 of 5).

Ten of 13 female bettongs carried pouch young (5 of 6 in the high diversity grids and 5 of 7 in the low diversity grids). There was no significant association between the proportion of females carrying pouch young and vegetation diversity, disturbance due to oil-field activities, or season of year (Table 5).

Discussion

Dynamics and Habitat Requirements of the Three Species

The impact of weather substantially outweighed the effects of both scale of vegetation mosaic and disturbance due to the oil field. Two hundred millimeters of rain fell in two days in mid-December 1988 (Cyclone Ilona), a month prior to the dry season trapping. This resulted in animals of the three species being in better condition in January than in June. Of female bandicoots captured in January, 82% carried pouch young (mostly small and unhaired), and many had been recruited into the population by June (indices of abundance suggested a 60% increase in population size between January and June). Possums and bettongs showed no significant change in either abundance or fecundity between the trapping periods. There was, however, a substantial increase in trap success for each of these species over the six months between trapping sessions (Table 2).

The impact of fire on a species or a community may be influenced by a variety of factors other than the spatial pattern of burn. These include the season, intensity, and frequency of burn and the season and amount of subsequent rainfall. Successional changes in plant composition and structure after fire in *Triodia* communities have been examined by Zimmer (1940), Burbidge (1944), Burbidge & Main (1971), Suijdendorp (1981), and Griffin (1991). Early stages in the succession are dominated by the germination of a suite of short-lived annual "fireweeds." On Barrow Island these include the dicots *Petalostylis labicheoides, Trichodesma zeylanicum, Stylobasium spathulatum, Adriana tomentosa, Corchorus parviflorus*, and *Heliotropium ovalifolium*. However, *Triodia* may again dominate the plant community within 2–3 years (Suijdendorp 1981). Burbidge and Main (1971) were unable to detect a known burn on Barrow Island from either the ground or a helicopter four years after the fire.

Short and Turner (1990) examined trends in abundance of mammals on Barrow Island over the period 1971–1988 (10–27 years after a fire) using spotlight data collected by Department of Conservation and Land Management (formerly Fisheries and Wildlife) and West Australian Petroleum. Bandicoots, possums, and bettongs showed no significant temporal trend. Short and Turner concluded that no major successional responses to fire had occurred over the mid-term (5–30 years) that might threaten survival of the species. This is in contrast to a study in forest in Tasmania (Driessen et al. 1991), where there was a significant increase in possums, *Trichosurus vulpecula*, and decrease in eastern bettongs, *Bettongia gaimardi*, in the six years following a fire. These changes were associated with substantial changes in the height and density of the understory that probably altered food availability, although the change for possums could have been due to a decrease in hunting pressure (Driessen et al. 1991).

At least one of the three species of *Isoodon, I. obesulus*, is reported to prefer early stages (6–9 years) of vegetation succession after clearing (Stoddart & Braithwaite 1979). Braithwaite (1983) maintained that this species required habitat that "burned fairly regularly." He attributed this to the abundant insect food in early successional vegetation. No such dependence on early successional stages after fire has been reported for *I. obesulus* in Tasmania (Heinsohn 1966) or for the northern brown bandicott *I. macrourus* (Gordon 1983).

Kerle (1983, 1985) concluded, from a study of habitat requirements, that numbers of the brush-tailed possum in northern Australia would be severely depleted by regular burning of its habitat. Regular fires reduced the growth of *Eucalyptus miniata* trees — reducing the likelihood of them having suitable nest hollows — and led to a decrease in the availability and diversity of food items. On Barrow Island possums shelter in cavities under cap rock, so fire at this site is unlikely to influence the availability of shelter and nest sites.

Hence, while the three species may be influenced to some degree by habitat change resulting from fire, there is no evidence that any of them requires a particular spatial pattern of burning to persist. The species have survived on Barrow Island through periods when much of the island would have been bare of vegetation after fire, and through long periods when the *Triodia* was unburnt.

Implications for the Vegetation Mosaic Hypothesis

The vegetation mosaic hypothesis was developed to explain the recent loss of species of medium-sized mammals from the spinifex deserts of mainland Australia. Its strength is that it explains the loss of mammals in northern deserts of Western and central Australia where other explanations — predators, rabbits, grazing by livestock, poisoning — appear inadequate. The loss of mammals from these northern deserts coincides with the timing of a major lifestyle and geographic shift by Aborigines (Johnson & Roff 1982; Johnson & Southgate 1990). Barrow Island retains several species that have been entirely lost or severely depleted on the mainland. Therefore, a comparison between Barrow Island and the mainland may point to factors important for the survival of these species. The climate and vegetation of Barrow Island are similar to that of much of the desert country to which the vegetation mosaic hypothesis is applied. There are, however, important differences between Barrow Island and the mainland.

One difference is the absence of Aboriginal occupation since Barrow was created as an island by rising sea level 8000 years ago (Abbott

1980). This, coupled with an oceanic "fire-break," has led to a fire regime that differs from that of the spinifex deserts of mainland Australia. Barrow appears to burn infrequently (at 50–100 year intervals?). In contrast, spinifex country on mainland Australia commonly burns at 5–30-year intervals in the absence of Aborigines (Hodgkinson et al. 1984) and considerably more frequently in their presence. The widespread presence of the fire-sensitive *Ficus platypoda* on Barrow Island is one indicator of the infrequent nature of fires there. On the mainland this species is confined to rocky slopes and gorges where fire is largely excluded.

Perhaps more obvious differences between island and mainland are the absence of introduced predators such as the dingo, fox, and cat, and of introduced herbivores such as sheep, cattle, rabbits, goats, donkeys, and camels.

The persistence of bettongs, possums, and bandicoots on Barrow Island (and of bettongs, hare-wallabies, and bandicoots on Bernier and Dorre Islands; Short & Turner 1992, 1993) in the absence of Aborigines and Aboriginal burning indicates that Aboriginal fire regimes per se are unnecessary for the survival of these species. The lack of any relationship between population parameters of the fauna of Barrow Island and vegetation diversity suggests that the scale of mosaic is similarly unimportant.

The persistence of species on Barrow and the pattern of their distribution across the island provide no support for the vegetation mosaic hypothesis as a general explanation of the decline of medium-sized native mammals in the spinifex deserts or in other major land-use zones. The experimental results warn against the uncritical acceptance of a hypothesis that has little or no empirical backing.

Changing fire regimes may have played a part in the demise of desert mammals, but through some mechanism other than the juxtaposition of seral stages or vegetation types. It may be that fire in central Australia acts through the greater predation (from both aerial and cursorial predators) that results from reduction in vegetation cover (Christensen & Maisey 1987), or through direct mortality of animals. The former effect would have been greatly exacerbated after the spread of foxes into the desert country between the 1930s and the 1950s (Finlayson 1958, 1961).

The absence of introduced herbivores and predators on Barrow Island is a simpler and more plausible explanation for the persistence of rare mammals there. Morton (1990) suggested that a widespread reduction of native mammals in the spinifex deserts resulted from the destruction of key refuge areas by rabbits and other introduced herbivores. This fragmentation of the populations of native mammals left them vulnerable to drought and elimination by introduced predators such as foxes and cats.

Morton's explanation may well reflect the historical changes that beset many species of Australiam mammal in the spinifex deserts over the last 100 years. It is likely that the combined impact of introduced herbivores and predators resulted in the decline and extinction of many native species. We suspect, however, that introduced predators—particularly foxes—could well have been effective in eliminating many species of native mammals from mainland Australia in the absence of the habitat change caused by the spread of introduced herbivores and without the accessory food source provided by rabbits. We are less certain that introduced herbivores, in the absence of exotic predators, would have had the same effect.

We view the survival of mammals on islands off the Western Australian coast to be due largely to the absence of predation from cursorial predators. This explanation is consistent with success in management of rare species on the mainland by controlling foxes (Kinnear et al. 1988), the differential pattern of survival of mammals on Australian islands to which Aborigines and their dingos had access (Abbott 1980) and the disparate success of reintroductions of macropods to islands without cursorial predators and to mainland and islands with such predators (Short et al. 1992). We conclude that management of relic and reintroduced populations of mammals in the spinifex deserts must include effective control of foxes and cats.

Acknowledgments

We would particularly like to thank Andrew Burbidge for introducing us to the vegetation mosaic hypothesis, encouraging us to think critically about it, and facilitating our field work on Barrow Island. Harry Butler introduced us to Barrow Island and willingly shared his vast accumulated knowledge of the island. All staff of West Australian Petroleum were highly supportive of the project. We would particularly like to thank Russell Lagdon, Peter Ford, and Gary Devenny. Financial support for the project was provided by the World Wide Fund for Nature, West Australian Petroleum, and CSIRO. Critical comment on an earlier draft of the manuscript was provided by Harry Butler, Ken Johnson, and Steve Morton.

References

Abbott, I. 1980. Aboriginal man as an exterminator of wallaby and kangaroo populations on islands around Australia. Oecologia (Berlin) **44**:347–354.

Allen H. 1983. Nineteenth century faunal changes in western New South Wales and Victoria. Working papers in anthropology, archaeology, linguistics and maori studies. Department of Anthropology, University of Auckland, New Zealand.

Australian Surveying and Land Information Group. 1990. Atlas of Australian resources. 6. Vegetation. Department of Administrative Services, Canberra, Australia.

Bolton, B. L., and P. K. Latz. 1978. The western hare-wallaby, *Lagorchestes hirsutus* (Gould) (Marcopodidae), in the Tanami Desert. Australian Wildlife Research **5**:285–293.

Braithwaite, R. W. 1983. Southern brown bandicoot, *Isoodon obesulus*. Page 94 in R. Strahan, editor. Complete book of Australian mammals. Angus and Robertson, Sydney, Australia.

Buckley, R. C. 1983. The flora and vegetation of Barrow Island, Western Australia. Journal of the Royal Society of Western Australia **66**:91–105.

Burbidge, N. T. 1944. Ecological succession observed during regeneration of *Triodia pungens* R. Br. after burning. Proceeding of the Royal Society of Western Australia **28**:149–156.

Burbidge, A. A. 1985. Fire and mammals in hummock grasslands of the arid zone. Pages 91–4 in J. R. Ford, editor. Fire ecology and management in Western Australian ecosystems. Western Australian Institute of Technology, Perth, Australia.

Burbidge, A. A., and A. R. Main. 1971. Report on a visit of inspection to Barrow Island November 1969. Department of Fisheries and Fauna, Western Australia.

Burbidge, A. A., and N. L. McKenzie. 1989. Patterns in the modern decline of Western Australia's vertebrate fauna: Causes and conservation implications. Biological Conservation **50**:143–198.

Burbidge, A. A., K. A. Johnson, P. J. Fuller, and R. I. Southgate. 1988. Aboriginal knowledge of the mammals of the central deserts of Australia. Australian Wildlife Research **15**:9–39.

Butler, W. H. 1970. A summary of the vertebrate fauna of Barrow Island, W. A. Western Australian Naturalist **11**:149–160.

Butler, W. H. 1975. Additions to the fauna of Barrow Island, W. A. Western Australian Naturalist **13**:78–80.

Caughley, G. 1977. Analysis of vertebrate populations. Wiley, London, England.

Christensen, P., and K. Maisey. 1987. The use of fire as a management tool in fauna conservation reserves. Pages 323–329 in D. A. Saunders, G. W. Arnold, A. A. Burbidge, and A. J. M. Hopkins, editors. Nature conservation: The role of remnants of native vegetation. Surrey Beatty, Sydney, Australia.

Copley, P. B., V. T. Read, A. C. Robinson, and C. H. S. Watts. 1990. Preliminary studies of the Nuyts Archipelago bandicoot *Isoodon obesulus nauticus* on the Franklin Islands, South Australia. Pages 345–356 in J. H. Seebeck, P. R. Brown, R. L. Wallis, and C. M. Kemper, editors. Bandicoots and bilbies. Surrey Beatty, New South Wales, Australia.

Cox, J. M. 1977. Barrow Island: An historical documentation. Report to West Australian Petroleum P/L, Perth, Western Australia.

Driessen, M. M., R. J. Taylor, and G. J. Hocking. 1991. Trends in the abundance of three marsupial species after fire. Australian Mammalogy **14**:121–124.

Finlayson, H. H. 1958. On central Australian mammals. Part III. The Potoroinae. Records of the South Australian Museum **13**:236–302.

Finlayson, H. H. 1961. On central Australiam mammals. Part IV. The distribution and status of Central Australian species. Records of the South Australiam Museum **14**:141–191.

Friend, J. A. 1990. Status of bandicoots in Western Australia. Pages 73–84 in J. H. Seebeck, P. R. Brown, R. L. Wallis, and C. M. Kemper, editors. Bandicoots and bilbies. Surrey Beatty, New South Wales, Australia.

Gentilli, J. 1972. Australian climatic patterns. Nelson, Sydney, Australia.

Gordon, G. 1974. Movements and activity of the shortnosed bandicoot *Isoodon macrourus* Gould (Marsupialia). Mammalia **38**:405–435.

Gordon, G. 1983. Northern Brown Bandicoot, *Isoodon macrourus*. Pages 96–97 in R. Strahan, editor. Complete book of Australian mammals. Angus and Robertson, Sydney, Australia.

Gould, R. A. 1971. Uses and effects of fire among the western desert Aborigines of Australia. Mankind **8**:14–24.

Griffin, G. F. 1991. Will it burn—should it burn? Management of the spinifex grasslands in inland Australia. Pages 63–76 in G. Chapman, editor. Desertified grasslands: Their biology and management. The Linnean Society of London, London, England.

Heinsohn, G. E. 1966. Ecology and reproduction of the Tasmanian bandicoots (*Perameles gunni* and *Isoodon obesulus*). University of California Publications in Zoology 80:1–96.

Hodgkinson, K. C., G. N. Harrington, G. F. Griffin, J. C. Noble, and M. D. Young. 1984. Management of vegetation with fire. Pages 141–156 in G. N. Harrington, A. D. Wilson, and M. D. Young, editors. Management of Australia's rangelands. CSIRO, Australia.

Johnson, K. A., and A. D. Roff. 1982. The western quoll, *Dasyurus geoffroii* (Dasyuridae, Marsupialia) in the Northern Territory: Historical records from venerable sources. Pages 211–226 in M. Archer, editor. Carnivorous marsupials. Royal Zoological Society of New South Wales, Sydney, Australia.

Johnson, K. A., and R. I. Southgate, 1990. Present and former status of bandicoots in the Northern Territory. Pages 85–92 in J. H. Seebeck, P. R. Brown, R. L. Wallis, and C. M. Kemper, editors. Bandicoots and bilbies. Surrey Beatty, New South Wales, Australia.

Kerle, J. A. 1983. Northern Brushtail Possum, *Trichosurus arnhemensis*. Page 149 in R. Strahan, editor. Complete book of Australian mammals, Angus and Robertson, Sydney, Australia.

Kerle, J. A. 1984. Variation in the ecology of *Trichosurus*: Its adaptive significance. Pages 115–128 in A. P. Smith and I. D. Hume, editors. Possums and Gliders. Australian Mammal Society, Sydney, Australia.

Kerle, J. A. 1985. Habitat preference and diet of the northern brush-tail possum *Trichosurus arnhemensis* in the Alligator Rivers regions, N. T. Proceedings of the Ecological Society of Australia 13:161–176.

Kerle, J. A., G. M. McKay, and G. B. Sharman. 1991. A systematic analysis of the brush-tailed possum, *Trichosurus vulpecula* (Kerr, 1792) (Marsupialia: Phalangeridae). Australian Journal of Zoology 39:313–331.

Kerle, J. A., J. N. Foulkes, R. G. Kimber, and D. Papenfus. 1992. The decline of the brushtail possum, *Trichosurus vulpecula* (Kerr 1798), in arid Australia. Rangeland Journal 14:107–127.

Kinnear, J. E., M. L. Onus, and R. N. Bromilow. 1988. Fox control and rock-wallaby dynamics. Australian Wildlife Research 15:435–450.

Kitchener, D. J., A. Chapman, B. G. Muir, and M.

Palmer. 1980. The conservation value for mammals of reserves in the Western Australian wheatbelt. Biological Conservation 18:179–207.

Krebs, C. J. 1985. Ecology. The experimental analysis of distribution and abundance. Harper and Row, Cambridge, England.

Lobert, B. 1990. Home range and activity period of the southern Brown Bandicoot (*Isoodon obesulus*) in a Victorian heathland. Pages 319–325 in J. H. Seebeck, P. R. Brown, R. L. Wallis, and C. M. Kemper, editors. Bandicoots and bilbies. Surry Beatty, New South Wales, Australia.

McKenzie, N. 1983. Golden Bandicott, *Isoodon auratus*. Page 98 in R. Strahan, editor, Complete book of Australian mammals. Angus and Robertson, Sydney, Australia.

Morton, S. R. 1990. The impact of European settlement on the vertebrate animals of arid Australia: A conceptual model. Proceedings of the Ecological Society of Australia 16:201–213.

Short, J., and B. Turner. 1990. The experimental reintroduction of the burrowing bettong to mainland Australia. Part II. The importance of a vegetation mosaic to the burrowing bettong. Report to World Wide Fund for Nature Australia, Sydney, Australia.

Short, J., and B. Turner. 1991. Distribution and abundance of spectacled hare-wallabies and euros on Barrow Island, Western Australia. Wildlife Research 18:421–429.

Short, J., and B. Turner. 1992. The distribution and abundance of the banded and rufous hare-wallabies, *Lagostrophus fasciatus* and *Lagorchestes hirsutus*. Biological Conservation 60:157–166.

Short, J., and B. Turner. 1993. The distribution and abundance of the burrowing bettong (Marsupialia: Macropodoidea). Wildlife Research 20:525–534.

Short, J., B. Turner, and C. Majors. 1989. The distribution, relative abundance, and habitat preferences of rare macropods and bandicoots on Barrow, Boodie, Bernier and Dorre Islands. Report to National Kangaroo Monitoring Unit. Australian National Parks and Wildlife Service, Canberra, Australia.

Short, J., S. D. Bradshaw, J. Giles, R. I. T. Prince, and G. Wilson. 1992. The reintroduction of macropods (Marsupialia: Macropodoidea) in Australia—a review. Biological Conservation 62:189–204.

Stoddart, D. M., and R. W. Braithwaite. 1979. A strategy for utilization of regenerating heathland habitat by the brown bandicoot (*Isoodon obesulus*; Marsupialia, Peramelidae). Journal of Animal Ecology 48:165–179.

Strahan, R. 1983. Complete book of Australian

mammals, Angus and Robertson, Sydney, Australia.

Suijdendorp, H. 1981. Responses of the hummock grasslands of northwestern Australia to fire. Pages 417–424 in A. M. Gill, R. H. Groves, and I. R. Noble, editors. Fire and the Australian biota. Australian Academy of Science, Canberra, Australia.

Winter, J. W. 1979. The status of endangered Australian Phalangeridae, Petauridae, Burramyidae, Tarsipedidae, and the koala. Pages 45–59 in M. J. Tyler, editor. The status of endangered Australian wildlife. Royal Zoological Society of South Australia.

Zimmer, W. J. 1940. Plant invasions in the mallee. Victorian Naturalist 56:143–147.

22
Fish Assemblage Recovery Along a Riverine Disturbance Gradient

Alan D. Kinsolving and Mark B. Bain

Introduction

Patterns of species diversity and community structure for a wide variety of stream organisms, including fish, have been related to streamflow patterns (reviewed in Poff and Ward 1989). Long-term streamflow characteristics (i.e., the magnitude of floods that recur less than annually) control channel morphology (Leopold et al. 1964, Dunne and Leopold 1978) and short-term changes in streamflow alter the immediate composition of the instream environment (depths, velocities, etc.). The product of streamflow and the physical structure of a stream channel can be regarded as a habitat template (sensu Southwood 1977, 1988) that constrains the number and types of species able to persist in a stream (Poff and Ward 1990).

Disturbance has been defined as any discrete event that kills, displaces, or damages organisms or populations (Sousa 1984), and as an event that disrupts the structure or function of a biological system (White and Pickett 1985, Sparks et al. 1990). Poff (1992) recently extended these definitions to include recurring events with ecological effects and varied attributes of frequency, intensity, duration, and predictability. The primary sources of natural environmental disturbance in lotic systems are streamflow events (e.g., droughts, floods; Resh et al. 1988) and hydrologic regimes (Ward and Stanford 1983, Poff and Ward 1989). Artificial fluctuations in streamflow and regulated flow regimes can be considered disturbances because empirical evidence clearly indicates that modified streamflows can alter stream communities (Petts 1984, Cushman 1985, Irvine 1985).

Bain et al. (1988) identified artificial flow fluctuations from hydroelectric dams as a disturbance that reduces fish community complexity, and Bain and Boltz (1989) hypothesized that rivers downstream of hydroelectric dams have a longitudinal (i.e., upstream-downstream) gradient of change in fish community characteristics. The hypothesized fish community gradient was regarded as a recovery gradient because disturbance effects would diminish with the downstream attenuation of flow fluctuations. Near hydroelectric dams with erratic water releases, shoreline fish assemblages are expected to be sparse and dominated by species that maintain populations in a wide variety of aquatic systems (macrohabitat generalists) because shoreline waters are continually relocated by fluctuating water levels. Fish using deep, midstream habitats are expected to be less affected by flow fluctuations. With increasing distance downstream, the extent of artificial flow fluctuation would decline because of the dynamics of pool storage and discharge (channel pondage, Dunne and Leopold 1978), and the shoreline fish assemblage was expected to become more abundant and diverse with the addition of species largely restricted to rivers and streams (fluvial specialists).

More than 150 studies of stream community recovery from disturbance have been analyzed recently (Niemi et al. 1990, Yount and Niemi 1990), but this information pertains almost entirely to recovery in time. Spatial patterns of stream recovery are poorly known with the exception of a few well-known cases such as the longitudinal recovery of stream biota from dissolved oxygen depletion (Hynes 1960). Spatial recovery in streams can be expected to display gradual change toward background conditions with increasing distance from a disturbance source or site. Natural gradients of fish community change have long been recognized (e.g., Burton and Odum 1945) in streams, with associated factors being habitat volume and diversity (Sheldon 1968, Schlosser 1982a, b, 1987) and environmental variability (Horwitz 1978, Schlosser 1990). Preservation and management of stream fish faunas will depend on incorporating spatial recovery information into decisions about permitting and siting of anthropogenic change.

In this study, we test the hypothesis that a gradient of fish community recovery occurs downstream of hydroelectric dams by attempting to develop fish assemblage gradients for macrohabitat generalists and fluvial specialists in shoreline waters of one highly flow-regulated river and one free-flowing river. This study focuses on the shallow, shoreline fish assemblage because these habitats are important refugia and nursery habitats for many river fishes (Schlosser 1985, 1987, Copp 1989), these habitats can contain the majority of fish and species in a river (Bain et al. 1988, Lobb and Orth 1991), shoreline habitats are the most sensitive habitats to fluctuating streamflow effects (Bain et al. 1988), and because obtaining highly quantitative samples of fish in deep, main-channel river habitats is difficult (Mahon 1980, Mann and Penczak 1984).

Study Area

The study was conducted on 66-km reaches of two medium-sized rivers (sensu Vannote et al. 1980) in the southeastern United States; the free-flowing Cahaba River and the highly flow-regulated Tallapoosa River (Fig. 1). These rivers are tributaries of the Alabama River, have similar annual flow regimes, and drain largely forested, moderate-relief basins with a mild climate. Both rivers are low-gradient, warmwater rivers with water that is circumneutral (acidity), soft to moderately soft, and low to moderate in specific conductance (Table 1). The Cahaba River study reach began in Centreville, Alabama (32°56′42″ N, 87°08′21″ W), at the base of the fall line rapids, and extended downstream to near Sprott, Alabama (32°37′41″ N, 87°15′29″ W). The Tallapoosa River study reach began 3 km below Thurlow Dam (32°30′45″ N, 85°53′21″ W), at the base of the fall line rapids, and extended downstream to near Wetumpka, Alabama (32°26′23″ N, 86°11′44″ W).

The fall line is a geologic and aquatic biogeographic boundary that delimits the range of many fish species in Alabama (Smith-Vaniz 1968) and other Atlantic coast states (Mayden 1987). The study rivers and their tributaries differ in physical structure above and below the fall line. Downstream of the fall line, stream channels tend to be wide, shallow, and meandering with most substrate composed of sand and clay. Above the fall line, the streams tend to have stable, rocky channels with well-developed and numerous pools, rapids, and riffles. Unlike other Eastern states, the fall line in Alabama is also the boundary of the Southwestern Appalachian and Southeast Plains ecoregions (Omernik 1987). Streams in different ecoregions typically vary in physicochemical attributes (Hughes and Larsen 1988) and fish fauna (Larsen et al. 1986, Hughes et al. 1987, Rohm et al. 1987, Lyons 1989).

Whereas the study reaches have similar instream habitats and many of the same fish species, the Cahaba River is smaller and drains some areas with limestone geology. The Cahaba River basin lacks any significant dams or water withdrawals. In contrast, the Tallapoosa River has been developed extensively for peak-demand, hydroelectric power production and hence it has dramatic, daily flow fluctuations (Fig. 2). Power production by Thurlow Dam (2 km upstream from the study reach) occurs

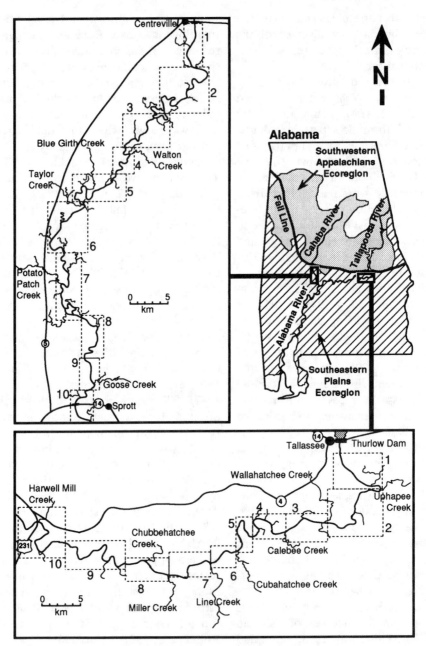

FIGURE 1. Location of the rivers and reaches used in this study with an inset for each reach showing the 10 sections used to group samples. The Alabama map shows the fall line (geologic and aquatic biogeographic boundary) and two ecoregions defined by Omernik (1987).

during some part of most weekdays and lasts various lengths of time. About 230 m^3/s of water is typically released during periods of electrical generation, and no water is released during nongenerating periods. Water is lost over the dam or through spillway gates only during unusually large floods (less than annu-

ally) and on other rare occasions (e.g., dam maintenance). On a few occasions, dam leakage has been measured at 2 m^3/s (Alabama Power Company, *unpublished data*).

As a consequence of hydroelectric dam operation, a gradient disturbance occurs downstream from Thurlow Dam. The upper end of

TABLE 1. Environmental attributes of the study reaches.*

	Cahaba River		Tallapoosa River	
Environmental attribute	Value	N	Value	N
Drainage area (km^2)				
Upstream end of study reach	2659		8620	
Downstream end of study reach	3556		12080	
Increase in drainage area over reach (%)	34		40	
Mean annual discharge (m^3, $\geq$60-yr record)	45.0		134.6	
Stream order	6		7	
Gradient (m/km)	0.48		0.65	
Water temperature (°C)				
Mean of sites at sampling	24.6	127	23.3	113
Standard deviation	4.99	127	4.03	113
Range	18–39	127	16–35	113
pH (mean)				
Upstream end of study reach	7.5	86	6.8	63
Downstream end of study reach	7.4	44	6.9	21
Total hardness (mg/L, mean)				
Upstream end of study reach	81.2	79	11.0	79
Downstream end of study reach	67.7	42	13.7	42
Conductivity (μS/cm, mean)				
Upstream end of study reach	196	155	41	63
Downstream end of study reach	146	137	48	22

*Hydrology and basin morphology statistics were computed from U.S. Geological Survey water quantity data and topographic maps. Water temperature statistics were computed from measurements at all sampling sites. All other water quality statistics were computed from all data in the STORET database maintained by the U.S. Environmental Protection Agency and the Alabama Department of Environmental Management.

the Tallapoosa River study reach alternates between lentic conditions and swift, turbulent water with a surface elevated >2 m in the channel. Gradually, tributary discharge adds water to the Tallapoosa River and at a point 38 km downstream river discharge does not fall below 5.7 m^3/s. Near the end of the study reach, fluctuations in water surface elevation are ≈1 m. Therefore, the Tallapoosa River study reach has a distinct flow regulation gradient (Fig. 2).

In addition to flow fluctuations, dams in the Tallapoosa River basin reduce river transport of coarse sediment and slightly reduce the seasonal variation in water temperature. However, thermal modification in the study reach is minor (Table 1) and inadequate to support coldwater fish species.

Methods

Fish and physical conditions were quantified at discrete 18-m^2 microhabitats along the shore-

lines of the Cahaba and Tallapoosa rivers at least 4 times from April through September 1988. Each river reach was delineated into 1.6-km (1 river mile) sections, and sample sites were selected in each section at random. A randomly selected site was replaced by the first suitable downstream location if the selected site was within 200 m of a permanent tributary or if water depth exceeded 0.75 m along the shoreline (e.g., vertical clay banks). Some sections of both rivers were not sampled during late summer because boat navigation was restricted by low discharge. During this study, 127 fish and habitat samples were collected in the Cahaba River and 113 samples in the Tallapoosa River.

Fish were collected at each site with one prepositioned area electrofisher (1.5 × 12 m, 18 m^2). This gear is effective for sampling a wide size range of fish in shallow water (Bain et al. 1985a, Bain and Finn 1991). The area electrofishers were connected to a custom power supply that produced pulsed direct current

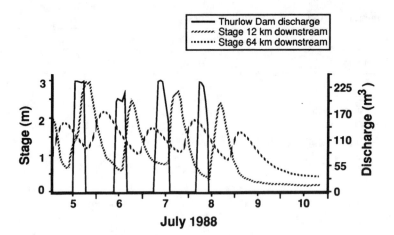

FIGURE 2. Daily hydrographs (stage or discharge) for a representative 6-d period on the Tallapoosa River showing the fluctuations in water release from the Thurlow hydroelectric dam (discharge data from the Alabama Power Company) and water surface elevations (stage data from the U.S. Geological Survey) at two points downstream of the dam. These hydrographs show the effect of discontinuous water releases on downstream water levels and the attenuation of water level fluctuations with distance downstream. Thurlow Dam discharge during generating periods is shown as peaks in the solid line, and there were no water releases during nongeneration periods (dam leakage estimated at 2 m^3/s).

(1–240 pulses/s with a scanning cycle of 0.7 s) between 2.5 and 4 kW. Each area electrofisher was positioned parallel to shore with the anode against the shore (within ≈ 0.3 m) and cathode towards the channel because fish are drawn towards the anode in a direct current field. A 30-m extension cord was connected to the end of the electrodes on each area electrofisher so they could be electrified without disturbing fish. Each area electrofisher was allowed to remain in place, undisturbed, for at least 0.5 h. After the field was energized, all fish in the area electrofisher were collected before turning off the power.

After collection of all fish from an area electrofisher, we measured water depth, current velocity, substrate type, and cover in the 18-m^2 sample site. Depth was measured at 1-m intervals along both the shoreline and channel sides of each site, resulting in 24 measurements per site. The mean water column velocity (recorded at 0.6 of the depth) was measured at four, evenly spaced points along the channel side of each site. Substrate was measured with a modification of the technique of Bain et al. (1985b). A 1.5-m lead-core rope was divided into 10 15-cm sections and within each section the dominant substrate code was recorded. Sub-

strate types were coded as (particle diameter in parentheses): 1 = smooth clay or bedrock, 2 = silt and sand (≤ 2 mm), 3 = gravel (>2–15 mm), 4 = pebble (>15–40 mm), 5 = cobble (>40–200 mm), 6 = boulder (>200 mm). A transect of substrate observations was made at 1.5-m intervals, for a total of 80 substrate observations per site. Non-substrate cover was quantified at the same 80 points as substrate using the technique of Kinsolving and Bain (1990). The presence or absence of nonsubstrate objects was noted for the water column above each 15-cm rope section. The total number of rope sections in which cover objects were present was used as a measure of cover density. When cover was present in a section, the number of surfaces (boundaries between the cover object and the water column) crossed by an imaginary 15 cm long plane projected upward from the bottom to the water surface were counted. This gave a measure of cover complexity, since more complex objects (e.g., a root wad) had many more surfaces than simpler objects like a single tree limb.

The raw physical habitat observations for each sampling site were used to compute site values for water depth, current velocity, substrate coarseness, substrate heterogeneity, cover density, and cover complexity. Water depth and current velocity were recorded as the mean of the site observations. Substrate coarseness was the mean of the 80 coded observations, and substrate heterogeneity was the standard deviation of those observations. Cover density had a minimum possible value of zero (no cover present) and a maximum possible value of 80 (cover present at all points in the site). Cover complexity for each site was calculated as the mean number of cover surfaces from those sampling points where cover was present.

For all statistical analyses and abundance plots, the number of fish (by species) at each sample site was transformed to an octave scale: <2 fish $= 1$, ≥ 2 to <4 fish $= 2$, ≥ 4 to <8 fish $= 3$, ≥ 8 to <16 fish $= 4$, ≥ 16 to <32 fish $= 5$, ≥ 32 to <64 fish $= 6$, etc. This transformation is similar to a logarithmic scale to the base 2, and it is effective for reducing typically high variability in organism counts to an intermediate range of abundance values (e.g., 0–10; Gauch 1982). The transformation is appropriate because community-level studies need to balance the overwhelming numerical influence of dominant species with the information from the occurrence of uncommon species. In addition, fish schooling behavior and variance associated with microhabitat-level samples contribute to great variations in species abundances.

All species were categorized as fluvial specialists or macrohabitat generalists using information on habitat use and distribution compiled from Scott and Crossman (1973), Pflieger (1975), Lee et al. (1980), Becker (1983), Robison and Buchanan (1984), and Burr and Warren (1986). Species classified as fluvial specialists were almost always reported from streams and rivers and were often described as requiring flowing-water habitats throughout life. Some information may have indicated that a species is occasionally found in lakes or reservoirs, but the vast majority of information pertained to streams. Macrohabitat generalists included those species that were commonly found in lakes, reservoirs, and streams and were able to complete their life cycle in any of these systems. For this study, the macrohabitat generalist category includes species that require

access to streams or flowing-water habitats for a specific life stage but otherwise are commonly found in lakes and reservoirs. Classifying these fluvial dependent species as macrohabitat generalists was unimportant in this study because only two species were of this type, blacktail redhorse (*Moxostoma poecilurum*) and white bass (*Morone chrysops*), and these fish use a wide variety of habitats at the life stage (juveniles) using shallow, shoreline waters. In most cases, the category for a species was obvious. However, at one time or another, even very strongly current-oriented species, such as many darters (Percidae), have been recorded in reservoirs; consequently, the distinction between groups is not always clear.

Sampling sites were pooled into 10 sections per river reach to develop fish assemblage gradients. Gauch (1982) states that by averaging out differences among samples, the formation of composites (equal to sections here) tends to raise the level of abstraction, emphasizing broader features of the data. Our primary interest was in the longitudinal changes in fish assemblages as a broad feature of each river reach. In addition, species occurrence and abundances were so highly variable that site by site data tended to become qualitative (presence or absence). The number of sections (10 per river) was chosen, a priori, following the general experience of Gauch (1982) that an intermediate range of number observations frequently captures most available information. Section boundaries were selected to minimize intersection differences in the number of samples and river length. Also, section boundaries were frequently set at the mouths of large tributaries (Fig. 1) so that changes in river size (drainage area) could be minimized within sections.

Longitudinal gradients of change in the characteristics of habitat and fish assemblages were tested with linear regression where river section was the independent variable. Characteristics of habitat were analyzed in the same manner as fish assemblage characteristics to determine if confounding gradients of habitat change occurred. Section means of habitat characteristics (e.g., mean depth, mean water velocity, etc.) and fish assemblage abundance (transformed numbers by habitat-use group) were the dependent variables. Changes in species diversity were tested using the number of species (species richness) in each river section. Abundance plots of all common species (recorded in five or more sections of the same river) by river section were used to inspect the overall pattern of change along the study reaches. For species abundance plots, the distribution of a species was considered incomplete if the species was not captured in two or more contiguous sections at either end of a study reach.

Results

The range and composition of sampled habitats were similar between the study reaches and along each river. We found little evidence that a gradient of change in shoreline habitat occurred on either river (Table 2). One exception was water depth in the Cahaba River where water depth decreased significantly downstream ($P = .003$).

A total of 16 183 fish of 44 species were collected. Almost all were <100 mm in total length and were either juveniles of large-bodied species (e.g., large-mouth bass, *Micropterus salmoides*) or juveniles and adults of small-bodied species (e.g., mosquitofish, *Gambusia affinis*). Similar numbers of macrohabitat generalists (20) and fluvial specialists (24) were recorded in the study (Table 3). Overall, fish densities were >6.5 times higher in the Cahaba River (112 fish/sample) than in the Tallapoosa River (17 fish/sample). Samples without fish were rare (2%; 2 of 127) in the Cahaba River and more common (12%, 13 of 113) in the Tallapoosa River. The number of species in shoreline waters was greater in the Tallapoosa River (36 species) than in the Cahaba River (30 species).

In the Cahaba River, a few macrohabitat generalists and fluvial specialists were consistently very abundant and all other species varied over a low range of abundances (Fig. 3). All generalist species were distributed throughout the 10 sections in the Cahaba River. Among the fluvial specialists, largescale stoneroller and

TABLE 2. Attributes of river sections sampled for fish and habitat. Habitat values are the mean of all samples from a section. Regression results are from univariate tests for a linear relation between river section and each habitat parameter.

Parameter	River section										Regression	
	1	2	3	4	5	6	7	8	9	10	R^2	P
	Cahaba River ($N = 127$ samples)											
Number of samples	9	22	15	12	12	6	13	14	14	10		
Section length (km)	6.8	9.6	7.1	5.5	6.7	6.1	6.5	6.4	6.5	4.8		
Increase in drainage (%)*	0.0	3.6	5.6	9.2	15.0	21.3	22.2	25.4	29.4	33.9		
Water velocity (cm/s)	11.01	8.09	16.48	18.29	12.60	8.48	15.45	12.87	7.84	26.33	0.12	.338
Water depth (cm)	25.23	20.63	21.63	22.74	16.46	19.92	16.31	17.92	16.72	16.46	0.68	.003
Substrate coarseness (index)	2.28	2.30	2.46	2.47	2.67	2.22	2.52	2.38	2.28	2.71	0.11	.351
Substrate heterogeneity	0.31	0.17	0.32	0.44	0.38	0.09	0.32	0.31	0.20	0.30	0.01	.752
Cover density	6.56	5.27	5.13	7.75	2.83	6.33	2.92	2.36	5.36	3.00	0.31	.097
Cover complexity	1.14	0.77	0.78	1.23	0.33	1.15	0.51	0.60	0.87	0.65	0.14	.297
	Tallapoosa River ($N = 113$ samples)											
Number of samples	15	9	6	13	13	8	12	12	11	14		
Section length (km)	6.4	6.5	5.4	7.4	4.9	4.2	5.4	6.5	9.6	9.7		
Increase in drainage (%)*	0.0	12.9	14.3	17.2	20.4	34.6	37.7	38.4	39.0	40.1		
Water velocity (cm/s)	3.58	7.95	13.36	14.48	12.00	7.73	8.87	13.69	11.94	15.07	0.31	.095
Water depth (cm)	32.34	20.73	26.93	20.57	22.41	21.65	26.91	18.64	23.08	20.15	0.26	.131
Substrate coarseness (index)	2.30	2.24	1.74	2.18	2.18	2.13	2.07	2.57	2.62	2.26	0.20	.193
Substrate heterogeneity	0.20	0.18	0.19	0.38	0.16	0.01	0.35	0.41	0.31	0.15	0.04	.591
Cover density	11.80	1.67	11.83	2.69	3.31	6.25	6.92	2.17	9.91	7.00	0.01	.813
Cover complexity	1.87	0.26	1.52	0.65	0.94	1.45	1.25	0.67	0.62	1.01	0.07	.459

*The reported percentage is the cumulative increase in drainage basin area from the upstream end of the study reach.

TABLE 3. Common and scientific names of fishes with the number of individuals, samples, and river sections in which each species occurred. The classification of each species into macrohabitat types is shown as G for macrohabitat generalists and FS for fluvial specialists.

Common name	Scientific name	Cahaba River			Tallapoosa River			Macro-habitat classification
		Fish	Samples	Sections	Fish	Samples	Sections	
Longnose gar	Lepisosteus osseus	...	...	...	1	1	1	G
Threadfin shad	Dorosoma petenense	4	1	1	2	1	1	G
Chain pickerel	Esox niger	1	1	1	1	1	1	G
Largescale stoneroller	Campostoma oligolepis	282	9	5	49	9	6	FS
Alabama shiner	Cyprinella callistia	3	1	1	...	...	...	FS
Blacktail shiner	Cyprinella venusta	4769	119	10	197	35	9	G
Silverjaw minnow	Ericymba buccata	266	28	9	52	10	7	FS
Speckled chub	Macrhybopsis aestivalis	19	5	4	22	7	6	FS
Bluehead chub	Nocomis leptocephalus	61	19	6	...	...	...	FS
Orangefin shiner	Notropis ammophilus	2918	105	10	96	23	8	FS
Emerald shiner	Notropis atherinoides	...	...	...	7	3	2	G
Rough shiner	Notropis baileyi	...	...	...	7	1	1	FS
Pretty shiner	Notropis bellus	...	...	...	40	5	3	FS
Striped shiner	Notropis chrysocephalus	8	2	1	...	...	...	FS
Fluvial shiner	Notropis edwardraneyi	...	...	...	233	16	8	FS
Pugnose minnow	Notropis emiliae	...	...	...	2	2	2	G
Silverstripe shiner	Notropis stilbius	61	10	5	...	...	...	FS
Weed shiner	Notropis texanus	29	3	3	97	18	8	G
Skygazer shiner	Notropis uranoscopus	723	46	10	229	22	7	FS
Mimic shiner	Notropis volucellus	1279	60	10	...	...	...	FS
Clear chub	Notropis winchelli	189	47	9	19	7	6	FS
Riffle minnow	Phenacobius catastomus	29	7	5	...	...	...	FS
Bullhead minnow	Pimephales vigilax	3018	108	10	54	18	8	G
Highfin carpsucker	Carpiodes velifer	33	11	6	13	5	4	FS
Alabama hog sucker	Hypentelium etowanum	1	1	1	6	3	3	FS
Blacktail redhorse*	Moxostoma poecilurum	12	9	5	22	5	4	G
Blackspotted topminnow	Fundulus olivaceus	72	31	10	39	22	9	G
Mosquitofish	Gambusia affinis	92	23	8	148	21	8	G
White bass*	Morone chrysops	...	...	...	1	1	1	G
Banded pygmy sunfish	Elassoma zonatum	...	...	...	1	1	1	G
Redbreast sunfish	Lepomis auritus	...	...	...	10	7	5	G
Bluegill	Lepomis macrochirus	86	26	9	246	35	10	G
Longear sunfish	Lepomis megalotis	253	54	10	214	45	10	G
Redear sunfish	Lepomis microlophus	5	1	1	2	2	2	G
Spotted sunfish	Lepomis punctatus	...	...	...	2	2	2	G
Spotted bass	Micropterus punctulatus	46	26	8	24	16	8	G
Largemouth bass	Micropterus salmoides	13	12	8	24	17	9	G
Naked sand darter	Ammocrypta beani	1	1	1	2	2	2	FS
Southern sand darter	Ammocrypta meridiana	14	9	6	...	...	...	FS
Johnny darter	Etheostoma nigrum	...	...	...	3	1	1	FS
Rock darter	Etheostoma rupestre	2	2	2	...	...	...	FS
Speckled darter	Etheostoma stigmaeum	...	...	...	11	8	5	FS
Blackbanded darter	Percina nigrofasciata	...	...	...	11	8	5	FS
Banded sculpin	Cottus carolinae	...	...	...	10	5	4	FS
	All species	14289	125	10	1894	100	10	

*These species require fluvial habitat for reproduction but use a wide variety of habitats during other life stages.

silver-stripe shiner were not recorded in the five downstream sections, and the southern sand darter was absent in the lower two sections.

Overall, the abundance and number of common species was consistent over the study area.

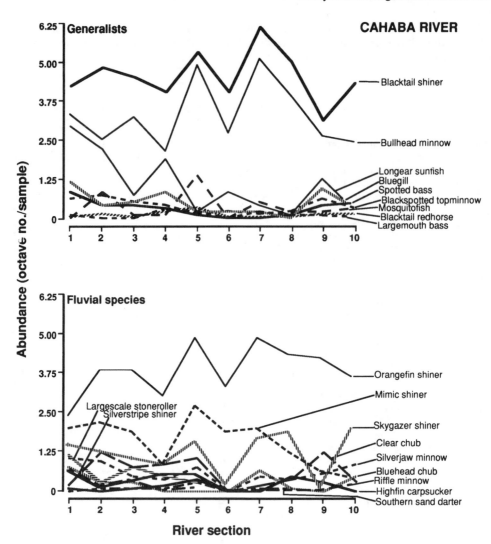

FIGURE 3. The mean abundance of each common species recorded in the Cahaba River grouped as macrohabitat generalists (top) and fluvial specialists (bottom). Species abundance lines were truncated if there were no captures in two or more terminal sections. Section number increases with distance downstream.

In the Tallapoosa River, generalist and fluvial specialist species were variable in abundance (Fig. 4) and consistently less abundant than in the Cahaba River. Numerically dominant species changed frequently and variations in the mean abundances of each species were large. In addition, many species were not distributed along the entire study reach. Of the generalists, mosquitofish and bullhead minnow were absent from the two upstream sections, and redbreast sunfish were not recorded in the last two downstream sections. Only three fluvial specialist species were re-

corded in the upper two sections. First appearance for three species occurred in section 3 and for three other species in section 4. Overall, the patterns of fish abundances and species distributions were highly variable on the Tallapoosa River, with only a few, sparsely collected fluvial species present in low numbers near the dam.

The mean abundance of macrohabitat generalists did not change as a function of river section in the Tallapoosa River ($P = .171$, Fig. 5), although there appeared to be a weak reduction in abundance in the Cahaba River ($P = $

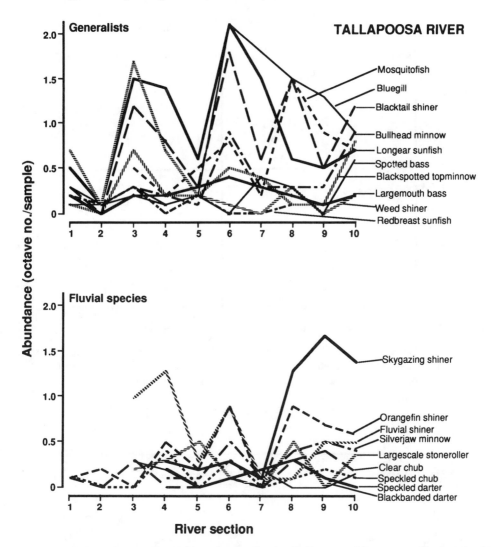

FIGURE 4. The mean abundance of common species in the Tallapoosa River grouped as macrohabitat generalists (top) and fluvial specialists (bottom). Species abundance lines were truncated if there were no captures in two or more terminal sections. Section number increases with distance downstream and section 1 begins near Thurlow Dam.

.057). For fluvial specialists, a linear trend of increasing abundance was significant along the Tallapoosa River ($P = .020$, $R^2 = 0.51$) but not along the Cahaba River ($P = .174$). Regression of the number of generalist species as a function of river section provided no evidence (Cahaba River $P = .713$, Tallapoosa River $P = .584$) that a linear trend existed along the Cahaba or Tallapoosa rivers (Fig. 6). The number of fluvial specialists increased along the Tallapoosa River ($P = .022$, $R^2 = 0.50$) but

may have decreased somewhat along the Cahaba River ($P = .070$).

Discussion

Shallow-water, shoreline habitat in the Cahaba and Tallapoosa rivers appeared physically comparable, but fish assemblages differed in abundance and composition. In general, both rivers had similar numbers of species and shared

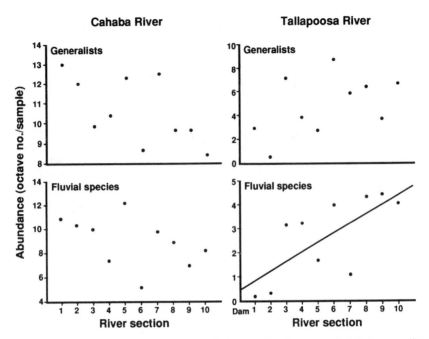

FIGURE 5. Relations between mean fish abundance and river section for macrohabitat generalists and fluvial specialists in the Cahaba and Tallapoosa rivers. A linear relation is shown for fluvial species on the Tallapoosa River because this was the only case with a clearly significant linear regression ($P = .020$, $R^2 = 0.51$).

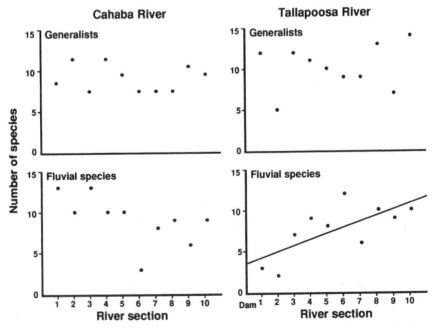

FIGURE 6. Relations between the number of recorded species and river section for macrohabitat generalists and fluvial specialists in the Cahaba and Tallapoosa rivers. A linear relation is shown for fluvial species on the Tallapoosa River because this was the only case with a clearly significant linear regression ($P = .022$, $R^2 = 0.50$).

many numerically dominant fishes. However, differences in overall fish density and the frequency of empty samples suggest that the shoreline assemblage was sparser in the Tallapoosa River than in the Cahaba River. Abundances of numerically dominant species were substantially less in the Tallapoosa than in the Cahaba. Although the water quality characteristics of the Cahaba indicate greater potential productivity, the magnitude of water quality differences between rivers does not seem adequate to explain the differences in shoreline fish abundance. Suppressed fish densities have been reported in systems experiencing chronic, anthropogenic streamflow fluctuations (Bain et al. 1988) and chronic, natural habitat fluctuations (e.g., Romer 1990).

The patterns of species abundances along each stream differed in several ways. In the Cahaba, the most abundant species tended to remain abundant throughout the study reach, and most less abundant species occurred throughout the reach. In the Tallapoosa, numerically dominant species changed over the study reach, and the abundance pattern for almost all species was highly variable. Many species in the Tallapoosa had discontinuous distributions, and fishless sections were most prevalent in the upper, more flow-regulated river sections. Also, in the Tallapoosa, high and low mean abundances of several species appeared synchronized, suggesting a pattern that could be attributed to tributary effects. The localized effects of tributaries on mainstream biotic properties and continuum patterns have been recognized (Vannote et al. 1980, Bruns et al. 1984, Minshall et al. 1985). If the shoreline fish assemblage of the Tallapoosa River was suppressed by adverse mainstream conditions, the influence of tributaries as sources of fish and as refugia from harsh mainstream conditions could have more apparent effects than in an undisturbed river.

Counter to our expectations was a weak trend towards a decline in the abundance of macrohabitat generalists and the number of fluvial species along the Cahaba River. Also, mean water depth was less in the most downstream sections. These suggested trends may be the residual effect of the transition in habitat and fish fauna associated with the river crossing

the fall line and the boundary of the two ecoregions (Fig. 1). Large-scale stoneroller and silverstripe shiner are abundant in streams and rivers upstream of the fall line, and these species gradually become less common in the lower Cahaba River watershed (Pierson et al. 1989). Also, the transition from a rocky, pool-riffle dominated river above the fall line to a sandy, meandering river downstream is not immediate. The gradual decrease in shoreline water depth with downstream distance on the Cahaba corresponds with the change to a broad, sandy river with numerous bars and shoals. Nevertheless, the suggested trends along the Cahaba further contrast the opposing and clear pattern in fish assemblage change along the Tallapoosa.

The lack of a distinct and statistically significant gradient in the abundance and diversity of macrohabitat generalists in both rivers was consistent with the hypothesis that flow fluctuations have little direct effect on species with broad habitat requirements. Habitat generalists varied in abundance and diversity in both rivers without any pattern corresponding to a flow regime. Studies on rainbow trout (*Oncorhynchus mykiss*, Irvine 1987) and adult smallmouth bass (*Micropterus dolomieui*, Bain et al. 1988) have indicated that highly regulated streamflows may not affect the density of species able to inhabit non-fluvial environments. Unlike the habitat generalists, a longitudinal gradient was evident for fluvial specialists in the Tallapoosa. The abundance and diversity of fluvial specialists increased as flow fluctuations became attenuated downstream of the Thurlow hydroelectric dam. Sections with the most variable flows had fluvial fish abundances so low that explanations aside from disturbance effects seem unlikely. No corresponding pattern was evident in the Cahaba, where the abundance of fluvial species remained relatively consistent.

The gradient in fluvial specialists on the Tallapoosa and the lack of a similar gradient on the Cahaba provided evidence that large, daily flow fluctuations cause a disturbance gradient in the Tallapoosa River. The most fluctuating river sections had a fish assemblage with fewer fish and species than less regulated sections and in the unregulated Cahaba River. Along such a disturbance gradient, habitat generalists can be expected to provide a base of fish and species that are variable in numbers and composition, but remain a relatively constant assemblage component. Fluvial specialists can be expected to supplement this base and gradually diversify and augment in numbers the fish assemblage. The differential response of generalists and specialists corresponds with a variety of other studies indicating that eurytopic species cope with environmental change better than stenotopic species (reviewed in Poff and Ward 1990). While the results of this study are specific to shallow, shoreline fishes, the overall pattern may reflect the nature of flow-regulated streams because shoreline areas can provide habitat to most fish and species in a river (Bain et al. 1988, Lobb and Orth 1991).

Although this study may be the first to document the recovery of a fish assemblage along a flow-regulation gradient, others have described similar patterns for macroinvertebrates. Voelz and Ward (1989, 1990, 1991) quantified invertebrate assemblages along a flow-regulation (unchanging within a season) gradient in the upper Colorado River. They concluded that the biota of regulated streams only partially resemble that of comparable free-flowing streams, and that sequential changes in faunas can indicate recovery potential. Voelz and Ward (1991) reported that their most downstream sampling site (12 km from a dam) had many similarities with an unmodified reference stream, and they reviewed other invertebrate studies of stream disturbance gradients. Although shoreline fish assemblages in the Tallapoosa change over a much greater length (at least 66 km) of river than invertebrates in Colorado, the form of flow regulation differed completely and other environmental modifications were important in the Voelz and Ward studies.

The modification of riverine habitat by impoundment and flow regulation is especially detrimental to species that are highly specialized for lotic environments (see Petts 1984 for an extensive review). This generalization has often been expressed using the labels obligate and facultative riverine fishes. However, obligate riverine species include some that require flowing-water conditions for a limited portion of their life cycle (e.g., reproduction in paddlefish, *Polyodon spathula*). Such species are not especially indicative to lotic disturbances because they persist or thrive in highly modified river environments that have accessible lotic habitats. Our classification groups species into those that are restricted to streams and rivers for all life stages (fluvial specialists), species that require riverine habitats for a part of their life cycle (riverine dependent [unimportant in this study]), and species that maintain populations in both lotic and lentic environments (macrohabitat generalists). This classification was intended as a refinement of the obligate and facultative categories so that species truly specialized for fluvial habitats could be isolated. Also, by applying specific classification rules, we could objectively categorize species, although a few classifications can be argued. Finally, the fluvial specialist category does not imply a preference for lotic conditions or a requirement for fluvial microhabitats. For example, striped shiners are restricted to streams and rivers, but they usually occupy very slow or still waters in large pools and backwaters (M. B. Bain, *unpublished data* for the study rivers; Becker 1983, Robison and Buchanan 1984).

Although we have identified a gradient of assemblage change on the Tallapoosa River, a recovery endpoint was not apparent. Yount and Niemi (1990) imply that recovery from disturbance occurs when a system returns to some normal bounds and is able to persist. Other recovery criteria such as original conditions, stable state, and an intact food web have been suggested, although Cairns (1990) argues that there is presently no clear basis for designating recovery. In our case, a variety of spatial recovery endpoints could be designated including the conclusion that the river does not recover in the study reach. The original condition of the fish assemblage is unknown and the Cahaba River cannot be used to precisely define one. In terms of fluvial fish density, the expected complement of species largely occurs by section 5 (≈ 30 km from the hydroelectric dam) but the trend for increasing density continues through the study reach. Similarly, fluvial species abundance continues to increase. A stable assemblage, in a longitudinal sense, does not appear until possibly the last three sections of the Tallapoosa River. Consequently, there appears to be no clear endpoint of recovery on the Tallapoosa River and our results suggest only that recovery is gradual and follows the gradient of attenuating flow fluctuations.

We hypothesized that a gradient of fish community recovery occurs downstream of a disturbance source, and for testing simplicity we assumed it would be a linear function of distance. Our findings suggest that spatial recovery may be non-linear because fluvial species appeared extremely depressed in the two river sections below Thurlow Dam, then greatly increased in diversity and abundance in the next few sections, and finally appeared to stabilize in the most downstream sections of the Tallapoosa River. There is little empirical evidence to support any particular form of recovery in space or time because essentially all studies of recovery have focused on the time to an endpoint. Studies of disturbed aquatic systems have identified colonization as the primary means of recovery rather than expansion of residual populations (Niemi et al. 1990). MacArthur and Wilson's (1967) species equilibrium model (theory of island biogeography) hypothesizes that colonization occurs at a non-linear rate relative to the number of resident species, and this theory has been used to explain recovery processes in a disturbed river (Minshall et al. 1983).

The species equilibrium model proposes that the rate of colonization (relative number of new species) will be maximum in empty environments, and that it will increasingly diminish as potential colonizers face competition with increasing numbers of resident species. While our findings suggest that disturbed river sections are populated through colonization, these areas were not devoid of fish and recovery was attributed to one component of the fish assemblage. Generalist fishes, and even disturbance-sensitive fishes such as fluvial specialists, include some species able to persist in the most degraded conditions. As disturbance severity declines, it is likely that conditions will be reached where many sensitive-type species appear at low abundances. As conditions approach normal, abundances will increase for the sensitive species already present, and further species additions would probably be rare.

Hence, a colonization curve could be expected for reasons similar to, but not precisely the same as those advanced by MacArthur and Wilson (1967). A non-linear recovery gradient may be appropriate as an hypothesis in future case studies, but detailed data on the mechanisms of recovery will be needed to model the process with confidence.

Yount and Niemi (1990) and Niemi et al. (1990) adapted the terminology of Bender et al. (1984) for their analysis of temporal recovery in aquatic systems. They distinguish between pulse disturbances that cause relatively instantaneous perturbation to aquatic biota, and press disturbances that cause sustained alteration. Pulse disturbances are localized in effect and followed by relatively rapid recovery through recolonization and immigration rather than expansion of a residual population. The proximity to unaffected habitats and refugia was a major correlate of rapid recovery. Press disturbances usually involve physical habitat modification with recovery possible only after hydraulic and hydrologic processes restore normal physical conditions. Also, press disturbances overwhelm the availability of adjacent refugia and colonization sources and the system moves to a new state dominated by tolerant species and lacking sensitive species. In other words, press disturbances tend to result in a new, biologically accommodated state rather than a recovered one.

There are spatial parallels for the temporal terms and concepts of Yount and Niemi (1990) and Niemi et al. (1990). Streamflow regulation is not a discrete event but a modification of a long-term regime, and hence it should be considered a press disturbance. As we documented near Thurlow Dam, fish assemblages at or near a press disturbance site (vicinity of regulating dam) may reflect biological accommodation. However, recovery does occur gradually in space (river distance) in the form of a gradient. In a spatial context, a disturbance with effects that persist beyond the immediate area, like flow regulation, can be termed a dissipating disturbance. Dissipating disturbances will be associated with a recovery gradient that spans distance and the downstream return of normal physicochemical conditions. Other examples of spatially dissipating disturbances would be large-scale, continuous point-source pollution (e.g., sewage discharge that depletes oxygen throughout a stream channel, Hynes 1960) and whole-discharge thermal change (e.g., Voelz and Ward 1989, 1990, 1991). While we did not study an analog to a pulse disturbance, a spatial equivalent would be a strictly localized disturbance with recovery so rapid it would appear like a step function (disturbed, non-disturbed) rather than a gradient. We expect some examples would be stream disturbances such as channelization of limited stream segments, road and pipeline crossings, and a typical power plant discharge (cooling water). We hypothesize that such localized disturbances would not extend beyond the immediately altered area due to the occurrence of an assemblage characteristic of adjacent habitats.

In this study, we hypothesized and tested for patterns in an unreplicated pair of disturbed and reference rivers. There are relatively few unregulated and unimpounded rivers in the United States (Benke 1990), and future demands for river development will continue to shrink the numbers. It may already be impossible to develop truly controlled, replicated, large-scale river experiments. However, unreplicated, paired-system experiments have provided extremely valuable data and understanding in ecology (Carpenter 1989). Although this approach provides evidence relative to a hypothesis, it is largely non-statistical and therefore lacks great confidence in conclusions. A series of such studies can act as replicates, staggered in time and location, and a series of studies can strengthen or weaken existing evidence and conclusions. Some new analyses have been proposed for large-scale experiments where minimal replication and controlled perturbation are feasible (see Carpenter 1990). While we could not incorporate these tools, we were able to go beyond the simplest paired-system approach by hypothesizing a system-wide pattern (recovery gradient) and testing for its occurrence. Our approach employs statistics for determining the presence of a gradient, but it does not circumvent the lack of system replication. Nevertheless, pattern detection in paired-systems does provide more convincing evidence than paired point estimates (e.g., mean density in two systems) because the probability of falsely identifying a system-wide pattern seems much less than the probability of obtaining an inaccurate point estimate.

The task of identifying recovery faces the same dilemma that Schindler (1987) described for identifying environmental stress: how do we determine when measurable variables vary outside a normal range? His recommended approach is to monitor sensitive ecosystem components for comparison with similar measurements in undisturbed, reference systems. Fluvial specialist fishes as a group appear sensitive to flow-related disturbance, and measures of their relative abundance and diversity are practical for bioassessment over large spatial scales (e.g., 20–100 km of river). More study of spatial recovery is needed to provide further evidence on the occurrence and form of gradients. If we had even general understanding of most factors that determine the rate of spatial recovery, environmental managers and regulators could use that knowledge to protect riverine fishes. For example, strict controls on dam discharges could be selectively required for cases involving long, unmodified river reaches without major tributaries. Conversely, restrictive discharge requirements may provide few benefits for riverine fishes in cases where recovery gradients are not possible (e.g., short river sections in a series of dams) or might be very short (e.g., dams on tributaries to much larger rivers). Finally, quantitative information on spatial recovery gradients for the range of common river disturbances is needed quickly to maintain much of the world's declining riverine fish faunas.

Acknowledgments

This study was sponsored by the Alabama Game and Fish Division. Critical comments on our work and suggestions for improvement were made by D. R. Bayne, D. R. DeVries, W. L. Fisher, M. S. Golden, N. R. Holler, B. L. Johnson, K. J. Scheidegger, and E. C. Webber. Cooperators of the Alabama Cooperative Fish and Wildlife Research Unit are the Alabama Game and Fish Division, Auburn University (Department of Fisheries and Allied Aquacultures, Department of Zoology and Wildlife Sciences, Alabama Agricultural Experiment Station), the U.S. Fish and Wildlife Service, and the Wildlife Management Institute.

References

Bain, M. B., and J. M. Boltz. 1989. Regulated streamflow and warmwater stream fish: a general hypothesis and research agenda. United States Fish and Wildlife Service Biological Report 89(18).

Bain, M. B., and J. T. Finn. 1991. Analysis of microhabitat use by fish: investigator effect and investigator bias. Rivers 2:57–65.

Bain, M. B., J. T. Finn, and H. E. Booke. 1985a. A quantitative method for sampling riverine microhabitats by electrofishing. North American Journal of Fisheries Management 5:489–493.

Bain, M. B., J. T. Finn, and H. E. Booke. 1985b. Quantifying stream substrate for habitat analysis studies. North American Journal of Fisheries Management 5:499–500.

Bain, M. B., J. T. Finn, and H. E. Booke. 1988. Streamflow regulation and fish community structure. Ecology 69:382–392.

Becker, G. C. 1983. Fishes of Wisconsin. University of Wisconsin Press, Madison, Wisconsin, USA.

Bender, E. A., T. J. Case, and M. E. Gilpin. 1984. Perturbation experiments in community ecology: theory and practice. Ecology 65:1–13.

Benke, A. C. 1990. A perspective on America's vanishing streams. Journal of the North American Benthological Society 9:77–88.

Bruns, D. A., G. W. Minshall, C. E. Cushing, K. W. Cummins, J. T. Brock, and R. L. Vannote. 1984. Tributaries as modifiers of the river continuum concept: analysis by polar ordination and regression models. Archiv für Hydrobiologie 99:208–220.

Burr, B. M., and M. C. Warren, Jr. 1986. A distributional atlas of Kentucky fishes. Kentucky Natural Preserves Commission Scientific and Technical Series 4, Frankfort, Kentucky, USA.

Burton, G. W., and E. P. Odum. 1945. The distribution of stream fish in the vicinity of Mountain Lake, Virginia. Ecology 26:182–194.

Cairns, J., Jr. 1990. Lack of theoretical basis for predicting rate and pathways of recovery. Environmental Management 14:517–526.

Carpenter, S. R. 1989. Replication and treatment strength in whole-lake experiments. Ecology 70:453–463.

———. 1990. Large-scale perturbations: opportunities for innovation. Ecology 71:2038–2043.

Copp, G. H. 1989. The habitat diversity and fish reproductive function of floodplain ecosystems. Environmental Biology of Fishes 26:1–27.

Cushman, R. M. 1985. Review of ecological effects of rapidly varying flows downstream of hydroelectric facilities. North American Journal of Fisheries Management 5:330–339.

Dunne, T., and L. B. Leopold. 1978. Water in environmental planning. W. H. Freeman, San Francisco, California, USA.

Gauch, H. G. 1982. Multivariate analysis in community ecology. Cambridge University Press, Cambridge, England.

Horwitz, R. J. 1978. Temporal variability patterns and the distributional patterns of stream fishes. Ecological Monographs 48:307–321.

Hughes, R. M., and D. P. Larsen. 1988. Ecoregions: an approach to surface water protection. Journal of the Water Pollution Control Federation 60:486–493.

Hughes, R. M., E. Rexstad, and C. E. Bond. 1987. The relationship of aquatic ecoregions, river basins, and physiographic provinces to the ichthyogeographic regions of Oregon. Copeia 1987:423–432.

Hynes, H. B. N. 1960. The biology of polluted waters. Liverpool University Press, Liverpool, England.

Irvine, J. R. 1985. Effects of successive flow perturbations on stream invertebrates. Canadian Journal of Fisheries and Aquatic Sciences 42:1922–1927.

———. 1987. Effects of varying flows in man-made streams on rainbow trout (Salmo gairdneri Richardson) fry. Pages 83–97 in J. F. Craig and J. B. Kemper, editors. Regulated streams. Plenum. New York, New York, USA.

Kinsolving, A. D., and M. B. Bain. 1990. A new approach to measuring cover in fish habitat studies. Journal of Freshwater Ecology 5:373–378.

Larsen, D. P., J. M. Omernik, R. M. Hughes, C. M. Rohm, T. R. Whittier, A. J. Kinney, A. L. Gallant, and D. R. Dudley. 1986. Correspondence between spatial patterns in fish assemblages in Ohio streams and aquatic ecoregions. Environmental Management 10:815–828.

Lee, D. S., C. R. Gilbert, C. H. Hocutt, R. E. Jenkins, D. E. Mcallister, and J. R. Stauffer, Jr. 1980. Atlas of North American freshwater fishes. North Carolina State Museum of Natural History, Raleigh, North Carolina, USA.

Leopold, L. B., M. G. Wolman, and J. P. Miller. 1964. Fluvial processes in geomorphology. W. H. Freeman, San Francisco, California, USA.

Lobb, M. D., III, and D. J. Orth. 1991. Habitat use by an assemblage of fish in a large warmwater stream. Transactions of the American Fisheries Society 120:65–78.

Lyons, J. 1989. Correspondence between the distribution of fish assemblages in Wisconsin streams and Omernik's ecoregions. American Midland Naturalist 122:163–182.

MacArthur, R. H., and E. O. Wilson. 1967. The theory of island biogeography. Princeton Univer sity Press, Princeton, New Jersey, USA.

Mahon, R. 1980. Accuracy of catch-effort methods for estimating fish density and biomass in streams. Environmental Biology of Fishes 5:343-360.

Mann, R. H., and T. Penczak. 1984. The efficiency of a new electrofishing technique in determining fish numbers in a large river in central Poland. Journal of Fish Biology 24:173-185.

Mayden, R. L. 1987. Historical ecology and North American highland fishes: a research program in community ecology. Pages 210-222 in W. J. Matthews and D. C. Heins, editors. Community and evolutionary ecology of North American stream fishes. University of Oklahoma Press, Norman, Oklahoma, USA.

Minshall, G. W., D. A. Andrews, and C. Y. Manuel-Faler. 1983. Application of island biogeographic theory to streams: macroinvertebrate recolonization of the Teton River, Idaho. Pages 279-297 in J. R. Barnes and G. W. Minshall, editors. Stream ecology: application and testing of general ecological theory. Plenum, New York, New York, USA.

Minshall, G. W., K. W. Cummins, R. C. Peterson, C. E. Cushing, D. A. Bruns, J. R. Sedell, and R. L. Vannote. 1985. Developments in stream ecosystem theory. Canadian Journal of Fisheries and Aquatic Sciences 42:1045-1055.

Niemi, G. J., P. DeVore, N. Detenbeck, D. Taylor, A. Lima, J. Pastor, J. D. Yount, and R. J. Naiman. 1990. Overview of case studies on recovery of aquatic systems from disturbance. Environmental Management 14:571-587.

Omernik, J. M. 1987. Ecoregions of the conterminous United States (with map supplement). Annals of the Association of American Geographers 77:118-125.

Petts, G. E. 1984. Impounded rivers. John Wiley & Sons, New York, New York, USA.

Pflieger, W. C. 1975. Fishes of Missouri. Missouri Department of Conservation, Jefferson City, Missouri, USA.

Pierson, J. M., W. M. Howell, R. A. Stiles, M. F. Mettee, P. E. O'Neil, R. D. Suttkus, and J. S. Ramsey. 1989. Fishes of the Cahaba River system in Alabama. Geological Survey of Alabama Bulletin 134.

Poff, N. L. 1992. Why disturbances can be predictable: a perspective on the definition of disturbance in streams. Journal of the North American Benthological Society 11:86-92.

Poff, N. L., and J. V. Ward. 1989. Implications of streamflow variability and predictability for lotic community structure: a regional analysis of streamflow patterns. Canadian Journal of Fisheries and Aquatic Sciences 46:1805-1818.

Poff, N. L., and J. V. Ward. 1990. Physical habitat template of lotic systems: recovery in the context of historical pattern of spatiotemporal heterogeneity. Environmental Management 14:629-645.

Resh, V. H., A. V. Brown, A. P. Covich, M. E. Gurtz, H. W. Li, G. W. Minshall, S. R. Reice, A. L. Sheldon, J. B. Wallace, and R. Wissmar. 1988. The role of disturbance in stream ecology. Journal of the North American Benthological Society 7:433-455.

Robison, H. W., and T. M. Buchanan. 1984. Fishes of Arkansas. University of Arkansas, Fayetteville, Arkansas, USA.

Rohm, C. M., J. W. Giese, and C. C. Bennett. 1987. Test of an aquatic ecoregion classification of streams in Arkansas. Journal of Freshwater Ecology 4:127-140.

Romer, G. S. 1990. Surf zone fish community and species response to a wave energy gradient. Journal of Fish Biology 36:279-287.

Schindler, D. W. 1987. Detecting ecosystem responses to anthropogenic stress. Canadian Journal of Fisheries and Aquatic Sciences 44(Supplement 1):6-25.

Schlosser, I. J. 1982a. Fish community structure and function along two habitat gradients in a headwater stream. Ecological Monographs 52:395-414.

_____. 1982b. Trophic structure, reproductive success, and growth rate of fishes in a natural and modified headwater stream. Canadian Journal of Fisheries and Aquatic Sciences 39:968-978.

_____. 1985. Flow regime, juvenile abundance, and the assemblage structure of stream fishes. Ecology 66:1484-1490.

_____. 1987. A conceptual framework for fish communities in small warmwater streams. Pages 17-24 in W. J. Matthews and D. J. Heins, editors. Community and evolutionary ecology of North American stream fishes. University of Oklahoma Press, Norman, Oklahoma, USA.

_____. 1990. Environmental variation, life history attributes, and community structure in stream fishes: implications for environmental management and assessment. Environmental Management 14:621-628.

Scott, W. B., and E. J. Crossman. 1973. Freshwater fishes of Canada. Fisheries Research Board of Canada Bulletin 184.

Sheldon, A. L. 1968. Species diversity and longitudinal succession in stream fishes. Ecology 49:193-198.

Smith-Vaniz, W. F. 1968. Freshwater fishes of Alabama. Alabama Agricultural Experiment Station, Auburn University, Alabama, USA.

Sousa, W. P. 1984. The role of disturbance in natural communities. Annual Review of Ecology and Systematics 15:353-391.

Southwood, T. R. E. 1977. Habitat, the template for ecological strategies? Journal of Animal Ecology 46:337–365.

———. 1988. Tactics, strategies and templets. Oikos 52:3–18.

Sparks, R. E., P. B. Bayley, S. L. Kohler, and L. L. Osborne. 1990. Disturbance and recovery of large floodplain rivers. Environmental Management 14:699–709.

Vannote, R. L., G. W. Minshall, K. W. Cummins, J. R. Sedell, and C. E. Cushing. 1980. The river continuum concept. Canadian Journal of Fisheries and Aquatic Sciences 37:130–137.

Voelz, N. J., and J. V. Ward. 1989. Biotic and abiotic gradients in a regulated high elevation Rocky Mountain river. Regulated Rivers 3:143–152.

Voelz, N. J., and J. V. Ward. 1990. Macroinvertebrate responses along a complex regulated stream environmental gradient. Regulated Rivers 5:365–374.

Voelz, N. J., and J. V. Ward. 1991. Biotic responses along the recovery gradient of a regulated stream. Canadian Journal of Fisheries and Aquatic Sciences 48:2477–2490.

Ward, J. V., and J. A. Stanford. 1983. The intermediate-disturbance hypothesis: an explanation for biotic diversity patterns in lotic ecosystems. Pages 347–356 in T. D. Fontaine, III, and S. M. Bartell, editors. Dynamics of lotic ecosystems. Ann Arbor Press, Ann Arbor, Michigan, USA.

White, P. S., and S. T. A. Pickett. 1985. Natural disturbance and patch dynamics: an introduction. Pages 3–9 in S. T. A. Pickett and P. S. White, editors. The ecology of natural disturbance and patch dynamics. Academic Press, New York, New York, USA.

Yount, J. D., and G. J. Niemi. 1990. Recovery of lotic communities and ecosystems from disturbance—a narrative review of case studies. Environmental Management 14:547–569.

23
Plant Invasions and the Role of Riparian Habitats: A Comparison of Four Species Alien to Central Europe

Petr Pyšek and Karel Prach

Introduction

In the process of plant invasion rivers may act as dispersal agents, supporting downstream movement of diaspores by water (e.g. Staniforth & Cavers, 1976; van der Pijl, 1982; Schneider & Sharitz, 1988; Skoglund, 1989). Moreover, periodic disturbances resulting from destructive flooding generally destroy or damage a large part of riparian vegetation (Ellenberg, 1988) thus creating openings that provide suitable habitats and favourable nutrient conditions for seedling establishment and subsequent colonization (Walker, Zasada & Chapin, 1986). Consequently, riparian habitats into which an alien species was successfully introduced and naturalized may serve as foci for subsequent spread into the adjacent landscape (Pyšek, 1991, 1993).

The present paper is based on historical reconstructions of spreading dynamics of four species alien to the flora of the Czech Republic, central Europe. All these species are extremely successful invaders in the present central European landscape (Kees & Krumrey, 1983; Pyšek, 1991; Pyšek & Prach, 1993a; Sukopp & Schick, 1993) and are the highest-growing central European representatives of the following life forms: annual (*Impatiens glandulifera* Royle), monocarpic perennial (*Heracleum mantegaz-*

zianum Somm. et Lev.), and polycarpic perennial (*Reynoutria japonica* Houtt. and *R. sachalinensis* (F. Schmidt Petropolit.) Nakai). Basic life history characteristics of the species under study are summarized in Table 1.

Based on floristic records covering more than a hundred years, the present paper addresses the following questions. (1) What are the between-species differences in the rate and timing of invasion? (2) Did the relationship of species to riparian habitats change during their invasions? Moreover, this paper aims to demonstrate how historical floristic records may be exploited for illustrating ecological phenomena.

Methods

Both published and unpublished floristic data including herbarium collections (Charles University Prague, National Museum Prague, and regional museums) were used in the reconstruction of invasion dynamics. For *Heracleum mantegazzianum* and *I. glandulifera*, complete referenced lists of localities have been published (Pyšek & Pyšek, 1993; Pyšek & Prach, 1993b). Information on the year of observation and habitat type was summarized. If the year of observation was not provided by the original author, the year of publication was used. As

TABLE 1. Characteristics of species investigated.

	Impatiens glandulifera	Heracleum mantegazzianum	Reynoutria japonica	Reynoutria sachalinensis
Area of origin	Himalayas	Caucasus	Far East	Far East
Life form	Annual	Monocarpic perennial (up to 4 years)	Polycarpic perennial	Polycarpic perennial
Maximum height	2.5 m	4–5 m	>3 m	4 m
Dispersal*	Water, man/ explosive seed capsule	Man, wind, water, animals	Man, water/ vegetative	Man, water/ vegetative
Main way of regeneration	Only by seeds	Seeds, tuberous root	Rhizomes (in Europe)	Rhizomes (in Europe)

Sources: Grime, Hodgson & Hunt, 1988; Perrins, Fitter & Williamson, 1990; Pyšek, 1991, personal observations.
*Listed according to decreasing importance (based on authors' field experience).

shown for *H. mantegazzianum* in a previous paper (Pyšek, 1991), the year of publication closely corresponds to the year of observation, since 81% of records are published within 5 years from the observation (the proportion of published localities significantly decreased with the interval between the observation and its publication, $Y = 1.22X^{-0.072}, r = -0.90, P < 0.0001$).

A previous paper (Pyšek, 1991) demonstrated how floristic data, systematically gathered over an area for a long time, may be used to reconstruct the pattern of invasion of a species on a large geographical scale. There are, however, some limitations to floristic data which should be emphasized. A sufficient intensity of floristic research within an area is necessary for a successful retrospective analysis of species spread. This is clearly kept because of the strong, long-term floristic tradition in the Czech Republic. If systematic recording of the flora is carried out, one can assume that the more common a species is, the more often it is recorded. The species itself should be currently (1) worthy of note, i.e. rare enough or otherwise interesting from the point of view of ecology, spreading dynamics etc., (2) conspicuous in order not to be overlooked and (3) taxonomically unproblematic, i.e. easily recognizable by amateur botanists who are the main producers of floristic data. These points may be considered reasonably fulfilled by the species studied.

Because of the data character (point records and a priori defined geographical area) it was not possible to use the size of the area occupied by a species as a measure of invasion (Hengeveld, 1989). Exponential regression models were thus fitted to the cumulative numbers of localities plotted against time (further termed as invasion curves) and the slope b of the regression line was used as a measure of the invasion rate (Trewick & Wade, 1986; Pyšek 1991; Pyšek and Prach, 1993a). Differences in slopes were tested using F-test (Snedecor & Cochran, 1967).

The beginning of exponential phase was defined as a year in which the parameters of the invasion curve are changing. This was detected using the maximum likelihood estimation of classic regression model parameters for partitioned model (Quandt, 1958, 1960). The variance of white noise was assumed to be equal in both parts of the partitioned model.

Results

History of Invasions

According to our current knowledge, all the species investigated were introduced to the Czech Republic at approximately the same time (Table 2), i.e. in the second half of the 19th century, having originally been planted as garden ornamental plants (see Pyšek, 1991 and Pyšek and Prach, 1993a for details on the history of introduction in *H. mantegazzianum* and *I. glandulifera*, respectively).

Considering the whole invasion process (i.e. since the time of introduction), the spreading rate was highest in *I. glandulifera*, followed by *R. japonica*. The spread of *H. mantegazzianum*

TABLE 2. Characteristics of the invasion process of the species studied.

	First appearance in Czech Republic		Duration of L	Beginning of E		Present number of localities	
	T	R		Loc. no.	Year	T	R
Impatiens glandulifera	1896	1900	40	12	1936	742	396
Heracleum mantegazzianum	1862	1900	80	4	1943	411	41
Reynoutria japonica	1892	1892	46	39	1938	515	128
R. sachalinensis	1869	1935	83	22	1952	152	36

See Methods for details of distinguishing the lag phase from the exponential one. T: in total; R: in riparian habitats; L: lag phase; E: exponential phase; Loc. no.: number of localities.

and *R. sachalinensis* was slower and the slopes *b* were not significantly different (Fig. 1, Table 3).

There was a lag phase of more than 80 years in *H. mantegazzianum* and *R. sachalinensis*. In two other species, the lag phase took about half of this period. The number of localities at the end of the lag phase was very low in *I. glandu-lifera* and *H. mantegazzianum*, indicating that the exponential phase began immediately after the species has established several foci in the area. On the other hand, both *Reynoutria* species did not begin to spread exponentially until having more than twenty localities in the region. The time at which the exponential phase of invasion began was similar in all species

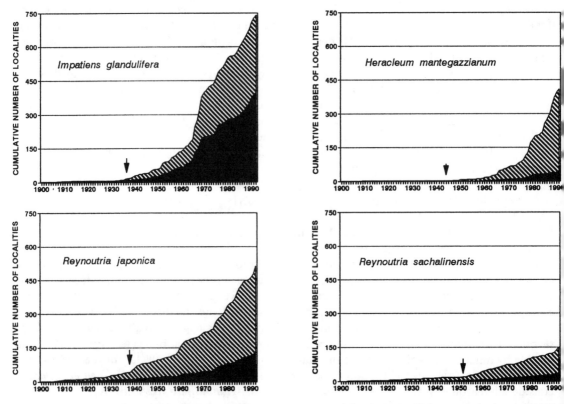

FIGURE 1. Invasion curves of the species studied in the Czech Republic. Only the 20th century is shown. Total cumulative number of localities is given on the *y*-axis and the part formed by riparian habitats is indicated by the solid area. The beginning of the exponential phase (see Methods for details on estimation) is indicated by an arrow. See Table 2 for summary of the number of localities. ■, Riparian; ◩, others.

TABLE 3. Summary of invasion rates of the species studied.

	Whole period of invasion		Exponential phase		
	T	R	T	R	O
Impatiens glandulifera	0.070 ± 0.0024a	0.064 ± 0.0027a	0.070 ± 0.0037a	0.074 ± 0.0040a	0.066 ± 0.0037a
Heracleum mantegazzianum	0.045 ± 0.0035b*	0.042 ± 0.0056bc	0.107 ± 0.0032b	0.096 ± 0.0062b	0.110 ± 0.0038b
Reynoutria japonica	0.059 ± 0.0026c*	0.050 ± 0.0015c	0.040 ± 0.0016c	0.043 ± 0.0013c	0.040 ± 0.0017c
R. sachalinensis	0.047 ± 0.0014b	0.058 ± 0.0028ab	0.043 ± 0.0034c	0.053 ± 0.0033d	0.043 ± 0.0024c

Invasion rate was expressed as a value of slope b from the regression equation CUMULATIVE NUMBER OF LOCALITIES $= \exp(a + b \times \text{YEAR})$. Slopes with their confidence intervals are given separately for all habitats (T), riparian (R) and other than riparian (O). Slopes that were not significantly different in F-test (according to Snedecor & Cochran, 1967) are bearing the same letter columnwise. See Methods for estimation of the beginning of the exponential phase.

*$P < 0.025$, otherwise $P < 0.001$.

(1936 in *I. glandulifera*, 1938 in *R. japonica*, 1943 in *H. mantegazzianum*) except *R. sachalinensis* in which it was delayed until 1952 (Table 2). *H. mantegazzianum* showed the highest rate of invasion during the exponential phase, followed by *I. glandulifera*. Both *Reynoutria* species were spreading at slower rates, the difference between both being not significant (Table 3).

Total numbers of localities reported up to 1992 are given in Table 2 for each species.

Role of Riparian Habitats

Treating riparian habitats separately, the highest rate of spread was found for *I. glandulifera*, followed by both *Reynoutria* species and *H. mantegazzianum* (Table 3). During the exponential phase, *I. glandulifera* ($F_{1,112} = 5.59$, $P < 0.05$) and *R. sachalinensis* ($F_{1,80} = 5.54$, $P < 0.05$) were spreading faster in riparian habitats than in the others. Conversely, the spread of *H. mantegazzianum* in other habitats was faster than in riparian ones ($F_{1,96} = 11.71$, $P < 0.01$) and no differences were found in *R. japonica* ($F_{1,108} = 3.70$, NS).

Assuming that different habitat types may play different parts in various periods of the invasion process, the lag and exponential phases of spread were analyzed separately for assessing habitat preferences in the species studied (Fig. 2). During the lag phase, the species differed significantly with respect to their habitat preferences (chi-square, $P < 0.0001$, Table 4). The closest relationship to

riparian habitats was found for *I. glandulifera*; this species was also successful in settlement sites. Similarly, *R. japonica* also preferred these two habitat types. Both species appeared in riparian sites immediately following their introduction to the Czech Republic (Table 2). In *H. mantegazzianum*, the number of localities reported for the lag phase (Table 2) was too low to draw conclusions; however, of the first four records, two were from parks. *R. sachalinensis* established successfully in semi-natural vegetation (forests, meadows) and parks in which it has originally been planted. There was a long period between the first recorded appearance of *R. sachalinensis* in the Czech Republic and its first appearance in riparian habitats (Table 2).

The between-species differences in habitat preferences during the exponential phase of spread were also significant (chi-square, $P < 0.0001$, Table 4). The habitat preferences were similar in both *Reynoutria* species (Fig. 2). In this period, the proportion of riparian localities (Fig. 3) increased significantly in *I. glandulifera*, *R. sachalinensis* and *R. japonica* and decreased in *H. mantegazzianum* (Table 5).

When the habitat preferences between lag and exponential phases were compared for particular species, significant differences (chi-square, $P < 0.01$) were found in all but *R. japonica* (Table 4).

Discussion

Although our results show that spreading along water courses played an important role in all

LAG PHASE

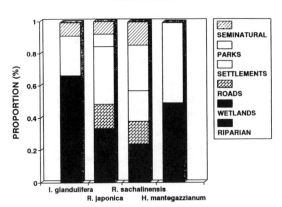

TABLE 4. Differences in habitat preferences between species and for each species between both phases of the invasion process (see Methods for distinguishing lag and exponential phases).

	Chi-square value	d.f.	P
Lag phase	205.78	15	<0.0001
Exponential phase	68.75	15	<0.0001
Lag v. exponential phase			
Impatiens glandulifera	17.80	5	0.0031
Heracleum mantegazzianum	151.82	5	<0.0001
Reynoutria japonica	3.03	5	0.6950
R. sachalinensis	25.48	5	<0.0001

Chi-square values are given.

EXPONENTIAL PHASE

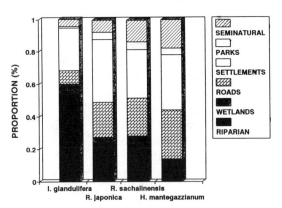

FIGURE 2. Distribution of localities with respect to habitat types. Localities recorded during the lag phase are compared with the distribution of those recorded during the exponential phase (see Methods for details on estimation). Semi-natural habitats (excluding riparian and wetlands) include meadows, forest margins, and forests. Roads and railways were grouped together.

the species investigated, at least in some period of their invasion, the behaviour of particular species with respect to riparian habitats is far from uniform (Table 5). *I. glandulifera* is an annual species producing large numbers of seeds which are effectively dispersed by stream and thus the water is, apart from man, the main dispersal factor. The seeds float for only a short period; however, they are transported along the bottom by stream and washed out and deposited on river banks during heavy floods (Lhotská & Kopecký, 1966). Moreover *I. glan-*

dulifera is, among those species studied, the one confined ecologically most closely to wet habitats (Trewick & Wade; 1986, Grime, Hodgson & Hunt, 1988; Perrins, Fitter & Williamson, 1990). Consequently, it is less frequently leaving riverbanks, brook shores or ponds and its spread into the landscape from these habitats was observed to be somewhat rare and/or temporary in the Czech Republic (Table 5). Generally, *I. glandulifera* is an example of the invasive species whose invasion abilities are, however, strongly limited to a certain habitat type (Pyšek and Prach, 1993a).

Water flow is also an important dispersal agent for vegetative fragments of both *Reynoutria* species (Conolly, 1977; Grime *et al.*, 1988;

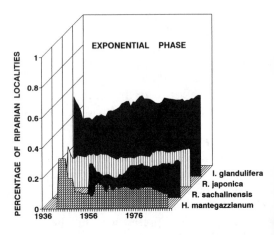

FIGURE 3. Changes in the proportion of riparian localities during the exponential phase of spread (see Methods for details on estimation). See Table 5 for testing the trends over time.

TABLE 5. Behaviour of the invasive species studied assessed with respect to riparian habitats.

	Affinity to riparian habitats during the phase		Expansion out of riparian habitats	%
	Lag	Exponential		
I. glandulifera	High	High, increasing (0.68***)	Occasional	0.47
H. mantegazzianum	Moderate	Low, decreasing (−0.30***)	Massive	0.90
R. japonica	Moderate	Moderate, increasing (0.53*)	Frequent	0.75
R. sachalinensis	Low	Moderate, increasing (0.65***)	Frequent	0.71

Changes in the percentage contribution of riparian localities to the total number of localities during the exponential phase were tested using Kendall correlation coefficient (values and significance levels are given in the parentheses. *$P < 0.05$, ***$P < 0.001$). Expansion from riparian habitats was assessed on the basis of the percentage of non-riparian localities reported up to the present, which is given in the last column.

Wade, 1992) but the spread by man (at present mostly unintentional, see Sukopp & Sukopp, 1988) still prevails in the studied region. Both *Reynoutria* species probably only rarely set viable seeds (Jehlík & Lhotská, 1970; Bailey, 1988) and therefore spreading by rhizomes plays a decisive part in species invasion. In the Czech Republic, floodplains areas are often used for allotments and both *Reynoutria* species are still being used as garden ornamental plants. Consequently, their rhizomes are often deposited with other garden waste close to rivers. Various building activities associated with the transport of soil containing rhizomes (Grime *et al.*, 1988) serve as another important vector transporting vegetative fragments in the study region (personal observation). Whereas *R. japonica* appears to be an extremely successful invader, the invasion of *R. sachalinensis* might recently have begun (note the magnitude of occurrence compared to the other species, Fig. 1) and the species seems to be in the period of 'river affinity'.

The lower proportion of riparian localities in *H. mantegazzianum* is due to its ability to expand massively to other habitat types (Table 5), namely settlements, road verges and forests. This fact may be explained by its competitive success over native flora in the wide range of habitats (Pyšek, 1993). Furthermore, the factors responsible for its dispersal are more variable than in other species studied. As well as water and man, animals and wind play an important role in dispersion (Jehlík & Lhotská, 1970; Williamson & Forbes, 1982; Lundström, 1984; Wyse Jackson, 1989; Pyšek 1991). However, even this species was found to spread first down the courses of great rivers during its invasion (Neiland *et al.*, 1972; Pyšek, 1991) and there was a period during which the riparian habitats were supporting its spread considerably (Pyšek, 1993). In general, however, the relatively low preferences of *H. mantegazzianum* for riparian habitats and, in spite of that, its high rate of spread indicates that other suitable 'transport habitats' (roads, railways, see Fig. 2) may serve as alternative routes for species spread.

In all species spreading in riparian habitats, the downstream movement of diaspores is assumed as demonstrated previously for *I. glandulifera* (Pyšek & Prach, 1993a) and *H. mantegazzianum* (Pyšek, 1993). However, the upstream movement cannot be excluded because of other dispersal vectors acting in river corridors not as unidirectionally as the water flow (Schneider & Sharitz, 1988).

Association with riparian habitats during the process of plant invasion seems also to be related to the character of the landscape in a given region. For all the species covered by our study, closer relationships to riparian habitats have been reported for the United Kingdom than is the case in the Czech Republic (e.g. Wyse Jackson, 1989; Wade, 1992). This may be explained by the fact that in the Czech landscape there are a number of various abandoned, disturbed and ill-managed sites outside rivers. These sites can be more easily colonized by invasive aliens. In the United Kingdom such suitable sites are more confined to river corridors, as the rest of the country is intensively managed.

Kornas (1990) reported thirty-three alien species established in Poland in semi-natural vegetation, of which seventeen are confined to disturbed riverside habitats. This author distinguished four stages of the naturalization process and suggested that the invasion of an alien into less disturbed sites should have been preceded by its establishment in heavily disturbed, so-called ruderal sites. Considering our results, this statement seems to be overgeneralized; at least *H. mantegazzianum* and *R. sachalinensis* were invading, although at a lower frequency than in man-made sites, into semi-natural forests and meadows immediately after introduction. Both these species were planted in parks and competed successfully with grass sward and herbs present in these sites as well as in adjacent semi-natural vegetation. On the other hand, *I. glandulifera* was planted as an ornamental plant in flower beds, i.e. kept from competition from spontaneous vegetation. This is reflected in the relatively high proportion of localities in settlement habitats where this species was present during the early period of invasion (40.3% in 1945) due to frequent escapes from cultivation; many of the earliest records refer to its occurrence as a garden weed. Correspondingly, during its invasion *I. glandulifera* was more confined to disturbed habitats, including river banks, and this is still true.

Considering the relationship between the life form of an invader and its invasion success, the present study indicates that the shorter the life span, the higher the rate of invasion. Both non-perennials included in the study, *I. glandulifera* and *H. mantegazzianum*, began to spread exponentially after having reached only a few localities in the area (although in the case of the latter species it was preceded by the long-lasting lag phase) subsequently their invasion continued at a higher rate than in the case of both perennials. Taking into account the characteristics generally suggested to encourage any species invasion (Baker, 1965; Newsome & Noble, 1986; Roy, 1990; Prach & Wade, 1992), some kind of trade-off between these traits may be suggested. Annual species profit from a large seed set and a rapid life cycle which is advantageous from the viewpoint of adaptations to changing environments (Grime, 1979 and others). On the other hand, long-lived species, especially the clonal ones as is the case with both *Reynoutria* species, profit from their ability to keep space once occupied, and spread

consistently within the close vicinity (Callaghan et al., 1992; Prach & Pyšek, 1993). The clonal life form is associated with high regeneration ability which may be considered another factor supporting species spread and dominance in a site. Regardless of the life form, the invasion success appears to be associated with high competitive ability (Noble, 1989; Roy, 1990) which is suggested to increase with plant height (Grime, 1979; Keddy, 1990). However, further research covering larger sets of species is required to test the relationship between a species' morphological and life-history traits and its invasion success. In addition, various specific species traits may be involved in determining the species' ability to spread. Different magnitudes of invasion found in *Reynoutria* species, whose life form, growth ability mode of dispersal and habitat preferences are very similar, may be explained by different density of rhizome systems in the soil (Sukopp & Schick, 1993). Another explanation for such difference may be that in the past, *R. japonica* was planted in Czech territory with a higher frequency than *R. sachalinensis* (V. Jehlík, personal communication).

Three methodological approaches may be distinguished in studies on plant invasions: (a) studies at the population level focused upon plant attributes determining invasion success, including manipulative experiments (Noble, 1989; Roy, 1990); (b) descriptive small-scale systematic studies of the invasion process (Williams, Hobbs & Hamburg, 1987; Thébaud & Debussche, 1991); and (c) landscape-viewed studies of invasions at large-scale level, up to continents, based upon decades-lasting continuous records or upon historical reconstruction from floristic or palaeological records (Woods & Davis, 1989; Trepl, 1984; Kornas, 1990; Pyšek, 1991); the former are unfortunately less frequent so far (Perring & Walters, 1962; Clegg & Grace, 1974; Williamson & Forbes, 1992; Haeupler & Schönfelder, 1988; Schönfelder, & Bresinsky, 1992; Hartl et al., 1992). Unfortunately, the studies at the landscape scale providing an insight into the history of species invasion on the time scale of centuries must be necessarily descriptive; an experimental approach is simply not applicable to such studies. When interpreting the results based on data sets such as those in the present study, one must be aware of the fact that some of them have only exploratory value and, consequently, some of the conclusions are necessarily speculative. However, the approach applied in the present paper may be considered useful for stating hypotheses on the species' ecological behaviour to be tested experimentally.

Conclusions

The following conclusions may be drawn to answer the questions put in the introduction.

(1) The rate of invasion differed in the species studied; it was highest in *I. glandulifera*, followed by *R. japonica*, *H. mantegazzianum*, and *R. sachalinensis*. The exponential phase began between the 1930s and 1950s in all species studied having been preceded by a lag phase of different duration. *H. mantegazzianum* and *I. glandulifera* began to increase exponentially after having reached only a small number of localities; subsequently their spread was faster than that of both *Reynoutria* species, which proceeded at a more even rate.

(2) Habitat preferences differed between the species investigated in both phases of the invasion process. The spread in riparian habitats was fastest in *I. glandulifera*, followed by *R. sachalinensis*, *R. japonica* and *H. mantegazzianum*. Riparian habitats played an important role in the invasion process of all species studied, at least in a certain period. The species dispersibility and frequency of habitats suitable for species establishment must be taken into account when evaluating such a role. Frequency of habitats available is closely related to land-use practices.

Acknowledgments

We thank two anonymous reviewers and Javier Puntieri for their comments on the previous version of the manuscript and for improving our English. Pavla Kotková and Tomáš Herben and are thanked for their comments on statistical treatment of data. Our thanks are also due to the following colleagues for providing their unpublished floristic data: J. Rydlo, Bohumil Slavík, J. Soják, V. Chán, K. Kubát, J. Hadinec, J. Sladký, R. Hlaváček, S. Kučera, F. Krahulec, J. Kolbek, N. Gutserová, M. Sandová and O. Roubínková. We thank L. Kučeravá for technical assistance.

References

Bailey, J.P. (1988) Putative *Reynoutria japonica* Houtt. × *Fallopia baldschuanica* (Regl) Holub hybrids discovered in Britain. *Watsonia*, **17**, 163–164.

Baker, H.G. (1965) Characteristics and modes of origin of weeds. *The genetics of colonizing species* (ed. by H. G. Baker and C. L. Stebbins), pp. 147–169, Academic Press, New York.

Callaghan, T.V., Carlsson, B.A., Jónsdóttir, I.S. Svenson, B.M. & Jonasson, S. (1992) Clonal plants and environmental change: introduction to the proceedings and summary. *Oikos*, **63**, 341–347.

Clegg, L.M. & Grace, J. (1974) The distribution of *Heracleum mantegazzianum* Somm. & Levier near Edinburgh. *Trans. Bot. Soc. Edinb.* **42**, 223–229.

Conolly, A.P. (1977) The distribution and history in the British isles of some alien species of *Polygonum* and *Reynoutria*. *Watsonia*, **11**, 291–311.

Ellenberg, H. (1988) *Vegetation ecology of central Europe*. Cambridge University Press, Cambridge.

Grime, J.P. (1979) *Plant strategies and vegetation processes*. J. Wiley, Chichester.

Grime, J.P., Hodgson, J., & Hunt, R. (1988) *Comparative plant ecology. A functional approach to common British species*. Unwin and Hyman, London.

Haeupler, H. & Schönfelder, P. (1988). *Farn- und Blütenpflanzen des Bundesrepublik Deutschland*. Eugen Ulmer, Stuttgart.

Hartl, H., Kniely, G., Leute, G.H., Niklfeld, H. & Perko, M. (1992) *Verbreitungsatlas der Farn- und Blütenpflanzen Kärntens*. Naturwiss. Ver. f. Kärnten, Klagenfurt.

Hengeveld, R. (1989) *Dynamics of biological invasions*. Chapman and Hall, London.

Jehlík, V. & Lhotská, M. (1970) *Contribution to the distribution and fruit biology of some synanthropic plants from the Pruho nice park, village of Pruhonica and Botic brook valley*. Studie CSAV, Praha 1970/7, 45–95 (in Czech).

Keddy, P.A. (1990) Competitive hierarchies and centrifugal organization in plant communities. *Perspectives on plant competition* (ed. by J. B. Grace and D. Tilman), pp. 266–290, Academic Press, San Diego.

Kees, H. & Krumrey, G. (1983) *Heracleum mantegazzianum*—Zierstaude, Unkraut und "Gifpflanze". *Gesunde Pflanzen*, **35**, 108–110.

Kornaś, J. (1990) Plant invasions in Central Europe: historical and ecological aspects. *Biological invasions in Europe and the Mediterranean Basin* (ed. by F. Di Castri F., A. J. Hansen and M. Debussche), pp. 19–36. Kluwer, Dordrecht.

Lhotská, M. & Kopecký, K. (1966) Zur Verbreitungsbiologie und Phytozönologie von *Impatiens glandulifera* Royle an den Flusssystemen der Svitava, Svratka und oberen Odra. *Preslia, Praha,* **38,** 376–385.

Lundström, H. (1984) Giant hogweed, *Heracleum mantegazzianum,* a threat to the Swedish countryside. *Weeds and weed control,* Vol. 1, pp 191–200. 25th Swedish Weed Conference, Uppsala.

Neiland, R. *et al.* (1987) Giant hogweed (*Heracleum mantegazzianum* Somm. & Lev.) by the river Allan and part of the river Forth. *Forth Nat. Hist.* **9,** 51–56.

Newsome, A.E. & Noble, I.R. (1986) Ecological and physiological characters of invading species. *Ecology of biological invasions: an Australian perspective* (ed. by R. H. Groves and J. J. Burden), pp. 1–20. Australian Academy of Science, Canberra.

Noble, I.R. (1989) Attributes of invaders and the invading process: terrestrial and vascular plants. *Biological invasions in Europe and the Mediterranean Basin* (ed. by F. Di Castri F., A. J. Hansen and M. Debussche) Kluwer, Dordrecht.

Perring, F.H. & Walters, S.M. (eds) (1962) *Atlas of the British flora.* Botanical Society of the British Isles, Nelson, London.

Perrins, J., Fitter, A. & Williamson, M. (1990) What makes *Impatiens glandulifera* invasive? *The biology and control of invasive plants* (ed. by J. Palmer), pp. 8–33. University of Wales, Cardiff.

Prach, K. & Pyšek, P. (1993b) Clonal species: what is their success in succession? *Folia Geobot. Phytotax. Praha,* **28** (in press).

Prach, K. & Wade, P.M. (1992) Population characteristics of expansive perennial herbs. *Preslia,* **64,** 45–51.

Pyšek, P. (1991) *Heracleum mantegazzianum* in the Czech Republic: the dynamics of spreading from the historical perspective. *Folia Geobot. Phytotax. Praha,* **26,** 439–454.

Pyšek, P. (1993) Ecological aspects of invasion of *Heracleum mantegazzianum* in the Czech Republic. *Invasive riparian weeds* (ed. by P.M. Wade, J. Brock and L. de Waal). J. Wiley, Chichester.

Pyšek, P. & Prach, K. (1993a) Invasion dynamics of *Impatiens glandulifera* in the Czech Republic—a century of spreading reconstructed on the regional scale. *J. Ecol* (subm.).

Pyšek, P. & Prach, K. (1993b) Survey of locations of *Impatiens glandulifera* in the Czech Republic and remarks on the dynamics of its invasion. *Zpr. Cs. Bot. Společ. Praha,* **28** (in Czech).

Pyšek, P. & Pyšek, A. (1993) Current occurrence of *Heracleum mantegazzianum* and survey of its localities in the Czech Republic. *Zpr. Cs. Bot. Společ. Praha.* **28** (in Czech).

Quandt, R.E. (1958) The estimation of the parameters of linear regression system obeying two separate regimes. *JASA,* **53,** 873–880.

Quandt, R.E. (1960) Tests of hypothesis that linear regression system obeys two separate regimes. *JASA,* **55,** 324–330.

Roy, J. (1990) In search of the characteristics of plant invaders. *Biological invasions in Europe and the Mediterranean Basin* (ed. by F. Di Castri F., A.J. Hansen and M. Debussche). Kluwer, Dordrecht.

Schneider, R.L. & Sharitz, R.R. (1988) Hydrochory and regeneration in a bald cypress-water tupelo swamp forest. *Ecology,* **69,** 1055–1063.

Schönfelder, P. & Bresinsky, A. (1990) *Verbreitungsatlas der Farn- und Blutenpflanzen Bayerns.* Eugen Ulmer, Stuttgart.

Skoglund, S.J. (1989) Seed dispersing agents in two regularly flooded river sites. *Can. J. Bot.* **68,** 754–760.

Snedecor, G.W. & Cochran, W.G. (1967) *Statistical methods.* Iowa University Press, Iowa.

Staniforth, R.J. & Cavers, P.B. (1976) An experimental study of water dispersal in *Polygonum* spp. *Can. J. Bot.* **54,** 2587–2596.

Sukopp, H. & Schick, B. (1993) Zur Biologie neophytischer *Reynoutria*-Arten in Mitteleuropa. II. Morphometrie der Sprosssysteme. Festschrift Zoller. *Diss. Bot.* **196,** 163–174.

Sukopp, H. & Sukopp, U. (1988) *Reynoutria japonica* Houtt. in Japan and in Europe. *Veröff. Geobot. Inst. ETH Stift. Rübel Zürich.* **98,** 354–372.

Thébaud, C. & Debussche, M. (1991) Rapid invasion of *Fraxinus ornus* L. along the Hérault River system in southern France: the importance of seed dispersal by water. *J. Biogeogr.* **18,** 7–12.

Trepl, L. (1984) Über *Impatiens parviflora* DC. als Agriophyt in Mitteleuropa. *Diss. Bot.* **73,** 1–400.

Trewick, S. & Wade, P.M. (1986) The distribution and dispersal of two alien species of *Impatiens,* waterway weeds in the British Isles. *Proceedings EWRS/AAB Symposium on Aquatic Weeds,* 351–356.

van der Pijl, L. (1982) *Principles of dispersal in higher plants.* Springer, New York.

Wade, P.M. (ed.) (1992) *The biology, ecology and management of the invasive riparian and aquatic plants Giant Hogweed, Himalayan Balsam, Japa-*

nese Knotweed and Swamp Stonecrop—a literature review. NRA Report, International Centre for Landscape Ecology, Loughborough.

Walker, L.R., Zasada, J.C. & Chapin, F.S. III. (1986) The role of life history processes in primary succession on an Alaskan floodplain. *Ecology,* **67,** 753–761.

Williams, K., Hobbs, R.J. & Hamburg, S.P. (1987) Invasion of an annual grassland in Northern California by *Baccharis pilularis* ssp. *consanguinea. Oecologia,* **72,** 464–465.

Williamson, J.A. & Forbes, J.C. (1982) Giant Hog-weed (*Heracleum mantegazzianum*): its spread and control with glyphosate in amenity areas. *Weeds. Proc. 1982 British Crop Protection Conference,* pp. 967–971.

Woods, K.D. & Davis, M.B. (1989) Palaeoecology of range limits: beech in the upper peninsula of Michigan. *Ecology,* **70,** 681–696.

Wyse Jackson, M. (1989) Observations on the Irish distribution of a plant with serious public health implications: Giant hogweed (*Heracleum mantegazzianum* Sommier & Levier). *Bull. Irish Biogeogr. Soc.* **12,** 94–112.

24
Biological Integrity Versus Biological Diversity as Policy Directives: Protecting Biotic Resources

Paul L. Angermeier and James R. Karr

Two phrases—*biological integrity* and *biological diversity*—have joined the lexicon of biologists and natural resource managers during the past two decades. The importance of these phrases is demonstrated by their influence on environmental research, regulatory, and policy agendas. The concepts behind the phrases are central to strategies being developed to sustain global resources (Lubchenco et al. 1991). Unfortunately, the phrases are widely used by the media, citizens, policy makers, and some biologists without adequate attention to the concepts they embody. Precise use of the terms *integrity* and *diversity* can help set and achieve societal goals for sustaining global resources; imprecise or inappropriate use may exacerbate biotic impoverishment—the systematic decline in biological resources (Woodwell 1990).

Although the related concepts of integrity and diversity were developed more or less independently (integrity in the study of aquatic systems, diversity in the study of terrestrial systems), both apply to all biotic systems. The US Clean Water Act and Canada's National Park Act enunciate the explicit goal of protecting biological integrity. No specific legislative mandate exists to protect biological diversity in the United States, but such protection became a central goal of the 1992 Earth Summit and the Global Biodiversity Protocol endorsed by many nations. The focal positions of the two concepts dictate that a clear understanding of their meanings is critical to developing effective resource policy.

Our review of current conceptions of integrity and diversity indicates that resource policy would be most effective if based on the more comprehensive goal of protecting biological integrity. Specific policy shifts related to that goal include a reliance on preventive rather than reactive management and a focus on landscapes rather than populations. We draw heavily from our experience with aquatic systems, but our conclusions apply equally to terrestrial systems.

Aquatic systems are appropriate models for illustrating general ecological consequences of anthropogenic impacts, because research is advanced on ecological impacts in these systems (e.g., Allan and Flecker 1993, Schindler 1990), and rates of extinction and endangerment for aquatic fauna exceed those for terrestrial fauna. Among North American animals, for example, Master (1990) reported that 20% of fishes, 36% of crayfishes, and 55% of mussels were extinct or imperiled, compared with 7% of mammals and birds. Similarly, only 4% of the federally protected aquatic species in the United States with recovery plans have recovered significantly, compared with 20% of protected terrestrial species (Williams and Neves 1992).

Defining Biological Diversity

The term *biological diversity* (or *biodiversity*) emerged as species extinction rates began to increase dramatically (Myers 1979). The specter of mass extinctions, combined with huge gaps in biological knowledge, has convinced many scientists that a global biological crisis exists (Wilson 1985). Moreover, because biological diversity provides important aesthetic, cultural, ecological, scientific, and utilitarian benefits to human society, the crisis is everyone's concern (Ehrlich and Wilson 1991).

One of the first formal definitions of biological diversity termed it "the variety and variability among living organisms and the ecological complexes in which they occur" (OTA 1987, p. 3). In addition, because "items are organized at many [biological] levels," biodiversity "encompasses different ecosystems, species, genes, and their relative abundance" (OTA 1987, p. 3). Other thorough discussions of biodiversity confirm that multiple organizational levels (e.g., genes, species, and ecosystems) are fundamental to the concept (Noss 1990, OTA 1987, Reid and Miller 1989), thereby distinguishing it from the much simpler concept of species diversity.

Hierarchies

Organizational hierarchies are useful tools for understanding complex biological phenomena. Several distinct hierarchies—taxonomic, genetic, and ecological (Table 1)—are relevant to biological diversity. We follow Reid and Miller (1989) in referring to biotic units at any level within a hierarchy as elements. Thus, species and classes are taxonomic elements, genes and chromosomes are genetic elements, and populations and biomes are ecological elements. Levels are nested within each hierarchy: a phylum comprises classes, a chromosome comprises genes, and a landscape comprises communities. The hierarchies in Table 1 are linked at the species-genome-population levels; any population of organisms has a taxonomic identity (species), which is characterized by a dis-

TABLE 1. Levels of organization in three hierarchies used to characterize biological diversity. These hierarchies are linked at the species-genome-population levels (see text for details) but not precisely at any other levels.

Taxonomic	Genetic	Ecological
Biota	Genome	Biosphere
Kingdom	Chromosome set	Biome
Division/Phylum	Chromosome	Landscape
Class	Gene	Ecosystem/Community
Order	Allele	Population
Family		
Genus		
Species		

tinct genome. However, taxa may share genetic elements, and ecological elements may share taxa.

Specifying levels within hierarchies and elements within levels may be arbitrary. For example, ecologists may add an ecological level for guilds, or taxonomists may debate the number of families within an order. Because each level and element contributes to biotic variety and value, all are appropriate targets of conservation. To focus assessment or conservation on a single hierarchy or level (e.g., species) is to arbitrarily ignore most biodiversity.

Spatiotemporal scale is not precisely defined by hierarchical level. Ecological elements are typically defined by spatial extent (e.g., a pond community or a desert landscape), yet most levels can correctly encompass a wide range of spatial scales (Allen and Hoekstra 1992). The dynamics of oak populations in a savannah landscape or of fungus populations in a stream-channel landscape, for example, may operate at vastly different spatial scales. The appropriateness of a spatiotemporal scale for studying a given element depends on the organisms and questions at issue (Levin 1992).

At ecological levels of organization above population (see Table 1), spatiotemporal bounds are often arbitrary, integration is often loose, and composition may be dynamic. However, these elements are not random assemblages, and they can be defined on the basis of ecological attributes and societal benefits. For example, the biota of the Chesapeake Bay basin is a legitimate element of biodiversity because it has objectively definable boundaries and confers societal benefits (e.g., fisheries) that would not exist if the component populations had not co-evolved.

We do not distinguish community and ecosystem as different hierarchical levels but rather as complementary ways of viewing the same system (Karr 1994, King 1993). Community perspectives are grounded in evolutionary biology and focus on the dynamics of organism distribution and abundance; ecosystem perspectives are grounded in thermodynamics and focus on the dynamics of energy and materials through and around organisms. Either perspective can be applied at any level in the ecological hierarchy.

Misconceptions

Because biological diversity is more comprehensive than species diversity, one must specify clearly the biological hierarchy and organizational level at issue in any discussion. In estimating biodiversity in a study area (e.g., a pond or continent), a researcher might count all the taxonomic elements present, all the genetic elements present, or all the ecological elements present. Even in the unlikely event that all the elements present are known, no accepted calculus permits integration of counts of elements across levels within a hierarchy (e.g., phyla and species) or across hierarchies (e.g., species and genes). Arguably, no such calculus should be sought.

Furthermore, the number of elements at different organizational levels need not be correlated. For example, there are more than twice as many marine phyla as terrestrial phyla, but fewer marine species (Ray and Grassle 1991). Similarly, Hoover and Parker (1991) found that species diversity and community diversity of overstory plants were inversely correlated among several Georgia landscapes. In neither example is it unequivocal which system has more biodiversity.

Failure to conceptually integrate the multiple aspects of biodiversity results in narrowly conceived comparisons. For example, Vane-Wright et al. (1991) measured biodiversity with an index of taxonomic diversity based on cladistics, which assesses distinctness of taxa. Similarly, Mares (1992) used a comparison of mammal diversity (at several taxonomic levels) among South American biomes to infer that biodiversity is greater in drylands than in lowland Amazon forest. These analyses are valuable, but they cannot be interpreted as comprehensive (or even representative) assessments of overall biodiversity because genetic and ecological hierarchies were ignored.

A common misuse of the term *biodiversity* makes it synonymous with *species diversity* (Redford and Sanderson 1992), a usage that trivializes the broader meaning of biodiversity and promotes misconceptions of conservation issues. Palmer (1992) takes this misconception to the extreme by depicting biodiversity loss as nothing more than species extinction. This incomplete view fails to recognize that elimination of extensive areas of old growth forest, dramatic declines in hundreds of genetically distinct salmonid stocks in the Pacific Northwest (Nehlsen et al. 1991), and the loss of chemically distinct populations from different portions of a species range (Eisner 1992) represent significant losses of biodiversity, regardless of whether any species become extinct. Other misuses of the term stem from inclusion of human-generated elements in assessments of an area's biodiversity (Angermeier 1994).

Defining Biological Integrity

Biological integrity refers to a system's wholeness, including presence of all appropriate elements and occurrence of all processes at appropriate rates. Whereas diversity is a collective property of system elements, integrity is a synthetic property of the system. Unlike diversity, which can be expressed simply as the number of kinds of items, integrity refers to conditions under little or no influence from human actions; a biota with high integrity reflects natural evolutionary and biogeographic processes.

The concept of biological integrity has played its largest policy role in the management of water resources where it first appeared in the 1972 reauthorization of the Water Pollution Control Act (now Clean Water Act; CWA).

The primary charge of the 1972 CWA and subsequent amendments was to "restore and maintain the chemical, physical, and biological integrity of the Nation's waters." This mandate has been the foundation for state and federal water-quality programs over the past two decades. Although implementation often has been ill-focused (Karr 1991), the concept of integrity is the primary directive for water policy in the United States.

The most influential definition of biological integrity was proposed by Frey (1975) and later applied by Karr and Dudley (1981). It defined the concept as "the capability of supporting and maintaining a balanced, integrated, adaptive community of organisms having a species composition, diversity, and functional organization comparable to that of natural habitat of the region" (Karr and Dudley, p. 56). Various forms of this definition now provide the basis for biotic assessment of surface waters by the US Environmental Protection Agency (EPA 1990) and numerous states (EPA 1991a).

Two important distinctions between integrity and diversity emerge from this definition. First, system integrity is reflected in both the biotic elements and the processes that generate and maintain those elements, whereas diversity describes only the elements. Again, following Reid and Miller (1989), we use processes to refer to a broad range of evolutionary, genetic, and ecological processes (Table 2). Integrity depends on processes occurring over many spatiotemporal scales, including cellular processes giving rise to genetic elements and ecosystem processes regulating the flow of energy and materials.

Although some authors (e.g., Noss 1990) explicitly include processes as components of diversity, we contend that processes are more appropriately considered as components of integrity. Process diversity is unlikely to provide an intuitive basis for distinguishing the biodiversity of different areas because areas vary in process rates rather than process occurrence. All areas support the processes of meiosis, speciation, disturbance, and predation, but rates vary dramatically. Moreover, changes in process rates cannot be interpreted as changes in diversity unless the number of participating elements also changes. For example, an increased rate of landscape disturbance need not produce more or fewer component communities and populations. Although processes clearly are essential to generate and maintain elements, their inclusion as components of biodiversity adds ambiguity without utility.

The second distinction between integrity and diversity is that only integrity is directly associated with evolutionary context. By definition, naturally evolved assemblages possess integrity but random assemblages do not. Adding exotic

TABLE 2. Elements, processes, and potential indicators of biological integrity for five levels of organization within three biological hierarchies. Assessing the integrity of a given area should incorporate indicators from multiple levels.

Hierarchy	Elements	Processes	Indicators
Taxonomic	Species	Range expansion or contraction	Range size
		Extinction	Number of populations
		Evolution	Isolating mechanisms
Genetic	Gene	Mutation	Number of alleles
		Recombination	Degree of linkage
		Selection	Inbreeding or outbreeding depression
Ecological	Population	Abundance fluctuation	Age or size structure
		Colonization or extinction	Dispersal behavior
		Evolution	Gene flow
	Assemblage	Competitive exclusion	Number of species
		Predation or parasitism	Species evenness
		Energy flow	Number of trophic links
		Nutrient cycling	Element redundancy
	Landscape	Disturbance	Fragmentation
		Succession	Number of communities
		Soil formation	Persistence

species or genes from distant populations may increase local diversity but it reduces integrity.

Most uses of the integrity concept focus on the community level of organization, but we suggest that integrity also applies to most other hierarchical levels in Table 1. Integrity of any biotic system can be assessed on the basis of attributes of elements and processes important to its genetic or ecological organization (see Table 2). However, the concept may not apply to taxa above the species level, because most taxonomic levels are artifacts of classification rather than functional biotic entities.

Because systems are hierarchical, an element is generated and maintained (in part) by processes occurring at organizational levels above and below its own level (O'Neill et al. 1989). For example, the integrity of a woodland community may depend on colonization dynamics of component populations as well as landscape-level disturbance dynamics. Thus, assessment of biological integrity should account for the influence of processes at multiple organizational levels and multiple spatiotemporal scales.

Selecting Benchmarks

The ability to recognize objectively and assess changes in integrity is critical for the concept's use in policy. The first hurdle in recognizing change in integrity is the selection of a benchmark state against which other states can be compared. Ecologists recognize that biological systems are not strictly deterministic but may develop (i.e., be organized) along multiple pathways as a result of different initial conditions, conditions in neighboring systems, and the sequence of influential events (Pickett et al. 1992). For example, marine intertidal communities are influenced by predation, disturbance, competition, physiological tolerances, and colonization from offshore. The relative importance of each process and the relative abundances of species at a given site depend on coastal circulation (Roughgarden et al. 1988). Variation in elements attributable to natural processes does not represent a variation in integrity, but variation caused by humans does.

Regier (1993) contends that states other than those evolved naturally can provide bench-

marks for integrity. Although unnatural states may be desirable for aesthetic, utilitarian, or other reasons, they cannot provide an objective basis for assessing biological integrity. Human-induced changes in biotic systems frequently are more rapid and severe than those occurring naturally. Thus, functional and evolutionary limits of the native biota provide objective bases for selecting appropriate integrity benchmarks (Pickett et al. 1992). For example, when forest harvest rates exceed regeneration rates, integrity is reduced, resulting in loss of late-successional communities. When a river is dammed, integrity is reduced, resulting in declines of populations adapted to the natural hydrological regime.

Evolutionary history should provide the primary basis for assessing biological integrity. Even the value of many artificial, human-generated elements (e.g., agricultural landscapes) depends on naturally evolved elements and processes, such as nitrogen-fixing bacteria and soil formation. Sadly, because of the pervasive effects of human actions, it is often difficult to characterize naturally evolved conditions. Because abilities to reconstruct historic scenarios of biotic conditions are likely to become even more impaired in the future, such efforts should proceed with the best information currently available.

Primacy of Integrity

Use of integrity as the primary management goal avoids the pitfalls of assuming that greater diversity or productivity is preferred. Knowledge of the couplings between biotic elements and processes is based largely on observations of stressed ecosystems. Experimental studies of whole lakes exposed to nutrient enrichment and acidification indicate that species composition responds more quickly and recovers more slowly than processes such as primary production, respiration, and nutrient cycling (Schindler 1990).

In a review of aquatic ecosystem and mesocosm responses to stress, Howarth (1991) found numerous examples of shifts in biotic elements that were unaccompanied by changes in process rates, but process changes were always accompanied by shifts in elements. These patterns are consistent with observations and predictions from forest ecosystems (Odum 1985) and support the hypothesis that ecological processes are buffered from perturbation by redundancy among elements (Bormann 1985). For example, multiple interchangeable elements (e.g., species) may drive a single process (e.g., nutrient cycle). Of course, given enough stress or element loss, any process can be impaired. As stress on system organization accumulates, nonlinear and threshold responses may result (see cases in Woodwell 1990).

Many changes in diversity can be evaluated objectively only on the basis of changes in integrity. For example, artificial nutrient enrichment of a naturally oligotrophic ecosystem may increase local species diversity yet eliminate a unique community. Such a change may be interpreted as either a gain or loss in diversity, but integrity is clearly reduced because of the shift away from native conditions. Human impacts in the Apalachicola River basin of the southeastern United States reduced freshwater flow into the estuary, resulting in elevated salinity and fish species diversity but loss of productivity and nursery function (Livingston 1991). Management for biological integrity would dictate maintenance of lower species diversity and higher productivity by restoring the original salinity dynamics.

Integrity goals also allow for natural fluctuation in element composition. Loss of a particular element (e.g., species) or replacement by a regionally appropriate one need not indicate a loss of integrity unless the processes associated with the element's maintenance become impaired. For example, natural metapopulation dynamics often include local, temporary extinctions balanced by recolonizations via dispersal (Hanski and Gilpin 1991). Such losses of populations do not indicate losses of integrity unless rates of extinction, dispersal, or recolonization are altered, as might occur in an artificially fragmented landscape.

The inadequacy of diversity as a policy directive is perhaps clearest in the evaluation of situations where humans add elements such as transferred genes, exotic species, or agricultural landscapes to natural systems (artificial biological diversity; Angermeier 1994). Artificial elements reduce integrity through widely documented effects on native elements and processes (Karr et al. 1986, Taylor et al. 1984, Vitousek 1990) and should be excluded from evaluations of biodiversity (Angermeier 1994).

Some (e.g., Palmer 1992) argue that artificial elements are components of biodiversity and therefore appropriate targets of biological conservation. We reject this argument for several reasons. First, culturally or technologically derived elements rarely perform life-support services as effectively as native elements (Ehrlich and Mooney 1983). Second, technology applied on massive spatial scales erodes biological integrity, ultimately leading to biotic impoverishment. And, third, including artificial diversity in conceptions of biodiversity wrongly legitimizes management strategies that erode native diversity.

Conceivably, through genetic engineering, species introduction, landscape modification, and other technologies, we could manufacture a biota with more elements, and thus more diversity, than the naturally evolved one, even to the exclusion of native elements. In contrast, the normative postulates of conservation biology (Soule 1985) were intended to protect products and processes of biogeography and evolution. Thus, current definitions of biodiversity should incorporate explicit native criteria.

In sum, biological integrity encompasses element composition (measured as number of items) and process performance (measured as rates) over multiple levels of organization; it is assessed in comparison with naturally evolved conditions within a given region. Biological integrity is thus generally defined as a system's ability to generate and maintain adaptive biotic elements through natural evolutionary processes. Current loss of biological diversity is tragic, but loss of biological integrity includes loss of diversity and breakdown in the processes necessary to generate future diversity.

Ecological Indicators

To assess biological integrity, one should be familiar with regional organizing processes and elements, including how they are influenced by human actions. A conceptual organization with five classes of interacting factors — physicochemical conditions, trophic base, habitat structure, temporal variation, and biotic interactions (Table 3) — has been useful in selecting ecological indicators to assess (Karr 1991, Karr et al. 1986) and tactics to restore (Gore 1985) integrity in aquatic systems.

Biological integrity can be assessed through diagnostic attributes or indicators, which ideally are sensitive to a range of stresses, able to distinguish stress-induced variation from natural variation, relevant to societal concerns, and easy to measure and interpret. Several authors (e.g., Karr 1991, Noss 1990, and Schaeffer et al. 1988) offer extensive lists of potential indicators of ecological integrity (also see Table 2); others have listed indicators of genetic integrity (Lande and Barrowclough 1987, Noss 1990). In practice, elements are used more frequently than processes as indicators of integrity because elements are typically more sensitive to degradation, more fully understood, and less expensive to monitor. Thus, biodiversity is an important indicator of biological integrity.

The complexity of biotic systems dictates that integrity assessments should incorporate a variety of indicators (including elements and processes) from multiple organizational levels and spatiotemporal scales. The index of biotic integrity (IBI) represents a successful approach for incorporating information from multiple indicators into a single numerical index (Karr 1991, Karr et al. 1986). Conditions observed in the system being assessed are compared to region-specific expectations for an undegraded system, (i.e., the reference condition).

The original IBI incorporated numerical criteria on species composition and diversity, trophic composition, population density, tolerance to human impacts, and individual health to assess integrity of lotic fish communities. The IBI has been used successfully in more than 20 states of the United States and in Canada, France, India, Poland, and Venezuela. Similar protocols (some using aquatic invertebrates) also have been developed for reservoirs, lakes, and estuaries (Deegan et al. 1993, EPA 1991b, Ohio EPA 1988). Efforts to apply such assessment approaches in terrestrial systems have lagged behind those in aquatic systems, but they can succeed if defensible criteria for appropriate indicators are developed.

Ecological Restoration

The goal of ecological restoration is to produce a self-sustaining system as similar as possible to

TABLE 3. Representatives of five classes of factors that organize ecological systems and provide a framework for assessing ecological integrity. Some factors are especially applicable in aquatic (A) or terrestrial (T) systems.

Class	Factors	
Physiochemical conditions	Temperature	Salinity
	pH	Precipitation (T)
	Insolation	Oxygen (A)
	Nutrients	Contaminants
Trophic base	Energy source	Energy content of food
	Productivity	Spatial distribution of food
	Food particle size	Energy transfer efficiency
Habitat structure	Spatial complexity	Vegetation form (T)
	Cover and refugia	Basin and channel form (A)
	Topography (T)	Substrate composition (A)
	Soil composition (T)	Water depth (A)
	Vegetation height (T)	Current velocity (A)
Temporal variation	Diurnal	Predictability
	Seasonal	Weather (T)
	Annual	Flow regime (A)
Biotic interactions	Competition	Disease
	Parasitism	Mutualism
	Predation	Coevolution

the native biota. But biological, socioeconomic, or technological constraints may limit our ability to attain that goal despite the best intentions. For example, past extinctions of many Great Lakes fish stocks prevent restoring integrity to those ecosystems even if exotic species and toxic chemicals could be removed. Similarly, as rangeland degradation progresses, the costs and time for restoration become increasingly prohibitive (Milton et al. 1994). Thus, restoration goals must be based on social and political constraints as well as biological potential. Once a goal (benchmark state) is selected, however, assessing restoration success is analogous to assessing integrity under other circumstances, which includes identifying organizing processes and selecting appropriate indicators.

Restoration methods usually mimic recovery from natural perturbations and reflect important organizational processes. Common approaches for aquatic systems include manipulating water quality, habitat structure, hydrology, riparian/watershed vegetation, and (less frequently) animal populations (Gore 1985, Osborne et al. 1993). Restoration of terrestrial systems typically focuses on establishing native vegetation and manipulating succession.

To maximize effectiveness, restoration efforts should employ and encourage natural ecological processes rather than technological fixes and should incorporate spatiotemporal scales large enough to maintain the full range of habitats necessary for the biota to persist under the expected disturbance regime. Failure to recognize important ecological relationships can result in counterproductive efforts. In the Mount St. Helens (Washington) blast area, for example, seeding slopes with grass and removing woody debris from streams actually hindered natural recovery processes (Franklin et al. 1988). On the other hand, many systems are remarkably responsive to appropriate restoration efforts. Years after a massive channelization project in the Kissimmee River in Florida, partial re-establishment of the flow regime quickly restored plant, invertebrate, fish, and bird assemblages (Toth 1993). As knowledge of ecological processes and the technology to mimic those processes advance, we expect ecological restoration to take its place as a successful discipline.

Policy Implications

Despite spending hundreds of millions of dollars on endangered species, the United States continues to lose biodiversity. We ascribe much of this loss to ineffective policy that emphasizes piecemeal conservation of the elements of diversity rather than comprehensive protection of the integrity of systems supporting those elements. Two major shifts are needed to produce more effective resource policy. First, goals of biological conservation and restoration should focus on protecting integrity (Karr 1993), especially the organizational processes that generate and maintain all elements, rather than focusing on the presence or absence of particular elements. Such an approach is more likely to prevent endangerment of elements and should be more cost-effective than emergency efforts to pull them back from the brink of extinction after serious degradation. Emergency tactics may be necessary where a focus on integrity fails to protect an element, but they should not be the primary basis of conservation, as in current policy.

Adoption of policy goals to protect integrity would help avoid difficult resource allocation problems such as estimating specific flows needed to sustain populations of endangered fishes in the Colorado River or endangered birds in Platte River wetlands. In fisheries, managing for integrity would not allow the widespread practice of stocking non-native fishes to be construed as enhancing biodiversity. In forestry, rather than using the range of stand ages in a forest or the range of tree ages in a stand (Lippke 1993) as measures of biodiversity, harvest schedules would mimic patterns of natural disturbance (Hunter 1990).

Appropriate roles of diversity in resource policy are in establishing conservation priorities, siting reserves, and indicating program success. However, policy makers must agree on which organizational levels and elements should be protected. Species and communities are commonly used to assess an area's conservation value, but genetic elements are rarely used. Gap analysis, for example, combines information on landscape-scale vegetation types with assumptions about the habitat associations of terrestrial vertebrates and butterflies to establish regional conservation priorities (Scott et al. 1993).

Policy effectiveness also could be improved by shifting focus from populations and species to landscapes. The organizational processes and ecological contexts that maintain populations typically operate at larger spatiotemporal scales than the populations themselves (Pickett et al. 1992). Because human impacts are applied at landscape scales, management prescriptions should be focused at the same scales. Landscape-scale approaches are especially important in managing aquatic systems, which can rarely rely on high-profile species (e.g., bald eagle or grizzly bear) to garner public support for protection.

Riparian zones and floodplains are critical landscape components linking aquatic and terrestrial systems; they regulate aquatic habitat formation, as well as entry of water, nutrients, and organic material into aquatic habitats (Gregory et al. 1991). Thus, management approaches focusing on strictly aquatic components (e.g., designation of a stream reach as wild and scenic or as critical habitat for an imperiled species) are unlikely to be effective over the long term. Application of integrity goals and landscape approaches are perhaps nowhere more important (or more politically challenging) than in estuaries or in anadromous fisheries, which depend on interactions among terrestrial, freshwater, marine, and even atmospheric systems. Implementation of integrity goals is likely to challenge the leadership of government agencies. Protection of biological integrity could be enhanced by restructuring tax and subsidy programs to eliminate conservation disincentives for private landowners and to distribute conservation costs and benefits equitably (Carlton 1986). Traditional agricultural, fisheries, forestry, game management, and mining agencies must replace their narrow, commodity and harvest-oriented philosophies with innovative perspectives founded on a broader range of social concerns, longer time frames, and more interagency cooperation (Salwasser 1991). Critical steps toward managing for biological integrity include establishing scientifically defensible benchmarks and assessment criteria.

Although these steps are potentially contentious, current uses of integrity goals indicate that success is attainable. Management programs for Kissimmee River, Ohio surface waters, and Canadian national parks are grounded in the goal of protecting or restoring biological integrity. The current shift in management of US national forests and parks should also involve a goal based on integrity. Emphasis on a method of management (i.e., ecosystem management) without a well-defined goal could be counterproductive.

Reserves alone are unlikely to sustain all biodiversity or even all species. Partnerships between government agencies and the public are essential to maintaining integrity and diversity across landscapes that include public and private lands. Noss and Harris (1986) proposed a promising conceptual approach in which interconnected networks of protected and multiple-use landscape components are managed to provide economic benefits yet protect ecological processes. Conservation biologists are exploring applications of this approach to regional landscapes such as the Pacific Northwest and Southern Appalachia (Mann and Plummer 1993). Similar management schemes could be effective in protecting the integrity of many ecosystems and landscapes, but where such preventive approaches fail, agencies should establish safety-net measures analogous to the Endangered Species Act to prevent important or unique ecosystems and landscapes from being destroyed.

Societal Choices

The causes of environmental degradation and loss of biodiversity are rooted in society's values and the ethical foundation from which values are pursued (Orr 1992). Solutions are likely to emerge only from a deep-seated will, not from better technology. Adopting biological integrity as a primary management goal provides a workable framework for sustainable resource use, but fostering integrity requires societal commitment well beyond government regulations and piecemeal protection. Such a commitment includes self-imposed limits on human population size and resource consumption, rethinking prevailing views of land stewardship and energy use, and viewing biological conservation as essential rather than as a luxury or nuisance.

Shifting our everyday thinking in this direction forces us to face the hard choices for which political rhetoric so often calls. Those choices are not likely to favor biological diversity unless people recognize the inherent value of unique biological elements and processes at all organizational levels. Conservation biologists should play a major role in articulating the value of biota, demonstrating links between biological integrity and economic stability, and dispelling the myth that technology can replace biodiversity or essential life-support services.

The decision to conserve or exhaust biotic resources is before us. It can be informed by science and influenced by government policy, but conservation primarily depends on a societal will grounded in recognition of its obligation to the future.

Acknowledgments

We thank P. D. Boersma, J. S. Edwards, W. E. Ensign, P. J. Jacobson, E. Serrano Karr, B. L. Kerans, I. J. Schlosser, M. L. Warren Jr., and four anonymous reviewers for helpful comments on earlier drafts. Two editors—Ellen W. Chu of *Illahee: Journal for the Northwest Environment* and Julie Ann Miller of *BioScience*—aided us immeasurably in expressing our ideas more clearly. The Cooperative Fish and Wildlife Research Unit is jointly supported by National Biological Survey, Virginia Department of Game and Inland Fisheries, and Virginia Polytechnic Institute and State University.

References

Allan, J. D., and A. S. Flecker. 1993. Biodiversity conservation in running waters. *BioScience* 43:32–43.

Allen, T. F. H., and T. W. Hoekstra. 1992. *Toward a Unified Ecology*. Columbia University Press, New York.

Angermeier, P. L. 1994. Does biodiversity include artificial diversity? *Conserv. Biol.* 8:600–602.

Bormann, F. H. 1985. Air pollution and forests: an ecosystem perspective. *BioScience* 35:434–441.

Carlton, R. L. 1986. Property rights and incentives in the preservation of species. Pages 255–267 in B. G. Norton, ed. *The Preservation of Species*. Princeton University Press, Princeton, NJ.

Deegan, L. A., J. T. Finn, S. G. Ayvazian, and C. Ryder. 1993. Feasibility and application of the index of biotic integrity in Massachusetts estuaries (EBI). Final Project Report, Ecosystem Center, Marine Biological Laboratory, Woods Hole, MA.

Ehrlich, P. R., and H. A. Mooney. 1983. Extinction, substitution, and ecosystem services. *BioScience* 33:248–254.

Ehrlich, P. R., and E. O. Wilson. 1991. Biodiversity studies: science and policy. *Science* 253:758–762.

Eisner, T. 1992. The hidden value of species diversity. *BioScience* 42:578.

Environmental Protection Agency (EPA). 1990. Biological criteria: national program guidance for surface waters. EPA-440/-90-004. EPA, Office of Water, Washington, DC.

———. 1991a. Biological criteria: state development and implementation efforts. EPA-440/5-91-003. EPA, Office of Water, Washington, DC.

———. 1991b. Biological criteria: research and regulation. EPA-440/5-91-005. EPA, Office of Water, Washington, DC.

Franklin, J. F., P. M. Frenzen, and F. J. Swanson. 1988. Re-creation of ecosystems at Mount St. Helens: contrasts in artificial and natural approaches. Pages 1–37 in J. Cairns Jr., ed. *Rehabilitating Damaged Ecosystems*. Vol. II. CRC Press, Boca Raton, FL.

Frey, D. 1975. Biological integrity of water: an historical perspective. Pages 127–139 in R. K. Ballentine and L. J. Guarraia, eds. *The Integrity of Water*. EPA, Washington, DC.

Gore, J. A., ed. 1985. *The Restoration of Rivers and Streams*. Butterworth, Boston, MA.

Gregory, S. V., F. J. Swanson, W. A. McKee, and K. W. Cummins, 1991. An ecosystem perspective of riparian zones. *BioScience* 41:540–551.

Hanski, I., and M. Gilpin. 1991. Meta-population dynamics: a brief history and conceptual domain. *Biol. J. Linn. Soc.* 42:3–16.

Hoover, S. R., and A. J. Parker. 1991. Spatial components of biotic diversity in landscapes of Georgia, USA. *Landscape Ecol.* 5:125–136.

Howarth, R. W. 1991. Comparative responses of aquatic ecosystems to toxic chemical stress. Pages 169–195 in J. Cole, G. Lovett, and S. Findlay, eds. *Comparative Analyses of Ecosystems*. Springer-Verlag, New York.

Hunter, M. L. Jr. 1990. *Wildlife, Forests, and Forestry*. Prentice-Hall, Englewood Cliffs, NJ.

Karr, J. R. 1991. Biological integrity: a long neglected aspect of water resource management. *Ecological Applications* 1:66–84.

———. 1993. Protecting ecological integrity: an urgent societal goal. *Yale Journal International Law* 18:297–306.

_____ . 1994. Landscapes and management for ecological integrity. Pages 229–251 in K. C. Kim and R. D. Weaver, eds. *Biodiversity and Landscapes.* Cambridge University Press, New York.

Karr, J. R., and D. R. Dudley. 1981. Ecological perspective on water quality goals. *Environ. Manage.* 5:55–68.

Karr, J. R., K. D. Fausch, P. L. Angermeier, P. R. Yant, and I. J. Schlosser. 1986. Assessing biological integrity in running waters: a method and its rationale. Special Publication 5. Illinois Natural History Survey, Champaign, IL.

King, A. W. 1993. Considerations of scale and hierarchy. Pages 19–45 in S. Woodley, J. Kay, and G. Francis, eds. *Ecological Integrity and the Management of Ecosystems.* St. Lucie Press, Delray Beach, FL.

Lande, R., and G. F. Barrowclough. 1987. Effective population size, genetic variation, and their use in population management. Pages 87–123 in M. E. Soule, ed. *Viable Populations for Conservation.* Cambridge University Press, New York.

Levin, S. A. 1992. The problem of pattern and scale in ecology. *Ecology* 73:1943–1967.

Lippke, B. 1993. Focus on preserving old growth is counterproductive to achieving biodiversity. *Northwest Environ. J.* 9:10–15.

Livingston, R. J. 1991. Historical relationships between research and resource management in the Apalachicola River estuary. *Ecological Applications* 1:361–382.

Lubchenco, J., et al. 1991. The sustainable biosphere initiative: an ecological research agenda. *Ecology* 72:371–412.

Mann, C. C., and M. L. Plummer. 1993. The high cost of biodiversity. *Science* 260:1868–1871.

Mares, M. A. 1992. Neotropical mammals and the myth of Amazonian biodiversity. *Science* 255:976–979.

Master, L. 1990. The imperiled status of North American aquatic animals. *Biodiversity Network News* 3(3):1–2, 7–8.

Milton, S. J., W. R. J. Dean, M. A. duPlessis, and W. R. Siegfried. 1994. A conceptual model of arid rangeland degradation: the escalating cost of declining productivity. *BioScience* 44:70–76.

Myers, N. 1979. *The Sinking Ark.* Pergamon Press, New York.

Nehlsen, W., J. E. Williams, and J. A. Lichatowich. 1991. Pacific salmon at the crossroads: stocks at risk from California, Oregon, Idaho, and Washington. *Fisheries (Bethesda)* 16(2):4–21.

Noss, R. F. 1990. Indicators for monitoring biodiversity: a hierarchical approach. *Conserv. Biol.* 4:355–364.

Noss, R. F., and L. D. Harris. 1986. Nodes, networks, and MUM's: preserving diversity at all scales. *Environ. Manage.* 10:299–309.

Odum, E. P. 1985. Trends expected in stressed ecosystems. *BioScience* 35:419–422.

Office of Technology Assessment (OTA). 1987. *Technologies to Maintain Biological Diversity.* Congress of the United States, OTA-F-330, Washington, DC.

Ohio Environmental Protection Agency (Ohio EPA). 1988. Biological criteria for the protection of aquatic life. Ohio EPA, Division of Water Quality Monitoring and Assessment, Surface Water Section, Columbus, OH.

O'Neill, R. V., A. R. Johnson, and A. W. King. 1989. A hierarchical framework for the analysis of scale. *Landscape Ecol.* 3:193–205.

Orr, D. W. 1992. *Ecological Literacy.* State University of New York Press, Albany, NY.

Osborne, L. L., P. B. Bayley, and L. W. Higler, eds. 1993. Lowland stream restoration: theory and practice. Special issue. *Freshwater Biol.* 29:187–342.

Palmer, T. 1992. The case for human beings. *Atlantic Monthly* 269(1):83–88.

Pickett, S. T. A., V. T. Parker, and P. L. Fiedler. 1992. The new paradigm in ecology: implications for conservation biology above the species level. Pages 65–88 in P. L. Fiedler and S. K. Jain, eds. *Conservation Biology.* Chapman & Hall, New York.

Ray, G. C., and J. F. Grassle. 1991. Marine biological diversity. *BioScience* 41:453–457.

Redford, K. H., and S. E. Sanderson. 1992. The brief barren marriage of biodiversity and sustainability? *Bull. Ecol. Soc. Am.* 73:36–39.

Reiger, H. A. 1993. The notion of natural and cultural integrity. Pages 3–18 in S. Woodley, J. Kay, and G. Francis, eds. *Ecological Integrity and the Management of Ecosystems.* St. Lucie Press, Delray Beach, FL.

Reid, W. V., and K. R. Miller. 1989. *Keeping Options Alive.* World Resources Institute, Washington, DC.

Roughgarden, J., S. Gaines, and H. Possingham. 1988. Recruitment dynamics in complex life cycles. *Science* 241:1460–1466.

Salwasser, H. 1991. In search of an ecosystem approach to endangered species conservation. Pages 247–265 in K. A. Kohm, ed. *Balancing on the Brink of Extinction.* Island Press, Washington, DC.

Schaeffer, D. J., E. E. Herricks, and H. W. Kerster. 1988. Ecosystem health: I. measuring ecosystem health. *Environ. Manage.* 12:445–455.

Schindler, D. W. 1990. Experimental perturbations of whole lakes as tests of hypotheses concerning ecosystem structure and function. *Oikos* 57:25–41.

Scott, J. M., et al. 1993. Gap analysis: a geographic approach to protection of biological diversity. Wildlife Monograph No. 123. The Wildlife Society, Bethesda, MD.

Soule, M. E. 1985. What is conservation biology? *BioScience* 35:727–734.

Taylor, J. N., W. R. Courtenay Jr., and J. A. McCann. 1984. Known impacts of exotic fishes in the continental United States. Pages 322–373 in W. R. Courtenay Jr. and J. R. Stauffer Jr., eds. *Distribution, Biology and Management of Exotic Fishes.* Johns Hopkins University Press, Baltimore, MD.

Toth, L. A. 1993. The ecological basis of the Kissimmee River restoration plan. *Fla. Sci.* 56(1):25–51.

Vane-Wright, R. I., C. J. Humphries, and P. H. Williams. 1991. What to protect? Systematics and the agony of choice. *Biol. Conserv.* 55:235–254.

Vitousek, P. M. 1990. Biological invasions and ecosystem processes: towards an integration of population biology and ecosystem studies. *Oikos* 57:7–13.

Williams, J. E., and R. J. Neves. 1992. Introducing the elements of biological diversity in the aquatic environment. *Trans. N. Am. Wildl. Nat. Resour. Conf.* 57:345–354.

Wilson, E. O. 1985. The biological diversity crisis. *BioScience* 35:700–706.

Woodwell, G. M., ed. 1990. *The Earth in Transition.* Cambridge University Press, New York.

PART 4
Promote Ecological Sustainability

25
Great Ideas in Ecology for the 1990s

Eugene P. Odum

In a commentary entitled "Science literacy" (Pool 1991), there is a table of "Science's top 20 greatest hits" as chosen by biologist Robert Hazen and physicist James Trefil. They suggested that these "great ideas" might be the basis for a course in general science, and comments from readers were invited.

In addition to the two laws of thermodynamics, Hazen and Trefil's list includes three other concepts that could be construed as ecological. These concepts are "everything on earth operates in cycles," "all forms of life evolved by natural selection," and "all life is connected."

For many years, I have contended that ecology is no longer a subdivision of biology but has emerged from its roots in biology to become a separate discipline that integrates organisms, the physical environment, and humans—in line with *oikos,* root of the word *ecology* (Odum 1977). From this view, the ecosystem level becomes the major focus. Populations are considered as ecosystem components and landscapes as associations of interacting ecosystems. This viewpoint is now generally accepted, as was indicated by a recent British Ecological Society survey in which members were asked to list what they considered the most important ecological concepts. The ecosystem was the concept most frequently listed (Cherrett 1989).

At the time that the science literacy article appeared, I was drawing up a list of basic concepts in ecology that might be included in courses designed to improve environmental literacy among undergraduates here at the University of Georgia. Here are 20 of my "great ideas" in ecology, as distinguished from "great ideas" in biology (e.g., DNA, genetic code, and general theory of natural selection). The last five items in my list relate to human ecology and the ecology-economics interface, which must be major foci in environmental literacy education in view of the increasingly serious global impacts resulting from human activities. The references I have selected for each concept may not be the best ones, and certainly they are not the only ones.

Concept 1. An ecosystem is a thermodynamically open, far from equilibrium, system. Input and output environments are an essential part of this concept. For example, in considering a forest tract, what is coming in and going out is as important as what is inside the tract. The same holds for a city. It is not a self-contained unit ecologically or economically; its future depends as much on the external life-support environment as on activities within city limits (Odum 1983, Patton 1972, Prigogine et al. 1972).

Concept 2. The source-sink concept: one area or population (the source) exports to another

area or population (the sink). This statement is a corollary to concept 1. It is applicable at ecosystem as well as population levels. At the ecosystem level, an area of high productivity (salt marsh, for example) may feed an area of low productivity (adjacent coastal waters). At the population level, a species in one area may have a higher reproduction rate than needed to sustain the population, and surplus individuals may provide recruitment for an adjacent area of low re-production. Food chains may also involve sources and sinks (see concept 12; Lewin 1989, Pulliam 1988).

Concept 3. In hierarchical organization of ecosystems, species interactions that tend to be unstable, nonequilibrium, or even chaotic are constrained by the slower interactions that characterize large systems. Short-term interactions, such as interspecific competition — the evolutionary arms race between a parasite and its host, herbivore-plant interactions, and predator-prey activities — tend to be oscillatory or cyclic. Large, complex systems — such as oceans, the atmosphere, soils, and large forests — tend to go from randomness to order and will tend to have more steady-state characteristics, for example, the atmosphere's gaseous balances.

Accordingly, large ecosystems tend to be more homeostatic than their components. This principle may be the most important of all, because it warns that what is true at one level may or may not be true at another level of organization. Also, if we are serious about sustainability, we must raise our focus in management and planning to large landscapes and beyond (Allen and Starr 1982, Kauffman 1990, O'Neill et al. 1986, Prigogine and Stengers 1984, Ulanowicz 1986).

Concept 4. The first signs of environmental stress usually occur at the population level, affecting especially sensitive species. If there is sufficient redundancy, other species may fill the functional niche occupied by the sensitive species. Even so, this early warning should not be ignored, because the backup components may not be as efficient. When the stress produces detectable ecosystem-level effects, the health and survival of the whole system is in jeopardy. This idea is a corollary of item 3: parts are less

stable than wholes (Odum 1985, 1990, Schindler 1990).

Concept 5. Feedback in an ecosystem is internal and has no fixed goal. There are no thermostats, chemostats, or other set-point controls in the biosphere. Cybernetics at the ecosystem level thus differs from that at the organism level (body temperature control, for example) or that of human-made mechanical systems (temperature control of a building, for example) where the control is external with a set point. Ecosystem control, where manifested, is the result of a network of internal feedback processes as yet little understood — another corollary of concept 3 (Patten and Odum 1981).

Concept 6. Natural selection may occur at more than one level. This idea is another corollary to concept 3. Accordingly, coevolution, group selection, and traditional Darwinism are all part of the hierarchical theory of evolution. Not only is the evolution of a species affected by the evolution of interacting species, but a species that benefits its community has survival value greater than a species that does not (Axelrod 1984, 1980, Axelrod and Hamilton 1981, Gould 1982, Wilson 1976, 1980).

Concept 7. There are two kinds of natural selection, or two aspects of the struggle for existence: organism versus organism, which leads to competition, and organism versus environment, which leads to mutualism. To survive, an organism does not compete with its environment as it might with another organism, but it must adapt to or modify its environment and its community in a cooperative manner. This concept was first suggested by Peter Kropotkin soon after Darwin. (Gould 1988, Kropotkin 1902).

Concept 8. Competition may lead to diversity rather than to extinction. Although competition plays a major role in shaping the species composition of biotic communities, competition exclusion (in which one species eliminates another, as in a flour beetle microcosm) is probably the exception rather than the rule in the open systems of nature. There, species are often able to shift their functional niches to avoid the deleterious effects of competition (den Boer 1986).

Concept 9. Evolution of mutualism increases when resources become scarce. Cooperation between species for mutual benefit has special survival value when resources become tied up in the biomass, as in mature forests, or when the soil or water is nutrient poor, as in some coral reefs or rainforests (Boucher et al. 1982, Odum and Biever 1984). The recent shift from confrontation to cooperation among the world's superpower nations may be a parallel in societal evolution (Kolodziej 1991).

Concept 10. Indirect effects may be as important as direct interactions in a food web and may contribute to network mutualism. When food chains function in food web networks, organisms at each end of a trophic series (for example, plankton and bass in a pond) do not interact directly but indirectly benefit each other. Bass benefit by eating planktiverous fish supported by the plankton, whereas plankton benefit when bass reduce the population of its predators. Accordingly, there are both negative (predator-prey) and positive (mutualistic) interactions in a food web network (Patton 1991, Wilson 1986).

Concept 11. Since the beginning of life on Earth, organisms have not only adapted to physical conditions but have modified the environment in ways that have proven to be beneficial to life in general (e.g., increase O_2 and reduce CO_2). This modified Gaia hypothesis is now accepted by many scientists. Especially important is the theory that microorganisms play major roles in vital nutrient cycles (especially the nitrogen cycle) and in atmospheric and oceanic homeostasis (Cloud 1988, Lovelock 1979, 1988, Kerr 1988, Margulis and Olendzenski 1991).

Concept 12. Heterotrophs may control energy flow in food webs. For example, in warm waters, bacteria may function as a sink in that they short-circuit energy flow so that less energy reaches the ocean bottom to support demersal fisheries. In cooler waters, bacteria are less active, allowing more of the fruits of primary production to reach the bottom (Pomeroy 1974, Pomeroy and Deibel 1986, Pomeroy and Wiebe 1988). Small heterotrophs may play similar controlling roles in terrestrial eco-

systems such as grasslands (Dyer et al. 1982, 1986, Seastadt and Crossley 1984). This concept is a corollary of concept 11.

Concept 13. An expanded approach to biodiversity should include genetic and landscape diversity, not just species diversity. The focus on preserving biodiversity must be at the landscape level, because the variety of species in any region depends on the size, variety, and dynamics of patches (ecosystems) and corridors (Odum 1982, Turner 1988, Wilson 1988).

Concept 14. Ecosystem development or autogenic ecological succession is a two-phase process. Early or pioneer stages tend to be stochastic as opportunistic species colonize, but later stages tend to be more self-organized (perhaps another corollary of concept 3; Odum 1989a).

Concept 15. Carrying capacity is a two-dimensional concept involving number of users and intensity of per capita use. These characteristics track in a reciprocal manner—as the intensity of per capita impact goes up, the number of individuals that can be supported by a given resource base goes down (Catton 1987). Recognition of this principle is important in estimating human carrying capacity at different quality-of-life levels and in determining how much buffer natural environment to set aside in land-use planning.

Concept 16. Input management is the only way to deal with nonpoint pollution. Reducing waste in developed countries by source reduction of the pollutants will not only reduce global-scale pollution but will spare resources needed to improve quality of life in undeveloped countries (Odum 1987, 1989b).

Concept 17. An expenditure of energy is always required to produce or maintain an energy flow or a material cycle. According to this net-energy concept, communities and systems, whether natural or human-made, as they become larger and more complex, require more of the available energy for maintenance (the so-called complexity theory). For example, when a city doubles in size, more than double the energy (and taxes) is required to maintain order (Odum and Odum 1981, Pippenger 1978).

Concept 18. There is an urgent need to bridge the gaps between human-made and natural life-support goods and services and between non-sustainable short-term and sustainable long-term management. Agroecosystems, tropical forests, and cities are of special concern. H. T. Odum's "emergy" concept and Daly and Cobb's index of sustainable economic welfare are examples of recent attempts to bridge these gaps (Daly and Cobb 1989, Folke and Kaberger 1991, Holden 1990, Odum 1988).

Concept 19. Transition costs are always associated with major changes in nature and in human affairs. Society has to decide who pays, for example, the cost of new equipment, procedures, and education in changing from high-input to low-input farming or in converting from air polluting to clean power plants (Renner 1991, Spencer et al. 1986).

Concept 20. A parasite-host model for man and the biosphere is a basis for turning from exploiting the earth to taking care of it (going from dominionship to stewardship, to use a biblical metaphor). Despite, or perhaps because of, technological achievements, humans remain parasitic on the biosphere for life support. Survival of a parasite depends on reducing virulence and establishing reward feedback that benefits the host (Alexander 1981, Anderson and May 1981, 1982, Levin and Pimentel 1981, Pimentel 1968, Pimentel and Stone 1968, Washburn et al. 1991). Similar relationships hold for herbivory and predation (Dyer et al. 1986, Lewin 1989, Owen and Wiegert 1976). In terms of human affairs, this concept involves reducing wastes and destruction of resources to reduce human virulence, promote the sustainability of renewable resources, and invest more in Earth care.

Concept Comparisons

In my list, I have covered the ecological items in Hazen and Trefil's "great ideas in science." Thermodynamics is represented in concept 1, natural selection in concepts 6 and 7 (and indirectly in most others), and cyclic behavior and connectiveness in concept 3.

The British Ecological Society survey listed approximately 50 wide-ranging items. The ed-

itor (Cherrett 1989) divided the 600 or so responding ecologists into two groups: practical holists and theoretical reductionists. I do not believe these dichotomies between holism and reductionism and that between theoretical and practical are very helpful. The most exciting of the concepts listed apply to all levels or to the interaction of levels, and any and all may have practical as well as theoretical aspects. I believe that my more comprehensive 20 concepts cover most of what is in the 50-item list.

I am sure that there are other concepts that might be added to my list, and I suspect that some of my choices may be considered by some ecologists as too hypothetical to have been included. I invite comments to further explore what concepts are most important to public knowledge of ecology (i.e., environmental literacy).

References

Alexander, M. 1981. Why microbial parasites and predators do not eliminate their prey and hosts. *Annu. Rev. Microbiol.* 35: 113–133.

Allen, T. F. H., and T. B. Starr. 1982. *Hierarchy: Perspectives for Ecological Complexity.* University of Chicago Press, Chicago.

Anderson, R. M., and R. M. May. 1981. The population dynamics of microparasites and their invertebrate hosts. *Philos. Trans. R. Soc. Lond. Biol. Sci. B* 291: 451–524.

———. 1982. Coevolution of hosts and parasites. *Parasitology* 85: 411–426.

Axelrod, R. 1984. *Evolution of Cooperation.* Basic Books, New York.

Axelrod, R., and W. D. Hamilton. 1981. The evolution of cooperation. *Science* 211: 1390–1396.

Boucher, D. S., S. James, and K. H. Keeler. 1982. The ecology of mutualism. *Annu. Rev. Ecol. Syst.* 13: 315–347.

Catton, W. R. 1987. The world's most polymorphic species: carrying capacity transgressed two ways. *BioScience* 37: 413–419.

Cherrett, J. M., ed. 1989. *Ecological Concepts: The Contribution of Ecology to an Understanding of the Natural World.* Blackwell Scientific Publ., London.

Cloud, P. E. 1988. Gaia modified. *Science* 240: 1716.

Daly, H. E., and B. J. Cobb. 1989. *For the Common Good: Redirecting the Economy Towards Community, the Environment, and a Sustainable Future.* Beacon Press, Boston.

den Boer, P. J. 1986. The present status of the competition exclusion principle. *Trends Ecol. Evol.* 1: 25–28.

Dyer, M. I., D. L. DeAngelis, and W. M. Post. 1986. A model of herbivore feedback in plant productivity. *Math Biosci.* 79: 171–184.

Dyer, M. I., J. K. Detling, J. K. Coleman, and D. W. Hilbert. 1982. The role of herbivores in grasslands. Pages 255–295 in J. R. Estes, R. J. Tyri, and J. N. Brunken, eds. *Grasses and Grasslands.* University of Oklahoma Press, Norman.

Folke, C., and T. Kaberger. 1991. *Linking the Natural Environment and the Economy: Essays from the Eco-Eco Group.* Kluvier Academic Publ., Boston.

Gould, S. J. 1982. Darwinism and the expansion of evolutionary theory. *Science* 216:380–387.

_____. 1988. Kropotkin was no crackpot. *Nat. Hist.* 97(7): 12–21.

Holden, C. 1990. Multidisciplinary look at a finite world. *Science* 249: 18–19.

Kauffman, S. 1990. Spontaneous order, evolution and life. *Science* 247: 1543–1544.

Kerr, R. A. 1988. No longer willful, Gaia becomes respectable. *Science* 240: 393–395.

Kolodziej, E. A. 1991. The cold war as cooperation. *Bulletin of the American Academy of Arts and Sciences* 44(7): 9–39.

Kropotkin, P. 1902. *Mutual Aid: A Factor of Evolution.* William Heinmann, London. Reprinted 1935, Extending Horizon Books, Boston.

Levin, S., and D. Pimentel. 1981. Selection of intermediate rates of increase in parasite-host systems. *Am. Nat.* 117: 308–315.

Lewin, R. 1989. Sources and sinks complicate ecology. *Science* 243: 477–478.

Lovelock, J. E. 1979. *Gaia: A New Look at Life on Earth.* Oxford University Press, New York.

_____. 1988. *The Ages of Gaia.* W.W. Norton, New York.

Margulis, L., and L. Olendzenski. 1991. *Environmental Evolution.* MIT Press, Cambridge, MA.

Odum, E. P. 1977. The emergence of ecology as a new integrative science. *Science* 195: 1289–1293.

_____. 1982. Diversity and the forest ecosystem. Pages 35–41 in J. L. and J. H. Cooley, eds. *Proceedings of a Workshop in Natural Diversity in Forest Ecosystems.* US Forest Service, SE Forest Experiment Station, Athens, GA.

_____. 1983. *Basic Ecology.* Saunders Publ., Philadelphia.

_____. 1985. Trends expected in stressed ecosystem. *BioScience* 35: 419–422.

_____. 1987. Reduced-input agriculture reduces nonpoint pollution. *J. Soil Water Conserv.* 42: 412–414.

_____. 1989a. *Ecology and Our Endangered Life-support systems.* Sinauer Assoc., Sunderland, MA.

_____. 1989b. Input management of production systems. *Science* 243: 177–182.

_____. 1990. Field experimental tests of ecosystem-level hypotheses. *Trends Ecol. Evol.* 5: 204–205.

Odum, E. P., and L. J. Biever. 1984. Resource quality, mutualism and energy partitioning in food chains. *Am. Nat.* 124: 360–376.

Odum, H. T. 1988. Self-organization, transformity and information. *Science* 242: 1132–1139.

Odum, H. T., and E. C. Odum. 1981. *Energy Basis for Man and Nature.* McGraw-Hill, New York.

O'Neill, R. V., D. L. DeAngelis, J. B. Waide, and T. F. H. Allen. 1986. *A Hierarchical Concept of Ecosystems.* Princeton University Press, Princeton, NJ.

Owen, D. F., and R. G. Weigert. 1976. Do consumers maximize plant fitness? *Oikos* 27:489–492.

Patten, B. C. 1978. Systems approach to the concept of environment. *Ohio J. Sci.* 78: 206–222.

_____. 1991. Network ecology: indirect determination of the life-environment relationship in ecosystems. Pages 288–351 in M. Higashi and T. P. Burns, eds. *Theoretical Studies of Ecosystems: The Network Perspective.* Cambridge University Press, New York.

Patten, B. C., and E. P. Odum. 1981. The cybernetic nature of ecosystems. *Am. Nat.* 118: 886–895.

Pimentel, D. 1968. Population regulation and genetic feedback. *Science* 159: 1432–1437.

Pimentel D., and F. A. Stone. 1968. Evolution and population ecology of parasite-host systems. *Can. Entomol.* 100: 655–662.

Pippenger, N. 1978. Complexity theory. *Sci. Am.* 238(6): 114–124.

Pomeroy, L. R. 1974. The ocean's food web: a changing paradigm. *BioScience* 24: 499–504.

Pomeroy, L. R., and D. Deibel. 1986. Temperature regulation of bacterial activity during the spring bloom in Newfoundland coastal waters. *Science* 233: 359–361.

Pomeroy, L. R., and W. J. Wiebe. 1988. Energetics of microbial food webs. *Hydrobiologia* 159: 7–18.

Pool, R. 1991. Science literacy: the enemy is us. *Science* 251: 266–267.

Prigogine, I., F. Nicoles, and A. Babloyantz. 1972. Thermodynamics and evolution. *Physics Today* 25(11): 23–38; 25(12): 138–142.

Prigogine, I., and I. Stengers. 1984. *Order Out of Chaos: Man's New Dialogue with Nature.* Bantam, New York.

Pulliam, H. R. 1988. Sources, sinks and population regulation. *Am. Nat.* 132: 652–661.

Renner, M. 1991. *Jobs in a Sustainable Economy.* Worldwatch paper 104, Worldwatch Institute, Washington, DC.

Schindler, D. W. 1990. Experimental perturbations of whole lakes as tests of hypotheses concerning ecosystem structure and function. *Oikos* 57: 25–41.

Seastadt, T. R., and D. A. Crossley. 1984. The influence of arthropods on ecosystems. *BioScience* 34: 157–161.

Spencer, D. F., S. B. Alpert, and H. H. Gilman. 1986. Cool water: demonstration of a clean and efficient new coal technology. *Science* 232: 609–612.

Turner, M. G. 1988. Landscape ecology: the effect of pattern on process. *Annu. Rev. Ecol. Syst.* 20: 171–197.

Ulanowicz, R. E. 1986. *Growth and Development: Ecosystem Phenomenology.* Springer-Verlag, New York.

Washburn, J. O., D. R. Mercer, and J. R. Anderson. 1991. Regulatory role of parasites: impact on host population shifts with resource availability. *Science* 253: 185–188.

Wilson, D. S. 1976. Evolution on the level of communities. *Science* 192: 1358–1360.

_____. 1980. *The Natural Selection of Populations and Communities.* Benjamin Cummings, Menlo Park, CA.

_____. 1986. Adaptive indirect effects. Pages 437–444 in J. Diamond and T.J. Case, eds. *Community Ecology.* Harper and Row, New York.

Wilson, E. O., ed. 1988. *Biodiversity.* National Academy Press, Washington, DC.

26

Can Extractive Reserves Save the Rain Forest? An Ecological and Socioeconomic Comparison of Nontimber Forest Product Extraction Systems in Petén, Guatemala and West Kalimantan, Indonesia

Nick Salafsky, Barbara L. Dugelby, and John W. Terborgh

Introduction

Beginning with the efforts of the National Council of Rubber Tappers in Brazil, considerable excitement has been generated over the past few years among the conservation and development community about the prospects of establishing extractive reserves that can maintain biodiversity while simultaneously providing a sustainable economic return to local peoples and governments (Fearnside 1989a; Schwartzman 1989; Allegretti 1990). This excitement was fueled by an article in *Nature* (Peters et al. 1989) proposing that the long-term financial return from the harvest of nontimber forest products found in a hectare of Amazonian rain forest far outweighed the net benefits of timber production or agricultural conversion of the same area of land. The findings from Peters et al. and similar studies (Alcorn 1989; Anderson & Jardim 1989; Prance 1989; Anderson 1990; Gómez-Pompa & Kaus 1990; Peters 1990) fueled the hopes of many conservationists, who found in these results an alluring mix of ecological, economic, and social justifications for preserving rainforest lands in a relatively pristine condition.

Several authors, including Peters et al. (1989), have attempted to temper this enthusiasm for extractive reserves by pointing out that hypothetical calculations of the income to be derived from an average hectare of extractive reserves have significant limitations (Fearnside 1989a; Vasquez & Gentry 1989; Browder 1990a, 1990b, 1992; Pinedo-Vasquez et al. 1990). The unique mix of ecological, socioeconomic, and political conditions existing at each potential reserve site makes generalization extremely risky. Only an analysis of a cross-section of tropical forest regions will support broad conclusions about the wide-scale applicability of the extractive reserve concept.

Unfortunately, the complex nature of nontimber forest product extraction systems currently precludes detailed quantitative comparisons of these systems. Nevertheless, the impending destruction of the world's remaining tropical forests, and the pressing needs of policy and decision makers for sound management guidelines, make it imperative that both the promise and limitations of the extractive reserve concept be fully understood.

In this paper, we compare nontimber forest product extraction systems in Guatemala and Indonesia to identify ecological, socioeconomic, and political parameters that need to be considered in the design and implementation of extractive reserves. For rhetorical purposes, we present evidence that these parameters facilitate sustained nontimber forest product extraction in the Petén and inhibit it in Kalimantan. It is not our intention, however, to imply that nontimber forest product extraction is perfectly suited to Petén or that it has no role to play in Kalimantan. Instead, our analysis seeks to underscore the importance of local conditions to the ultimate success or failure of an extractive reserve, while simultaneously recognizing the influence of national and international economic and political factors. Although this paper focuses primarily on market-oriented extractive reserves, many of the issues we discuss in the context of commercial nontimber forest products may apply as well to other land uses, such as collection of nontimber forest products for household consumption or small-scale timber extraction.

Study Sites, Methods, and Background Descriptions

Study Sites and Methods

Descriptions of the two extractive systems in this paper are based on research visits to Guatemala and Indonesia in 1990, supplemented by previous research experience in both these regions. In Guatemala, field work was conducted in the Department of Petén, in the villages of Uaxactun and Carmelita and in harvesting camps near El Mirador, all of which are located in the newly created Maya Biosphere Reserve (Fig. 1). In Indonesia, field work was conducted in the province of West Kalimantan, focusing on the villages of Keranji and Benawai Agung, both of which border Gunung Palung National Park (Fig. 2).

We spent several weeks in each country, living in the study villages, accompanying workers into the forest on harvesting trips, and conducting standardized interviews with the members of households engaged in nontimber forest product extraction, processing, and trade. We also spoke extensively in each country with storekeepers and traders, with local, regional, and national governmental authorities, and with scientists conducting research on these topics.

Background Descriptions of Extractive Systems

Petén

Nontimber forest products have been exploited in the Petén since at least the time of the Mayan civilization. According to recent evidence, the Mayas relied heavily on numerous nontimber forest products for subsistence and trade during much of the first millennium A.D. (Nations & Nigh 1980; Voorhies 1982; Gómez-Pompa & Kaus 1990). Today, many *Peteneros* still rely on a highly diversified portfolio of forest products to meet food, fuel, fodder, construction, and medicinal needs (Heinzman & Reining 1989).

Three primary products—chicle (latex from *Manikara zapota* trees), *xate* (pronounced *shá'te*—fronds from *Chamaedorea spp.* palms), and allspice (fruits from *Pimenta dioica* trees)—form the basis of an extensive export-oriented extraction system (Table 1). Chicle extraction began in the late nineteenth century and peaked in the 1930s and 1940s, when P. K. Wrigley built airstrips in Petén to harvest chicle, used in the production of chewing gum. In recent years, with the advent of synthetic latex, chicle extraction has declined to some degree. Allspice and xate extraction, however, have grown substantially since 1960 (Schwartz 1990). Today, extraction of these three products has developed into a major industry, employing harvesters who collect the products from the forest, contractors who supply the harvesters and bring the products to regional transportation hubs, and processors who prepare the products for export. It has been estimated that this extractive industry provides part- or full-time employment for over 7000 people in Petén and an annual export revenue of US $4–7 million to the Guatemalan economy (Nations 1989).

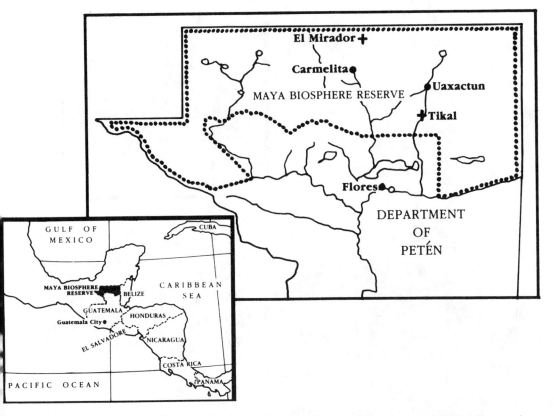

FIGURE 1. Location of study villages in the Maya Biosphere Reserve, Department of Petén, Guatemala.

West Kalimantan

Nontimber forest products have undoubtedly been exploited for centuries in Kalimantan by indigenous Dayak tribes and various immigrant settlers. As in Guatemala, local villagers harvest many different products from the forest (Burkhill 1935; deBeer & McDermott 1989). The extraction systems employed in West Kalimantan are less formally organized than those in the Petén. Nevertheless, a number of products are collected on a part-time basis by villagers who go into the forest for periods lasting from a half-day to a month. Harvested products are generally sold to local shopkeepers and middlemen. No figures are currently available concerning the value of extractive industries in the region. Government statistics for Indonesia as a whole, however, estimate that the annual value of nontimber forest products exported in the years 1983 to 1987 ranged between US $18 and $48 million excluding rattan, and US $127 and $238 million including rattan (deBeer &

McDermott 1989). The value of products consumed domestically, although undocumented, is probably even higher.

Ecological Factors

Density of Exploited Species

A primary factor in determining the success of an extractive system is the density at which the desired species occur in the forest (Peters 1990). The density of a given plant (or animal) species tends to be inversely related to the overall diversity of the ecosystem. Species density affects the *search time* necessary to locate target individuals, the *travel time* needed to move between these individuals, and the *carrying time* needed to bring the gathered product back to a central collecting point. A basic tenet of optimal foraging models of behavior holds that as the density of a given species decreases (and thus search, travel, and carrying times in-

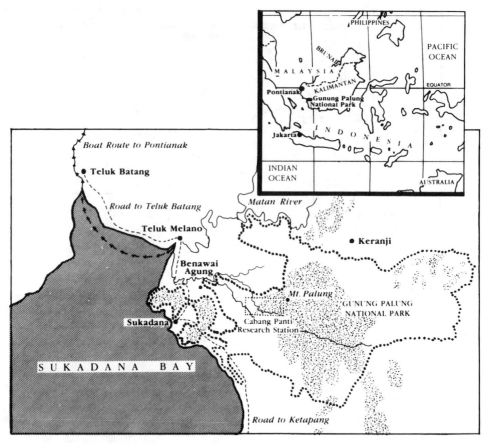

FIGURE 2. Location of study villages near Gunung Palung National Park, West Kalimantan, Indonesia.

crease), the overall return from the product decreases and hence the product should be less preferred (Charnov 1976; Krebs & McCleery 1984).

Petén

Subtropical forests such as those of the Petén contain a relatively low tree species diversity (on the order of 50–100 species per ha) and a correspondingly higher mean density of individuals per species (Lindell 1937; Foster & Brokaw 1982). A recent survey of three sites in Petén found densities of 47, 33, and 23 chicle trees (>10 cm dbh) per ha; 9, 31, and 12 allspice trees (>10 cm dbh) per ha; and 2279 and 2479 xate plants per ha in active harvesting zones (only the latter two sites have numbers reported for *xate*) (Cabrera Madrid et al. 1990). The high densities of these plants enable harvesters to minimize search, travel, and carrying

times and thus to collect large amounts of the product over short periods of time. We followed one xate harvester who was on average able to cut fronds (1–5 per plant) from 211 plants per hour over the course of a morning's work.

Kalimantan

The tropical forests of Kalimantan, by contrast, have high species diversity (on the order of 150–225 species per ha) and a correspondingly low density of individual taxa (Wallace 1878; Whitmore 1984; Gentry 1988). Low densities of producing plants have a profoundly negative impact on the ability of villagers to harvest many potential forest products efficiently. In general, to overcome the high costs of search, travel, and carrying time, harvesters are restricted to the most valuable products, or those that have a predictable and/or patchy distribution. For example, gaharu, a resin that

TABLE 1. Selected forest products with a major role in local extraction systems.

Common Name	Scientific Name	Form	Part	Uses	Harvest Season	Final Market	Harvesting Method
Petén							
Chicle	*Manikara zapota* (Sapotaceae)	canopy tree	latex	chewing gum & glue	Sep–Feb	export to Japan, US, Europe	tapped from tree in rainy season
Xate	*Chamaedorea* spp. (Arecaceae)	understory palm	leaf	floral arrangements for weddings	Mar–Jun	export to US, Europe	cut from plant
Allspice	*Pimenta dioica* (Myrtaceae)	mid-canopy tree	ripe fruit	culinary & pickling spice	Jul–Aug	export to US, Europe, Russia	picked from felled branches
West Kalimantan							
Gaharu	*Aquilaria* spp. (Thymeliaceae)	mid-canopy tree	resin from infected heartwood	perfume for weddings & funeral rituals	continuous?	export to Saudi Arabia, Asia	cut from felled trees and then boiled out
Medang	*Litsea* spp. (Lauraceae)	mid-canopy tree	bark	aromatic mosquito repellent coils	continuous?	domestic urban areas, export to S.E. Asia	stripped from felled trees, cut, dried and pressed
Illipe nuts	*Shorea* spp. (Diptercarpaceae)	emergent tree	seed	cooking oil & chocolate	masting periods (3–7 yrs)	domestic & export to S.E. Asia	gathered from ground, dried and pressed
Ironwood	*Eusideroxylon zwageri* (Lauraceae)	canopy tree	timber	termite & rot resistant wood	continuous?	local & regional	hand felled, floated out on rafts
Damar	*Dipterocarpus* spp. (Dipterocarpaceae)	emergent tree	exuded resin	caulk for ship construction	continuous?	local & regional	collected from cuts in bark
Rattan	*Calamus* spp. (Arecaceae)	climbing liana	stem	furniture & construction	continuous?	regional & export	cut and pulled down from canopy
Wild fruits	various	understory & canopy trees	fruits & seeds	food	masting & seasonal	local & regional	picked or collected off the ground

is derived from the diseased heartwood of several *Aquilaria* species and used in perfume production, can bring as much as several thousand dollars per tree in local markets (deBeer & McDermott 1989; Dixon et al. 1991). This product has a sufficient value to induce harvesters to travel throughout the forest in search of infected trees (Jessup & Peluso 1986). For the majority of potential products, however, the economic return may not justify the time required to find them in the forest.

Temporal Availability of Products

Another important factor in determining the success of an extractive system is the temporal availability of products. Ecological constraints such as fruiting phenology and economic constraints such as seasonal demand can dictate that a given product be only periodically harvestable. An ideal extractive system should be based on a mix of products whose availability and demand periods are staggered so as to sustain harvest activities throughout the year.

Petén

In Petén, the phenology of and demand for the three main nontimber forest products allows them to be exploited at different times of the year in a complementary fashion. Chicle can be harvested only during the rainy season (August to January) when relative moisture, temperature, and rate of transpiration enhances the flow of latex through the trees (Karling 1934). Although xate palm fronds are harvested year-round, the peak demand period (in part created by the spring wedding season in the U.S. and Europe) occurs between March and June. Finally, allspice fruits can be harvested only in July and August when the crop is fully ripened, a period that falls between the end of peak xate demand and the onset of chicle collection. The sequential nature of these harvest seasons enables harvesters to earn steady cash income throughout the year while moderating the demands placed on continually available resources such as *xate*.

Kalimantan

In contrast, the forests of Kalimantan and Southeast Asia as a whole are characterized by extreme phenological cycles. Many species of trees produce fruit only once every 3–5 years during a *masting season,* at which time many taxa fruit simultaneously. Masting is hypothesized to be proximally cued by brief periods of warm, dry days (or associated cool, clear nights) and ultimately driven by the need to satiate seed predators (Janzen 1974; Ashton et al. 1988). According to this hypothesis, masting limits the numbers of fruit- and seed-eating animals in the forest by restricting food availability to brief periods, when the animals are swamped by more fruit than they can possibly consume.

This phenological pattern severely affects human fruit and seed "predators" in much the same way it does animals. Extraction systems and markets for many perishable fruit products are extremely difficult to maintain, given that the fruit may be available for only one month every several years. To make matters worse, during these periods of availability, local markets soften because everyone can easily obtain their own supply from the forest. In addition, the glut causes prices to bottom out in urban markets (Pinedo-Vasquez et al. 1990). The harvest of wild fruits also has to compete with that of related domestic fruits that follow the same phenological cycle. For example, wild durian (*Durio* spp.) are occasionally harvested by villagers but are rarely transported to urban markets, since more valuable domestic durians are available at the same time.

This phenological dilemma can be avoided to some degree by gathering nonperishable fruits. One example can be seen in the harvest of illipe nuts, seeds of *Shorea* spp., which are harvested during intense masting periods and pressed for their oils. Because the seeds can be converted into an unperishable commodity, a large industry has developed that provides an important though sporadic source of cash income to Dayak villagers. The large temporally (and spatially) clumped patches of seeds produced by the masting trees also offset some of the problems arising from low population densities of target species.

Product and Ecosystem Sustainability

Finally, it is important to consider the long-term sustainability both of populations of harvested species and of the overall ecosystem. Organisms in a complex ecosystem such as a tropical forest are interlinked in ways that may frustrate naive efforts at management. The removal of a given product from a forest may upset delicate natural balances, which could then reverberate throughout the ecosystem (Gilbert 1980; Terborgh 1988).

Extractive activities are most likely to be unsustainable if they result in killing or damaging individuals of the target species, especially if it is a long-lived organism such as a canopy tree. Extraction may also be unsustainable if it affects reproduction, thus depressing regeneration rates. For example, harvesting of dispersed seeds may inhibit seedling survival and population regeneration (Janzen 1970).

Harvesting in areas of high seed density (such as beneath the crown of the mother tree), however, may not have as negative an impact on regeneration, especially if seedling survival is largely density dependent (seedling survival is inhibited under conditions of high seed density). Fortunately, human and animal seed predators generally prefer these high density patches. Recent empirical work in Madre de Dios, Peru, reveals that Brazil nut (*Bertholletia excelsa*) harvesters collect nuts only under the most productive individual trees, ignoring many others (E. Oritz, personal communication). Moreover, they search only under the crown and ignore more peripheral zones where the seeds most important to recruitment might be expected to lie. If these widely dispersed seeds escape harvesters, then extractive activities may have little effect on regeneration (Peters 1990).

On the level of the overall ecosystem, extractive activities will ultimately be unsustainable if critical nutrients, such as nitrogen and phosphorus, are removed at rates greater than that at which they are being replenished (Likens et al. 1977). The prevailing paradigm is that tropical forests on very poor soils are delicately balanced ecosystems sustained by highly efficient nutrient recycling mechanisms (Jordan 1985; Stallard 1988). Many nontimber forest products contain substantial concentrations of limiting nutrients, and over the long term it is possible that the export of these nutrients may be unsustainable.

Petén

The extractive industries of the Petén have prospered for decades because the target resources have so far not been overexploited. Harvesting practices do not appear to damage plants or jeopardize population recruitment if done correctly. Individual chicle trees have been tapped for decades with seemingly minimal harmful effects, although inexperienced harvesters may fatally injure trees by cutting too deep (Karling 1934). Likewise, xate harvest, which does not take reproductive parts of plants and leaves enough fronds to support continued growth, theoretically seems sustainable. Cabrera-Madrid et al. (1990), however, compared xate palm densities in harvested areas with the protected forests of Tikal National Park and found that there was an average of 2279 and 2478 stems per ha in two harvested areas, whereas there was an average of 4506 stems per ha in the protected forest. It is more difficult to determine the effect of harvesting practices on the ecosystem as a whole, but harvested products probably contain minimal concentrations of critical nutrients.

Kalimantan

In Kalimantan, there is little or no available evidence concerning the long-term sustainability of nontimber forest product harvest practices. One might suspect, however, that some of the current harvest practices may be unsustainable, as they either destroy the harvested individual or remove reproductive parts. For example, harvests of both gaharu and medang (bark from *Litsea* spp. used in the production of mosquito repellent coils), as currently practiced, involve cutting down the tree to obtain the desired product. Likewise, the illipe nut harvest could be adversely impacting *Shorea* reproduction. Although reproductively active *Shorea* trees produce vast crops of seeds, only a few survive insect and vertebrate seed predators, even fewer are successfully dispersed to suitable germination sites, and fewer yet survive to become saplings and then adults (Howe & Smallwood 1982). Removal of illipe nuts may reduce the number of seeds going though these natural evolutionary bottlenecks to a level that may diminish effective replacement. Harvested products in Kalimantan, such as illipe nuts (which like other seeds are nutrient rich), could also conceivably constitute an unsustainable drain on the overall ecosystem over the long term.

Socioeconomic and Political Factors

Resource Tenure and Conservation Incentives

An important problem in many extractive systems is the lack of incentives for individuals to conserve available resources for long-term use. Instead, individuals respond to perverse incentives to overharvest resources that are often rooted in existing land and resource tenure regimes (McNeely 1988). Many forest products are open-access resources that are publicly owned, yet their use is not governed by formal or even informal rules (Ciriacy-Wantrup & Bishop 1985). In such situations, individual harvesters have little or no incentive to conserve or manage the resource. In contrast, traditional societies with established rules governing resource use operate under a managed system of common property that can potentially provide incentives for conservation (Berkes et al. 1989).

Petén

Although most of the land and resources in the Maya Biosphere Reserve are publicly owned, there is some precedent for resource tenure within the nontimber forest product industries. This tenure appears to be expressed in informal rules or understandings among contractors and harvesters that guide their interactions and harvesting practices. The limited number of contractors, and their informal monopolies over traditional harvesting areas, seem to give them a long-term interest in preserving the resource base. Likewise, many of the harvesters work with one another and with one contractor on a long-term basis, and therefore have incentives to cooperate in maintaining productivity (Schwartz 1990). As an example, several *xateros* mentioned to us that they would not generally harvest leaves within 15 minutes' walk of the camp; instead, these were to be reserved for harvesters who needed a few leaves to make up a full bundle at the end of the day.

Kalimantan

Parts of Kalimantan have well-documented common property regimes (see Jessup & Peluso 1986). In the natural forests surrounding Gunung Palung National Park, however, there appears to be little or no evidence of resource tenure or common property rules governing nontimber forest product extraction. Instead, resource extraction undergoes pronounced boom and bust cycles in which a resource is discovered, exploited, and then driven to near extinction (Gentry & Vasquez 1988).

The gaharu wood industry provides one of the best examples of this cycle. Gaharu was first harvested in the more densely populated regions of Southeast Asia (deBeer & McDermott 1989). As the stock of trees became depleted, the market shifted to more remote areas, such as Kalimantan. When a market developed in the Gunung Palung region in the mid 1980s, villagers suddenly began searching for gaharu. Most of the best quality wood was harvested by teams of professionals who swept the forest on month-long harvesting trips. Many others undertook harvesting trips as well, which generally resulted in the harvest of lower quality wood and even the felling of undiseased trees. As a result, the regional populations of gaharu were depleted within a few years (even in Gunung Palung National Park, where harvesting was illegal), and the industry moved on to other even more remote areas. The over-harvesting of gaharu, driven by its high value, provides an extreme example of the boom and bust cycles to which the extractive industry is prone. Similar yet less precipitous cases of resource depletion have occurred, however, with many other extractive products such as ironwood (a termite-resistant silca-impregnated wood from the species *Eusideroxylon zwageri*) and medang.

Physical and Social Infrastructure

Extraction of products for both household consumption and market sale requires that harvesters be able to transport products from the forest to the point of consumption or sale. This need for transport is especially important with regard to commercially-oriented products, since markets for these goods normally exist only in large towns or regional trading centers. Accordingly, the success of extractive systems often depends on the availability of developed physical (roads, trading boats) and social (middlemen, export companies) infrastructures that enable harvesters to transport and sell their products.

Physical infrastructure is of greatest importance when harvested products are delicate or perishable. Many potentially marketable forest fruits, for example, are not found in regional or international markets, in part because spoilage during transport cannot be avoided. Latexes and fibers, on the other hand, can survive rough handling and storage for long periods if transport is not readily available.

Social infrastructure, consisting of middlemen and traders, is also crucial (Jessup & Peluso 1986). It is difficult for many harvesters who live in remote villages to travel to regional markets. Accordingly, the presence of traders and storekeepers enables harvesters to sell their products at or near the village or harvesting site. In addition, the middlemen may provide credit and/or supplies, which enable villages living a hand-to-mouth existence to go on extended harvesting trips. Although the typical combination of low product prices and high interest rates on supplies ensures that economic rents accrue mostly to middlemen, in their absence the harvesters would earn nothing. This is not to say that such debt peonage is necessarily a desirable situation (Fearnside 1989a). Harvesters may nonetheless prefer its conditions to those offered by other more risky subsistence alternatives.

Petén

The extractive industries of Petén are based on commodities that are relatively durable and easily transported. In their processed form, chicle, allspice, and xate all last over three months in a marketable condition and are not very susceptible to transportation damage. In addition, the processing and transportation infrastructures established during the heyday of chicle extraction, consisting of forest camps, warehouses, airstrips, roads, and air and ship transportation routes, are still used today for the movement of all nontimber forest products.

Moreover, a well-established patron-client system exists that facilitates the organization and financing of thousands of harvesters and the transportation of their products. Middlemen contractors serve as patrons to teams of harvesters, whom they transport to harvest camps within the forest. The patrons loan harvesters money to pay for food and other supplies needed by their families during their absence. Harvesters also buy food on credit from the patron while in the forest, often at a 100% to 500% mark-up (Heinzman & Reining 1988). At the end of each week, harvesters are paid for their production, less loaned money. Over the past few years, competition among contractors for laborers has enabled harvesters to gain more leverage and, accordingly, better wages and working conditions.

Kalimantan

In Kalimantan, by contrast, there are few roads, and most transport is by water. Products harvested from the forest generally have to be small and light enough to be transported from the forest in small canoes that may have to be carried around obstructions in the streams. For some highly valuable products such as ironwood, it becomes worthwhile to expend enormous efforts to transport the wood (which is so heavy that it sinks and has to be lashed to rafts of floating wood). For many other products, however, the expected return may not compensate for the transport costs.

Although arrangements with middlemen exist in the Gunung Palung region, they tend to be somewhat informal. For example, medang harvesters generally go to the forest on their own for up to ten days at a time to harvest the bark, having first obtained supplies from a local shopkeeper at high interest rates. Although it is difficult to document, the lack of competition among middlemen for harvesting territories may in part be responsible for restricting extractive activities.

Product Demand

For many extractive industries, perhaps the most important impediment is the lack of demand for harvested products either in commercial markets or for household consumption

(Pinedo-Vasquez et al. 1990). This problem is compounded by the presence of synthetic substitutes for many products and the fact that many products that seem ideal in theory may conflict with established cultural taboos and preferences (deBeer & McDermott 1989). In general, products with a low demand elasticity (a unit-free measurement of change in demand in relation to change in price) will have steady markets, whereas products with a high elasticity will tend to be in high demand only when they can be obtained at low cost. It is also important to distinguish between demand originating in ephemeral fads from that based on long-term needs (Browder 1992).

Petén

The extractive reserve system in Petén has been successful because it was developed to meet an existing market demand. Chicle extraction, for example, was directly driven by the demand for chewing gum. When petroleum-based latex substitutes became available, the bottom fell out of the market and has only recently revived as some gum manufacturers have resumed using natural latex, and as other industrial uses for chicle (such as glue) have been developed. Similarly, annual extraction of allspice reportedly depends on (of all things) the success of the Russian herring catch, since the spice is a major component of the pickling brine.

Kalimantan

In Kalimantan as well, products such as gaharu that have a significant demand have been extensively exploited. Other readily available products, however, have been less exploited due to insufficient demand. For example, damar resin, an exudate collected from *Dipterocarpus* spp., is sold in many urban and village markets for use as a shipping caulk, but the overall demand is not great. Accordingly, the price paid to collectors makes extensive harvesting uneconomical. Furthermore, the price of damar is ultimately constrained by the price of synthetic caulking materials, which may be more abundant in large urban areas. Similarly, game animals could potentially be sustainably (or unsustainably) harvested in the Gunung Palung region (Caldecott 1988), but they are

not because Islamic law prohibits consumption of most wild animals. Animals such as the rusa deer (*Cervus unicolor*) that are not taboo, however, are readily eaten when available.

Pressures for Alternative Land Uses and Political Power

The final socioeconomic factor to consider in the design of extractive systems is the extent to which other land uses, such as timber production or conversion to agriculture, may economically and politically outcompete the extractive industries. Economic benefit-cost models can be used to compare the long-term net present value of maintaining extractive reserves versus using the land for timber production or agriculture (Peters et al. 1989). These models have limitations, however, in that they may make unrealistic assumptions about the various ecological and economic factors outlined above; they do not always account for consumption of products by local peoples and they may not be appropriate to forestry situations (Fearnside 1989b). Furthermore, there may be many cases in which the models demonstrate the exact opposite of what conservationists want to hear, namely that timber production may be the most profitable use of forest lands, especially in the face of high discount rates and logging incentives created by distorted tax systems and capital markets (Repetto 1988; Fearnside 1989b). Finally, extractive industries may be unable to compete with plantations or agroforestry systems that can produce similar products through a more intensive use of the land.

The limitation of the economic benefit-cost approach is especially significant in light of current political realities. It may be possible to show that a given extractive system would have a higher net present value than a logging operation on the same area of land. Nonetheless, if the benefits of the extractive reserve accrue to local villagers and small-scale middlemen, whereas the economic rents from logging would be captured by urban elite with strong political connections, it is not hard to guess which system would win out. Similarly, land-hungry agricultural transmigrants from densely populated areas generally have greater political influence than harvesters living in remote forest villages. Although recent events in Eastern Europe graphically demonstrate the potential for rapid political reform, extractive reserve design should take into account existing political realities.

Extractive reserves can generally support only low human population densities. Fearnside (1989a) found that experienced brazil-nut harvesters in the Amazon used 300–500 ha of forest for each family (a population density of 1.0–1.7 persons per km^2). Needless to say, it may be extremely difficult for land planners to justify extractive reserves that require tens or hundreds of hectares of forest to support one household in countries that have extremely high population densities.

Ultimately, even if extractive reserves are more economically sensible than competing land-use alternatives, there is no guarantee that the society as a whole will use economic efficiency as a decision-making criterion.

Petén

In the Petén, the extractive system has persisted in part because some of the most powerful people in the region (including members of the military) capture the majority of rents from extracted products. Logging has had some impact on deforestation rates over the past decades. There is also increasing pressure to convert forest to agricultural lands. Nonetheless, the government has temporarily halted timber concessions and banned immigration into the newly created Maya Biosphere Reserve. In part, this decision indicates that maintaining the forest as an extractive reserve is in the interests of politically powerful individuals. The extent of this commitment may be tested in the near future, given recent reports that the loggers, who have exhausted the mahogany and cedar trees that were their primary timber sources, are now beginning to cut chicle trees illegally.

Kalimantan

In Kalimantan, by contrast, extractive reserve rents accrue to Dayak and Melayu villagers and to Chinese traders and storekeepers, who can be very wealthy but have little or no political power. Instead, political power is held by the owners of timber companies and by the Javanese and Balinese transmigrants. A stark example of the effects of these power dynamics can be seen in one of our study villages, which is located in the midst of a recently exploited timber concession. Over the past decade, the forest surrounding the Dayak village has been stripped away, effectively eliminating many extractive activities that were occurring there. In the initial years, villagers were minimally employed by the timber company, but today the few jobs that are available are held by Javanese immigrants.

Logging in Kalimantan is big business, generating enormous revenues (Gillis 1988; Repetto 1988). Wood is the major export in the region; the value of Indonesian timber exports in the early 1980s was roughly an order of magnitude greater than that of nontimber forest products (Gillis 1988). Although no explicit study of the potential value of extractive reserves in the region has been performed, it seems unlikely that current levels of nontimber forest product extraction will be comparable, even over the long term. Over time, it may be possible for indigenous groups to organize and obtain power as the Brazilian rubber tappers union has done, but this is a difficult process that is compounded by the Southeast Asian cultural reluctance to defy existing authorities. Overall, it does not seem realistic to imagine that Dayak villagers will be able to compete effectively with urban elite and multinational timber corporations for access to the forest.

Perhaps the ultimate constraint on the extraction of nontimber forest products from natural forest in the Gunung Palung region is that village residents have an alternative source for many forest products. Over the past centuries, residents of this area have developed extensive forest gardens, a multi-species agroforestry system that produces a wide variety of products similar to those harvested from natural forest. Harvest of products from these managed agroforestry systems has a number of advantages over extraction from natural forests, in that the forest gardens (1) offer enhanced densities of desired species, (2) have phenological cycles released from the masting pattern (and thus produce fruit more often), (3) are to some degree protected against animal competitors, (4) enjoy clearly defined property rights, (5) are located near villages, and (6) provide products that are culturally preferred and of a higher quality. For example, although many fruits such as the langsat (*Lansium* spp.) grow in both forest and gardens, there is little or no incentive to harvest the wild fruits for either market sale or household consumption since they generally ripen at the same time as those in the forest gardens. Similarly, the secure tenure rights of the gardens make them the only place in the region near settlements in which relatively large ironwood trees can be found.

Discussion

Table 2 presents a summary of the ecological and socioeconomic factors that determine the potential and limitations of a nontimber forest product extraction system. As demonstrated by the above examples, the ecological and socioeconomic conditions of Petén generally favor the creation of extractive reserves. Extractive systems in Petén have the potential to be economically sustainable while preserving relatively intact forest. In Kalimantan, on the other hand, prevailing ecological and socioeconomic conditions make it unlikely that extractive reserves will play a major part in saving the rain forest. This is not to say that nontimber forest

TABLE 2. Summary of extraction systems in Petén and W. Kalimantan (see text for detailed explanations of rankings).

Parameter	Petén	Kalimantan	Significance	Potential management options
Ecological Factors				
density of exploited species	high	low	determines search and travel times	enrichment plantings
temporal availability of products	regular	irregular (masting)	determines steadiness of harvesting income	artificial selection
sustainability (species level)	moderate (?)	low (?)	determines long-term viability of exploited species	seedling plantings, bottleneck removal
sustainability (overall system)	moderate (?)	?	determines long-term viability of overall system	reduction of nutrient loss
Socioeconomic and Political Factors				
resource tenure	partially specified	open-access	determines incentives for resource conservation	legalize and enforce tenure rights
physical infrastructure	moderately developed	minimally developed	determines ability to transport products to market	build infrastructure
social infrastructure	highly developed	moderately developed	determines opportunities for harvesters to find work	co-ops and incentives for fair middlemen
market demand	moderate to high	low to high	determines degree to which products will be harvested	marketing techniques
political power of industry participants	moderate	minimal	determines degree to which extractive system can be maintained	political action and education
pressure for alternative land use	minimal	strong	determines competition to convert forest lands	holistic land management strategies
existence of alternative agroforestry system	no	yes	determines labor and resources invested into forest system	holistic land management strategies

product extraction has no role to play in this region, but rather that it will have to be a part of a diversified multiple-use management plan.

As outlined in Table 2, many of the ecological limitations on extractive reserves could ultimately be overcome by the design and implementation of careful management regimes. For example, low species densities in a forest could be artificially enhanced by enrichment plantings of desired species (Gómez-Pompa & Kaus 1990; Viana 1990; Leighton et al. 1991). Phenological limitations could be offset by developing new products that can be harvested in off-periods (Heinzman & Reining 1988; Dixon et al. 1991) or by finding genetic strains of existing products that produce fruit at desirable times (Peters 1991). Maintenance of population levels could be enhanced through management techniques and educational efforts ensuring that sufficient reproduction occurs (Leighton et al. 1991; Robinson & Redford 1991). And, finally, ecosystem sustainability could be enhanced through reduction of nutrient removals by techniques such as husking fruits while still in the forest (Jordan 1985).

Socioeconomic factors limiting extractive reserves can also be mitigated through various management regimes. For example, resource tenure could be granted to open-access resources, or existing traditional rights could be recognized (Bromley & Cernea 1989). Infrastructure could be enhanced to improve transportation (Browder 1990b). Cooperatives could be organized to allow harvesters to transport their products to market without losing most of the rents to middlemen (Jessup & Peluso 1986). Marketing techniques could be employed to increase both domestic and export demand for nontimber forest products (deBeer & McDermott 1989; Dixon et al. 1991). Political action and education could be used to obtain political power for disenfranchised forest villagers and to formally recognize the benefits they may be receiving from the forest (Allegretti 1990). And, finally, holistic land management strategies could be used to obtain the maximum possible sustained economic return from a region (McNeely & Miller 1984; McNeely et al. 1990).

In considering any one of these potential management strategies, it is very important to distinguish between what is technically feasible and what is politically and economically feasible. For example, an enrichment planting of a given species may be biologically feasible, but if the trees require decades to mature and produce a salable product, the incentive to plant the trees will be minimal. It is also important to consider that the relevant factors may be interlinked in complex ways. For instance, building roads into a forest to enhance access to products may lead to overharvesting and also enhance rates of deforestation by loggers and agriculturalists. Accordingly, management strategies will have to be carefully designed and implemented to suit the unique ecological and cultural context of each local situation.

Ultimately, any proposed land use, such as an extractive reserve, needs to be considered as part of a comprehensive regional strategy. Extractive reserves are not stand-alone solutions, but they can effectively complement other land uses. Existing land-use patterns in the study villages in Kalimantan provide one model of such a land-use spectrum that encompasses farms, home gardens, forest gardens, timber and nontimber extraction areas, and natural forest (Fig. 3). This diverse arrangement of natural and managed habitats provides a wide range of products and environmental services to local human inhabitants, while buffering core conservation areas. Likewise, in Guatemala, the Maya Biosphere Reserve (Nations 1989) provides another model of holistic land-use planning that encompasses agriculture, timber and nontimber forest product extraction, ecotourism, and protected forest.

Conclusions

Based on our observations in Petén and Kalimantan, we offer the following conclusions concerning the extractive reserve concept and appropriate uses for tropical forest lands.

- Extractive reserves are not the panacea that some people would have them be and by themselves will not save the world's rain forests.

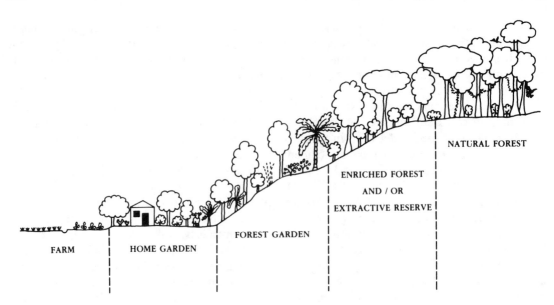

FIGURE 3. The land-use spectrum in the Gunung Palung region. The following is a list of some products and services generated by the different components of the land-use spectrum. Products are listed by English name or, where no common English name exits, by regional and/or local name with scientific name, following Sastrapradja et al. (1977a, 1977b, 1977c, 1978).

Farms: rice, soybeans, cassava, various vegetables, grass (for tethered cattle).

Home gardens: bananas, papaya, pineapple, hot peppers, rambutan (*Nephilium lappaceum*), ornamental flowers and shrubs, chickens, ducks, goats.

Forest gardens: durian (*Durio zibenthinus*), rubber, coffee, bamboo, sugar palm (*Arenga sacharifera*), pakawai (*Durio* spp.), mangosteen (*Garcinia mangostana*), langsat and duku (*Lansium* spp.), jambu (*Szygium* spp.), terap (*Artocarpus elasticus*), nangka (*Artocarpus heterophyllus*), cempadak (*Artocarpus integer*), jeluti (*Artocarpus* spp.), machang (*Mangifera foetida*), kuweni (*Mangifera* spp.), belimbing (*Averhoa* spp.), lengkeng (*Euphoria longen*), rambai (*Baccaurea motleyana*), gandaria (*Bouea macrophylla*), tengkawang (*Shorea* spp.), and more.

Enriched forests and/or extractive reserves (potentially): various fruits, timber, rattan, gaharu, damar, medang, illipe nuts, medicinal plants, game animals.

Natural forests: new fruit species, new strains of existing fruits, pollinators, ground water recharge, carbon sink, eco-tourism, and option, existence and spiritual values.

- Extractive systems can, however, be an important component of a broader land-use spectrum that includes agricultural plots, forest gardens, timber and nontimber extraction areas, and conservation forest and provides a variety of products and services.
- Local ecological, economic, political, and cultural conditions need to be understood and incorporated into decisions concerning the feasibility, role, location, and extent of extractive reserves, as well as the design, implementation, and management of such systems.
- Various management strategies and policy decisions can facilitate the development of successful extractive systems, but they need to

be carefully selected so as not to cause unintended negative effects.

Acknowledgments

N. Salafsky and B. L. Dugelby would like to thank the National Council of Protected Areas (CONAP), the National Commission on the Environment (CONAMA), the Center for Conservation Studies, University of San Carlos (CECON), and the Tropical Agronomic Research Center (CATIE) for sponsoring and supporting our work in the Petén, and the Indonesian National MAB Committee-LIPI, the Center for Research and Development in Biology-LIPI (Puslitbang Biology-LIPI), and the Forest Conservation Service (PHPA) for sponsoring our visit to Kalimantan. We would like especially to thank the residents of the villages of Uaxactun, Carmelita, Keranji, and Benawai Agung, who took us into their homes and put up with our questions.

In addition, we are grateful for the assistance and advice that we received from all of those who helped us along the way, including A. Lehnhoff, H. Rivera, I. Ponciano, B. Heinzman, M. Cabrera Madrid, S. López, C. Reining, A. Solórzano, and M. McLemore in Guatemala, and T. Heinald, D. Sastrapradja, Soetikno W., J. Sugardjito, E. Sumaradja, H. Prayitno, Amir, M. Leighton, Ah-Yen, C. Knab, M. Kusneti, H. Makinuddin, Tadyn, C. Webb, N. Yonkow, S. Zens, I. Koswara, E. Hammond, C. Peters, and E. Saffran in Indonesia.

Funding for our travels was provided by the Sally Hughes-Schrader Travel Grant of the Duke Sigma-Xi Chapter and the Duke Graduate School, the Lazar Fellowship Fund of the Duke School of Forestry and Environmental Studies International Group, the Duke-University of North Carolina Latin American Studies Program, the Josiah Charles Trent Memorial Foundation, and the Explorer's Club. Support for writing this paper was provided by the Duke University Center for Tropical Conservation under a cooperative agreement with the U.S. Agency for International Development.

We thank J. Browder, L. Curren, C. Danks, C. Knab, M. McKean, R. Oren, G. Paoli, D. Peart, F. Putz, N. Schwartz, and three anonymous referees for reading and providing helpful comments on various drafts of this manuscript, and V. Salafsky for drawing and typesetting the figures and maps.

References

Alcorn, J. B. 1989. An economic analysis of Huastec Mayan forest management. Pages 182–206 in J. O. Browder, editor. Fragile lands of Latin America: strategies for sustainable development. Westview Press, Boulder, Colorado.

Allegretti, M. H. 1990. Extractive reserves: an alternative for reconciling development and environmental conservation in Amazonia. Pages 252–264 in A. B. Anderson, editor. Alternatives to deforestation: steps toward sustainable use of the Amazonian rain forest. Columbia University Press, New York, New York.

Anderson, A. B. 1990. Extraction and forest management by rural inhabitants in the Amazon estuary. Pages 65–85 in A. B. Anderson, editor. Alternatives to deforestation: steps toward sustainable use of the Amazonian rain forest. Columbia University Press, New York, New York.

Anderson, A. B., and M. A. G. Jardim. 1989. Costs and benefits of floodplain forest management by rural inhabitants in the Amazon estuary: a case study of Açai palm production. Pages 114–129 in J. O. Browder, editor. Fragile lands of Latin America: strategies for sustainable development. Westview Press, Boulder, Colorado.

Ashton, P. S., T. J. Givinish, and S. Appanah. 1988. Staggered flowering in the Dipterocarpaceae: new insights into floral induction and the evolution of mast fruiting in the seasonal tropics. American Naturalist 132:44–66.

Berkes, F., D. Feeny, B. J. McCay, and J. M. Acheson. 1989. The benefits of the commons. Nature 340:91–93.

Bromley, D. W., and M. M. Cernea. 1989. Management of common property natural resources: some conceptual and operational fallacies. World Bank, Washington, D.C.

Browder, J. O. 1990a. Extractive reserves will not save tropics. BioScience 40:626.

Browder, J. O. 1990b. Beyond the limits of extraction: tropical forest alternatives to extractive reserves. The Rainforest Harvest Conference, London, England.

Browder, J. O. 1992. The limits of extractivism. BioScience 42:174–182.

Burkhill, I. H. 1935. A dictionary of the economic products of the Malay Peninsula. Crown Agents for the Colonies, London, England.

Cabrera Madrid, M., R. Heinzman, S. López, C. Reining, and A. Solórzano. 1990. Non-timber forest products in the Maya Biosphere Reserve: results of ecological and socioeconomic surveys and recommendations for management and investigations. Unpublished draft report.

Caldecott, J. 1988. Hunting and wildlife management in Sarawak. International Union for the Conservation of Nature and Natural Resources, Gland, Switzerland.

Charnov, E. L. 1976. Optimal foraging: the marginal value theorem. Theoretical Population Biology 9:129–136.

Ciriacy-Wantrup, S., and R. Bishop. 1985. Common property as a concept in natural resource policy. Natural Resources Journal 4:173–727.

deBeer, J. H., and M. J. McDermott. 1989. The economic value of non-timber forest products in Southeast Asia with an emphasis on Indonesia, Malaysia, and Thailand. Netherlands Committee for the International Union for the Conservation of Nature and Natural Resources, Amsterdam, The Netherlands.

Dixon, A., H. Roditi, and L. Silverman. 1991. From

forest to market: a feasibility study of the development of selected nontimber forest products from Borneo for the U.S. market. Project Borneo/Harvard Business School, Cambridge, Massachusetts.

Fearnside, P. M. 1989a. Extractive reserves in Brazilian Amazonia. BioScience 39:387-393.

Fearnside, P. M. 1989b. Forest management in Amazonia: the need for new criteria in evaluating development options. Forest Ecology and Management 27:61-79.

Foster, R. B., and N. Brokaw. 1982. Structure and history of the vegetation of Barro Colorado Island. Pages 67-82 in E. G. Leigh, A. S. Rand, and D. M. Windsor, editors. The ecology of a tropical forest: seasonal rhythms and long-term changes. Smithsonian Institution Press, Washington, D.C.

Gentry, A. H. 1988. Changes in plant community diversity and floristic composition on evolutionary and geographic gradients. Annals of the Missouri Botanical Garden 75:1-34.

Gentry, A. H., and R. Vasquez. 1988. Where have all the Ceibas gone? A case history of mismanagement of a tropical forest resource. Forest Ecology and Management 23:73-76.

Gilbert, L. 1980. Food web organization and the conservation of neotropical diversity. Pages 11-34 in M. E. Soulé and B. A. Wilcox, editors. Conservation biology: an evolutionary-ecological perspective. Sinauer Associates, Sunderland, Massachusetts.

Gillis, M. 1988. Indonesia: public policies, resource management, and the tropical forest. Pages 43-113 in R. Repetto and M. Gillis, editors. Public policies and the misuse of forest resources. Cambridge University Press, Cambridge, England.

Gómez-Pompa, A., and A. Kaus. 1990. Traditional management of tropical forests in Mexico. Pages 45-64 in A. B. Anderson, editor. Alternatives to deforestation: steps toward sustainable use of the Amazonian rain forest. Columbia University Press, New York, New York.

Heinzman, R. M., and C. S. Reining. 1988. Sustained rural development: extractive forest reserves in the Northern Petén of Guatemala. United States Agency for International Development, Guatemala City, Guatemala.

Heinzman, R. M., and C. S. Reining. 1989. Nontimber forest products in Belize and their role in a biosphere reserve model. Unpublished report. Institute of Economic Botany, New York Botanical Garden, Bronx, New York.

Howe, H. F., and J. Smallwood. 1982. Ecology of seed dispersal. Annual Review of Ecology and Systematics 13:201-28.

Janzen, D. H. 1970. Herbivores and the number of species in tropical forests. American Naturalist 104:501-528.

Janzen, D. H. 1974. Tropical blackwater rivers, animals, and mast fruiting by the Dipterocarpaceae. Biotropica 6:69-103.

Jessup, T. C., and N. L. Peluso. 1986. Minor forest products as common property resources in East Kalimantan, Indonesia. Common Property Resource Management, National Academy Press, Washington, D.C.

Jordan, C. F. 1985. Nutrient cycling in tropical forest ecosystems. John Wiley and Sons, Chichester, England.

Karling, J. S. 1934. Dendrograph studies on Achras zapota in relation to the optimum conditions for tapping. American Journal of Botany 21:161-193.

Krebs, J. R., and R. H. McCleery. 1984. Optimization in behavioral ecology. Pages 91-121 in J. R. Krebs and N. B. Davies, editors. Behavioral ecology: an ecological approach. Sinauer Associates, Sunderland, Massachusetts.

Leighton, M., P. C. Schulze, and D. R. Peart. 1991. Appraisals of enrichment planting in selectively logged forest in Kalimantan: preliminary analysis of economic and ecological variables. Conference on Interactions of People and Forests in Kalimantan. New York Botanical Garden, Bronx, New York.

Likens, G. E., F. H. Bormann, R. S. Pierce, J. S. Eaton, and N. M. Johnson. 1977. Biogeochemistry of a forested ecosystem. Springer-Verlag, New York, New York.

Lindell, C. L. 1937. The vegetation of Petén. Carnegie Institute of Washington, Washington, D.C.

McNeely, J. A. 1988. Economics and biological diversity: using economic incentives to conserve biological resources. International Union for the Conservation of Nature and Natural Resources, Gland, Switzerland.

McNeely, J. A., and K. A. Miller (eds.) 1984. National parks, conservation, and development: the role of protected areas in sustaining society. Smithsonian Institution Press, Washington, D.C.

McNeely, J. A., K. R. Miller, W. V. Reid, R. A. Mittermeier, and T. B. Werner. 1990. Conserving the world's biodiversity. International Union for the Conservation of Nature and Natural Resources, Gland, Switzerland.

Nations, J. D. 1989. La Reserva del al Biosfera Maya, Petén: estudio tecnico. Report to the Guatemalan National Council of Protected Areas (CONAP), Guatemala City, Guatemala.

Nations, J. D., and R. B. Nigh. 1980. The evolutionary potential of Lacondon Maya sustained-yield tropical forest agriculture. Journal of Anthropological Research 36:1-30.

Peters, C. M. 1990. Population ecology and management of forest fruit trees in Peruvian Amazonia. Pages 86–98 in A. B. Anderson, editor. Alternatives to deforestation: steps toward sustainable use of the Amazonian rain forest. Columbia University Press, New York, New York.

Peters, C. M. 1991. Population ecology and management of illipe nut in a mixed dipterocarp hill forest. Conference on Interactions of People and Forests in Kalimantan. New York Botanical Garden, Bronx, New York.

Peters, C. M., A. H. Gentry, and R. O. Mendelsohn. 1989. Valuation of an Amazonian rain forest. Nature 339:655–656.

Pinedo-Vasquez, M., D. Zarin, P. Jipp, and J. Chota-Inuma. 1990. Use-values of tree species in a communal forest reserve in Northeast Peru. Conservation Biology 4:405–416.

Prance, G. T. 1989. Economic prospects from tropical rainforest ethnobotany. Pages 61–74 in J. O. Browder, editor. Fragile lands of Latin America: strategies for sustainable development. Westview Press, Boulder, Colorado.

Repetto, R. 1988. The forest for the trees: government policy and the misuse of forest resources. World Resources Institute, Washington, D.C.

Robinson, J. G., and K. H. Redford. 1991. Sustainable harvest of neotropical forest animals. Pages 415–429 in J. G. Robinson and K. H. Redford, editors. Neotropical wildlife use and conservation. University of Chicago Press, Chicago, Illinois.

Sastrapradja, S., S. H. Lubis, E., Djajasukma, H. Soetarno, and I. Lubis. 1977a. Sayur-sayuran (Vegetables). Lembaga Biologi Nasional (LBN-LIPI), Bogor, Indonesia.

Sastrapradja, S., N. Wulijarni-Soetijptu, S. Danimihardja, and R. Soejono. 1977b. Ubi-ubian (Root and Tuber Crops). Lembaga Biologi Nasional (LBN-LIPI), Bogor, Indonesia.

Sastrapradja, S., U. Sutisna, G. Panggabea, J. P. Mogea, S. Sukardjo, and A. T. Sumarto. 1977c. Buah-buahan (Fruits). Lembaga Biologi Nasional (LBN-LIPI), Bogor, Indonesia.

Sastrapradja, S., S. Danimihardja, R. Soejono, N. W. Soetjipto, and M. S. Prana. 1978. Tanaman Industri (Industrial Plants). Lembaga Biologi Nasional (LBN-LIPI), Bogor, Indonesia.

Schwartz, N. B. 1990. Forest society: a social history of Petén, Guatemala. University of Pennsylvania Press, Philadelphia, Pennsylvania.

Schwartzman, S. 1989. Extractive reserves: the rubber tappers' strategy for sustainable use of the Amazon rainforest. Pages 150–165 in J. O. Browder, editor. Fragile lands of Latin America: strategies for sustainable development. Westview Press, Boulder, Colorado.

Stallard, R. F. 1988. Weathering and erosion in the humid tropics. Pages 225–246 in A. Lerman and M. Meybeck, editors. Physical and chemical weathering in geochemical cycles. Kluwer Academic Publishers, Dordrecht, The Netherlands.

Terborgh, J. W. 1988. The big things that run the world—a sequel to E. O. Wilson. Conservation Biology 2:402–403.

Vasquez, R., and A. H. Gentry. 1989. Use and misuse of forest harvested fruits in the Iquitos area. Conservation Biology 3:350–361.

Viana, V. M. 1990. Seed and seedling availability as a basis for management of natural forest regeneration. Pages 99–115 in A. B. Anderson, editor. Alternatives to deforestation: steps toward sustainable use of the Amazonian rain forest. Columbia University Press, New York, New York.

Voorhies, B. 1982. An ecological model of the Early Maya of the Central Lowlands. In K. V. Flannery, editor. Maya subsistence. Academic Press, New York, New York.

Wallace, A. R. 1878. Tropical nature and other essays. Macmillan, London, England.

Whitmore, T. C. 1984. Tropical rain forests of the Far East. Clarendon Press, Oxford, England.

27
Sustainable Use of the Tropical Rain Forest: Evidence from the Avifauna in a Shifting-Cultivation Habitat Mosaic in the Colombian Amazon

Germán I. Andrade and Heidi Rubio-Torgler

Introduction

Many bird species of Amazonian forests are adapted to habitat mosaics produced by natural disturbance regimes (Haffer 1991), among which the most important are large-magnitude and high-intensity disturbances (*sensu* Connell & Slatyer 1977) produced by the action of the rivers on the upland forests (Foster 1980; Salo et al. 1986). The river-created habitats support as much as 15% of the avifauna in the Amazon river banks (Remsen & Parker 1983). These habitats are also the source of many species that invade disturbed areas in the upland forests (Terborgh & Weske 1969). Within the "terra firme" forest, tree-fall gaps are the most common natural disturbances (small magnitude and low intensity; *sensu* Connell & Slatyer 1977) and are integral to forest dynamics (Denslow 1987). Schemske and Brokaw (1981) observed in Panama that natural tree-fall gaps sustain 26% of the bird species that inhabit the forest interior. Similar results were obtained in Puerto Rico and Costa Rica (Wunderle et al. 1987; Levey 1988*a*). More recently, attention has been drawn to the study of the effect of anthropogenic disturbance, especially on the composition and diversity of avifaunas and the ability of bird species to survive in fragmented and isolated habitats (Willis 1979; Bierregaard 1986; Lovejoy & Bierregaard 1990).

The study of anthropogenic disturbances that lead to secondary growth has received less attention, even though this is the most widespread kind of disturbance in extensive regions of western Amazonia (Johns 1992). Blake et al. (1990) described the avifauna of secondary habitats of different age at La Selva (Costa Rica), showing that some invader species originated in open land, whereas others were canopy birds that follow the foliage-air interface of the forest (*sensu* Stiles 1983). In human-made habitat mosaics in Pará (Brazil), Novaes (1980) found that 42% of bird species are associated with disturbed areas. Finally, in Tefé (Brazilian Amazon), Johns (1992) compared the avifauna of undisturbed forest, slightly logged forest, secondary growth, and crop fields, and he concluded that many species are found in most habitat types, although the similarity decreases with increasing disturbance intensity. Johns reported that the overlap index between early secondary growth and undisturbed understories is almost zero (see also Lovejoy 1974). Johns (1992) emphasized that there are no data available to determine at which stage of the regeneration process the original understory avifauna is restored. In this paper we address this question, by comparing the composition of the avifauna in a chronoseries of known-age regenerating plots created by shifting agriculture and undisturbed understories in the Colombian Amazon.

Reprinted with permission from Conservation Biology vol. 8, pp. 545–554. Copyright 1994 The Society for Conservation Biology and Blackwell Science, Inc.

Study Area and Methods

This study was carried out in the "terra firme" forest (*sensu* Pires & Prace 1985) at the Mmtí-Paraná river in the Colombian Amazon (Figure 1.). The climate is warm and wet (>3500 mm annual precipitation, mean annual temperature 24°C; Walschburger & von Hildebrand 1989). Shifting agricultural practices in the area include the slash and burn of areas of about one hectare, which are planted, harvested and abandoned within five years, with decreasing intensity of use over this period. Secondary growth is then allowed to continue until the original structure of the forest is regenerated (Walschburger & von Hildebrand 1990). The secondary areas studied are described in detail by Walschburger and von Hildebrand (1990), who determined, based on physiognomy and floristics, two distinct phases of vegetation development: "young secondary growth" from abandonment to five years, composed of weeds and short-lived shrubs; and "old secondary growth" from five to about 20 years of abandonment. This phase is characterized by typical fast-growing trees. Between July 1985 and April 1986, bimonthly captures of birds with mist nets (12 m

by 2.5 m, 35 mm mesh) located at ground level (0–2.5 m) were carried out in four primary forest and six secondary areas that had been abandoned from one to 17 years previously. Mist nets were opened before dawn and closed at midday. Sampling effort was similar for all sites (mean: 422.4 net hours, CV: 1.47%), for a total of 4323 net hours (Table 1). Captured birds were identified, measured, banded, and released at their original capture site. In this article we present data only on the composition of the avifauna along the forest regeneration gradient. Percentage similarity indexes (Beals 1960) between pairs of stations were calculated following Schemske & Brokaw (1981) and Wunderle et al. (1987), in which

$$PS\ 1\ 100 - \sum_{1/2} |ai - bi| = \sum_{min} (ai,\ bi)$$

where ai is the percentage of individuals of species i in the sample a and bi is the percentage of individuals of the species i in the sample b. A matrix of similarity indices between pairs of sampled sites was prepared. Trophic structure was defined according to the modified categories of Robinson and Terborgh (1990), based on diet and strata. Species were assigned to 11

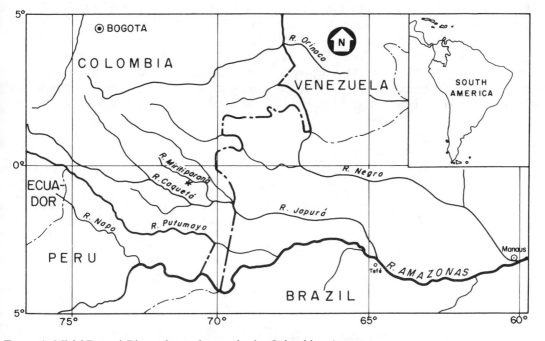

FIGURE 1. Mmtí Paraná River, the study area in the Colombian Amazon.

TABLE 1. Summary of captures.

Age (years)	YSG			OSG			U			
	1–2	2–3	5	7–10	13–15	15–17	32	33	29	34
Species	36	5	45	27	38	34	32	33	29	34
Captures	115	71	114	45	90	105	79	82	75	102
Net hours	419	425	430	429	426	425	423	423	412	412
Captures/net hours	0.27	0.17	0.26	0.10	0.21	0.25	0.19	0.19	0.18	0.25

YSG = young secondary growth; OSG = old secondary growth; U = undisturbed understories (forest).

trophic groups based on Sick (1985) and Johns (1992), on field observations, and on a preliminary analysis of fecal samples (Appendix).

Results

Overall at ground level, 878 captures of 103 species belonging to 20 families were obtained (Appendix). Similarity indices between pairs of stations are presented in Table 2. Comparing similarity indices within each habitat type, there are no consistent differences among the sites (Kruskal Wallis one-way ANOVA, KW = 4.4, $p > 0.1$), suggesting similar levels of local heterogeneity. On the other hand, similarity indices between habitats showed rank values of 13.6 for young secondary growth and old secondary growth, 11.7 for young secondary growth and understories, and 24.9 for old secondary growth and understories (KW statistic = 12.7, $p < 0.0017$). These values indicating that young secondary growth avifauna was least similar to other sites, whereas old

secondary growth and primary forest had the most similar avifauna.

In order to avoid artifacts produced by single captures, only species captured more than once were taken into account when community-wide comparisons were made. The low capture frequency of many species also limits the possibility of an analysis of habitat selection. In fact, 38 species (36.8%) were represented by single captures.

Four species (4.8%) were exclusive of understories; among them only *Picummnus* spp, *Cercomacra tyrannina*, *Myrrmeciza athrotorax*, *Thamnophilus murinus*, *Empidonax traillii*, *Thryothorus coraya*, *Troglodytes aedon*, *Ramphocelus carbo*, *Oryzoborus angolensis*, and *Saltator maximus* are true second-growth species. Associated with disturbed areas were 49 species (47.5%), among which 10 (10.6%) were exclusive of young plots, five (4.8%) of old secondary growth, and three (2.9%) of both types of secondary growth. Species exclusively found in young secondary growth and with single captures were *Manacus manacus*, *Myrmotherula obscura*, and *Microbates collaris*.

TABLE 2. Similarity index between pairs of sites.

Sites (years)	1–2	2–3	5	7–10	13–15	15–17	U	U	U	U
1–2	–	31.3	45.7	36.6	40.8	32.1	38.7	14.5	25.8	43.8
2–3	–	–	38.9	42.7	47.6	41.3	39.4	42.2	45.9	39.5
5	–	–	–	37.3	44.6	38.0	41.9	24.3	32.4	42.8
7–10	–	–	–	–	53.9	57.7	57.6	33.2	45.1	51.8
13–15	–	–	–	–	–	51.8	50.4	39.5	44.4	58.5
15–17	–	–	–	–	–	–	60.5	50.8	54.9	53.8
U	–	–	–	–	–	–	–	45.0	56.9	52.2
U	–	–	–	–	–	–	–	–	58.6	37.7
U	–	–	–	–	–	–	–	–	–	46.1
U	–	–	–	–	–	–	–	–	–	–

U = undisturbed understories (forest sites).

Distribution of abundance (percentage of species in different abundance classes) for the three major habitat types has a typical pattern (Figure 2). Significant differences, however, were found for distributions between young secondary growth and old secondary growth (two-sample Kolmogorov-Smirnow test, $p <$ 0.05) and between old secondary growth and forest (Kolmogorov-Smirnow, $p < 0.01$). These differences are attributable mainly to a higher proportion of rarely captured species (0.15%) in old secondary growth originating mainly from forest edges or the canopy, all of which were represented by single captures. These include *Jacamerops aurea, Nonnula brunnea, Selenidera reinwardtii, Attila spadiceus, Laniocera hypopirrha, Pseudattila phoenicurus, Euphonia chlorotica, Cyanerpes ceruleus,* and *Tangara callophrys.* Another cause of between-habitat differences is the higher proportion of common species (capture proportion more than 5%) in the understory, such as *Pithys albifrons* and *Pipra coronata.*

For some common species several tendencies are clear. *Pipra coronata* did not show a preference for a particular habitat. Others, such as *Pithys albifrons* and *Hylophylax poecilonota,* for unknown reasons concentrated in young secondary habitats and in forest. *Pipra erythrocephala,* found in all habitat types, was captured more often in old secondary growth. Among the 11 species found only in early regenerating areas, *Tyrannulus elatus* and *Saltator maximus*—the former a foliage-air interface specialist and the latter found at edges and openings—were captured at high rates. The remaining nine were lower in abundance, coinciding with old secondary growth, where all five species were of the low abundance type. Relative to the total avifauna, species that use the foliage-air interface do not contribute substantially to the young second-growth samples (about 5% of the captures and six species).

There were small differences in the number of species captured per trophic class in the three habitats (Figure 3). All trophic classes were present in all habitats, except the canopy and shrub insectivores (*Myiopagis caniceps* and *Empidonax traillii* respectively) and one transient piscivore (*Chloroceryle india*) only found in the understory. Insectivores were present in all habitat types; ant-followers, however, were

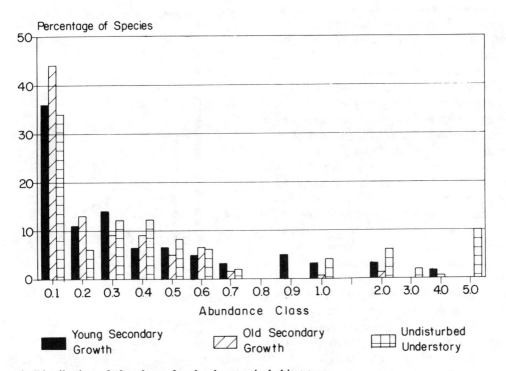

FIGURE 2. Distribution of abundance for the three main habitat types.

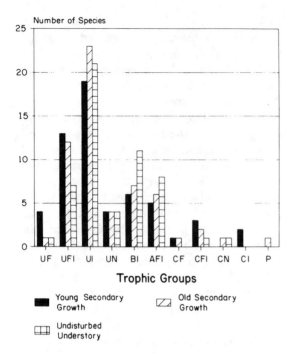

FIGURE 3. Species richness of birds of different trophic classes in three main habitat types.

more diverse in the understory. In young second growth they were represented by *Gymnopithys leucaspis*.

Larger differences were found in the abundance of birds of different trophic classes (Figure 4). Almost all classes were represented by more individuals in secondary growth areas. Bark-gleaning insectivores were more frequently found in both young and old second growth; *Xiphorhynchus ocellatus* was more frequently captured in secondary growth. *Picumnus pumilus* and *Picumnus exilis* also seemed to prefer secondary growth areas. Understory insectivores and nectarivores did not present major differences in abundance in the three habitat types. By contrast, understory frugivore-insectivores were more abundant in young second growth, while old second growth and undisturbed understories had similar abundances. Five nectarivore species were found in the secondary growth, among which only *Polyplancta aurescens* was absent in the understory. *Glaucis hirsuta* was observed in riparian forests where large clumps of Heliconias (Heliconiaceae) were abundant. In general, as in other lowland sites, a relative scarcity of frugivores

(28%, 29 species) and nectarivores (6.7%, 7 species) was observed.

Samples of avifauna in secondary growth and primary forest are slightly distinct, especially due to the occurrence in the former of canopy birds and natural secondary-growth species that follow the foliage-air interface (*sensu* Stiles 1983). This avifauna persists during the first 10 years of regeneration, after which the foliage-air interface is high above the ground and the environmental conditions of the understory are restored; then most of the birds caught near ground are typical to those of the forest interior.

Discussion

Higher capture rates of individuals and species in secondary habitats seem to be a general pattern in tropical forest areas (Wong 1986; Levey 1988*b*; Bierregaard 1990; Blake et al. 1990; Johns 1992), presumably due to high interhabitat species movement (Terborgh et al. 1990) or to the availability at edges of more

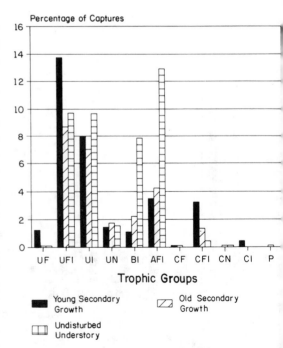

FIGURE 4. Abundance of birds of different trophic classes in three main habitat types.

resources, and not necessarily to higher number of specialized edge species (Johns 1992). In our study site we did not find great differences in capture rates, nor in the composition of the avifaunas between sites. Also, comparison of similarity indices among sites within the same type of vegetation showed similar degrees of heterogeneity.

An important proportion of species reported was captured in the regenerating habitat mosaic (49 species, 47.5%). Out of these, 19 species (18.4%) were found only in disturbed areas. These figures are higher than those of Manu (32%; Terborgh 1986) and Pará (42%, Novaes 1980). This may be due to the fact that, even though shifting cultivation is a high-intensity disturbance, more birds used these patches because regenerating plots are interspersed with forest and are followed by a vigorous regeneration (Walschburger & von Hildebrand 1990).

The lack or low abundance of some typical secondary-growth species suggests that these small-scale shifting agricultural plots, and the limited extent and isolation of the man-made habitat mosaics in the area, do not provide enough suitable habitat for invader species. The doves *Columbina* spp., common in samples of early secondary vegetation at Tefé (Johns 1992), were not present in our samples at Mirití-Paraná, and *Crotophaga ani* and *Muscivora tyrannus* were not as abundant there as at other localities in the Caquetá river 100 km downstream, where these species are common. Similarly, the avifauna of secondary vegetation at Mirití-Paraná is less rich in species originating in natural secondary vegetation and river edges. In Manu, such species comprise a large proportion of the avifauna in mature forest samples (69 species of 250; Terborgh et al. 1990). Their scarcity at Mirití-Paraná is probably due to the limited extension of habitats associated with alluvial plains in interfluvial terra firme forests and along small rivers and forest streams.

As in other studies (Johns 1992), the frequency of trophic groups in the regeneration process was a better indicator of the disturbance-regeneration regime. In the regenerating plots studied, the restoration of the environmental conditions of the understory seems to occur when the foliage-air interface is located farther

from the ground in regenerating areas of at least 10 years since abandonment. Understory insectivores increase as regeneration proceeds, which is explained by their physiological and ecological dependence on the interior of the forest (Orians 1969; Karr & Freemark 1983). At Mirití-Paraná, as in other places, most ant-followers avoided disturbed areas and were more common in primary forest sites, being especially sensitive to disturbances (Bierregaard 1990; Johns 1992).

Canopy insectivores present greater variation in the use of lower strata, which limits their use as indicators of regeneration. Some Tyrannidae were captured only in young secondary growth, and none was present in the understory samples. At Manaus they move along the forest's-vertical profile during the dry season, presumably as a response to microclimatic changes (Bierregaard 1990); Tyrannidae are also generally more diverse in disturbed patches (Johns 1990). Even less sensitive to the disturbance produced by man-made gaps were bark-gleaning insectivores. In Mirití-Paraná they were captured in regenerating fields more frequently than in Tefé (Johns 1990). At Manu some of these species were encountered mainly in mature habitat (Robinson & Terborgh 1990). This difference is apparently an artifact of the limited portion of the vegetation profile that is sampled with ground-level mist nets.

Frugivores represent 10% of the species at Mirití-Paraná, a lower figure than in Central America (17.2% at La Selva; Blake et al. 1990) and Panama (15.2%; Karr et al. 1982), but similar to other Amazonian sites (14% at Manu; Robinson & Terborgh 1990). The higher abundance in the sample of forest of some small frugivores (*Pipra coronata, Mionectes oleaginea,* and *Ramphocelus carbo*) that depend on the fruit of small plants of the Melastomataceae family, or of those associated with secondary habitat (Levey 1988b; Thiollay 1992), may result from the high occurrence of shifting agriculture in this area, occupied 20 years before this study began. Their higher abundance in the forest may also be an artifact of placing the understory nets near the forest edge. Small frugivores, however, seemed more common in the study area than in large tracks of completely undisturbed forest.

Nectarivores at Mirití-Paraná were caught in the forest understory and, more frequently, in regenerating areas due to the abundance of large clumps of *Heliconia*. These species are always slightly more abundant in secondary habitat. The relative scarcity of nectarivores and frugivores in our sample, as well as in other upland Amazonian sites (Bierregaard 1990), is probably due to the lower productivity of undisturbed Amazonian understories (Gentry & Emmons 1987).

Granivores may indicate severe, large-scale disturbance in the forest ecosystem because of their dependence on grasses. Their absence from mist-netting data at Mirití-Paraná (even though *Volatinia jacarina* was observed) suggests that, even under high-intensity disturbance regimes such as shifting agriculture, the small scale of disturbance does not allow invasion of these species. By contrast, granivores were very common in the samples of early secondary vegetation at Tefé (Johns 1990). The absence from the regeneration areas of colonizer species originating in riparian habitats, such as *Monasa* spp, *Chelidoptera tenebrosa, Paroaria gularis,* and *Crotophaga major* (see also Terborgh & Weske 1969) may be the result of the limited extension of riparian habitat in the interfluvial upland forests, or a manifestation of lower species diversity in areas far from large rivers (Haffer 1990). Data on avifaunal composition on chronoseries of habitat mosaics on the alluvial plains of white-water rivers are needed to test this hypothesis.

Finally, with growing concerns about the real role that forest-dwelling native amazonias can play in biodiversity conservation (Redford & Stearman 1993), our data suggest that—at least for terra firme forests—small-scale, shifting agriculture mimics from the avifauna standpoint the natural gap-phase dynamics of the forest when crop fields are interspersed in the forest and second growth is allowed to occur.

Acknowledgments

This work was part of the study of the Yukuna-Tanimuka indigenous agriculture system carried out by Colombian Fundación Puerto Rastrojo. Field work was supported by the Colombian Institute of Natural Resources and the Environment (G. I. Andrade) and by Fundación Puerto Rastrojo (H. Rubio-Torgler). The Center for the Study of Bird Migration of Brazil kindly provided rings. The authors recognize A. Lieberman, whose work at Mirití resulted in the preliminary inventory of the avifauna from which this project started. We are grateful to T. Walschburger, A. Hurtado, M. Quiñones, and R. Ortiz for their help in the field, and J. Echeverri and T. Walschburger in data analysis. This paper greatly benefited from comments by L. M. Renjifo, L. Roselli, F. G. Stiles, A. Lieberman, P. Feinsinger, D. Levy, E. Raez, and two anonymous reviewers.

References

Beals, E. W. 1960. Forest bird communities in the Apostle island of Wisconsin. Wilson Bulletin **72**:156–181.

Bierregaard, R. O., Jr. 1986. Changes in bird communities in virgin forest and isolated Amazonian forest fragments. Ibis **128**:166–167.

Bierregaard, R. O., Jr. 1990. Species composition and trophic organization of the understory bird communities in a Central Amazonian terra firme forest. In A. H. Gentry, editor. Four Neotropical rainforests. Yale University Press, New Haven, Connecticut.

Blake, J. G., F. G. Stiles, and B. A. Loiselle. 1990. Birds of la Selva biological station: Habitat use, trophic composition, and migrants. In A. H. Gentry, editor. Four Neotropical rainforests. Yale University Press, New Haven, Connecticut.

Connell, J. H., and R. O. Slatyer. 1977. Mechanisms of succession in natural communities and their role in community stability and organization. American Naturalist **111(4)**:1114–1119.

Denslow, J. S. 1987. Tropical rainforest gaps and tree species diversity. Annual Review of Ecology and Systematics 1987 **18**:431–451.

Foster, R. 1980. Heterogeneity and disturbance in tropical vegetation. In M. E. Soulé & B. A. Wilcox, editors. Conservation biology: An evolutionary-ecological perspective. Sinauer Associates, Sunderland, Massachusetts.

Gentry, A. H., and L. Emmons. 1987. Geographical variation in fertility, phenology, and composition of the understory of Neotropical forests. Biotropica **19**:216–227.

Haffer, J. 1990. Avian species richness in tropical South America. Studies on Neotropical Fauna and Environment **25(3)**:157–183.

Haffer, J. 1991. Mosaic distribution patterns of Neotropical forest birds and underlying cyclic disturbance processes. In H. Remmert, editors. The mosaic-cycle concept of ecosystems. Ecological

Studies, vol. 85. Springer Verlag, New York.

Johns, A. D. 1992. Response of Amazonian rain forest birds to habitat modification. Journal of Tropical Ecology 7:417–437.

Karr, J. R., and K. E. Freemark. 1983. Habitat selection and environmental gradients: Dynamics in the "stable" tropics. Ecology 64:1481–1494.

Karr, J. R., D. W. Schemske, and N. L. Brokaw. 1982. Temporal variation in the understory bird community of a tropical forest. In E. G. Leigh, A. Stanley Rand, and D. M. Windsor, editors. The ecology of a tropical forest: Seasonal rhythms and long-term changes. Smithsonian Institution Press, Washington, D.C.

Levey, D. J. 1988a. Tropical wet forest treefall gaps and distributions of understory birds and plants. Ecology 69(4):1076–1089.

Levey, D. J. 1988b. Spatial and temporal variation in Costa Rican fruit and fruit-eating bird abundance. Ecological Monographs 58:251–269.

Lovejoy, T. E. 1974. Bird diversity and abundance in Amazon forest communities. The Living Bird 13:127–191.

Lovejoy, T. E., and R. O. Bierregaard, Jr. 1990. Central Amazonian forests and the minimum critical size of ecosystems project. In A. H. Gentry, editor. Four Neotropical rainforests. Yale University Press, New Haven, Connecticut.

Novaes, F. C. 1980. Observaçoes sobre a avifauna do alto curso do Rio Paru de Leste, Estado do Pará. Boletim do Museu Paraense Emilio Goeldi, Nova Serie 100:1–58.

Orians, G. H. 1969. The number of bird species in some tropical forests. Ecology 50:783–801.

Pires, J. M., and G. T. Prance. 1985. The vegetation types of the Brazilian Amazon. In G. T. Prance and T. E. Lovejoy, editors. Key environments: Amazonia. Pergamon Press, New York.

Redford, K. H., and A. Maclean. 1993. Forest-dwelling native amazonians and the conservation of biodiversity: Interests in common or in collision? Conservation Biology 7(2):248–255.

Remsen, J. V., Jr., and T. A. Parker, III. 1983. Contribution of river-created habitats to bird species richness in Amazonia. Biotropica 12:23–30.

Robinson, S. K., and J. Terborgh. 1990. Bird community of the Cocha Cashu biological station in Amazonian Peru. In A. H. Gentry, editor. Four Neotropical rainforests. Yale University Press, New Haven, Connecticut.

Salo, J., R. Kalliola, I. Hakkinen, Y. Makinen, P. Nimela, M. Puhakka, and P. D. Coley. 1986. River dynamics and the diversity of Amazon lowland forest. Nature 322:254–258.

Schemske, D. W., N. Brokaw. 1981. Treefalls and the distribution of understory birds in a tropical forest. Ecology 62:938–945.

Sick, H. 1985. Ornitologia Brasileira, 2 vols. Editora Universidade de Brasilia, Brasilia, Brazil.

Stiles, F. G. 1983. Birds. Introduction. Pages 502–530 in D. H. Janzen, editor. Costa Rican natural history. University of Chicago Press, Chicago, Illinois.

Terborgh, J. W., and J. S. Weske. 1969. Colonization of secondary habitats by Peruvian Birds. Ecology 50:765–781.

Terborgh, J. W., S. T. Robinson, T. A. Parker, III, C. A. Munn, and N. Pierpont. 1990. Structure and organization of an amazonian forest bird community. Ecological Monographs 60(2):213–238.

Thiollay, J. M. 1992. Influence of selective logging on bird species diversity in a Guianan rain forest. Conservation Biology 6(1):47–63.

Walschburger, T., and P. von Hildebrand. 1989. Observaciones sobre la utilización estacional del bosque húmedo tropical por los indígenas del río Mirití. Colombia Amazonica 3(1):51–74.

Walschburger, T., and P. von Hildebrand. 1990. The first 26 years of forest regeneration in natural and man-made gaps in the Colombian Amazon. Page 457 in A. Gomez-Pompa, T. C., Withmore, and M. Hadley, editors. Rain forest regeneration and management. Man and the Biosphere Series, vol. 6. United Nations Education.

Willis, E. O. 1979. The composition of avian communities in remanescent woodlots in southern Brazil. Papeis Avulsos Zoologia 33(1):1–25.

Wong, M. 1986. Trophic organization of understory birds in a Malaysian dipterocarp forest. The Auk 103:100–116.

Wunderle, J. M., A. Diaz, I. Velazques, and R. Scaharron. 1987. Forest openings and the distribution of understory birds in a Puerto Rican rain forest. Wilson Bulletin 99:22–37.

Appendix

Species, trophic groups, capture rates, and total captures in three major habitat types.

Habitat Codes: YSG = young secondary growth (<10 years); OSG = old secondary growth (10–17 years); U = undisturbed understory. Trophic Groups Codes: UF = understory frugivores; UFI = understory frugivores-insectivores; UI = understory insectivores; UN = understory nectarivores; BI = bark-gleaning insectivores; AFI = ant-follower insectivores; CF = canopy frugivores; CFI = canopy frugivores-insectivores; CN = canopy nectivores; CI = canopy insectivores; P = piscivores.

| FAMILY and Species | Trophic Group | Capture Rates (100 net hours) | | | Total Captures |
		YSG	OSG	U	
COLUMBIDAE (1)					
Geotrygon montana	UF	0.23			3
TROCHILIDAE (5)					
Phaethornis bourcieri	UN	0.08	0.46	0.06	8
Phaethornis superciliosus	UN	0.54	0.46	0.48	2
Polyplancta aurescens	UN		0.08		1
Thalurania furcata	UN	0.23	0.15		5
Threnetes leucurus	UN	0.15		0.06	3
ALCEDINIDAE (1)					
Chloroceryle inda	P		0.08		1
MOMOTIDAE (1)					
Momotus momota	UFI			0.06	1
GALBULIDAE (2)					
Galbula albirostris	UI	0.08	0.08	0.06	3
Jacamerops auréa	UI		0.08		1
BUCCONIDAE (2)					
Nonnula brunnea	UI		0.08		1
Notharcus ordii	UI	0.08	0.08		2
CAPITONIDAE (1)					
Capito niger	CF			0.06	1
RAMPHASTIDAE (1)					
Selenidera reinwardtii	CF		0.08		1
PICIDAE (2)					
Picumnus pumilus	BI	0.15	0.08		3
Picumnus exilis	BI		0.08		1
DENDROCOLAPTIDAE (8)					
Deconychura strictolaema	BI	0.08	0.23	0.23	8
Deconychura longicauda	BI			0.06	1
Dendrocincla fuliginosa	BI	0.31	0.31	0.17	12
Dendrocincla merula	AFI		0.23	0.17	6
Glyphorhynchus spirurus	BI	0.08	0.70	1.19	30
Lepidocolaptes albolineatus	BI			0.06	1
Sittasomus griseicapillus	BI	0.08			1
Xiphorhynchus ocellatus	BI	0.08		0.17	4
Xiphorhynchus guttatus	BI		0.23	0.17	6
FURNARIDAE (8)					
Automolus infuscatus	UI	0.08	0.31	0.41	12
Automolus rubiginosis	UI			0.17	3
Hyloctistes subulatus	UI	0.08		0.06	2
Philydor pyrrhodes	UI	0.15	0.46	0.23	12
Sclerurus mexicanus	UI		0.23	0.35	9
Sclerurus caudatus	UI			0.06	1
Synallaxis rutilans	UI			0.06	1
Xenops minutus	BI		0.15	0.29	7
FORMICARDIAE (25)					
Ceromacra tyrannina	UI	1.98	0.39		19
Dichrozona cincta	UI			0.29	5
Formicarius colma	UI	0.15	0.08	0.23	7
Formicarius sp.	UI	0.39		0.89	20

(continued)

Appendix (*continued*)

FAMILY and Species	Trophic Group	Capture Rates (100 net hours)			Total Captures
		YSG	OSG	U	
Gymnophithys leucaspis	AFI	0.86	0.31	1.43	49
Hylophylax naevia	AFI			0.29	5
Hylophylax poecilonota	AFI	0.39	1.40	1.61	50
Hypocnemis cantator	UI	0.31	0.85	0.35	21
Myrmothera haematonota	UI		0.15	0.17	5
Myrmeciza athrotorax	UI	0.15			2
Myrmoborus myotherinus	UI	0.31		0.11	6
Myrmothera campanisa	UI			0.06	1
Myrmotherula axilaris	UI		0.08		1
Myrmotherula menetriesii	UI	0.08	0.08		2
Myromotherula obscura	UI		0.08		1
Percnostola caurensis	AFI			0.06	1
Percnostola leucosticta	AFI			0.06	1
Percnostola rufifrons	AFI	0.31	0.23	0.29	12
Pithys albifrons	AFI	0.62	1.25	2.75	70
Phlegopsis erythroptera	AFI	0.23	0.15	0.17	8
Pygiptila stelaris	UI		0.23		3
Thamnomanes caesius	UI	0.08	0.46	0.35	13
Thamnomanes ardesiacus	UI			0.06	1
Thamnophilus aethiops	UI		0.15		2
Thamnophilus murinus	UI		0.15		2
PIPRIDAE (8)					
Manacus manacus	UFI		0.08		1
Neopipo cinnamonea	UFI	0.08			1
Pipra coronata	UFI	3.06	2.81	2.93	124
Pipra erythrocephala	UFI	1.56	1.71	0.53	51
Pipra pipra	UFI	0.08	0.08		2
Schiffornis turdinus	UFI	0.39	0.15	0.23	11
Teleonema filicauda	UFI	0.70	0.31	0.06	14
Tyranneutes stolzmanni	UFI	0.23	0.08		4
TYRANNIDAE (15)					
Attila spadiceus	UFI		0.08		1
Corythopis torquata	UI		0.08		1
Empidonax traillii	CI	0.08			1
Laniocera hypopirrha	UI		0.08		1
Mionectes macconnelli	UFI	0.23			3
Mionectes oleaginea	UFI	1.64	0.62	1.07	47
Myiornis eucaudatus	UI		0.15	0.11	4
Myiopagis caniceps	CI	0.23			3
Onychorynchus coronatus	UI	0.15			2
Platyrinchus platyrhynchos	UI			0.06	1
Platyrinchus saturatus	UI			0.06	1
Pseudattila phoenicurus	UI		0.08		1
Ramphotrigon ruficauda	UI		0.23		3
Terenotriccus erythrurus	UI		0.23		3
Tyrannulus elatus	CI	0.62			8
TROGLODYTIDAE (5)					
Henicorhina leucosticta	UI		0.08	0.17	4
Microcerculus marginatus	UI	1.02	0.08	0.47	22
Thryothorus coraya	UI	0.39	0.15		7
Troglodytes aedon	UI	0.23			3
Cyphorinus arada	UI	0.08			1
TURDIDAE (3)					
Catharus maximus	UFI	0.54			7
Turdus albicollis	UFI	0.08	0.08		2
Turdus lawrencii	UFI	0.47	0.31	0.17	13

(continued)

Appendix (*continued*)

FAMILY and Species	Trophic Group	Capture Rates (100 net hours)			Total Captures
		YSG	OSG	U	
SYLVIDAE (1)					
Microbates collaris	UI		0.08		1
VIREONIDAE (1)					
Vireo olivaceus	UF	0.08			1
THRAUPIDAE (8)					
Cyanerpe's caeruleus	CN		0.08		1
Dacnis cayana	CN	0.08			1
Euphonia chlorotica	CF	0.08			1
Ramphocelus carbo	CFI	1.49	0.31		23
Tachyphonus surinamus	CFI	0.62	0.54	0.29	20
Tachyphonus cristatus	CFI	0.08			1
Tangara callophrys	CFI		0.08		1
Thraupis palmarum	CFI	0.08			1
FRINGILIDAE (4)					
Arremon taciturnus	UF	0.08			1
Cyanocompsa cyanoides	UF	0.47	0.08	0.06	8
Oryzoborus angolensis	UF	0.08			1
Saltator maximus	UFI	0.47			6

28
The Effects of Management Systems on Ground-Foraging Ant Diversity in Costa Rica

Dana S. Roth, Ivette Perfecto, and Beverly Rathcke

Introduction

The loss of biological diversity is an ecological crisis of profound and universal impact (Wilson 1988). Most of this loss is occurring in tropical regions as a result of conversion of forest to agriculture and pasture (Myers 1984). The "traditional" solution to reduce loss of species has been to establish national parks and other protected areas and attempt to exclude local populations of people from them. While these pristine parks are critical to protecting many species that cannot survive habitat modification, small islands of pristine forest may not be optimal for the protection of many species on a long-term basis, particularly if they are surrounded by huge expanses of disturbed habitats that restrict migration (Lovejoy et al. 1986, Bierregaard et al. 1992, but see Wu et al. 1993). More recently, it has been suggested that diverse, locally managed systems may best protect species diversity (Harris and Eisenberg 1989, McNeely 1989). Comparative studies on the effects of different land-use systems on biological diversity (e.g., Wilson and Johns 1982, Johns 1985) could provide preliminary data useful in making management decisions for these systems.

To assess the effects of management systems on species diversity, we chose to study ants for the following reasons: Ants are frequently the most abundant insects in tropical forests (Wilson 1987, 1992). In samples from a Peruvian rain forest, ants comprised > 70% of individual insects (Wilson 1992). Likewise, Holloway and Stork (1991) demonstrated that the abundance of the family Formicidae was second only to the order Diptera in diversity studies in Borneo. Their wide distribution throughout the world in diverse habitats makes ants strong indicators of biological diversity.

As indicators of ecosystem condition (Majer 1990, Perfecto 1991a, b), ant assemblages often reflect the degree of habitat disturbance and/or succession (Torres 1984a, b) in a community. Ants play important roles in structuring communities, from nutrient cycling in the soil (Lal 1988) and roles in seed dispersal (Kleinfeldt 1978, Bond and Slingsby 1984, Beattie 1985, Majer 1990) to influencing the faunal (Andersen 1992, De Kock et al. 1992) and floral (Davidson et al. 1978, Carroll 1983, Janzen 1983) communities. Their involvement in a variety of often symbiotic interactions with other species of animals and plants (Carroll and Janzen 1973, Kleinfeldt 1978, Huxley 1980, Beattie and Culver 1982, Roberts and Heithaus 1986, Hölldobler and Wilson 1990) illustrates such importance. In managed ecosystems, such as agricultural systems, where ants are both pests (Samways 1983) and pest control agents

(Perfecto 1990, Perfect 1991, Waage 1991, Perfecto and Sediles 1992), they play a role in shaping communities through species interactions (Majer 1972, Room 1975, Jackson 1984), including competition (Haering and Fox 1987).

The objective of this study is to consider ants as general indicators of biodiversity to examine the potential for conservation in a mosaic that includes human-influenced agricultural and/or forestry systems as well as parks and reserves (e.g., Pimentel et al. 1992). In using ground-foraging ants as indicators, we are not suggesting that they are a surrogate for other biodiversity, but rather that, as a component of biodiversity, they could reflect patterns of loss of diversity of other species. We compare the effects of anthropogenic disturbances on ant diversity in lowland tropical rain forest and nearby agricultural systems in Costa Rica. Specifically, we investigate the number and heterogeneity of species in different land use systems based on a gradient of increased levels of disturbance (Fig. 1). We investigate the hypothesis that as the level of disturbance increases, species richness and evenness within sites is reduced, while a few species become more dominant. We also test the hypothesis that β-diversity (the change in species diversity along the gradient of different management systems) is higher between different land uses than between different sites of similar land uses.

Methods

Study Area

This study was conducted in undisturbed lowland wet rain forest (Hartshorn 1983) and 24-yr-old abandoned cacao plantations in the 1510-ha La Selva Biological Station (10°26′ N and 83°59′ W; Clark 1990), and nearby productive cacao and banana plantations, important agri-

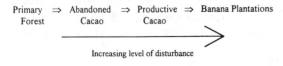

FIGURE 1. The gradient of anthropogenic disturbances used in this study.

cultural systems in the Sarapiquí region of Costa Rica (Boucher et al. 1983). Cacao plantations in the area are relatively small (2–3 ha) with significant numbers of large trees remaining for shade, and few agricultural inputs (D. Roth, *personal observation and conversations with farmers;* Table 1). Banana plantations are immense, some exceeding 20 000 ha and requiring large amounts of agricultural inputs. At the time of the study, this zone was undergoing rapid changes in land use as extensive plantations of export bananas were being planted, replacing small farms, pastures, and forests (D. Roth, *personal observations*).

Three undisturbed (primary) forest sites were located on Camino Circular Lejano (CCL), Sendero Holdridge (SHO), and Camino Experimental Sur (CES), and two abandoned cacao sites (AC-1, AC-2) were located on the Sendero Occidental (SOC), where cacao plantations were abandoned in 1968 (Clark 1990); both of these habitats were located in La Selva Biological Station. Four cacao plantations in the vicinity of the reserve were selected (Table 1). The first plantation, located in Chilamate (Chilamate) was bordered by the road from Puerto Viejo to San José (via Vara Blanca) on one side, pasture on two sides, and secondary forest and a stream on the fourth. The second plantation, in the town of La Virgén de Sarapiquí (LV-1) was adjacent to the Puerto Viejo-San José road on one side, a pasture and houses on two sides, and secondary forest and stream on the fourth. Cacao trees in this plantation were irregularly spaced, and there were more open areas. The third cacao plantation (LV-2), located 3 km south of LV-1 on the west side of the road, ran lengthwise parallel to the Puerto Viejo-San José road and was surrounded on the other long side by secondary forest. Mixed stands of primarily papaya trees and pasture were adjacent on the other two sides, respectively. The fourth plantation, located in El Tigre de Sarapiquí (El Tigre) was located along the dirt road (now paved, 1993) from Puerto Viejo to Rio Frio. The plantation was bordered by agricultural fields and secondary forest at the back and by houses and a clearing on the other two sides. Diversity of vegetation in each of the cacao plantations varied, but plants such as palms, *Musa* spp., *Erythrina* spp., *Cordia* spp., and a

TABLE 1. A description of study sites within the Sarapiquí region of Costa Rica.

Habitat	Site	Age (yr)	Size (ha)	Distance to La Selva Reserve (km)	Management system
Primary forest	CCL	. . . *	. . .	. . .	In 1510 ha. La Selva reserve
	SHO	. . .	. . .	. . .	. . .
	CES	. . .	. . .	. . .	. . .
Abandoned cacao	AC-1	24	. . .	. . .	In 1510 ha. La Selva reserve
	AC-2	24	. . .	. . .	. . .
Productive cacao	Chilamate	12	3	8	No pesticide applied except poison administered directly onto *Paraponera clavata* nests
	LV-1	6	3	18	No pesticides applied; irregularly spaced cacao trees, leaving open areas with grass
	LV-2	12	2	21	No pesticides applied; high diversity of non-cacao plants
	El Tigre	6	2 to 3	4	Fungicide applied every 4 mo; few noncacao plants
Banana	Pto. Viejo	<1	>10^4	15	Nematicides and fungicides regularly applied on ground and aerially; recently established plantation (10-mo-old plants)
	Rio Frio	24	1800	30	Nematicides and fungicides regularly applied; ground covered by loose dirt and banana leaves and stems; older banana site

* . . . = not applicable.

variety of herbaceous vegetation were commonly found (Table 1). The two banana plantations sampled, one near Puerto Viejo (PV) and one in Rio Frio (RF), were large in comparison to all of the other sites (Table 1). We sampled at a site where the first bananas had been planted in April to June 1991, so at the time of sampling, they were between 7 and 10 mo old. In Rio Frio, we sampled in an older site located in "Finca 4," across from the packing plant near the entrance to the plantations from Puerto Viejo (Table 1).

All sites were situated on alluvial terraces of the Sarapiquí or Puerto Viejo rivers or their tributaries. The agricultural systems studied normally occur on alluvial soils; for the purposes of controlling for altitudinal, climatic, and edaphic factors affecting species distributions, sites within La Selva Biological Station were also chosen on alluvial soils.

To sample ants within each site, we placed 10 parallel transects separated by 5–15 m perpendicular to a 100 m long edge. Along each transect, we placed 20 tuna baits (3–4 g each) on the ground. Baits were separated from each other by ≈1.5 m. Tuna baits have often been used in studies of ant communities and attract a generalist assemblage of ants (e.g., Levings and Franks 1982, Torres 1984a, b, Nestel and Dickschen 1990, Perfecto 1991 a, b). After baits were laid, we waited 15 min to allow ants to find the baits and recruit other workers. We then proceeded to check baits in the order in which they were laid, collected specimens of each new ant species with an aspirator, and recorded our observations of all ant species at the bait. Ants were sampled mornings between 0730 and 1200. Sampling was conducted in the dry season from January through March 1992.

We counted all ants that were found at or in the close vicinity (within 10 cm) of the baits. We attempted to equalize sampling effort at each bait by counting only those individuals seen at the bait or moving to or from the bait for the fixed time of observation, ≈30 s. Ants were sampled once in each transect.

Individual ants from each sample were pointed, pinned, and sorted according to species. Ants were later identified by Stefan P. Cover, at Harvard University Museum of Comparative Zoology, and John T. Longino, at the Evergreen State College. A list of ant species was tabulated for each bait in every transect, site, and habitat so that the relative distribution of each species of ant could be determined. A complete list of species can be found in the Appendix. From these data, indices of diversity were calculated and compared across sites and habitats.

Voucher specimens were deposited at the Museum of Comparative Zoology at Harvard University in the United States, and at the Instituto Nacional de Biodiversidad (INBio) in Costa Rica.

Data were collected on soil temperature, canopy cover, and leaf litter for each transect and averages were calculated for each site. Soil temperature was measured by a thermometer placed in the soil for 2 min before recording a reading. Canopy cover was measured using a spherical densiometer, a hand-held, concave mirror with gridlines, held level at 1 m from the ground. Openings in the canopy were manually counted within the grid and a conversion factor yielded the canopy cover value. Six samples of leaf litter were collected from within each of the sampling sites by randomly choosing grid coordinates, and collecting leaf litter in a 0.25 × 0.25 m square. Litter was air dried in a funnel apparatus (Southwood 1987, with modifications by R. K. Colwell, *personal communication*) and later weighed.

Many diversity indices exist, each with its own strengths and weaknesses. No single index encompasses all of the characteristics of an ideal index including high discriminant ability, low sensitivity to sample size, and ease in calculation (Magurran 1988). Therefore, it is best to use a combination of them. We chose indices that reflect a combination of richness, dominance, evenness, and relative abundance, the latter three of which reflect heterogeneity. These indices helped us to interpret what has happened to biological diversity in the community of terrestrial ants that come to tuna baits in different land-use systems. Based on literature and recommendations (Samways 1984, Southwood 1987, Magurran 1988, Krebs 1989; G. Fowler, *personal communication*), we chose to

use S (species richness), H (Shannon index), d (Berger-Parker index), and E (Shannon evenness index) to examine α-diversity within all of the sites and habitats, and PS (Percent Similarity) to examine β-diversity among the habitats. A brief discussion of each of the indices follows: S: Species richness equals the total number of species in the community. We chose this index because it provides a great deal of information about the community, especially with respect to relative stage of succession (Samways 1984). As an index, S is easily conceptualized and comparable across habitats. Because all species in a community cannot usually be enumerated (Krebs 1989), species-area relationships were also calculated. H: Shannon's index of diversity reflects both evenness and richness (Colwell and Huston 1991) and is commonly used in diversity studies (Krebs 1989). It was calculated by the equation:

$$H = - \sum_{i=1}^{s} p_i \ln p_i,$$

where p_i is the proportion of individuals of the ith species, and S is the total number of species. d: The Berger-Parker dominance measure expresses the proportional importance of the most abundant species (Magurran 1988). Low values indicate lowered dominance by any one species in a system and are generally accompanied by increased evenness of species (Magurran 1988). The Berger-Parker index, d, equals $p_{i(max)}$, the proportion of the most abundant species. E: Shannon's evenness index indicates relative abundances of species in terms of evenness and is based on the Shannon index of diversity. Both the Berger-Parker dominance index and the Shannon evenness index are important measures of heterogeneity. The Shannon evenness index is calculated by the equation:

$$E = H/H_{max} = H/\ln S,$$

where H is Shannon's diversity index and $\ln S$ is the natural logarithm of the number of species. PS: Percent Similarity shows the proportions of species in common between sites; this index ranges from zero if there is no species overlap to 100 if there is complete species overlap (Krebs 1989). The Percent Similarity index is relatively insensitive to sample size and species diversity (Krebs 1989). Percent similarity is calculated by the equation:

$$P = \Sigma \ \text{minimum}(p_{1i}, p_{2i}) \times 100,$$

where P equals the percentage similarity between sites 1 and 2, p_{1i} equals the proportion of species i in community sample 1 and p_{2i} equals the proportion of species i in community sample 2 (Krebs 1989). β-diversity, discussed in this study, is proportional to the inverse of similarity.

Diversity was statistically examined using two approaches. In the first, we pooled data from all of the transects to get one value per site for each site for each index (S, H, d, E); since individual transects are not considered separately, this method did not account for differences among transects within the sites. This approach was used so that diversity of ants could be examined relative to soil temperature, leaf litter amount, and canopy cover, for which observations were pooled to give one value per variable per site. One-way analyses of covariance (ANCOVA) were conducted to compare habitats using diversity indices as the dependent variable and soil temperature, canopy cover, and leaf litter separately as the covariates. The objective of these analyses was to test for differences between habitats and for correlation between diversity and abiotic factors, as has been suggested in the literature (e.g., Levings 1983, Torres 1984a, b).

In the second approach, we analyzed diversity data by comparing measures of α-diversity analyzed for each transect. Analyses at this level examined heterogeneity on a finer scale (i.e., within each transect). A two-factor (sites, habitats) unbalanced nested design was used for an ANOVA on each of the dependent variables (S, H, d, E) using SAS General Linear Models Procedures (SAS 1990). The unbalanced design arose as a result of different numbers of sites for each habitat type; there were the same number of observations within each site. Each variable was calculated on the transect level, with 10 observations made for each site. A Type III (SAS 1990) ANOVA model was used to compare diversity within sites and sites nested within habitats. When the model found statistically significant differences in habitats and sites, least significant difference (LSD)

multiple comparison tests were used to determine which were significantly different.

Parametric statistical tests were used as there was no major deviation from the assumptions of normality of the data. Transformations of the data were unnecessary except for Shannon's evenness index, E, where an arcsine transformation of the data resulted in a normal distribution.

Because all species commonly cannot be sampled, we used species-area relationships as a tool to predict the maximum number of species in an area larger than that sampled (Connor and McCoy 1979). In this study, area was defined as the number of transects, with more transects covering more area. Total species number was estimated using a bootstrap technique where the original data were resampled (Stout and Vandermeer 1975; A. Kaufmann and J. H. Vandermeer, *unpublished program*). The procedure involved sampling n values with replacement, and a species-area curve was obtained by repeatedly resampling 10 times for increasing area. The predicted number of species for each site was calculated from the slope of a linear equation fit to the reciprocal of the data points (i.e., where the curve levels off).

Results

A total of 109 species comprising 31 genera, and five of the eight neotropical subfamilies, were sampled in this study (Appendix). Most of the ant species belonged to the subfamily Myrmicinae (78), followed by the subfamily Ponerinae (24), Formicinae (5), and Pseudomyrmecinae (1) and Dolichoderinae (1). Primary forest sites had the highest number of species, followed by abandoned cacao sites (Table 2). Productive cacao and banana sites had significantly fewer species ($P < .0001$; Table 2).

Some species were found in only one habitat type (Table 3). On average, 7.67 species per primary forest site were restricted to forest, 5.5 species per site were restricted to abandoned cacao, three species per site were restricted to cacao plantations and five species per site were restricted to banana (Table 3).

Leaf litter and canopy cover were not significantly different in any of the sites ($P > .05$,

TABLE 2. Diversity indices for ant diversity in each site based on pooled data from transects. S = species richness, H = Shannon's index of diversity, E = Shannon's evenness index, and d = Berger-Parker dominance index.

Habitat	Site	Index			
		S	H	E	d
Primary forest	CCL	37	2.986	0.827	0.163
	SHO	39	3.256	0.889	0.118
	CES	43	3.094	0.823	0.135
Abandoned	AC-1	36	3.045	0.850	0.157
cacao	AC-2	36	2.803	0.782	0.19
Productive	Chilamate	16	1.093	0.394	0.753
cacao	LV-1	17	1.731	0.611	0.45
	LV-2	26	2.126	0.660	0.473
	El Tigre	11	1.535	0.640	0.421
Banana	Pto. Viejo	14	0.823	0.315	0.83
	Rio Frio	13	1.927	0.751	0.305

Table 4), but soil temperature was significantly different between sites (one-way ANOVA; P = .016). Soil temperature was slightly higher in banana plantations than in the other habitats. None of the diversity indices covaried significantly with any of these microsite characteristics.

The major differences in diversity were between primary forest and abandoned cacao habitats on the one hand and productive cacao

TABLE 3. List of ant species found exclusively in the indicated habitats.

Primary forest	Abandoned cacao	Cacao plantations	Banana plantations
Apterostigma sp. 1	*Cyphomyrmex* sp. 2	*Acromyrmex* sp. 1	*Cardiocondyla* sp. 1
Carabarella sp. 1	*Eurhopalothrix* sp. 1	*A. volcanus*	*Cyphomyrmex* sp. 1
Pheidole sp. 8	*Hylomyrma* sp. 1	*Aphaenogaster* sp. 1	*Pheidole* sp. 13
Pheidole sp. 9	*Octostruma* sp. 1	*Crematogaster* sp. 3	*Pheidole* sp. 26
Pheidole sp. 10	*Pheidole* sp. 27	*Monomorium floricola*	*Pheidole* sp. 34
Pheidole sp. 14	*Pheidole* sp. 35	*Pheidole* sp. 33	*P. punctatissima*
Pheidole sp. 17	*Smithistruma* sp. 1	*Trachymyrmex* sp. 2	*Solenopsis* sp. 2
Pheidole sp. 21	*Solenopsis* sp. 7	*Pseudomyrmex* sp. 1	*Tetramorium*
Pheidole sp. 22	*Solenopsis* sp. 9	*Gnamptogenys* sp. 2	*bicarinatum*
Pheidole sp. 32	*Gnamptogenys bispinosa*	*Pachycondyla constricta*	*Hypoponera* sp. 3
P. cephalica	*Brachymyrmex* sp. 2	*Brachymyrmex* sp. 1	*Hypoponera* sp. 4
P. fiorii		*Tapinoma melano-*	
Solenopsis unid.		*cephalum*	
Solenopsis sp. 5			
Solenopsis sp. 6			
Ectatomma gibbum			
Gnamptogenys sp. 4			
Odontomachus			
brunneus			
O. hastatus			
O. opaciventris			
Pachycondyla villosa			
Paraponera clavata			
Paratrechina sp. 2			
Total 23	11	12	10
Number of sites 3	2	4	2
Average number of unique species per site 7.67	5.5	3	5

TABLE 4. Site characteristics (soil temperature, leaf litter, and canopy cover) for each site. Values indicated are means for each site ± 1 standard deviation. Sample size, n, equals 6 samples. (Leaf litter units measure dry mass.)

Habitat	Site	Soil temperature (°C)	Leaf litter (g/0.0625 m^2)	Canopy cover (%)
Primary forest	CCL	23.4 ± 0.55	38.3 ± 14.3	91.5 ± 3.6
	SHO	23.3 ± 0.05	53.2 ± 28.6	92.6 ± 3.62
	CES	23.0 ± 0.63	27.4 ± 13.4	92.7 ± 4.20
Abandoned cacao	AC-1	23.0 ± 0.0	58.9 ± 44.93	87.9 ± 6.64
	AC-2	22.7 ± 1.15	33.2 ± 10.25	90.3 ± 5.60
Productive cacao	Chilamate	23.5 ± 0.71	83.1 ± 18.57	90.0 ± 5.80
	LV-1	23.8 ± 0.84	57.9 ± 12.77	94.4 ± 3.92
	LV-2	22.8 ± 0.05	59.5 ± 11.23	91.6 ± 7.90
	El Tigre	23.7 ± 0.58	53.8 ± 11.54	94.8 ± 2.73
Banana	PV	24.2 ± 0.75	25.3 ± 24.27	95.0 ± 4.80
	RF	25.2 ± 0.75	37.3 ± 40.25	88.0 ± 7.29

and banana habitats on the other. Based on an ANOVA, there were significant differences among habitats for every diversity index at the habitat level ($P < .0001$, Tables 5 and 6). However, least significant difference (LSD) multiple comparison tests showed that primary forest and abandoned cacao were not significantly different from each other except in evenness, E (Table 7). Cacao and banana plantations were also not significantly different from each other.

Sites nested within habitats showed significant differences for all indices except richness, S (Table 6) demonstrating variability among sites in similar habitats. Because the sites within

habitats varied, we used one-way ANOVA within each habitat type; we showed significant differences between the two banana sites ($P < .05$) and between the four productive cacao sites ($P < .05$; Table 8). LSD tests determined that the cacao sites LV-1 and LV-2 were not significantly different from each other for any of the diversity indices, and that the Chilamate site and the El Tigre site were significantly different only in terms of evenness, E ($P < .05$), but not for any of the other variables. Chilamate and El Tigre were consistently significantly different from LV-1 and LV-2 for all indices ($P < .05$) except E. Chilamate had a significantly lower evenness index than all other

TABLE 5. Ant diversity indices based on average values ± 1 standard deviation within the 10 transects nested within each site ($n = 10$). S = species richness, H = Shannon's index of diversity, E = Shannon's evenness index, and d = Berger-Parker dominance index.

Habitat	Site	Diversity indices			
		S	H	E	d
Primary forest	CCL	10.9 ± 2.9	2.197 ± 0.3	0.937 ± 0.03	0.202 ± 0.4
	SHO	10.7 ± 2.9	2.174 ± 0.3	0.928 ± 0.03	0.242 ± 0.08
	CES	11.9 ± 2.2	2.233 ± 0.2	0.908 ± 0.04	0.238 ± 0.05
Abandoned cacao	AC-1	11.0 ± 4.4	2.014 ± 0.5	0.870 ± 0.06	0.312 ± 0.1
	AC-2	10.3 ± 2.1	2.034 ± 0.3	0.876 ± 0.05	0.284 ± 0.08
Mean ± 1 SD		10.96 ± 0.59	2.13 ± 0.1	0.904 ± 0.03	0.256 ± 0.04
Productive cacao	Chilamate	3.70 ± 2.1	0.706 ± 0.4	0.555 ± 0.1	0.799 ± 0.1
	LV-1	5.9 ± 1.4	1.358 ± 0.3	0.772 ± 0.1	0.491 ± 0.1
	LV-2	5.8 ± 2.1	1.378 ± 0.4	0.804 ± 0.1	0.488 ± 0.2
	El Tigre	3.8 ± 2.0	0.805 ± 0.5	0.681 ± 0.2	0.712 ± 0.2
Banana	PV	2.8 ± 0.8	0.518 ± 0.3	0.516 ± 0.2	0.829 ± 0.1
	RF	4.5 ± 1.4	1.325 ± 0.3	0.920 ± 0.07	0.404 ± 0.1
Mean ± 1 SD		4.41 ± 1.2	1.02 ± 0.39	0.71 ± 0.15	0.62 ± 0.18

TABLE 6. Level of significance based on analysis of variance (ANOVA) comparing habitats and sites nested within habitats. Site data are based on ant diversity of 10 transects measured in each site.

	Diversity indices			
	S	H	E	d
Habitat	$P = .0001$	$P = .0001$	$P = .0001$	$P = .0001$
Site (habitat)	$P = .1073$	$P = .0001$	$P = .0001$	$P = .0001$

cacao sites, reflecting high dominance by one species, *Ectatomma ruidum*. In addition, LV-2 had significantly higher degree of evenness than El Tigre (Table 8). There were only two banana sites sampled and they differed from each other with respect to all indices ($P < .05$).

With the exception of banana plantation sites, percent similarity of ant assemblages was highest in sites of similar land uses (Table 9); therefore, β-diversity (inversely related to similarity) was lower among sites in similar land use systems, as hypothesized. The high overlap of species in the disturbed sites (i.e., productive cacao and Rio Frio banana sites) seemed primarily a result of high abundances of dominant species, particularly *Ectatomma ruidum* and *Pheidole subarmata*. Sites within La Selva Biological station (abandoned cacao and primary forest) had more species in common but at much lower abundances (Table 9; Appendix).

The predicted number (according to species-area curves) of species in the community at each site shows much higher species richness in the primary forest and abandoned cacao sites than in the agricultural sites (Table 10 and Fig. 2), as would be expected given the actual results on species richness mentioned above. Banana and productive cacao sites level off at a much lower number of samples than do forest sites. Like the forest and abandoned cacao sites, the productive cacao site (LV-2) with the highest noncacao plant diversity levels off at a much later point.

Discussion

There were clear changes in ant diversity over the gradient of anthropogenic disturbances studied. The major trend showed that with greater disturbance there was lowered richness and evenness and increased dominance of fewer species. The abandoned cacao sites had 91% as many species as primary forest sites, whereas cacao and banana plantation sites had 44 and 34% as many species as the forest sites, respectively. Each of the productive agricultural systems showed greatly reduced ant diversity by all of the indices, each of which emphasizes a different aspect of diversity. Therefore, it is possible to qualitatively assess what has happened as a result of the conversion of forest land into agricultural land.

The most dramatic differences in diversity along the gradient of disturbance from primary forest to banana plantations were between sites on La Selva Biological Station (forest and abandoned cacao) and those outside of the station (cacao and banana plantations; Fig. 1). Ant diversity (based on Shannon's Index, H) decreased in more heavily disturbed areas. In addition, the decrease in the total number of ant species corresponded to increased domination by one or two species in the community, as has been observed previously (Majer 1972, Samways 1983, Jackson 1984). Two cacao plantations, LV-1 and LV-2, had lower levels of

TABLE 7. Least significant difference (LSD) multiple comparison tests comparing ant diversity in all habitats.

Habitat	Abandoned cacao				Cacao				Banana			
	S	H	E	d	S	H	E	d	S	H	E	d
Forest	NS	NS	*	NS	*	*	*	*	*	*	*	*
Abandoned cacao					*	*	*	*	*	*	*	*
Productive cacao									NS	NS	NS	NS

*Differences significant at $P < .05$; NS = not significant.

TABLE 8. Least significant differences comparisons for ant diversity in productive cacao plantations.

Site	Chilamate				LV-1				LV-2				El Tigre			
	S	H	E	d	S	H	E	d	S	H	E	d	S	H	E	d
Chilamate	. . .	. . .	. . .	. . .	*	*	*	*	*	*	*	*	NS	NS	*	NS
LV-1					. . .	. . .	. . .	. . .	NS	NS	NS	NS	*	*	NS	*
LV-2									. . .	. . .	. . .	. . .	*	*	*	*

* Differences significant at $P < .005$.

dominant species (Berger-Parker dominance Index, d) than the reserve sites but higher than the other two cacao (Chilamate and El Tigre) and banana sites. Similarly, these sites (LV-1 and LV-2) with lowered dominance by any one species also contained a more even distribution of ant species (E), and are thus considered more diverse. The sites with fewer dominant species were also more vegetatively diverse and heterogeneous than the more highly disturbed sites (D. Roth, *personal observations*). Higher ant diversity in forest and abandoned cacao sites than in banana and cacao sites (Tables 5 and 6) could be attributed to changes in successional stage of the habitat (Castro et al. 1990), or to vegetational diversity and heterogeneity, microclimate, and soil and litter characteristics, as has been demonstrated previously (Southwood et al. 1979, Majer 1990, Neumann 1991).

Factors such as age of the habitat and proximity to colonizing sources may also affect diversity. Our banana results support previous studies that show that the age of the system affects the level of dominant species in a habitat

(Lévieux and Diomande 1985, Majer and Camer-Pesci 1991), although there were no replicate sites for each banana site to better support this result. For example, the older banana plantation in Rio Frio had fewer dominant species than the younger Puerto Viejo plantation, which was heavily dominated by *Solenopsis geminata* (index d, in Table 2, per site; in Table 5, per transect; Appendix), a species known to be an early recolonizer after disturbance (Perfecto 1991*b*).

Soil temperature, leaf litter amount, and canopy cover that have previously been demonstrated to affect diversity of insects (e.g., Levings 1983, Torres 1984*a*, *b*, Lynch et al. 1988) were not correlated to ant diversity in this study. The failure to find correlations could reflect difficulty in making accurate measurements in highly variable habitats and sites. We observed higher heterogeneity in canopy cover in the reserve and cacao sites compared to the banana sites, which may be reflected by the higher soil temperature in the banana plantations. The quality of leaf litter in the different

TABLE 9. Percent similarity in ant species for all sites. CHIL = Chilamate cacao plantation. All other abbreviations for sites defined in *Methods: Study sites*. Superscripts indicate comparisons between sites of similar land uses: 1 = primary forest, 2 = abandoned cacao, 3 = productive cacao, and 4 = banana.

		Primary forest			Abandoned cacao		Productive cacao				Banana	
		CCL	SHO	CES	AC-1	AC-2	CHIL	LV-1	LV-2	ET	B-PV	B-RF
Primary forest	CCL		41.75[1]	47.6[1]	38.85	31.6	5.95	3.6	14.1	6.5	3.3	3.35
	SHO			44.0[1]	39.1	46.8	15.9	10.1	29.5	15.9	0.5	15.1
	CES				35.3	33.7	4.5	4.5	14.5	7.0	0.5	1.7
Abandoned cacao	AC-1					37.9[2]	12.6	12.1	24.4	14.5	3.6	11.1
	AC-2						23.3	10.3	30.9	22.8	3.0	21.0
Productive cacao	Chilamate							17.9[3]	53.5[3]	39.4[3]	10.7	33.7
	LV-1								29.0[3]	64.2[3]	12.4	25.6
	LV-2									57.2[3]	5.4	46.8
	El Tigre										8.4	47.3
Banana	Pto. Viejo											6.1[4]

TABLE 10. Observed and predicted maximum number of ant species for each site. The predicted number of ant species was based on species-area relationships (Fig. 2).

	Site	Observed number	Predicted number of species
Primary forest	CCL	37	44
	SHO	39	54
	CES	43	58
Abandoned cacao	AC-1	36	46
	AC-2	36	42
Productive cacao	Chilamate	16	21
	LV-1	17	20
	LV-2	25	37
	ET	11	14
Banana	PV	14	19
	RF	13	16

sites was also highly variable, making sampling consistently difficult. Our failure to demonstrate correlations between microsite characteristics and ant diversity must, therefore, be interpreted cautiously.

An important result in this study was the apparent recovery (based on comparison with primary forest) of ground-foraging ant diversity in the abandoned cacao plantations. This resilience in abandoned cacao sites may, in part, be due to the close proximity of primary forest, which serves as a source from which plant and animals can colonize a disturbed site. Whether land use systems such as agroforestry plantations are able to recover biological diversity if they are surrounded by pasture or large-scale agricultural systems presents an important question. With the current state of highly modified tropical landscapes, this is a difficult question to resolve — especially when source populations have already been lost in the modifications. Based on observations in this study, we suggest that proximity to a colonizing source is important. Three of the four cacao plantations (Chilamate, LV-1, El Tigre, all similar in size) were surrounded

Species-Area Curves for All Sites

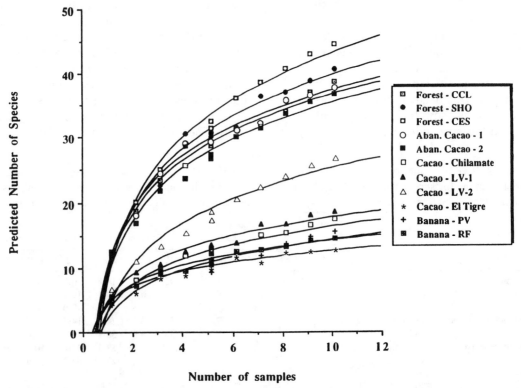

FIGURE 2. Species-area relationship for all sites. Forest and abandoned cacao sites are predicted to have the most ant species. A cacao plantation (LV-2), adjacent to secondary forest, has an intermediate number of species. The fewest species are found in the other cacao and banana plantations.

primarily by road and pasture with small pockets of secondary forest (for LV-1 and Chilamate) at one end. These three had fewer ant species than did the fourth (LV-2), which was bordered lengthwise by a secondary forest; this plantation (LV-2) had considerably more vegetational diversity, including native trees, than the others. This site was not significantly different from LV-1 for all of the diversity indices based on pooled transect data, but the overall diversity of this site was higher, suggesting higher fine-scale heterogeneity within the site.

Diversity of insects seems to be highly influenced by isolation, the spatial scale of the land use system, and surrounding land use systems, although studies supporting this latter idea are scarce. Klein's (1989) study of carabid beetles in forest fragments in the Brazilian Amazon shows that forest beetles did not move across clear-cut areas between forest fragments and were only found between fragments after secondary growth had invaded certain areas. It is apparent that isolation may present insurmountable barriers for some species. In reference to ants, this could be an important factor governing their ability to arrive at sites suitable for colonization. In heavily grazed or intensive agricultural areas, the possibility of mortality for dispersing insects increases because of use of chemical pesticides (e.g., Risch et al. 1986) or because of dessication (Klein 1989). Thus, it seems that a variety of abiotic and biotic factors, individual species' biology, and landscape pattern all play a role in shaping insect diversity in a community.

Even though indices of α-diversity stayed relatively the same in or across habitats (e.g., in the banana sites and certain cacao sites), we demonstrated that β-diversity was greatest in heterogeneous habitats separated by significant distances. β-diversity along the gradient of the management systems included in this study was higher than that across different sites under similar management (except for the banana sites). The exception, the two banana plantations that were separated by > 35 km, had only 6.1% species in common. In this instance, the age of the plantation may also have affected species assemblages, since the species in the younger plantation are rapid recolonizers of disturbed sites. β-diversity was lower in productive cacao and banana sites, since similar species were found in both, but these species were,

in general, different from those found in the reserve (Table 9). β-diversity was lowest between forest and abandoned cacao sites, which were in close proximity to each other (Table 9). Nevertheless, each reserve site had a higher number of ants unique to that site. We found a total of 17 ant species that occurred in only one site of the primary forest, and nowhere else. (A total of 23 species were encountered in only one site from all of the other habitats combined.) Species found in only one habitat type may be good indicators of the effects of habitat disturbance if their range is restricted by a certain level of disturbance. For example, ants that do not come to tuna baits, such as army ants or other specialized predators, may be more sensitive to disturbance. In the absence of data addressing qualitative differences in diversity, it may be useful to use these species as indicators of disturbance. It is also important to note that although the total number of species was highest in primary forest, each habitat site had a considerable number of ants not found in any other habitat type (Table 3), thus increasing β-diversity. The increase in β-diversity has important implications for conservation and supports the idea that biodiversity may be conserved in a mosaic of management systems.

The results of this study, i.e., the significant reduction of ant diversity from forest to agricultural systems, may underestimate differences in diversity among the habitats. Based on the ant community sampled — ants that come to tuna baits — the loss is most likely a conservative estimate. The results may be biased because tuna baits tend to attract ants that are generalists: such species may be more prevalent in modified habitats since they usually are more common and less likely to be affected by disturbance or more likely to recolonize after a disturbance (Connell 1978, Denslow 1985, Hansson 1991). On the contrary, less disturbed habitats may favor more specialized species: in the forest and abandoned cacao, we observed many terrestrial ants nearby (e.g., leaf-cutter ants) that never arrived at tuna baits and thus were not counted. For more specialized organisms, ants and otherwise, the loss of biodiversity in the sites we studied is likely to be more severe.

Because our choice of research sites was limited by what was available, and there were no productive cacao plantations as close to the

primary forest as the abandoned cacao plantations, our results were confounded by the effect
of proximity to primary forest. The results were
also confounded by the vast differences in area
of each land use system. The difference in
diversity could be a consequence of proximity
to primary forest or a consequence of the scale
and/or actual land use system. Our data do not
allow us to distinguish among these alternatives.

If the results reflect true differences in the
system of management, then we can conclude
that ant species diversity seems able to recover
after a fallow period following disturbance
such as conversion to cacao plantation. The
results of our study, demonstrated by comparing diversity of ants in forest to that in
abandoned cacao and cacao plantations, suggest that 91% of the species richness of ground-
foraging ants that come to tuna baits is regained within 25 yr (the time since
abandonment of cacao plantations in La
Selva). Similar results were found in a selectively logged area in Indonesia where no significant differences in diversity of some birds and
mammals were found between undisturbed
forest and selectively logged areas 3–5 yr after
logging (Wilson and Johns 1982), except that
the same study found a dramatic reduction in
diversity of these vertebrates when the forests
were replaced by plantations. A study in abandoned cacao of different ages and at different
distances from primary forest would be required to estimate if and when the habitat starts
to recover in terms of species richness. Heinen
(1992) found that herpetofaunal diversity in an
old abandoned cacao plantation was intermediate between that of the forest and a more
recently abandoned cacao plantation, supporting our results. Even though our results
indicate that cacao may not irreparably damage
biological diversity over the long term, the
effect of proximity to primary forest (affecting
colonization by organisms) needs to be better
elucidated. In addition, recolonization of many
plants may also be important in providing a
more diverse environment for insects, as suggested above.

As indicators of biodiversity, ground-
foraging ants represent a group that is likely to
reflect what happens to other generalist organisms, and underestimate what would happen to

more specialized organisms in disturbed habitats. If they, as indicators, provide insight into
the relationship between land use systems and
biodiversity, they could be useful in land use
planning and management of habitat. Such
management could address needs of local people, and at the same time, the need to preserve
biological diversity. In this study, we aimed to
examine such potential for conservation of
biological diversity in systems that may have
less aesthetic appeal, but which may, over the
long term, ensure that more natural areas with
their intrinsic wealth of species will be protected. If organisms are able to migrate across
or inhabit a mosaic of managed systems (including forest reserves), the chance of their
extinction diminishes.

Acknowledgments

We gratefully acknowledge R. K. Colwell, J. H.
Vandermeer, and two anonymous reviewers for their
detailed comments on earlier versions of this manuscript, S. P. Cover and J. T. Longino for their
tremendous help in identifying ant specimens, G.
Fowler and P. Park for statistical advice, I. Alvarez,
A. Martinez, C. Salazar, Standard Fruit, and Cobal
for their time and the use of their plantations, and
the staff at La Selva Biological Station for scientific
advice and logistical support. Funds for a pilot study
were provided through the Organization for Tropical
Studies, and partial funding for the study was provided by Rackham Graduate School of the University of Michigan to D. S. Roth.

References

Andersen, A. N. 1992. Regulation of "momentary"
    diversity by dominant species in exceptionally rich
    ant communities of the Australian seasonal trop-
    ics. American Naturalist **140**:401–420.
Beattie, A. J. 1985. The evolutionary ecology of
    ant-plant mutualisms. Cambridge University
    Press, New York, New York, USA.
Beattie, A. J., and D. C. Culver. 1982. Inhumation:
    how ants and other invertebrates help seeds. Na-
    ture **297**:627.
Bierregaard, R. O., Jr., T. E. Lovejoy, V. Kapos, A.
    A. dos Santos, and R. W. Hutchings. 1992. The
    biological dynamics of tropical rainforest frag-
    ments: a prospective comparison of fragments and
    continuous forest. BioScience **42**:859–866.
Bond, W., and P. Slingsby. 1984. Collapse of an
    ant-plant mutualism: the Argentine ant (*Iri-*

domyrmex humilis) and myrmecochorous Proteaceae. Ecology **65**:1031–1037.

Boucher, D. H., M. Hansen, S. Risch, and J. H. Vandermeer. 1983. Agriculture. Pages 66–73 *in* D. H. Janzen, editor. Costa Rican natural history. University of Chicago Press, Chicago, Illinois, USA.

Carroll, C. R. 1983. *Azteca* (Hormiga Azteca, Azteca ants, *Cecropia* ants). Pages 691–693 *in* D. H. Janzen, editor. Costa Rican natural history. University of Chicago Press, Chicago, Illinois, USA.

Carroll, C. R., and D. H. Janzen. 1973. Ecology of foraging by ants. Annual Review of Ecology and Systematics **4**:231–257.

Castro, A. G., M. V. B. Quieroz, and L. M. Araújo. 1990. O papel do distúrbio na estructura de comunidades de formigas (Hymenoptera: Formicidae). Revista Brasileira Entomologia **34**(1): 201–213.

Clark, D. B. 1990. La Selva Biological Station: a blueprint for stimulating tropical research. Pages 9–27 *in* A. Gentry, editor. Four neotropical rainforests. Yale University Press, New Haven, Connecticut, USA.

Colwell, R. K., and M. A. Huston. 1991. Conceptual framework and research issues for species diversity at the community level. Pages 37–71 *in* O. T. Solbrig, editor. From genes to ecosystems: a research agenda for biodiversity. International Union of Biological Sciences, Paris, France.

Connell, J. H. 1978. Diversity in tropical rain forests and coral reefs. Science **199**(24 March):1302–1310.

Connor, E. F., and E. D. McCoy. 1979. The statistics and biology of the species-area relationship. American Naturalist **113**:791–833.

Davidson, D. W., J. T. Longino, and R. R. Snelling. 1978. Pruning of host plant neighbors by ants: an experimental approach. Ecology **69**:801–808.

De Kock, A. E., J. H. Giliomee, K. L. Pringle, and J. D. Majer. 1992. The influence of fire, vegetation age and Argentine ants (*Iridomyrmex humilis*) on ant communities in Swartboskloof. Pages 203–215 *in* B. W. van Wilgen, D. M. Richardson, E. J. Kruger, and H. J. van Hensbergen, editors. Fire in South African mountain fynbos: ecosystem, community, and species response of Swartboskloof. Ecological Studies 93. Springer-Verlag, Berlin, Germany.

Denslow, J. S. 1985. Disturbance-mediated coexistence of species. Pages 307–323 *in* S. T. A. Pickett and P. S. White, editors. The ecology of natural disturbance and patch dynamics. Academic Press, Tallahassee, Florida, USA.

Haering, R., and B. J. Fox. 1987. Short-term coexistence and long-term competitive displacement of two dominant species of *Iridomyrmex*: the succes-

sional response of ants to regenerating habitats. Journal of Animal Ecology **56**:495–507.

Hansson, L. 1991. Dispersal and connectivity in metapopulations. Biological Journal of the Linnean Society **42**:89–103.

Harris, L., and J. Eisenberg. 1989. Enhanced linkages: necessary steps for success in conservation of faunal diversity. Pages 166–181 *in* D. Western and M. C. Pearl, editors. Conservation for the twenty-first century. Oxford University Press, New York, New York, USA.

Hartshorn, G. S. 1983. Plants. Pages 118–183 *in* D. H. Janzen, editor. Costa Rican natural history. University of Chicago Press, Chicago, Illinois, USA.

Heinen, J. T. 1992. Comparisons of the leaflitter herpetofauna in abandoned cacao plantations and primary rain forest in Costa Rica: some implications for faunal restoration. Biotropica **24**:431–439.

Hölldobler, B., and E. O. Wilson. 1990. The ants. Belknap Press of Harvard University Press, Cambridge, Massachusetts, USA.

Holloway, J. D., and N. E. Stork. 1991. The dimensions of biodiversity: the use of invertebrates as indicators of human impact. Pages 37–62 *in* D. L. Hawksworth, editor. The biodiversity of microorganisms and invertebrates: its role in sustainable agriculture. C. A. B. International, Wallingford, U.K.

Huxley, C. 1980. Symbiosis between ants and epiphytes. Biological Reviews of the Cambridge Philosophical Society **55**:321–340.

Jackson, D. A. 1984. Ant distribution patterns in a Cameroonian cocoa plantation: investigation of the ant mosaic hypothesis. Oecologia (Berlin) **62**:318–324.

Janzen, D. H., editor. 1983. Costa Rican natural history. University of Chicago Press, Chicago, Illinois, USA.

Johns, A. D. 1985. Selective logging and wildlife conservation in tropical rain-forest: problems and recommendations. Biological Conservation **31**:355–375.

Klein, B. 1989. Effects of forest fragmentation on dung and carrion beetle communities in central Amazonia. Ecology **70**:1715–1725.

Kleinfeldt, S. E. 1978. Ant-gardens: the interaction of *Codonanthe crassifolia* (Gesneriaceae) and *Crematogaster longispina* (Formicidae). Ecology **59**:449–456.

Krebs, C. J. 1989. Ecological methodology. Harper and Row, New York, New York, USA.

Lal, R. 1988. Effects of macrofauna on soil properties in tropical ecosystems. Agriculture, Ecosystems and Environment **24**:101–116.

Lévieux, J., and T. Diomande. 1985. Évolution des peuplements de fourmis terricoles selon l'age de la végétation dans une forêt de Cote d'Ivoire intacte ou soumise a l'action humaine. Insectes Sociaux 32(2):128–139.

Levings, S. C. 1983. Seasonal, annual, and among-site variation in the ground ant community of a deciduous tropical forest: some causes of patchy species distributions. Ecological Monographs 53:435–455.

Levings, S. C., and N. R. Franks. 1982. Patterns of nest dispersion in a tropical ground ant community. Ecology 63:338–344.

Lovejoy, T. E., R. O. Bierregaard, Jr., A. B. Rylands, J. R. Malcolm, C. E. Quintela, L. H. Harper, K. S. Brown, Jr., A. H. Powell, G. V. N. Powell, H. O. R. Schubart, and M. B. Hays. 1986. Edge and other effects of isolation on Amazon forest fragments. Pages 257–285 in M. E. Soulé, editor. Conservation biology: the science of scarcity and diversity. Sinauer Associates, Sunderland, Massachusetts, USA.

Lynch, J. F., A. R. Johnson, and E. C. Balinsky. 1988. Spatial and temporal variation in the abundance and diversity of ants (Hymenoptera: Formicidae) in the soil and litter layers of a Maryland forest. American Midland Naturalist 119:31–40.

Magurran, A. E. 1988. Ecological diversity and its measurement. Princeton University Press, Princeton, New Jersey, USA.

Majer, J. D. 1972. The ant mosaic in Ghana cocoa farms. Bulletin of Entomological Research 62:151–160.

———. 1990. Rehabilitation of disturbed land: long-term prospects for the recolonization of fauna. Proceedings of the Ecological Society of Australia 16:509–519.

Majer, J. D., and P. Camer-Pesci. 1991. Ant species in tropical Australian tree crops and native ecosystems—is there a mosaic? Biotropica 23(2): 173–181.

McNeely, J. A. 1989. Protected areas and human ecology: how national parks can contribute to sustaining societies in the twenty-first century. Pages 150–157 in D. Western and M. C. Pearl, editors. Conservation for the twenty-first century. Oxford University Press, New York, New York, USA.

Myers, N. 1984. The primary source: tropical forests and our future. W. W. Norton, New York, New York, USA.

Nestel, D., and F. Dickschen. 1990. The foraging kinetics of ground ant communities in different Mexican coffee agroecosystems. Oecologia 84:58–63.

Neumann, F. G. 1991. Responses of litter arthro-pods to major natural or artificial ecological disturbances in mountain ash forest. Australian Journal of Ecology 16:19–32.

Perfect, T. J. 1991. Biodiversity and tropical pest management. Pages 145–148 in D. L. Hawksworth, editor. The biodiversity of microorganisms and invertebrates: its role in sustainable agriculture. C.A.B. International, Wallingford, U.K.

Perfecto, I. 1990. Indirect and direct effects in a tropical agroecosystem: the maize-pest-ant system in Nicaragua. Ecology 71:2125–2134.

———. 1991a. Ants (Hymenoptera: Formicidae) as natural control agents of pests in irrigated maize in Nicaragua. Journal of Economic Entomology 84(1):65–70.

———. 1991b. Dynamics of Solenopsis geminata in a tropical fallow field after ploughing. Oikos (Copenhagen) 62:139–144.

Perfecto, I., and A. Sediles. 1992. Vegetational diversity, ants (Hymenoptera: Formicidae), and herbivorous pests in a neotropical agroecosystem. Environmental Entomology 21(1):61–67.

Pimentel, D., U. Stachow, D. A. Takacs, H. W. Brubaker, A. R. Dumas, J. J. Meaney, J. A. S. O'Neill, D. E. Onsi, and D. B. Corzilius. 1992. Conserving biological diversity in agricultural/forestry systems: most biological diversity exists in human-managed ecosystems. BioScience 42: 354–362.

Risch, S. J., D. Pimentel, and H. Grover. 1986. Corn monoculture versus old field: effects of low levels of insecticides. Ecology 67:505–515.

Roberts, J. T., and E. R. Heithaus. 1986. Ants rearrange the vertebrate-generated seed shadow of a neotropical fig tree. Ecology 67:1046–1051.

Room, P. M. 1975. Relative distributions of ant species in cocoa plantations in Papua New Guinea. Journal of Applied Ecology 12:47–61.

Samways, M. J. 1983. Community structure of ants (Hymenoptera: Formicidae) in a series of habitats associated with citrus. Journal of Applied Ecology 20:833–847.

———. 1984. A practical comparison of diversity indices based on a series of small agricultural ant communities. Phytophylactica 16:275–278.

SAS/STAT. 1990. Version 6, Fourth edition. SAS Institute, Cary, North Carolina, USA.

Southwood, T. R. E. 1987. Diversity, species packing and habitat description. Pages 420–455 in Ecological methods with particular reference to the study of insect populations. Chapman and Hall, Cambridge, U.K.

Southwood, T. R. E., V. K. Brown, and P. M. Reader. 1979. The relationships of plant and insect diversities in succession. Biological Journal of the Linnean Society 12:327–348.

Stout, J., and J. H. Vandermeer. 1975. Comparison of species-richness for stream-inhabiting insects in tropical and mid-latitude streams. American Naturalist **109**:263–280.

Torres, J. A. 1984*a*. Diversity and distribution of ant communities in Puerto Rico. Biotropica **16**(4): 296–303.

_____. 1984*b*. Niches and coexistence of ant communities in Puerto Rico: repeated patterns. Biotropica **16**(4):284–295.

Waage, J. K. 1991. Biodiversity as a resource for biological control. Pages 149–163 *in* D. L. Hawksworth, editor. The biodiversity of microorganisms and invertebrates: its role in sustainable agriculture. C.A.B. International, Wallingford, U.K.

Wilson, E. O. 1987. The arboreal ant fauna of Peruvian Amazon forests: a first assessment. Biotropica **19**(3):245–251.

_____, editor. 1988. Biodiversity. National Academy Press, Washington, D.C., USA.

_____. 1992. The effects of complex social life on evolution and biodiversity. Oikos **63**(1):13–18.

Wilson, W. L., and A. D. Johns. 1982. Diversity and abundance of selected animal species in undisturbed forest, selectively logged forest and plantations in East Kalimantan, Indonesia. Biological Conservation **24**:205–218.

Wu, J., J. L. Vankat, and Y. Barlas. 1993. Effects of patch connectivity and arrangement on animal metapopulation dynamics: a simulation study. Ecological Modelling **65**:221–254.

Appendix

List of all ants collected from primary forest, abandoned cacao plantations, productive cacao plantations, and banana in this study. Numbers indicate relative proportions of species in each site. The relative proportion of species $A = N_A/N_{all}$ where N_A = the number of times species A was found at baits within a site, and N_{all} = the sum of the number of all ant species found at all baits within the same site.

Formicidae	Primary forest			Abandoned cacao		Productive cacao plantations				Banana plantations	
	CCL	SHO	CES	AC-1	AC-2	Chila-mate	LV-1	LV-2	El Tigre	Pto. Viejo	Rio Frio
Myrmicinae											
Acromyrmex sp. 1						0.013					
A. volcanus						0.006		0.006			
Aphaenogaster sp. 1								0.006			
A. araneoides	0.117	0.054	0.11	0.14	0.081						
Apterostigma sp. 1		0.005									
Atta cephalotes			0.004	0.01				0.012			
Carabarella sp. 1	0.004	0.005									
Cardiocondyla sp. 1										0.011	
Crematogaster sp. 1			0.004		0.022						
Crematogaster sp. 2	0.021	0.025	0.02	0.024							
Crematogaster sp. 3							0.00				
Cyphomyrmex sp. 1										0.011	
Cyphomyrmex sp. 2					0.004						
Erebomyrma nevermanni	0.004	0.039	0.039	0.021	0.018						
Eurhopalothrix sp. 1					0.004						
Hylomyrma sp. 1					0.004						
Monomorium floricola								0.012			
Octostruma sp. 1					0.004						
Pheidole sp. 1	0.021	0.005	0.012	0.035	0.018	0.019		0.03	0.07		
Pheidole sp. 2	0.008	0.054	0.045	0.045	0.033	0.013	0.005	0.048			
Pheidole sp. 3	0.058		0.004	0.021				0.012			
Pheidole sp. 4			0.004							0.006	
Pheidole sp. 5	0.054	0.025	0.135	0.01	0.007		0.005			0.012	
Pheidole sp. 7						0.006	0.099	0.084	0.082		0.166
Pheidole sp. 8			0.016								
Pheidole sp. 9	0.021	0.039	0.098								
Pheidole sp. 10	0.004	0.01	0.024								
Pheidole sp. 11			0.041	0.024							
Pheidole sp. 12	0.004	0.029	0.004	0.003	0.037						
Pheidole sp. 13											0.033
Pheidole sp. 14			0.004								
Pheidole sp. 15	0.163	0.078	0.053		0.026						
Pheidole sp. 16		0.039		0.003	0.007						
Pheidole sp. 17			0.102								
Pheidole sp. 19	0.004	0.005	0.004	0.017	0.015						
Pheidole sp. 20		0.01			0.022						
Pheidole sp. 21		0.025									
Pheidole sp. 22	0.067										
Pheidole sp. 23	0.004				0.004						
Pheidole sp. 24	0.017										0.011
Pheidole sp. 25		0.005		0.157	0.019						
Pheidole sp. 26										0.005	
Pheidole sp. 27				0.003	0.055						
Pheidole sp. 28			0.012	0.003				0.012			
Pheidole sp. 30			0.008								
Pheidole sp. 31		0.005		0.01							
Pheidole sp. 32		0.005									
Pheidole sp. 33								0.006			

(continued)

Appendix (*continued*)

Formicidae	Primary forest			Abandoned cacao		Productive cacao plantations				Banana plantations	
	CCL	SHO	CES	AC-1	AC-2	Chila-mate	LV-1	LV-2	El Tigre	Pto. Viejo	Rio Frio
Pheidole sp. 34										0.005	
Pheidole sp. 35					0.007						
Pheidole sp. 36	0.125	0.005		0.007	0.007						
P. annectans	0.079	0.039	0.073	0.028	0.084			0.006		0.005	
P. cephalica		0.005									
P. longiscapa	0.05	0.015	0.024	0.003							
P. nr. mamore		0.004			0.007						
P. punctatissima										0.005	0.04
P. nr. subarmata				0.003							0.258
P. subarmata							0.45	0.096	0.421	0.043	0.06
P. fiori	0.075	0.01	0.008								
Sericomyrmex sp. 1			0.008			0.006	0.005				
Smithistruma sp. 1					0.004						
Solenopsis unid.		0.005									
Solenopsis sp. 1	0.004					0.006		0.012			
Solenopsis sp. 2											0.007
Solenopsis sp. 3	0.004			0.077	0.007		0.018				0.013
Solenopsis sp. 4	0.017	0.029			0.011						
Solenopsis sp. 5	0.004										
Solenopsis sp. 6	0.004										
Solenopsis sp. 7				0.038							
Solenopsis sp. 8		0.005	0.004	0.042	0.004			0.005			
Solenopsis sp. 9				0.035							
S. geminata				0.031		0.101	0.081	0.006	0.041	0.83	0.013
Tetramorium bicarinatum										0.005	
Trachymyrmex sp. 1				0.003		0.019					
Trachymyrmex sp. 2									0.006		
Trachymyrmex sp. 3	0.004		0.02	0.007			0.005	0.012	0.018		
Wasmannia sp. 1					0.007					0.011	
W. auropunctata	0.05	0.039	0.004	0.021	0.007	0.006	0.005	0.024	0.006		
Pseudomymicinae											
Pseudomyrmex sp. 1							0.005				
Ponerinae											
Ectatomma gibbum	0.004		0.02								
E. ruidum	0.004	0.118	0.004	0.038	0.179	0.753	0.068	0.473	0.322		0.305
E. tuberculatum			0.004					0.012			
Gnamptogenys sp. 2								0.006			
Gnamptogenys sp. 3							0.212	0.012			0.007
Gnamptogenys sp. 4			0.004								
G. bispinosa					0.004						
Hypoponera sp. 1	0.004	0.015	0.01	0.01	0.011			0.006			
Hypoponera sp. 2		0.005	0.004	0.007							
Hypoponera sp. 3										0.005	
Hypoponera sp. 4										0.011	
Odontomachus bauri								0.006			0.04
O. brunneus	0.004										
O. chelifer	0.021	0.059	0.012					0.024			
O. erythrocephalus	0.017			0.018		0.006				0.043	
O. hastatus			0.004								
O. laticeps	0.008	0.054	0.016	0.01	0.004						
O. opaciventris		0.005									
Pachycondyla apicalis	0.017	0.025	0.02	0.007							
P. constricta						0.006					

(*continued*)

Appendix (*continued*)

Formicidae	Primary forest			Abandoned cacao		Productive cacao plantations				Banana plantations	
	CCL	SHO	CES	AC-1	AC-2	Chila-mate	LV-1	LV-2	El Tigre	Pto. Viejo	Rio Frio
P. harpax	0.0125	0.02		0.031	0.011	0.013					0.046
P. obscuricornis			0.008	0.003		0.006					
P. villosa		0.005									
Paraponera clavata			0.004								
Formicinae											
Brachymyrmex sp. 1						0.019	0.009				
Brachymyrmex sp. 2					0.007						
Campanotus sp. 1			0.004				0.014		0.006		
Paratrechina sp. 1	0.033	0.059	0.016	0.063	0.077		0.009	0.06	0.018		0.013
Paratrechina sp. 2		0.025									
Dolichoderinae											
Tapinoma melanocephalum								0.018			

29

An Approach for Managing Vertebrate Diversity Across Multiple-Use Landscapes

Andrew J. Hansen, Steven L. Garman, Barbara Marks, and Dean L. Urban

Introduction

There is widespread agreement that biological diversity is valuable and that it is rapidly being lost (Myers 1979, Wilson 1988, Soulé 1991). Consequently, the conservation of biodiversity has emerged as a major international issue, and numerous laws, research initiatives, and management strategies have been enacted. Yet, biological diversity continues to decline even in wealthy and technologically advanced countries (Ehrlich and Wilson 1991). The "endangered species" approach of protecting species after they are at risk is insufficient for several reasons (Rohlf 1991, Mann and Plummer 1992). Nature preservation is also failing because reserves are often too few, too small, and too isolated to maintain natural processes and species (Pickett and Thompson 1978, Noss and Harris 1986, Newmark 1987, Hunter 1991).

Many ecologists now recommend complementing these traditional approaches with rigorous efforts to maintain biodiversity in human-dominated landscapes (Noss and Harris 1986, Brussard 1991, Hansen et al. 1991). In the United States the expanse of public lands in a "semi-natural" condition provides an opportunity to use ecological principles to manage for both commodity production and biodiversity (Wilcove 1989, Westman 1990). Land stewards are increasingly embracing this approach (Gillis 1990), and many management plans now call for resource production and for the maintenance of biological diversity at several organizational levels.

Unfortunately, no one knows how to protect "genetic, species, ecosystem, and landscape diversity" (Salwasser 1991) on lands intensively managed for wood, forage, and other products. Land managers are now wrestling with such difficult questions as: How can we manage effectively without adequate knowledge of the distribution and ecology of biodiversity? What elements of biodiversity can realistically be maintained and over what spatial and temporal scales in a managed ecosystem? What predictive approaches can best identify the likely impacts of alternative management scenarios on biodiversity and other resources? How can biological diversity be monitored to ensure that management strategies are successful?

We present here an approach for managing vertebrate species diversity in multiple-use lands at the landscape scale (e.g., 1000–20 000 ha). Vertebrates were selected because they are better known than most other organisms. Our underlying conceptual model is that animal community response to landscape change can be explained by (1) the suite of life histories represented in the local community and (2) the local trajectory of landscape change (Urban et al. 1988, Hansen and Urban 1992, Hansen et al. 1992, Urban et al. 1992). The essence of the approach is to use data on the life history and habitat use of each species in a community to classify habitat suitability across the planning area. Computer models are then used to project habitat abundance for each species under different management regimes. With such information, land managers can choose the regime that best meets their objectives.

The approach is described in five steps and illustrated with an analysis of a watershed in western Oregon. Local objectives, data bases, and expertise will dictate how the approach can best be implemented in other settings.

Step 1: Set Clear Objectives

Objectives that clearly state the desired resource and conservation priorities are critical for successful management. Well-focused objectives facilitate the development of precise landscape designs for achieving the objectives. They also provide a basis for evaluating the success of the strategies that are implemented. Factors that should be considered when establishing objectives include the level of specification required and the regional context of the planning unit.

Minimum Specifications

Setting clear objectives relative to biodiversity is especially difficult because the term is so all encompassing. Management plans often list nebulous objectives such as maintaining "biodiversity," "ecosystem health," or "ecosystem sustainability." These broad goals are too general for building specific management prescriptions or for assessing whether the prescriptions are successful.

A biodiversity management plan should specify at a minimum: response variables, target levels, and spatial/temporal domains. *Response variables* are the entities being managed. Both the organizational levels(s) of interest (e.g., deme, species, community, landscape) and the specific entities within each level need be clearly elucidated (e.g., species level: all vertebrate species, species requiring late-seral habitats, or sensitive species). *Target level* spec-

ifies the relative or absolute abundance of the response variables that is considered sufficient (e.g., minimizing the number of species that fall below minimum viable population sizes or maintaining a specific ratio of seral stages). *Spatial/temporal domain* indicates the area and time period over which the target levels of the response variables are to be maintained (e.g., over at least 80% of the planning area for at least 100 yr). Objective criteria should be used for defining these variables, levels, and domains to facilitate evaluation of whether the objectives are met.

Hierarchical Planning

Resource patterns within a planning area both influence and are influenced by factors at broader spatial scales (Noss and Harris 1986). This necessitates a hierarchical planning framework where objectives for a particular spatial scale are set with respect to broader-scale constraints and finer-scale mechanisms (Allen and Starr 1982). The importance of a planning area for a particular species, for example, can best be determined with knowledge of the regional or continental distribution of the species. The planning area may play an important role in the larger system by providing key habitats for regionally rare species or offering strategic dispersal routes under climate change. At the same time, fine-scale patterns and processes in the planning area need be considered inasmuch as they influence local animal populations. For example, the patterning of microhabitats such as canopy layering or coarse woody debris may strongly affect the demography of local populations.

Hierarchical planning is difficult because organisms and processes differ in the characteristic scales over which they operate (O'Neill et al. 1986). Thus, it is challenging to identify one set of levels in the hierarchy (planning levels) that are meaningful for all the resources of interest. It is also difficult to obtain and integrate data across a range of scales. Coarse-scale data bases such as those being generated by the U.S. Fish and Wildlife Service's Breeding Bird Survey (Droege 1990) and Gap Analysis Project (Scott et al. 1987, 1991) are useful for establishing the regional context. Data at increasingly fine spatial scales are helpful for setting local objectives. The Interagency Scientific Committee on the Northern Spotted Owl (Thomas et al. 1990) provided an excellent example of integrating data across scales and carrying out hierarchical planning for a single species.

Our example deals primarily with one level in the planning hierarchy, the landscape or watershed level. However, the principles can, and ultimately should be applied at several spatial scales.

Example

The approach was applied to a 3318-ha section of the Cook-Quentin watershed in the Willamette National Forest of the western Cascades of Oregon. This area was selected because sufficient data were readily available and land-use patterns there are typical of multiple-use federal forest lands in the region. Only vegetation and bird habitat patterns are considered. Topography, geomorphology, hydrology, and roads are ignored in this example. These factors can exert strong influence over forest productivity and/or vertebrate habitat suitability and should be considered where local data allow. The watershed is within the western hemlock (*Tsuga heterophylla*) and the Pacific silver fir (*Abies amabilis*) vegetation zones (Franklin and Dyrness 1973). Approximately 22% of the area has been clear-cut under a staggered-setting design (i.e., dispersed harvest units) and reforested. The remaining area supports natural young, mature, and old-growth stands (Fig. 1).

A regional analysis was not performed. We assumed for the example that the basin is rather unusual in the area in having large patches of mature and old-growth forests that provide habitat for late-serial bird species. We also assumed that habitats for bird species requiring large trees, snags, and/or fallen trees in open-canopy stands are rare in the watershed and in the region due to the suppression of natural disturbance and past clear-cutting (Hansen et al. 1991).

Objectives were set as follows:

1) Maximize across the planning area, in perpetuity, habitat diversity for bird species

COOK-QUENTIN WATERSHED
Present Vegetation Patterns

☐ OC-N	(0-20 yr)	▦ OC-M	(0-20 yr)
▤ Y-N	(30-70 yr)	▥ Y-M	(30-70 yr)
▨ M-N	(80-190 yr)	▦ M-M	(80-190 yr)
■ OG	(200+ yr)		

FIGURE 1. Map of current seral stage distribution in the Cook-Quentin watershed (Oregon, USA). Abbreviations are as follows and definitions of the age-classes are listed in Table 2. OC-N = open-canopy natural; Y-N = young natural; M-N = mature natural; OG = old growth; OC-M = open canopy managed; Y-M = young managed; M-M = mature managed.

requiring late-seral (mature and old-growth) habitats.

2) Maximize across the planning area, in perpetuity, habitat diversity for bird species requiring structurally rich, open-canopy habitats.

3) Produce saw timber (trees > 30 cm diameter at breast height) at levels compatible with objectives 1 and 2.

Step 2: Associate Target Species with Specific Habitat Configurations

Management goals often direct maintaining viable populations of all native vertebrates. Where this is the case, we recommend an approach intermediate between the "coarse-filter" and "fine-filter" approaches described by Noss (1987) and Hunter (1991). The coarse filter approach of maintaining communities or ecosystems in hopes of maintaining the species within them is appropriate where knowledge is lacking on the ecologies of species of interest (as is usually the case with taxonomic groups other than vascular plants, butterflies, and vertebrates). Without explicit reference to the habitat requirements of individual species, however, it is difficult to establish landscape design criteria to maintain these species and to evaluate how well an implemented design conserves species. On the other hand, intensive management of individual species (fine-filter approach) is usually not possible for many species in a community because of limitations in detailed demographic data and in financial resources needed to acquire such data.

We suggest an intermediate approach when the goal is to maintain most or all members of a vertebrate community. We recommend that habitat suitability and life-history attributes be used as surrogates for detailed demographic data for the vertebrate species in the planning area. Objective analysis of the patterning of suitable habitats across the planning area and the life-history attributes of individual species can then be used to select the subset of species that are likely to be sensitive to landscape change and merit additional demographic research see *Step 3*).

There is a strong theoretical basis (James et al. 1984, Pulliam 1988, Urban and Smith 1989) and empirical evidence (Capen 1981, Cody 1985, Verner et al. 1986) for using habitat as an indicator of demography. It is important that the appropriate habitat attributes and scales of habitat be considered. Most studies have focused on vegetation and documented the importance of vegetation structure and spatial patterning in explaining animal distributions (see Hunter 1990 and Rodiek and Bolen 1991). Habitat characteristics involving primary productivity, geomorphology, hydrology, soils, and disturbance history are less often considered in vertebrate studies. It is likely, however, that such factors are also extremely important and should be included in habitat analyses.

Animal habitat relationships should be measured over a range of spatial and temporal scales (Urban et al. 1992). Individual species probably operate over a characteristic range of scales, and the habitat elements explaining most of the variation in species abundance may differ among scales (Harris 1984, Wiens et al. 1986, Neilson et al. 1992) (Fig. 2). Differences in scales of habitat use among species are probably a function of differences in life-history attributes such as body size, metabolic rates, home range size, and vagility (Hansen and Urban 1992). Thus it is desirable to measure animal habitat relations at multiple scales and to perform objective analyses of the type and strength of habitat association at each scale. These data can then be used to evaluate the response of each species to changes in habitat patterning involving one or more scales.

Several analytical methods exist for quantifying animal habitat relationships and for classifying habitat suitability in independent field plots. Common methods include simple seral-stage associations (e.g., Thomas 1979, Verner and Boss 1980, and Brown 1985), multivariate statistical methods (Capen 1981, Verner et al. 1986), and Habitat Suitability Models (HSI) (USFWS 1981, Schamberger and O'Neil 1986).

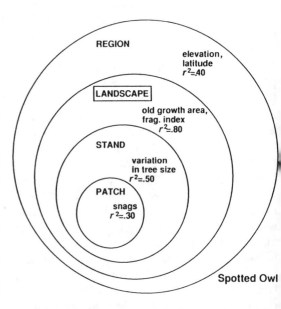

FIGURE 2. Hypothetical example illustrating that the types and strengths of animal habitat relationships may differ among spatial scales. Listed for each scale are the habitat features explaining most of the variation in Spotted Owl abundance and the strength of the association. Note in this example that the strongest correlation occurs at the landscape level.

These methods differ in the resolution of data needed for calibration and in accuracy in classifying habitat suitability. Approaches based on seral-stage association or discriminant function analysis typically identify habitats simply as suitable or unsuitable. Logistic regression generates a probability that a plot is suitable. HSI models produce output of a relative ranking from 0 (least suitable) to 1 (most suitable). And regression analyses can rate habitat quality in terms of population abundance. Which methods are most appropriate for a given application depends upon local expertise, data availability, and desired predictive capability.

While habitat may serve as a useful indicator of animal demography, it is important to point out that the relationship is seldom perfect. Biotic interactions (e.g., predation, competition, etc.), disturbances, chance demographic events, and other factors may all complicate the species-habitat associations. For this reason it is very important to validate animal-habitat models to determine if the error level is acceptable for the application at hand. Well-designed monitoring programs are useful for providing data for validation (see *Step 5*).

Land managers will typically find that information is insufficient to produce rigorous habitat models for local species. Given that management activities will proceed anyway, we recommend in the short term that managers use the best available data to good advantage. As little as we know about animal habitat associations in most regions, much more information is generally available than is currently being used for management. Over the longer term, managers should implement field studies to provide data to develop and validate habitat models.

Example

For the Cook-Quentin application we required a means of assessing bird habitat relationships that could be derived from existing studies, that considered habitat factors at two or three spatial scales, and that could interface with the habitat simulation model chosen for the application.

We settled on a simple habitat-classification scheme that considers just four variables: seral-stage association, microhabitat association, response to edge, and minimum territory size (Table 1). The initial species list included those birds identified by Brown (1985) as having primary habitats in low- and mid-elevation conifer and conifer-hardwood forests in Oregon and Washington west of the Cascade Mountain crest. We supplemented Brown's habitat evaluation with data from the other sources listed in Table 1. Included in the analysis were the 51 species for which sufficient data were available.

Step 3: Assess the Potential Viability of Species

Objectively ranking species in terms of sensitivity to landscape change can determine which species most merit additional research or special management consideration. One approach is to map the abundance of suitable habitats across the planning area. A second approach is to examine the life-history attributes of the species.

Population viability is strongly related to area of suitable habitat (Laurance 1991) and to population size (Pimm et al. 1988), which is often a function of habitat area. Mapping habitat suitability across the planning area can identify species that may be at risk because of habitat shortages. This mapping can be done by cross-tabulating the distribution of habitat attributes over the planning area with the habitat classification functions developed in Step 2. Geographical information systems (GIS) are especially useful for managing, analyzing, and displaying these data.

Knowledge of species life histories is also useful for evaluating the potential viability of a population in a given habitat configuration. Several studies have found that certain life-history traits are strongly correlated with proneness to extinction, including short longevity, low reproductive rate, constrained dispersal, specialization on particular foods or habitats, and large home-range size (Whitcomb et al. 1981, Pimm et al. 1988, Laurance 1991). A systematic evaluation of key life-history traits of each species can identify those species

TABLE 1. Bird species, life-history traits, projected sensitivity to landscape change, and habitat suitability in the planning areas.*

Species	Reproductive effort	Nest type†	Nest height (m)†	Minimum territory size (ha)	Seral-stage assoc.‡	Micro-habitat assoc.	Response to edge§‖	Response to patch size§	Sensitivity score	Area suitable (%)
Neotropical migrants										
Hermit Warbler										
Dendroica occidentalis	4	O	17.7	0.0	M, OG	G	G	...	15	36
Solitary Vireo										
Vireo solitarius	4	O	11.3	1.7	M, OG	G	G	...	16	36
Hammond's Flycatcher										
Empidonax hammondii	4	O	7.6	0.0	M, OG	G	G	G	16	36
Western Wood-Pewee										
Contopus sordidulus	3	O	7.6	1.2	M, OG	G	G	...	17	36
Western Flycatcher										
Empidonax difficilis	6	O	4.6	0.0	M, OG	G	G	G	15	36
Olive-sided Flycatcher										
Contopus borealis	4	O	12.2	0.0	OC	G	E	G	17	16
Townsend's Warbler										
Dendroica townsendi	4	O	3.7	0.0	M, OG	G	G	...	15	36
Orange-crowned Warbler										
Vermivora celata	5	O	0.6	0.0	OC	G	I	...	19	10
Vaux's Swift										
Chaetura vauxi	5	H	1.2	0.0	OG	N	G	...	15	26
Wilson's Warbler										
Wilsonia pusilla	5	O	0.	0.2	G	G	G	...	17	100
Black-throated Gray Warbler										
Dendroica nigrescens	4	O	7.0	0.0	G	G	G	...	15	100
Western Tanager										
Piranga ludoviciana	4	O	11.0	0.0	OC	G	E	...	16	29
Tree Swallow										
Tachycineta bicolor	5	H	3.1	0.0	OC, M, OG	N	G	...	13	16
Swainson's Thrush										
Catharus ustulatus	8	O	3.7	0.0	Y, M, OG	G	I	...	16	48
Short-distance migrants										
American Goldfinch										
Carduelis tristis	5	O	4.6	0.0	OC	G	I	...	16	10
American Robin										
Turdus migratorius	8	O	4.6	0.0	OC	G	G	P	16	22
Western Bluebird										
Sialia mexicana	5	H	7.6	0.3	OC	N	G	...	13	0
Hermit Thrush										
Catharus guttatus	5	O	1.2	0.6	Y, M, OG	G	G	G	15	78
Rufous Hummingbird										
Selasphorus rufus	4	O	2.4	0.0	OC, OG	G	G	...	14	58

*Bird species were drawn from those listed by Brown (1985) as having primary habitats in low- to mid-elevation conifer and conifer-hardwood forests in western Oregon and Washington. Sensitivity score is a relative index of sensitivity to landscape change based on life-history traits. Area suitable is the percentage of the Cook-Quentin planning area that was rated as suitable habitat by the LSPA model. Migration strategy is from Ehrlich et al. 1988 and Love 1990. All other data are from Brown 1985 unless otherwise noted. ... denotes missing data.

    Character variables are coded; for all variables: G = Generalist; for Nest type: O = Open, H = Hole, P = Parasite; for Seral-stage association: OC = Open Canopy (<30 yr), Y = Young (30–70 yr), M = Mature (80–190 yr), OG = Old Growth (>190 yr); for Microhabitat association: N = Natural (large trees, snags, fallen trees); for Response to edge: E = Edge specialist, I = Interior specialist; for Response to patch size: P = Positive.

    †From Ehrlich *et al.* 1988.

    ‡Serving as primary habitat as defined by Brown 1985.

    §From Rosenburg and Raphael 1986.

    ‖A. J. Hansen, J. Peterson, and E. Howarth, *unpublished data.*

TABLE 1. (*continued*)

Species	Repro-ductive effort	Nest type†	Nest height (m)†	Minimum territory size (ha)	Seral-stage assoc.‡	Micro-habitat assoc.	Response to edge§‖	Response to patch size§	Sensi-tivity score	Area suitable (%)
				Residents						
Brown Creeper										
Certhia americana	6	H	8.0	1.7	M, OG	N	G	G	13	36
Northern Goshawk										
Accipiter gentilis	3	O	12.2	100.0	M, OG	G	G	. . .	15	36
Winter Wren										
Troglodytes troglodytes	6	H	0.8	0.3	M, OG	N	I	P	17	19
Hairy Woodpecker										
Picoides villosus	4	H	9.8	0.0	G	N	G	G	11	78
Cooper's Hawk										
Accipiter cooperi	4	O	12.2	100.0	Y, M, OG	G	G	. . .	18	78
Blue Grouse										
Dendragapus obscurus	9	O	0.0	0.0	G	g	G	G	18	100
Chestnut-backed Chickadee										
Parus rufescens	7	H	2.1	1.3	M, OG	N	G	P	14	36
Sharp-shinned Hawk										
Accipiter straitus	6	O	10.7	100.0	Y, M, OG	G	G	P	18	78
Varied Thrush										
Ixoreus naevius	4	O	8.3	20.0	M, OG	G	I	. . .	17	19
Golden-crowned Kinglet										
Regulus satrapa	16	O	9.8	0.3	Y, M, OG	G	G	G	11	78
Pileated Woodpecker										
Dryocopus pileatus	4	H	13.8	128.0	M, OG	N	G	P	15	36
Red Crossbill										
Loxia curvirosta	4	O	7.0	0.0	M, OG	G	G	. . .	13	36
Red-breasted Nuthatch										
Sitta canadensis	8	H	6.7	0.9	M, OG	N	G	G	11	36
Gray Jay										
Perisoreus canadensis	4	O	5.2	64.0	Y, M, OG	G	G	. . .	18	77
Barred Owl										
Strix varia	3	H	10.7	0.0	M, OG	N	G	. . .	13	36
North Pygmy Owl										
Glaucidium gnoma	5	H	4.3	0.0	M, OG	N	G	G	12	36
American Kestrel										
Falco sparverius	5	H	14.1	100.0	OC	N	G	. . .	15	0
White-crowned Sparrow										
Zonotrichia leucophrys	8	O	0.8	0.0	OC	G	I	. . .	18	10
Rufous-sided Towhee										
Pipilo erythropthalmus	8	O	0.8	0.0	OC	N	I	. . .	16	100
Song Sparrow										
Melospiza melodia	8	O	0.6	0.3	OC	G	I	. . .	16	10
Mountain Quail										
Oreortyx pictus	10	O	0.0	2.0	OC	G	G	G	18	22
Spotted Owl										
Strix occidentalis	2	O	6.1	100.0	M, OG	N	I	P	20	0
Northern Saw-whet Owl										
Aegolius acadicus	5	H	11.3	0.0	G	N	G	. . .	13	100
Northern Flicker										
Colaptes auratus	9	H	3.4	16.0	OC, M, OG	N	G	. . .	10	36
American Crow										
Corvus brachyrhynchos	6	O	10.7	0.0	G	G	G	. . .	13	100
Great Horned Owl										
Bubo virginianus	4	O	12.2	25.0	OC, M, OG	G	E	. . .	16	29
Red-tailed Hawk										
Buteo jamaicensis	3	O	13.1	100.0	OC, M, OG	G	G	. . .	14	51

(*continued*)

TABLE 1. (*continued*)

Species	Repro-ductive effort	Nest type†	Nest height (m)†	Minimum territory size (ha)	Seral-stage assoc.‡	Micro-habitat assoc.	Response to edge§‖	Response to patch size§	Sensi-tivity score	Area suitable (%)
Steller's Jay										
Cyanocitta stelleri	4	O	5.2	0.0	G	G	G	G	13	100
Pine Siskin										
Carduelis pinus	8	O	8.6	0.0	G	G	G	. . .	12	100
Purple Finch										
Carpodacus purpureus	8	O	7.0	0.0	G	G	G	. . .	13	100
Common Raven										
Corvus corax	8	O	6.1	0.0	G	G	G	. . .	13	100
Dark-eyed Junco										
Junco hyemalis	10	O	3.1	0.0	G	G	G	. . .	11	100

most likely to be at risk. Life-history data for vertebrates can be derived from field guides and primary ecological literature.

By assessing both habitat availability and life-history attributes, managers can identify the species that are especially sensitive to management. Such species may merit special management approaches and/or more detailed demographic studies.

Example

Habitat mapping. — Current vegetation patterns in the Cook-Quentin watershed were described using two USDA Forest Service data bases. The Mature and Over-Mature (MOM's) inventory used aerial photographs to classify stands (≥ 2 ha in size) according to tree diameter and height. These data were not validated for the Cook-Quentin landscape. Field surveys in the Fall Creek watershed on the nearby Lowell Ranger District, Willamette National Forest, indicated that the MOM's survey classified overstory size class correctly in 78% of the stands sampled (G. Marsh, *unpublished report* to the Willamette National Forest).

We reclassified the MOM's data for Cook-Quentin by seral stage and stand age (Table 2) for compatibility with the habitat classification functions and our landscape model. A second data set, derived from aerial photographs, delineated the location and the harvest date of all stands that were clear-cut in the past. These stands were labeled as managed, and were assumed to be devoid of the large trees, snags, and fallen trees that are known to be important

microhabitat elements for several vertebrate species, are typical in natural forests, and are generally absent in traditionally managed plantations in the region (Hansen et al. 1991). Hence, the vegetation map included three seral stages of managed forest (open canopy, young, and mature) and four seral stages of natural forest (open canopy, young, mature, and old growth). Nonvegetational features such as streams and rock outcroppings were not considered.

Habitat suitability across the basin was determined using a computer program that crosstabulated the habitat requirements for each species with the vegetation characteristics of each 2.5-ha cell. For those species responding to edges, the zone of attraction or avoidance was assumed to be within 160 m of the edge. Also, the area requirement had to be met within a patch; use of two or more neighboring patches was not considered.

Relatively little of the planning area was rated as suitable for several of the bird species (Table 1). No habitat was available for Western Bluebird, American Kestrel, and Spotted Owl (scientific names of all bird species are listed in Table 1). The open-canopy patches with natural microhabitats required by Western Bluebird and American Kestrel were not present. Also unavailable were patches of mature and oldgrowth forest large enough for Spotted Owl. (In reality, this species does exist in the Cook-Quentin watershed. Individual breeding pairs likely make use of several neighboring patches of older forest, a strategy not considered in our model.) Only 10% of the landscape was suit

TABLE 2. Convention used to convert USDA Forest Service vegetation data for the Cook-Quentin watershed to seral stages used in the landscape modeling application. dbh = diameter at breast height.

| MOMS* vegetation classes | Landscape model | |
	Seral stage	Stand age (yr)
Seedling (≤1.4 m height)	Open canopy	10
Sapling (1.4–6.0 m height)	Open canopy	20
Pole (>6.0 m height, <20 cm dbh)	Open canopy	30
Small (20–53 cm dbh)†	Young	55
Large (>53 cm dbh)†	Mature	140
Old growth†	Old growth	250

*Mature and Over-Mature inventory (see *Step 3: Assess . . . . Example: Habitat mapping*).

†Criteria for old growth described in Old Growth Definition Task Group (1986).

able for four species that require the interiors of open-canopy stands: Orange-crowned Warbler, Song Sparrow, White-crowned Sparrow, and American Goldfinch. The Olive-sided Fly-catcher, an edge species, found only 16% of the landscape suitable. Nineteen percent of the area was available to Varied Thrush and Winter Wren, birds found primarily in the interiors of mature and old-growth forest. In addition to being relatively rare, habitat for many of these species was fragmented (Fig. 3), a fact that could further jeopardize local populations.

Field data were not available to validate the habitat suitability maps. As mentioned above, this is an important step in real-world applications.

Life-history traits. We also used information on several life-history characteristics to derive a "sensitivity" index (Hansen and Urban 1992) of the potential responsiveness of each species to landscape change. Species were rated from 1 (least sensitive) to 3 (most sensitive) for each of the eight life-history traits (Table 1). A total score for a species was derived by summing the scores across traits. The rationales for the criteria generally follow the findings of Whitcomb et al. (1981). The validity of this approach was supported by a significant correlation between the sensitivity scores of Pacific Northwest bird species and their regional population trends over the past 20 yr (Hansen and Urban 1992).

Scores ranged from 11 (least sensitive) to 20 (most sensitive) (Table 1). Among those with the highest scores were Spotted Owl (20), Orange-crowned Warbler (19), Olive-sided Flycatcher (17), Winter Wren (17), and Varied Thrush (17). All of these also have limited habitat in the planning area (see above), hence they are among the species that may be most vulnerable under some management activities.

The mapping of habitats within the planning area and the life-history analyses identified several bird species potentially sensitive to landscape change that had not previously been recognized by conservationists (e.g., some early-successional species: Orange-crowned Warbler, White-crowned Sparrow). This fact emphasizes the value of objective approaches for rating species viability. Such species may merit additional research and management attention.

We emphasize that attention to spatial scale is important in assessing species sensitivity. The data we used in the life-history analysis repre-

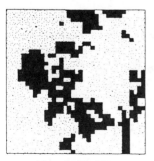

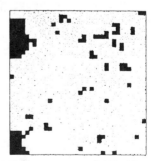

Vaux's Swift   Olive-sided Flycatcher   Orange-crowned Warbler

FIGURE 3. Maps of suitable habitat (■) in the Cook-Quentin planning area for bird species associated respectively (left to right) with old-growth, edge, and open-canopy interior habitats.

sent approximations across the range of each species, making this analysis more or less regional in scale. The habitat mapping, in contrast, only considered the Cook-Quentin landscape. A similar mapping effort at larger spatial scales (regional or continental) is required to place the local results in a context for evaluation. If, for example, the Orange-crowned Warbler has abundant habitats or large populations elsewhere in the region, managers of the Cook-Quentin landscape may choose not to be concerned about the shortage of habitat for this species in the planning area. As mentioned under *Step 1*, local objectives need to be derived based on information at several spatial scales.

We are now attempting to expand our approach for habitat mapping to the subregional scale. Species range maps generated by the Gap Analysis Project (Scott et al. 1987, 1991) and various continental surveys of bird population trends (e.g., Droege 1990) are also useful in establishing a broader-scale context for evaluating the sensitivity of local species.

Step 4: Project Future Habitat Patterns Under Alternative Management Prescriptions Using Simulation Models

Land managers have a long history of trying to assess the likely future consequences of differing management strategies. Key challenges for landscape management are to develop a comprehensive set of alternative landscape designs and to perform objective trade-off analyses of resource response under each design.

Designing landscapes for biodiversity is a topic currently attracting much attention, but there are few good examples or comprehensive guidelines. General principles are presented in Harris (1984) and Hunter (1991). Thomas et al. (1990) and K. N. Johnson, J. R. Franklin, J. W. Thomas, and J. Gordon (*unpublished report* [1991] to the Committee on Agriculture, U.S. House of Representatives, Washington, D.C.: available from Forest Research Laboratory Publications, College of Forestry, Oregon State University, Corvallis, Oregon, USA) offer case studies of regional-scale designs for late-seral and riparian species in the Pacific Northwest (PNW) of the United States.

Knowledge of landscape dynamics in presettlement times may sometimes offer guidance for modern landscape design. Information on the relationships among disturbance regimes, habitat patterns, and vertebrates in natural landscapes can provide a context for understanding and managing current landscapes (e.g., Romme and Despain 1989, Hansen et al. 1991). However, we caution against using a snapshot of spatial patterns from the past as a literal guide for a desired future condition. The high level of spatial and temporal variation in many presettlement landscapes may be unacceptable today. Also, modern landscapes are often rescaled and bounded in such a way that the movements of disturbance and organisms typical in the past are not now possible (Urban et al. 1987).

Presettlement fire regimes in the PNW, for example, were extremely variable spatially and temporally (Morrison and Swanson 1990). During periods when wildfire was intense over large areas, forest-dwelling species probably persisted in small refugial areas. Some of these species likely recolonized burned habitats slowly during the decades to centuries following the disturbance. Unless such relaxation periods and suitable dispersal corridors are provided, modern populations would likely be lost under these patterns.

We conclude that there is no alternative but to use ecological principles to design landscape patterns deliberately to meet management objectives. Much work is needed on how this can best be done.

Once a set of management alternatives has been designated, trade-off analyses of resource responses can help to determine which alternatives best meet the management objectives. Computer simulation models can be extremely useful for projecting the responses of several resource variables over long time periods and large areas. A variety of models have been developed to simulate vegetation dynamics and vertebrate habitats. These models differ in the degree of biological realism in their formulations, the spatial scale at which vegetation is considered, and extent of spatial interaction among neighboring cells. Reviews can be found in Shugart (1984), Verner et al. (1986), and

Huston et al. (1988). The choice of which type of model to use for vertebrate habitat applications depends upon the questions being addressed, the data sets available, the prediction accuracy required, and the programming expertise available. Unfortunately, "easy to use" packages are not generally available. A substantial investment is usually required to adapt a general model to a particular location.

Example

Models. We used the landscape model LSPA (Li 1989, Hansen et al. 1992) and the gap model ZELIG.PNW (Urban 1990) to simulate four disturbance-management regimes in the Cook-Quentin watershed. LSPA is a geometric model that simulates change in a gridded landscape according to a user-specified timber harvest regime involving cutting-unit size, spatial distribution of cutting units, and harvest rate. Vegetation dynamics are not modeled directly; stands are assigned to seral stages based on the time elapsed since disturbance. The model calculates several landscape metrics at each time-step and classifies habitat suitability for each bird species according to the criteria described under *Step 2.*

ZELIG is a generic version of the gap model FORET (Shugart 1984) that was designed to be adapted to diverse forest types. These models simulate the establishment, growth, and death of individual trees on small plots equivalent to the area shaded by a canopy tree. Tree demography is stochastically constrained by tree life-history traits and local environmental conditions. Output from several simulated gaps is aggregated for a statistical description of stand dynamics. ZELIG.PNW is a version developed for forests in the western Oregon Cascades. It incorporates several updates to ZELIG since the model was introduced (Urban 1990), involving tree height-diameter relationships, leaf area, tree growth, and soil moisture (D. L. Urban and S. L. Garman, *unpublished manuscript*). ZELIG.PNW additionally includes subroutines to simulate diverse silvicultural prescriptions and snag and fallen-log dynamics (Garman et al. 1992). The model has performed well in validations for chronosequences of natural and managed forests at elevations of about 1000 m in the western Oregon Cascades (Garman et al. 1992. D. L. Urban and S. L. Garman, *unpublished manuscript*) (Fig. 4).

For this demonstration ZELIG.PNW was run using environmental conditions and tree

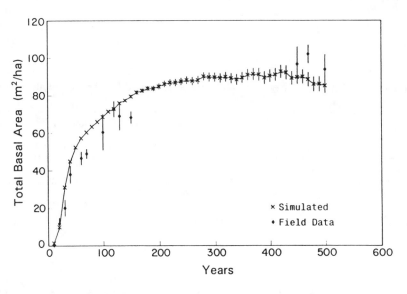

FIGURE 4. Trends in total basal area under natural succession at 1000 m elevation, western slope of the Oregon Cascades mountains as described by field data and as simulated with the gap model ZELIG.PNW (see *Step 4 . . . : Example: Models*). Simulated data are based on 30 model plots. Sample sizes of field data varied from 5 to 42 reference stands. Data shown are means ± 1 SE.

species for the 1000-m elevation in the west-central Oregon Cascades. Thirty 0.1-ha plots were modeled for each simulation and the results averaged to represent the stand. ZELIG. PNW was initialized with a 250-yr-old simulated stand for all four prescriptions. The response variables derived from ZELIG were standing basal area of saw timber (trees > 30 cm in diameter at breast height) and basal area of saw timber harvested.

Wood production across the Cook-Quentin landscape under each disturbance-management scenario was determined by first calculating the area occupied by each seral stage at each time-step. The amounts of standing saw timber and saw timber harvested for each seral stage were derived from the ZELIG runs for the time-steps equivalent to the median ages of the seral stages. These values were then summed across seral stages to obtain totals for the landscape at that time step. In the wood production run, no trees were retained during harvest, and the harvest level for each seral stage was assumed to be equal to the simulated basal area of saw timber for the mean age class within the seral stage. The same procedure was used for the multiple-use run except that basal area of saw timber harvested was reduced by 9.3% for each seral stage. This level is equivalent to the amount retained (9.8 trees/ha) at the start of the ZELIG simulation of multiple use.

The variable reported here as total wood production was calculated as the cumulative basal area of saw timber harvested up until a time-step plus the basal area of saw timber standing at that time-step.

Management alternatives. The four disturbance-management regimes simulated are described in Table 3. A presettlement fire regime was modeled to offer a point of reference for the management prescriptions. We previously simulated the regime of high-severity fires documented for the Cook-Quentin watershed using a model similar to LSPA (Hansen et al. 1992). We generated a starting landscape for the present application by initializing the fire model with 200-yr-old forest and simulating 220 yr of forest succession to reduce the effects of initial vegetation conditions. A run for an additional 140 yr is reported here.

The other three prescriptions were initialized with present vegetation patterns in the Cook-Quentin planning area. The wood production run is typical of that used on private forest lands in the PNW today. Cutting units were maximally dispersed under a staggered-setting design. One 70-yr rotation was simulated, and the results at year 70 were considered steady state for an additional rotation. The multiple-use prescription, based on principles advanced as ecological forestry (Franklin 1992), had larger harvest units, a longer rotation, and a higher level of tree, snag, and log retention than the wood-production run. Also, the units were

TABLE 3. Natural disturbance and management alternatives simulated for the Cook-Quentin watershed (Oregon).

| Model* | Variable | Prescription | | | |
		Natural fire	Wood production	Multi-use forestry	No action
LSPA	Disturbance patch size (ha)	8.6†	22.5	40.0	NA‡
	Disturbance pattern	Random	Maximum dispersal	Maximum aggregation	NA
	Rotation length (yr)	114§	70	140	NA
	Minimum harvest or burn age (yr)	20	55	55	NA
ZELIG	Retention level (no. of trees/ha)	9.8 PSME/ha	none	9.8 PSME/ha	NA
	Inseeding rate (seeds/ha)	Natural	988 PSME	988 PSME	988 PSME
	Thinning yr 15 and 30 (stems/ha)	None	543 PSME	380 PSME	None
				163 TSHE£	

*The landscape simulation model LSPA and the gap model ZELIG.PNW are described in *Step 4: Project future habitat patterns . . . : Example: Models.* † Patch size is modeled as an exponential function with a mean of 8.6 ha.

‡NA = not applicable.

§ Fire rotation interval is an exponential function with a mean of 8.8% of the landscape per decade.

‖PSME = *Pseudotsuga menziesii.*

¶TSHE = *Tsuga heterophylla.*

maximally aggregated using a quarter-strip cutting design (Li 1989). The final (no action) prescription had no management activities over the 140-yr simulation period.

We assumed for the purposes of bird habitat classification that all stands in the natural fire and multiple-use runs contained sufficient levels of live trees, snags, and fallen trees to support the bird species requiring these features. These features are removed under traditional clear-cutting, and thus we assumed that all harvest units in the wood-production run and harvest units under age 110 in the no-action run were unsuitable habitat for species requiring such microhabitat features.

Results. Landscape geometry and the distribution of seral stages differed substantially among the four scenarios (Fig. 5). The wood-production run lost all natural microhabitats and late seral stages by year 70 (Fig. 6a and 6b), while early seral stages remained abundant (Fig. 6c). Results of the no-action run were somewhat the inverse of those under wood production. The multiple-use run maintained high levels of natural microhabitats and moderate levels of early and late seral stages, patterns also generated by the natural-fire run. Total density of edges between patches of different seral stage, a measure of landscape fragmentation, was highest under the natural-fire regime, intermediate under no action and wood production, and lowest under multiple use over much of the simulation (Fig. 7). We did not differentiate between the habitat suitability of forest edges created by wildfire and those created by timber harvest. In reality, fire-generated edges likely differ from the edges of harvest units in structure and likely result in less extreme changes in forest interior microclimate and vertebrate habitat suitability. Even so, the fire run suggests that patch shape and seral-stage distribution were complex under the pre-settlement fire regime in this basin (see also Morrison and Swanson 1990). This finding counters the notion that presettlement landscapes in this portion of the Cascades were continuous expanses of old growth.

Habitat diversity for all bird species was substantially lower under wood production than under the other three scenarios (Fig. 8a), due mostly to the loss of natural microhabitats

and late seral stages. None of the 18 species primarily associated with late seral stages had habitat after year 70 under wood production (Fig. 8b); all but one of these species (Spotted Owl) had some suitable habitat in the other three runs. Habitat richness for early seral species was highest under natural fire and multiple use, slightly lower under wood production, and went to 0 under no action (Fig. 8c).

Production of saw timber was substantially greater under wood production than under multiple use. By year 140, total wood production under the wood production run was 49% greater than that under multiple use (Fig. 9). This difference was due to: (1) modeled tree-growth rates being reduced in the multiple-use prescription by overstory retention and by thinning for mixed species, and (2) more rapid conversion of older, slower-growing natural stands to younger, faster-growing plantations under the wood-production run.

This exercise is, to our knowledge, the first attempt to quantify the consequences for wildlife habitat and wood production of alternative management scenarios that considers both stand- and landscape-level factors. The simulations predict that the multiple-use prescription would maintain bird habitat diversity for all species, late-seral species, and early-seral species at levels similar to the natural fire regime. Wood production, however, is substantially reduced under the multiple-use run compared with the wood-production run. Land managers, after weighing the assumptions and limitations of the methodology, can use this information in choosing the alternative that is most likely to promote management objectives. The multiple-use run would be the obvious choice based on the stated objectives of this demonstration and the spatial and temporal scale at which the analysis was conducted.

Step 5: Implement Preferred or Experimental Strategies and Monitor the Responses of Habitats and Species

Implementation of a preferred strategy (or strategies) is an experiment in itself that can

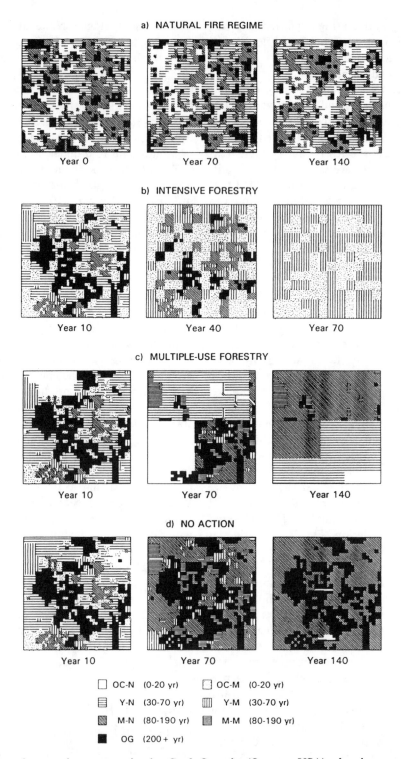

FIGURE 5. Maps of vegetation pattern in the Cook-Quentin (Oregon, USA) planning area for selected time-steps as simulated by the landscape model LSPA for four disturbance-management scenarios: (a) natural fire; (b) wood production; (c) multiple use; and (d) no additional management intervention. OC-N = open-canopy natural; Y-N = young natural; M-N = mature natural; OG = old growth; OC-M = open-canopy managed; Y-M = young managed; M-M = mature managed.

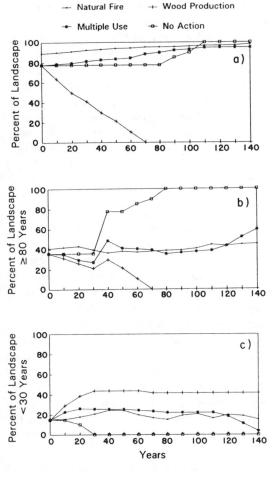

FIGURE 6. Proportion of the simulated landscape under each of four management scenarios containing: (a) levels of canopy trees, snags, and coarse woody debris sufficient for bird species requiring these features; (b) mature and old-growth forest (≥ 80 yr); and (c) early seral stands (< 30 yr).

sive. This fact is in evidence in the Pacific Northwest where even the most basic information on habitat distributions in management units (e.g., snag levels in harvest units) has not been successfully assembled, let alone the more comprehensive information needed to manage vertebrate habitat diversity.

We suggest that managers of vertebrate habitats monitor the effectiveness of implementing a prescription, the responses of habitat to the management action, and the population responses of select species. Monitoring programs should consider multiple temporal and spatial scales. See Noss (1990) for a thoughtful approach to monitoring biodiversity.

Creative approaches are needed to collect these data in an efficient and cost-effective fashion. Remote sensing offers promise for sampling habitats from microsite to landscape and regional levels (e.g., Cohen and Spies 1992). Monitoring wildlife species abundance still requires field sampling. Ultimately, interagency cooperation may offer the best hope of designing and implementing appropriate monitoring protocols. Expanded funding levels will also be essential. The payoffs of rigorous monitoring are apt to be considerable. These data are needed both for evaluating the extent to which management objectives are met and for testing and updating wildlife habitat and computer simulation models.

Completion of Step 5 leads back to Step 1. An iterative process of reevaluating management objectives, performing trade-off analyses, and conducting management experiments of-

reveal a great deal about resource response to manipulation (see Walters [1986] and Walters and Holling [1990] for reviews). In fact, land managers may sometimes wish to subdivide the planning area and implement two or more management alternatives, in a replicated fashion if possible, and compare results.

A well-designed monitoring program is critical for learning from any management experiment. This view is widely held among federal forest managers. Most forest plans in our region call for some level of monitoring, but designing and implementing effective monitoring programs is neither simple nor inexpen-

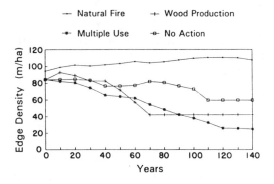

FIGURE 7. Density of edges among seven seral stages in the planning area under the four management scenarios.

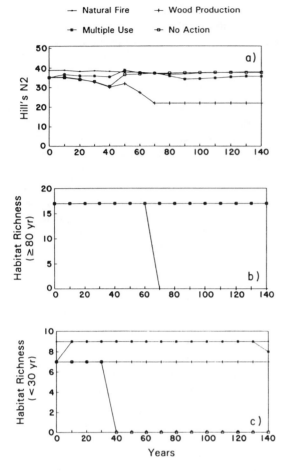

FIGURE 8. Bird habitat diversity in the planning area under the four management scenarios. (a) Diversity (Hill's N2) for all bird species. (b) Number of species primarily associated with mature and old-growth stands with some suitable habitat (habitat richness) (total n = 18). (c) Habitat richness for species primarily associated with early seral habitats (total n = 9).

fers hope for successfully managing vertebrate diversity on multiple-use lands.

Conclusion

Society and the courts are increasingly demanding objective and effective management strategies that balance complex resource demands. The approach we describe for managing vertebrates on multiple-use lands is a logical step towards more effective management of biological diversity.

One limitation of the approach is that it is restricted to organisms for which taxonomy, habitat requirements, and life-history attributes are described. This is the case for vertebrates and vascular plants in many areas, but not for most invertebrates and nonvascular plants (which represent the great majority of species). The coarse-filter approach (Noss 1987, Hunter 1991) may be the best alternative for such groups until more information becomes available. Efforts such as this one on the better known taxa are useful for delineating the types of information that are needed on groups yet to be studied in any detail.

Another limitation of the approach is that habitat suitability is evaluated rather than vertebrate demography. As mentioned earlier, habitat is known to be an imperfect indicator of demography. Consequently, other approaches have been developed that simulate reproduction, dispersal, and survival of individual animals across complex landscapes (e.g., Pulliam et al. 1992). Unfortunately, data for parameterizing these models is available for relatively few vertebrate species and these species may not necessarily be the ones most sensitive to human activities. For this reason we advocate using the habitat-based approach as a way to consider most vertebrates in a community. The demographic approaches should additionally be used for those species that are sufficiently well known.

The demonstration for the Cook-Quentin

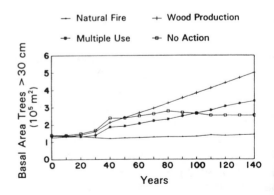

FIGURE 9. Total wood production (cumulative basal area harvested up to each time-step plus basal area of live trees) of saw timber (trees > 30 cm in diameter at breast height) in the planning area under the four management scenarios.

landscape also has various limitations. The vertebrate habitat models used have not been validated, and thus we were unable to assess the accuracy of the predictions. In reality, current data are insufficient to validate the habitat functions. We believe it is better to base management decisions on objectively derived habitat functions, even if unvalidated, than not to consider vertebrates at all. In the longer term, obtaining data for validation should be a high priority among those charged with managing vertebrate diversity.

Also, our simulations did not consider landscape attributes such as topography, stream courses, and road networks. These factors are likely important to forest productivity, logging feasibility, and animal habitat suitability in many landscapes (e.g., Li 1989). Some important ecological processes were also not considered, including propagation of natural disturbance, long-term site productivity, seed dispersal, and the role of animals in altering habitats. Knowledge of such processes is important for predicting habitat response to management. For example, traditional timber management in the region may reduce long-term site productivity and result in declining timber yields in successive rotations (Franklin 1992), an outcome not considered under our wood-production run. Some of these factors are now being incorporated into our models (topography, streams, site index, and nutrient cycling) to increase the realism of the simulations.

A final limitation is that this demonstration did not consider the regional context of the planning area. Applying the approach across neighboring landscapes would be useful for setting objectives, evaluating vertebrate species sensitivity to landscape change, and interpreting the trade-off analyses.

On the other hand, the approach is attractive in offering a logical set of steps that, when implemented iteratively, should improve both our knowledge base and the effectiveness of management. The approach allows for most vertebrates in a planning area to be considered and for their potential sensitivity to land use to be evaluated objectively. Such assessment of species sensitivity is important because species that are abundant in a location can become rare in relatively short time periods in changing landscapes (Hansen et al. 1992). Another strength of the approach is the tandem use of simulation models and management experiments. Given that stand- and landscape-scale experiments on forest dynamics will require decades to complete, computer modeling can be used now to perform trade-off analyses of resource response under differing management strategies. Modeling also is useful for identifying information gaps. Management experiments complement modeling by providing data for parameterizing the models and for assessing the effectiveness of various landscape designs. A final strength of the approach is that it encourages use of the best information currently available. Rather than ignoring ecological data because they are incomplete, it encourages that management decisions be based on the best information that is available at the time and advocates implementation of mechanisms to improve the knowledge base.

Acknowledgments

The framework for the approach evolved from discussions with Hank Shugart, Tom Smith, Fred Swanson, and numerous forest managers. We thank Mark Harmon, Charlie Halprin, and Phil Sollins for collaboration in adapting ZELIG to the PNW. Joe Lint, David Mladenoff, Roger Monthey, Jim Morrison, and Dave Perry reviewed the manuscript. This work was supported by the COPE Program, College of Forestry, Oregon State University, the USDA Forest Service Pacific Northwest Forest Experiment Station, the USDA Forest Service New Perspectives Program, and the Environmental Protection Agency.

References

Allen, T. F. H., and T. B. Starr. 1982. Hierarchy: perspectives for ecological complexity. University of Chicago Press. Chicago, Illinois, USA.

Brown, E. R., technical editor. 1985. Management of wildlife and fish habitats in forests of western Oregon and Washington. USDA Forest Service Report RG-F&WL-192-1985. USDA Forest Service, Portland, Oregon, USA.

Brussard, P. F. 1991. The role of ecology in biological conservation. Ecological Applications 1:6–12.

Capen, D. E., editor. The use of multivariate statistics in studies of wildlife habitat. USDA Forest

Service General Technical Report RM-87. Rocky Mountain Forest and Range Experiment Station, Fort Collins, Colorado, USA.

Cody, M. L. 1985. An introduction to habitat selection in birds. Pages 4–46 in M. L. Cody, editor. Habitat selection in birds. Academic Press, Orlando, Florida, USA.

Cohen, W. B., and T. A. Spies. 1992. Estimating structural attributes of Douglas-Fir/Western Hemlock forest stands from landsat and SPOT imagery. Remote Sensing on the Environment 41:1–17.

Droege, S. 1990. The North American breeding bird survey. In J. R. Sauer and S. Droege, editors. Survey designs and statistical methods for the estimation of avian population trends. U.S. Fish and Wildlife Service, Biological Report 90:1–4.

Ehrlich, P. R., D. S. Dobkin, and D. Wheye. 1988. The birder's handbook. Simon & Schuster, New York, New York, USA.

Ehrlich, P. R., and E. O. Wilson. 1991. Biodiversity studies: science and policy. Science 253:758–762.

Franklin, J. F. 1992. Scientific basis for new perspectives in forests and streams. Pages 25–72 in R. Naiman, editor. Watershed management: balancing sustainability and environmental change. Springer-Verlag, New York, New York, USA.

Franklin, J. F., and C. T. Dyrness. 1973. Natural vegetation of Oregon and Washington. Oregon State University Press, Corvallis, Oregon, USA.

Garman, S. L., A. J. Hansen, D. L. Urban, and P. F. Lee. 1992. Alternative silvicultural practices and diversity of animal habitat in western Oregon: a computer simulation approach. Pages 777–781 in P. Luker, editor. Proceedings of the 1992 Summer Simulation Conference. The Society for Computer Simulation, Reno, Nevada, USA.

Gillis, A. M. 1990. The new forestry: an ecosystem approach to land management. BioScience 40:558–562.

Hansen, A. J., T. A. Spies, F. J. Swanson, and J. L. Ohmann. 1991. Lessons from natural forests: implications for conserving biodiversity in natural forests. BioScience 41:382–392.

Hansen, A. J. and D. L. Urban. 1992. Avian response to landscape pattern: the role of species life histories. Landscape Ecology 7:163–180.

Hansen, A. J., D. L. Urban, and B. Marks. 1992. Avian community dynamics: the interplay of human landscape trajectories and species life histories. Pages 170–195 in A. J. Hansen and F. di Castri, editors. Landscape boundaries: consequences for biological diversity and ecological flows. Ecological Studies Series. Springer-Verlag, New York, New York, USA.

Harris, L. D. 1984. The fragmented forest. University of Chicago Press, Chicago, Illinois, USA.

Hunter, M. L., Jr. 1990. Wildlife, forests, and forestry. Prentice Hall, Englewood Cliffs, New Jersey, USA.

———. 1991. Coping with ignorance: the coarse-filter strategy for maintaining biodiversity. Pages 266–281 in K. A. Kohm, editor. Balancing on the brink of extinction. Island Press, Washington, D.C., USA.

Huston, M., D. DeAngelis, and W. Post. 1988. New computer models unify ecological theory. BioScience 38:682–691.

James, F. C., R. F. Johnston, G. J. Niemi, and W. J. Boecklen. 1984. The Grinnellian niche of the Wood Thrush. American Naturalist 124:17–47.

Laurance, W. F. 1991. Ecological correlates of extinction proneness in Australian tropical rain forest mammals. Conservation Biology 5:1–11.

Li, H. 1989. Spatio-temporal pattern analysis of managed forest landscapes: a simulation approach. Dissertation. Oregon State University, Corvallis, Oregon, USA.

Love, T. 1990. Distribution maps of Pacific Northwest birds using neotropical habitats. Oregon Birds 16:56–85.

Mann, C. C., and M. L. Plummer. 1992. The butterfly problem. The Atlantic 269(1):47–70.

Morrison, P. H., and F. J. Swanson. 1990. Fire history and pattern in a Cascade Range landscape. USDA Forest Service General Technical Report PNW-GTR-254, USDA Forest Service, Portland, Oregon, USA.

Myers, N. 1979. The sinking ark. A new look at the problem of disappearing species. Pergamon, Elmsford, United Kingdom.

Neilson, R. P., G. A. King, R. L. De Velice, and J. M. Lenihan. 1992. Regional and local vegetation patterns: the responses of vegetation diversity to subcontinental air masses. Pages 129–149 in A. J. Hansen and F. di Castri, editors. Landscape boundaries: consequences for biological diversity and ecological flows. Ecological Studies Series. Springer-Verlag, New York, New York, USA.

Newmark, W. D. 1987. A land-bridge island perspective on mammalian extinctions in western North American parks. Nature 325:430–432.

Noss, R. F. 1987. From plant communities to landscapes in conservation inventories: a look at the Nature Conservancy (USA). Biological Conservation 41:11–37.

———. 1990. Indicators for monitoring biodiversity: a hierarchical approach. Conservation Biology 4:355–364.

Noss, R. F., and L. D. Harris. 1986. Nodes, networks, and MUMs: preserving diversity at all scales. Environmental Management 10:299–309.

Old-Growth Definition Task Group. 1986. Interim definitions for old-growth Douglas-fir and mixed-conifer forests in the Pacific Northwest and California. Forest Service Research Note PNW-447. USDA Forest Service, Portland, Oregon, USA.

O'Neill, R. V., D. L. DeAngelis, J. B. Waide, and T. F. H. Allen. 1986. A hierarchical concept of the ecosystem. Princeton University Press, Princeton, New Jersey, USA.

Pickett, S. T. A., and J. N. Thompson. 1978. Patch dynamics and the size of nature reserves. Biological Conservation 13:27–37.

Pimm, S. L., H. L. Jones, and J. Diamond. 1988. On the risk of extinction. American Naturalist 132:757–785.

Pulliam, H. R. 1988. Sources, sinks, and population regulation. American Naturalist 132:652–661.

Pulliam, H. R., J. B. Dunning, Jr., and J. Liu. 1992. Population dynamics in complex landscapes: a case study. Ecological Applications 2:165–177.

Rodiek, J. E., and E. G. Bolen. 1991. Wildlife and habitats in managed landscapes. Island Press, Washington, D.C., USA.

Rohlf, D. J. 1991. Six biological reasons why the endangered species act doesn't work—and what to do about it. Conservation Biology 5:273–282.

Romme, W. H., and D. G. Despain. 1989. Historical perspective on the Yellowstone fires of 1988. BioScience 39:695–699.

Rosenberg, K. V., and M. G. Raphael. 1986. Effects of forest fragmentation on vertebrates in Douglas-fir forests. Pages 263–272 in J. Verner, M. L. Morrison, and C. J. Ralph, editors. Wildlife 2000: modeling habitat relationships of terrestrial vertebrates. University of Wisconsin Press, Madison, Wisconsin, USA.

Salwasser, Hal. 1991. In search of an ecosystem approach to endangered species conservation. Pages 247–265 in K. A. Kohm, editor. Balancing on the brink of extinction. Island Press, Washington, D.C., USA.

Schamberger, M. L., and L. J. O'Neil. 1986. Concepts and constraints of habitat-model testing. Pages 5–10 in J. Verner, M. L. Morrison, and C. J. Ralph, editors. Wildlife 2000: modeling habitat relationships of terrestrial vertebrates. University of Wisconsin Press, Madison, Wisconsin, USA.

Scott, J. M., B. Csuti, J. D. Jacobi, and J. E. Estes. 1987. Species richness: a geographic approach to protecting future biological diversity. BioScience 37:782–788.

Scott, J. M., B. Csuti, K. Smith, J. E. Estes, and S. Caicco. 1991. Gap analysis of species richness and vegetation cover: an integrated biodiversity conservation strategy. Pages 282–297 in K. A. Kohm, editor. Balancing on the brink of extinction. Is-

land Press, Washington, D.C., USA.

Shugart, H. H. 1984. A theory of forest dynamics. Springer-Verlag, New York, New York, USA.

Soulé, M. E. 1991. Conservation: tactics for a constant crisis. Science 253:744–750.

Thomas, J. W., editor. 1979. Wildlife habitats in managed forests of the Blue Mountains of Oregon and Washington. Agriculture Handbook 553. U.S. Department of Agriculture, Forest Service, Washington, D.C., USA.

Thomas, J. W., E. D. Forsman, J. B. Lint, E. C. Meslow, B. R. Noon, and J. Verner. 1990. A conservation strategy for the Northern Spotted Owl. Interagency Scientific Committee to Address the Conservation of the Northern Spotted Owl (USDA: Forest Service, USDI: Bureau of Land Management, Fish and Wildlife Service, and National Park Service). 1990-791-171/20026. United States Government Printing Office, Washington, D.C., USA.

Urban, D. L. 1990. A versatile model to simulate forest pattern. A user's guide to ZELIG version 1.0. University of Virginia, Charlottesville, Virginia, USA.

Urban, D. L., A. J. Hansen, D. O. Wallin and P. J. Halpin. 1992. Life-history attributes and biodiversity: scaling implications for global change. Pages 173–195 in O. T. Solbrig, H. M. van Emdem, and P. G. van Oordt, editors. Biological diversity and global change. Monograph 8. International Union of Biological Sciences. Paris, France.

Urban, D. L., R. V. O'Neill, and H. H. Shugart. 1987. Landscape ecology: a hierarchical perspective can help scientists understand spatial patterns. BioScience 37:119–127.

Urban, D. L., H. H. Shugart, Jr., D. L. DeAngelis, and R. V. O'Neill. 1988. Forest bird demography in a landscape mosaic. Publication Number 2853. Oak Ridge National Laboratory, Oak Ridge, Tennessee, USA.

Urban, D. L., and T. M. Smith. 1989. Microhabitat pattern and the structure of forest bird communities. American Naturalist 133:811–829.

USFWS [U.S. Fish and Wildlife Service]. 1981. Standards for the development of suitability index models. Ecological Services Manual 103. USDI Fish and Wildlife Service, Division of Ecological Services. U.S. Government Printing Office, Washington, D.C., USA.

Verner, J., and A. S. Boss, technical coordinators. 1980. California wildlife and their habitats: western Sierra Nevada. USDA Forest Service, General Technical Report PSW-37. Pacific Southwest Forest and Range Experiment Station, Berkeley, California, USA.

Verner, J., M. L. Morrison, and C. J. Ralph, editors. 1986. Wildlife 2000: modeling habitat relationships of terrestrial vertebrates. University of Wisconsin Press, Madison, Wisconsin, USA.

Walters, C. J. 1986. Adaptive management of renewable resources. McGraw-Hill, New York, New York, USA.

Walters, C. J., and C. S. Holling. 1990. Large-scale management experiments and learning by doing. Ecology 71:2060–2068.

Westman, W. E. 1990. Managing for biodiversity. BioScience 40:26–33.

Whitcomb, R. F., C. S. Robbins, J. F. Lynch, B. L. Whitcomb, K. Klimkiewicz, and D. Bystrak. 1981. Effects of forest fragmentation on avifauna of the eastern deciduous forest. Pages 125–205 *in* R. L. Burgess and D. M. Sharpe, editors. Forest island dynamics in man-dominated landscapes. Springer-Verlag, New York, New York, USA.

Wiens, J. A., J. T. Rotenberry, and B. Van Horne. 1986. A lesson in the limitations of field experiments: Shrub-steppe birds and habitat alteration. Ecology 67:365–376.

Wilcove, D. S. 1989. Protecting biodiversity in multiple-use lands: lessons from the U.S. Forest Service. Trends in Ecology and Evolution 4:385–388.

Wilson, E. O., editor. 1988. Biodiversity. National Academy Press, Washington, D.C., USA.

30
Cross-Scale Morphology, Geometry, and Dynamics of Ecosystems

C. S. Holling

Introduction

Community ecology and ecosystem ecology seem to have existed in different worlds. Levin (1989) suggests that the gulf between the two is the consequence of the different historical traditions in each. Community ecology, for example, emerged from basic studies, where generalized patterns were sought in the natural interactions among the biota. From the outset, the goal has been to deduce general and simple theory. On the other hand, many of the modelling approaches developed to understand ecosystem dynamics emerged from specific applied problems, where not only biotic but abiotic and human disturbances transformed ecosystem function. That tradition, therefore, is often more complete, but at the price of producing a collection of complex specific examples from which generalization is difficult.

So long as the image of these two fields concentrates on caricatures, they will never come together. There are, for example, ecosystem models that can be legitimately criticized for presuming that ever-increasing detail improves prediction, but such examples are not at the heart of advances in the field (Holling 1978, Walters 1986). Similarly, community studies that do not provide testable connections to reality and that ignore scale are not at the

leading edge of that field, either (Strong et al. 1984, Brown and Maurer 1989, Wiens 1989).

Both fields have matured sufficiently that the differences have begun to blur. Community and population biologists have begun to address applied problems—e.g., of multi-species fisheries (May et al. 1979), and of conservation (Soulé and Wilcox 1980). They have joined ecosystem ecologists in grappling with the challenge of continental-scale shifts in climate and land use, reflected, for example, in Brown and Maurer's (1989) call for "macroecology." And ecosystem ecologists have become active in formulating theory by utilizing applied studies to provide focus and relevance—e.g., deriving simplified theoretical abstractions of complex models (Ludwig et al. 1978), or developing a thermodynamic theory for disturbed and transformed ecosystems (Odum 1985).

There is now recognition of the need for a new contract between community and ecosystem ecology, reflected in the recognition that animals shape their ecosystems (McNaughton et al. 1988, Naiman 1988) and that species dynamics can be more sensitive to ecosystem stress than ecosystem process rates (Schindler 1990, Vitousek 1990). This paper crystallizes and focuses these concerns by providing evidence that a relatively few processes, having distinct frequencies in space and time, structure terrestrial ecosystems, entrain other variables, and set the rhythm of ecosystem dynamics. Moreover, the signature of that rhythm and its discontinuous, lumpy character is echoed in the discontinuous nature of the body-mass distributions of animal communities.

This paper presents the results of an effort to test the reality of a consistent pattern that emerged from a comparative study of 23 examples of managed ecosystems (summarized in Holling 1986). Those studies were in the traditions of systems ecology and suggested that the way ecosystems are structured in time and space could be described, simplified, and generalized. The process of testing the generalizations, however, led to approaches more consistent with community ecology. Therefore this paper evolved into one that also became a formal effort to blend the two fields by utilizing theories, techniques, and examples from each.

Since the paper is complex, Table 1 is provided as a reader's guide and summary.

Inferring the Pattern

The 23 examples of managed ecosystems fall into four groups: forest insect pests, forest fire, semi-arid grasslands, and fisheries. The essence of the dynamic behavior of the models developed for these situations proved to be generated by the actions and interactions of 3-4 sets of variables and associated processes, each of which operated at distinctly different speeds (as reviewed in Holling 1986). The speeds were therefore discontinuously distributed, and differed from their neighbors often by as much as an order of magnitude. A summary of the conclusions is presented in Table 2.

But are those features the consequence of the way modellers make decisions rather than the results of ecosystem organization? It does not help, particularly, that the models were based upon extensive knowledge of biological processes and, in most instances, used parameters that were independently estimated in the field. But some support is provided by critical studies like the forest-budworm one, in which the model continued to faithfully represent very different behaviors in the field occurring because of different climates in different geographical regions or because of different pest

management and harvesting strategies (Clark et al. 1979). Nevertheless, the uneasy feeling that such conclusions are a figment of the way we think led to the series of different tests that forms the foundations for this paper. For the moment, then, the discontinuous nature of time dynamics should be considered a hypothesis.

For the models, at least, this structure organizes the time behavior of variables into a small number of cycles, presumably abstracted from a larger set that continue at smaller and larger scales than the range selected. As an example, we predict four dominant cycles for insect defoliator-forest dynamics in boreal forest ecosystems (Clark et al. 1979, McNamee et al. 1981, Holling 1988, 1992): those with periods of 3-5 yr, 10-15 yr, 35-40 yr, and > 80 yr. Each is controlled by a specific key variable together with lagged interactions with other variables. For example, the 3-5 yr cycle is generated by an interaction among needles, insect defoliators, and parasites (McNamee 1979). All are fast variables, with generation times of one year, and the lags between them generate the periodicity.

The intermediate periodicity of 10-15 yr is dominated by interactions between defoliation and the time for recovery of foliage quantity and quality. Since growth is inhibited and there is little tree mortality, the basic timing is set by the generation time for crown foliage — for conifers, typically 8-12 yr. This periodicity is one of the most commonly monitored in the boreal forest (Myers 1988). In some cases, it is associated with outbreaks that spread to cover tens of thousands of square kilometres, and in other cases it is associated with localized non-spreading outbreaks (Berryman 1987). In both cases tree growth is slowed, but tree mortality is infrequent.

The 35-40 yr cycle is classically represented by spruce budworm in the eastern area of the distribution of spruce-balsam fir forests in North America (Morris 1963, Clark et al. 1979). Its periodicity is the consequence of the interaction between the fast variables associated with the insects, and the slowest variable, the trees. Essentially, agents like insectivorous birds (Holling 1988) inhibit outbreaks whenever budworm populations are very low and the crowns are sufficiently small that search-

TABLE 1. A reader's guide.

I. Inferring the Pattern

A review of 23 examples of managed ecosystems suggests that a small number of clusters of biotic and abiotic variables impose structure over a very wide range of scales in all ecosystems i.e.,

1) *The Extended Keystone Hypothesis:* All terrestrial ecosystems are controlled and organized by a small set of key plant, animal, and abiotic processes. As a corollary:

2) *The Entrainment Hypothesis:* Within any one ecosystem, time-series data for biotic variables should have periodicities that cluster into a small number of sets, reflecting the generation time of one of the critical structuring variables.

3) These propositions need refinement to become testable, and so a search starts to identify a relevant set of testable hypotheses. A lumpy distribution of morphological attributes of animals emerges as a possibility.

II. Are There Size Clumps?

Adult body mass is a robust morphological measurement that might provide a direction.

1) If ecosystems have a discontinuous architecture, then animals should demonstrate attributes of size and behavior that are scaled by the discontinuous architecture of the landscapes in which they live. Specifically, there will be gaps in the distribution of body masses.

2) Data for mammals and birds of the boreal forest and short-grass prairie and for birds of North America show that such body-mass gaps exist. Body-mass distributions are fundamentally discontinuous, but what causes the gaps?

III. What Are the Size Gaps?

The multi-causal world of ecology makes it likely that single hypotheses, even if valid, will be rejected. A set of alternatives is therefore needed.

1) Four alternative hypotheses could explain these body-mass gaps:

a) *The Textural-Discontinuity Hypothesis:* Animals demonstrate the existence of a hierarchical structure and discontinuous texture of the landscape they inhabit by having a discontinuous distribution of their sizes, searching scales, and behavioral choices.

b) *The Limited-Morph Hypothesis:* There are only a limited number of life-forms (e.g., locomotory modes) possible for animals, each of which is constrained to function effectively only over a limited range of sizes.

c) *The "Urtier" Historical Hypothesis:* The species of animals are drawn from a limited number of ancestral forms whose organizational constraints preclude evolution of intermediate sizes.

d) *The Trophic-Trough Hypothesis:* Size-dependent trophic interactions initiate and maintain lumpy patterns of body-mass distributions independent of any other mechanism.

2) Those alternatives are tested by comparing body-mass clumps in ecosystems having different spatial structures (forest, prairie, and pelagic) and in animal groups having fundamentally different body plans (birds vs. mammals) or feeding habits (carnivores vs. herbivores).

3) All but the Textural-Discontinuity Hypothesis are rejected as an initiating cause. A review of pelagic, benthic, and African savanna communities substantiates the conclusion that discontinuities in body-mass distributions are universal in terrestrial landscapes.

4) Thus the geometry of landscapes and ecosystems is organized into a small number of quanta with distinct architectural attributes, and these quanta shape the morphology of animals.

IV. Resource Utilization in Discontinuous Landscapes

The analysis now shifts from the qualitative identification of structure to measurement of its quantity. Both the number of species with different body masses and the size of their home ranges reflect the amount of resources available at different scales. Moreover, the relation between home-range size and body mass provides a way to convert body masses into absolute measures of spatial attributes in the landscape. The result:

1) The amount of resources in each ecosystem quantum is different in each ecosystem and landscape.

2) The overall way that resources are distributed among animals of different sizes is similar in all landscapes, except, perhaps, in those subject to rapid transition.

3) The spatial extent of home ranges of mammals of the same body mass is determined by the productivity of the landscape and by their trophic status. The less productive the landscape and the farther up the trophic chain, the larger the spatial extent of their foraging areas.

4) All animals measure elements in the landscape with a spatial grain defined as a function of their size. As a consequence, there is a unified response of all trophic levels and of mammals and birds to the geometric structure of resources available in each quanta. The sampling grain of the animal interacts with the texture of the ecosystem to define an invariant feature of home ranges and of body-mass clump structure that unifies all mammals and birds and all trophic groups.

V. Hierarchy of Decisions and Hierarchy of Opportunities

Those conclusions now set the stage to analyze how animals make decisions in a discontinuous environment and how those decisions can result in the discontinuous structure of body-mass distributions.

1) There is a hierarchy of decisions made by all animals that increases in scale from food, to forageable patch, to habitats, to home range, and to region. The mass-specific scale of the hierarchy of decisions is calculated by using allometric relationships to convert the body-mass clump limits into absolute space and time equivalents.

2) The spatial extent of decisions made by animals of the largest clump categories ranges from roughly tens of centimetres to thousands of kilometres. In contrast, those of the smallest category exploit a smaller range from millimetres

(continued)

TABLE 1. (*continued*)

to kilometres. As a consequence, animals of different body-mass clumps encounter different discontinuities in their environment. Those discontinuities are potentially hierarchical and fractal.

3) A hierarchical organization in a landscape generates abrupt shifts in the kind, distances, and sizes of objects when the grain of measurement reaches a value that aggregates objects into a new set of objects in a new hierarchical level. Since the size of an animal defines the grain of its measurement, then a body-mass gap exists at those sizes and sampling grains where there is an abrupt transition in the attributes of objects (their size and inter-object distances) between two hierarchical levels or between two domains of different fractal dimension.

4) Scale-specific impacts of disturbances on vegetation therefore have scale-specific impacts on the body-mass groups most affected. Landscape fragmentation will have differential effects on different sizes of animals depending on the spatial grain of the fragmentation.

VI. Revisiting Ecosystem Dynamics

The circle of the argument is now closed by turning from the structure of animal morphology and of ecosystem geometry back to the original issues raised by ecosystem dynamics.

1) The landscape is hierarchically structured by a small number of structuring processes into a small number of nested levels, each of which has its own physical textures and temporal frequencies. That is, the processes that generate discontinuous time dynamics also generate discontinuous physical structure.

2) At least six hierarchical levels can be identified, each of which is dominated by one category of structuring processes. The smaller and faster scales are dominated by vegetative processes, the intermediate by disturbance and environmental processes, and the largest and slowest by geomorphological and evolutionary processes.

3) The intermediate, mesoscale processes transfer local events at the scale of patch into large-scale consequences because they have a spreading, or spatially contagious character. Mesoscale spatial dynamics and the distribution of vegetation are dominated by animals, water, wind, fire, and people. They represent the processes that mediate the kind and speed of response of vegetated landscapes to possible climate change.

4) Those organisms that are members of the set of structuring processes make their own environment by producing some of the structures within the nested set. At the same time, these structures create niches at a variety of scales for a larger set of entrained species.

5) The structuring processes establish a cycle of birth, growth and storage, and death and renewal as a nested set of such cycles, each with its own range of scales. For the microscales, fresh needles cycle yearly, the crown of foliage cycles with a decadal period, and trees or gaps cycle at close to a century or longer period.

6) Those cycles are organized by four functions: exploitation, conservation, release, and reorganization. Stability and productivity are determined by the exploitation and conservation sequence. Resilience and recovery are determined by the release and reorganization sequence.

7) There are two points in the cycle where slower and larger levels in ecosystems become briefly vulnerable to transformation because of small events and fast processes. One is at release, when the system becomes overconnected and brittle as it moves toward maturity. The other is at reorganization, when the system is underconnected, with weak organization and weak regulation. It is at those points that alternative futures can suddenly emerge for devolution or evolution.

VII. The Human Animal

One animal, the human, provides a specific perspective into the sources of and responses to lumpy geometry and dynamics.

1) Human activities transform the texture of landscapes at mesoscales and provide a specific example of how a discontinuous hierarchy of decisions intersects with a landscape hierarchy to produce the opportunities that animals exploit.

2) The spatial grain and ambit of the human animal is restricted not by body-mass class, but by technological innovation. It is that process that expands both human choices and human impacts to regional and planetary scales.

3) The paper closes with a research agenda to link processes across scales and to integrate ecosystem dynamics with community structure.

---

ing by birds is concentrated within a small volume of foliage per hectare. Gradual growth of the trees and closure of crowns forces the searching activity by birds to be diluted over such large volumes of foliage that predation mortality declines and budworm populations escape to generate a spreading outbreak. In these cases, tree mortality can be up to 80% over extensive areas. It is an example of multi-stable behavior (Holling 1973, Sinclair

et al. 1990) that is described in more detail elsewhere (Ludwig et al. 1978, Clark et al. 1979).

Finally, the ≥ 80-yr cycle represents the longest successional period and is ultimately set by tree longevity. The proximate cause of tree death can be any of a number of mortality agents—wind (Sprugel and Bormann 1981), fire (Heinselman 1981a, b), and insect or disease (Davis 1981, 1989).

TABLE 2. Key variables and speeds in four groups of managed ecosystems.

| The system | Variables | | | Key reference |
	Fast	Intermediate	Slow	
Forest insect	Insect, needles	Foliage crown	Trees	NcNamee et al. 1981, Holling 1991
Forest fire	Intensity	Fuel	Trees	Holling 1980
Savanna	Annual grasses	Perennial grasses	Shrubs	Walker et al. 1969
Aquatic	Phytoplankton	Zooplankton	Fish	Steele 1985

These four periodicities are almost certainly abstracted from a larger set. Still-slower nutrient and successional dynamics could generate cycles of multiple centuries, and the faster dynamics of soil organisms can generate periodicities of months (Pastor and Mladenoff 1992). Nevertheless, although these models are only partial representations of ecosystem function, they do provide a basis for inferring the features of more complete ecosystem patterns.

This inference is based in part upon the observation that the behavior of all of the models was organized by the action and interaction of 3–4 variables each having a qualitatively different speed (as summarized earlier in Table 2). Although the specific processes and variables that caused the four periods of cyclic behavior were unique to each situation, the primary determinant was the trees themselves, with new needles having a 1-yr generation time, crowns of foliage an 8–12 yr generation time, and the trees themselves a generation time of >70 yr. The question now is how the different cyclic patterns interrelate to each other.

In one example, the jack pine sawfly-jack pine system (McLeod 1979), three of the cycles could occur in the same simulation and were generated as a function of the age of the forest and stochastic weather effects. Both the simulations and field data showed periods of fast cycles at low insect densities, periods when populations abruptly increased to cause significant defoliation in a 10–15 yr cycle, and periods where populations could abruptly jump still higher and trigger tree mortality and a renewal of the 60–80 yr successional cycle. Similarly, the budworm model generated not only the 35–40 yr boom and bust cycle over the whole 70 000-km^2 region modelled, but also an ≈15-yr cycle with little tree mortality in a few sites in that region. Such "anomalous" sites are known in nature, and one such site provided the principal source of data for one of the major analyses of budworm population dynamics that reached conclusions very different (Royama 1984) from those described above.

That suggests that the occurrence of these cycles of different periods is not a matter of either one or the other, but that they all occur as a nested set within an ecosystem. Moreover, each will have both a temporal and spatial aspect. Local areas with fast cycles are contained within stands where there will be synchronous slower ones (or less-frequent disturbances), which in turn are contained within still larger regions with still-slower cycles, and so on.

This strongly suggests that ecosystems could be hierarchically structured by a small set of processes into discontinuous domains of influence or "lumps," each nested within another and each defined by a specific periodicity/ frequency and spatial architecture. Such an argument is an extension of Paine's revealing demonstration of the effects of keystone variables. In his original example (Paine 1966, 1974), experimental manipulations of littoral communities showed clearly that starfish predation of mussels structured the spatial distribution and sizes of patches of species in the community. That keystone argument can be extended to include macroscopic plants and their associated disturbance mechanisms as obvious contributors to spatial architecture and to abiotic disturbances like fire. Thus it seems possible that a small number of clusters of biotic and abiotic variables will impose structure over a very wide range of scales in all ecosystems. Formally, the hypothesis can be stated as follows:

The Extended Keystone Hypothesis: All terrestrial ecosystems are controlled and organized by a small set of key plant, animal, and abiotic processes. They form interacting clusters of relationships, each of which determines the temporal and spatial structure over a constrained range of scales. The overall extent of these influences covers at least centimetres to hundreds of kilometres in space and months to centuries in time.

But again, does this generalized conclusion simply reflect how we impose order on complex systems rather than how ecosystems are in fact structured? Independent tests of the proposition are necessary.

The first test concerns the nested cycles. If ecosystems are structured by a small number of dominant cycles whose frequencies (or periods) are discontinuously distributed, then we could propose that all ecosystem variables will track one of those nested cycles. Each periodicity will act as an attractor, entraining other variables. This can occur because the basic time constants of some variables are in the neighborhood of the structuring cycle, or because evolutionary adaptations lead to the occupation of each of these specific, frequency-defined niches. An example is the way a tree species, like jack pine, can develop adaptations to fire that tune the life-span, seed release, and the fire cycle as an interacting set. The formal hypothesis can be expressed as follows:

The Entrainment Hypothesis: Within any one ecosystem, time-series data for biotic variables should have periodicities that cluster into a small number of sets, reflecting the generation time of one of the critical structuring variables, i.e., periodicities (or frequencies) should be discontinuously distributed in a predictable way.

Since the shortest period predicted by the models for the eastern boreal forest is 3–5 yr and the longest period is >80 yr, a test would ideally require data sets with a resolution of one year and a duration that can contain at least three of the longer cycles—i.e., 250 yr. Such series are available from data on tree rings (Blais 1965, 1968) and annually varved lake-sediment cores (Clark 1988), but typical durations are shorter and biased toward major events like extensive insect outbreaks and major fires. I am working on such a survey, which will become part of a later, companion paper. For the moment, it is significant that Carpenter and Leavitt (1991) proposed a similar "variance transmission model" and confirmed a pattern

of entrainment they predicted using a 93-yr sediment core from a lake in Wisconsin. Since other time-series data are sparse and potentially biased, however, a more rigorous test is desirable. That led to the proposition that if ecosystems are hierarchically structured in time they will also be in space. Spatial attributes such as object size and physical texture should also be discontinuously distributed into a small number of categories.

In a general way, we perceive the landscape as a nested hierarchy (Allen and Starr 1982, Urban et al. 1987, Steele 1989), with each level having its own range of scales in both space and time (Fig. 1). The forest-insect models described above represent a subset of that hierarchy confined to the finely textured fast variables of needles and their associated defoliating insects and parasites, the slower and coarser crown and its associated suite of insectivorous birds, and the still slower and coarser patches associated with plant species competing for nutrients, light, and water.

But again, how real is that perception of hierarchy and of discontinuous structures? How fundamental are they to the function of the ecosystem and the organization of communities? The proposition needs refinement to become a testable hypothesis.

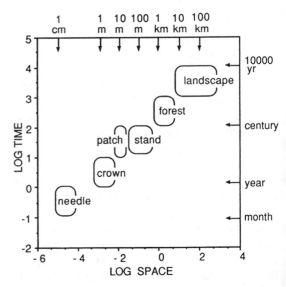

FIGURE 1. Logarithmic time and space scales of landscape elements and ecosystems of the boreal forest. The temporal scale is based on dimensions in years, the spatial scale on dimensions in kilometres.

Testable hypotheses are trivially easy to develop if the question is defined narrowly enough. The challenge is to develop alternative sets of testable hypotheses having real power for generalization. What hypotheses can be developed in this case that are both relevant and testable?

Community ecology has been shaped by a search for simple order that might underlie the apparent diversity and complexity of nature. It was R. H. MacArthur's compelling intuition that the physical architecture of the environment shaped the structure of biotic components of communities and the morphology and behavior of the associated animal species. A variety of studies (MacArthur and MacArthur, 1961, MacArthur et al. 1966, Willson 1974, Rotenberry and Wiens 1980) have shown that, at some scales, patterns of species distribution and community structure are associated with vegetation structure and habitat configuration or "physiognomy" (Whittaker 1975).

But other studies challenge those conclusions. Wiens and Rotenberry (1981), for example, show that plant species composition (floristics), rather than habitat architecture, determines the community composition of a guild of ground-foraging birds in local subsets of steppe habitat. When data from several such different habitats are aggregated over a continental scale, however, those floristic effects are masked and the physical architecture of the habitat correlates well with measures of bird community structure, such as species diversity.

Much of the controversy focuses on the question of whether the assemblages of organisms occupying a specific locality, region, or continent represent a stable and persistent set of relationships that have achieved an equilibrium species composition. The counterargument is that while such equilibria might exist in principle, the actual communities are continually on a transient trajectory, often far from equilibria, and crudely tracking ever-changing fast and slow environmental variables and ever-changing attractors. Such an argument is more consistent with the present understanding of ecosystem dynamics (Holling 1973, 1986). Hence when a heterogeneous collection of various "non-equilibrial" local assemblages are summed over continental areas, the patterns of community structure that emerge might simply be artifacts of the summation procedure (Wiens 1984).

The issue concerns as much the precision and scale of the postulated pattern as it does the precision and scale of the data to demonstrate that pattern. A robust analysis should match the two. The coarser or more aggregated the pattern, the more likely it will be general and the more likely a robust test can be devised.

As an example, a more specific hypothesis to test the reality of hierarchical organization of ecosystems into a set of discontinuous spatial and temporal properties is that animals will have attributes of size and behavior that are scaled by the discontinuous architecture of the landscapes in which they live. These attributes will therefore also be discontinuously distributed into a small number of clumps. This predicted pattern is a qualitative one. The test is independent of the kind of distribution that might describe or explain the distribution. It is dependent only on whether it is discontinuous or not. As I shall show later (near the end of *Are there size clumps?*), that pattern is robust and is insensitive to assumptions of whether there are equilibrial communities or not.

In order to match the proposed pattern with the appropriate scale and precision of data, it is important to choose data whose variability does not mask the pattern itself. Biomass/size distributions, for example, are appropriate for developing models of community productivity in aquatic ecosystems (Dickie et al. 1987), but would likely obscure any discontinuities because of the variability introduced by age and density. Such data are too broad and variable. At the other extreme, data from specific guilds of organisms are appropriate for analyzing questions of competitive displacement in terrestrial communities (MacArthur and MacArthur 1961), but the range of scales in such data is too narrow for exploring questions of cross-scale discontinuities. We need a "Goldilocks Solution" (as suggested by Wiens' [1984] commentary on MacArthur [1972]), that avoids too broad and blunt a measuring instrument on the one hand and too fine and scale-specific a one on the other.

That led me to choose the sizes of adults of

species with determinate growth representing a major taxonomic group that occupies a major landscape or biome. Similar data from other taxonomic groups with different morphologies can then provide tests of consistency. More variable fine-scaled data concerning, for example, search behavior and home-range size, can then be used to evaluate sources of discontinuities and to define absolute spatial scales.

Are There Size Clumps?

I first collected data on body masses of adult birds of all species that breed in the boreal forest of North America east of the foothills of the Rocky Mountains. Only species breeding in pure conifer or mixed conifer-deciduous forests

were included (Appendix 1). As alternative hypotheses were proposed, additional, similarly defined data were assembled for the mammals of the boreal forest (Appendix 2) and the birds and mammals of the short-grass prairie of southern Alberta (Appendices 3 and 4, respectively).

The body masses of the species are shown as cumulative distributions in Fig. 2. In such plots, discontinuities would produce a staircase effect, with the steeply sloping "risers" representing species in clumps of similar body mass and the "treads" representing the breaks where body masses abruptly jump into another clump. The plots of Fig. 2 do suggest such a structure, although the eye tends to smooth out bumps on rising curves of this sort. A different and more precise way is therefore needed to

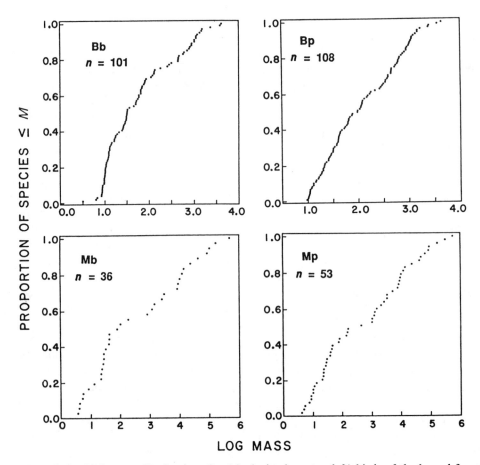

FIGURE 2. Cumulative body-mass distributions for (clockwise from top left) birds of the boreal forest (Bb), birds of the short-grass prairies (Bp), mammals of the short-grass prairies (Mp) and mammals of the boreal forest (Mb).

present the data by focusing less on clumps than on the gaps between them. That is, since clumps exist because of gaps between them, we need a gap detector, not a clump detector!

That can be devised from the difference in body masses between neighboring species that are arranged in rank order of increasing body mass. Simple body-mass ratios do expose a strong discontinuous structure, but with a trend that increases with body mass. The following body-mass difference index was therefore designed to remove that trend while still retaining the structure:

$$\text{Body-Mass Difference Index} = (M_{n+1} - M_{n-1})/(M_n)^\gamma$$

where M_n is the body mass of the nth species in rank order of increasing size and γ is an exponent sufficient to detrend the data. I found that the data were satisfactorily detrended if γ was 1.3 for birds and 1.1 for mammals. I note in passing: it is possible that those values, determined by using an iterative procedure, might have some biological significance, since later (see *Resource utilization in discontinuous landscapes: Home range*) the same values appear as exponents relating home-range size to body mass.

The results from the boreal region forest and prairies are shown for the birds in Fig. 3 and for the mammals in Fig. 4. All show distinct size ranges of clumps characterized by small body-

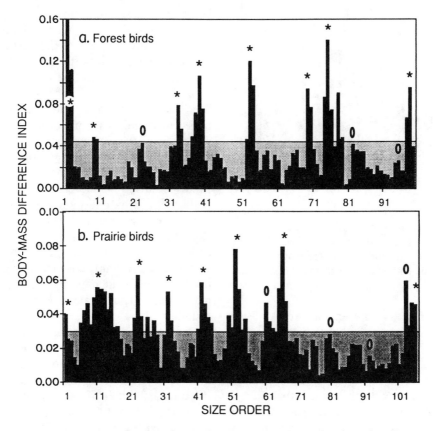

FIGURE 3. Distribution of body-mass gaps for birds of the boreal forest and prairie region as their size increases. The body-mass difference index = $(M_{n+1} - M_{n-1})/(M_n)^\gamma$, where M_n is the body mass of the nth species in rank order of increasing size and γ is an exponent sufficient to detrend the data. The horizontal line is 1 SE above the average. Asterisks show the body-mass gaps that were identified. The criterion was to choose the maximum index value that occurs once the index exceeds the horizontal line, followed by at least two values below that line. Those same breaks were identified independently by a hierarchical clustering technique, which identified additional ones as indicated by small open ovals.

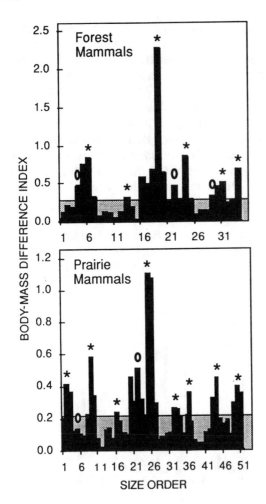

FIGURE 4. Distribution of body-mass gaps for mammals of the boreal region as their size increases. The horizontal line is 1 se below the average. Asterisks and ovals show the body-mass gaps that were identified using the criteria described in the Fig. 3 legend.

mass differences between neighbors. Each range is interrupted by an abrupt increase in the body-mass difference index. Those abrupt increases identify body-mass gaps or discontinuities. As can be seen from the figures, most of the mass gaps are extreme enough that there is considerable latitude for placing a horizontal criterion line that identifies gaps as those that exceed that line. If a trial line starts by being placed midway through the large gaps, as the line is lowered, the same large gaps continue to be identified until a position is reached where a class of smaller gaps abruptly is encountered. Those smaller gaps might well be real, but in

order to produce consistent counts of gaps I chose a final criterion line that included the first of these smaller gaps but stopped short of intersecting others. Although that line happened to end up being positioned at the mean + 1 se for all the bird data sets (Fig. 3), and at the mean − 1 se for the mammal (Fig. 4), I do not place any significance upon those particular values. The asterisks identify the gaps intersected by this line.

As a further check, a hierarchical clustering technique independently identified the clumps of similar body masses by computing Euclidian distances between farthest neighbors (Chambers et al. 1983). That approach identified the same discontinuities marked by asterisks, and also identified a few other breaks, as indicated by the open ovals in Figs. 3 and 4. Most of the latter noticeably stand out, and I suspect they are just as real as the others, though not as obvious. More data sets of this sort are needed to evolve the most robust technique. Until that is done, I chose to use the more conservative identification of gaps shown by the asterisks. That more conservative choice affects the details of what follows in minor ways, but not the substance of the conclusions.

Could the breaks be artifacts of rounding off the numbers? That would be the case if the jumps occurred whenever the last significant digit changed by one integer. The body-mass data have anywhere from three to five significant figures, depending on the size of the animals. In every instance, however, the gaps exceeded changes introduced by integer shifts in the last significant digit by several orders of magnitude. They cannot be artifacts of rounding off.

Could the discontinuities simply occur by chance variability in small samples taken from fundamentally continuous distributions? Continuous distributions have traditionally been fitted to such kinds of data, typically using the log-normal (May 1978) or, with better success, a truncated log-normal distribution (Schoener and Jansen 1968). It is basically irrelevant for the purposes of this study, however, what quantitative form of distribution best describes the data. It is important only whether the distribution is discontinuous or not. Nevertheless, in passing, the log-normal distribution proved to provide a very poor fit, with the probabilities of it describing the distributions ranging from 0.046 to 2.0×10^{-5}.

A bootstrap statistical technique was used to determine the likelihood of the distribution being so lumpy by chance. The approach draws samples from a "master" distribution that is clearly not discontinuous, each sample having the same mean, variance, and sample size as those of the observed distribution. Initially I intended to use the distribution of the body masses of the birds in North America (from Dunning 1984) as the master distribution, assuming any geographically local discontinuities would be averaged out in a continent-size data base. Although these data showed no obvious lumpy structure (Fig. 5a), they showed a strong discontinuous pattern when they were filtered by taking a running average ($n = 11$). This pattern was retained even when the species of the boreal forest and the boreal short-grass prairie were removed (as in Fig. 5b), showing that the strong lumpy quality I had seen for the distributions of these two groups did not impose the structure on the whole.

Therefore I rejected the North American birds as the source for a master distribution and turned to the log-normal, since clearly it is a continuous distribution. One thousand mock distributions were generated for each of the four base data sets. The number of discontinuities was compared for each distribution using a particularly conservative criterion chosen because it was simple to program. A break or gap was recorded if at least two consecutive values of the body-mass difference index exceeded the mean plus 2 SE, followed by at least four values below that. Note that the purpose was not to fit the log-normal, but simply to determine how often by chance a continuous distribution like the log-normal would produce the same degree of lumpiness as the data.

A typical result is shown in Fig. 6 for the birds of the short-grass prairies, where such a criterion identifies six clumps in the real data (Fig. 6a). Curve A in the figure is the average body-mass difference index of all 1000 samples drawn from the log-normal distribution. Fig. 6b shows how frequently breaks occurred in the mock log-normal samples by this criterion (from 0 breaks to 7, with a mode of 2). The

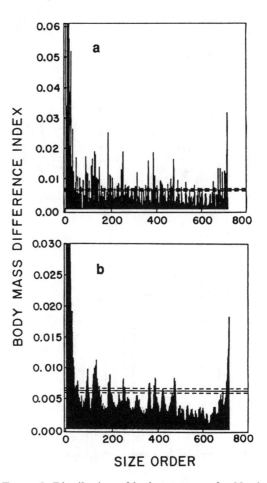

FIGURE 5. Distribution of body-mass gaps for North American birds, less the species in the boreal forest and short-grass prairie. Body-mass difference index as in Fig. 3 legend. Top graph (a) includes data for individual species arranged in increasing size; bottom graph (b) is a running average ($n = 11$) of those individual species data. The horizontal lines are the average (——) with 1 SE (- - -) above and below it.

actual bird data have six gaps defined by the conservative criterion, a number that occurred in only 14 of the 1000 mock distributions. Hence there is a probability of .014 for such lumpiness to occur by chance. Similar probabilities for the remaining three data sets ranged from .01 to .005. Therefore, it is highly unlikely that the lumpiness observed occurs by chance. The gaps are real, and the distributions are fundamentally discontinuous.

Moreover, the ability to detect discontinuities is insensitive to assumptions that the animal communities represent equilibrial associations.

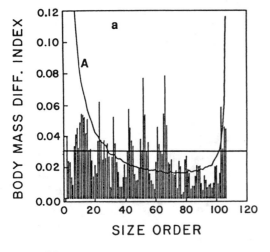

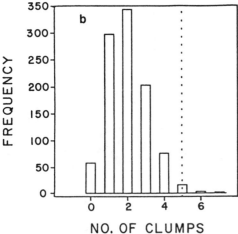

FIGURE 6. An example of the bootstrap statistical test for clumpiness using the birds of the short-grass prairie. (a) Observed distribution of body-mass gaps vs. rank order of increasing size. The horizontal line is the average + 2 SE. The curve (A) is the average value of the body-mass difference index (described in Fig. 3 legend) of all the 1000 samples drawn from the master log-normal distribution. (b) Frequency distribution of number of clumps found in the 1000 samples drawn from the log normal. A clump was arbitrarily defined as two departures above the horizontal line shown in the top graph followed by four below. The vertical dotted line is the number of clumps in the observed data as defined by the same criterion. Only 14 of the 1000 samples drawn from the log-normal distribution have the same clumpiness as the observed data.

Consider the cumulative distributions of the kind shown earlier in Fig. 2. At the extreme, discontinuities would appear as a series of steps in a staircase. If one species was removed arbitrarily, this would simply lower the height of the rise and, in some cases, also increase the width of the tread. The break itself would be detected so long as the clump (the rise) was represented by at least two species. As a consequence, even incomplete species sets can reveal the discontinuities. Only the precision of the quantitative estimate of the body mass where the break occurs would be affected.

But what do discontinuities measure?

What Are the Size Clumps?

Discontinuities define the gaps that separate clumps. The discontinuities for boreal bird body masses form seven distinct body-mass groups or clumps, with a variable proportion of the species represented in each clump (Fig. 7a). In this section I first want to discover clues as to what causes those gaps and clumps, so that formal hypotheses might be designed and tested. That requires a general survey of biological attributes that might correlate with the clump structure.

Developing Useful Hypotheses

The aquatic birds—ducks, herons, and shore birds—are concentrated in the larger clump, as a consequence of larger size associated with their aquatic habit, whereas the terrestrial species are spread throughout all clumps. Each clump is very much a "mixed bag" of terrestrial species, with the hummingbirds, kinglets, warblers, and chickadees dominating the first two clumps and the hawks and owls the two largest. Except for the families having only one species, each family is represented in ≥2 clumps. The family with the most species (the Emberizidae with 29 species: the warblers, sparrows, and blackbirds) is represented in six of the seven clumps. The next largest families, the woodpeckers and owls, each contain seven species and are spread among four and three clumps, respectively. Hence there is little taxonomic correlation with the clumps.

When I reviewed a tally of the habitats noted for the birds of Canada in Godfrey (1986), I could not discover any consistent association of

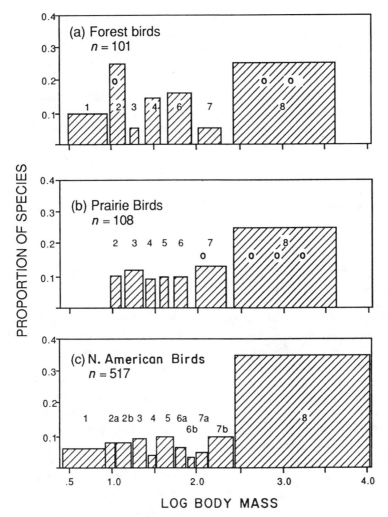

FIGURE 7. Proportion of bird species in each body-mass clump category in the (a) boreal region forest, (b) boreal region short-grass prairie, and (c) for all remaining North American species. Body mass was measured in grams. Small open ovals indicate additional breaks identified with the hierarchical clustering technique. The position and width of each bar represents the range of masses in a particular body-mass clump.

clumps with coniferous, deciduous, or mixed woods, let alone more specific botanical classifications. Similarly, there was no correlation of clumps with seral stages of the plant communities—mature or young forest, recently burned or unforested. Finally, each clump contained more than one guild foraging style as identified by Ehrlich et al. (1988)—nectar feeders and gleaners of foliage, bark, and ground are represented in the first two groups, for example, and gleaners of ground and bark as well as patrollers are represented in the three largest clumps.

This set of plant, seral stage, and foraging classifications therefore does not uniquely distinguish the clumps. They reflect a different set of attributes by which species partition resources. They have little to do with characterizing the clumps themselves.

But the clumps may be distinguished by the scales at which birds make decisions to start, continue, or abandon activities associated with survival and opportunity. Any animal makes such decisions at a number of scales, depending on the motivation. For example, Gass and Montgomerie (1981) describe a sequence for hummingbirds that has general applications. During their southward migration, hummingbirds choose mountainous over other terrain, then choose red-looking meadows on mountainsides if they can find them and thick brushfields likely to be patrolled by sapsuckers if they cannot, then densely flowered parts of meadows over sparsely flowered, then nectar-containing flowers over ones without, and so on. I shall concentrate on such hierarchies of decision and hierarchies of opportunity in the section of the same name, but for now this representation provides a focus for identifying a narrower set of observations that point to a useful direction for forming hypotheses.

Consider, for example, one level in the hierarchy of decisions, i.e., the size of patches within which bouts of foraging are concentrated. Although each species among the kinglets, Brown Creeper, Winter Wren, warblers, and chickadees of the first two clumps forage in different parts of tree crowns (MacArthur 1958), on tree trunks, in thickets, and on the ground, they all concentrate foraging bouts among small elements within those strata, such as, for example, single branches or closely associated groups of branches. Similarly, the species in the third clump can be associated with wet, shrubby, or bog areas, like Lincoln's Sparrow, or the Northern Waterthrush, or with the purely terrestrial understory of forests, like the Ovenbird. But again these species share a similar scale for their concentrated bouts of foraging—in this instance within habitat patches measured in metres (e.g., Zach and Falls 1979).

The still larger birds of clump category 4 include a number of species like the White-throated Sparrow and thrushes, typically perceived as being ground feeders, but which also forage opportunistically in tree crowns when budworm or other insect defoliators are abundant. Some of these bird species are more common in stands of young trees (e.g., White-throated Sparrow and Hermit Thrush) and others in mature forests (e.g., Swainson's Thrush), but all concentrate bouts of searching in patches in the forest measured in tens of metres.

Pine Grosbeaks and Evening Grosbeaks, typical of the next largest clump, are non-territorial flocking birds that at times are found in stands of young trees, feeding on buds, and at times in stands of mature trees, feeding on seeds or insects like budworm larvae. The basic scale for their searching bouts is at the scale of a stand of trees of a few hundred metres.

Finally, clumps that contain the largest species include the owls, hawks, and water birds, which concentrate among landscape elements— between parts of water and land formations, for example, or between clearings and forested areas. Some require a combination of those elements. The Great Gray Owl and the Long-eared Owl, for example, nest in wooded areas but use large areas of meadows in their search for voles. That suggests that there are interwoven mosaics of different kinds of attributes in the landscape that are utilized for different purposes. All sizes of animals are likely to utilize, for different purposes, such multi-attribute mosaics. The reality only becomes obvious for the larger birds, however, because our human scale of observation matches the scale of their alternative uses. There is clearly a scale-dependent subjective bias in such natural-history observations.

Nevertheless, if those behavioral observations have any consistency, then there should be a correlation between the coarseness of the grain or texture of the habitats chosen and the clump group. In order to evaluate the possible significance of that correlation, a texture index was developed from the habitat descriptions of birds described in Godfrey (1986), as supplemented by Ehrlich et al. (1988). Each of the habitat descriptors was first assigned to one of three categories defined by the size of the objects mentioned in the descriptions of each species' habits and habitats. The fine category (F) included objects measured in metres, such as shrubs, small trees, thickets, pools, small streams, etc. The intermediate category (I) included habitat objects measured in tens of metres, such as bogs, clearings, forest edges, fields, bays, rivers, etc. The coarse category (C) included objects measured in hundreds of metres and more, such as savannas, deserts, tundras, extensive forests, prairies, etc. The presence or absence of these three categories was tallied for each species, and a scale of five grades of increasing coarseness was given to that tally as follows:

$$F < (F + I) < I < (I + C) \quad \text{or}$$
$$(F + I + C) < (F + C) \quad \text{or} \quad C.$$

Each of those five categories was given a weighting from 1 (the finest) to 5 (the coarsest) and each species was tallied accordingly. A texture index was then calculated as an average within each clump. The results are shown in Fig. 8.

The index increases regularly from the clump group with the smallest birds to the one with the largest. Because of the potential subjectivity of the base information, however, in no sense is this a rigorous test of the proposition that the clumps are uniquely characterized by scale of decisions. But it does something equally important. It establishes a non-trivial basis to develop a set of alternative hypotheses that are both testable and potentially generalizable.

Four Hypotheses

A discontinuous distribution of body masses is a logical consequence of a correlation between textural choice and body mass on the one hand (as shown in Fig. 8) combined with a hierarchically organized landscape of the kind suggested in Fig. 1 on the other. The breaks in the body-mass distribution would reflect the abrupt shift of species into a different scale range for decisions that exploit different scales of grain or texture of another hierarchical level. The distribution of body masses, therefore, could be a kind of bioassay of landscape structure. These remarks reinforce the original proposi-

tion that ecosystems are organized into a small number of qualitatively discrete temporal and spatial quanta, and lead to a formal hypothesis for testing:

The Textural-Discontinuity Hypothesis. Animals should demonstrate the existence of a hierarchical structure and of the discontinuous texture of the landscape they inhabit by having a discontinuous distribution of their sizes, searching scales, and behavioral choices. Landscapes with different hierarchical structures should have corresponding differences in the clumps identified by such a bioassay.

Three viable alternatives to this hypothesis, however, can explain the discontinuous distribution of body masses by referring to discontinuous features of developmental morphology, evolutionary history, or trophic relationships rather than to those of ecosystem dynamics and structure.

One alternative hypothesis is that mammals and birds have a limited number of life-forms, e.g., locomotory modes, each of which is constrained to animals within distinct ranges of size. The hovering flight mode of hummingbirds, for example, is constrained to lower size ranges; the soaring flight of albatrosses is constrained to larger size ranges. If there are few enough possible life forms, a discontinuous distribution of sizes could result even in the presence of a continuous distribution of spatial attributes. Its formal expression is as follows:

The Limited-Morph Hypothesis. There are only a limited number of life-forms (e.g., locomotory modes) possible for animals, each of which is constrained to function effectively only over a limited range of sizes. There are so few of these life-forms possible, that sizes of animals cluster into a small number of clumps even if the spatial attributes of their habitats are continuously distributed.

Another alternative hypothesis emerges from the historical character of the evolution and behavior of any complex systems. That is, rare events can shape structure and behavior. In this case, past events could have led to the survival of a small number of ancestral forms ("Urtier") whose further adaptation is constrained to a small number of discrete size ranges, i.e.,

The "Urtier" Historical Hypothesis. The species of animals have been drawn from a limited number of ancestral forms whose organizational constraints preclude evolution of intermediate sizes even though

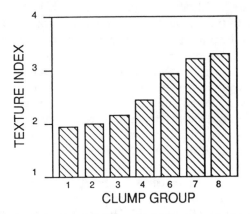

FIGURE 8. Index of choices of habitat texture by birds of the boreal forest assorted into body-mass groupings (clumps). High values indicate coarse environmental texture. (Note that there is no Group 5 in boreal forest birds.)

such sizes would find adaptive possibilities in a landscape with continuously distributed opportunities.

The final alternative is a trophic hypothesis. There is a rough relationship between the size of some animals and the size range of the items they eat (Vezina 1985). As a consequence, this selection could produce "standing" waves in the spectrum of animal sizes. That is:

The Trophic-Trough Hypothesis. Size-dependent trophic interactions will initiate and maintain lumpy patterns of body-mass distributions independent of any other mechanism.

Note that all of these hypotheses are expressed as single causes that could alone initiate discontinuities in body-mass distributions. If there is, in fact, only one such initiating cause, then it is inevitable that causes implied in other hypotheses would reinforce pre-determined discontinuities.

I tested the alternative hypotheses in two ways, first by comparing clump sizes in ecosystems having different spatial structures and second by comparing clump sizes in animal groups having fundamentally different body plans (birds vs. mammals) or feeding habits (carnivores vs. herbivores). The first set of tests challenges, in particular, the relevance of the Textural-Discontinuity Hypothesis. The second set puts the Limited-Morph, Urtier, and Trophic-Trough Hypotheses more at risk.

Tests Contrasting Different Ecosystems

One of the most obvious contrasts between boreal forests and the short-grass prairie is that forests have a large amount of finely textured material distributed through a large volume of aboveground crown, whereas the prairies have a much smaller amount of fine material distributed over an essentially two-dimensional layer just above the ground. Since the amount of above-ground fine-textured material is much less in the prairies than in the boreal forest, the Textural-Discontinuity Hypothesis suggests that the number of species in the smaller clump categories should be sharply reduced or even eliminated in the prairies. In order to test that, I collected species and body-mass information

for the birds of the short-grass prairie in the same latitude bands as the boreal forest, i.e., an area that includes much of the Canadian Province of Alberta as well as the southern portion of Saskatchewan. If such a reduction does not occur, the hypothesis is disproved. Note that restricting the latitude chosen is critical. More-southern grasslands could conceivably make up for any such reduction because of the increased productivity associated with more solar radiation in a year.

The species and body masses of birds of the short-grass prairie are shown in Appendix 3 and the body-mass difference index in Fig. 3. Just as with boreal bird species, there is a pronounced discontinuous pattern that defines the set of body-mass clumps shown in Fig. 7b.

As predicted, clump category 1 is eliminated entirely from the short-grass prairie and category 2 has sharply fewer species. If one of the Limited-Morph, Urtier, or Trophic-Trough Hypotheses provided the only explanation, it is difficult to understand why only body-mass clump category 1 is absent from the boreal prairie, but present in the boreal forest. In addition, a surprising new category appeared (clump category 5 in Fig. 7b) that was absent in the boreal forest. Its existence in particular brings the Urtier Hypothesis into question, since if an ancestral type could evolve into such size ranges in the prairies, there seems no obvious reason why this would not also have happened in the boreal forest.

Some of the species of this new clump 5 are associated with isolated perches and trees and with the scattered shrubs typical of some parts of the prairie landscape (e.g., Brown Thrasher, Loggerhead Shrike, kingbirds, and Sage Thrasher). The appearance of clump category 5 therefore suggests a possible connection with the scattered and isolated (from a human perspective) shrubs, trees, and promontories in the prairie — attributes that are not matched by similar dispersion of utilized objects in the forest. The elimination of clump category 1, the reduction in category 2, and the apparent creation of category 5, are each consistent with the Textural-Discontinuity Hypothesis, but weakly contrary to the Limited-Morph, Urtier, and Trophic-Trough Hypotheses.

But those differences between boreal and

prairie bird size distributions are not so dramatic as the otherwise remarkable similarity in clump categories in each of these very different landscapes. If, for the moment, we ignore the discontinuities absent in one of the landscapes, but present in the other, the remaining pairs can be plotted against each other (Fig. 9). The very tight one-on-one relation indicates that both communities of birds either share many of the same species, respond to very similar architectural features, or are formed by identical morphological forces of organismic development or trophic organization.

Concerning the first possibility, the species complex in the two landscapes is quite different, for example, only 17 of the 108 prairie species are found in the boreal forest. Twelve of these are in the smallest (four species) or largest (eight species) clump categories. The remaining five species are scattered among the other clumps. Moreover, if the body masses of those species that are shared are removed from each distribution, the location of the mass gaps remains unchanged, demonstrating again the robust nature of the analysis. The explanation therefore is limited either to one concerning response to ecosystem architecture, or to one concerning either shared morphology or history. At first thought, the closeness of the one-on-one fit seems a more likely consequence of one of the morphological or historical expla-

nations than of the ecosystem architecture one. Is it possible for different ecosystems to have sufficiently invariant architectural structure to generate such similarities in clump structure?

The issue raised by such strikingly similar, but not identical, clumps in different landscapes is whether there are ecological and landscape processes that have the same consequence on architecture, irrespective of the ecosystem concerned. To explore that further, the clump structure of the body masses of the birds of North America presented by Dunning (1984) was analyzed—after first removing those species found in the boreal forest and short-grass prairie in order to develop a data set uncontaminated by the structure of these bird communities. The resulting clump structure (Figs. 5b and 7c) shows some striking similarities with the categories identified in the boreal and prairie data sets. All of the first five categories in Fig. 7c are seen in one or both of Fig. 7a and b. The apparent new subdivision of the clump-2 category in the North American birds is even suggested in the forest data set by the discontinuity detected by the hierarchical cluster analysis (Fig. 3a). Although clump category 8 contains considerable structure, as shown by the body-mass index figure (Fig. 3), it is not analyzed in detail here since, as I shall show shortly, the presence of water birds masks structure.

The basic conclusion (Fig. 7) is that from clump 6 and above, clumps additional to those of the forest and short-grass prairie appear, and below that some clumps may be present or absent. This suggests some unique properties do exist in different ecosystems on top of some universal properties shared by all landscapes. Hence it seems possible that only a few processes over a limited range of scales uniquely characterize any one ecosystem and landscape and that they do so by determining whether a particular range of textural scales is present or absent. Other properties are universal characteristics of all landscapes.

This possibility smacks of earlier work on ecological processes in which it was shown that there were certain universal components of a process like predation that underlay all examples. Other, subsidiary components could be present or absent, and the possible combina-

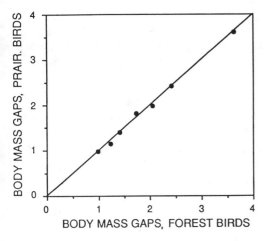

FIGURE 9. Relationship between gaps in the distribution of log of the body mass of birds of the boreal region prairies and forests. Body mass was measured in grams. $Y = 1.00X - 0.009$, $r^2 = 0.996$.

tions generated the great diversity of different types (Holling 1963, 1965). If so, it would provide a powerful way to simplify and generalize the structural characteristics of landscapes and their effects across scales—by identifying those attributes and processes that are universal and another small set that determine the uniqueness of a landscape. The importance of that possibility suggests the need for further tests.

One way is to search for habitats that are extremely different from the prairie or boreal, in order to test the limits of similarity that are possible for the clump structure of birds. The ocean environment is one such. Although there are structural discontinuities produced by physical forces that form a nested set that increase in scale from turbulence, to waves, to eddies, to gyres (Steele 1985, 1989), such structural features are spatially less fixed and, at least above a few centimetres, generate less-extreme differences in textural properties at different scale ranges. Time sequences of remotely sensed images of algal biomass in oceans, for example, show continually changing patterns where the flows are turbulent at all scales (Steele 1989). It seems possible, therefore, that size distributions of pelagic organisms, in contrast to terrestrial ones, might be influenced more by trophic structure than by physical architecture. To test that, species and body masses of pelagic birds were obtained and analyzed, again from the same latitude band as the boreal forest and short-grass prairie, i.e., the coast and near ocean of British Columbia and Alaska (Appendix 5).

Figs. 10 and 11 compare the structure of the pelagic bird body masses with those of the terrestrial and water components of the prairie bird fauna. First, removing the water birds from the prairie data set leaves the clump structure of the purely terrestrial component essentially unchanged (i.e., compare Fig. 3b with Fig. 10a and b for the body-mass difference indices and Fig. 7b with Fig. 11a and b for the clump structure). The only difference is a weak subdivision of clump category 7 and a stronger hint of additional discontinuities in category 8. Second, as predicted, the structure of both sets of water bird data (Fig. 11b and c) is very different from that of the terrestrial sets.

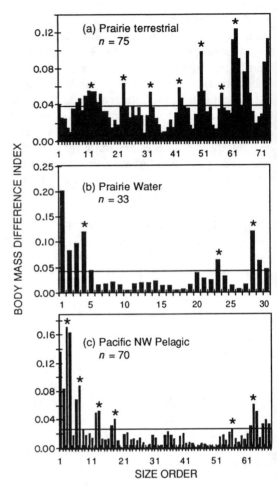

FIGURE 10. Distribution of body-mass gaps as size increases for the terrestrial and water birds of the prairies and the pelagic birds of the Pacific Northwest of North America. Asterisks show the body-mass gaps identified. Body mass difference index is $(M_{n+1} - M_{n-1})(M_n)^\gamma$. The horizontal lines are 1 SE above the average, for each group of birds.

The pattern shown by the body-mass difference index is much less pronounced for the water birds (Fig. 10b and c). The bulk of species is concentrated in one clump category (8) (Fig. 11b and c) as defined by the criteria used for defining clumps in the terrestrial data sets, just as the Textural-Discontinuity Hypothesis would suggest. Hence the extreme difference between the physical architecture of water and terrestrial habitats seems to be reflected in predictable differences in the body-mass clumps. The Textural-Discontinuity Hypothesis again resists disproof.

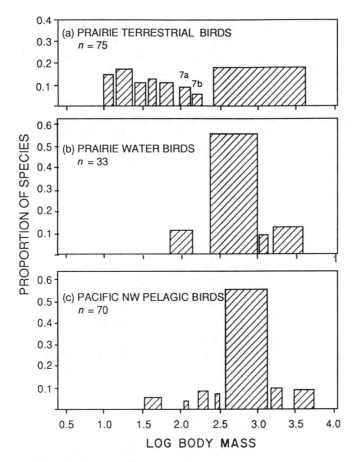

FIGURE 11. Proportion of bird species in each body-mass clump category for the terrestrial and water birds of the boreal prairies and the pelagic birds of the Pacific Northwest of North America. The width of each bar represents the range of masses in a particular body-mass clump.

The tests to this point challenged the Textural-Discontinuity Hypothesis in particular by comparing the body-mass distributions of the same Class of animals, the birds (Class Aves), occupying very different landscapes. The Limited-Morph, Urtier and Trophic-Trough Hypotheses, however, still remain largely untested. In the next section, therefore, the emphasis shifts to challenging these hypotheses directly by comparing profoundly different life-forms, body plans, and trophic classes of animals that occupy the same landscapes.

Tests Contrasting Different Body Plans

Three comparisons are made in this section — between birds and mammals of the boreal forest, between birds and mammals of the prairie, and between carnivorous and herbivorous/omnivorous mammals of the prairie. The life-form of each member of those pairs is dramatically different from the other — birds fly and hop while mammals creep, walk, and run; predators stalk and pursue while herbivores browse and flee. As a consequence, the resulting body plans and organizational constraints of each member of the pairs are so different that there should be little similarity in clump structure between the members if either the Limited-Morph, Urtier, or Trophic-Trough Hypotheses contain the sole causation of clump structure. Alternatively, if the Textural-Discontinuity Hypothesis is the leading cause, then the clump structure of the members of each pair should have strong similarities.

The species and body masses of the mammals of the boreal forest and short-grass prairie are shown in Appendices 2 and 4. The body-mass difference index was shown earlier in Fig. 4, and the clump structure of the mammals of the boreal forest and short-grass prairie is shown in Fig. 12a and b. Because there were more prairie mammals than boreal, it also proved possible to further subdivide the prairie mammals into approximately equal numbers of carnivores on the one hand and herbivores and omnivores on the other. Their clump structure is shown in Fig. 12c and d.

There are striking similarities between the clump structure of birds and those of mammals of the boreal (compare Fig. 7a with Fig. 12a) and of the prairies (compare Fig. 7b with Fig. 12b). Even clump category 5, which appeared in the prairie bird data set but was absent in the boreal, is similarly present in the prairie mammal data set, but not in the boreal mammal set. That reinforces the earlier conclusion that the difference in the architecture of the prairie landscape occurs at a defined and narrow range of spatial scales. The only qualitative difference is that clump category 1, present for the boreal birds, is absent for the boreal mammals. Other than that single excep-

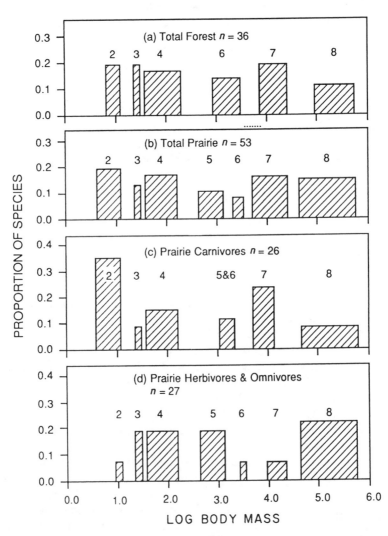

FIGURE 12. Proportion of species of mammals in body-mass clump categories for boreal region forest and prairie. Body masses were measured in grams. The width of each bar represents the range of masses in a particular body-mass clump.

tion, which I shall return to after discussing the similarities, the qualitative pattern for birds in the two landscapes is identical to that of mammals.

That remarkable similarity makes the equal similarity between the carnivore clump pattern and the herbivore/omnivore pattern (Fig. 12c and d) almost an anti-climax. The three pairs of comparisons present a strong case against the Limited-Morph, Urtier, and Trophic-Trough Hypotheses. I have not been able to conjure an explanation that constraints of morphological possibilities alone could lead to identical patterns in animals with such diverse body plans and evolutionary histories. Nor can I imagine how trophic processes on their own could produce identical patterns in both herbivores and carnivores. But the similarity in pattern is entirely consistent with the argument that the set of species in each of the paired comparisons is responding to the same landscape and ecosystem structures over a very wide range of scales.

Although the qualitative patterns are almost identical, there is a pronounced quantitative difference between the body mass of birds and that of mammals at each discontinuity. Since the largest bird in the prairie data set is the Golden Eagle at ≈ 4 kg and the largest mammal is the American bison at ≈ 560 kg, the body-mass range covered by birds is radically narrower than that for mammals. If the two classes of animal are viewed as two "measuring instruments," providing, in their body-mass discontinuities, a bioassay of the same landscape structures, then each is calibrated differently. The cross-calibration of the two can be obtained by plotting the discontinuities identified by bird body-mass clumps against those identified by mammal body-mass clumps (Fig. 13).

The plot shows a very tight linear relationship ($r^2 = 0.975$), with a slope close to 1/3 (0.327). In addition, the regression coefficients for the boreal data set are indistinguishable from those for the prairie. Finally, the only differences are that the prairie data set has one additional point (from clump category 5) that is absent in the boreal, and the boreal birds show the smallest clump category 1, while the boreal mammals do not.

I shall first deal with the differences. As I argued earlier, the presence of clump category 5

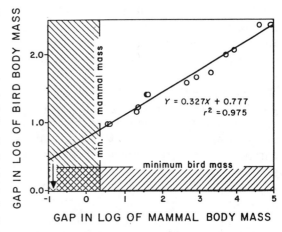

FIGURE 13. Calibration of bird body-mass gaps to mammal body-mass gaps, with minimum masses permitted for physiological reasons. The vertical arrow identifies the mammal body-mass gap that is missing because of physiological constraints on size. Body masses were measured in grams.

in the prairies and its absence in the boreal, suggests an architectural feature unique to the prairie landscape. It further reinforces the suggestion that there are invariant processes and architectural structures shared by most landscapes, and a few other processes and structures that uniquely characterize specific landscapes.

For the remaining difference, the boreal mammals lack clump category 1 because of the way mammals are physiologically constrained in minimum animal size. That minimum is $\approx 2–3$ g for endotherms, and one explanation is that it is determined by limitations in delivery of sufficient oxygen to tissues set by heart size and blood viscosity (Schmidt-Nielson 1984). The lower body-mass limit for clump category 1 is set by the smallest boreal bird, the Ruby-throated Hummingbird. Its body mass of 3.15 g is just above the minimum allowable size. But if there were a comparably sized mammal at that discontinuity, the body mass would have to be well below that permitted for physiological reasons, as shown by the vertical arrow in the figure. Hence the way mammals respond to structure makes it impossible for them to have a viable size small enough to produce clump category 1.

The most obvious differences between body-mass gaps of birds and those of mammals is the big difference in their values. One explanation might be that aerodynamic constraints on size force birds to utilize a different and narrower range of scales in the landscape. But, countering that, is the ability of birds to cover more space in a shorter time. A small bird can monitor the same range of spatial scales that only a much larger mammal can. Finally, the closeness of fit of the plot of bird body-mass gaps against those of mammals (Fig. 13) suggests that both groups are affected by the same attributes over exactly the same scale range. They use different measuring units to "bioassay" architectural features simply because flying birds and pedestrian mammals have fundamentally different relationships to space.

Following the Textural-Discontinuity Hypothesis, let us suppose that there is a textural attribute of landscape elements, T, that each mammal and bird measures over a particular scale range. Then we may assume that T is measured by a scaling relationship in which the parameters for the birds differ from those for the mammals. That is,

$$T = a_b M_b^{sb} \text{ for birds, and}$$

$$T = a_m M_m^{sm} \text{ for mammals,}$$

where a_b and a_m = weighting coefficients for birds and mammals, respectively, M_b and M_m = body mass (in grams) of birds and mammals, respectively, and s_b and s_m = exponent for birds and mammals, respectively.

Because of the equality of both body-mass functions with T,

$$a_b M_b^{sb} = a_m M_m^{sm}, \quad \text{or}$$

$$(\log M_b)_c = (s_m/s_b(\log M_m)_c) + [\log(a_m/a_b)]/s_b,$$

where c = clump category.

Hence, a plot of $\log M_b$ against $\log M_m$ at comparable discontinuities should yield a straight line, just as it does in Fig. 13. The slope is not significantly different from 1/3, suggesting that mammals measure the lineal size of landscape elements such as radii, perimeters, and nearest-neighbor distances whereas birds measure the same attributes in volumetric terms. That is not entirely unreasonable, since mammals live on surfaces and birds utilize three-dimensional space.

I conclude from these tests comparing different landscapes and different body plans that the discontinuous distribution of animal body masses reflects a similarly discontinuous structure in the landscape over a wide range of scales. The striking similarity, but not identity, between the clump structure of prairie animals and boreal ones indicates that many of the processes that form structure are common to any landscape or ecosystem, but a few are landscape specific, particularly over the coarser ranges of scale. In contrast, the great differences in clump structure between landscapes and "waterscapes" indicate that fundamentally different processes shape structure in the two systems and that the distribution of morphological attributes of organisms reflect those differences.

In brief, the geometry of landscapes and ecosystems shapes the morphology of animals.

Are Clumps the Rule or the Exception?

If physical architecture does determine patterns of size in all ecosystems, then aquatic benthic communities that develop on sediment should reflect the very obvious and simple textural categories of that substrate by having equally obvious and simple patterns in their size distributions. That is what Schwinghamer and his colleagues (Schwinghamer 1981, 1985) demonstrated for a number of benthic communities. The biomass distributions of near-shore, abyssal, and continental-shelf communities have two very obvious troughs that separate the communities into three distinct body-mass clumps. The clump with the smallest particle sizes includes bacteria and other microflora attached to the surface of silt grains. The clump containing intermediate-size animals is associated with meiofauna that live in the interstices between particles. And in the clump containing the largest macrofauna, burrowing and surface animals occupy the substrate as an integral whole.

These studies conclude that the physical character of sediments establishes three scale ranges for evolving three distinct size-dependent strategies, an argument that I have expressed here as reflecting distinct textural features of the habitat over different scale ranges. Hence the smallest scale range is characterized by the texture and size of particle surfaces, the intermediate scale range by the texture and size of water-laden pores between particles, and the largest scale range by the texture and size of the whole substrate. Because the physical structure of sediments at these three scales is so discrete and clear, it is a convincing argument that benthic organisms would be unable to evolve to exploit textures that bridged more than one of these scale ranges of these different textures.

In contrast, the size distributions of the biomass of pelagic communities seem to be flatter (Kerr 1974, Sheldon et al. 1977), although the data are so variable that the averaging techniques used are likely to mask discontinuities. That is why I chose body masses of only adults and avoided the averaging effects of lumping data into arbitrary size categories. Nevertheless, the same biomass data and averaging techniques do not mask the very strong structure of aquatic benthic communities. That is the same conclusion I presented earlier (see *Tests contrasting different ecosystems*) when I compared the adult body-mass gaps of pelagic bird communities with those of terrestrial birds. Open water lacks fixed physical structures, and although there are discontinuities (Steele 1989) it seems likely that the turbulent nature of water movement will generate broader ranges of scale-invariant structures interrupted by changes in texture over constrained ranges of scale. That, at least, is what Steele's analysis of ocean data suggests. As an example, there is an overall trend in the power spectrum of temperature variance that is interrupted by sharp peaks. One is at 50 d and 100 km, associated with the dynamics of eddy movement, and one is at 20 d and a few kilometres, associated with internal waves. As a consequence, the size distributions of pelagic organisms are likely to be dominated more by trophic than by architectural attributes. A model of plankton dynamics, for example (Silvert and Platt 1978), shows that production peaks for one size class of plankton would become evened out by opportunistic feeding and stochastic processes. Where architectural discontinuities establish size clumping, however, trophic interrelations then could certainly reinforce them. Trophic relationships by themselves, however, seem unlikely to be able to initiate the clump structure observed.

A few scattered examples of published body-mass discontinuities in terrestrial ecosystems have either been mentioned in passing or explained differently. Caughley and Krebs (1983) explicitly proposed an interesting hypothesis that would generate one such discontinuity. That hypothesis is a broader extension of the Trophic-Trough Hypothesis suggested earlier. They proposed that body masses of mammals would be distributed bimodally if populations of small mammals with body masses < 25–35 kg are regulated by processes intrinsic to the population (e.g., genetic and behavioral) and if populations of larger mammals are regulated primarily by extrinsic trophic processes (e.g., predation and resource limitation). They compiled body-mass data for herbivorous and omnivorous mammals, pooled from communities in Europe, Canada, East Africa, and Thailand, and presented frequency histograms that revealed not just one but six troughs, in ranges similar to the ones described in this paper. They did not explain or evaluate the multiple clumps but concluded that these data do not provide particularly convincing evidence for the bimodality they predicted, although one of the troughs is in the range they expected.

Subsequently Caughley (1987) discovered that the appearance of multimodality was an artifact of an error in converting body-mass classes to logarithms. This is one of the biases that makes histograms inappropriate for analysis of distribution patterns. Since they did not present their original body-mass data, however, I could not apply the techniques I have used in this paper to interpret their data further.

I could discover only two other examples in the literature that described discontinuities in body-mass distributions of terrestrial species. Wiens and Rotenberry (1980) observed that there was a gap in the body-mass distributions of 12 species of ground-foraging birds in the shrub steppe that they studied, and I observed that the 36 species of insectivorous birds of the spruce-fir forests of eastern North America were organized into three distinct size groups (Holling 1988). Both studies mentioned the gaps in body mass as an interesting and unexplained curiosity, and I used the fact as a convenient way to stratify the sample of birds so that parameters of their functional responses need be estimated only for the three groups and not for each of the 36 species. But at that time I certainly did not suspect any general property.

Although I could not discover further examples of body-mass discontinuities in the literature, Jarman's classic behavioral study of the social organization of African antelope species (Jarman 1974) did conclude that there were five distinct behavioral classes, characterized by the selectivity and scale of their feeding choices, by the size of groups that formed, by mobility, and by anti-predator behavior. Those are precisely the attributes that suggested the Textural-Discontinuity Hypothesis, and yet the body mass of animals in Jarman's five classes seem to overlap, with little indication of discontinuities. Because of Jarman's breadth and depth of behavioral experience and insight, these data are potentially a significant set to disprove that hypothesis. I therefore reanalyzed his data using the techniques used here.

I first obtained as clean a set as possible by using only those species occupying grasslands and savannas and by eliminating species that Jarman was unable to assign unambiguously to one of the five groups because of incomplete information. These data are presented in Appendix 6. There proved to be significant body-mass discontinuities, and the resulting body-mass clumps are compared to Jarman's behavioral classes in Fig. 14. With the exception of his class C, each behavioral class is unambiguously defined by the body mass clump structure. In fact, the body-mass gap analysis suggests that class A might well subdivide into two groups. The one exception is that class C exactly overlaps the two size classes that define classes B and D. The latter two are very selective feeders. Class B includes species that feed entirely on grass or entirely on browse, and class D includes species that feed very selectively on grasses. In contrast, Jarman describes the feeding behavior of Class C as being more flexible. Animals of this class feed on a variety of both grasses and browse, in a range of vegetation types, and shift their diet seasonally. Therefore, with a sufficient quantity of resources at different scales and with a seasonal

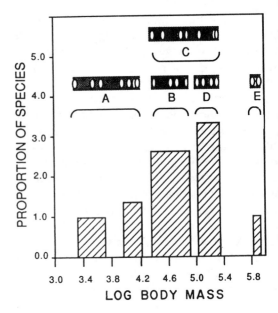

FIGURE 14. Proportion of species of African antelope in each body-mass clump category. The horizontal bars and letters A–E are the behavioral classes identified by Jarman (1974), and the open ovals are the body masses of animals in each class. Body masses were measured in grams. The width of each bar represents the range of masses in a particular body-mass clump.

shift in specialization, a strategy can evolve that competes effectively at two textural ranges of habitat with species that have evolved to be specialists in either one of the two. These data therefore further support the Textural-Discontinuity Hypothesis and extend the demonstration to include a specific guild on another continent.

But if this conclusion is so general, why have there not been more demonstrations of discontinuities? Certainly many examples in the literature explore the characteristics of body-mass distributions, and yet none emphasize the existence of discontinuities except as noted above. My own struggles with the propositions described here provide, I believe, the explanation.

Ecology, and indeed all of science, has been dominated since Newton by simplifications that explicitly or implicitly assume that processes connecting state variables generate one stable state, that those variables typically exist near an equilibrium, and that continuous, smooth behaviors and distributions are the rule. I became convinced almost 20 yr ago that theory, causal analysis, and example demonstrated that ecological time dynamics were organized by more than one stable state, that abrupt shifts between stability domains were common, and that behavior far from equilibria was critical for maintenance of diversity and resilience (Holling 1973). Since then, those conclusions have been deepened and extended by the 23 case studies described at the beginning of this paper. Moreover, the importance of non-linear dynamics of complex systems and of discontinuous properties now pervades the sciences—e.g., evolution (Gould and Eldredge 1977), developmental biology (Kauffman 1983), ecology (May 1977, Schaffer 1985), chemistry and physics (Prigogine and Nicolas 1971), and economics (Arthur 1990). It seems a small step to move from that view of discontinuous dynamics in time and multiple stable states to similar properties in space. And yet it took me an uncomfortably long time to make the connection. The paradigm of continuous phenomena is deeply rooted.

We are also conditioned in biology to present data on distributions as frequency histograms. That is what I first attempted in testing the discontinuity hypothesis, using data on birds of the boreal forests, and saw little evidence for discontinuities. And yet I knew such discontinuities existed because I had remarked on them, in passing, during years of studying insect populations, birds, and mammals in forests. It quickly became obvious that histograms themselves can both mask and generate pattern. If one expects a continuous distribution, then any jaggedness in the plots is viewed as the noise from small sample sizes. The normal reaction is to broaden the interval so as to increase sample size. The result is that any discontinuous pattern is masked. Similarly, if one's expectation is for discontinuities, the interval can be shortened until any desirable level of discontinuity seems to emerge. Some investigators, such as Schoener (1968), perceive the weakness of frequency histograms and use cumulative body-mass distributions. That uses the data more effectively, but again expectations help the eye to smooth out the bumps along a rising line. That is why I shifted the analysis to the body-mass difference index in order to detect body-mass gaps rather than body-mass clumps, and

to a hierarchical cluster method that used body-mass dissimilarities rather than body-mass similarities to define groups.

Hence the reason why discontinuities have been rarely noted is because expectations not only dictate the models and theories we derive, they also determine the patterns in data we see and remark. And the dominant expectation has been for continuous distributions.

I conclude, therefore, that discontinuities in body-mass distributions are universal in landscapes associated with fixed physical structures. Moreover, I expect that discontinuities shown earlier at a biome level will also occur for birds and mammals in site-specific ecosystems of at least a few hundred square kilometres in size— i.e., the size that contains the home range of the largest mammals or birds. There might, however, be fewer discontinuities because at some scales the spatial structures will be ecosystem specific. The discontinuities are caused by scale-limited strategies of behavioral decisions and life-styles that evolve in response to habitats that have different architectural attributes over different scale ranges. Such discontinuous architecture should be produced wherever fixed structures are generated by biological or physical processes. Hence the communities in terrestrial, benthic, shallow lake, littoral, and coral ecosystems should all show strong discontinuities in the distribution of adult body masses of species. I speculate that pelagic communities in open oceans are likely to demonstrate weaker discontinuities because of the different spectral properties of land and water. Since the structure of ecological systems rarely has single causes, however, it is likely that once the discontinuities are initiated by the discontinuous architecture, they become reinforced by morphological adaptations, by life-history traits, and by trophic relations. The previous tests of hypotheses, however, suggest that these latter processes are insufficient in themselves to explain the body-mass gaps.

Resource Utilization in Discontinuous Landscapes

In this final analytical section I shall draw upon additional data to infer the quantity of resources available to each clump category and how they are utilized. The intent is to determine if such quantitative data can provide a more precise explanation and quantification of the specific attributes of landscape elements and of the measurement units animals use to measure those attributes.

To this point in the analysis I have concluded that the discontinuities in the body-mass distributions are a bioassay of the qualitative, not quantitative, structure of the landscapes. That is, they indicate the presence of structure, but not its amount. To make clear the distinction between the body-mass clumps, which bioassay qualitative structure, and the ecosystem and landscape categories, which have quantitative properties of amount as well as qualitative ones of structure, I shall call the latter "ecosystem quanta."

I shall explore these properties of ecosystem quanta by comparing different landscapes and life-forms in three ways. First, the distribution of species by body mass provides information about the way resources are shared among species. Second, extensive data on home-range sizes of mammals and birds provide information on how resources are utilized and the absolute scales of their utilization. Both data sets help define invariant properties of landscapes and the way they are utilized by mammals and birds. Finally, that information provides the basis to demonstrate how the behavioral decisions animals make at different scales interact with the hierarchical nature of the quanta. This makes it possible to convert the body-mass units of the clump categories into an absolute linear scale that measures quanta over ranges from centimetres to hundreds of kilometres.

Species-Body-Mass Distributions

The number of species in each clump category is likely, in part at least, to reflect the quantity of resources utilizable at the appropriate scales in the landscape. Hence the percentage of the species in the different body-mass clump categories should reflect the amount of usable resources in the associated habitat quantal categories. The percentage of species in different clump categories has already been shown for

birds in forests and prairies (Fig. 7a and b), for birds of North America as a whole (Fig. 7c), and for mammals in forest and prairie (Fig. 12a and b). When I plotted those values for every paired combination of these five data sets there clearly was no correlation.

Hence, as reflected by the number of species in a clump category, each different landscape distributes the amount of usable resources differently among the clumps. That is precisely what should be expected. Earlier the bioassay suggested that different landscapes had very similar textural categories, at least at the lower scale ranges. That is, the texture of a tree in the river valleys of the prairies is not going to be much different from that of a tree in the forests. But the amount of resources within that textural category will be clearly very different in each ecosystem. Hence the number of species in each clump category should be landscape specific. Moreover, since there was also little correlation between the bird and mammal percentage-of-species-per-clump comparisons, the species in each of those different classes must utilize the resources differently within any one landscape. Again, that is expected, because different amounts of resources are available in each textural category because of the different spatial geometries of birds and mammals.

I conclude, therefore, that the amount of resources in each quantum is unique in each ecosystem and landscape, and that within any landscape the amount of resources available to birds is different from that available to mammals.

Another way to compare resource distribution is to ignore the clump structure and plot the body-mass distributions in quantile-quantile plots (Chambers et al. 1983). Such plots compare two distributions by plotting the log of body masses of one against the log of body masses of the other at the same quantile values, as in Fig. 15a–d. Fig. 15a–d makes comparisons within the same taxonomic group, i.e., between landscapes. The three bird-to-bird comparisons and the one mammal-to-mammal comparison all show an overall regression line with a slope close to 1 and an intercept close to 0. That is, all bird distributions have the same overall shape—independent of landscape type, as do all the mammal data sets.

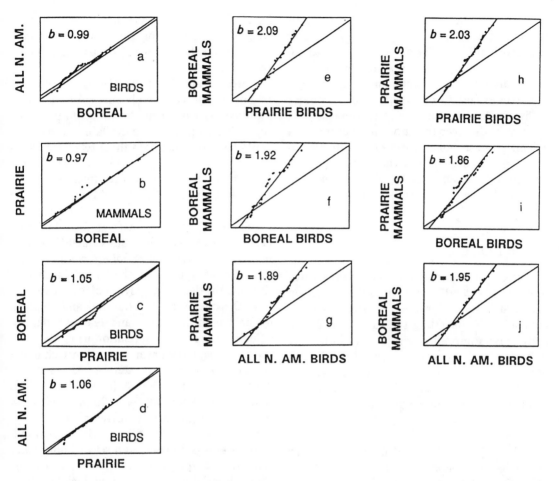

FIGURE 15. (a–d) Cross-ecosystem comparison of the body masses of a single taxon at the same quantile; (e–j) Cross-taxon comparison of the body masses of mammals (Y axis) and birds (X axis) at the same quantile. (Quantiles, here, scale a rank order of species by body mass into proportions of the total number of species, identifying the mass value at each proportion.) b = slope; 45° line added for comparison.

This observation provides another example of an invariant property of different ecosystems. That is, although the total productivity or magnitude of resources available in each landscape is certainly different, the overall way that resources are distributed among animals of different sizes is identical in all landscapes, except, perhaps, in those subject to rapid transition.

There are, however, noticeable deviations about each line that are different from landscape to landscape. They either reflect the unique quantal structure of each ecosystem, the incompleteness of the species lists, the non-equilibrium character of the communities, or some combination of all those.

Each of the remaining six comparisons (Fig. 15e–j) are between taxonomic groups within the same landscape type. The regression lines are very different from those just discussed. All have slopes close to 2.0 and intercepts averaging about −1.3. That is,

$$(\log M_m)_i = 2.0(\log M_b)_i - 1.3,$$

where M_m = body mass of mammals, M_b = body mass of birds, and i = the quantile value.

The negative intercept indicates that the body-mass distributions of the mammals are shifted to lower body masses than those of the birds, perhaps because pedestrian mammals have a lower energy demand than flying birds do. A slope > 1.0 indicates that the body-mass distributions of mammals are spread more widely than those of birds. The slope is so close to 2.0 that it is tempting to use scaling arguments as an explanation. That is, mammals exploit their resources essentially by sweeping out encounter paths of a certain width, i.e., with dimension 1, whereas birds sweep out encounter paths with a certain cross-sectional area, i.e., with dimension 2. As a consequence, the resources available in a landscape can be divided among a narrower range of bird sizes than mammal sizes.

Home Range

The effect of animal size on home-range area provides a second source of information on quantitative resource availability that opens the possibility of converting the values for clump body masses into an absolute scale within the landscape. In particular, it proved possible to use those data to identify the invariant feature that generates the same clump structure for carnivores as for omnivores and herbivores (Fig. 12c and d) and for mammals as for birds (Fig. 13).

McNab (1963) started a small industry when he published a paper showing that the home ranges (H) of mammals scaled to body mass ($M[W$ in McNab 1963]) in the classic allometric relation:

$$H = aM^b, \quad \text{or}$$

$$\log H = b \log M + \log a.$$

For the data set he had at the time, the slope, b, was not significantly different from 0.75, the value to be expected if metabolic needs were the sole determinant of the size of the foraging area. Since then, data have been obtained for many more species, for birds as well as mammals and for different trophic groups (Schoener 1968, Harestad and Bunnell 1979, Jenkins 1981, Lindstedt et al. 1986). Several different values have been derived for the value of the slope, each of which has drawn upon a different explanation. There is now agreement that the slope is >0.75, suggesting that some relationship exists in addition to metabolic needs. But it is not clear whether the slope for

birds differs from that for mammals, nor whether the slope for carnivorous mammals differs from that for herbivorous and omnivorous ones.

Harestad and Bunnell's (1979) reevaluation of home range and body mass for mammals significantly expanded the number of species for which estimates were available so that they could be partitioned into three sets — for carnivores, omnivores, and herbivores. The slopes for herbivores and omnivores (means $\pm$ 1 SE) were close to one (1.02 ± 0.11, $n = 28$, and 0.92 ± 0.13, $n = 7$, respectively). The slope for carnivores, however, was significantly larger (1.36 ± 0.16, $n = 20$). Subsequently, Gittleman and Harvey (1982) collected home-range data for 45 species of the order Carnivora, drawing the data largely from different sources than the ones used by Harestad and Bunnell. When I used Gittleman and Harvey's data to calculate the regression relation using the ordinary least-squares linear regression techniques used by all of the above authors, the slope for carnivores was 1.07 ± 0.20 ($n = 26$, $r^2 = 0.55$) and for omnivores 0.86 ± 0.20 ($n = 19$, $r^2 = 0.52$), not significantly different from each other or from a slope of 1.0. Similarly, Lindstedt et al. (1986) collated a set of home-range data for 15 species of carnivores, restricting themselves largely to data that had been collected using radiotelemetry tracking. They quote a slope of 1.03 ($r^2 = 0.66$), whose 90% confidence interval contains neither 0.74 nor 1.36.

It seems likely, therefore, that the higher slope quoted for carnivores by Harestad and Bunnell (1979) occurred because of some bias. There are two sources of bias in all these analyses. One occurs because both home-range size and animal size have an associated error term. LaBarbera (1989) has shown that in such cases the simple least-squares linear regression techniques used by all these authors is biased, and a bivariate regression technique yields less-biased estimates. It is not clear which of a number of alternative techniques of bivariate analysis are most appropriate, so I chose to use both simple least-squares regression and principal axis correlation analysis (Sokal and Rohlf 1969) in order to bound estimates of parameters within a likely range.

The other source of bias is biological, caused by the mixed nature of such data sets. McNab (1983), for example, shows that metabolic relations are affected by variables other than body size, of which specifics of food type are one of the major influences. In addition, home-range sizes have been shown to be affected by latitude (Harestad and Bunnell 1979, Lindstedt et al. 1986), habitat productivity (Gass et al. 1976, Ward and Krebs 1985), group size (Gittleman and Harvey 1982), and foraging style (Swihart et al. 1988). Hence it is easy for data sets with small numbers of species to be biased, particularly if species at smaller or larger extremes are differentially affected. That is, it is not only a matter of mixing apples and oranges, it is a matter of mixing small grapes and big grapefruit!

In order to moderate those biases, I therefore combined all the data sets noted above together with a number of others from the literature (Appendix 7), and used both simple least-squares linear regression and principal axis correlation analysis to calculate coefficients. I did not include any data for shrews because their small size is correlated with unique metabolic rates. I eliminated redundancies and any data points for which there were fewer than five replicates, and calculated the average body masses and home-range sizes for each species in each of the references noted. That produced data for 33 species of carnivores, 28 species of omnivores, and 34 species of herbivores. Since some species had more than one estimate from different geographic locations and habitat conditions, the total sample sizes were 53, 37, and 40 respectively. The results are shown in Fig. 16 and the parameter estimates in Table 3.

None of the slopes for carnivores, omnivores, and herbivores are significantly different from each other. The higher slope presumed for carnivores described in the literature therefore seems to have been caused by small sample sizes and the bias from mixed samples. As expected, the intercepts are very different from each other, resulting in home-range sizes that increase from herbivores to omnivores to carnivores over all body masses. Carnivore home ranges, for example, are ≈20 times the size of those of herbivores, reflecting the lower energy available for that higher trophic level. Hence

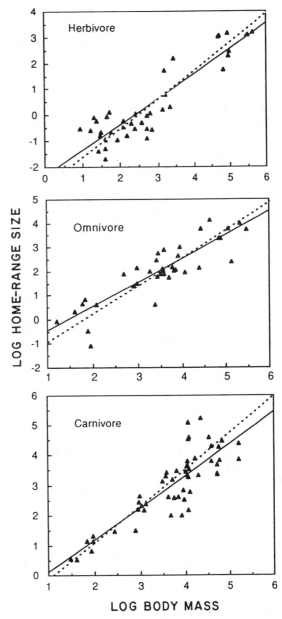

FIGURE 16. Home-range size (measured in hectares) vs. body mass (measured in grams) of mammals. — = simple least-squares regression; - - - = principal axis correlation.

home-range sizes increase from herbivores to omnivores to carnivores as a consequence of the reduced energy available to progressively higher trophic levels.

Slopes calculated using the bivariate technique are consistently higher than those using simple regression, as LaBarbera (1989) has noted. Since the errors associated with estimates of home range are likely to be greater than those associated with estimates of body mass, the least biased estimates are likely to be bounded by the two sets of estimates presented in Table 3. An estimate of around 1.1 seems reasonable.

These relationships not only provide a good reference point for converting body masses into their spatial equivalents, they also give insight into the textural characteristics of landscapes. Harestad and Bunnell (1979) provided a useful starting point when they suggested that an animal utilizes the minimum area that can sustain its energy requirements. That is,

$$H \propto R/P, \qquad (1)$$

where R = energy requirements of an animal of body mass M (in kilojoules per day per unit of area), and P = energy in the environment utilizable by the animal (in kilojoules per day per unit of area). R is proportional to metabolic rate so that

$$R \propto M^{0.75}.$$

Lindstedt et al. (1986) argue, reasonably, that the energy requirements of an animal also depend upon biological time, i.e., on the duration of such critical stages as gestation and maturation. These scale to $M^{0.25}$, so that a more complete expression for total energy demand is given as

$$R \propto M^{0.75} \times M^{0.25} \propto M^{1.00}. \qquad (2)$$

P, on the other hand, reflects the availability of resources in the habitat that affect survival and reproduction. The amount of those resources available to an animal depends on the textural roughness of the environment and upon the spatial grain at which that environment is sampled by the animal (Morse et al. 1985). A small animal will be exposed to textures at fine grains and a large animal at coarse ones.

Mandelbrot (1982) has formalized the description of such textural attributes using fractal geometry, and the concept has begun to be applied in ecology (Burrough 1981, Kent and Wong 1982, Bradbury et al. 1984, Frontier 1986, Krummel et al. 1987, Milne 1988, 1991, Wiens and Milne 1989, Loehle 1990). One of Mandelbrot's more intuitively revealing examples considers the length of a coastline as measured by animals with various step lengths. If the roughness of the coastline's texture is self-similar, then its complexity will be the same at any scale. That is, an enlarged 10-cm sample of the coastline, for example, will look identical to a 1-km sample if both are displayed at the same size. If an ant paces out the length of such a coastline by following its contours between two points, separated by, say, 1 km "as the crow flies," then the distance measured, or the amount of resources experienced, will be greater than if a mouse did the pacing, or still greater than if a moose did the pacing. That is, relative to the 1-km sample, the ant follows a more tortuous and longer path than the moose.

The same effect is produced if the lengths are measured by different step lengths or by bands

TABLE 3. Parameter values for the regression and correlation of log(home range, in hectares) vs. log(mass, in grams) for mammals (data from Appendix 7).

Trophic status	n	Simple linear regression				Principal axis correlation		
		Intercept, t, at 1.0 g ($\bar{X} \pm 1$ SE)	Slope, b ($\bar{X} \pm 1$ SE)	r^2		Intercept, t, at 1.0 g	Slope, b, and 95% confidence interval	Correlation coefficient
Carnivore	53	-1.06 ± 0.38	1.09 ± 0.10	0.71		-1.85	1.31 1.10–1.57	0.85
Omnivore	37	1.48 ± 0.34	0.99 ± 0.09	0.77		-2.04	1.15 0.95–1.40	0.87
Herbivore	40	-2.38 ± 0.23	0.99 ± 0.07	0.82		-2.69	1.10 0.942–1.28	0.90

of different widths that, as they unroll, are centered on the sinuous line. For a fractal curve of this sort, the length experienced, l, with a step length or band width u, is given by

$$l(u) \propto u^{(1-D)}, \qquad (3)$$

where D = the fractal dimension.

D lies in the range $1 \le D \le 2$ for a line that is a transect swept out by a mammal as it moves across a surface. If D is 1, then the line is straight and smooth. As D increases, the line becomes rougher and more sinuous, until, as D approaches 2, the line fills up a two-dimensional area. For a surface, D lies in the range $2 \le D \le 3$.

The amount of resources, or the usable energy available per unit area, will be proportional to the grain of sampling, or step length, so that,

$$P \propto l(u), \quad \text{i.e.,} \quad \text{from Eq. 3,}$$
$$P \propto u^{(1-D)}.$$

Since u, the step length, is a lineal measure, it scales to body mass as $M^{0.33}$; then

$$P \propto M^{0.33(1-D)}. \qquad (4)$$

Substituting Eq. 4 and Eq. 2 in Eq. 1,

$$H \propto M^{1.00}/M^{0.33(1-D)}, \quad \text{or} \qquad (5)$$
$$H \propto M^{(0.67 + 0.33D)}; \quad \text{i.e.,} \qquad (6)$$
$$H = tM^{(0.67 + 0.33D)}, \qquad (7)$$

where t is a constant combined from energy metabolic efficiency constants and from the value for the energy available to the animal as a consequence of overall ecosystem productivity and the animal's trophic status. It will therefore assume a different value in landscapes of different productivity and for animals of different trophic status.

Earlier, when discussing Table 3, I suggested that the least-biased estimate for the exponent in Eq. 7 likely lies around 1.1. If this value is substituted as the exponent expression of that equation, the fractal dimension is estimated to be 1.3. When the actual fractal dimension of vegetation is directly measured, it is found to lie between 1.2 and 1.85, with an average of 1.4 (Morse et al. 1985, Krummel et al. 1987). As a

consequence, the value of the fractal dimension computed from the home-range relationship is not unreasonable, lying well within the range directly measured at different scales in landscapes.

In review, this analysis of the home range of mammals leads to the following conclusions. Home ranges or foraging areas, like any spatial information, can be viewed as having not only a spatial extent, analogous, for example, to the overall size of a satellite image, but a grain size as well, analogous to the pixel size of an image (Wiens 1990). Eq. 7 indicates that mammals of the same body mass, with the same longevity and metabolic rates and efficiencies, will have foraging areas with a spatial extent determined by the productivity of the landscape and by their trophic status. The less productive the landscape and the farther up the trophic chain, the larger the spatial extent of their foraging areas. That is why the literature demonstrates strong effects of latitude and habitat productivity on home-range size, and why carnivores have home-range sizes 20 times those of herbivores, independent of their body mass (Fig. 16).

In contrast, the spatial grain at which the landscape is sampled by animals is largely independent of the landscape and of the animal's trophic status. It is a function of body size, and, presumably, of body form. Therefore the slopes of the home range against body size log-log relationships are similar for all trophic levels. Differences in the productivity and fractal dimension of different landscapes contribute to the variability about each line, but, over the range of body masses considered, that variability is swamped by purely size-dependent attributes affecting longevity, metabolism, and step length.

This latter conclusion now connects back to the earlier analysis of clump body-mass structure. It provides the explanation for why mammalian herbivores, omnivores, and carnivores show body-mass gaps at the same values. They are all measuring the geometry of elements in the landscape with a spatial grain defined by their step length or size alone.

How, then, do birds utilize quanta, when the body-mass values that define the clump gaps are so different? Schoener (1968) collated data on bird territory sizes and foraging areas and found the slope of the regression for bird carnivores was 1.31 ± 0.06 and 1.39 ± 0.08 (means $\pm$ 1 SE), for territory and foraging areas, respectively. More than anything else, the coincidence of those values with the slope that Harestad and Bunnell (1979) originally found for carnivorous mammals perpetuated the idea that carnivores are somehow fundamentally different from herbivores in the exponent that converts body mass to home-range size. And yet my analysis earlier led to the conclusion that carnivores are not significantly different in this respect, and that the slopes for mammalian carnivores, omnivores, and herbivores are all ≈ 1.1.

Schoener's data set for home-range and territory sizes of birds was therefore also expanded and analyzed in the same manner as for mammals (Appendix 8). Data were obtained for an additional 15 species and, following Schoener, all species were classified as carnivore (90–100% animal food), omnivore (10–90% animal food), or herbivore (0–10% animal food). Since there were only five herbivores, those data were pooled with the omnivores. The results for territory and home range were combined and are presented in Fig. 17.

The slope for the bird carnivore regression (Fig. 17a) is 1.36 ± 0.10 (mean $\pm$ 1 SE, $n = 43$, $r^2 = 0.81$), not at all different from the values that Schoener calculated, and higher, although not significantly so, than the comparable figures for mammals. The intercept is -1.66 ± 0.177. The comparable values calculated using principal axis correlation analysis were 1.59 for the slope (95% confidence interval between 1.36 and 1.86) and -2.00 for the intercept. These data, however, are much less reliable than those for mammals largely because there is more clumping of replicates within narrow size/feeding-class ranges.

That is what makes the regression relationship for the herbivores and omnivores almost meaningless (the dotted line in Fig. 17b). Again it is a problem of mixing grapes and grapefruit. For example, there is a cloud of points concentrated at a low range of body masses and a few scattered points at higher body masses. All the species in the lower cloud of points are formally

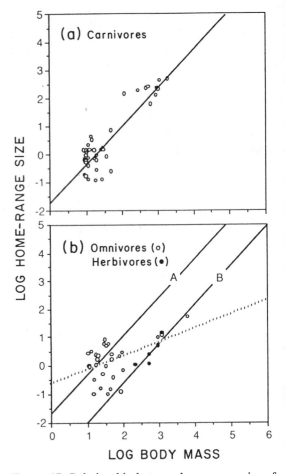

FIGURE 17. Relationship between home-range size of birds and their body masses: (a) carnivores; (b) herbivores and omnivores. The dotted line in the bottom graph is fitted through all points. Line A is the same as that in the top graph. Line B has the same slope and is drawn through the five pure herbivore points (●).

classified as omnivore because they feed on both invertebrates and seeds. But this hardly qualifies them as omnivores in the mammalian sense, since foliage is a vanishingly small part of their diet. Only the few points at higher body masses are dominated by true foliage-eating herbivores.

Therefore an alternative, but speculative, treatment seems appropriate in order to bound the regimes of the possible. The upper solid line (A) in Fig. 17b is the same as the one plotted in Fig. 17a, showing that those seed- and insect-eating species in the lower cloud of points could well have home ranges that conform to the overall carnivore pattern. The lower solid line (B) is drawn parallel to that, but through the five herbivore points. The latter represent data for the Blue Grouse, Willow Ptarmigan, Prairie-Chicken, Sage Grouse, and Northern Bobwhite, all species that consume vegetation, much like herbivorous mammals. Moreover, the two solid lines are displaced from each other about the same amount as were the mammalian carnivore and herbivore lines, suggesting that mass-specific territory sizes of bird carnivores could also be ≈ 20 times the size of those of herbivores. Therefore, despite the ambiguities, it is possible that all trophic classes of birds have the same exponent relating foraging area size to body mass, and that the exponent might have a higher value than the one for mammals.

It is therefore important to reevaluate the previous conclusion that size only defines the spatial grain for both resource utilization and body-mass clump structure. Is it possible that the exponent of the home-range against body-size relationship of flying birds could be higher than that of pedestrian mammals because of their different relationship to three-dimensional space?

The argument that home-range sizes of both birds and mammals reflect the same attributes of the landscape therefore has to deal with the possibility that the exponent relating home range to body mass can be above at least 1.3 for birds rather than near 1.1 as presumed for mammals. Earlier, another difference was encountered between birds and mammals that concerned body-mass distributions (Fig. 15). The explanation involved differences in the geometry of search for resources. That is, mammals exploit their resources by sweeping out encounter paths of a certain width, i.e., with dimension 1, whereas birds sweep out encounter paths with a certain cross-sectional area, i.e., with dimension 2. Hence the sampling grain or unit step for birds should be an area, and scale to $M^{0.67}$, rather than a lineal measure that scales to $M^{0.33}$, as in the case of mammals. Hence, P, the amount of utilizable energy in the environment, should change from

$$P \propto M^{0.33(1-D)} \quad \text{for mammals (Eq. 4), to}$$
$$P \propto (M^{0.67(1-D)} \quad \text{for birds,}$$

and the equations for size of home range of birds would therefore change to

$$H \propto M^{1.00}/M^{0.67(1-D)}, \quad \text{or}$$
$$H \propto M^{(0.33+0.67D)}.$$

If the actual exponent lies between 1.2 and 1.5, a quite reasonable fractal dimension of between 1.3 and 1.7 is calculated when substituted for the exponent portion of Eq. 8. Thus the home-range vs. body-mass relation of birds as well as of mammals can generate similar estimates of fractal dimension (D) that lie well within the range of values for vegetation (1.2 to 1.85) that are directly measured in nature (Morse et al. 1985, Krummel et al. 1987).

The key point I want to emphasize is not those details, however, but that both mammals and birds measure the same discontinuities in the landscape, but might be using different units for measurement. Earlier in this section, when I compared mammalian herbivores, omnivores, and carnivores, I concluded that the body masses defining clump categories were the same for all trophic types of mammals because their size uniquely defines a lineal step length or spatial grain for measurement of landscape elements. The same can be true of birds if the geometry of their unit step is a unit area rather than a unit length.

I conclude, therefore, that animals of all taxonomic classes and all trophic levels measure attributes of objects at the same mass-specific spatial grain. The only difference is in the dimensionality that defines the unit step of a mammal from that of a bird. And that reflects the fact that mammals walk and birds fly!

In summary, there are four conclusions in this section on habitat quanta and their utilization:

(1) The amount of resources in each quantum is different in each ecosystem and landscape.

(2) The overall way that resources are distributed among animals of different sizes is similar in all landscapes, except, perhaps, in those subject to rapid transition.

(3) The spatial extent of home ranges of mammals of the same body mass is determined

by the productivity of the landscape and by their trophic status. The less productive the landscape and the longer the trophic chain, the larger the spatial extent of their foraging areas.

(4) All animals measure elements in the landscape with a spatial grain defined as a function of their size. As a consequence, there is a unified response of all trophic levels and of mammals and birds to the geometric structure of resources available in each quantum. The sampling grain of the animal interacts with the texture of the ecosystem to define an invariant feature of home ranges and of body-mass clump structure that unifies all mammals and birds and all trophic groups.

Those conclusions now set the stage in the next section to analyze how animals make decisions in the environment and how those decisions can result in the discontinuous structure of body-mass distributions.

Hierarchy of Decisions and Hierarchy of Opportunity

The objects encountered by animals are either edible, frightful, lovable, ignorable, or novel. The first three define the resources needed in the landscape to provide food, protection, and conditions for reproduction. Their utilization has come to be well described by optimal foraging theory (Charnov 1976, Krebs and Davies 1978, Mangel and Clark 1986), where some objective function—e.g., net energy or available time—is maximized in the short term under constraints. The latter two define resources that are essential for long-term persistence of a species in a fluctuating environment. I argue that they have much more to do with adaptive choices in changing circumstances and rules of thumb that minimize information acquisition and processing. The rules of thumb that persist are those with the least demand on information while still permitting survival and reproduction consistently over long periods.

Consider the set of decisions made when an animal locates itself in an area during a breeding season. In so doing, it reacts to past experience and to proximate stimuli that corre-

late with future availability of necessary resources. There is a nested set of decisions.

As an example, the large wading birds of the Florida peninsula and Cuba, like the ibis, Great and Snowy Egrets, and Wood Storks (Ciconiiformes), potentially have available a number of different wetland areas, ranging from the extensive Everglade system in the south, to the marshes, ponds, and lakes of north-central Florida and even on to those in the Carolinas (Palmer 1962). Tagging records (Byrd 1978) indicate that birds can locate in areas from up to several hundred to one or two thousand kilometres from their birth place. Such scales describe the spatial extent for decisions affecting location. But just as the information in a satellite image of a landscape has not only a spatial extent, but also a resolution, or pixel size, so the information utilized at each level of decision also has a resolution or grain for that decision (Wiens 1989, 1990). In this case of locating an area, the grain is the size of the smallest region of wetlands that can be consistently occupied in good and bad times—probably measured in the order of 100 km or so.

The decision is strongly influenced by past experience during previous breeding periods. As a consequence of such fidelity to place, changed environmental or landscape attributes can leave a population "trapped" for some years in an environment adequate for survival, but inadequate for reproduction. That is what has been happening in the Florida Everglades as a consequence of changes in water delivery. Nesting success of wading birds has declined by 90% since the mid 1960s (Frederick and Collopy 1988). Thus there is not only a spatial scale defining a decision, there is also a temporal scale whose extent, or time horizon, can be defined as the time taken to extinguish the fidelity for place, and whose resolution can be defined as the time since the last experience. In this example, that is measured as decades for the time horizon and a year for the time resolution.

Areal decisions are only one of a nested set, each of which has its own spatial extent and grain and its own time horizon and resolution. The levels and their scales are suggested for this illustrative example in Fig. 18. In developing

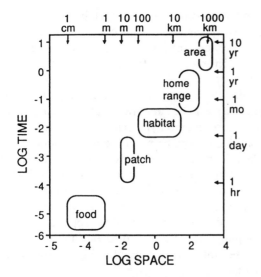

FIGURE 18. Scales in the hierarchy of decisions made by large wading birds. Time is measured in years, space in kilometres.

this example I have emphasized feeding as the principal need, but the same hierarchy applies to the decisions for protection, nesting, mate selection, exploration, etc.

At the next-finer scale, locations of foraging areas or home ranges are decided within the spatial extent of the few hundred kilometres of area chosen. The decisions are made on the basis of the immediate success in acquiring food and undisturbed roosts, and presumably on the basis of other attributes of water bodies that have become associated with future resource availability through a rule-of-thumb decision rule. A reasonable foraging radius for many of these birds is ≈ 20 km (range: 3–30 km), so that the spatial grain for deciding territory is ≈ 20 km and the spatial extent is a few hundreds of kilometres. Some birds have been tracked for up to 60 km of unswerving flight in their search for new foraging areas when success within the normal foraging area is poor (G. T. Bancroft, *personal communication*). Such decisions are made within a year and are likely based on foraging success as integrated over one to a few weeks by some internal measure of nutritional status. Hence the time horizon can be estimated as a year and the time resolution as a week.

The next-finer set of decisions is for locating habitats within the foraging area where

searching for food is concentrated. The extent is now defined by foraging distance—about 20 km. Each day decisions are made within this area for wetlands (shallow ponds, lake edges, marshes) at depths suitable for foraging and containing prey in adequate concentrations. The grain is set by the minimum size of pond utilized, 50 to 100 m according to my observations of Great Egrets (*Casmerodius albus*). The decisions are again determined by some immediate signals, such as the presence of other feeding birds, but also by success in the previous days. The time horizon is therefore a few days to perhaps a week, and the resolution one day.

Food is rarely distributed evenly within a pond, nor is ease of access. The next decisions therefore are for forageable patches whose extent and grain is about 50 m and 10 m, respectively. The time for abandonment can be the length of a foraging bout—at the most, a few hours, and the time resolution perhaps a few tens of minutes.

The final decisions are for the prey themselves, where the extent is set by the reaction distance of the predator, about a metre, and the grain by the prey size, about a centimetre. The time horizon is in the tens of minutes and the resolution under a minute.

This example describes for a particular species what Orians (1980) and Gass and Montgomerie (1981) identify as the hierarchy of decisions made by all animals. The same levels and objects of decision are universal for all birds or mammals, although the scale ranges for each are specific to the species. They are also specific to the size of animal. As a consequence, it should be possible to develop a generalized hierarchy of decisions by converting the body-mass clump limits into their space and time equivalents.

I did this by using allometric relations to "anchor" the space–time domain of each of the clump categories at three points. At the smallest scale, I used equations that related prey size and prey capture rate to predator size in order to anchor the food-choice box. At an intermediate scale, I used the home-range equations developed in the previous subsection entitled *Home range* and equations concerning rates of habitat utilization to anchor the habitat-choice box. Finally, at the largest scale, I used estimates of body-mass-dependent dispersal distance and longevity to anchor the areal-choice box. The equations are derived and the anchor points calculated in Appendix 9.

As an example, the spatial domains for decisions by carnivorous mammals of the boreal forest are shown in Fig. 19 and the temporal domains in Fig. 21. The hierarchy for the boreal landscape, shown earlier in Fig. 1, and the domains of atmospheric variation are added for comparison, drawing inspiration from Clark (1985). Although the quantitative positions for each of the hierarchies are approximate, three main conclusions can be made.

The first conclusion concerns the role of spatial structure in forming discontinuous body-mass distributions. The second concerns differences in the time domain of ecosystems that

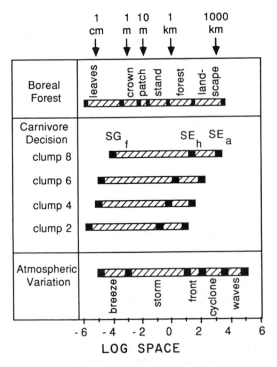

FIGURE 19. Comparison of spatial scales: of elements of the boreal forest landscape, of decisions by carnivore mammals in four of the clump categories, and of atmospheric variation. The spatial scale is based on distances in kilometres. SG_r = spatial grain for food decisions; SE_h = spatial extent for home-range decisions; SE_a = spatial extent for areal decisions.

control decisions and, ultimately, the distribution of species. The third concerns the interaction between the landscape hierarchy, the decision hierarchy, and the scales of atmospheric variation.

Fig. 19 shows that the spatial range for decisions covers the same range as the ecosystem-landscape hierarchy. That is, there is a tight spatial coupling between these two hierarchies. That is precisely what is expected if spatial discontinuities are the primary source of body-mass discontinuities, and this confirmation simply helps substantiate the reality of the spatial conversions derived in Appendix 9. The coupling transforms distribution of body masses into distinct clumps in the following manner.

The ecosystem from which opportunities are sought is itself discontinuously structured. Animals of each clump category exploit objects within a different range of scales. The largest animals exploit those in a range from roughly tens of centimetres to thousands of kilometres. In contrast, those of the smallest category shown in Fig. 19 exploit a smaller range, from millimetres to kilometres. As a consequence, animals of different body-mass clumps encounter different discontinuities in their environment. Those discontinuities are potentially of two sorts—hierarchical, as suggested in the stylized landscape of Fig. 20, and fractal.

As shown in Fig. 20, a hierarchical organization generates abrupt shifts in the kinds, distances, and sizes of objects when the grain of measurement reaches a value that aggregates objects into a new set of objects in a new hierarchical level. Since the size of an animal defines the grain of its measurement, then a body-mass gap exists at those sizes and sampling grains where there is an abrupt transition in the attributes of objects (their size and inter-object distances) between two hierarchical levels or between two domains of different fractal dimension.

As a consequence, different-sized animals will be affected by different scales of disturbance of resource objects in different ways. For example, Morton (1990) explains the extinction of all middle-sized Australian mammals since European settlement with arguments entirely consistent with the clump-structure explanation

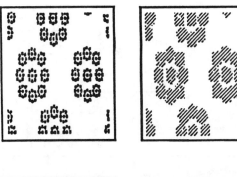

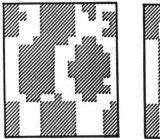
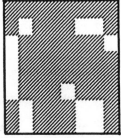

FIGURE 20. Example of a stylized landscape (upper left) with a four-level hierarchy of oval objects designed to be self-similar at all scales. The smallest oval solid dots can be imagined as feeding patches within a pond (a), the group of such dots as a pond (b), the group of those ponds as wetland region (c), and the full set of wetland regions as an area (d). In each of the remaining three versions the pixel size becomes progressively greater, and those can be imagined as being produced by the grain a wading bird uses for choice of ponds (upper right), for wetland region (lower left) or for area (lower right). It is an example of a hierarchical structure that has the same fractal dimension at all scales.

presented here. Changed patterns of fire combined with the impacts of rabbits changed the pattern of productivity on the Australian landscape. Fig. 20 provides a stylized example to explain the effect. The changes Morton describes would still leave patches of productivity, but at certain spatial grains they would be scattered at greater distances. It is as if, in Fig. 20, every other of the groups of dots (imagined as ponds in the example) was removed but everything else remained unchanged. Very small animals could still find resources among the patches within one of the existing "ponds" [see (a) in Fig. 20 legend] just as before. Very big animals have a grain of such a size that those separate ponds fuse into single forageable

units [see (c) in Fig. 20] and have an ambit large enough to make them accessible irrespective of whether every other one is removed or not. Intermediate-sized animals would be the only ones faced with potentially serious increases in inter-patch distances [see (b) in Fig. 20] that could threaten existence in bad times. Scale-specific impacts of disturbances on vegetation therefore have scale-specific impacts on the body-mass groups most affected. Landscape fragmentation (Harris 1984) will have differential effects on different sizes of animals depending on the spatial grain of the fragmentation. That is potentially predictable.

In addition to hierarchical breaks in object kind, size, and distance, there are specific scales where the fractal dimension, D, can suddenly change, as different processes act to determine texture. Some of those are coincident with a change in hierarchical level, but that need not be so as is shown in the scale-invariant example of Fig. 20. As an example of discontinuous changes in fractal structure, Morse et al. (1985) analyzed the fractal nature of woody plant parts and demonstrated that the fractal dimension of Virginia creeper (*Parthenocissus quinquefolia*) over small scale ranges was 1.28, but suddenly changed to 1.55 as the scale increased. In another example over a larger range of scales, Bradbury et al. (1984) demonstrated that coral reefs have three ranges of scale invariance, each with a different fractal dimension. The roughly textured individual branches are in a range below 10 cm ($D = 1.1$), the smoother adult colonies are in the range of 20 to 200 cm ($D = 1.02$–1.08) and the again-more-textured grooves and buttresses are in the range of 5 to 30 m ($D = 1.15$–1.17). Over a still larger range of scales, Kent and Wong (1982) showed that the shoreline of lakes in the boreal region is formed by two distinct geomorphological processes. Erosional processes dominate below ≈ 350 m and produce a coast line of $D = 1.14$. Glacial corrosive processes dominate above 350 m and produce a rougher coast line of $D = 1.44$. Hence changes in hierarchical levels and of fractal dimension across ecosystem and landscape grains impose a discontinuous structure on the opportunities for satisfying needs.

The earlier analysis of home ranges demonstrated that the spatial extent for decisions such as those involving home ranges varies, depending on trophic status, food type, and habitat productivity. In contrast, the spatial grain for decisions, or the step unit, is an invariant function of body mass and the dimensional geometry alone. Hence the trade-off that determines the size of an animal has two features. On the one hand, to be smaller in a textured environment means a smaller grain for selection of resources that exposes the animal to more surface or resources per unit area. To be larger, on the other hand, means less resources per unit area because of a larger sampling grain, but the ability to move greater distances so as to balance that reduction. Hence each scale range that contains sufficient resources to support populations will be exploited by species whose body masses are ones that will satisfy this trade-off. If the environment were not hierarchically structured and were, in addition, self-similar at all scales, animal body masses could be distributed continuously. But that is not the case.

I conclude, therefore, that the discontinuities in body-mass distribution occur because grain-specific breaks in spatial architecture exclude certain sizes. If there were animals with those sizes, they would be ill adapted to exploit objects in both the smaller quantal ranges as well as those in the next larger quantal range. Therefore body-mass clump categories exist because the gaps between clump categories represent a worst possible solution. Once the worst is eliminated, considerable adaptive opportunities are therefore left for diverse and flexible solutions within a quantal range, some of which can be sub-optimal and still persist.

The second major conclusion concerns the very different relationship that the decision and ecosystem hierarchies have in time (Fig. 21). They do not match in time the way they do in space. The overall decision hierarchy operates at a speed three to four orders of magnitude faster than the ecosystem hierarchy. That means that the slower dynamics of the ecosystem and landscape constrain and control the variability experienced for animal decisions. The upper end of the decision hierarchy, however, does overlap the same time ranges as the lower end of the ecosystem hierarchy. Hence the larger scale decisions for area become cou-

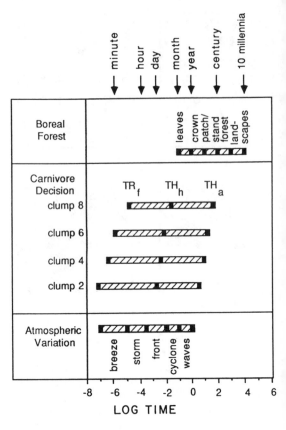

FIGURE 21. Comparison of temporal scales of elements of the boreal forest landscape, of decisions by carnivore mammals in four of the clump categories, and of atmospheric variation. The logarithmic scale is based on times in years. TR_f = temporal resolution for food decisions; TH_h = temporal horizon for home range decisions; TH_a = temporal horizon for areal decisions.

pled to the variations in vegetation texture and quantity over a scale range from needles to stands (Fig. 21). Those attributes are not slow enough simply to be dealt with as constants. Instead, they become determinants of where to locate and when to abandon areas. Therefore vegetation texture at those scales is important for questions of extinctions, invasions, and animal distributions.

The third and final conclusion concerns the interaction among the landscape hierarchy, the decision hierarchy, and the scales of temperature and moisture variation. The band of atmospheric variation overlaps the decision hierarchy, but not the ecosystem hierarchy (Fig. 21). Because of the matching of time dynamics between decisions and weather, variation mediated by the atmosphere directly influences behavioral decisions at all scales. It is for that reason that animals have evolved ways to partially control the effects of such physical variation, through a variety of behavioral and physiological homeostatic mechanisms that regulate, for example, internal body temperature.

The ecosystem hierarchy, however, does not overlap the domain of atmospheric variation in a significant range of scales. That is, the ecosystem can operate as if weather is constant, at least at less-than-seasonal scales of variation. As a consequence, the physical architecture of the ecosystem itself mediates variation by moderating the fast dynamics of air through the slower dynamics of vegetation. For example, the daily temperature variation within a stand of trees is noticeably less extreme than in clearcut areas (Heckert 1959). Similarly, at a larger scale, the onset of spring and snow melt is accelerated by heat accumulation and reradiation by low-albedo coniferous foliage in the boreal forest (Hare and Ritchie 1972). To a degree, therefore, the boreal forest "makes its own weather" and the animals living therein are exposed to a more moderate and slower variation in temperature and moisture than they would otherwise be.

These features of the interaction between different space and time domains of variation of the ecosystem, of behavior, and of the physical environment are summarized in Fig. 22. It provides the point of departure needed to connect community structure with ecosystem dynamics. That will be dealt with in the next section.

In summary, a species persists in a landscape if its needs match resources in both good and bad times. In many cases it is probably the times of crises that determine species compositions, as Wiens and Rotenberry (1980) suggest for birds of the shrub steppe. Those times are dictated by environmental fluctuations, which themselves occur at various scales in space and time. The internal consequences of those fluctuations to the individual animal are regulated by physiological homeostatic mechanisms. And their external consequences are moderated by the physical architecture of the ecosystem and landscape. As a consequence, species persis-

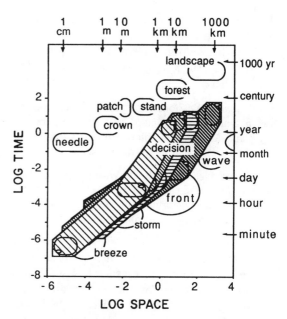

FIGURE 22. Comparison of time and space scales: of the forest ecosystem hierarchy, of decisions by carnivore mammals in four of the clump categories (differently shaded bands), and of atmospheric variation. The open ovals in the decision bands are the anchor points for food decisions, home-range decisions, and area decisions calculated in Appendix 9.

tence is determined by the match between the hierarchy of decisions, the hierarchy of ecosystem attributes, and the way that landscapes/ecosystems transform climatic variability at different scales. The latter two combine to make a hierarchy of opportunity in space and time. They form what Southwood (1977) has called the habitat templet. Since opportunities are distributed discontinuously across scales, gaps in the body-mass distributions of animals appear because competitive solutions become infeasible at those body masses.

Revisiting Ecosystem Dynamics

This study was launched by conclusions derived from a comparative study of ecosystem dynamics (Holling 1986). Most of the analyses presented here, however, concern static structure and not dynamics. This final section closes the circle to return to that theme in order to connect community structure with ecosystem dynamics. The key issue is how the observed structure is produced.

An earlier section, *Inferring the pattern*, set the stage by suggesting two hypotheses, summarized here:

The Extended Keystone Hypothesis: All ecosystems are controlled and organized by a small number of key plant, animal, and abiotic processes that structure the landscape at different scales.

The Entrainment Hypothesis: Within any one ecosystem, the periodicities and architectural attributes of the critical structuring processes will establish a nested set of periodicities and spatial features that become attractors for other variables.

The specific hypotheses and analyses that evolved from these more general propositions provided the focus for the earlier analytical sections and led to conclusions consistent with them. Moreover, the extended keystone hypothesis is similar to other views that have assigned ecosystem processes to characteristic scales (e.g., Harris 1980, Clark 1985, Steele 1989). The difference is that I argue that there are relatively few structuring processes and their influence is expressed at a relatively few scale ranges. The entertainment hypothesis flows directly from and can be tested with time-series data, as Carpenter and Leavitt (1991) have done with time-series data from lakes.

I conclude, therefore, that the landscape is hierarchically structured by a small number of structuring processes into a small number of nested levels, and that those form physical textures and temporal frequencies specific to each level. Each of the habitat quanta exists within a specific range of scales. The physical architecture within each level has specific scale-invariant properties, such as object kind, size, distance, and fractal dimension. The temporal frequencies are similarly clustered about unique speeds. Consequently, I conclude that the processes that generate discontinuous time dynamics also generate discontinuous physical structure.

I will explore the consequences of those conclusions by first describing some specific examples of structuring processes in the boreal forest, and second, by emphasizing how hierarchies themselves are dynamically changing organizations.

Structuring Processes

The structuring processes of different hierarchical levels in the boreal forest are shown in Table 4. Note that these processes are defined as those that both produce structure and are affected by that structure. As a consequence, climatic processes are included at only two scales, where vegetation and climate interact. One is at the scale of the patch and below, where vegetation biomass can moderate weather to produce a slower and less-extreme microclimate. The other is at the scale of the landscape, where mesoclimate can be affected by the albedo, roughness, and evapotranspiration rates from vegetation. Otherwise climatic variables can be dealt with as exogenous driving variables that can be decomposed from the system because of their much greater speed (see Fig. 21).

Those organisms that are members of the set of structuring variables make their own environment by producing some of the structures within the nested set. At the same time, these structures create niches at a variety of scales for a larger set of species than the structuring ones. Species of organisms in that larger set are affected by and dependent on the structure, but contribute little to it qualitatively. In that sense they can be termed "entrained" species. The number of these species depends on the productivity and accessibility of ecosystem resources, particularly during critical life-history stages such as those related to reproduction and juvenile survival.

The entrained species, however, are not simply "along for the ride." In the evolutionary theater (Hutchinson 1965) they also can be the bit actors who are understudies of the principals. Changed circumstances can lead them to assume structuring functions—e.g., species of insect defoliators, whose populations are normally confined to small scales, that become outbreak species when agricultural or forestry practices produce monocultures of subdominant species; or grasses, common only early in forest succession, that become dominant under a shift in climate and an increase in fire frequency. They are, in theatrical terms, truly "fifth business" (Davies 1970), functioning to bridge crisis moments so that the plot can advance. Over evolutionary time scales, the

TABLE 4. Hierarchical levels in the boreal forested landscape with scales, structuring variables, and structuring processes. The scales and levels were earlier presented graphically in Figs. 1 and 22.

Level	Scales		Categories of structuring processes	Structuring variables	Structuring processes
	Time (yr)	Space (km)			
Needle	0.1 to 1	0.01×10^{-3} to 0.1×10^{-3}	Physiological processes	Leaves, herbs, grass by age and condition; detritus; nutrients	Photosynthesis and respiration; decomposition; nutrient uptake
Crown	1 to 10	1×10^{-3} to 10×10^{-3}	Autecological processes	Crown volume; bush and log density; herbivore and predator density; soil structure, nutrients; seed density	Plant growth; seed production; foraging on vegetation, seeds, and herbivores; animal population dynamics; decomposition
Patch/Gap	10 to 100	10×10^{-3} to 30×10^{-3}	Plant competitive processes of gap dynamics	Dominant and subdominant tree and bush sizes; fuel; soil structure; seed density	Tree growth, competition, and mortality; soil immobilization and mobilization; vegetation effects on microclimate
Tree stand	10 to 100	0.03 to 1	Mesoscale disturbance and dispersal processes	Insect, disease, and fuel distribution; tree and plant age, density, and condition	Disturbance dynamics (fire, insect, and disease); tree harvesting; seed dispersal
Landscape	100 to 1000	1 to 100	Watershed processes	Topography; forest, aquatic, and open, ecosystems	Erosion; watershed hydrology; mesoclimatic interactions with vegetation
Boreal zone	1000 to 10 000	100 to >3000	Planetary and evolutionary processes	Precipitation; temperature; bedrock	Evolution; geomorphology; planetary dynamics

pool of entrained species is critical for providing alternative opportunities. Over ecological time scales, the structuring species dominate.

Three approximate scale ranges exist, each defined by a broad class of processes that dominate over those ranges of scale. The microscales are dominated by vegetative processes, the mesoscales by disturbance and environmental processes, and the macroscales by geomorphological and evolutionary processes.

Those three different classes of processes determine the architecture over the relevant scale range. At the microscale the physiognomy of trees, bushes, and smaller plants introduces discontinuous structures and textures that are likely to be the same irrespective of the ecosystem concerned. That is, to a significant

degree a needle, a leaf, a grass or forb, a bush, or a tree have similar geometrical attributes whether they grow in the boreal forests of Canada, the river valleys of the Great Central Plains of North America, the tree islands of the Everglades, or the savannas of Africa. Hence, as observed earlier, the smaller animal body-mass clump categories in one landscape, like the boreal zone prairies, can be the same as those in another, like the boreal forests.

At the other extreme of scale, much slower and larger geomorphological processes shape topographic, hydrologic, and edaphic properties. Some, like postglacial retreat, are still in the process of affecting species distributions in northern regions. The architectural patterns produced by macroscale processes, however, cover a wider variety of possibilities than those produced by microscale processes. The mountains of the Rockies, the mosaic of granitic lakes and hills on the Canadian Shield, the rolling landscape of the pot-hole prairies, and the flat topography and extent of the Everglades river of grass contain spatial patterns sufficiently different as to affect presence or absence of body-mass clump categories. Those macroscale processes are slow enough, however, that it is likely that body-mass clump structure, as distinct from species-specific composition, has stabilized within present ecosystems.

It is at the intervening mesoscale where a number of quite different biotic and abiotic processes operate that have a speed and a variety of spatial consequences that can affect clump structure within decades.

These mesoscale processes transfer local events at the scale of patch into large-scale consequences because they have a spreading, or spatially contagious, character. Some of them are abiotic—water, fire, and wind. But others are zootic, and are mediated by movements of animals as they disperse, migrate, and move over hundreds to thousands of kilometres—outbreak insects that defoliate, ungulates that graze, large mammals and birds that prey. Still others are the consequences of human transformations of regional landscapes. Mesoscale spatial dynamics and the distribution of vegetation are dominated by animals, people, water, wind, and fire. It is at this scale, for example, that forest insects and fire can exert a role in facilitating a rapid transformation of boreal ecosystems as a consequence of climate change (Holling 1992).

Such mesoscale processes are different in different landscapes and, as a consequence, the body-mass clump structure and textural categories at intermediate to large scales should be landscape specific. They may well provide measures of ecosystem responses to change.

Dynamics of Hierarchies

Hierarchies of this sort are not static structures. The levels are maintained by the processes described, but only within a limited domain. Consider, for example, the microscale processes. Those processes aggregate to establish a

cycle of birth, growth and storage, death, and renewal, as suggested in Fig. 23. Moreover, there is a nested set of such cycles, each with its own range of scales. For the microscales, fresh needles cycle yearly, the crown of foliage cycles with a decadal period, and trees, or gaps, cycle at close to a century or longer period. Hence views of succession that perceive forests moving to a sustained equilibrium are limited.

Over the last decade the literature on ecosystems has led to major revisions in the original Clementsian view of succession as being a highly ordered sequence of species assemblages moving toward a sustained climax whose characteristics are determined by climate and edaphic conditions. This revision comes from extensive comparative field studies (West et al. 1981), from critical experimental manipulations of watersheds (Bormann and Likens 1981, Vitousek and Matson 1984), from paleoecological reconstructions (Delcourt et al. 1983, Davis 1986), and from studies that link systems models and field research (West et al. 1981).

The revisions include four principal points. First, invasion of persistent species after disturbance and during succession can be highly probabilistic. Second, both early- and late-successional species can be present continuously. Third, large and small disturbances triggered by events like fire, wind, and herbivores are an inherent part of the internal dynamics, and in many cases set the timing of successional cycles. Fourth, some disturbances can carry the ecosystem into quite different stability domains—mixed grass and tree savannas into shrub-dominated semi-deserts, for example (Walker 1981). In summary, therefore, the notion of a sustained climax is a useful, but essentially static, and incomplete equilibrium view. The combination of these advances in ecosystem understanding with studies of population systems has led to one version of a synthesis that emphasizes four primary stages in an ecosystem cycle (Holling 1986).

The traditional view of ecosystem succession has been usefully seen as being controlled by two functions: *exploitation*, in which rapid colonization of recently disturbed areas is emphasized, and *conservation*, in which slow accumulation and storage of energy and material is emphasized. But the revisions in under-

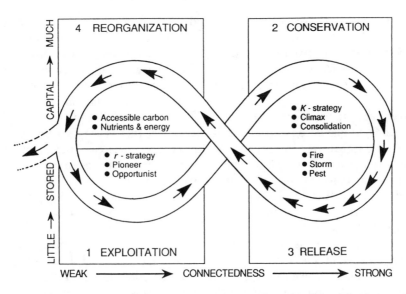

FIGURE 23. The four ecosystem functions and the flow of events between them. The arrows show the speed of that flow in the ecosystem cycle, where arrows close to each other indicate a rapidly changing situation and arrows far from each other indicate a slowly changing situation. The cycle reflects changes in two attributes, i.e., (1) Y axis: the amount of accumulated capital (nutrients, carbon) stored in variables that are the dominant keystone variables at the moment and (2) X axis: the degree of connectedness among variables. The exit from the cycle indicated at the left of the figure indicates the stage where a flip is most likely into a less or more productive and organized system, i.e., devolution or evolution as revolution!

standing indicate that two additional functions are needed. One is that of *release*, or "creative destruction," a term borrowed from the economist Schumpeter (1950, as reviewed in Elliott [1980]), in which the tightly bound accumulation of biomass and nutrients becomes increasingly fragile (overconnected) until it is suddenly released by agents such as forest fires, insect pests, or intense pulses of grazing. The second is one of *reorganization*, in which soil processes of mobilization and immobilization minimize nutrient loss and reorganize nutrients to become available for the next phase of exploitation.

During this cycle, biological time flows unevenly. The progression in the ecosystem cycle proceeds from the exploitation phase (Fig. 23: box 1) slowly to conservation (box 2), very rapidly to release (box 3), rapidly to reorganization (box 4), and rapidly back to exploitation. Connectedness and stability increase, and a "capital" of nutrients and biomass is slowly accumulated during the sequence from exploitation to conservation. The system eventually becomes overconnected, so that rapid change is

triggered. The agents of disturbance might be wind, fire, disease, insect outbreak, or a combination of these. The stored capital is then released and the system loses its tight organization, to permit renewal of the same stable state.

That pattern is discontinuous and is dependent on the existence of changing multi-stable states that trigger and organize the release and reorganization functions. Resilience and recovery are determined by the release and reorganization sequence, whereas stability and productivity are determined by the exploitation and conservation sequence.

A critical feature of hierarchies is the asymmetric interactions between levels (Allen and Starr 1982, O'Neill et al. 1986). In particular, the larger, slower levels maintain constraints within which faster levels operate. In that sense, therefore, slower levels control faster ones. If that was the only asymmetry, however, then it would be impossible for organisms to exert control over slower environmental variables. That is the criticism that many geologists make of the Gaia theory (Lovelock 1988)—how could slow geomorphic processes possibly be affected by fast biological ones? However, the birth, death, and renewal cycle shown in Fig. 23 transforms hierarchies from fixed static structures to dynamic entities whose vulnerability to small disturbances changes at different points in the cycle.

There are two key states where slower and larger levels in ecosystems become briefly vulnerable to dramatic transformation because of small events and fast processes. One is when the system becomes overconnected and brittle as it moves toward maturity (Fig. 23: box 2). At these stages, there are tight competitive relations among the plant species. From an equilibrium perspective, the system is highly stable (i.e., fast return times in the face of small disturbances), but from a resilience perspective, sensu Holling (1987), the domain over which stabilizing forces can operate becomes increasingly small. Brittleness comes from loss of resilience. Hence the system becomes an accident waiting to happen. In the boreal forest, for example, the accident might be a contagious fire that becomes increasingly likely as the amount, extent, and flammability of fuel accumulates. Or it could be a spreading insect outbreak triggered as the increasing amount of foliage both increases food and habitat for defoliating insects and decreases the efficiency of search by their vertebrate predators (Holling 1988).

Small and fast variables can also dominate slow and large ones at the stage of reorganization (Fig. 23: box 4). At this stage, the system is underconnected, with weak organization and weak regulation. As a consequence, it is the stage most affected by probabilistic events that allow a diversity of entrained species, as well as exotic invaders, to become established. On the one hand, it is the stage most vulnerable to erosion and to the loss of accumulated capital. On the other hand, it is the stage from which jumps to unexpectedly different and more-productive systems are possible. At this stage, instability comes because of loss of regulation rather than from the brittleness of reduced resilience.

The degree to which small, fast events influence larger, slower ones is critically dependent upon the accumulation, cycling, and conservation of accumulated capital. And that in turn depends upon the mesoscale disturbance processes. If, because of human management or macroscale changes, the disturbance destroys too much accumulated capital over too large an area, the system can flip into a qualitatively different stable state that can persist until there is explicit rehabilitation by management. As an example, that is why grazing at sustained but moderate levels can transform productive savannas into less-productive systems dominated by woody shrubs (Walker et al. 1969). The question for issues of human transformation from the scale of fields to the planet, therefore, is how much change does it take to release disturbances whose intensity and extent are so great that the renewal capital is destroyed or regeneration of the existing plant species is prevented.

In summary, the earlier tests of possible causes for body-mass gaps led to the conclusion that the geometry of landscapes shapes the morphology of animals. This section on ecosystem dynamics and structure now adds the conclusion that key organisms themselves, together with abiotic disturbances, in turn shape the geometry of ecosystems.

The Human Animal

It is often useful to view a body of developed argument from a fresh perspective. The impact of humans provides that opportunity, because we are an animal faced by our own resource decisions and because we are an ecological force.

Homo sapiens exists at a unique position in the hierarchy of body-mass clump categories. The human male's average body mass of 82 kg lies in the gap between the second largest body-mass clump category for mammals and the largest one. Within the boreal forest (Appendix 2), for example, the wolf is on one side of that gap at 43 kg and the white-tailed deer on the other side at 86 kg. Within the boreal zone prairies, the cougar is on one side at 66 kg and the white-tailed deer again on the other side at 86 kg. Even among the antelopes of the savannas and grasslands of East Africa, the lesser kudu is on one side at 79 kg and the topi on the other at 115 kg (Appendix 6).

Earlier I argued that those gaps exist because

they represent a "forbidden" range of sizes where animals would be ill adapted to exploit resources of the next-smaller or next-larger textural ranges. It is tempting to speculate that this exception occurs because humans' generalized morphology combines with an ability to develop and adapt specialized technology to turn a least feasible body-mass solution into the best possible one. The specialized tools, habitation, and defense of hunters and gatherers, for example, together with domestication of hunting companions, open up efficient ways to utilize opportunities over a very wide range of scales. In addition, the use of fire places early humans in the role of a structuring process, capable, in temperate North America, for example, of transforming mosaics of grassland and woods with a spatial grain of ≈1 km into extensive regions of contiguous grass or forest. Both hunting pressure and such landscape transformations could clearly affect the larger body-mass categories.

As an example, evidence has accumulated to suggest that human invasion into North America ≈11 000 yr BP account for the sudden extinction of a large proportion of the megaherbivores during the Pleistocene (Martin 1967). In much less than 1000 yr, 73% of the genera of herbivores weighing >44 kg became extinct. Note that this definition of megaherbivore by paleontologists includes only the largest size clump category I identified in the boreal forested and prairie landscapes. South America and Australia were also affected, with, respectively, 80 and 86% of the megaherbivores becoming extinct. Those regions with long histories of habitation, like Africa and tropical Asia, did not experience massive extinctions at that time (Martin 1984, Webb 1984).

Simulations using a "blitzkrieg" model developed by Mosimann and Martin (1975) suggest it would have been possible to achieve such extinctions in North and South America through hunting alone as new human invaders swept in a wave front, typical of invading species, from Alaska to Patagonia at an average rate of 160 km per decade. The timing of those extinctions is coincident with the appearance of peoples of the Clovis hunting cultures in the Americas.

In addition, all species of the very large grazing and browsing herbivores >1000 kg

(mammoths, gomphotheres, ground sloths, and mastodons) became extinct (Owen-Smith 1989), thereby eliminating structuring herbivores whose presence would have maintained a habitat mosaic appropriate for their size category. Fire, as suggested earlier, could then have led rapidly to vegetation cover of one type over extensive areas, thereby accelerating declines. Since the largest size clump category now living in the boreal forest ranges from 86 kg (white-tailed deer) to 483 kg (moose), it seems likely that the extinction of the even-larger herbivores >1000 kg eliminated a distinct clump of body masses associated with a unique mosaic of grasses and woods in North America 11 000 yr ago. This category still exists in Africa and includes the elands and water buffalo (Fig. 14 and Appendix 6) as well as elephants and rhino. They are well known for their maintenance of a mosaic structure of vegetation at spatial grains of vegetation measured in tens of kilometres (Owen-Smith 1989).

As techniques for agriculture and for permanent settlements became established in Europe, humans developed a more pervasive ability to shape textures in the landscape for human use. For example, the present mosaic of fields, villages, and woodlots in long-established rural areas in Europe, such as those of southern Austria, reflects a transformation from extensively wooded conditions to a mosaic of fields, woods, and villages with a grain measured in the 100 m to kilometre range. By this stage of human development, fire now became an agent to be curtailed to very circumscribed areas, rather than to be used to transform large areas.

The locations of the villages in that region still reflect the scale of human choices for early means of food production. Villages are spaced ≈7 km apart, making the most distant fields within 3 or 4 km, or a one-half hour walk, leaving the remainder of the daylight hours for work. Just as 3–4 km fixes the spatial grain for choosing farming sites around villages, a day defines the time resolution for the farming activities associated with the village.

Once in settled locations, excess capital could be accumulated to provide opportunities for regional trade in surplus goods. Ken Follett's historical novel (1990) of the development of Gothic cathedrals during the twelfth century describes an English King's edict that a village could establish a market, and thereby accumulate even further wealth, only if it was at least 14 miles (22 km) from any other market. The goal was to assure that no one had to spend more than one third of a day each week or month travelling to the nearest market (≈ 11 km) and one third for the return, leaving one third for market business. For this level in the hierarchy of choices, 11 km describes the spatial grain for choice of a market location and a week or month the temporal resolution for its use.

I have presented these brief examples to emphasize three points. First, these human examples support the generality of the argument that a discontinuous hierarchy of decisions intersects with a landscape hierarchy to produce the opportunities that animals exploit. Moreover, human activities demonstrate the direct and indirect effect of structuring processes that transform landscapes and affect the presence of body-mass clump categories over a meso- to macroscale range.

Second, the emergence of local markets provides an example that adds an additional insight into the dynamics of hierarchies. Not only can hierarchies lose levels and degrade because of loss of accumulated capital, as described earlier, they can also evolve new, larger, and slower levels when excess capital accumulates beyond a critical threshold. Again that is a discontinuous process that promotes a rapid transition to an added and expanded scale of influence. Finally, cultural evolution of new social organizations is accompanied by technological innovation, combined with novel procedures and institutions. That creates new classes of opportunity and the evolution of new systems.

A key feature of such technological change is the development of new modes of locomotion that open new scale ranges for opportunity and influence. For example, the horse allowed the half-hour to 1-hr trip between work and home to expand from 3 or 4 km to perhaps a dozen. The automobile, and associated infrastructure, expanded trips of the same duration to 40 km. Spatial scale increased without necessarily changing temporal scale for locating home and work. Trips between work and home are still

largely daily ones. The consequence has been patterns of urbanization that, on the one hand, provide unique economic and cultural opportunities and, on the other, produce environmental and ecological impacts that expand to a regional scale of hundreds of kilometres. The spatial grain and ambit of the human animal is restricted not by body-mass class, but by technological innovation. It is that process that expands both human choices and human impacts to regional and planetary scales.

Conclusions

The increasing consequence of regional and global changes induced by humans has stimulated advances in understanding, in methods, and in techniques for both science and management of natural and managed environments at regional and even planetary scales (Clark and Munn 1986, Holling 1989, Walters and Holling 1990). I hope that this paper further contributes to that by providing a context for linking processes across scales and by providing a way to integrate ecosystem dynamics with community structure.

I base that hope on four principal conclusions from this study that help define a research agenda for analyzing cross-scale dynamics of ecosystems in space and time.

1) The landscape is structured hierarchically by a small number of structuring processes into a small number of levels, each characterized by a distinct scale of "architectural" texture and of temporal speed of variables.

This structure organizes the behavior of ecosystems into a nested set of cycles, each with its own time constant and spatial texture. It is important to quantify these temporal and spatial quanta because they define specific regions of scale invariance at which different scale-limited models are needed. Such scale-limited models are present state-of-the-art practice, and, at the minimum, a family of such models needs to be developed or adapted to predict systems dynamics in time and space. Examples of such scale-constrained time/space models include the spruce-fir-budworm system of eastern North America (Clark and Holling 1979, Clark et al. 1979), estuary dynamics in

Louisiana (Costanza et al. 1986), and the Everglades ecosystem of south Florida (Walters et al. 1992). But—in a more intriguing way—I believe that these hierarchical attributes, together with the key processes that impose structure, will make it possible to also develop a "meta-model" that will focus explicitly on cross-scale interactions.

2) Each of the small number of processes that influence structure does so over limited scale ranges. The temporal and architectural structure of ecosystem quanta are determined by three broad groups of processes, each dominating over different ranges of scale.

Vegetative processes that determine plant growth and form define a discontinuous texture at fine microscales of centimetres to tens of metres in space and days to decades in time. As a consequence, the textures are largely the same in different ecosystems containing plants of the same physiognomy. The amount of vegetation and the successional pattern over these scales, however, vary in different ecosystems.

At the other, macroscale extreme, slow geomorphological processes define a basic topographic and edaphic structure at large scales of hundreds to thousands of kilometres and centuries to millennia. They are slow enough that their influence can be dealt with as constants when issues are ones of human impact.

In between, contagious disturbance processes such as fire, insect outbreak, plant disease, and water flow form patterns over spatial scales of tens of metres to hundreds of kilometres. In addition, the direct impacts of grazing by large herbivores and the indirect effects of large predators and animal disease further transform spatial patterns over these mesoscales. These processes have time scales of years to decades, making them critically important in determining whether present human influences, such as climate change induced by accumulating greenhouse gases, trigger a transformation of ecosystems, and, if so, how rapidly.

Vegetation amount is affected by weather and soils, and the vegetation in turn modifies climate and soil structure at two scale ranges. One is at the scale of the patch and below, where vegetation biomass can moderate weather to produce a slower and less extreme microclimate. The other is at the scale of the landscape, where mesoclimate can be affected by the albedo, roughness, and evapotranspiration rates from vegetation. Otherwise, climatic variables can be dealt with as exogenous driving variables that can be decomposed from the system because of their much greater speed.

The ranges over which any set of processes dominates results in regions of self-similarity or scale invariance. Satellite imagery can be used to identify those regions using fractal and hierarchical techniques of analysis. Those data and techniques provide a basis for developing a monitoring scheme to assess vulnerabilities of ecosystems to environmental changes.

3) Because of the non-linear nature of mesoscale disturbance processes, fine-scale knowledge of autecology cannot simply be aggregated to represent behavior at scales beyond the scale of a patch or gap.

At the critical mesoscales, distinct disturbance processes are triggered at thresholds of tens of metres to kilometres. These structuring variables of disturbance form the kind and amount of structure found at mesoscales by causing local events to cascade upward in scale to affect landscape patterns. They are both abiotic—e.g., fire, storm, and flood (Sprugel and Bormann 1981)—and zootic—e.g., insect outbreaks (Berryman 1987, Holling 1992), beaver influences, pulsed grazing by large ungulates and predation by animals that both exert "top-down" control (McNaughton et al. 1988, Naiman 1988), contagious plant diseases, and human harvesting activities (Clark et al. 1979).

Analyses of the function of mesoscale disturbance processes can provide a bridge between analyses of patch dynamics on the one hand and those of global atmospheric dynamics on the other. The former provide an understanding of ecosystem dynamics up to a spatial scale of tens of metres (Botkin et al. 1972, Pickett et al. 1986). The latter (Clark and Munn 1986, Menabe and Wetherald 1986, Dickinson 1989) have led to the development of General Circulation Models that are beginning to provide understanding of planetary atmospheric dynamics down to a scale of hundreds of kilometres. Non-linear mesoscale processes function in between those two. Their analysis requires new methods for measurement and experiment at unfamiliarly large scales (Carpenter 1990).

4) Behavioral and morphological attributes of animals can be used as a bioassay of existing landscape structure or as a predictor of the impacts of changes in vegetation pattern on animal community structure.

Variables other than the keystone ones, such as the entrained fauna of ecosystems, are affected by and "track" the structure of the hierarchical levels, but do not themselves significantly affect that structure. Since the hierarchical levels have discontinuously distributed spatial and temporal properties, therefore the sizes, behavioral decisions, and foraging ranges of animals are distributed into discontinuous groupings. The characteristics of body-mass clump structure therefore can be used as a succinct bioassay of existing or past ecosystem structure. Changes in that structure will change the groupings in a predictable way.

Hence the impacts of regional development or of climate change on faunal community structure can be predicted from observed (e.g., by remote sensing and ground monitoring networks) or predicted (by models) changes in the structure of vegetation.

In conclusion, the speed and expansion of human impacts on the planet's environment and biota require fresh approaches to deal with cross-scale phenomena. Increasingly, unexpected local events can have their source half a world away. In order to develop policies for sustainable development, a biosphere theory that embraces all scales is essential. This paper is a step in that direction. But such a theory can only come inductively from specific examples of cross-scale analysis and synthesis. Nature is sufficiently surprising that I, at least, am not smart enough to have deduced the patterns described herein and their causes without a framework that allowed nature to point the direction for me.

Acknowledgments

This paper has been greatly enriched and improved by a large number of colleagues, relatives, and friends. Throughout its evolution, discussions with Carl Walters, Lee Gass, and Lance Gunderson helped deepen, extend, and clarify the ideas and tests. At an early stage, Chris Holling (an economist, no less) suggested and named the Urtier Hypothesis as one of the alternatives. Ilse Holling spent weeks precisely collating natural-history information on birds and analyzed the possible patterns that revealed the correlation of habitat texture with body clump category. Ian McTaggart Cowan and Dick Canning gave unstinting help filling in my gaps in knowledge of the mammals and birds of the short-grass prairies. Tony Sinclair and Bryan Walker did the same for the ungulates of East Africa. Lee Gass, Peter Fredericks, and Steve Carpenter spent an astonishing amount of time and care reviewing the manuscript in a way that helped me greatly improve its clarity and focus. Gordon Orians and Peter Feinsinger shoe-horned reviews during trips and gave me the assurance to dare to dabble as I did in community ecology. I hope they do not regret it. Finally, Candace Lane and Theresa Patterson nurtured me and the paper through the data collection, analysis, and proof reading stages. I am indebted to all. This work was supported by a NASA/EOS grant number NAGW 2524 as part of the Interdisciplinary Scientific Investigations of the Earth Observing Systems program.

References

Allen, T. F. H., and T. B. Starr. 1982. Hierarchy: perspectives for ecological complexity. University of Chicago Press, Chicago, Illinois, USA.

Arthur, B. 1990. Positive feedbacks in the economy. Scientific American **262**:92–99.

Balch, R. E. 1933/34. The balsam woolley aphid, *Adelges piceae* (Ratz.) in Canada. Scientific Agriculture **14**:374–383.

Banfield, A. W. F. 1974. The mammals of Canada. University of Toronto Press, Toronto, Ontario, Canada.

Bekoff, M. 1977. *Canis latrans*. Mammalian Species **79**:1–9.

Berryman, A. A. 1987. The theory and classification of outbreaks. Pages 3–30 *in* P. Barbosa and J. C. Schultz, editors. Insect outbreaks. Academic Press, San Diego, California, USA.

Blais, J. R. 1965. Spruce budworm outbreaks in the past three centuries in the Laurentide Park, Quebec. Forest Science **11**:130–138.

———. 1968. Regional variation in susceptibility of eastern North American forests to budworm attack based upon history of outbreaks. Forestry Chronical **44**:17–23.

Bormann, F. H., and G. E. Likens. 1981. Patterns and process in a forested ecosystem. Springer-Verlag, New York, New York, USA.

Botkin, D. B., J. F. Janak, and J. R. Wallis. 1972. Some ecological consequences of a computer model of forest growth. Journal of Ecology **60**:849–872.

Bradbury, R. H., R. E. Reichelt, and D. G. Green. 1984. Fractals in ecology: methods and interpretation. Marine Ecology **14**:295–296.

Burrough, P. A. 1981. Fractal dimensions of landscapes and other environmental data. Nature **294**:240–242.

Brown, J. H., and B. A. Maurer. 1989. Macroecology: the division of food and space among species on continents. Science **243**:1145–1150.

Byrd, M. A. 1978. Dispersal and movements of six North American ciconiiforms. Pages 161–185 *in* A. Sprunt IV, J. C. Ogden, and S. Winkler, editors. Wading birds. National Audubon Society, New York, New York, USA.

Carpenter, S. R. 1990. Large-scale perturbations: opportunities for innovation. Ecology **71**:2038–2043.

Carpenter, S. R., and P. R. Leavitt. 1991. Temporal variation in paleoimnological record arising from a tropic cascade. Ecology **72**:277–285.

Caughley, G. 1987. The distribution of eutherian body weights. Oecologia (Berlin) **74**:317–320.

Caughley, G., and C. J. Krebs. 1983. Are big mammals simply little mammals writ large? Oecologia (Berlin) **59**:7–17.

Chambers, J. M., W. S. Cleveland, B. Kleiner, and P. Tukey. 1983. Graphical methods for data analysis. Duxbury, Boston, Massachusetts, USA.

Charnov, E. L. 1976. Optimal foraging: the marginal value theorem. Theoretical Population Biology **9**:129–136.

Clark, W. C. 1985. Scales of climate impacts. Climate Change **7**:5–27.

Clark, J. S. 1988. Effect of climate change on fire regimes in northwestern Minnesota. Nature **334**:233–235.

Clark, W. C., and C. S. Holling. 1979. Process models, equilibrium structures, and population dynamics: on the formulation and testing of realistic theory in ecology. Population Ecology **25**:29–52.

Clark, W. C., D. D. Jones, and C. S. Holling. 1979. Lessons for ecological policy design: a case study of ecosystems management. Ecological Modelling **7**:1–53.

Clark, W. C., and R. E. Munn. 1986. Sustainable development of the biosphere. Cambridge University Press, Cambridge, England.

Costanza, R., F. H. Sklar, and J. W. Day, Jr. 1986. Modelling spatial and temporal succession in the Atchafalya/Terrebonne marsh/estuarine complex in South Louisiana. Pages 387–404 *in* D. A. Wolfe, editor. Estuarine variability. Academic Press, New York, New York, USA.

Dagg, A. I. 1974. Mammals of Ontario. Otter Press, Waterloo, Ontario, Canada.

Davies, R. 1970. Fifth business. Macmillan of Canada, Toronto, Ontario, Canada.

Davis, M. B. 1981. Quaternary history and the stability of forest communities. Pages 132–153 *in* D. C. West, H. H. Shugart, and D. B. Botkin, editors. Forest succession: concepts and application. Springer-Verlag, New York, New York, USA.

———. 1986. Climatic instability, time lags, and community disequilibrium. Pages 269–284 *in* J. Diamond and T. Case, editors. Community ecology. Harper & Row, New York, New York, USA.

———. 1989. Retrospective studies. Pages 71–89 *in* G. E. Likens, editor. Long-term studies in ecology: approaches and alternatives. Springer-Verlag New York, New York, USA.

Delcourt, H. R., P. A. Delcourt, and T. I. Webb. 1983. Dynamic plant ecology: the spectrum of vegetational change in space and time. Quaternary Science Reviews **1**:153–175.

Dickie, L. M., S. R. Kerr, and P. R. Boudreau. 1987. Size-dependent processes underlying regularities in ecosystem structure. Ecological Monographs **57**:233–250.

Dickinson, R. E. 1989. Predicting climate effects. Nature **342**:343–344.

Dunning, J. B., Jr. 1984. Body weights of 686 species of North American birds. Monograph Number 1. Western Bird Banding Association, Cave Creek, Arizona, USA.

Ehrlich, P., D. S. Doblin, and D. Wheye. 1988. The birder's handbook. Simon and Schuster, New York, New York, USA.

Elliott, J. E. 1980. Marx and Schumpeter on capitalism's creative destruction: a comparative restatement. Quarterly Journal of Economics **95**:46–58.

Erskine, A. J. 1977. Birds in boreal Canada: communities, densities and adaptations. Canadian Wildlife Service Report Series Number **41**.

Farlow, J. O. 1976. A consideration of the tropic dynamics of a late Cretaceous large-scale dinosaur community (Old-man Formation). Ecology **57**:841–857.

Follett, K. 1990. The pillars of the Earth. Penguin Books USA, New York, New York, USA.

Forsyth, Adrian. 1985. Mammals of the Canadian wild. Camden House, Charlotte, Vermont, USA.

Frederick, P. C., and M. W. Collopy. 1988. Reproductive ecology of wading birds in relation to water conditions in the Florida Everglades. Technical report number 30. Florida Cooperative Fish and Wildlife Research Unit, School of Forestry Research and Conservation, University of Florida, Gainesville, Florida, USA.

Frontier, S. 1986. Applications of fractal theory to ecology. Pages 335–378 in P. Legendre, editor. Development in numerical ecology. NATO Advanced Research Workshop. NATO, Berlin, Germany.

Fuller, T. K., W. E. Berg, and D. W. Kuehn. 1985. Bobcat home range size and daytime cover-type use in northcentral Minnesota. Journal of Mammalogy 66:568–571.

Gass, C. L., G. Angehr, and J. Centa. 1976. Regulations of food supply by feeding territoriality in the rufous hummingbird. Canadian Journal of Zoology 54:2046–2054.

Gass, C. L., and R. D. Montgomerie. 1981. Hummingbird foraging behavior: decision-making and energy regulation. Pages 159–194 in A. C. Kamil and T. D. Sargent, editors. Foraging behavior: ecological, ethological, and psychological approaches. Garland STPM Press, New York, New York, USA.

Gittleman, J. L., and P. H. Harvey. 1982. Carnivore home-range size, metabolic needs and ecology. Behavioral Ecology and Sociobiology 10:57–63.

Godfrey, W. E. 1986. The birds of Canada. National Museum of Canada. Queen's Printer, Ottawa, Ontario, Canada.

Gould, S. J., and N. Eldredge. 1977. Punctuated equilibria: the tempo and mode of evolution reconsidered. Paleobiology 3:115–151.

Hare, K. F., and J. C. Ritchie. 1972. The boreal bioclimates. Geographical Review 62:633–665.

Harestad, A. S., and F. L. Bunnell. 1979. Home range and body weight — a reevaluation. Ecology 60:389–402.

Harris, G. P. 1980. Temporal and spatial scales in phyto-plankton ecology. Mechanisms, methods, models, and management. Canadian Journal of Fisheries and Aquatic Sciences 37:877–900.

Harris, L. D. 1984. The fragmented forest. University of Chicago Press, Chicago, Illinois, USA.

Heckert, L. 1959. Die klimatischen verhaltnisse in laubwaldern. Zeitschrift für Meteorologie 13:211–233.

Heinselman, M. L. 1981a. Fire and succession in the conifer forest of northern North America. Pages 374–405 in Forest succession: concepts and application. Springer-Verlag, New York, New York, USA.

_____ . 1981b. Fire and succession in the conifer forests of northern North America. USDA Forest Service General Technical Report WO-26.

Holling, C. S. 1963. An experimental component analysis of population processes. Memoirs of the Entomological Society of Canada 32:22–32.

_____ . 1965. The functional response of predators to prey density and its role in mimicry and population regulation. Memoirs of the Entomological Society of Canada 45:1–60.

_____ . 1973. Resilience and stability of ecological systems. Annual Review of Ecology and Systematics 4:1–23.

_____ . 1978. Adaptive environment assessment and management. EIA [Environmental Impact Assessment] Review 2:24–25.

_____ . 1980. Forest insects, forest fires and resilience. Pages 445–464 in H. Mooney, J. M. Bonnicksen, N. L. Christensen, J. E. Lotan, and W. A. Reiners, editors. USDA Forest Service General Technical Report WO-26.

_____ . 1986. Resilience of ecosystems; local surprise and global change. Pages 292–317 in W. C. Clark and R. E. Munn, editors. Sustainable development of the biosphere. Cambridge University Press, Cambridge, England.

_____ . 1987. Simplifying the complex: the paradigms of ecological function and structure. European Journal of Operational Research 30:139–146.

_____ . 1988. Temperate forest insect outbreaks, tropical deforestation and migratory birds. Memoirs of the Entomological Society of Canada 146:21–32.

_____ . 1989. Integrating science for sustainable development. Journal of Business Administration 19:73–83.

_____ . 1992. The role of forest insects in structuring the boreal landscape. Pages 170–191 in H. H. Shugart, R. Leemans, and G. B. Bonan, editors. A systems analysis of the global boreal forest. Cambridge University Press, Cambridge, England.

Hutchinson, G. E. 1965. Ecological theory and the evolutionary play. Yale University Press, New Haven, Connecticut, USA.

Jarman, P. J. 1974. The social organisation of antelope in relation to their ecology. Behaviour 48:215–267.

Jenkins, S. H. 1981. Common patterns in home range-body size relationships of birds and mammals. American Naturalist 118:126–128.

Jenkins, S. H., and P. E. Busher. 1979. Castor canadensis. Mammalian Species 120:1–8.

Kauffman, S. A. 1983. Developmental constraints: internal factors in evolution. Pages 195–225 in N. H. B. C. Goodwin and C. G. Wylie, editors. Development and evolution. Cambridge University Press, Cambridge, England.

Kent, C., and J. Wong. 1982. An index of littoral zone complexity and its measurement. Canadian Journal of Fisheries and Aquatic Sciences 39:847–853.

Kerr, S. R. 1974. Theory of size distribution in

ecological communities. Journal of the Fisheries Research Board of Canada 31:1859–1862.

King, C. M. 1983. *Mustela erminea.* Mammalian Species 195:1–8.

Kirkland, G. L., Jr. 1981. *Sorex dispar* and *Sorex gaspensis.* Mammalian Species 155:1–4.

Krebs, J. R., and N. B. Davies. 1978. Behavioral ecology: an evolutionary approach. Blackwell Scientific, Oxford, England.

Krummel, J. R., R. H. Gardner, G. Sugihara, R. V. O'Neill, and P. R. Coleman. 1987. Landscape patterns in a disturbed environment. Oikos 48:321–324.

LaBarbera, M. 1989. Analyzing body size as a factor in ecology and evolution. Annual Review of Ecology and Systematics 20:97–117.

Landsdowne, J. F., and J. A. Livingston. 1966. Birds of the northern forest. McClelland and Stewart, Atheneum, New York, USA.

Levin, S. A. 1989. Challenges in the development of a theory of community and ecosystem structure and function. Pages 1–394 in J. Roughgarden, R. M. May, and S. A. Levin, editors. Perspectives in ecological theory. Princeton University Press, Princeton, New Jersey, USA.

Lindstedt, S. L., B. J. Miller, and S. W. Buskirk. 1986. Home range, time, and body size in mammals. Ecology 67:413–418.

Loehle, C. 1990. Home range: a fractal approach. Landscape Ecology 5:39–52.

Long, C. A. 1974. *Microsorex hoyi* and *Microsorex thompsoni.* Mammalian Species 33:1–4.

Lovelock, J. 1988. The ages of Gaia. W. W. Norton, New York, New York, USA.

Ludwig, D., D. D. Jones, and C. S. Holling. 1978. Qualitative analysis of insect outbreak systems: the spruce budworm and forest. Journal of Animal Ecology 47:315–332.

MacArthur, R. H. 1958. Population ecology of some warblers of northeastern forests. Ecology 39:599–619.

———. 1972. Geographical ecology. Princeton University Press, Princeton, New Jersey, USA.

MacArthur, R. W., and J. W. MacArthur. 1961. On bird species diversity. Ecology 42:594–598.

MacArthur, R. W., H. Recher, and M. L. Cody. 1966. On the relation between habitat selection and species diversity. American Naturalist 100:319–332.

Macdonald D., editor. 1984. The encyclopedia of mammals. 2. George Allen and Unwin, London, England.

Mace, G. M., and P. H. Harvey. 1983. Energetic constraints on home-range size. American Naturalist 121:120–132.

Mandelbrot, B. B. 1982. The fractal geometry of nature. W. H. Freeman, New York, New York, USA.

Mangel, M., and C. W. Clark. 1986. Towards a unified foraging theory. Ecology 67:1127–1138.

Martin, P. S. 1967. Prehistoric overkill. Pages 75–120 in P. S. Martin and H. E. Wright, Jr., editors. Pleistocene extinctions. Yale University Press, New Haven, Connecticut, USA.

———. 1984. Prehistoric overkill: the global model. Pages 354–403 in P. S. Martin and R. G. Klein, editors. Quaternary extinctions. University of Arizona Press, Tucson, Arizona, USA.

May, R. M. 1977. Thresholds and breakpoints in ecosystems with a multiplicity of stable states. Nature 269:471–477.

———. 1978. The dynamics and diversity of insect faunas. Pages 188–204 in L. A. Mound and N. Waloff, editors. Diversity of insect faunas. Blackwell Scientific, Oxford, England.

May, R. M., J. R. Beddington, C. W. Clark, S. J. Holt, and R. M. Laws. 1979. Management of multispecies fisheries. Science 205:267–277.

McLeod, J. M. 1979. Discontinuous stability in a sawfly life system and its relevance to pest management strategies. Pages 68–81 in W. E. Walters, editor. Selected papers in forest entomology from the XV International Congress of Entomology, USDA Forest Service General Technical Report WO-8.

McNab, B. K. 1963. Bioenergetics and the determination of home range size. American Naturalist 97:133–140.

———. 1983. Energetics, body size, and the limits to endothermy. Journal of Zoology 199:1–29.

McNamee, P. J. 1979. A process model for eastern blackheaded budworm. Canadian Entomologist 115:55–66.

McNamee, P. J., J. M. McLeod, and C. S. Holling. 1981. The structure and behavior of defoliating insect-forest systems. Researches on Population Ecology 23:280–298.

McNaughton, S. J., R. W. Ruess, and S. W. Seagle. 1988. Large mammals and process dynamics in African ecosystems. BioScience 38:794–800.

Mech, D. L. 1974. *Canis lupus.* Mammalian Species 37:1–6.

Menabe, S., and R. T. Wetherald. 1986. Reduction in summer soil wetness induced by an increase in atmospheric carbon dioxide. Science 232:626–628.

Merritt, J. F. 1981. *Clethrionomys gapperi.* Mammalian Species 146:1–9.

Milne, B. T. 1988. Measuring the fractal geometry of landscapes. Applied Mathematics and Computation 27:67–79.

———. 1991. Lessons from applying fractal models to landscape patterns. Pages 199–235 in M. G.

Turner and R. H. Gardner, editors. Ecological studies. Springer-Verlag, New York, New York, USA.

Miller, G. L., and B. W. Carroll. 1989. Modeling vertebrate dispersal distances: alternatives to the geometric distribution. Ecology 70:977–986.

Morris, R. F. 1963. The dynamics of epidemic spruce budworm populations. Memoirs of the Entomological Society of Canada 21.

Morse, D. R., J. H. Lawton, and M. M. Dodson. 1985. Fractal dimension of vegetation and the distribution of arthropod body lengths. Nature 314:731–733.

Morton, S. R. 1990. The impact of European settlement on the vertebrate animals of arid Australia: a conceptual model. Proceedings of the Ecological Society 16:201–213.

Mosimann, J. E., and P. S. Martin. 1975. Simulating overkill by Paleoindians. American Scientist 63:304–313.

Myers, J. H. 1988. Can a general hypothesis explain population cycles of forest Lepidoptera. Pages 179–242 in J. B. Cragg, editor. Advances in Ecological Research, Academic Press, London, England.

Naiman, R. J. 1988. Animal influences on ecosystem dynamics. BioScience 38:750–752.

O'Neill, R. V., D. L. DeAngelis, J. B. Waide, and T. F. H. Allen. 1986. A hierarchical concept of ecosystems. Princeton University Press, Princeton, New Jersey, USA.

Odum, E. P. 1985. Trends expected in stressed ecosystems. BioScience 35:419–422.

Orians, G. H. 1980. General theory and applications to human behavior. Pages 49–66 in J. S. Lockard, editor. Evolution of human social behavior. Elsevier, New York, New York, USA.

Owen-Smith, N. 1989. Megafaunal extinctions: the conservation message from 11,000 years B.C. Conservation Biology 3:405–411.

Paine, R. T. 1966. Food web complexity and species diversity. American Naturalist 100:65–75.

_____ . 1974. Intertidal community structure: experimental studies on the relationship between a dominant competitor and its principal predator. Oecologia (Berlin) 15:93–120.

Palmer, R. S. 1962. Handbook of North American birds. Yale University Press, New Haven, Connecticut, USA.

Pastor, J., and D. J. Mladenoff. 1992. The southern boreal northern-hardwood forest border. Pages 216–240 in H. H. Shugart, R. Leemans, and G. B. Bonan, editors. A systems analysis of the global boreal forest. Cambridge University Press, Cambridge, England.

Peters, R. H. 1983. The ecological implications of body size. Cambridge University Press, Cambridge, England.

Peterson, R. L. 1955. North American moose. University of Toronto Press, Toronto, Ontario, Canada.

_____ . 1966. The mammals of Eastern Canada. Oxford University Press, Toronto, Ontario, Canada.

Pickett, S. T. A., J. Kolasa, J. J. Armesto, and S. L. Collins. 1986. Disturbance and ecological hierarchy. Journal of Theoretical Biology 54:1–26.

Powell, R. A. 1981. Martes pennanti. Mammalian Species 156:1–6.

Prigogine, I., and G. Nicolas. 1971. Biological order, structure and instabilities. Quarterly Review of Biophysics 4:107–148.

Radinsky, L. 1978. Evolution of brain size in carnivores and ungulates. American Naturalist 112:815–831.

Rotenberry, J. T., and J. A. Wiens. 1980. Habitat structure, patchiness, and avian communities in North American steppe vegetation: a multivariate analysis. Ecology 61:1228–1250.

Royama, T. 1984. Population dynamics of the spruce budworm Choristoneura fumiferana. Ecological Monographs 54:429–462.

Schaffer, W. M. 1985. Order and chaos in ecological systems. Ecology 66:93–106.

Schindler, D. W. 1990. Experimental perturbations of whole lakes as tests of hypotheses concerning ecosystem structure and function. Oikos 57:25–41.

Schmidt-Nielson, K. 1984. Scaling: why is animal size so important? Cambridge University Press, Cambridge, England.

Schoener, T. W. 1968. Sizes of feeding territories among birds. Ecology 49:123–141.

Schoener, T. W., and D. H. Jansen. 1968. Notes on environmental determinants of tropical versus temperate insect size patterns. American Naturalist 102:207–224.

Schumpeter, J. A. 1950. Capitalism, socialism and democracy. Harper & Row, New York, New York, USA.

Schwinghamer, P. 1981. Characteristic size distributions of integral benthic communities. Canadian Journal of Fisheries and Aquatic Sciences 38:1255–1263.

_____ . 1985. Observations on size-structure and pelagic coupling of some shelf and abyssal benthic communities. Pages 347–359 in P. E. Gibbs, editor. Proceedings from the Marine Biology Symposium, Plymouth, UK, 19 September 1984. Cambridge University Press, Cambridge, England.

Sheldon, R. W., W. H. Sutcliffe, Jr., and M. A. Paranjape. 1977. Structure of pelagic food chain and relationship between plankton and fish production. Journal of the Fisheries Research Board of Canada **34**:2344–2353.

Silvert, W., and T. Platt. 1978. Energy flux in the pelagic ecosystem: a time dependent equation. Limnology and Oceanography **23**:813–816.

Sinclair, A. R. E., P. D. Olsen, and T. D. Redhead. 1990. Can predators regulate small mammal populations? Oikos **59**:382–392.

Smith, T. M., and H. H. Shugart. 1987. Territory size variation in the ovenbird: the role of habitat structure. Ecology **68**:695–704.

Sokal, R. R., and J. F. Rohlf. 1969. Biometry. W. H. Freeman, New York, New York, USA.

Soulé, M. E., and B. A. Wilcox. 1980. Conservation biology: an evolutionary-ecological perspective. Sinauer, Sunderland, Massachusetts, USA.

Southwood, T. R. E. 1977. Habitat, the templet for ecological strategies? Journal of Animal Ecology **46**:337–365.

Speirs, J. M. 1985. Birds of Ontario. Volumes I and II. Natural Heritage/Natural History Inc., Toronto, Ontario, Canada.

Sprugel, D. G., and F. H. Bormann. 1981. Natural disturbance and the steady state in high-altitude balsam fir forests. Science **211**:390–393.

Steele, J. H. 1985. A comparison of terrestrial and marine systems. Nature **313**:355–358.

_____ . 1989. The ocean 'landscape'. Landscape Ecology **3**:185–192.

Strong, D. R. J., D. Simberloff, L. G. Abele, and A. B. Thistle, 1984. Ecological communities: conceptual issues and the evidence. Princeton University Press, Princeton, New Jersey, USA.

Swihart, R. K. 1986. Home range-body mass allometry in rabbits and hares (Leporidae). Acta Theriologica **31**:139–148.

Swihart, R. K., N. A. Slade, and B. J. Bergstrom. 1988. Relating body size to the rate of home range use in mammals. Ecology **69**:393–399.

Terres, J. K. 1980. The Audubon Society encyclopedia of North American birds. Alfred A. Knopf, New York, New York, USA.

Tranzmann, A. W. 1981. *Alces alces*. Mammalian Species **154**:1–7.

Urban, D. L., R. V. O'Neill, and H. H. J. Shugart. 1987. Landscape ecology. BioScience **37**:119–127.

Vezina, A. F. 1985. Empirical relationships between predator and prey size among terrestrial vertebrate predators. Oecologia (Berlin) **67**:555–565.

Vitousek, P. M. 1990. Biological invasions and ecosystem processes: towards an integration of population biology and ecosystem studies. Oikos **57**:7–13.

Vitousek, P. M., and P. A. Matson. 1984. Mechanisms of nitrogen retention in forest ecosystems: a field experiment. Science **225**:51–52.

Walker, B. H. 1981. Is succession a viable concept in African savanna ecosystems? Pages 431–447 *in* D. C. West, H. H. Shugart, and D. B. Botkin, editors. Forest succession: concepts and application. Springer-Verlag, New York, New York, USA.

Walker, B. H., D. Ludwig, C. S. Holling, and R. M. Peterman. 1969. Stability of semi-arid savanna grazing systems. Journal of Ecology **69**:473–498.

Walters, C. J. 1986. Adaptive management of renewable resources. McGraw Hill, New York, New York, USA.

Walters, C., L. Gunderson, and C. S. Holling. 1992. Experimental policies for water management in the Everglades. Ecological Applications **2**:189–202.

Walters, C. J., and C. S. Holling. 1990. Large-scale management experiments and learning by doing. Ecology **71**:2060–2068.

Ward, R. M. P., and C. J. Krebs. 1985. Behavioural responses of lynx to declining snowshoe hare abundance. Canadian Journal of Zoology **63**:2817–2824.

Wasserman, F. E. 1980. Territorial behavior in a pair of White-Throated Sparrows. Wilson Bulletin **92**:74–87.

Webb, S. D. 1984. Ten million years of mammal extinctions in North America. Pages 189–210 *in* P. S. Martin and R. G. Klein, editors. Quaternary extinctions. University of Arizona Press, Tucson, Arizona, USA.

Wells-Gosling, N., and L. R. Heaney. 1984. *Glaucomys sabrinus*. Mammalian Species **229**:1–8.

West, D. C., H. H. Shugart, and D. B. Botkin. 1981. Forest succession: concepts and application. Springer-Verlag, New York, New York, USA.

Whitaker, J. O., Jr., and R. E. Wrigley. 1972. *Napaeozapus insignis*. Mammalian Species **14**:1–6.

Whittaker, R. H. 1975. Communities and ecosystems. Macmillan, New York, New York, USA.

Wiens, J. A. 1984. On understanding a nonequilibrium world: myth and reality in community patterns and processes. Pages 439–457 *in* D. R. Strong, Jr., D. Simberloff, L. Abele, and A. R. Thistle, editors. Ecological communities: conceptual issues and the evidence. Princeton University Press, Princeton, New Jersey, USA.

_____ . 1989. Spatial scaling in ecology. Functional Ecology **3**:385–397.

_____ . 1990. On the use of "grain size" in ecology. Functional Ecology **4**:720.

Wiens, J. A., and B. T. Milne. 1989. Scaling of "landscapes" in landscape ecology, or, landscape ecology from a beetle's perspective. Landscape Ecology **3**:87–96.

Wiens, J. A., and J. T Rotenberry. 1980. Patterns of morphology and ecology in grassland and shrub-steppe bird populations. Ecological Monographs **50**:287–308.

Wiens, J. A., J. T. Rotenberry, and B. Van Horne. 1985. Territory size variations in shrubsteppe birds. Auk **102**:500–505.

Wiens, J. A., and J. T. Rotenberry. 1981. Habitat associations and community structure of birds in shrubsteppe environments. Ecological Monographs **51**:21–41.

Willson, M. F. 1974. Avian community organization and habitat structure. Ecology **55**:1017–1029.

Wooding, F. H. 1982. Wild mammals of Canada. McGraw-Hill Ryerson, New York, New York, USA.

Woods, C. A. 1973. *Erethizon dorsatum*. Mammalian Species **29**:1–6.

Zach, R., and B. Falls. 1979. Foraging and territoriality of male ovenbirds (Aves: Parulidae) in a heterogeneous habitat. Journal of Animal Ecology **48**:33–52.

BIRDS OF THE BOREAL FOREST

Species and body masses of boreal forest birds found east of the Manitoba-Ontario border in pure conifer or mixed conifer stands.* (Entries are listed in order of increasing body mass.)

Common name	Family	Scientific name	Species no.	Body mass (g)†	Log_{10} mass	Body mass difference index‡ $[(M_{n+1} - M_{n-1})/M_n^{1.3}]$
Ruby-throated Hummingbird	Trochilidae	*Archilochus colubris*	1	3.15	0.498	. . .
Golden-crowned Kinglet	Muscicapidae	*Regulus satrapa*	2	6.20	0.792	0.3266
Ruby-crowned Kinglet	Muscicapidae	*Regulus calendula*	3	6.65	0.823	0.1789
American Redstart	Emberizidae	*Setophaga ruticilla*	4	8.30	0.919	0.1117
Brown Creeper	Certhiidae	*Certhia americana*	5	8.40	0.924	0.0189
Parula Warbler	Emberizidae	*Parula americana*	6	8.60	0.934	0.0183
Magnolia Warbler	Emberizidae	*Dendroica magnolia*	7	8.70	0.940	0.0090
Nashville Warbler	Emberizidae	*Vermivora ruficapilla*	8	8.75	0.942	0.0060
Black-throated Green Warbler	Emberizidae	*Dendroica virens*	9	8.80	0.944	0.0089
Winter Wren	Troglodytidae	*Troglodytes troglodytes*	10	8.90	0.949	0.0467
Chestnut-sided Warbler	Emberizidae	*Dendroica pensylvanica*	11	9.60	0.982	0.0449
Blackburnian Warbler	Emberizidae	*Dendroica fusca*	12	9.75	0.989	0.0104
Boreal Chickadee	Paridae	*Parus hudsonicus*	13	9.80	0.991	0.0026
Red-breasted Nuthatch	Sittidae	*Sitta canadensis*	14	9.80	0.991	0.0103
Tennessee Warbler	Emberizidae	*Vermivora peregrina*	15	10.00	1.000	0.0150
Common Yellowthroat	Emberizidae	*Geothlypis trichas*	16	10.10	1.004	0.0074
Black-throated Blue Warbler	Emberizidae	*Dendroica caerulescens*	17	10.15	1.006	0.0098
Least Flycatcher	Tyrannidae	*Empidonax minimus*	18	10.30	1.013	0.0072
Palm Warbler	Emberizidae	*Dendroica palmarum*	19	10.30	1.013	0.0048
Canada Warbler	Emberizidae	*Wilsonia canadensis*	20	10.40	1.017	0.0238
Black-capped Chickadee	Paridae	*Parus atricapillus*	21	10.80	1.033	0.0181
Black and White Warbler	Emberizidae	*Mniotilta varia*	22	10.80	1.033	0.0091
Cape May Warbler	Emberizidae	*Dendroica tigrina*	23	11.00	1.041	0.0354
Yellow-bellied Flycatcher	Tyrannidae	*Empidonax flaviventris*	24	11.60	1.064	0.0413
Chipping Sparrow	Emberizidae	*Spizella passerina*	25	12.00	1.079	0.0237
Philadelphia Vireo	Vireonidae	*Vireo philadelphicus*	26	12.20	10.86	0.0194
Mourning Warbler	Emberizidae	*Oporonis philadelphia*	27	12.50	1.097	0.0131
Myrtle Warbler/Yellow-rumped Warbler	Emberizidae	*Dendroica coronata*	28	12.55	1.099	0.0019
Bay-breasted Warbler	Emberizidae	*Dendroica castanea*	29	12.55	1.099	0.0168
Blackpoll Warbler	Emberizidae	*Dendroica striata*	30	13.00	1.114	0.0160
Common Redpoll	Fringillidae	*Carduelis flammea*	31	13.00	1.114	0.0143
Traill's Flycatcher/Adler Flycatcher	Tyrannidae	*Empidonax alnorum*	32	13.40	1.127	0.0377
Eastern Wood-Pewee	Tyrannidae	*Contopus virens*	33	14.10	1.149	0.0385
Pine Siskin		*Carduelis pinus*	34	14.60	1.164	0.0766
Solitary Vireo	Fringillidae	*Vireo solitarus*	35	16.60	1.220	0.0545
Red-eyed Vireo	Vireonidae	*Vireo olivaceus*	36	16.70	1.223	0.0206
Lincoln's Sparrow	Vireonidae	*Melospiza lincolnii*	37	17.40	1.241	0.0268
Northern Waterthrush	Emberizidae	*Seirus noveboracensis*	38	17.80	1.250	0.0474
Ovenbird	Emberizidae	*Seirus aurocapillus*	39	19.40	1.288	0.0699
White-breasted Nuthatch	Emberizidae	*Sitta carolinensis*	40	21.10	1.324	0.1044
Purple Finch	Sittidae	*Carpodacus purpureus*	41	24.90	1.396	0.0735

*The following references were used to produce the above species list: Godfrey 1966, Landsdowne and Livingston 1966, Erskine 1977, Speirs 1985.

  † All body masses, with the exception of the Clay-colored Sparrow and the Chipping Sparrow, were taken from Dunning (1984). When body masses were listed for both males and females, an average of the two was used. The body mass for the Clay-colored Sparrow was taken from Terres (1980). The body mass for the Chipping Sparrow was obtained from a specimen caught by Richard J. Cannings, Department of Zoology, University of British Columbia, in May 1988.

  ‡In this index, M_n is the body mass of the nth species in rank order of increasing size and 1.3 is an exponent sufficient to detrend the data.

APPENDIX 1 (*continued*)

Common name	Family	Scientific name	Species no.	Body mass (g)†	Log_{10} mass	Body mass difference index‡ $[(M_{n+1} - M_{n-1})/M_n^{1.3}]$
White-throated Sparrow	Fringillidae	*Zonotrichia albicollis*	42	25.90	1.413	0.0247
White-winged Crossbill	Emberizidae	*Loxia leucoptera*	43	26.60	1.425	0.0155
Downy Woodpecker	Fringillidae	*Picoides pubescens*	44	27.00	1.431	0.0276
Scarlet Tanager	Picidae	*Piranga olivacea*	45	28.60	1.456	0.0307
White-crowned Sparrow	Emberizidae	*Zonotrichia leucophrys*	46	29.40	1.468	0.0271
Swainson's Thrush	Muscicapidae	*Catharus ustulatus*	47	30.80	1.489	0.0186
Hermit Thrush	Muscicapidae	*Catharus guttatus*	48	31.00	1.491	0.0046
Veery	Muscicapidae	*Catharus fuscescens*	49	31.20	1.494	0.0097
Cedar Waxwing	Bombycillidae	*Bombycilla cedrorum*	50	31.85	1.503	0.0100
Olive-sided Flycatcher	Tyrannidae	*Contopus borealis*	51	32.10	1.507	0.0050
Fox Sparrow	Emberizidae	*Passerella iliaca*	52	32.30	1.509	0.0076
Gray-cheeked Thrush	Muscicapidae	*Catharus minimus*	53	32.80	1.516	0.0449
Red Crossbill	Fringillidae	*Loxia curvirostra*	54	36.50	1.562	0.1192
Rose-breasted Grosbeak	Emberizidae	*Pheucticus ludovicianus*	55	45.60	1.659	0.0962
Yellow-bellied Sapsucker	Picidae	*Sphyrapicus varius*	56	50.30	1.702	0.0344
Solitary Sandpiper	Scolopacidae	*Tringa solitaria*	57	51.20	1.709	0.0159
Whip-poor-will	Caprimulgidae	*Caprimulgus vociferous*	58	52.95	1.724	0.0299
Pine Grosbeak	Fringillidae	*Pinicola enucleator*	59	56.40	1.751	0.0341
Evening Grosbeak	Fringillidae	*Coccothraustes vespertinus*	60	59.40	1.774	0.0166
Rusty Blackbird	Emberizidae	*Euphagus carolinus*	61	59.75	1.776	0.0304
Northern Shrike	Laniidae	*Lanius excubitor*	62	65.60	1.817	0.0256
Northern Three-toed Woodpecker	Picidae	*Picoides tridactylus*	63	65.65	1.817	0.0028
Hairy Woodpecker	Picidae	*Picoides villosus*	64	66.25	1.821	0.0157
Black-backed Three-toed Woodpecker	Picidae	*Picoides arcticus*	65	69.30	1.841	0.0196
Canada Jay	Corvidae	*Perisoreus canadensis*	66	71.10	1.852	0.0313
American Robin	Muscicapidae	*Turdus migratorius*	67	77.30	1.888	0.348
Lesser Yellowlegs	Scolopacidae	*Tringa flavipes*	68	81.00	1.908	0.0183
Saw-whet Owl	Strigidae	*Aegolius acadicus*	69	82.85	1.918	0.0186
Blue Jay	Corvidae	*Cyanocitta cristata*	70	86.80	1.939	0.0925
Common Grackle	Emberizidae	*Quiscalus quiscula*	71	113.50	2.055	0.0750
Common Snipe	Scolopacidae	*Gallinago gallinago*	72	122.00	2.086	0.0359
Yellow-shafted Flicker/Northern Flicker	Picidae	*Colaptes auratus*	73	132.00	2.121	0.0210
Boreal Owl	Strigidae	*Aegolius funereus*	74	134.00	2.127	0.0112
Sharp-shinned Hawk	Accipitridae	*Accipiter striatus*	75	138.50	2.141	0.0847
Pigeon Hawk/Merlin	Falconidae	*Falco columbarius*	76	185.50	2.268	0.1389
Long-eared Owl	Strigidae	*Asio otus*	77	262.00	2.418	0.0729
Pileated Woodpecker	Picidae	*Dryocopus pileatus*	78	287.00	2.458	0.0383
Hawk-Owl	Strigidae	*Surnia ulula*	79	322.00	2.508	0.0884
Common Crow	Corvidae	*Corvus brachyrhynchos*	80	448.00	2.651	0.0468
Horned Grebe	Podicipedidae	*Podiceps auritus*	81	453.00	2.656	0.0025
Broad-winged Hawk	Accipitridae	*Buteo platypterus*	82	455.00	2.658	0.0074
Spruce Grouse	Phasianidae	*Dendragapus canadensis*	83	474.00	2.676	0.0404
Ruffed Grouse	Phasianidae	*Bonasa umbellus*	84	576.50	2.761	0.0350
Hooded Merganser	Anatidae	*Lophodytes cucullatus*	85	610.00	2.785	0.0335
Barred Owl	Strigidae	*Strix varia*	86	716.50	2.855	0.0333
Peregrine Falcon	Falconidae	*Falco peregrinus*	87	781.50	2.893	0.0180
Lesser Scaup	Anatidae	*Aythya affinis*	88	820.00	2.914	0.0169
Sharp-tailed Grouse	Phasianidae	*Tympanuchus phasianellus*	89	885.00	2.947	0.0118
Common Goldeneye	Anatidae	*Bucephala clangula*	90	900.00	2.954	0.0197
Red-breasted Merganser	Anatidae	*Mergus serrator*	91	1021.50	3.009	0.0152
Goshawk	Accipitridae	*Accipiter gentilis*	92	1024.50	3.011	0.0116
Great Gray Owl	Strigidae	*Strix nebulosa*	93	1116.50	3.048	0.0111

(continued)

APPENDIX 1 (*continued*)

Common name	Family	Scientific name	Species no.	Body mass (g)†	Log$_{10}$ mass	Body mass difference index‡ [$(M_{n+1} - M_{n-1})/M_n^{1.3}$]
Red-tailed Hawk	Accipitridae	*Buteo jamaicensis*	94	1126.00	3.052	0.0089
Common Raven	Corvidae	*Corvus corax*	95	1199.00	3.079	0.0223
White-winged Scoter	Anatidae	*Melanitta fusca*	96	1350.00	3.130	0.0244
Osprey	Accipitridae	*Pandion haliaetus*	97	1485.50	3.172	0.0146
Great Horned Owl	Strigidae	*Bubo virginianus*	98	1543.50	3.189	0.0648
Great Blue Heron	Ardeidae	*Ardea herodias*	99	2390.00	3.378	0.0934
Canada Goose	Anatidae	*Branta canadensis*	100	3847.50	3.585	0.0381
Common Loon	Gaviidae	*Gavia immer*	101	4134.00	3.616	. . .

APPENDIX 2

MAMMALS OF THE BOREAL FOREST

Species and body masses of boreal forest mammals found east of the Manitoba-Ontario border in pure conifer or mixed conifer forests.*

Common name	Scientific name	Species no.	Body mass (g)	Log_{10} mass	Body mass difference index‡ $[(M_{n+1} - M_{n-1}/M_n^{1.1}]$
Gaspe shrew	*Sorex gaspensis*	1	3.685	0.566	. . .
Pygmy shrew	*Microsorex hoyi*	2	4.110	0.614	0.106
Northern long-eared bat	*Myotis septentrionalis*	3	4.252	0.629	0.176
Common shrew	*Sorex cinereus*	4	5.244	0.720	0.161
Gray long-tailed shrew	*Sorex dispar*	5	5.500	0.740	0.282
Smoky shrew	*Sorex fumeus*	6	8.363	0.922	0.415
Water shrew	*Sorex palustris*	7	13.182	1.120	0.379
Woodland jumping mouse	*Napaeozapus insignis*	8	22.450	1.351	0.173
Deer mouse	*Peromyscus maniculatus*	9	22.963	1.361	0.027
Short-tailed shrew	*Blarina brevicauda*	10	24.522	1.390	0.048
Gapper's red-backed vole	*Clethrionomys gapperi*	11	26.932	1.430	0.042
Southern bog lemming	*Synaptomys cooperi*	12	28.300	1.452	0.015
Hoary bat	*Lasiurus cinereus*	13	28.350	1.453	0.044
Northern bog lemming	*Synaptomys borealis*	14	33.000	1.519	0.111
Yellownose vole	*Microtus chrotorrhinus*	15	42.525	1.629	0.073
Least chipmunk	*Eutamias minimus*	16	43.942	1.643	0.011
Least weasel	*Mustela nivalis*	17	44.509	1.648	0.153
Ermine	*Mustela erminea*	18	80.797	1.907	0.183
Northern flying squirrel	*Glaucomys sabrinus*	19	104.895	2.021	0.173
Red squirrel	*Tamiasciurus hudsonicus*	20	191.362	2.282	0.364
Marten	*Martes americana*	21	839.145	2.924	0.248
Mink	*Mustela vison*	22	1224.698	3.088	0.073
Snowshoe hare	*Lepus americanus*	23	1496.853	3.175	0.094
Striped skunk	*Mephitis mephitis*	24	2642.173	3.422	0.082
Fisher	*Martes pennanti*	25	3118.445	3.494	0.128
Porcupine	*Erethizon dorsatum*	26	8504.850	3.930	0.103
River otter	*Lutra canadensis*	27	9071.840	3.958	0.017
Lynx	*Lynx canadensis*	28	10149.121	4.006	0.029
Wolverine	*Gulo gulo*	29	12303.683	4.090	0.030
Coyote	*Canis latrans*	30	14061.352	4.148	0.061
Beaver	*Castor canadensis*	31	23995.016	4.380	0.096
Wolf	*Canis lupus*	32	43204.638	4.636	0.103
White-tailed deer	*Odocoileus virginianus*	33	86416.533	4.937	0.067
Woodland caribou	*Rangifer tarandus caribou*	34	105686.930	5.024	0.050
Black bear	*Ursus americanus*	35	169643.400	5.230	0.107
Moose	*Alces alces andersoni*	36	481828.100	5.683	. . .

*The following references were used to produce the above species list: Balch 1933/34, Peterson 1955, 1966, Whitaker and Wrigley 1972, Woods 1973, Banfield 1974, Dagg 1974, Long 1974, Mech 1974, Bekoff 1977, Jenkins and Busher 1979, Kirkland 1981, Merritt 1981, Tranzmann 1981, Wooding 1982, King 1983, Macdonald 1984, Wells-Gosling and Heaney 1984, Forsyth 1985.

†Most body masses were taken from the Forsyth (1985) reference above and confirmed by other references where available. The exceptions were the gray long-tailed shrew, woodland jumping mouse, southern bog lemming, and northern bog lemming, which were taken from the Banfield (1974) and Peterson (1955, 1966) references. When two different masses were given for females and for males an average mass was used.

‡Index as in Appendix 1.

APPENDIX 3

BIRDS OF THE BOREAL PRAIRIE

Species and body masses of boreal prairie birds found east of the Alberta short-grass prairie.*

Common name	Family name	Scientific name	Species no.	Body mass (g)†	Log₁₀ mass	Body mass difference index‡ [$(M_{n+1} - M_{n-1})/M_n^{1.3}$]
Yellow Warbler	Emberizidae	*Dendroica petechia*	1	9.500	0.978	. . .
Common Yellowthroat	Emberizidae	*Geothlypis trichas*	2	10.100	1.004	0.0396
Least Flycatcher	Tyrannidae	*Empidonax minimus*	3	10.300	1.013	0.0241
Clay-colored Sparrow	Emberizidae	*Spizella pallida*	4	10.600	1.025	0.0232
Black-capped Chickadee	Paridae	*Parus atricapillus*	5	10.800	1.033	0.0136
House Wren	Troglodytidae	*Troglodytes aedon*	6	10.900	1.037	0.0090
Brewer's Sparrow	Emberizidae	*Spizella breweri*	7	11.000	1.041	0.0337
Long-billed Marsh Wren/ Marsh Wren	Cinclidae	*Cistothorus palustris*	8	11.660	1.067	0.0410
Chipping Sparrow	Emberizidae	*Spizella passerina*	9	12.000	1.079	0.0451
Western Wood Pewee	Tyrannidae	*Contopus sordidulus*	10	12.800	1.107	0.0327
American Goldfinch	Fringillidae	*Carduelis tristis*	11	12.900	1.111	0.0486
Violet-green Swallow	Hirundinidae	*Tachycineta thallassina*	12	14.150	1.151	0.0543
Bank Swallow	Hirundinidae	*Riparia riparia*	13	14.600	1.164	0.0536
Rough-winged Swallow	Hirundinidae	*Stelgidopteryx serripennis*	14	15.900	1.201	0.0521
Rock Wren	Troglodytidae	*Salpinctes obsoletus*	15	16.500	1.217	0.0418
Baird's Sparrow	Emberizidae	*Ammodramus bairdii*	16	17.500	1.243	0.0508
Barn Swallow	Hirundinidae	*Hirundo rustica*	17	18.600	1.270	0.0313
Chestnut-collared Longspur	Emberizidae	*Calcarius ornatus*	18	18.900	1.276	0.0318
Savannah Sparrow	Emberizidae	*Passerculus sandwichensis*	19	20.050	1.302	0.0243
Tree Swallow	Hirundinidae	*Tachycineta bicolor*	20	20.100	1.303	0.0142
Song Sparrow	Emberizidae	*Melospiza melodia*	21	20.750	1.317	0.0213
Say's Phoebe	Tyrannidae	*Sayornis saya*	22	21.200	1.326	0.0160
Cliff Swallow	Hirundinidae	*Hirundo pyrrhonota*	23	21.600	1.334	0.0368
McCown's Longspur	Emberizidae	*Calcarius mccownii*	24	23.200	1.365	0.0621
Sprague's Pipit	Bombycillidae	*Anthus spragueii*	25	25.300	1.403	0.0375
Vesper Sparrow	Emberizidae	*Pooecetes gramineus*	26	25.700	1.410	0.0250
Downy Woodpecker	Picidae	*Picoides pubescens*	27	27.000	1.431	0.0372
Mountain Bluebird	Muscicapidae	*Sialia currucoides*	28	28.400	1.453	0.0258
Veery	Muscicapidae	*Catharus fuscescens*	30	31.200	1.494	0.0268
Horned Lark	Alaudidae	*Eremophila alpestris*	31	31.350	1.496	0.0074
Cedar Waxwing	Bombycillidae	*Bombycilla cedrorum*	32	31.850	1.503	0.0267
Northern Oriole	Emberizidae	*Icterus galbula*	33	33.750	1.528	0.0521
Gray Catbird	Mimidae	*Dumetella carolinensis*	34	36.900	1.567	0.0353
Lark Bunting	Emberizidae	*Calamospiza melanocorys*	35	37.600	1.575	0.0233
Eastern Kingbird	Tyrannidae	*Tyrannus tyrannus*	36	39.500	1.597	0.0168
Western Kingbird	Tyrannidae	*Tyrannus verticalis*	37	39.600	1.598	0.0075
Spotted Sandpiper	Scolopacidae	*Actitis macularia*	38	40.400	1.606	0.0073
Rufous-sided Towhee	Emberizidae	*Pipilo erythrophthalmus*	39	40.500	1.607	0.0134
Bobolink	Emberizidae	*Dolichonyx oryzivorus*	40	42.050	1.624	0.0217
Sage Thrasher	Mimidae	*Oreoscoptes montanus*	41	43.300	1.636	0.0138
Brown-headed Cowbird	Emberizidae	*Molothrus ater*	42	43.900	1.642	0.0300
Loggerhead Shrike	Laniidae	*Lanius ludovicianus*	43	47.400	1.676	0.0573
Red-winged Blackbird	Emberizidae	*Agelaius phoeniceus*	44	52.550	1.721	0.0452
Piping Plover	Charadriidae	*Charadrius melodus*	45	55.200	1.742	0.0370
Wilson's Phalarope	Scolopacidae	*Phalaropus tricolor*	46	59.350	1.773	0.0337
Common Nighthawk	Caprimulgidae	*Chordeiles minor*	47	62.000	1.792	0.0154
Brewer's Blackbird	Emberizidae	*Euphagus cyanocephalus*	48	62.650	1.797	0.0115
Yellow-headed Blackbird	Emberizidae	*Xanthocephalus xanthocephalus*	49	64.500	1.810	0.0118
Black Tern	Laridae	*Chlidonias niger*	50	65.300	1.815	0.0188

(*continued*)

APPENDIX 3 *(continued)*

Common name	Family name	Scientific name	Species no.	Body mass (g)†	Log$_{10}$ mass	Body mass difference index‡ [$(M_{n+1} - M_{n-1})/M_n^{1.3}$]
Brown Thrasher	Mimidae	*Toxostoma rufum*	51	68.800	1.838	0.0380
Sora	Rallidae	*Porzana caroina*	52	74.600	1.873	0.0313
American Robin	Muscicapidae	*Turdus migratorius*	53	77.300	1.888	0.0771
Killdeer	Charadriidae	*Charadrius vociferus*	54	96.550	1.985	0.0536
Western Meadowlark	Emberizidae	*Sturnella neglecta*	55	97.700	1.990	0.0296
Mountain Plover	Haematopodidae	*Charadrius montanus*	56	108.000	2.033	0.0359
Common Grackle	Emberizidae	*Quiscalus quiscula*	57	113.500	2.055	0.0160
American Kestrel	Falconidae	*Falco sparverius*	58	115.500	2.063	0.0115
Mourning Dove	Columbidae	*Zenaida macroura*	59	119.000	2.076	0.0090
Common Tern	Laridae	*Sterna hirundo*	60	120.000	2.079	0.0059
Common Snipe	Scolopacidae	*Gallinago gallinago*	61	122.000	2.086	0.0233
Northern Flicker	Picidae	*Colaptes auratus*	62	132.000	2.121	0.0455
Belted Kingfisher	Alcedinidae	*Ceryle alcyon*	63	148.000	2.170	0.0347
Burrowing Owl	Strigidae	*Athene cunicularia*	64	155.000	2.190	0.0305
Upland Sandpiper	Scolopacidae	*Bartramia longicauda*	65	169.500	2.229	0.0285
Black-billed Magpie	Corvidae	*Pica pica*	66	177.500	2.249	0.0542
Willet	Scolopacidae	*Catoptrophorus semipalmatus*	67	215.000	2.332	0.0785
Long-eared Owl	Strigidae	*Asio otus*	68	262.000	2.418	0.0467
Franklin's Gull	Laridae	*Larus pipixcan*	69	280.000	2.447	0.0231
Eared Grebe	Podicipedidae	*Podiceps nigricollis*	70	297.000	2.473	0.0220
American Avocet	Recurvirostridae	*Recurvirostra americana*	71	316.000	2.500	0.0248
American Green-winged Teal	Anatidae	*Anas crecca*	72	341.000	2.533	0.0155
Short-eared Owl	Strigidae	*Asio flammeus*	73	346.500	2.540	0.0150
Marbled Godwit	Scolopacidae	*Limosa fedoa*	74	371.000	2.569	0.0178
Cinnamon Teal	Anatidae	*Anas cyanoptera*	75	385.500	2.586	0.0065
Blue-winged Teal	Anatidae	*Anas discors*	76	386.000	2.587	0.0239
Marsh Hawk (Northern Harrier)	Accipitridae	*Circus cyaneus*	77	440.500	2.644	0.0205
Pied-billed Grebe	Podicipedidae	*Podilymbus podiceps*	78	442.000	2.645	0.0027
Common Crow	Corvidae	*Corvus brachyrhynchos*	79	448.000	2.651	0.0039
Horned Grebe	Podicipedidae	*Podiceps auritus*	80	453.000	2.656	0.0248
Ring-billed Gull	Laridae	*Larus delawarensis*	81	518.500	2.715	0.0271
Ruddy Duck	Anatidae	*Oxyura jamaicensis*	82	544.500	2.736	0.0189
Long-billed Curlew	Scolopacidae	*Numenius americanus*	83	586.500	2.768	0.0162
California Gull	Laridae	*Larus californicus*	84	609.000	2.785	0.0064
Northern Shoveler	Anatidae	*Anas clypeata*	85	613.000	2.787	0.0078
American Coot	Rallidae	*Fulica americana*	86	642.000	2.808	0.0080
American Bittern	Ardeidae	*Botaurus lentiginosus*	88	706.000	2.849	0.0212
American Wigeon	Anatidae	*Anas americana*	89	755.500	2.878	0.0207
Lesser Scaup	Anatidae	*Aythya affinis*	90	820.000	2.914	0.0208
Black-crowned Night Heron	Ardeidae	*Nycticorax nycticorax*	91	883.000	2.946	0.0096
Sharp-tailed Grouse	Phasianidae	*Tympanuchus phasianellus*	92	885.00	2.947	0.0054
Gadwall	Anatidae	*Anas strepera*	93	919.500	2.964	0.0145
Swainson's Hawk	Accipitridae	*Buteo swainsoni*	94	988.500	2.995	0.0116
Pintail	Anatidae	*Anas acuta*	95	1010.500	3.005	0.0070
Redhead	Anatidae	*Aythya americana*	96	1045.000	3.019	0.0085
Mallard	Anatidae	*Anas platyrhynchos*	97	1082.000	3.034	0.0092
Red-tailed Hawk	Accipitridae	*Buteo jamaicensis*	98	1126.000	3.052	0.0068
Ferruginous Hawk	Accipitridae	*Buteo regalis*	99	1145.000	3.059	0.0098
Canvasback	Anatidae	*Aythya valisineria*	100	1219.000	3.086	0.0204
Great Horned Owl	Strigidae	*Bubo virginianus occidentalis*	101	1354.500	3.132	0.0210
Turkey Vulture	Cathartidae	*Cathartes aura*	102	1467.000	3.166	0.0094

(continued)

APPENDIX 3 (*continued*)

Common name	Family name	Scientific name	Species no.	Body mass (g)†	Log$_{10}$ mass	Body mass difference index‡ $[(M_{n+1} - M_{n-1})/M_n^{1.3}]$
Western Grebe	Podicipedidae	*Aechmophorus occidentalis*	103	1477.000	3.169	0.0161
Double-crested Cormorant	Phalacrocoracidae	*Phalacrocorax auritus*	104	1679.000	3.225	0.0586
Great Blue Heron	Ardeidae	*Ardea herodias*	105	2390.000	3.378	0.0320
Sage Grouse	Phasianidae	*Centrocercus urophasianus*	106	2467.500	3.392	0.0457
Canada Goose	Anatidae	*Branta canadensis canadensis*	107	3564.000	3.552	0.0444
Golden Eagle	Accipitridae	*Aquila chrysaetos*	108	4308.000	3.634	. . .

*The species list was developed from a check-list of Albertan Birds produced by the Provincial Museum of Alberta in 1981. The species recorded were those that had not been introduced and that fell into the categories: breeds, resident, fairly common, common, uncommon, and occasional.

† All body masses, with the exception of the Chipping Sparrow, Long-billed Marsh Wren, and Brewer's Sparrow, were taken from Dunning (1984). The body mass for the Chipping Sparrow was obtained from a specimen caught by Richard J. Cannings, Department of Zoology, University of British Columbia, in May 1988. The body masses for the Long-billed Marsh Wren and Brewer's Sparrow were taken from Terres (1980).

‡Index defined as in Appendix 1.

APPENDIX 4

MAMMALS OF THE BOREAL PRAIRIE

Species and body masses of boreal prairie mammals found in the short-grass prairie of the boreal region.*

Common name	Scientific name	Species no.	Body mass† (g)	Log_{10} mass	Body mass difference index‡ $[(M_{n+1} - M_{n-1})/M_n^{1.1}]$
Pygmy shrew	*Sorex hoyi hoyi*	1	4.110	0.614	. . .
Western small-footed bat	*Myotis ciliolabrum*	2	4.819	0.683	0.201
Prairie shrew	*Sorex haydeni*	3	5.244	0.720	0.412
Dusky shrew	*Sorex obscurus soperi*	4	7.371	0.868	0.362
Olive-backed pocket mouse	*Perognathus fasciatus*	5	8.505	0.930	0.121
Little brown bat	*Myotis lucifugus*	6	8.646	0.937	0.132
Long-eared bat	*Myotis evotis*	7	9.922	0.997	0.102
Silver-haired bat	*Lasionycteris noctivagans*	8	9.922	0.997	0.091
Red bat	*Lasiurus borealis*	9	11.056	1.044	0.222
Western harvest mouse	*Reithrodontomys megalotis*	10	13.041	1.115	0.576
Western jumping mouse	*Zapus princeps*	11	20.766	1.317	0.338
Big brown bat	*Eptesicus fuscus*	12	22.538	1.353	0.071
Deer mouse	*Peromyscus maniculatus*	13	22.963	1.361	0.018
White-footed mouse	*Peromyscus leucopus*	14	23.105	1.364	0.125
Southern red-backed vole	*Clethrionomys gapperi*	15	26.932	1.430	0.140
Hoary bat	*Lasiurus cinereus*	16	28.350	1.453	0.072
Sagebrush vole	*Lagurus crutatus*	17	29.767	1.474	0.237
Northern grasshopper mouse	*Onychomys leucogaster*	18	38.272	1.583	0.180
Meadow vole	*Microtus pennsylvanicus*	19	39.690	1.599	0.109
Least weasel	*Mustela nivalis*	20	44.509	1.648	0.098
Prairie vole	*Microtus ochrogaster*	21	46.068	1.663	0.453
Ord's kangaroo rat	*Dipodomys ordii terrosus*	22	75.127	1.876	0.300
Ermine	*Mustela erminea*	23	80.797	1.907	0.509
Northern pocket gopher	*Thomomys talpoides*	24	138.915	2.143	0.312
Thirteen-lined ground squirrel	*Spermophilus tridecemlineatus*	25	151.672	2.181	0.082
Long-tailed weasel	*Mustela frenata*	26	159.468	2.203	1.103
Richardson's ground squirrel	*Spermophilus richardsonii*	27	443.677	2.647	1.069
Muskrat	*Ondatra zibethicus*	28	1031.921	3.014	0.290
Black-footed ferret	*Mustela nigripes*	29	1043.261	3.018	0.060
Nuttall's cottontail	*Sylvilagus nuttallii*	30	1156.659	3.063	0.077
Mink	*Mustela vison*	31	1224.698	3.088	0.050
Black-tailed prairie dog	*Cynomys ludovicianus*	32	1281.397	3.108	0.104
Snowshoe hare	*Lepus americanus*	33	1496.853	3.175	0.259
Swift fox	*Vulpes velox*	34	2086.523	3.319	0.256
Striped skunk	*Mephitis mephitis*	35	2642.173	3.422	0.219
White-tailed jack rabbit	*Lepus townsendii*	36	3356.580	3.526	0.338
Red fox	*Vulpes vulpes*	37	5193.628	3.715	0.364
American badger	*Taxidea taxus*	38	7801.782	3.892	0.173
American procupine	*Erethizon dorsatum*	39	8504.850	3.930	0.060
River otter	*Lutra canadensis*	40	9071.840	3.958	0.045
Raccoon	*Procyon lotor*	41	9525.432	3.979	0.045
Lynx	*Lynx canadensis*	42	10149.121	4.006	0.109
Wolverine	*Gulo gulo*	43	12303.683	4.090	0.124
Coyote	*Canis latrans*	44	14061.352	4.148	0.320
Beaver	*Castor canadensis*	45	23995.016	4.380	0.443
Wolf	*Canis lupus*	46	43204.638	4.636	0.193

*The following references were used to produce the species list: Banfield 1974, Forsyth 1985. Note: Ian McTaggart-Cowan provided information that helped extend the species list developed using the above references.

†All body masses were taken from the Forsyth (1985) reference. The body mass used for the prairie shrew, however, is that of the common shrew, and the body mass used for Nuttall's cottontail is that of the Eastern cottontail. When two different body masses were given for females and for males an average mass was used.

‡Index as in Appendix 1.

(continued)

APPENDIX 4 (*CONTINUED*)

Common name	Scientific name	Species no.	Body mass† (g)	Log_{10} mass	Body mass difference index‡ [$(M_{n+1} - M_{n-1})/M_n^{1.1}$]
Pronghorn antelope	*Antilocapra americana*	47	48194.150	4.683	0.164
Cougar	*Felis concolor*	48	66507.927	4.823	0.189
White-tailed deer	*Odocoileus virginianus*	49	86409.276	4.937	0.095
Mule deer	*Odocoileus hemionus*	50	91965.778	4.964	0.289
Black bear	*Ursus americanus*	51	169643.400	5.230	0.390
American elk	*Cervus canadensis*	52	312751.680	5.495	0.349
Bison	*Bison bison*	53	556557.380	5.746	. . .

Pelagic Birds of Northwest North America

Species and body masses of marine birds that feed in the waters off the coast of British Columbia (Canada) and Washington state (USA).*

Common name	Scientific name	Species no.	Body mass (g)	$\log_{10}$ mass	Body mass difference index‡ $[(M_{n+1} - M_{n-1})/M_n^{1.3}]$
Red-necked Phalarope	*Phalaropus lobatus*	1	33.500	1.5250	. . .
Leach's Storm-Petrel	*Oceanodroma leucorhoa*	2	39.800	1.5999	0.1361
Fork-tailed Storm-Petrel	*Oceanodroma furcata*	3	55.300	1.7427	0.0835
Red Phalarope	*Phalropus fulicaria*	4	55.650	1.7455	0.1711
Arctic Tern	*Sterna paradisaea*	5	110.000	2.0414	0.1635
Common Tern	*Sterna hirundo*	6	120.000	2.0792	0.0182
Aleutian Tern	*Sterna aleutica*	7	120.000	2.0792	0.0690
Xantus' Murrelet	*Synthliboramphus hypoleucus*	8	167.000	2.2227	0.0877
Cassin's Auklet	*Ptychoramphus aleuticus*	9	188.000	2.2742	0.0256
Sabine's Gull	*Xema sabini*	10	191.000	2.2810	0.0174
Ancient Murrelet	*Synthliboramphus antiquus*	11	206.000	2.3139	0.0196
Bonaparte's Gull	*Larus philadelphia*	12	212.000	2.3263	0.0140
Marbled Murrelet	*Brachyramphus marmoratus*	13	222.000	2.3464	0.0488
Black-vented Shearwater	*Puffinus opisthomelas*	14	276.000	2.4409	0.0515
Long-tailed Jaeger	*Stercorarius longicaudus*	15	296.500	2.4720	0.0129
Eared Grebe	*Podiceps nigrocolis*	16	297.000	2.4728	0.0112
Mottled Petrel	*Pterodroma inexpectata*	17	316.000	2.4997	0.0119
Parakeet Auklet	*Cyclorrhynchus psittacula*	18	318.000	2.5024	0.0320
Buller's Shearwater	*Puffinus bulleri*	19	380.000	2.5798	0.0401
Bufflehead	*Bucephala albeola*	20	403.500	2.6058	0.0100
Mew Gull	*Larus canus*	21	403.500	2.6058	0.0014
Black-legged Kittiwake	*Rissa tridacytla*	22	407.000	2.6096	0.0193
Horned Grebe	*Podiceps auritus*	23	453.000	2.6561	0.0216
Parasitic Jaeger	*Stercorarius parasiticus*	24	464.500	2.6670	0.0118
Pigeon Guillemot	*Cepphus columba*	25	487.000	2.6875	0.0119
Heerman's Gull	*Larus heermanni*	26	500.000	2.6990	0.0106
Rhinoceros Auklet	*Cerorhinca monocerata*	27	520.000	2.7160	0.0132
Short-tailed Shearwater	*Puffinus tenuirostris*	28	543.000	2.7348	0.0072
Northern Fulmar	*Fulmarus glacialis*	29	544.000	2.7356	0.0004
Ruddy Duck	*Oxyura jamaicensis*	30	544.500	2.7360	0.0046
Flesh-footed Shearwater	*Puffinus carneipes*	31	560.000	2.7482	0.0177
California Gull	*Larus californicus*	32	609.000	2.7846	0.0133
Hooded Merganser	*Lophodytes cucullatus*	33	610.000	2.7853	0.0025
Horned Puffin	*Fratercula corniculata*	34	619.000	2.7917	0.0032
Harlequin Duck	*Histrionicus histrionicus*	35	622.500	2.7941	0.0178
Pomarine Jaeger	*Stercorarius pomarinus*	36	694.000	2.8414	0.0225
Pink-footed Shearwater	*Puffinus creatopus*	37	721.000	2.8579	0.0176
Tufted Puffin	*Fratercula cirrhata*	38	779.000	2.8915	0.0121
Peregrine Falcon	*Falco peregrinus*	39	781.500	2.8929	0.0015
Sooty Shearwater	*Puffinus griseus*	40	787.000	2.8960	0.0166
Oldsquaw	*Clangula hyemalis*	41	873.000	2.9410	0.0198
Common Goldeneye	*Bucephala clangula*	42	900.000	2.9542	0.0061
Barrow's Goldeneye	*Bucephala islandica*	43	910.000	2.9590	0.0071
Greater Scaup	*Aythya marila*	44	944.500	2.9752	0.0063
Black Scoter	*Melanitta nigra*	45	950.000	2.9777	0.0008
Surf Scoter	*Melanitta perspicillata*	46	950.000	2.9777	0.0021
Thick-billed Murre	*Uria lomvia*	47	964.000	2.9841	0.0064
Common Murre	*Uria aalge*	48	992.500	2.9967	0.0047
Thayer's Gull	*Larus thayeri*	49	996.000	2.9983	0.0025
Glaucous-winged Gull	*Larus glaucescens*	50	1010.000	3.0043	0.0022

(continued)

APPENDIX 5 (*continued*)

Common name	Scientific name	Species no.	Body mass (g)	Log_{10} mass	Body mass difference index‡ $[(M_{n+1} - M_{n-1})/M_n^{1.3}]$
Western Gull	*Larus occidentalis*	51	1011.000	3.0048	0.0016
Red-breasted Merganser	*Mergus serrator*	52	1021.500	3.0092	0.0017
Red-necked Grebe	*Podiceps grisegena*	53	1023.000	3.0099	0.0152
Herring Gull	*Larus argentatus*	54	1135.000	3.0550	0.0174
South Pole Skua	*Catharacta maccormicki*	55	1156.000	3.0630	0.0101
Canvasback	*Aythya valisineria*	56	1219.000	3.0860	0.0218
White-winged Scoter	*Melanitta fusca*	57	1350.000	3.1303	0.0260
Common Merganser	*Mergus merganser*	58	1470.500	3.1675	0.0123
Western Grebe	*Aechmophorus occidentalis*	59	1477.000	3.1694	0.0073
Red-throated Loon	*Gavia stellata*	60	1551.000	3.1906	0.0158
Pacific Loon	*Gavia pacifica*	61	1659.000	3.2198	0.0107
Double-crested Cormorant	*Phalacrocorax auritus*	62	1679.000	3.2251	0.0193
Pelagic Cormorant	*Phalacrocorax pelagicus*	63	1915.000	3.2822	0.0298
Brandt's Cormorant	*Phalacrocorax penicillatus*	64	2103.000	3.3228	0.0605
Laysan Albatross	*Diomedea immutabilis*	65	3041.500	3.4831	0.0503
Black-footed Abaltross	*Diomedea nigripes*	66	3148.000	3.4980	0.0135
Brown Pelican	*Pelecanus occidentalis*	67	3392.000	3.5305	0.0335
Common Loon	*Gavia immer*	68	4134.000	3.6164	0.0387
Bald Eagle	*Haliaeetus leucocephalus*	69	4683.500	3.6706	0.0324
Yellow-billed Loon	*Gavia adamsii*	70	5438.000	3.7354	. . .

*The list was developed by Dick Cannings, Department of Zoology, University of British Columbia, and masses were obtained from Dunning (1984).

APPENDIX 6

ANTELOPES OF THE AFRICAN SAVANNA

Species and body masses of African antelopes of grasslands and savannas.*

Common name	Scientific name	Species no.	Body mass (g)	Log$_{10}$ mass	Body mass difference index [$(M_{n+1} - M_{n-1})/M_n^{1.1}$]
Guenther's long-snouted dik-dik	*Rhynchotragus guentheri*	1	2000	3.3010	. . .
Salt's and Phillip's dik-dik	*Madoqua saltiana*	2	3500	3.5441	0.0247
Kirk's and Damaraland long snouted dik-dik	*Rhynchotragus kirki*	3	4500	3.6532	0.0802
Sharpe's grysbok	*Raphicerus sharpei*	4	9000	3.9542	0.0217
Grimm's, grey, or common duiker	*Sylvicapra grimmia*	5	12000	4.0792	0.0100
Steinbok	*Raphicerus campestris*	6	14000	4.1461	0.0081
Klipspringer	*Oreotragus*	7	16000	4.2041	0.0223
Thomson's gazelle	*Gazella thomsoni*	8	22500	4.3522	0.0055
Mountain reedbuck	*Redunca fulvorufula*	9	25000	4.3979	0.0173
Springbok, springbuck	*Antidorcas marsupialis*	10	34000	4.5315	0.0116
Bohor reedbuck	*Redunca redunca*	11	43000	4.6335	0.0009
Gerenuk	*Litocranius walleri*	12	44000	4.6435	0.0097
Bushbuck	*Tragelaphus scriptus*	13	54500	4.7364	0.0052
Grant's gazelle	*Gazella (Nanger) granti*	14	62000	4.7924	0.0009
Impala	*Aepyceros melampus*	15	63500	4.8028	0.0091
Kob	*Kobus (Adenota) kob*	16	79500	4.9004	0.0000
Lesser kudu	*Tragelaphus imberbis*	17	79500	4.9004	0.0145
Topi, tiang, korrigum	*Damaliscus korrigum*	18	113500	5.0550	0.0004
Nyala	*Tragelaphus angasi*	19	115000	5.0607	0.0044
Lichtenstein's hartebeest	*Alcelaphus lichtensteini*	20	131500	5.1189	0.0035
Tsessbe	*Damaliscus lunatus*	21	147500	5.1688	0.0024
Black wildebeest or white-tailed gnu	*Connochaetes gnou*	22	160000	5.2041	0.0009
Bubal hartebeest	*Alcelaphus buselaphus*	23	165500	5.2188	0.0008
Red hartebeest	*Alcelpahus caama*	24	170500	5.2317	0.0053
Common waterbuck	*Kobus ellipsiprymnus*	25	204000	5.3096	0.0014
Defassa waterbuck	*Kobus defassa*	26	215500	5.3334	0.0000
Blue wildebeest or brindled gnu	*Connochaetesta urinus*	27	215500	5.3334	0.0506
Cape eland	*Taurotragus oryx*	28	650000	5.8129	0.0014
Giant eland	*Taurotragus derbianus*	29	700000	5.8451	0.0030
African buffalo (plains)	*Syncerus caffer*	30	820000	5.9138	. . .

*The list of species that are known to inhabit the savanna and grasslands was chosen by Brian Walker (CSIRO, Division of Wildlife and Rangeland Research, P.O. Box 84, Lyneham, A.C.T. 2602, Australia) and Tony Sinclair (Department of Zoology, University of British Columbia, Vancouver, British Columbia, Canada V6R 3W9) from Jarman (1974).

APPENDIX 7

MAMMAL HOME RANGES

Shrews and fossorial mammals were eliminated, as were any data with < 5 replicates. C = carnivores, H = herbivores, and O = omnivores.

Scientific name	Diet	Body mass (g)	Log₁₀ mass	Home range (ha)	Log₁₀ home range	Author
Onychomys leucogaster	C	31.	1.491	3.62	0.559	Mace and Harvey 1983
Mustela rixosa	C	42.	1.623	3.28	0.516	Harestad and Bunnell 1979
Mustela erminea	C	70.	1.845	14.00	1.146	Lindstedt et al. 1986
Mustela nivalis	C	85.	1.929	6.75	0.829	Lindstedt et al. 1986
Mustela erminea	C	93.	1.968	20.64	1.315	Harestad and Bunnell 1979
Mustela erminea	C	95.	1.978	14.00	1.146	Gittleman and Harvey 1982
Helogale parvula	C	270.	2.431	30.00	1.477	Gittleman and Harvey 1982
Herpestes auropunctatus	C	780.	2.892	31.00	1.491	Gittleman and Harvey 1982
Martes americana	C	870.	2.940	160.00	2.204	Gittleman and Harvey 1982
Mustela vison	C	910.	2.959	270.00	2.431	Gittleman and Harvey 1982
Martes americana	C	963.	2.983	432.50	2.636	Lindstedt et al. 1986
Martes americana	C	1043.	3.018	209.31	2.321	Harestad and Bunnell 1979
Martes martes	C	1200.	3.079	150.00	2.176	Gittleman and Harvey 1982
Mungos mungo	C	1260.	3.100	240.00	2.380	Gittleman and Harvey 1982
Martes pennanti	C	3075.	3.488	1356.00	3.132	Lindstedt et al. 1986
Alopex lagopus	C	3600.	3.556	2080.00	3.318	Lindstedt et al. 1986
Martes pennanti	C	3750.	3.574	2590.00	3.413	Gittleman and Harvey 1982
Vulpes vulpes	C	4100.	3.613	410.00	2.613	Gittleman and Harvey 1982
Taxidea taxus	C	4100.	3.613	410.00	2.613	Gittleman and Harvey 1982
Felis silvestris	C	4700.	3.672	101.00	2.004	Gittleman and Harvey 1982
Vulpes vulpes	C	5050.	3.703	1552.00	3.191	Lindstedt et al. 1986
Vulpes vulpes	C	5448.	3.736	387.34	2.588	Harestad and Bunnell 1979
Lynx rufus	C	6200.	3.792	3070.00	3.487	Gittleman and Harvey 1982
Taxidea taxus	C	6950.	3.842	410.00	2.613	Lindstedt et al. 1986
Proteles cristatus	C	8300.	3.919	100.00	2.000	Gittleman and Harvey 1982
Lutrogale perspicillata	C	8800.	3.944	700.00	2.845	Gittleman and Harvey 1982
Lynx rufus	C	9072.	3.958	320.82	2.506	Harestad and Bunnell 1979
Lynx canadensis	C	10149.	4.006	2791.70	3.446	Ward and Krebs 1985
Canis latrans	C	10600.	4.025	4200.00	3.623	Gittleman and Harvey 1982
Lynx rufus	C	11000.	4.041	2211.33	3.345	Lindstedt et al. 1986
Lynx rufus	C	11011.	4.042	6225.00	3.794	Litvaitis et al. 1986
Lynx rufus	C	11011.	4.042	4700.00	3.672	Fuller et al. 1985
Lynx canadensis	C	11567.	4.063	1852.40	3.268	Harested and Bunnell 1979
Lynx lynx	C	11567.	4.063	35546.15	4.551	Harestad and Bunnell 1979
Gulo gulo	C	11620.	4.065	122500.00	5.088	Gittleman and Harvey 1982
Leptailurus serval	C	11700.	4.068	150.00	2.176	Gittleman and Harvey 1982
Canis latrans	C	12000.	4.079	3410.00	3.533	Lindstedt et al. 1986
Gulo gulo	C	12000.	4.079	40500.00	4.607	Lindstedt et al. 1986
Canis latrans	C	12600.	4.100	618.50	2.791	Swihart et al. 1988
Canis latrans	C	15890.	4.201	7597.57	3.881	Harestad and Bunnell 1979
Lynx lynx	C	19300.	4.286	2240.00	3.350	Gittleman and Harvey 1982
Lycaon pictus	C	22000.	4.342	175000.00	5.243	Gittleman and Harvey 1982
Canis lupus	C	33200.	4.521	39160.00	4.593	Gittleman and Harvey 1982
Canis lupus	C	37422.	4.573	20276.88	4.307	Harestad and Bunnell 1979
Canis lupus	C	39000.	4.591	6250.00	3.796	Lindstedt et al. 1986
Felis concolor	C	51800.	4.714	4860.00	3.687	Gittleman and Harvey 1982
Crocuta crocuta	C	52000.	4.716	2500.00	3.398	Gittleman and Harvey 1982
Panthera pardus	C	52400.	4.719	2360.00	3.373	Gittleman and Harvey 1982
Felis concolor	C	56000.	4.748	19365.00	4.287	Lindstedt et al. 1986
Acinonyx jubatus	C	58800.	4.769	6750.00	3.829	Gittleman and Harvey 1982
Felis concolor	C	67000.	4.826	29733.33	4.473	Harestad and Bunnell 1979
Panthera leo	C	155800.	5.193	24000.00	4.380	Gittleman and Harvey 1982

(continued)

APPENDIX 7 (*continued*)

Scientific name	Diet	Body mass (g)	Log$_{10}$ mass	Home range (ha)	Log$_{10}$ home range	Author
Panthera tigris	C	161000.	5.207	7140.00	3.854	Gittleman and Harvey 1982
Peromyscus maniculatus	O	16.	1.204	0.81	−0.092	Harestad and Bunnell 1979
Tamias minimus	O	40.	1.602	2.10	0.322	Swihart et al. 1988
Tamias umbrinus	O	60.	1.778	4.55	0.658	Swihart et al. 1988
Tamias quadrivittatus	O	70.	1.845	6.73	0.828	Swihart et al. 1988
Sigmodon hispidus	O	80.	1.903	0.34	−0.469	Mace and Harvey 1983
Tamias striatus	O	90.	1.954	0.08	−1.097	Swihart et al. 1988
Glaucomys (volans, sabrinus)	O	123.	2.091	4.14	0.617	Mace and Harvey 1983
Herpestes sanguineus	O	490.	2.690	75.00	1.875	Gittleman and Harvey 1982
Galidea elegans	O	810.	2.908	23.00	1.362	Gittleman and Harvey 1982
Bassariscus astutus	O	950.	2.978	31.50	1.498	Lindstedt et al. 1986
Bassariscus astutus	O	950.	2.978	140.00	2.146	Gittleman and Harvey 1982
Fossa fossa	O	1800.	3.255	100.00	2.000	Gittleman and Harvey 1982
Mephitis mephitis	O	2400.	3.380	4.00	0.602	Gittleman and Harvey 1982
Mephitis mephitis	O	2586.	3.413	294.67	2.469	Harestad and Bunnell 1979
Didelphis marsupialis	O	2724.	3.435	59.88	1.777	Harestad and Bunnell 1979
Urocyon cinereoargenteus	O	2800.	3.447	548.00	2.739	Swihart et al. 1988
Nandinia binotata	O	3200.	3.505	73.00	1.863	Swihart et al. 1988
Urocyon cinereoargenteus	O	3600.	3.556	122.00	2.086	Lindstedt et al. 1986
Urocyon cinereoargenteus	O	3700.	3.568	110.00	2.041	Gittleman and Harvey 1982
Ichneumia albicauda	O	3900.	3.591	800.00	2.903	Gittleman and Harvey 1982
Otocyon megalotis	O	3900.	3.591	73.00	1.863	Gittleman and Harvey 1982
Nasua narica	O	5000.	3.699	55.00	1.740	Gittleman and Harvey 1982
Cerdocyon thous	O	6000.	3.778	150.00	2.176	Gittleman and Harvey 1982
Procyon lotor	O	6400.	3.806	110.00	2.041	Gittleman and Harvey 1982
Procyon lotor	O	7264.	3.861	113.73	2.056	Harestad and Bunnell 1979
Procyon lotor	O	8000.	3.903	430.20	2.634	Swihart et al. 1988
Canis aureus	O	8800.	3.944	1000.00	3.000	Gittleman and Harvey 1982
Meles meles	O	11600.	4.064	87.00	1.940	Gittleman and Harvey 1982
Pecari tajacu	O	23814.	4.377	135.21	2.131	Harestad and Bunnell 1979
Hyaena hyaena	O	26800.	4.428	5800.00	3.763	Gittleman and Harvey 1982
Hyaena brunnea	O	43300.	4.636	13250.00	4.122	Gittleman and Harvey 1982
Ursus americanus	O	69200.	4.840	2285.00	3.359	Swihart et al. 1988
Ursus americanus	O	76204.	4.882	2413.09	3.383	Harestad and Bunnell 1979
Ursus americanus	O	110500.	5.043	5630.00	3.751	Gittleman and Harvey 1982
Ailuropoda melanoleuca	O	135000.	5.130	250.00	2.398	Gittleman and Harvey 1982
Ursus arctos	O	204120.	5.310	9283.13	3.968	Harestad and Bunnell 1979
Ursus arctos	O	298500.	5.475	5310.00	3.725	Gittleman and Harvey 1982
Perognathus longimembris	H	8.	0.919	0.31	−0.509	Mace and Harvey 1983
Clethrionomys gapperi	H	16.	1.204	0.25	−0.602	Harestad and Bunnell 1979
Microtus oregoni	H	20.	1.301	0.81	−0.092	Harestad and Bunnell 1979
Pitymys pinetorum	H	23.	1.362	0.58	−0.237	Mace and Harvey 1983
Synaptomys cooperi	H	26.	1.415	0.04	−1.398	Mace and Harvey 1983
Microtus (arvalis, oeconomous, agrestis)	H	29.	1.467	0.16	−0.796	Mace and Harvey 1983
Apodemus flavicollis	H	31.	1.490	0.22	−0.658	Mace and Harvey 1983
Microtus pennsylvanicus	H	40.	1.602	0.12	−0.921	Harestad and Bunnell 1979
Microtus pennsylvanicus	H	40.	1.602	0.02	−1.699	Swihart et al. 1988
Microtus ochragaster	H	40.	1.602	0.11	−0.959	Swihart et al. 1988
Synaptomys cooperi	H	40.	1.602	0.05	−1.301	Swihart et al. 1988
Oryzomys capita	H	45.	1.653	0.89	−0.051	Mace and Harvey 1983
Dipodomys (agilis, merriami, ordii)	H	51.	1.708	1.29	0.111	Mace and Harvey 1983
Dicrostonyx groenlandicus	H	58.	1.763	0.20	−0.699	Mace and Harvey 1983
Tamias striatus	H	85.	1.929	0.11	−0.959	Harestad and Bunnell 1979
Sigmodon hispidus	H	120.	2.079	0.59	−0.229	Swihart et al. 1988
Ochotona princeps	H	122.	2.086	0.35	−0.456	Harestad and Bunnell 1979
Ochotona princeps	H	150.	2.176	0.16	−0.796	Swihart et al. 1988

(*continued*)

APPENDIX 7 (continued)

Scientific name	Diet	Body mass (g)	Log$_{10}$ mass	Home range (ha)	Log$_{10}$ home range	Author
Tamiasciurus hudsonicus	H	200	2.301	0.49	−0.310	Swihart et al. 1988
Microcavia australis	H	240.	2.380	0.31	−0.509	Mace and Harvey 1983
Tamiasciurus hudsonicus	H	254.	2.405	1.10	0.041	Harestad and Bunnell 1979
Proechimys (guyanensis, semispinosus)	H	383.	2.583	0.53	−0.276	Mace and Harvey 1983
Sciurus carolinensis	H	500.	2.699	0.95	−0.022	Harestad and Bunnell 1979
Sciurus griseus	H	500.	2.699	0.30	−0.523	Harestad and Bunnell 1979
Cavia aperea	H	525.	2.720	0.13	−0.886	Mace and Harvey 1983
Sciurus carolinensis	H	625.	2.796	1.18	0.072	Mace and Harvey 1983
Sylvilagus bachmani	H	695.	2.842	0.28	−0.553	Harestad and Bunnell 1979
Sylvilagus floridanus	H	1322.	3.121	1.62	0.210	Harestad and Bunnell 1979
Lepus americanus	H	1500.	3.176	53.80	1.731	Swihart et al. 1988
Lepus americanus	H	1543.	3.188	5.93	0.773	Harestad and Bunnell 1979
Sylvilagus aquaticus	H	2198.	3.342	2.12	0.326	Harestad and Bunnell 1979
Lepus californicus	H	2700.	3.431	156.00	2.193	Swihart et al. 1988
Antilocapra americana	H	48000.	4.681	1060.47	3.025	Harestad and Bunnell 1979
Antilocapra americana	H	50300.	4.702	1142.00	3.058	Swihart et al. 1988
Odocoileus hemionus columbianus	H	64638.	4.810	58.85	1.770	Harestad and Bunnell 1979
Ovis canadensis	H	87000.	4.940	1433.40	3.156	Harestad and Bunnell 1979
Odocoileus virginianus	H	90720.	4.958	196.06	2.292	Harestad and Bunnell 1979
Odocoileus hemionus hemionus	H	97524.	4.989	285.27	2.455	Harestad and Bunnell 1979
Cervus canadensis	H	300510.	5.478	1292.54	3.111	Harestad and Bunnell 1979
Alces alces	H	411075.	5.614	1609.53	3.207	Harestad and Bunnell 1979

APPENDIX 8
BIRD HOME RANGES
Data with < 5 replicates were eliminated. C = carnivores, H = herbivores, and O = omnivores.

Scientific name	Diet	Body mass (g)	Log₁₀ body mass	Home range (ha)	Log₁₀ home range (ha)	Author
Vireo atricapillus	C	8.500	0.929	1.497	0.175	Schoener 1968
Dendroica magnolia	C	8.600	0.934	0.728	−0.138	Schoener 1968
Setophaga ruticilla	C	9.000	0.9054	0.194	−0.712	Schoener 1968
Dendroica virens	C	9.000	0.954	0.648	−0.189	Schoener 1968
Troglodytes troglodytes	C	9.500	0.978	1.012	0.005	Schoener 1968
Dendroica petechia	C	9.500	0.978	0.170	−0.770	Schoener 1968
Dendroica pensylvanica	C	9.600	0.982	0.607	−0.217	Schoener 1968
Dendroica fusca	C	9.700	0.987	0.526	−0.279	Schoener 1968
Geothylpis trichas	C	9.800	0.991	0.526	−0.279	Schoener 1968
Empidonax minimus	C	9.900	0.996	0.178	−0.749	Schoener 1968
Vireo belli	C	10.000	1.000	1.174	0.070	Schoener 1968
Parus carolinensis	C	10.100	1.004	1.497	0.175	Schoener 1968
Parus atricapillus	C	11.000	1.041	1.457	0.163	Schoener 1968
Parus palustris	C	11.000	1.041	2.266	0.355	Schoener 1968
Troglodytes aedon	C	11.200	1.049	0.405	−0.393	Schoener 1968
Oporornis philadelphia	C	11.300	1.053	0.769	−0.114	Schoener 1968
Vireo griseus	C	11.400	1.057	0.134	−0.874	Schoener 1968
Empidonax wrighti	C	12.300	1.090	1.578	0.198	Schoener 1968
Contopus virens	C	13.800	1.140	4.371	0.641	Schoener 1968
Dendroica kirtlandi	C	14.000	1.146	3.399	0.531	Schoener 1968
Protonotaria citrea	C	16.100	1.207	1.497	0.175	Schoener 1968
Vireo olivaceus	C	17.600	1.246	0.728	−0.138	Schoener 1968
Thyrothorus ludovicianus	C	18.500	1.267	0.121	−0.916	Schoener 1968
Seiurus aurocapillus	C	18.900	1.276	1.012	0.005	Schoener 1968
Seiurus aurocapillus	C	19.400	1.288	0.635	−0.197	Zach and Falls 1979
Seiurus aurocapillus	C	19.400	1.288	0.290	−0.538	Smith and Shugart 1987
Oenanthe oenanthe	C	25.200	1.401	1.538	0.187	Schoener 1968
Icteria virens	C	27.000	1.431	0.134	−0.874	Schoener 1968
Calcarius laponicus	C	28.600	1.456	1.760	0.246	Mace and Harvey 1983
Lanius collurio	C	30.000	1.477	1.578	0.198	Schoener 1968
Spiza americana	C	35.000	1.544	0.890	−0.051	Mace and Harvey 1983
Scardofella inca	C	48.000	1.681	0.260	−0.585	Mace and Harvey 1983
Lanius ludovicianus	C	48.100	1.682	7.568	0.879	Schoener 1968
Falco sparverius	C	112.000	2.049	141.645	2.151	Schoener 1968
Elanus caerulus	C	270.000	2.431	182.200	2.261	Mace and Harvey 1983
Accipiter cooperi	C	469.000	2.671	225.418	2.353	Schoener 1968
Circus cyaneus	C	521.000	2.717	252.128	2.402	Schoener 1968
Buteo lineatus	C	626.000	2.797	63.943	1.806	Schoener 1968
Buteo buteo	C	846.000	2.927	130.718	2.116	Schoener 1968
Buteo sawinsoni	C	971.000	2.987	246.462	2.392	Schoener 1968
Accipiter gentilis	C	978.000	2.990	212.468	2.327	Schoener 1968
Buteo jamaicensis	C	1126.000	3.052	424.935	2.628	Schoener 1968
Nyctea scandiaca	C	1920.000	3.283	493.734	2.693	Schoener 1968
Sitta pusilla	O	10.200	1.009	2.752	0.440	Schoener 1968
Sitta pygmaea	O	10.600	1.025	0.971	−0.013	Schoener 1968
Spizella breweri	O	11.000	1.041	0.918	−0.037	Weins et al. 1985
Spizella passerina	O	12.200	1.086	3.076	0.488	Schoener 1968
Passerina cyanea	O	14.300	1.155	0.105	−0.978	Schoener 1968
Chamaea fasciata	O	14.800	1.170	0.324	−0.490	Schoener 1968
Parus inornatus	O	16.600	1.220	2.428	0.385	Schoener 1968
Ammodramus savannarum	O	16.700	1.223	1.093	0.038	Schoener 1968
Spizella arborea	O	18.100	1.258	1.700	0.230	Schoener 1968
Amphispiza belli	O	18.900	1.276	2.278	0.358	Weins et al. 1985

(continued)

APPENDIX 8 (*continued*)

Scientific name	Diet	Body mass (g)	Log_{10} body mass	Home range (ha)	Log_{10} home range (ha)	Author
Melospiza melodia	O	21.600	1.334	0.162	−0.791	Schoener 1968
Zonotrichcia albicollis	O	25.900	1.413	0.508	−0.294	Wasserman 1980
Guiraca caerulea	O	27.900	1.446	6.192	0.792	Schoener 1968
Lululla arborea	O	30.000	1.477	8.296	0.919	Schoener 1968
Sialia sialis	O	30.800	1.489	1.012	0.005	Schoener 1968
Habia rubica	O	32.800	1.516	4.856	0.686	Schoener 1968
Dumatella carolinensis	O	35.900	1.555	0.105	−0.978	Schoener 1968
Habia fuscicauda	O	37.700	1.576	6.071	0.783	Schoener 1968
Richmondena cardinalis	O	41.200	1.615	0.150	−0.825	Schoener 1968
Pipilio fuscus	O	44.700	1.650	2.590	0.413	Schoener 1968
Pipilio aberti	O	46.300	1.666	1.619	0.209	Schoener 1968
Mimus polyglottos	O	50.100	1.700	0.405	−0.393	Schoener 1968
Aphelocoma coerulescens	O	72.400	1.860	2.145	0.331	Schoener 1968
Turdus migratorius	O	80.200	1.904	0.121	−0.916	Schoener 1968
Sturnella neglecta	O	89.000	1.949	3.035	0.482	Schoener 1968
Sutrnella magna	O	89.000	1.949	3.035	0.482	Schoener 1968
Turdus merula	O	92.000	1.964	0.728	−0.138	Schoener 1968
Colinus virginianus	H	200.000	2.301	1.100	0.041	Mace and Harvey 1983
Lagopus lagopus	H	510.000	2.708	2.570	0.410	Mace and Harvey 1983
Dendragapus obscura	H	510.000	2.708	1.210	0.083	Mace and Harvey 1983
Bonasa umbellus	O	908.000	2.958	10.000	1.000	Mace and Harvey 1983
Tymphanuchus cupido	H	908.000	2.958	5.320	0.726	Mace and Harvey 1983
Centrocercus urophasianus	H	1135.000	3.055	14.600	1.164	Mace and Harvey 1983
Phasiana colchicus	O	1136.000	3.055	12.500	1.097	Mace and Harvey 1983
Meleagris gallopavo	O	6023.000	3.780	54.960	1.740	Mace and Harvey 1983

APPENDIX 9
DERIVATION OF EQUATIONS CONVERTING BODY MASSES TO SCALES OF CHOICE

Abbreviations used throughout this appendix: SE = spatial extent (in kilometres); SG = spatial grain (in kilometres); TH = time horizon (in years); TR = time resolution (in years). Subscripts refer to choices for area (a), habitat (h), and food (f).

Scaling conversions for habitat choice

SE_h, for carnivores: $H = 8.71 \times 10^{-4}M^{1.02}$ (from Table 2), where H = home range (in square kilometres), M = body mass (in grams), and $SE_h = H^{0.5}$; therefore

$$SE_h = 0.0295M^{0.51} \qquad (9.1)$$

SE_h, for herbivores: $H = 4.17 \times 10^{-5}M^{1.02}$ (from Table 2), and

$$SE_h = 0.00646M^{0.51}. \qquad (9.2)$$

TR_h and TH_h, for carnivores and herbivores, respectively (from Swihart et al. 1988): TTI = $354M_k^{0.25}$, where TTI = time to independence (in minutes), and M_k = body mass (in kilograms). The "time to independence" is the time interval at which an animal's current position is influenced only by its pattern of home-range use, not by its position Δt minutes earlier. It therefore is the same as TR, or temporal resolution.

Converting to years and grams,

$$TR_h = 1.20 \times 10^{-4}M^{0.22}.$$

Assuming that $TH_h = 10TR_h$, then

$$TH_h = 1.20 \times 10^{-3}M^{0.22}. \qquad (9.3)$$

Scaling conversion for area choice

$SE_a = 50 SE_h$ (from Miller and Carrol 1989), or for carnivores:

$$SE_a = 1.47M^{0.51} \qquad (9.4)$$

and, for herbivores:

$$SE_a = 0.323 M^{0.51}. \qquad (9.5)$$

TH_a is equal to life span. Averaging parameters presented in Peters (1983),

$$TH_a = 2.95M^{0.21}, \qquad (9.6)$$

where TH_a = life span (in years) or time resolution for area, and M = body mass (in grams).

Scaling conversion for food choice

SF_f and TR_f are the spatial grain and time resolution for food choice, respectively. For predators they are expressed as the length of the minimum-sized prey and the time between captures, respectively.

SG_f for predators: From Peters (1983:277) the minimum prey size is given by:

$$w_{min} = 2.10 \times 10^{-5}M^{1.12}, \qquad (9.7)$$

where w_{min} = minimum prey size (in grams). Prey body mass can be converted to length, from Radinsky (1978),

$$w = 25 \times 10^{12/3} \qquad (9.8)$$

where w = body mass of prey animal (in grams) and l = length (in kilometres). Substituting Eq. 9.8 in Eq. 9.7,

$$SG_f = 9.44 \times 10^{-7}M^{0.37} \qquad (9.9)$$

for clump categories $c \le 3$. This same equation is used for herbivores, as well.

TR_f, for predators: The time resolution for food choice is expressed as the time between successive captures of the minimum-sized prey, when the full diet is made up of prey of that size.

TABLE A.1. Summary of equations converting body mass to spatial and temporal equivalents.*

Level of choice	Scale	Carnivore	Herbivore
Area	SE	1.47 $M^{0.51}$	0.323$M^{0.51}$
	TH	2.95$M^{0.21}$	2.95$M^{0.51}$
Habitat	SE	0.0295$M^{0.51}$	0.00646$M^{0.51}$
	TH	$(1.20 \times 10^{-3})M^{0.22}$	$(1.20 \times 10^{-3})M^{0.23}$
Food	SG	$(9.44 \times 10^{-7})M^{0.37}$	$(9.44 \times 10^{-7})M^{0.37}$
	TR	$(5.19 \times 10^{-8})M^{0.43}$	$(5.19 \times 10^{-8})M^{0.43}$

*SE = spatial extent (in kilometres); SG = spatial grain (in kilometres); TH = time horizon (in years); TR = time resolution (in years); and M = body mass (in grams).

From Holling (1965), the time interval between captures at satiation is given as

$$T_I = T_t/N_{ak}.\qquad(9.10)$$

where T_I = time spent handling and searching for one prey (in days), N_{ak} = maximum number of prey eaten per day, $T_t = 1$ d, and

$$N_{ak} = I/w,\qquad(9.11)$$

where I = no. of grams required to be eaten per predator per day, and w = body mass of prey.

From Farlow (1976)

$$I = 1.11\,M^{0.69}.\qquad(9.12)$$

Substituting Eq. 9.12 in Eq. 9.11 in Eq. 9.10, $T_I = w/1.11M^{0.69}$, or $T_I = 0.9wM^{-0.69}$. With TR_f expressed in years, therefore,

$$TR_f = 0.00247\ w_{min}M^{-0.69}.\qquad(9.13)$$

Substituting Eq. 9.5 in Eq. 9.13,

$$TR_f = 5.19 \times 10^{-8}M^{0.43}.\qquad(9.14)$$

31
Scale and Biodiversity Policy: A Hierarchical Approach

Bryan G. Norton and Robert E. Ulanowicz

Introduction

There exists a broad consensus supporting the protection of biological diversity; but the exact meaning of this consensus for policy is not clear. In the United States, for example, the *Endangered Species Act* emphasizes protection of species. But this emphasis has led to the question: Since approximately 99% of all species that have existed on earth are now extinct, how can it be so urgent that we reduce anthropogenic species extinctions? The standard answer to this question—that extinction itself is not bad, but rather that the accelerated *rate* and broadened *scale* of extinctions is unacceptable—likewise raises more questions than answers. One might ask, what would be an "acceptable" rate of extinctions? If species are not sacrosanct, what then *is* the proper target of protection? These questions are important because our inability to answer them indicates huge gaps in our understanding of environmental management and of biodiversity protection: It is not clear at what scale the problem of biodiversity loss should be addressed; Nor is it clear that measuring rates of species loss is the only or best criterion for measuring the success or failure of protection efforts.

In this paper we explore the policy implications of a hierarchical approach to protecting biological diversity. The hierarchical approach, which represents a specific application of general systems theory,[1,2] models natural complexity as a hierarchy of embedded systems represented on different *scales*. A major assumption of hierarchy theory is that smaller subsystems change according to a more rapid dynamic than do larger systems.[1,3-5] We believe that this correlation between system size (hierarchical level) and rate of change introduces some conceptual order into discussions of the proper scale on which to address environmental policy goals, and we illustrate our approach by applying it to biological diversity policy.

While much of our conceptual apparatus is adapted from theoretical ecology, we do not consider our work to be scientific in the narrow sense that it consists of value-free descriptions and explanatory hypotheses. We, on the contrary, believe that conservation biology is a normative science—like medicine it is guided most basically by a commitment to important social values. Just as medical research must fulfill both a criterion of methodological rigor *and* a criterion of relevance—usefulness in healing patients—conservation biologists are likewise obligated to characterize ecological systems in ways that are not only accurate, but useful in protection and recovery programs. The goal of conservation biology should therefore be to examine dynamics that affect environmentally important goals. Social values, and our attempts to understand how to protect them, direct conservation biology by pinpointing crucial natural dynamics that should be understood and protected. Thus, while species are of course important because species are essential participants in natural dynamics, we intend to shift the focus of biodiversity policy to protecting the *health* of socially important natural processes.

This approach eschews purely scientific delineation of goals for conservation biology, and departs from the pure science paradigm. But this approach can be regarded as value-free in another and more realistic sense. Whether elements of nature are valued for themselves (intrinsically) or for future humans (instrumentally), we can provide a scientific argument that it is multigenerational, ecosystem-level dynamics that should be the target of protection policy. Because protecting ecological processes that unfold across multiple generations is the only way to sustain species diversity for future generations, and because we are committed to this policy goal, the question of whether species or future humans are ultimately valued is rendered moot.

A Theory of Scale for Biodiversity Protection

Contemporary philosophy and physical theory have converged to show that there exist many consistent and coherent accounts of reality as we experience it.[6,7] One manifestation of this more general result directly affects scalar questions. Newtonian physics assumed that the world could be understood on a single, unified scale, there being no universal constants in the Newtonian system of description. "Scale therefore becomes all-important," in the words of Ilya Prigogene and Isabelle Stengers, "because the universe is no longer homogeneous, and the synoptic perspective is abandoned in favor of a hierarchically organized, multiscalar and dynamic world."[6]

Choice of system boundaries and scale are therefore an essential part of describing a

system that is to be managed for a given purpose, and thus the best description of a system is one that describes dynamic processes on a scale determinative of priority social goals. We are therefore not bothered by the recognition that ecologists use the concept of an ecosystem loosely and variably. We recognize choices to bound a given system in time and space as decisions based broadly on the usefulness of certain models in understanding targeted physical processes. Since there are many useful ways to understand a system, articulated social goals must direct choices as to how natural systems are described. Choice of the proper scale on which to address an environmental problem such as species loss is therefore an interactive process in which definitions of policy goals guide choices of system boundaries, even as scientific descriptions of processes, and human impacts on them, help us to refine our understanding of policy goals. Determining the correct scale and perspective from which to address environmental problems therefore involves a complex interaction of value definition, concept formation, and scientific description—an interaction in which the articulation of environmental goals drives science (Fig. 1).

We emphasize the development of a physical scale for conservation biology, and assume a high social value on protecting biological diversity. We proceed to combine this assumption with hierarchical principles and to explore the implications of this combination for biodiversity policy. We believe that the emerging concept of ecosystem health, understood in conjunction with hierarchy theory, should guide policy debate. The outcome of that debate, admittedly a political affair (as is any process of value articulation), should in turn guide biological diversity policy.

The difficult theoretical problem we have posed for ourselves is as follows: Given that the scale of ecosystem description is relative to choices regarding the concepts and values we operate with—and these, in turn, are relative to goals and value determinations—how can ecosystem scale and boundaries be constructed on a rational basis? Implicit in this question is the recognition that a choice can be *relative* to certain factors, including public values, without

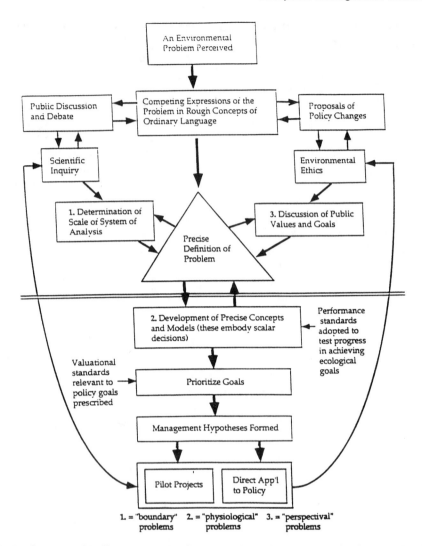

FIGURE 1. The Environmental Policy Process. Environmental problems are not clearly formulated when they first emerge in public discourse. Determination of the proper scale at which a problem should be "modelled" requires an interactive, public process in which public values guide scientific development of models. Once the problem is precisely defined and models developed, the process of experimentation with solutions can begin.

thereby becoming *subjective* (not amenable to rational analysis). Choices to employ certain concepts to describe an ecosystem, and choices to view it on a particular scale, involve tremendous latitude and depend on the goals of the researcher. Nevertheless, these decisions are in fact constrained by the goals of managers as well as those of the researcher. To understand a natural dynamic in order to protect it requires that the dynamic be modelled at a scale relevant to social values.

Scale and Biodiversity Policy

The goal of conserving biological diversity for the benefit of future generations determines the temporal horizon of biodiversity policy. Relying on hierarchy theory we reason that since the policy horizon for this social value is many generations, we must concentrate on a large-scale dynamic such as the dynamic that determines total diversity over landscape level systems. This approach squares with what we know, biologically: landscapes are essentially patchy. Many populations of plants and animals are ephemeral and we cannot save every population of every species, nor can we save every species. As the scale of human activities on the earth increases and human-dominated landscapes prevail more and more, it is inevitable that the rate of ecological change will accelerate. The goal of policy should be to maintain the health of the dynamics that support and retain diversity on a large geographical scale. The proposed approach therefore agrees with advocates of species protection in most cases, although for different reasons. Policies should usually protect species because an accelerating rate of species loss is the best available benchmark of illness in ecological systems. According to our approach, however, the value of species is mainly in their contribution to a larger dynamic, and we do not believe huge expenditures are always justified to save ecologically marginal species. The problem, of course, is to specify what is too large an expenditure and to define "ecologically marginal". On the approach developed here, these definitions must be built upon a theoretically adequate conception of system scale, one that is also useful in guiding protection and restoration efforts and in communicating with the public as it articulates goals and values.

The limits to any dimensional description of a system have been formalized in the discipline of *dimensional analysis,* for which the Buckingham-Pi Theorem provides the foundation.[8] This theorem states that there are a limited number of dimensionless groupings of the physical parameters of a system (expressed in terms of fundamental dimensions of the system) that are sufficient to control the dynamics of the system. A corollary of the Buckingham-Pi theorem, which states that only those dimensional groupings near order unity are important to the dynamical system description, is of special relevance here. That is, if any of the characteristic parameters becomes disproportionate with respect to the scale of the other system phenomena, then it becomes irrelevant to the system description. Either it is so slow as to appear constant or so fast as always to be in equilibrium with other limiting factors. Thus, in a real and quantitative

way, the Buckingham-Pi theorem allows us to circumscribe the domain of applicability—the focal level in a hierarchy—for any given system feature.

Let us, for example, look at how the Buckingham-Pi theorem would apply to a specific management problem. German foresters of the 19th century emphasized production of timber and converted huge areas of the German forest to monocultural spruce. Initially, yields of high-quality timber increased. After three or four iterations, however, yields plummeted. Young trees could not penetrate the soil with their roots and a condition called soil sickness developed.[9] Analysis showed that soil composition had been altered because essential microorganisms had been lost. In this example, descriptions at a particular scale—the scale of economic forestry—had been assumed to provide a complete and unique description of reality. Failure to recognize that timber production is a process that exists as a part of a system that has evolved over centuries and that that system is supported by processes existing on a longer scale than is registered in the language of production forestry resulted in a serious management failure.

The Buckingham-Pi theorem provides a tool by which we can pinpoint the proper scale at which to formulate policy in cases such as this. Suppose at the outset that we are ignorant of the actual dynamics of litter decay and wish to determine which parameters of the system, at what scale, are pertinent. We could list the following parameters as candidates: D_o, the initial density of litter on one square meter of forest floor; F, the rate of litter fall per year; B_o, the initial density of bacteria among the litter; r, the instantaneous rate of litter decay; L, the characteristic length of the patch we are observing; and h, Planck's constant. There are three fundamental dimensions (mass, length, and time) among the six parameters. The Buckingham-Pi theorem says that there will be $3 (= 6 - 3)$ dimensionless groupings that characterize the system dynamics. Without going into the details about how they are determined, those three groupings (pi numbers) may be taken as B_o/D_o, $D_o r/F$ and h/FL^4.

Now we go into the field and laboratory and

actually measure these parameters. Our (hypothetical) estimates are:

$D_o = 1.2$ Kg m^{-2}

$F = 0.5$ Kg m^{-2}y^{-1}

$B = 0.00009$ Kg m^{-2}

$r = 0.41$ y^{-1}

$L = 1$ m

$h = 2.09 \times 10^{-26}$ Kg m^{-2} y^{-1}, Planck's constant which means that:

$B_o/D = 7.5 \times 10^{-5}$

$D_o r/F = 0.99$

$h/FL^4 = 4.18 \times 10^{-26}$

As one could have guessed, the grouping that contains Planck's constant, h, is much less than one, and the corollary to Buckingham-Pi says that this grouping will not influence the dynamics we are observing to any visible extent. That is, phenomena at the submolecular level occur so fast and over such a small space that we need not concern ourselves with those details. What is less obvious is that the first pi number, B_o/D_o, is also very small, so that one need not be concerned with following bacterial concentrations. What this result implies is that some unknown factor (e.g., nitrogen concentration, soil moisture, soil temperature, aeration rate, etc.) is limiting the breakdown of the litter. The bacteria themselves grow very quickly, and their densities will rise and fall in very short order to track the unknown limiting factor.

The dynamics of forest litter are best described in this case by the second grouping, $D_o r/F$, which characterizes the ratio of the decay rate to that of the supply. That the two are very comparable indicates that litter buildup should be a slow process. In fact, one can calculate a characteristic time for the process by dividing the difference between the supply and decay rates by the stock of litter present, i.e.

$$(F - D_o r)/F = 0.00667 \text{y}^{-1}$$

The reciprocal of this accretion rate is 150 years, which accords with our intuition that soil buildup is a very slow process. The slowness of that process explains how, by concentrating on forest production of timber in a single cycle of planting and harvest (8–40 years), the relevant scale for economics, the German foresters ignored a crucial factor in sustainable forest use

and, in the process of simplifying the system, actually impoverished it on a longer scale of time. The scale relevant to a policy of intergenerational sustainability correlates very closely with the ecological parameter of soil build-up because this parameter is associated with maintaining both the diversity and the productivity of the system over multiple generations. According to this hypothetical example, a public concern for long-term sustainability of forest products and for maintenance of forest health implies an approximate horizon of 150 years. A policy horizon of this length is suggested by the recovery time from damage to soil composition and we arrive at a scale based on the production function for leaf litter, the crucial variable if our public concern is indefinite intergenerational sustainability of forest productivity. The characteristic dynamic of forest development is that of carbon retention and the rate of soil build-up is probably the rate-limiting factor most relevant to maintaining and encouraging long-term forest productivity. In this way, an understanding of the production function affecting a public value of intergenerational sustainability, when coupled with a hierarchical understanding of ecosystem structure and functions, determines the proper scale for addressing an environmental problem.

Whole Ecosystem Management

The goal of biological diversity policy should be, given the long time horizon of the policy of intergenerational sustainability, to protect *total diversity at the landscape level of ecological organization*. While we do not intend to reify any given system-description as complete and uniquely correct, we do believe that one can scientifically determine ecosystem boundaries and membranes *provided a priority social goal such as protecting biodiversity* is specified. Ecosystem-level management is distinguished by its concern for characteristics of whole systems—characteristics that cannot be reduced to aggregated characteristics of its parts. The decision to emphasize whole ecosystem management is a decision to employ hierarchical, rather than aggregative models—it is to seek models that integrate policy goals on multiple levels or scales, rather than simply counting bottom-line costs and benefits.[5,10,11] To add a whole eco-system level to a management plan is to resolve to manage, in addition to managing components of the system for resource production, the system as a system. Successful ecosystem management will necessarily be management that has conceptualized the system in a way that focuses attention on the central features of the system—features that are important to supporting important public values. The intuitive idea of ecosystem health is valuable because it focuses attention on the larger systems in nature and away from the special interests of individuals and groups. Competing and special interests, and the goals they articulate, must be integrated into the larger-scale goal of protecting the health and integrity of the larger ecological system.[12] While decisions regarding particular elements of the landscape, especially those in the private sector, will be managed according to economic goals and criteria, biodiversity policy focuses on the larger scale. The regulative idea of a healthy ecological system organizes, tests, and integrates these special interests on a landscape level of organization. A priority goal of conservation biologists must be educating the public toward a better understanding of ecological management, and in helping citizens to articulate their values and to express those values in management decisions.

But the analogy of ecosystem health/integrity is best understood as an intuitive guide, rather than as a specific determinant of policy choices,[13] because, like all analogies, it eventually breaks down. The strength of the medical analogy is that it focuses attention on the overall organization of the system: just as a good physician would not treat a specific organ without paying attention to impacts on the health of the entire organism, whole ecosystem managers must constantly monitor impacts of human activities on the larger ecosystems that form the human environment. The medical analogy is important in emphasizing the importance of systems thinking, and of a recognition of multiple levels and scales on which systems change dynamically.[14] But the medical analogy has an important drawback.[5] Whereas human medicine and veterinary medicine focus on individual organisms and are guided by the un-

questioned goal of protecting the health of patients, ecosystems are multi-scalar and have no obvious identity. No prior, overriding consideration like the Hippocratic oath determines which level of the complex hierarchies of nature should be considered the "whole", organismic level of the system. Managers have considerable latitude in choosing the boundaries, and hence the scale, of the systems they monitor and manage. We believe that choices within this latitude will remain indeterminate until a viable consensus regarding management goals has been articulated. In cases where public goals have been clearly formulated, scientific description of the internal functioning of ecological systems will provide guidance regarding the location of boundaries, and regarding which internal compartments/membranes of the system to emphasize (Fig. 1). A whole ecosystem is a system whose boundaries include essential elements of a dynamic relevant to important social values for a region.

Ecosystem health/integrity therefore stands as the central policy concept to guide ecologically understood environmental management; and, we are arguing, public values—aesthetic, economic and moral—all depend on protecting the processes that support the health of larger-scaled ecological systems. These systems create the context for those activities, and in this sense are crucial elements in their value.[5,11,15] The local, cultural goal of protecting the capacity of systems to react creatively and productively to disturbance, whether footprints of hikers or harvests by oystermen, therefore can sometimes take precedence over the short-term goals of individuals and economic interest groups.

Autopoietic Systems

Whole ecosystem management must be understood as management of a self-organizing system—a system that creates and maintains itself by homeostatic and homeorhetic responses to changing conditions. We describe the creative feature of ecosystems as *autopoiesis* (from the Greek term meaning "self-making").[16–18] Emphasis on autopoiesis implies that the macroscalar boundaries separating the system from its surroundings as well as the smaller-scaled boundaries that separate the system into subsystems or "organs" are chosen to accentuate dynamics essential to sustaining biodiversity. It is not claimed that the features we emphasize are the only features of ecosystems that could be spotlighted, it is claimed only that the scalar choices (boundaries and membranes) represent conceptualizations (models) of the system that are *managerially relevant* and *naturally appropriate* given the goal of protecting healthy ecosystems and their elements over many generations.

While we agree with those, such as Botkin,[14] who emphasize the dynamism of natural systems, we recognize also that dynamically creative change requires a certain amount of stability in the form of larger, slower-changing systems that provide "stable" backgrounds for the processes of iteration and reiteration that allow evolutionary development. Evolutionary creativity on long scales requires creative solutions to environmental constraints that are essentially "fixed" on the scale of individual specimens.

A commitment to ecological sustainability, the resolution to protect complex and creative ecological systems for future generations, assumes the possibility of stability across multiple generations. Stability, here, is treated as a "well-founded illusion of scale." It is an illusion because the system on all levels is constantly dynamic. But this illusion, from a human perspective, is nevertheless well founded because large-scale ecosystems have historically changed sufficiently slowly that there existed continuity of landscape across human generations.

From the environmental standpoint a most important attribute of self-supporting units is their ability to adapt to new circumstances in *creative* ways. This creativity supports the ability of natural systems to rebound in response to heavy economic exploitation and also explains their ability to absorb human wastes. As human activities become ever more intrusive in the systems of nature, these creative adaptations will become even more crucial. Creativity has been perceived as relevant only to conscious, goal-forming agents. But as new developments in physical theory have made clear, the process of creation is ubiquitous in the universe

and at times can even transpire in systems not containing living members.[6] Ulanowicz[19] has argued that the capacity for creativity constitutes the crux of what is normally referred to as *ecosystem health*. But the capacity for creativity is too often misperceived, which comes as no surprise, given the difficulties in describing it in semantic, much less quantitative terms. The emerging consensus[19–21] indicates that creative action is contingent upon two mutually exclusive properties of the performing system.

First, it is necessary that any system capable of solving a novel problem possess a requisite amount of ordered complexity. Order implies constraints — events impinging upon the system or subsystem must initiate a channeled sequence of reactions (which may be and probably must be reflexive to some extent) that culminate in the response of the system to that input, e.g. compensation, indifference, counteraction, co-option, etc. Without such coherence, creativity is impossible, and Atlan[20] demonstrates how thresholds in ordered complexity must be surpassed before a system is capable of creative action. This side of creativity is widely understood. It is unquestioned that an organism or system must possess enough "apparatus" before creativity is possible. But some of the most tightly ordered objects in the universe are machines — artifacts that are incapable of truly creative actions, primarily because they lack an adequate degree of inherent disorder.

It is not so universally acknowledged (or, in many cases, even suspected) that incoherence is also a prerequisite for creative action. Before creativity is possible, a system must possess a potential "reservoir" of stochastic, disconnected, inefficient features that constitute the raw building blocks of effective innovation. In the course of normal functioning such disutility appears as an "overhead" or an encumbrance. However, when faced with a perturbation or problem, it is this background of dysfunctional repertoires that is utilized to meet the exigency. Background species or marginally extant trophic pathways — system redundancies — can be activated in response to a disruption of the normally dominant means an ecosystem employs to process material and energy. If the disturbance is recurrent or persistent, the new response eventually will be incorporated into permanent coherent structure. This idea of freedom resonates in public values with the emphasis on wilderness protection and with the importance placed on protecting wildness wherever possible and appropriate.[5,22]

The concepts of order and incoherence may seem subjective to some, but Ulanowicz[23] has suggested that it is possible to employ results from information theory (quantitative epistemology) to estimate the relative amounts of each of these attributes possessed by a given system. To attach numbers to these system properties, it is necessary first to describe the system as a collection of subunits linked together by processes that can be quantified. For example, ecosystems are often described as a collection of species or other aggregations of organisms linked one to another by exchanges of material or energy. These exchanges can be assigned physical units and measured or otherwise estimated in the field or laboratory.

Once the ecosystem has been bounded and then characterized as a network of palpable flows, one can employ information theory to quantify the diversity of flows in this ensemble as if each flow were independent of all others. Of course, the exchanges do not occur in random, unconnected fashion. There is an order in the pattern of trophic connections and temporal sequences. Such order gives rise to a component of the overall diversity of flows as computed by a variable called the average mutual information of the network topology.[24] Ulanowicz[5] has given the name *ascendancy* to a scaled version of the mutual information. Systems with more clearly defined pathways of cause and effect will exhibit higher values of ascendancy. One can rigorously prove that the mutual information and linkages can never exceed the measure for the diversity of flows. This condition has led Ulanowicz[25,26] to call the latter term the system *capacity* for growth and development. System capacity obviously is tightly coupled with the biodiversity of the system. We are here hypothesizing that this idea may also serve as the link between the intuitively understood policy concept of ecosystem health and the more precise, quantitative disciplines of systems theory and information analysis.

The amount by which the capacity exceeds the mutual information has been called the system *overhead*. All those system features which contribute nothing to its order and coherence by definition add to its overhead. These include redundant and inefficient pathways, stochastic and ill-phased events, etc.

In terms of these three concepts — ascendancy, capacity, and overhead — one can enumerate the requirements for a system to act creatively in response to a novel circumstance: (i) The system must have a high capacity for growth and development, i.e. its biodiversity and complexity must remain high. (ii) Most of this capacity needs to be expressed as ordered and coherent ascendancy. (iii) Some capacity must remain as unstructured and incoherent overhead to afford the system the degrees of freedom necessary to respond to novel environmental stimuli.

A biotic system satisfying all three requirements can be termed "healthy."[19] Thus, we can suggest a definition of ecosystem health for public policy consideration: "An ecological system is healthy and free from 'distress syndrome' if it is stable and sustainable, i.e. if it is active and maintains its organization and autonomy over time."[27] The goal of sustaining ecosystem health so defined therefore involves maintaining a capacity for autopoietic activity on the scale relevant to many human generations.

The Value of Biodiversity

One advantage of the approach to scale and policy goals sketched here is that it bypasses intransigent value questions and focuses attention on concrete and achievable goals. It does so by reversing the usual valuational methods of utilitarians and economists, who place a price-value on species and then aggregate toward a total value for ecosystems. On our approach, the policy-driving values are ecosystem-level processes; we save species *both* because we value them directly (at least in many cases) *and* because of their roles in ecosystem processes. But since the processes must in the long run protect the species, the question of ultimate value, species or ecosystems, will arise only in those cases where large expenditures are required to save an ecologically marginal species.

If we are committed to saving species/ biodiversity for future generations and wish to introduce dollar figures into policy debates, we should estimate the total value of the ecosystem dynamic that protects species to be equivalent to costs that would be incurred to maintain individual species in alternative ways. If we do not protect species in the wild, they must then be protected in zoos or other artificially managed areas. The cost of artificial protection would be prohibitive for more than a few species. We therefore adopt the intermediate goal — which is instrumental to the goal of protecting ecosystem processes — of protecting as many species as possible. But this is not to say that one would never declare a particular species too expensive to save, given its ecological role. The obligation to protect species is therefore best understood in the terms of the *Safe Minimum Standard* as formulated by Ciriacy-Wantrup and developed by Richard Bishop.[28,29] Endangered species policy should be governed by the rule: protect all species, as long as the costs are bearable.

This approach to valuation also suggests a new sort of partnership between biologists, economists, and the public. Emphasis on the self-perpetuating features of ecological systems and their role in achieving social goals such as species preservation implies highest priority for studies that promise to characterize the structures, functions, and processes that make an ecological system a habitat capable of perpetuating species for many generations. Economists also have important roles. By developing new methods of valuation for deciding policy priorities and by determining costs of various alternatives for maintaining functioning habitats, economists can make protection efforts more efficient. Especially, they must develop incentive systems that will encourage healthy economies that are compatible with protecting ecosystem health. Since ecosystem health is as much an evaluation as it is a descriptive concept, both economists and ecologists must work to inform the public about management options and work to develop scientific models that both express and, through an interactive process, improve values. It is therefore a high priority to develop new methods of valuation that are sufficiently interactive to contribute to the dynamic process of defining and protecting ecosystem health.

Conclusion

Ecosystematic, hierarchical management recognizes that many choices we make, both individually and collectively, will introduce disturbances on a local scale; as when a field is plowed, a fire set, or a forest plot harvested. The recommendation that we manage for ecosystem health as well as for productivity in the various cells of the system implies that, when we disturb a wetland, for example, we will look also at the impacts of the disturbance on the larger level of the landscape. This approach would recognize that, in managing *particular* fields or wetlands, we usually seek to maximize productivity and economic efficiency. One might call management on this cellular level *resource management*. But the hierarchical approach also recognizes more inclusive levels of management, levels where we are concerned about the healthy functioning of the creative systems of nature and about the continued existence, across the landscape, of indigenous species and distinctive ecological communities — what is popularly called "biodiversity". We have emphasized that a hierarchical, whole ecosystem approach to management recognizes multiple levels of system organization; the scale on which an environmental problem is addressed must depend on the public goals that are given prominence.

Because the public derives many values — economic, cultural, and aesthetic — from the landscape, no single ranking of environmental goals can be adequate to guide public policy. Hierarchical thinking helps us to avoid policy gridlock, however, if we recognize that successful policy will encourage a patchy landscape. On the level of field or farm, economic criteria will predominate, while on the ecosystem level we must manage for total diversity and complexity. Here, macroscalar criteria must guide the development of incentives that protect ecosystem health. This general approach seeks integration of levels; it places priority on finding new and various methods and procedures, and on arranging economic incentives to encourage economic development that has minimal negative impact on large ecological systems. This process will be political. A variety of economically efficient policies will be delineated; simultaneously, expectations will be set for maintenance of the health of the larger system that perpetuates complexity and total diversity. Good policies will be those that fulfill key criteria on both levels (Fig. 2).

The political process of developing a biodiversity policy for any given region should be guided by three central principles.

I. Efforts at maintaining biodiversity should be directed at maintaining the total diversity of the landscape over multiple generations. Landscape-level goals must be defined more precisely by increased articulation of biodiversity values within a locale. Good management will require public dialogue as much as expert opinion because the definition of goals and development of scientific understanding is an interactive and experimental process.

II. Diversity must be understood dynamically, in terms of healthy processes, rather than merely as maintenance of current elements of the system. The development of landscape-level models will involve choosing a scale and a perspective from which to both understand and manage large ecological systems. Dimensional analysis, combined with information theory and applied to hierarchical models, can provide techniques to help pinpoint dynamics associated with important public values and their support.

III. Economic activities that complement and enhance, rather than oppose and degrade, ecological processes are to be preferred and encouraged. Recognizing that natural systems will react creatively to change, we should develop economic incentives to encourage economic development that mimics natural disturbances.[30]

References

1. Allen, T.F.H. and Starr, T. B. 1982. *Hierarchy: Perspectives for Ecological Complexity*. University of Chicago Press, Chicago.
2. Hayden, F.G. 1989. *Survey of Methodologies for Valuing Externalities and Public Goods*. Prepared for the U.S. Environmental Protection Agency, Office of Environmental Planning. Contract number: 68-01-7363.
3. O'Neil, R.V. 1988. Hierarchy theory and global change. In: *Scales and Global Change*. Rosswall, T., Woodmansee, R.G. and Risser, P.G. (eds). John Wiley and Sons, Ltd., New York.
4. Norton, B.G. 1990. Context and hierarchy in Aldo Leopold's theory of environmental management. *Ecol. Econ.*
5. Norton, B.G. 1991. *Toward Unity Among Environmentalists*. Oxford University Press, New York.
6. Prigogine, I. and Stengers, I. 1984. *Order Out of Chaos: Man's New Dialogue with Nature*. Bantam, New York.
7. Quine, W.V.O. 1960. *Word and Object*. MIT Press, Cambridge, MA.
8. Long, R.R. 1963. Dimensional analysis. In: *Engineering Science Mechanics*. Ch. 9, Prentice-Hall, Englewood Cliffs, NJ.
9. Meine, C. 1988. *Aldo Leopold: His Life and Work*. University of Wisconsin Press, Madison, WI.
10. Page T. 1977. *Conservation and Economic Efficiency*. Johns Hopkins University Press.
11. Page, T. 1991. Sustainability and the problem of valuation. In: *The Ecological Economics of Sustainability*. Costanza, R. (ed.). Columbia University Press.
12. Leopold, A. 1949. *Sand County Almanac*. Oxford University Press, London.
13. Ehrenfeld, D. 1992. Ecosystem health and eco-

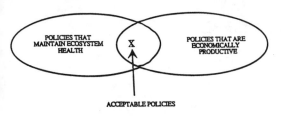

FIGURE 2. A hierachical system of analysis does not seek a single, synoptic "bottom-line" determination of the best policy from every perspective, but rather defines (through public debate and political processes) policies that will protect ecosystem health and policies that fulfill essential ecological and economic criteria. If no such policies have been proposed, there should be an intense effort to develop them.

logical theories. In: *Ecosystem Health: New Goals in Environmental Management*. Costanza, R., Norton, B. and Haskell, B. (eds). Island Press, Covelo, CA.

14. Botkin, D.B. 1990. *Discordant Harmonies*. Oxford University Press, New York.

15. Page, T. 1992. Environmental existentialism. In: *Ecosystem Health: New Goals in Environmental Management*. Costanza, R., Norton, B. and Haskell B. (eds). Island Press, Covelo, CA.

16. Maturna, H.R. and Varela, F.J. 1980. Autopoiesis: The organization of the living. In: *Autopoiesis and Cognition*, Maturna, H.R. and Varela, F.J. (eds). D. Reidel Publishing Co., Boston.

17. Rees, W.R. 1990. The ecology of sustainable development. *The Ecologist 20*, 18–23.

18. Callicott, J.B. 1989. *In Defense of the Land Ethic: Essays in Environmental Philosophy*. State University of New York Press, Albany.

19. Ulanowicz, R.E. 1986. A phenomenological perspective of ecological development. In: *Aquatic Toxicology and Environmental Fate: Ninth Volume*. Posten, T.M. and Purdy, R. (eds). American Society for Testing and Materials, Philadelphia, ASTM STP 921, p. 73–81.

20. Atlan, H. 1974. On a formal definition of organization. *J. Theor. Biol. 45*, 295–304.

21. Wagensberg, J., Garcia, A. and Sole, R.V. 1990. Connectivity and information transfer in flow networks: Two magic numbers in ecology? *Bull. Math. Biol. 52*, 733–740.

22. McNamee, T.M. 1986. Putting nature first: A proposal for whole ecosystem management. *Orion Nature Quarterly 5*, 5–19.

23. Ulanowicz, R.E. 1986. *Growth and Development: Ecosystems Phenomenology*. Springer-Verlag, New York.

24. Rutledge, R.W., Basorre, B.L. and Mulholland, R.J. 1976. Ecological stability: An information theory viewpoint. *J. Theor. Biol. 57*, 355–371.

25. Ulanowicz, R.E. 1980. An hypothesis on the development of natural communities. *J. Theor. Biol. 85*, 233–245.

26. Ulanowicz, R.E. and Norden, J.S. 1990. Symmetrical overhead in flow networks. *Int. J. Sys. Sci. 21*, 429–437.

27. Haskell, B.D., Norton, B.G. and Costanza, R. (eds). 1992. *Ecosystem Health: New Goals for Environmental Management*. Island Press, Covelo, CA.

28. Ciriarcy-Wantrup, S.V. 1959. *Resource Conservation: Economics and Politics*. University of California Division of Agricultural Services, Berkeley and Los Angeles.

29. Bishop, R. 1978. Endangered species and uncertainty: the economics of safe minimum standards. *Am. J. Agric. Econ. 60*.

30. This research was supported by the U.S. Environmental Protection Agency, Office of Policy, Planning and Evaluation, through a cooperative agreement with Chesapeake Biological Laboratory. Benjamin Haskell contributed to the research for this paper.

32
Population, Sustainability, and Earth's Carrying Capacity

Gretchen C. Daily and Paul R. Ehrlich

The twentieth century has been marked by a profound historical development: an unwitting evolution of the power to seriously impair human life-support systems. Nuclear weapons represent one source of this power. Yet, even the complexities of global arms control are dwarfed by those inherent in restraining runaway growth of the scale of the human enterprise, the second source of possible disaster. Diminishing the nuclear threat involves relatively few parties, well-established international protocols, alternate strategies that carry easily assessed costs and benefits, short- and long-term incentives that are largely congruent, and widespread recognition of the severity of the threat. In contrast, just the opposite applies to curbing the increasingly devastating impact of the human population. In particular, the most personal life decisions of every inhabitant of the planet are involved and these are controlled by socioeconomic systems in which the incentives for sacrificing the future for the present are often overwhelming.

This article provides a framework for estimating the population sizes and lifestyles that could be sustained without undermining the potential of the planet to support future generations. We also investigate how human activity may increase or reduce Earth's carrying capacity for *Homo sapiens*. We first describe the current demographic situation and then examine various biophysical and social dimensions of carrying capacity.

Our analysis is necessarily preliminary and relatively simple; we anticipate that it will undergo revision. Nonetheless, it provides ample basis for policy formulation. Uncertainty about the exact dimensions of future carrying capacity should not constitute an excuse to postpone action. Consider the costs being incurred today of doing so little to halt the population explosion, whose basic dimensions were understood decades ago.

The Current Population Situation

The human population is now so large and growing so rapidly that even popular magazines are referring to the possibility of a "demographic winter" (*Time* 1991). The current population of 5.5 billion, growing at an annual rate of 1.7%, will add approximately 93 million people this year, equivalent to more than the population of Mexico (unless otherwise noted, demographic statistics are from, or projected from, PRB 1991).

Growth rates vary greatly from region to region. The combined population of less-developed nations (excluding China) is growing

Reprinted with permission from Bioscience vol. 42, pp. 761–771. Copyright 1992 American Institute of Biological Sciences.

at approximately 2.4% annually and will double in 30 years if no changes in fertility or mortality rates occur. The average annual rate of increase in more-developed nations is 0.5%, with an associated doubling time of 137 years. Many of those countries have slowed their population growth to a near halt or have stopped growing altogether.

The regional contrast in age structures is even more striking. The mean fraction of the population under 15 years of age in more-developed countries is 21%. In less-developed countries (excluding China) it is 39%; in Kenya it is fully 50%. Age structures so heavily skewed toward young people generate tremendous demographic momentum. For example, suppose the total fertility rate (average completed family size) of India plummets over the next 33 years from 3.9 to 2.2 children (replacement fertility). Under that optimistic scenario (assuming no rise in death rates), India's population, today some 870 million, would continue to grow until near the end of the next century, topping out at approximately 2 billion people.

The slow progress in reducing fertility in recent years is reflected in the repeated upward revisions of United Nations projections (UNFPA 1991). The current estimate for the 2025 population is 8.5 billion, with growth eventually leveling off at approximately 11.6 billion around 2150. These projections are based on optimistic assumptions of continued declines in population growth rates.

Despite the tremendous uncertainty inherent in any population projections, it is clear that in the next century Earth will be faced with having to support at least twice its current human population. Whether the life-support systems of the planet can sustain the impact of so many people is not at all certain.

Environmental Impact

One measure of the impact of the global population is the fraction of the terrestrial net primary productivity (the basic energy supply of all terrestrial animals) directly consumed, co-opted, or eliminated by human activity. This figure has reached approximately 40% (Vitousek et al. 1986). Projected increases in pop-

ulation alone could double this level of exploitation, causing the demise of many ecosystems on whose services human beings depend.

The impact (I) of any population can be expressed as a product of three characteristics: the population's size (P), its affluence or per-capita consumption (A), and the environmental damage (T) inflicted by the technologies used to supply each unit of consumption (Ehrlich and Ehrlich 1990, Ehrlich and Holdren 1971, Holdren and Ehrlich 1974).

$$I = PAT$$

These factors are not independent. For example, T varies as a nonlinear function of P, A, and rates of change in both of these. This dependence is evident in the influence of population density and economic activity on the choice of local and regional energy-supply technologies (Holdren 1991a) and on land management practices. Per-capita impact is generally higher in very poor as well as in affluent societies.

Demographic statistics give a misleading impression of the population problem because of the vast regional differences in impact. Although less-developed nations contain almost four-fifths of the world's population and are growing very rapidly, high per-capita rates of consumption and the large-scale use of environmentally damaging technologies greatly magnify the impact of industrialized countries.

Because of the difficulty in estimating the A and T factors in isolation, per-capita energy use is sometimes employed as an imperfect surrogate for their product. Using that crude measure, and dividing the rich and poor nations at a per-capita gross national product of $4000 (1990 dollars), each inhabitant of the former does roughly 7.5 times more damage to Earth's life-support systems than does an inhabitant of the latter (Holdren 1991a). At the extremes, the impact of a typical person in a desperately poor country is roughly a thirtieth that of an average citizen of the United States. The US population has a larger impact than that of any other nation in the world (Ehrlich and Ehrlich 1991, Holdren 1991a,b).

The population projections and estimates of total and relative impact bring into sharp focus a question that should be the concern of every biologist, if not every human being: how many people can the planet support in the long run?

The Concept of Carrying Capacity

Ecologists define carrying capacity as the maximal population size of a given species that an area can support without reducing its ability to support the same species in the future. Specifically, it is "a measure of the amount of renewable resources in the environment in units of the number of organisms these resources can support" (Roughgarden 1979, p. 305) and is specified as K in the biological literature. *Carrying capacity* is a function of characteristics of both the area and the organism. A larger or richer area will, *ceteris paribus*, have a higher carrying capacity. Similarly, a given area will be able to support a larger population of a species with relatively low energetic requirements (e.g., lizards) than one at the same trophic level with high energetic requirements (e.g., birds of the same individual body mass as the lizards). The carrying capacity of an area with constant size and richness would be expected to change only as fast as organisms evolve different resource requirements. Though the concept is clear, carrying capacity is usually difficult to estimate.

For human beings, the matter is complicated by two factors: substantial individual differences in types and quantities of resources consumed and rapid cultural (including technological) evolution of the types and quantities of resources supplying each unit of consumption. Thus, carrying capacity varies markedly with culture and level of economic development.

We therefore distinguish between biophysical carrying capacity, the maximal population size that could be sustained biophysically under given technological capabilities, and social carrying capacities, the maxima that could be sustained under various social systems (and, especially, the associated patterns of resource consumption). At any level of technological development, social carrying capacities are necessarily less than biophysical carrying capacity, because the latter implies a human factory-farm lifestyle that would be not only universally undesirable but also unattainable because of inefficiencies inherent in social resource distribution systems (Hardin 1986). Human ingenuity has enabled dramatic increases in both biophysical and social carrying capacities for *H. sapiens*, and potential exists for further increases.

Carrying capacity today. Given current technologies, levels of consumption, and socioeconomic organization, has ingenuity made today's population sustainable? The answer to this question is clearly no, by a simple standard. The current population of 5.5 billion is being maintained only through the exhaustion and dispersion of a one-time inheritance of natural capital (Ehrlich and Ehrlich 1990), including topsoil, groundwater, and biodiversity. The rapid depletion of these essential resources, coupled with a worldwide degradation of land (Jacobs 1991, Myers 1984, Postel 1989) and atmospheric quality (Jones and Wigley 1989, Schneider 1990), indicate that the human enterprise has not only exceeded its current social carrying capacity, but it is actually reducing future potential biophysical carrying capacities by depleting essential natural capital stocks.

The usual consequence for an animal population that exceeds its local biophysical carrying capacity is a population decline, brought about by a combination of increased mortality, reduced fecundity, and emigration where possible (Klein 1968, Mech 1966, Scheffer 1951). A classic example is that of 29 reindeer introduced to St. Matthew Island, which propagated to 6000, destroyed their resource base, and declined to fewer than 50 individuals (Klein 1968). Can human beings lower their per-capita impact at a rate sufficiently high to counterbalance their explosive increases in population?

Carrying capacity for saints. Two general assertions could support a claim that today's overshoot of social carrying capacity is temporary. The first is that people will alter their lifestyles (lower consumption, A in the $I = PAT$ equation) and thereby reduce their impact. Although we strongly encourage such changes in lifestyle, we believe the development of policies to bring the population to (or below) social carrying capacity requires defining human beings as the animals now in existence. Planning a world for highly cooperative, anti-materialistic, ecologically sensitive vegetarians would be of little value in correcting today's situation. Indeed, a statement by demographer Nathan Keyfitz (1991) puts into perspective the view that behavioral changes will keep *H. sapiens* below social carrying capacity:

If we have one point of empirically backed knowledge, it is that *bad policies are widespread and persistent*. Social science has to take account of them [our emphasis].

In short, it seems prudent to evaluate the problem of sustainability for selfish, myopic people who are poorly organized politically, socially, and economically.

Technological optimism. The second assertion is that technological advances will sufficiently lower per-capita impacts through reductions in T that no major changes in lifestyle will be necessary. This assertion represents a level of optimism held primarily by nonscientists. (A 1992 joint statement by the US National Academy of Sciences and the British Royal Society expresses a distinct lack of such optimism.) Technical progress will undoubtedly lead to efficiency improvements, resource substitutions, and other innovations that are currently unimaginable. Different estimates of future rates of technical progress are the crux of much of the disagreement between ecologists and economists regarding the state of the world. Nonetheless, the costs of planning development under incorrect assumptions are much higher with overestimates of such rates than with underestimates (Costanza 1989).

A few simple calculations show why we believe it imprudent to count on technological innovation to reduce the scale of future human activities to remain within carrying capacity. Employing energy use as an imperfect surrogate for per-capita impact, in 1990 1.2 billion rich people were using an average of 7.5 kilowatts (kW) per person, for a total energy use of 9.0 terawatts (TW; 10^{12} watts). In contrast, 4.1 billion poor people were using 1 kW per person, and 4.1 TW in aggregate (Holdren 1991a). The total environmental impact was thus 13.1 TW.

Suppose that human population growth were eventually halted at 12 billion people and that development succeeded in raising global per capita energy use to 7.5 kW (approximately 4 kW below current US use). Then, total impact would be 90 TW. Because there is mounting evidence that 13.1 TW usage is too large for Earth to sustain, one needs little imagination to picture the environmental results of energy expenditures some sevenfold greater. Neither physicists nor ecologists are sanguine about improving technological performance sevenfold in the time available.

There is, indeed, little justification for counting on technological miracles to accommodate the billions more people soon to crowd the planet when the vast majority of the current population subsists under conditions that no one reading this article would voluntarily accept. Past expectations of the rate of development and penetration of improved technologies have not been fulfilled. In the 1960s, for example, it was widely claimed that technological advances, such as nuclear agroindustrial complexes (e.g., ORNL 1968), would provide 5.5 billion people with food, health care, education, and opportunity. Although the Green Revolution did increase food production more rapidly than some pessimists (e.g., Paddock and Paddock 1967) predicted, the gains were not generally made on a sustainable basis and are thus unlikely to continue (Ehrlich et al. 1992). At present, approximately a billion people do not obtain enough dietary energy to carry out normal work activities.

Furthermore, as many nonscientists fail to grasp, technological achievements cannot make biophysical carrying capacity infinite. Consider food production, for example. Soil can be made more productive by adding nutrients and irrigation; yields could possibly be increased further if it were economically feasible to grow crops hydroponically and sunlight were supplemented by artificial light. However, biophysical limits would be reached by the maximal possible photosynthetic efficiency. Even if a method were found to manufacture carbohydrates that was more efficient than photosynthesis, that efficiency, too, would have a maximum. The bottom line is that the laws of thermodynamics inevitably limit biophysical carrying capacity (Fremlin 1964) if shortages of inputs or ecological collapse do not intervene first.

Sustainability

A sustainable process is one that can be maintained without interruption, weakening, or loss of valued qualities. Sustainability is a necessary and sufficient condition for a population to be at or below any carrying capacity. Sustainable development has thus been defined as "development that meets the needs and aspirations of the present without compromising the ability of future generations to meet their own needs" (Brundtland 1987, p. 43). Implicit in the desire for sustainability is the moral conviction that the current generation should pass on its inheritance of natural wealth, not unchanged, but undiminished in potential to support future generations.

In any discussion of sustainability, it is clearly necessary to establish relevant temporal and spatial scales. The time scale that will be considered here is tens of human generations—that is, hundreds of years to a millenium. The spatial scale is obviously constrained by the size of the planet, a closed system for most purposes. Though trade enables populations to sustainably exceed local and regional carrying capacities, all accounts must balance for Earth as a whole.

Classification of resources. How does one determine a sustainable level of consumption? To address this question, we start by specifying several resource types and analyzing the con-

straints on their use independently. Then, the paramount importance of interactions deriving from the simultaneous use of a resource required for multiple activities is considered. We also highlight throughout means by which humanity could increase the maximum sustainable levels of resource consumption (dimensions of biophysical carrying capacity).

Our scheme involves the somewhat arbitrary classification of continuously distributed elements into discrete units to bring into focus key aspects of sustainability. First, there are the resources that provide free services to humanity without necessarily undergoing depletion or degradation (Table 1, first column). These resources include microbial nutrient cyclers and soil generators, natural pest-control agents, and pollinators of crops. Of special importance are the forests, which help to maintain a balance of gases in the atmosphere, to ameliorate local climate, to provide habitat for wildlife, to control erosion, and to run the hydrologic cycle. Other resources, such as food, drinking water, energy, and the capacity of the environment to absorb pollutants, are necessarily consumed, dispersed, or degraded as the benefits are derived from them.

Second, there is an important distinction in practice between renewable and nonrenewable resources, although renewal rates are continuously distributed. Renewable resources tend to

TABLE 1. Resource classification scheme with some examples. This classification scheme makes explicit three key parameters that determine the nature of the maximum sustainable level of use for a particular resource. The examples provided are by no means exhaustive.

Resource type		Not necessarily degraded or dispersed in use	Necessarily degraded or dispersed in use
Nonrenewable (at current use rates)	Essential	Stratospheric ozone, tropical forests, biodiversity	Time or opportunity
	Substitutable	Materials that supply some services (e.g., diamonds and gold for aesthetic and wealth repository purposes)	Nonrenewable energy sources (e.g., fossil fuels), some other minerals
Renewable	Essential	Ecosystem elements that supply services (e.g., soil microbes, some temperate forests, pollinators)	Solar energy, fresh water; some soil used for agriculture
	Substitutable	Species that supply some services (e.g., animals for power, transport, insulin, and vaccines; trees for cooling buildings)	Wood for construction, any particular food type

be flow-limited and are reconstituted after human consumption or dispersion through natural processes driven by solar energy (which may be enhanced by human investment, as when trees are planted). Nonrenewable resources are generally stock-limited and have either very low or no renewal rates and prohibitive reconstitution costs (though one or more recyclings before ultimate discard may be possible; Ehrlich et al. 1977). The rate of degradation and erosion of topsoil (according to one estimate a net 25 billion tons erosion loss per year; Brown and Wolf 1984) is so much in excess of its rate of creation that soil has been turned into an essentially nonrenewable resource on any relevant time scale. The same can be said of groundwater in many aquifers (e.g., Wittwer 1989) and biodiversity (Ehrlich and Wilson 1991).

Last, resources may be further classified into two types: those for which substitutes are either currently or foreseeably available (substitutable resources) and those for which complete substitution at the required scale is currently and foreseeably impossible (essential resources). Substitutable resources include fossil fuels, some metals and minerals, and some natural fibers. Essential resources include fertile soils, fresh water, and biodiversity. The classification of some resources may vary depending on the manner in which they are used; for example, forests as sources of wood are substitutable resources because wood has substitutes for most purposes, whereas forests as sources of ecosystem services generally constitute essential resources.

Maximum sustainable use. The maximum sustainable level of use (MSU) of a resource depends on how it is classified with respect to the preceding attributes and on socioeconomic factors. Using the classification scheme of Table 1, we may now specify a theoretical MSU for each resource type, which represent dimensions of biophysical carrying capacity.

In the case of resources from which humanity *may* benefit without causing their depletion or degradation, MSU is proportional to the total extent of the resource: the greater the forested area, for example, the greater the scale of ecosystem services provided by forest. In this general case, sustaining maximal use is a matter

of safeguarding the ability of such resources to provide humanity with services. For uses that necessarily alter resources in the process of deriving benefits from them, MSU depends on the resource's renewability and substitutability.

Let us consider nonrenewables first. No resources that are absolutely essential for human life have been classically considered nonrenewable except those for which supplies are so large (e.g., calcium) as to make worrying about them pointless. A notable exception may be time or opportunities to prevent irreversible, possibly catastrophic consequences of anthropogenic impacts on the environment.

Numerous nonrenewable substitutable resources are critical to maintaining certain features of today's civilization, although their disappearance would not threaten human existence. Iron, for example, is used heavily in the production and transport of energy and goods in industrial societies. By definition, there is no sustainable rate of consumption of nonrenewables; the closest approximation is a quasisustainable consumption rate equivalent to (or lower than) the rate of generation of substitutes. The primary difficulty in the use of nonrenewables is not exhaustion per se (because quantities are generally gigantic), but rather the technical, economic, environmental, and sociopolitical difficulties associated with declining quality of the resources (with respect to, for example, distance, depth, and concentration) and with the transitions to substitutes (e.g., Holdren 1991a).

At first glance, it might seem that stocks and flows of renewable resources would require the least effort to maintain simply because they are regenerated for us. However, increasing human demands on the biophysical environment make it difficult to limit the use of many renewable resources to a sustainable rate. It is therefore critical to consider how MSUs of renewable resources vary as a function of those stocks, that is, how human activity may increase or reduce those elements of biophysical carrying capacity.

For a renewable essential resource that is necessarily consumed, degraded, or dispersed in the extraction of value from it, the MSU is equivalent to its renewal rate. MSU (and maximum sustainable yield) increases monotonically with the global extent of resource stocks (e.g., agricultural soils, harvested forest, and groundwater) above a critical point. As the land area covered with productive agricultural soil, supporting intact forest, or underlaid by freshwater aquifers is reduced, the MSU of these resources is proportionally diminished. The minimum represents the point below which the constituent stocks are so small that the resource cannot be used sustainably. For example, very thin soils are agriculturally unproductive (UNEP 1984), and regeneration of trees may fail in small remnant forest patches subject to deleterious edge and isolation effects.

Interestingly, surface water also features a linear relationship between MSU and stock, and it illustrates a case where humanity may increase MSU by altering the spatial and temporal distribution of the resource. Although humanity exercises substantial control over the distribution of water among different (natural or artificial) channels and reservoirs (White 1988), it has relatively little direct control of the total stock. Furthermore, silting of dams and salinization of agricultural water may represent barriers to increasing the long-term MSU of water through anthropogenic manipulation. Recently, humanity has unwittingly reduced the total annual input to some surface water systems through deforestation and desertification (Myers 1989). More dramatic changes in regional stocks of surface water are expected as a consequence of global warming (Gleick 1989, Schneider 1990, Tegart et al. 1990).

The extraction of resources is generally managed not at the global spatial scale but at local or regional levels. Several functional relationships between MSU and a single local resource stock are possible. The curve in Figure 1a describes a general relationship between MSU of agricultural soil and the stock (soil depth). While soil depth remains sufficiently greater than the rooting depth of crops or other plants, soil loss has little or no negative effect on productivity, but productivity decreases with soil depth below this threshold. Initially negligible costs of losing soil to erosion may become steep as soil thins below this threshold (called the critical point, C*). The soil depth on most of the cropland in Haiti appears to be substantially below C* (Terborgh 1989, WRI 1992a).

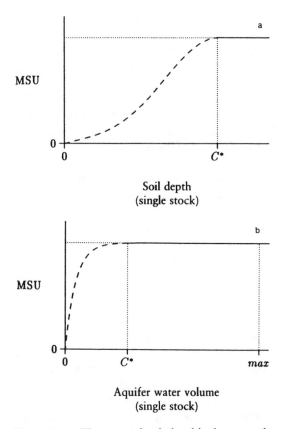

Figure 1. **a.** The general relationship between the maximum sustainable level of use (MSU) of soil and its depth. **b.** The general relationship between the MSU of an aquifer and its stock. The precise functional forms below C* are uncertain.

Agricultural productivity worldwide is suffering because of such land degradation (UNEP 1984).

The local depletion of aquifers also exemplifies this general relationship between a single stock of a renewable resource and its MSU (Figure 1b). MSU is equivalent to the rate of recharge at any stock above C*. MSU is constant across nearly all values of stock because the renewal rate is largely stock independent. At stock levels below C*, aquifers may suffer from salinization or collapse (Dunne and Leopold 1978), reducing MSU.

There are two important differences between the management of soils and aquifers, although the functional forms below the critical point are uncertain. First, many aquifers contain orders of magnitude more water than the critical volume, whereas soils are rarely more than a few times deeper than the critical depth. Second, MSU of water from aquifers may decline more rapidly below C* than that of many soils (NAS 1989).

A hypothetical relationship between MSU and a forest harvestable at maximum sustainable yield is depicted in Figure 2. Though the precise functional form depends on forest type and harvesting method, the rate of forest regeneration is highest at a biomass density below the maximum attainable. At extremely high densities, trees suffer from overcrowding; at very low densities, microclimatic and other conditions may become unfavorable for germination and sapling recruitment.

Where resources in high demand and in short supply are overharvested, a positive feedback cycle is established, thereby sequentially depleting the stocks and lowering the MSUs. For example, overharvesting of fuelwood, the primary source of energy for more than half of the world's population, has created severe local and regional shortages. To supply domestic energy, these shortages are countered by overharvesting increasingly distant supplies and by burning animal dung and crop residues, important inputs to the maintenance of soil productivity (WRI 1992b). For any essential resources that may limit the size of the human population (e.g., fertile soil, forest products and services, and fresh water), depletion constitutes a reduction in biophysical carrying capacity of the planet.

MSUs of renewable substitutable resources

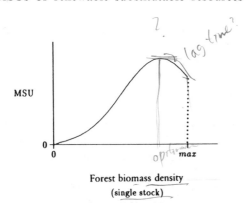

FIGURE 2. The general relationship between the maximum sustainable level of use (MSU) of a forest under harvest and its biomass density.

that are necessarily consumed, degraded, or dispersed are also equivalent to their renewal rates (that may be enhanced by human investment). Maintenance of the function served by such resources could also be sustained if the supply were exhausted at a rate less than or, at most, equivalent to the rate of generation of substitutes. Thus, coal and then petroleum and other substitutes replaced wood as a primary source of industrial energy.

Maximum sustainable abuse. We next consider the passive use of natural biogeochemical processes to absorb waste and to reconstitute component resources therein, also elements of biophysical carrying capacity. This analysis on sustaining output rates complements the foregoing one that concerns sustaining input rates. The maximal sustainable emission rate of a pollutant into the environment (maximum sustainable level of abuse; MSA) is defined as the rate above which unacceptable damage is caused. Specifying levels of damage that are unacceptable is a subject of a complex literature on risk analysis (see for example, Ehrlich and Ehrlich 1991, Kates et al. 1985).

Humanity exercises some control over four parameters relating to MSA: the type of pollutant released, the spatial distribution of the pollutant, the total stock of pollutant in the environment, and the scale and health of natural (or human-made) ecosystems that are meant to absorb the pollutant. In this article, we explore the following two relationships: first, that between MSA and the scale and health of the ecosystem(s) into which the waste is released; and, second, that between the total stock of accumulated pollutant and the ability of the environment to buffer *H. sapiens* from harmful effects.

Pollutants whose rates of removal are limited, at least in part, by biological processes differ from those whose removal rates are not biolimited. Removal may be achieved by degradation into benign products, dilution to harmless levels, or transfer into sinks. Virtually all organic wastes (e.g., sewage and pulp mill effluents) are biolimited. Examples of pollutants whose removal rates are not biolimited include asbestos and radioactive materials.

MSA is a function of the pollutant's distribution and rate of removal and of the sensitivity of the affected systems to its concentration. For a given spatiotemporal distribution of pollutant, MSA is the level of emission that produces the highest concentration of pollutant that can be tolerated by the most sensitive system element. If the removal mechanism is the most sensitive, then MSA is equivalent to the maximal sustainable average rate of removal. For example, MSA for organic waste flushed into an aquatic system is equal to the maximal emission rate that does not lead to eutrophication. System elements other than those involved in removal may be most sensitive. Thus, for a toxic waste that can be degraded by specialized bacteria, MSA may be limited by the sensitivity of components of the recipient ecosystems other than the bacteria (e.g., shellfish, fishes, seabirds, and marine mammals in the case of oil spilled into the oceans).

Variation in the emission or removal rates must be incorporated into the calculation of MSA. Although the average removal rate may be sufficient to prevent long-term buildup of a pollutant, variation in the rate may allow temporary but harmful concentrations to develop, as in the cases of air pollution in city basins that are only periodically swept clean by winds or of acid pulses associated with the spring melt of acid snow.

MSA may be increased in two ways. The first is by manipulation of the distribution of pollutant into concentrations that maximize the removal rate or buffering capacity of the environment. The second is by enhancement of the removal rate by increasing the extent and capacity of systems involved in its removal, be they natural ecosystems or sewage treatment plants.

The same analysis applies, to pollutants whose rates of degradation or uptake by sinks are not biolimited. Although their removal rates are independent of the scale and capacity of ecosystems, their MSAs may depend on these factors to the extent that ecosystems buffer humanity and other life-forms from negative impacts by, for instance, dilution. Any level of waste generation could be considered quasisustainable (even for pollutants with no degradation rates, such as asbestos) until the capacity of the environment to buffer humanity and its life support systems from unacceptably harmful effects is exceeded.

Interactions. The preceding analysis enables calculation of upper bounds on carrying capacities by dividing each MSU and MSA by the minimal or desired average per-capita use or abuse and finding the minimum among all those resources. However, the simultaneous use of different resources usually involves complex, indirect interactions that constrain MSUs and MSAs of a resource required for multiple activities (e.g., forests).

A systems approach is required to keep account of how one activity may impinge on another. To determine a sustainable use of coal, for example, one must account for the damage (e.g., in the form of acid precipitation, strip mining, and global warming) done to natural systems that reduces MSUs and MSAs of those systems. Sustainable farming requires similar comparison of all marginal costs (including decreases in MSAs of soils and water supplies) of applying pesticides and fertilizers to the marginal benefits derived in short-term increases in yield.

Furthermore, a given activity may cause perturbations that have unintended, indirect effects on other system elements. In the case of marine systems, for example, the MSU of a harvested species may depend not only on its own population dynamics (stock-dependent renewal rate), but on the importance of that species in controlling the population dynamics of other species. Harvesting high on the food chain may trigger undesirable population explosions of species lower down. Similarly, harvesting organisms low on the food chain (e.g., krill) may result in the collapse of populations of valued species that consume them (Orians 1990).

The resolution of conflicting demands on interdependent resources involves a complex set of social and economic considerations. Biologists can contribute by describing quantitatively alternative patterns of sustainable use and the relative magnitudes of the carrying capacities resulting from each.

Lag times. A crucial difficulty in assessing whether a given human activity is sustainable is the time that passes between the onset of the activity and human perception of its impact. A delay in perceiving the impact may result from either an actual lag time before its manifestation or from an inability to detect the impact under routine monitoring.

In the case of CFC-catalyzed ozone depletion, there is an actual lag time of approximately a decade between the release of an average CFC molecule and its arrival to the upper atmosphere where it is active. Yet, ozone thinning was only predicted and then detected approximately half a century after freons first came into use. The delay between predicting (Arrhenius 1896) and detecting global warming with certainty is apparently more than a century (Tegart et al. 1990, Schneider 1990); by the time the effects are manifest, irreversible deleterious changes may have occurred (Daily et al. 1991).

Social Dimensions of Carrying Capacity

Social dimensions of carrying capacity include lifestyle aspirations, epidemiological factors, patterns of socially controlled resource distribution, the disparity between private and social costs, the difficulty in formulating rational policy in the face of uncertainty, and various other features of human sociopolitical and economic organization. Although the full complexity of such social dimensions requires investigation beyond the scope of this article, as illustrations, we briefly outline some of the issues surrounding discounting, the global commons, international trade, and prices.

Discounting over time. There are numerous situations (sometimes called social traps), in which the immediate, local incentives are inconsistent with the long-run, global best interest of both the individual and society, and with the maintenance of carrying capacity (Costanza 1987, Cross and Guyer 1980, Platt 1973). One of the most pervasive causes of social traps is the natural human tendency to discount costs that appear remote, either in time or space.

The most straightforward reason for discounting is to adjust for the time value of money: the value of $1000 delivered today is higher than that of $1000 to be delivered in ten years because of benefits that can be derived from investing the money over the decade. Discounting is done routinely in the context of cost/benefit analysis and has enormous influence on fiscal policy in every arena (e.g., Lind 1982).

Although, in principle, discounting is valid, two problems make discounting over a substantial time horizon (several decades or more) a gamble with the welfare of future generations. Estimates of future costs and benefits are uncertain, and there is both subjectivity and uncertainty in the selection of an appropriate discount rate.

Economists have great difficulty assigning monetary value to many of today's environmental amenities (e.g., clean air and national parks) and risks (e.g., global warming and ozone depletion), much less those of the future. When future costs are uncertain, a risk-averse policy would require discounting less than if they could be predicted with certainty. However, when analysts cannot agree on the uncertainties, too often they make no adjustment at all in the discount rate. The result is an underestimate of potential future costs, such that projects that imperil future generations appear more favorable than they should. These uncertainties are compounded over the period for which the calculation is made; the longer the time horizon, the greater the gamble. And when essential resources are involved, that gamble is with future carrying capacities.

The problem with discounting is not simply that decision makers often fail to apply it appropriately. The very process of discounting (especially at rates as high as 10%) encourages the public to underestimate the importance of future costs and defer their payment. Consider the problem of determining whether society would profit by taking measures now to deter the onset of global warming. Suppose that inaction will result in a known and certain cost of $100 billion to be incurred in 100 years. Discounted at 10% (on an annual basis), the present value of that cost is reduced to a mere $7.2 million. In a cost/benefit framework, investment in any deterrent whose net immediate cost exceeded $7.2 million seems irrational. But that discounted cost is so deceptively small that society may foolishly fail to invest even that minimal amount to solve a potentially serious future problem.

Choosing not to take action now presumes that posterity will be richer than we are, easily able to pay the $100 billion. In the recent past, successive generations have indeed enjoyed ever greater average wealth, but this trend may not continue until the time comes to pay for these deferred costs (Lind 1982; see also, for example, Fuchs and Reklis 1992). In short, this method of analysis should not be applied to long-term resource management because it constitutes a recipe for a growing burden of environmental debt, resulting in lower future carrying capacities.

Discounting by distance. Another form of discounting, also important and innate in policy judgements relevant to carrying capacity, is discounting over distance. The significance of events (including the magnitude of benefits and costs) occurring at a distance is discounted. The distance may be measured in strictly geographic terms, or it may be remoteness in a social, economic, or political sense.

Discounting over distance is reflected in several dimensions of human behavior and judgment. Consider how societies value domestic environmental health relative to that abroad. Japan is using timber stripped from virgin forests in several nations (including the United States) for low-quality products such as concrete forms, while carefully protecting its own forests. Twenty-five percent of all pesticides exported from the United States are heavily restricted or banned by the United States and other industrialized nations (Weir and Schapiro 1981). The German government made little effort to control industrial emissions until the effects of acid precipitation were manifest in its own forests and soils (to the tune of costing $1.4 billion per year). By then, approximately 18,000 Swedish lakes had acidified to the point that fish stocks were severely reduced, in part due to German emissions (Myers 1984).

In some instances, discounting by distance is clearly in the best interest of the discounters, but misjudgment of the relevant distance may exact a penalty. Overestimation of distance contributes to the extraction and sale, at below-market values, of natural resources (such as timber) from regions that are geographically and socioeconomically remote from policy centers in Washington, DC (e.g., Alaska and Colorado), and clearly confers a net cost to the United States (Wirth and Heinz 1991).

Overestimates of the relevant distance have led to profound environmental problems with direct implications for carrying capacity. For example, until recently, the upper atmosphere was considered so remote as to encourage emission of airborne pollutants that did not cause local or regional smog problems. It came as a surprise that the connections between the gaseous composition of the seemingly distant stratosphere and our day-to-day lives are actually very tight (Daily et al 1991). Similarly, the ability of humanity to vastly alter global biogeochemical cycles through local and regional habitat conversion has only become apparent in recent decades.

Currently, the many indications that human society has exceeded social carrying capacity and is paying a price for it are barely noticed. The negative impact of human activity on the planet usually manifests itself first to those whose lives are tightly dependent on the health of fragile, local ecosystems. Yet, by the time many current environmental problems directly affect decision-makers, whose lives are buffered by distance and economic well-being, it will be far too late to correct them. Ecologist Thomas Lovejoy's program of taking policy-makers and celebrities to tropical forests has helped make apparent the intimate connections to parts of the biosphere that are often misperceived as remote.

For different reasons, discounting over time and distance both encourage behavior that may reduce carrying capacity for future generations. Pressing economic problems often cause developing nations to apply higher discount rates to the future cost of depleting essential resources (as in accepting toxic wastes and environmentally damaging industries rejected by rich countries). Discounting over distance fosters the illusion that wealthy nations and individuals can afford to ignore the increasingly desperate plight of the poor.

The global commons. There are several reasons why it is in the selfish best interest of developed nations to narrow the gap between rich and poor. First, it will help the developing nations to protect their vast reservoirs of biodiversity, whose destruction affects at least two major elements of carrying capacity. The need for wild plants and microorganisms, which already supply the active ingredients in more than 25% of modern pharmaceuticals, may become acute as the human population grows more susceptible to disease (Ehrlich and Ehrlich

1990). Biodiversity is also critical to maintaining crop resistance to pests and drought, supplying the raw materials for genetic engineering and thus hopefully permitting the future phenomenal boost in agricultural yields required to feed an exponentially growing population (Ehrlich et al. 1992).

Second, developing nations have the power to degrade severely the entire planet's life support systems simply by following development paths taken by the rich. Elementary calculations indicate that the mobilization of coal reserves (e.g., in China or India) to fuel even a modest increment of development could overwhelm any efforts by industrialized nations to compensate by reducing their own greenhouse gas emissions (Ehrlich and Ehrlich 1989). Similarly, large increases in methane and nitrous oxide fluxes would accompany planned expansion of agriculture and the continued destruction of tropical forest. The rapid deployment of less-damaging technologies (such as solar-hydrogen energy technologies) in developed nations and their transfer to the rest of the world is required to secure just this atmospheric element of the global commons.

Third, the ever-growing disparity between rich and poor carries forbidding implications for social carrying capacity, including intensifying economic dislocation and social strife as the transfer of capital, labor, and refugees across steepening gradients accelerates. Political challenges also loom large as the ranks of those with little to lose increase, nuclear capability proliferates in the developing world, and vulnerability to terrorism increases (e.g., Schneider and Mesirow 1976).

In short, there is no lifeboat escape possibility for the rich. All nations will have to come to grips with the limits to carrying capacity. Unless measures are taken by the rich to facilitate sustainable development, the continued destruction of humanity's life support systems (and a reduction in biophysical carrying capacity) is virtually guaranteed.

International trade. Trade may increase global biophysical carrying capacity by lifting regional constraints arising from the naturally heterogeneous distribution of resources. If there were no trade at all, then global biophysical carrying capacity would equal the sum of all local biophysical carrying capacities. Trade may also increase global biophysical carrying capacity through the increased efficiency that results from regional specialization in the production of goods.

Exceeding local and regional carrying capacities on a sustainable basis through trade has the unfortunate effect of encouraging the "Netherlands fallacy" (Ehrlich and Holdren 1971): the idea that all regions could simultaneously sustain populations that sum to more than global carrying capacity. Regional and local development plans need to account for the global balance of trade in resources.

The optimal size of resource catchment areas needs consideration with respect to economies of scale and the incentives for sustainable resource management. Empirical evidence suggests that economic incentives favor better management of natural resources by local communities with long-term stakes in sustainability than by distant parties driven to maximize short-term profit (see examples in Ehrlich and Ehrlich 1991). A better understanding is needed of the trade-offs between the efficiency associated with large industries and the better quality of local resource management.

Finally, the organization and regulation of international commerce is extremely important to evaluation of carrying capacity, but it is also complex and poorly understood (see, e.g., Culbertson 1991, Daly and Cobb 1989, Keynes 1933). For example, standard economic thought tends to support free trade. However, completely unregulated international trade could reduce carrying capacity by tending to diminish international diversity, thereby increasing the vulnerability of nations to disasters in other regions (e.g., droughts in distant grain belts) and limiting their ability to learn lessons from their own successes and failures (e.g., Culbertson 1991).

Prices. Prices relate to both biophysical and social carrying capacities in at least two important ways. First, underpricing of resources encourages unsustainable management. Underpricing often occurs because future generations have no means of making their demands for a resource known. The future demand for the water in the Ogalalla aquifer clearly is not reflected in its current price. One solution would be to regulate prices of essential resources to keep their use sustainable.

Prices also play an important role in the rates of innovation. High prices constitute incentives for research and development of technologies that are more efficient or that substitute more abundant for scarce resources. Such price-induced innovation appears to be the rule and can be seen clearly in the development of agriculture (Hayami and Ruttan 1985). The price of food is obviously related to the agricultural dimension of biophysical and social carrying capacities.

Achieving Sustainability

We wish to reemphasize that our analyses are necessarily preliminary, intended to provide a framework for subsequent more-detailed and quantitative studies. In particular, central determinants of social carrying capacity lie in the domain of interactions among resources, among sociopolitical and economic factors, and between biophysical and social constraints. However, the complexity of these interactions makes it unlikely that they will be sufficiently well evaluated in the next several decades to allow firm calculations of any carrying capacity. From a policy perspective, the current great uncertainty in future social carrying capacity is irrelevant because the human population is likely to remain above that carrying capacity for decades at least.

Global assessments of MSUs and MSAs of critical resources such as forests and the atmosphere should be undertaken immediately, in the tradition already established for greenhouse gases. Such assessments would provide measures of relative contributions of nations to the preservation or destruction of the global commons. They could thus form the basis for international treaties and possible control schemes, such as the issuing of tradable permits for consumption of fractions of global MSUs and MSAs.

Nations and regions should evaluate MSUs and MSAs for their key resources. Even cursory examinations can be informative (e.g., Daly 1990). Fresh water, both surface and underground, is an obvious top candidate for evaluation in many regions, including the United States, Mexico, much of Africa and China, and the Middle East. Other inputs to agriculture, especially topsoils, require examination everywhere, in a context of revised natural resource accounting (Repetto et al. 1987). MSUs and MSAs that pose the greatest constraints will determine the carrying capacities of any region in the absence of imports. Especially careful consideration must be given to assumptions about maintaining access to limiting resources through trade, because the last frontiers for acquiring cheap and plentiful resources are closing (e.g., Folke et al. 1991).

Because further degradation of the global environment is inevitable, interdisciplinary evaluations of the relative costs of alternative evils and their communication to the public is necessary. Some provision of insurance should be taken in proportion to the level of uncertainty and the severity of possible deleterious effects of given activities. In the meantime, no further net loss of essential elements of natural capital should be incurred.

Several potentially effective social (especially market) mechanisms have been suggested to make short-term incentives consistent with long-term sustainability. These mechanisms include fees for use of common-property resources, taxes on the depletion of natural capital, and flexible environmental assurance bonding systems for regulating activity that may be environmentally damaging, but whose effects are uncertain (Costanza 1987, Costanza and Daly 1992, Costanza and Perrings 1990). Implementation and further development of such methods of avoiding social traps is essential.

Frequently lacking, however, is a vision of a desired world that would establish a basic social carrying capacity for human beings. In the short run, efforts must be made to minimize the damage to Earth's systems, while providing the requisites of a decent life to the entire global population. In the long run, however, public discussions should be encouraged to guide policy on sustainable resource management. Sound science is central to the estimation of carrying capacities and the development and evaluation of technologies, but it can give minimal guidance at best regarding the issues surrounding the question of the kinds of lives people would choose to live.

The current decade is crucial, marking a

window of environmental and political opportunity that may soon close. Environmentally, each moment of inaction further entrains irreversible trends, such as the global extinction of biodiversity and alteration of the gaseous composition of the atmosphere. Though it is certainly possible that intensifying human impact on the planet will precipitate a sudden disaster, it seems more likely that humanity will just gradually erode Earth's life-support capabilities over the next few decades. The more important window may thus be a political one for laying the institutional foundations for desired change. Right now, in the wake of the United Nations Conference on Environment and Development, citizens and national governments may be at a peak in receptivity to acknowledging environmental problems and tackling their solutions. Let us seize the day.

Acknowledgments

We wish to thank S. Alexander, A. Ehrlich, M. Feldman, N. Haddad, H. Mooney, J. Roughgarden, T. Sisk, and P. Vitousek (Department of Biology, Stanford); L. Daniel and T. Daniel (Bureau of Economics, Federal Trade Commission); W. Falcon and R. Naylor (Institute for International Studies, Stanford); L. Goulder (Department of Economics, Stanford); J. Harte, J. Holdren, and A. Kinzig (Energy and Resources Group, University of California, Berkeley); D. Murphy, A. Launer, and J. McLaughlin (Center for Conservation Biology, Stanford); P. Matson (NASA, Ames); S. Schneider (National Center for Atmospheric Research); and V. Valdivia (Siemens A.G.) for helpful comments on the manuscript. Holdren, Matson, and Vitousek have been closely involved with us in the Stanford Carrying Capacity Project. Reviewers for *BioScience*, including R. Costanza, H. Daly, J. Miller, N. Myers, and one anonymous referee, were also extremely helpful. This project was made possible by the generous support of the W. Alton Jones Foundation and several private donors.

References

Arrhenius, S. 1896. On the influence of carbonic acid in the air upon the temperature on the ground. *London, Edinburgh, and Dublin Philosophical Magazine and Journal of Science* 1896: 237–276.

Brown, L. R., and E. C. Wolf. 1984. Soil erosion:

quiet crisis in the world economy. Worldwatch Paper 60, Worldwatch Institute, Washington, DC.

Brundtland, G. H., chair. 1987. *Our Common Future*. Oxford University Press, New York.

Costanza, R. 1987. Social traps and environmental policy. *BioScience* 37:407–412.

———. 1989. What is ecological economics? *Ecol. Econ.* 1:1–7.

Costanza, R., and H. E. Daly. 1992. Natural capital and sustainable development. *Cons. Biol.* 6(1): 1–10.

Costanza, R., and C. Perrings. 1990. A flexible assurance bonding system for improved environmental management. *Ecol. Econ.* 2:57–75.

Cross, J. G., and M. J. Guyer. 1980. *Social Traps*. University of Michigan Press, Ann Arbor.

Culbertson, J. M. 1991. US "free trade" with Mexico: progress or self-destruction? *Social Contract* 2:7–11.

Daily, G. C., P. R. Ehrlich, H. A. Mooney, and A. H. Ehrlich. 1991. Greenhouse economics: learn before you leap. *Ecol. Econ.* 4:1–10.

Daly, H. E. 1990. Carrying capacity as a tool of development policy: the Ecuadoran Amazon and the Paraguayan Chaco. *Ecol. Econ.* 2:187–195.

Daly, H. E., and J. B. Cobb Jr. 1989. *For the Common Good*. Beacon Press, Boston.

Dunne, T., and L. B. Leopold. 1978. *Water in Environmental Planning*. W. H. Freeman, San Francisco.

Ehrlich, P. R., and A. H. Ehrlich. 1989. How the rich can save the poor and themselves: lessons from the global warming. Pages 287–294 in S. Gupta and R. Pachauri, eds. *Proceedings of the International Conference on Global Warming and Climate Change: Perspectives from Developing Countries*. Tata Energy Research Institute, New Delhi, 21–23 February.

———. 1990. *The Population Explosion*. Simon and Schuster, New York.

———. 1991. *Healing the Planet*. Addison-Wesley, New York.

Ehrlich, P. R., A. H. Ehrlich, and G. C. Daily. 1992. Population, ecosystem services, and the human food supply. Morrison Institute for Population and Resource Studies Working Paper No. 44, Stanford University, Stanford, CA.

Ehrlich, P. R., A. H. Ehrlich, and J. P. Holdren. 1977. *Ecoscience: Population, Resources, Environment*. W. H. Freeman, San Francisco.

Ehrlich, P. R., and J. P. Holdren. 1971. Impact of population growth. *Science* 171:1212–1217.

Ehrlich, P. R., and E. O. Wilson. 1991. Biodiversity studies: science and policy. *Science* 253:758–762.

Folke, C., M. Hammer, and A. Jansson. 1991. Life-support value of ecosystems: a case study of the Baltic Sea Region. *Ecol. Econ.* 3:123–137.

Fremlin, T. 1964. How many people can the world support? *New Scientist* 29 October.

Fuchs, V. R., and D. M. Reklis. 1992. America's children: economic perspectives and policy options. *Science* 255:41–46.

Gleick, P. H. 1989. Global climatic changes and geopolitics: pressures on developed and developing countries. Pages 603–621 in A. Berger, S. Schneider, and J. C. Duplessy, eds. *Climate and Geo-Sciences: A Challenge for Science and Society in the 21st Century.* Kluwer Academic Publ., Dordrecht, The Netherlands.

Hardin, G. 1986. Cultural carrying capacity: a biological approach to human problems. *BioScience* 36:599–606.

Hayami, Y., and V. W. Ruttan. 1985. *Agricultural Development: An International Perspective.* Johns Hopkins University Press, Baltimore.

Holdren, J. P. 1991a. Population and the energy problem. *Popul. Environ.* 12:231–255.

———. 1991b. Testimony before the Committee on Science, Space, and Technology (Hearing on Technologies and Strategies for Addressing Global Climate Change), US House of Representatives, 17 July.

Holdren, J. P., and P. R. Ehrlich. 1974. Human population and the global environment. *Am. Sci.* 62:282–292.

Jacobs, L. 1991. *Waste of the West: Public Lands Ranching.* Jacobs, Tucson, AZ.

Jones, R. R., and T. Wigley. 1989. *Ozone Depletion: Health and Environmental Consequences.* John Wiley & Sons, New York.

Kates, R., C. Hohenemser, and J. Kasperson, eds. 1985. *Perilous Progress: Managing the Hazards of Technology.* Westview Press, Boulder, CO.

Keyfitz, N. 1991. Population and development within the ecosphere: one view of the literature. *Population Index* 57:5–22.

Keynes, J. M. 1933. National self-sufficiency. Yale Review, Summer.

Klein, D. R. 1968. The introduction, increase, and crash of reindeer on St. Matthew Island. *J. Wildl. Manage.* 32:350–367.

Lind, R. C. 1982. *Discounting for Time and Risk in Energy Policy.* Johns Hopkins University Press, Baltimore.

Mech, L. D. 1966. *The Wolves of Isle Royale.* Fauna of the National Parks of the United States, Fauna Series 7.

Myers, N. 1984. *Gaia: An Atlas of Planet Management.* Anchor Press, New York.

———. 1989. Tropical deforestation and climate change. Pages 341–353 in A. Berger, S. Schneider, and J. C. Duplessy, eds. *Climate and Geo-*

Sciences: A Challenge for Science and Society in the 21st Century. Kluwer Academic Publ., Dordrecht, The Netherlands.

National Academy of Sciences (NAS). 1989. *Alternative Agriculture.* National Academy Press, Washington, DC.

Oak Ridge National Laboratory (ORNL). 1968. Nuclear energy centers, industrial and agroindustrial complexes. Summary Report ORNL-4291.

Orians, G. H. 1990. Ecological sustainability. *Environment* 32(9):10–15, 34–39.

Paddock, W., and P. Paddock. 1967. *Famine 1975!* Little, Brown, Boston.

Platt, J. 1973. Social traps. *Am. Psychol.* 28:642–651.

Population Reference Bureau (PRB). 1991. *World Population Data Sheet.* Washington, DC.

Postel, S. 1989. Halting land degradation. Pages 21–40 in L. Brown, A. Durning, C. Flavin, L. Heise, J. Jacobson, S. Postel, M. Renner, C. P. Shea, and L. Starke, eds. *State of the World 1989.* Norton, New York.

Repetto, R., M. Wells, C. Beer, and F. Rossini. 1987. *Natural Resource Accounting for Indonesia.* World Resources Institute, Washington, DC.

Roughgarden, J. 1979. *Theory of Population Genetics and Evolutionary Ecology: An Introduction.* Macmillan, New York.

Scheffer, V. B. 1951. The rise and fall of a reindeer herd. *Science Monthly* 73:356–362.

Schneider, S. H. 1990. *Global Warming.* Random House, New York.

Schneider, S. H., and L. E. Mesirow. 1976. *The Genesis Strategy: Climate and Global Survival.* Plenum, New York.

Tegart, W. J. M., G. W. Sheldon, and D. C. Griffiths, eds. 1990. *Climate Change: The IPCC Impacts Assessment.* Australian Government Publication Service, Canberra.

Terborgh, J. 1989. *Where Have All the Birds Gone?* Princeton University Press, Princeton, NJ.

Time Magazine. 1991. 24 June, p. 60.

United Nations Environment Programme (UNEP). 1984. *Environmental Data Report.* Blackwell, Ltd., Oxford, UK.

United Nations Population Fund (UNFPA). 1991. *The State of World Population 1991.* UNFPA, New York.

Vitousek, P. M., P. R. Ehrlich, A. H. Ehrlich, and P. A. Matson. 1986. Human appropriation of the products of photosynthesis. *BioScience* 36:368–373.

Weir, D., and R. Schapiro. 1981. *Circle of Poison.* Institute for Food Development and Policy, San Francisco.

White, G. F. 1988. A century of change in world

water management. Pages 248–259 in *Earth '88: Changing Geographic Perspectives*. National Geographic Society, Washington, DC.

Wilson, E. 1989. The value of biodiversity. *Sci. Am.* September, p. 108–116.

Wittwer, S. H. 1989. Food problems in the next decades. Pages 119–134 in D. Botkin, M. Caswell, J. Estes, and A. Orio. *Changing the Global Environment: Perspectives on Human Involvement.*

Academic Press, Boston.

Wirth, T., and J. Heinz. 1991. Project 88—Round II. Incentives for Action: Designing Market-Based Environmental Strategies, Washington, DC.

World Resources Institute (WRI). 1992a. *Environmental Almanac 1992*. Houghton Mifflin, Boston.

_____ . 1992b. *World Resources 1992–93*. Oxford University Press, New York.

33
Biodiversity and Ecosystem Function

Paul G. Risser

Introduction

Biological diversity occurs at several hierarchical levels, from genes to individuals, from populations, species, communities, and ecosystems to landscapes. At each of these levels there are important relationships between biodiversity and ecosystem processes and between biodiversity and the ways in which ecosystems respond to disturbance and changing global conditions.

Much of the current consideration about the importance of biodiversity in ecosystem processes, however, is based on experiments and observations at the species level. For example, Tilman and Downing (1994) experimentally increased the biodiversity of North American grasslands with added nitrogen. When these grasslands were subsequently confronted with drought (an important controlling variable in prairie ecosystems), those with higher plant-species diversity had a much faster recovery of primary production than those with lower species diversity. Thus, biodiversity had a direct relationship with ecosystem processes and with the response of the ecosystem to disturbance.

Relationships between biological diversity and ecosystem function are inherently complex and operate at many spatial and temporal scales. The current state of the science involves collecting and aggregating examples of these relationships at various scales, but it is impractical to study and measure all the potential relationships. There is a need for guidelines to identify the conditions under which particularly important connections between biodiversity and ecosystem function are likely to occur. Specifically, we need a strategy to simplify the myriad potential relationships between biodiversity and ecosystem function and to identify the most important of these. Such a strategy is analogous to developing strategies for selecting geographical areas of high biodiversity and/or rich and rare species for preservation (Pendergast et al. 1993).

Simplification Rationale

Holling (1992) offers important ideas that can form the basis of a simplifying approach by arguing that a relatively small set of plant, animal, and abiotic processes structure ecosystems across scales of time and space. In addition, he postulates that in the temporal dimension these key structuring processes dictate a few dominant temporal frequencies that drive other processes. Finally, he hypothesizes that these structuring processes should also generate a discontinuous distribution of spatial struc-

Reprinted with permission from Conservation Biology vol. 9, pp. 742–746. Copyright 1995 The Society for Conservation Biology and Blackwell Science, Inc.

tures coupled with discontinuous temporal frequencies.

The landscape forms a hierarchy that contains breaks in object sizes, object proximities, and textures at particular scales. Holling (1992) tested this contention that ecosystems should have this "lumpy" architecture by examining discontinuities in body-mass structural characteristics and found these discontinuous distributions in mammals (using three trophic levels) and birds. The breaks in geometry in the landscape occur because structuring processes exert their influence over defined ranges of scale in three dominating classes of processes. At small spatial scales of centimeters to tens of meters and time scales of days to decades, the structure of the ecosystem is driven by vegetative processes that determine plant growth, plant form, and soil structure. At very long and broad scales, slow geomorphological processes dominate the formation of a topographic and edaphic structure. These processes operate at the scales of hundreds of thousands of kilometers and centuries to millennia.

At a third, intermediate scale, contagious disturbance processes such as fire, insect outbreak, plant disease, and water flow dominate the formation of patterns over spatial scales of hundreds of meters to hundreds of kilometers and on time scales of years to decades. Although Holling did not differentiate the relative importance of these three scales, it is arguable that it is at this intermediate scale where both direct and indirect effects have their greatest consequences for biodiversity and ecosystem behavior. Indeed, the example of the relationship of biodiversity with drought recovery (Tilman & Downing 1994) fits this model. The spatial and temporal scales are of intermediate dimensions in Holling's construct, and drought, along with grazing and fire, is clearly one of the few dominating processes that structure grasslands in the central plains of North America (Risser et al. 1981). In general, the direct effects of grazing, forestry, land-use change (the products of human activities) and the indirect effects of large predators and animal disease further transform spatial patterns over these intermediate scales, determining whether present local, regional, and global human influences will trigger changes in the relationship between bio-

diversity and ecosystem function — and, if so, how rapidly (Holling 1992).

Strategy

At this point, it is useful to recognize the fundamental components of the proposed strategy for parsimoniously identifying the most significant relationships between biodiversity and ecosystem function. From Holling's work, it can be argued that much of the behavior of ecosystems can be understood from relatively few dominating processes. These processes can be categorized into a small number of general classes, and, moreover, the focus on identifying these processes can be placed at intermediate scales of space and time. In accepting this simplification, it is important to realize that ecosystems are inherently dynamic over time (Sprugel & Bormann 1981; Davis & Zabinski 1992) and that they interact across landscape spatial scales (Risser et al. 1984). From a practical view, therefore, the most useful strategy is first to identify and focus on the key dominating ecosystem processes, recognizing that they may change over time and that ecosystems interact over the landscape and cannot be considered in isolation. This strategy does not require that all possible interactions between biodiversity and ecosystem process at many spatial and temporal scales be evaluated. Rather, it argues that ecosystem processes, through their internal dynamics and operations over the landscape, integrate the most important spatial and temporal scales — and, by focusing first on the dominating ecosystem processes (Holling 1992), we are led to the most important relationships between ecosystem and biodiversity.

Further Simplification

The strategy described above greatly simplifies the challenge of integrating biodiversity and ecosystem function over many scales of space and time. As a further refinement of this proposed strategy, however, there are other possible simplifying principles. The focus on the key processes described (Holling 1992) can be placed at intermediate scales of space and time for identifying the conditions under which interactions between biodiversity and ecosystem are particularly intense. Also, the greatest biodiversity is frequently found in those ecosystems in which the disturbance levels are of intermediate intensity (Connell 1978).

Moist tropical forests are generally rich in tree species, and it has been assumed that at least a portion of this diversity can be attributed to long periods of environmental stability (Huston 1993). We know, however, that these forests continuously undergo localized change, and there are now some indications of a more widespread increase in tree turnover rates since the 1950s (Phillips & Gentry 1994; Pimm & Sugden 1994). If this trend exists, shifts in composition would be expected toward more climbing, early-successional, fast-growing, gap-dependent tree species. Over time, there would be changes in the spatial distribution of forest types, in primary production, and in carbon sequestering, and these spatial patterns would become more heterogeneous. The relationship between biodiversity (forest types) and ecosystem processes (carbon sequestering) would change relatively rapidly during succession, until the final species composition was established. Thus, the early to midsuccessional status can be just as useful for evaluating biodiversity and ecosystem function in tropical terrestrial ecosystems as it is in temperate ecosystems.

Evaluation of invading species offers another means of identifying important relationships between biodiversity and ecosystem function. In the nitrogen-limited, open-canopied forests of Hawaii, the exotic, actinorrhizal nitrogen-fixing *Myrica faya* was able to invade young volcanic soils (Vitousek 1990). Invasion by *Myrica* altered the availability as well as the quantity of nitrogen in these sites. Although there may be relatively few cases in which changes in biodiversity via invasions alter large-scale ecosystem properties, there are some cases, as evidenced here. Vitousek (1990) suggests that invaders can change ecosystems when they (1) differ substantially from native species in resource acquisition or utilization, (2) alter the trophic structure of the invaded area, or (3) alter disturbance frequency and/or intensity.

In attempting to identify conditions under which biodiversity and ecosystem relationships are particularly significant, it is useful to consider the idea that the highest plant-species diversity is found on poor soils and the lowest plant diversity on the best soils (Huston 1993). The presumed mechanism is that, under the most favorable conditions, diversity among species can be reduced by competition, although abiotic disturbances such as fires or hurricanes, as well as predators and parasites, can prevent competitive dominance by eliminating some of the dominant competitors. Huston argues that this inverse association between plant diversity and soil fertility is found at local, regional, and global scales. The pattern does not always exist, and there are exceptions: the highest diversity of large marine vertebrates and predatory birds and mammals is usually found in productive habitats. Nevertheless, for an initial approximation, soil fertility may focus initial consideration on plant biodiversity and ecosystem processes.

The importance of species interactions in structuring ecosystem processes has been known for many years (Paine 1971). In some systems, changes in the biodiversity of a relatively small number of key species can have major effects on ecosystem behavior over relatively short time scales. In addition, species dynamics can be more sensitive than ecosystem processes to disturbance and stress (Schindler 1990). There may be many cases, however, in which ecosystem processes and their relationship to biodiversity are primarily controlled by abiotic rather than biotic interactions. Rotenberry and Wiens (1980) studied bird communities for five years throughout the structurally simple steppe vegetation in North America. The abundance of many bird species was correlated with the physiognomic features of the habitat, but there was little apparent interaction among the dynamics of coexisting species. Thus, although biotic interactions (such as pollinators in tropical forests and consumers in rocky intertidal habitats) may be important in relationships between biodiversity and ecosystem function (Paine 1971), in some cases the abiotic interactions are far more important and can explain most of the ecosystem dynamics (Rotenberry & Wiens 1980).

Ecotones, or transitional areas between ecosystems or vegetation types, are frequently high in biodiversity. A part of this increased biodiversity arises because the ecotone contains species characteristic of both adjacent communities as well as some species adapted to the ecotone itself. In addition, habitats in the ecotonal region are likely to be fragmented, thus leading to high biodiversity (Whitcomb et al. 1981; Grover & Musick 1990). Recently, there has been an increased appreciation for the ways in which ecotones can modify the flows of water, nutrients, and other materials in addition to the movement of plants and animals across the landscape (Peterjohn & Correll 1984; Wiens et al. 1985). Because ecotones frequently support high biodiversity and may also control important ecosystem processes, they should be among the first conditions examined when the relationships between biodiversity and ecosystem processes are considered.

Summary

Ecosystems are internally variable over time and space due to inherent properties; they interact at landscape levels, so the condition of one may affect others, and, on many scales, disturbance affects interactions between biodiversity and the behavior of ecosystems. Given this enormous complexity, it is easy to see why science frequently fails to provide policy makers with clear advice about priorities for conservation and about the relative importance among possible explanations of the relationships between biodiversity and ecosystem functions. Ironically, the better understood the ecological system, the less need there is for simplifying constructs; less well-known ecological systems require simplified initial strategies. Therefore, this simplifying strategy will not specify the integrated scales of time and space of all ecological complexity within the best-known systems; rather, it offers a set of guidelines for initial consideration of relationships between biodiversity and ecosystem function.

Instead of constructing an artificial framework of representative hierarchical scales of space and time, I propose that the most effective strategy is first to focus on the identification of the few dominating ecological processes that structure the ecosystem (Holling 1992) and

then to seek to understand the relationships between biodiversity and these ecosystem processes. These key processes will integrate across space and time scales of most importance.

The key relationships can be further refined by examining several additional situations that apparently lead to important interactions between biodiversity and ecosystem processes: (1) early to midsuccessional status (Phillips & Gentry 1994); (2) low soil fertility (Huston 1993); (3) intermediate levels of disturbance (Connell 1978); (4) biotic interactions only when there is collaborative evidence of importance (Rotenberry & Wiens 1980), (5) invading species that differ significantly from native species in resource acquisition or utilization (Vitousek 1990); and (6) ecotones (Grover & Musick 1990). This simplifying and integrating strategy requires further testing, and its domain of applicability needs to be firmly established. In the meantime, this strategy will be useful for organizing information and for guiding those who need to formulate policy and set priorities for research and natural resources.

References

Connell, J. H. 1978. Diversity in tropical rain forests and coral reefs. High diversity of trees and corals is maintained only in a non-equilibrium state. Science 119:1302–1310.

Davis, M. B., and C. Zabinski. 1992. Changes in geographical range resulting from greenhouse warming: effects of biodiversity in forests. Pages 297–308 in R. Peters, and T. Lovejoys, editors. Global warming and biological diversity. Yale University Press, New Haven, Connecticut.

Grover, H. D., and H. B. Musick. 1990. Shrubland encroachment in southern New Mexico. USA: An analysis of desertification processes in the American Southwest. Climate Change 16:165–190.

Holling, C. S. 1992. Cross-scale morphology, geometry, and dynamics of ecosystems. Ecological Monographs 62:447–502.

Huston, M. 1993. Biological diversity, soils, and economics. Science 262:1676–1680.

Paine, R. T. 1971. A short-term experimental investigation of resource partitioning in a New Zealand rocky intertidal habitat. Ecology 52:1096–1106.

Pendergast, J. R., R. M. Quinn, J. H. Lawton, B. C. Eversham, and D. W. Gibbons. 1993. Rare species, the coincidence of diversity hotspots, and conservation strategies. Nature 365:335–337.

Peterjohn, W. T., and D. L. Correll. 1984. Nutrient

dynamics in an agricultural watershed: observation on the role of a riparian forest. Ecology 65:1466–1475.

Phillips, O. L. and A. H. Gentry. 1994. Increasing turnover through time in tropical forests. Science 263:954–958.

Pimm, S. L. and A. M. Sugden. 1994. Tropical diversity and global change. Science 263:264.

Risser, P. G., E. C. Birney. H. D. Blocker, W. S. May, W. J. Parton, and J. A. Wiens. 1981. The true prairie ecosystem. Hutchinson and Ross, Stroudsburg, Pennsylvania.

Risser, P. G., J. R. Karr, and R. T. T. Forman. 1984. Landscape ecology. Directions and approaches. Special Publication no. 2. Illinois Natural History Survey, Champaign.

Rotenberry, J. T., and J. A. Wiens. 1980. Habitat structure, patchiness, and avian communities in North American steppe vegetation: a multivariate analysis. Ecology 61:1238–1250.

Schindler, D. W. 1990. Experimental perturbations of whole lakes as tests of hypotheses concerning ecosystem structure and function. Oikos 57:24–41.

Sprugel, D. G., and F. H. Bormann. 1981. Natural disturbance and the steady state in high-altitude balsam fir forests. Science 211:390–393.

Tilman, D., and J. A. Downing. 1994. Biodiversity and stability in grasslands. Nature 367:363–365.

Vitousek, P. M. 1990. Biological invasions and ecosystem processes. Towards an integration of population biology and ecosystem studies. Oikos 57:7–13.

Whitcomb, R. F., C. S. Robbins, J. F. Lynch, B. L. Whitcomb, K. Klimikiewicz, and D. Bystrak 1981. Effects of forest fragmentation on avifauna of the eastern deciduous forest. Pages 125–205 in R. L. Burgess and D. M. Sharpe, editors. Forest island dynamics in man-dominated landscapes. Springer Verlag, New York.

Wiens, J. A., C. S. Crawford, and J. R. Gosz. 1985. Boundary dynamics. A conceptual framework for studying landscape ecosystems. Oikos. 45:421–427.

Index